T0344795

EX LIBRIS

EVIDENCE AND PROCEDURES FOR BOUNDARY LOCATION

EVIDENCE AND PROCEDURES FOR BOUNDARY LOCATION

Seventh Edition

Donald A. Wilson, BSF, MSF, LLS, PLS, RPF
Land Boundary Consultant, Newfields, New Hampshire

Charles A. Nettleman, III, PhD, PLS, Esq.
Walter G. Robillard

Based on the original concept of Curtis M. Brown

This book is printed on acid-free paper.

Published by John Wiley & Sons, Inc., Hoboken, New Jersey

Published simultaneously in Canada

Library of Congress Cataloging-in-Publication Data

Names: Robillard, Walter G. (Walter George), 1930- | Wilson, Donald A., 1941- author. | Nettleman, Charles A., III, author.
Title: Evidence and procedures for boundary location / Donald A. Wilson, BSF, MSF, LLS, PLS, RPF, Land Boundary Consultant, Newfields, New Hampshire; Charles A. Nettleman III; Walter G. Robillard, Esq., BSF, MAPA, JD, LLM, RLS, FSMS, Attorney at Law and Boundary Consultant, Atlanta, Georgia.
Description: Seventh edition. | Hoboken : Wiley, 2021. | Includes index.
Identifiers: LCCN 2021012087 (print) | LCCN 2021012088 (ebook) | ISBN 9781119719397 (cloth) | ISBN 9781119719427 (adobe pdf) | ISBN 9781119719434 (epub)
Subjects: LCSH: Boundaries (Estates)—United States.
Classification: LCC KF639 .B73 2021 (print) | LCC KF639 (ebook) | DDC 346.7304/32--dc23
LC record available at https://lccn.loc.gov/2021012087
LC ebook record available at https://lccn.loc.gov/2021012088

Cover Design: Wiley
Cover Image: © Dietmar Rauscher/iStockphoto

SKY10075181_051424

CONTENTS

Preface ix

1 Introduction 1
2 Definition, Scope, and Nature of Evidence 13
3 Words as Evidence 51
4 Evidence and Technology 77
5 Other Types or Species of Evidence 91
6 Calculations and Measurements as Evidence 139
7 Plats as Evidence 189
8 Evidence of Water Boundaries 213
9 Using Historical Knowledge as Evidence 243
10 Recording and Preserving Evidence 319
11 Procedures for Locating Boundaries Described by Words 339
12 Original Surveys and Related Platting Laws 379
13 Unwritten Transfers of Land Ownership 397
14 Guarantees of Title and Location 441
15 Using and Understanding Words in Boundary Descriptions 459
16 Professional Liability 513
17 Professional Stature 533
18 The Surveyor in Court 551
19 The Surveyor, the Law, and Evidence: A Professional Relationship 581

Appendix A The Surveyor's Report 633

Appendix B Wooden Evidence 643

Appendix C **The (Quasi-)Judicial Function of Surveyors** **649**

Appendix D ***Geodaesia*** **659**

Appendix E **Land Acts that Created the Public Lands** **673**

Index **685**

PREFACE

Now in its seventh edition, *Evidence and Procedures for Boundary Location* has evolved over the years into an expanded version of Curtis Brown's original vision. Based on a comment made at the 1948 annual meeting of the American Congress on Surveying and Mapping, and joined by Winfield Eldridge, the authors wrote in the Preface to the first edition in 1962: "It is strange and sad that ours is probably the only profession in the country that has no book devoted specifically to the profession. There are a number of textbooks on surveying, but there is no textbook that is keyed directly to the practicing property surveyor. Such a book has not as yet been published though recent works have partially closed the gap. Much still needs to the done to provide the profession with comprehensive practical coverage of the field of property surveying" (*Surveying and Mapping*, vol. IX, no. 4 (1949), p. 291).

In 1958, the annual conference of the California Council of Civil Engineers and Land Surveyors made a request to Curtis Brown, a licensed California land surveyor, to prepare text material directed toward professional surveyors. It became advisable to increase the scope so as to have a text applicable to all areas of the United States, so in 1960, Winfield Eldridge of the University of Illinois was invited to be a co-author. Together, they accomplished their task, and in 1962 saw the first edition come to print. After the untimely death of Professor Eldridge in 1966, Curtis Brown invited Walter Robillard of Georgia and Donald Wilson of New Hampshire as co-authors, so that, in his words, the eastern part of the United States with its differences from the public land states, would be well represented. As the three authors worked to refine the overall text to include new principles and take advantage of changes in technology, the book has now arrived at a revision of the sixth edition, in this seventh edition, incorporating new ideas and challenges. In a profession such as ours, with continual changes and improvements, a technical publication soon becomes outdated and must constantly be revised to remain current. Technology advances and the law evolves accordingly.

The aims of this title have always been threefold: (1) the recognition and presentation of evidence in accordance with legal principles; (2) a study manual to aid the student in learning the details of property surveying; and (3) a reference manual for the practicing surveyor to offer guidance when faced with conflicts and difficult situations. As a testament to its usefulness, the book has been consulted and cited by numerous courts at all levels throughout the United States. This edition, following in the footsteps of Brown, Eldridge, and Robillard, endeavors to keep true to the existing goals and bring the text material to meet the demands of a new frontier in property surveying.

It would also be remiss not to thank George Cole for his help with the water boundaries chapter in previous editions. In keeping with the original vision, all authors and contributors to the various editions have been licensed land surveyors, writing for other land surveyors and those in related disciplines who rely on surveying and surveyors.

Dr. Tony Nettleman
Georgetown, Texas

Don Wilson
Newfields, New Hampshire

1

INTRODUCTION

1.1 Scope of the Book

In writing any book, authors should have a goal or philosophy as to what it is they wish to accomplish. We have such a philosophy: It is our desire to help the student and the practicing surveyor, as well as those aspiring young people who wish to enter the surveying profession, appreciate some of what we feel are the finer aspects of surveying, that is, the legal aspects as well as that area of the law that relates to and encompasses evidence in all of its aspects.

This is a book about survey evidence and, as such, it is not a book on "how to survey" but what to do with your survey when it is completed and what your clients can expect from your survey. Many consider the surveyor as the individual standing behind a transit or theodolite turning angles or collecting data for a new road or measuring for a new subdivision. This book will have a double focus: First, it is aimed at those students who wish to enter the surveying profession and at those surveyors who locate boundary lines and land parcels or utilize evidence in searching for and locating the footsteps of the earlier surveyors who originally created the parcels or property.

Second, we wish to provide a reference textbook that future professional land surveyors may use as a study guide to prepare for their Fundamentals, Principles & Practice, and state-specific examinations. In writing this book, the authors have attempted to modernize the format to be more in keeping with the legal texts of today. The authors have attempted to make the references and the terminology consistent with legal courses to be in keeping with the modern approach. In this we have tried to modernize the land surveyor's responsibilities and liabilities.

The functions of land surveyors should be considered as varied and different in each state. For simplicity, boundary surveying can be divided into two general areas or disciplines: (1) locating or relocating originally described parcels of land; and (2) creating new parcels. In keeping with recent legal decisions, we have somewhat modified some of the terminology. For instance, seldom is the term *property line* or *property boundary* used. It is our belief that property rights, including property boundaries, are legal questions and as such are not addressed by land surveyors. Surveyors locate boundaries, or land boundaries or deed lines. They do not and cannot locate property rights.

Corners and the lines connecting those corners define the extent of property rights that are described in the land description. These lines may coincide with property rights or they may be separate and independent. These originally created lines remain fixed and unalterable except as provided by law, such as the loss or gaining of property rights or the addition or elimination of contiguous parcels of land in the same ownership.

This concept is what gives rise to the phrase "following the footsteps" because a parcel, once established, should remain fixed through any series of conveyances. This phrase is historical in that many earlier decisions coined the phrase, which now has been adopted and adapted by modern-day courts in rendering decisions. A grantee who acquires a parcel owned by a grantor determines the boundaries of his/her purchase at the time that the particular parcel was carved out of some larger tract. He or she takes to the bounds of the estate of his or her grantor, who, in turn, took to the limits of that grantor's estate, to the time of the creation of the boundaries [1].

The first portion of the book (Chapters 1–11) covers resurveys or retracements of former surveys based on the record; the latter part (Chapters 12–19) covers the creation of new parcels or the division of land in which the surveyor may create the record. The importance of evidence in both phases will be stressed in the following pages.

The first problem one encounters is in the definition of *surveying*. No two states or the federal government or the international community describe surveying in the same terms; nor do they have the same requirements for one to become qualified to create, identify, retrace, and then testify in judicial tribunals as to boundaries or to survey real property.

How original surveys are made is subject to control by the legislative branch of the government, and since there are 50 states and the federal government, land subdivision laws and regulations of original surveys are extremely variable and usually are regulated by standards, statutes, rules, and regulations. However, after parcels have been created and the land has been divided and described, it is left to the courts to interpret the position of the original boundaries. Today, practically all original surveys must begin from a survey of an existing parcel. For this reason, the first portion of the book pertains to the location of previously described parcels. It deals with the evidence and methods for locating these corners, lines, and parcels.

In attempting to survey and locate a described parcel of land, the only permanent and correct location of its boundaries is where a court of competent jurisdiction would locate them. To know where a court would locate property boundaries, the surveyor must have expert knowledge and understanding of the laws of boundaries and evidence. Yet regardless of where the surveyor would locate the boundaries of the parcel, the final location of any boundary is nothing more than an opinion of the evidence recovered, evaluated, and then interpreted, and this is always subject to review by the courts and by other subsequent surveyors.

Once a boundary is questioned or litigated and the parties seek determination by a jury, the trial is usually divided into two parts: It is left to the jury to decide what the facts are and it is left to the judge to apply the law to the facts. This has led many courts to hold to the old saying: "What boundaries are is a matter of law; where boundaries are is a matter of fact."

Thus, in a trial, the jury decides where an original monument position was located based on evidence the surveyor used to formulate an opinion. The judge decides whether the monument or measurement is controlling as a matter of evidence presented at the trial and not necessarily as a matter of law. In a survey based on the record, the surveyor may be asked to determine both of these questions, either knowingly or unknowingly. Chapter 2 examines the laws of evidence necessary to prove facts and the order of importance of discovered evidence.

According to the *Statute of Frauds* first enacted in England in 1677 and later adopted both in statute law and common law in the United States, land ownership must be proven by some form of written evidence. To prove the right of legal possession, a written document must be produced witnessing such right. In early decisions, courts found that the requirement for written deeds sometimes caused an individual to commit fraud; thus, the concept of title passing without a writing was created, yet English courts recognized a concept from Roman law, adverse possession, by calling it an unwritten title. Since a legally created unwritten title is legally superior to a written title, one may say that the written title is either extinguished or reduced in status to a junior interest.

In recent years, a marked change has occurred in the courts' thinking on the subject of professional responsibility and liability. The concepts of privity of contract and the time of commencement of the running of any statute of limitations are vastly different from what they were 50 years ago. Chapter 16 and a portion of Chapter 17 discuss these subjects with the hope that student will be able to limit or possibly escape professional exposure for liability when entering the profession.

According to decisions reported in court cases, surveyors, in retracing old boundary lines, are directed and obligated to follow the "footsteps of the original or creating surveyor"; therefore, it is essential that any surveyor who practices in the area of property or boundary identification has knowledge of the historical background of land surveys in general and the geographic area specifically and that he or she know under which laws they were originally performed. The authors believe that the term *footsteps* equates to a question of *evidence*: evidence created and then evidence recovered. Evidence will be discussed to include many facets. The purpose of discussing surveying history in Chapter 9 is to aid present-day surveyors in understanding why we must follow certain procedures when locating property boundaries. Even this history must be used as a tool for the present and not as a device to present the romance of the past. Exploration of historical background and the study of the development of various survey systems and equipment will provide the needed background of laws and customs governing property owners' rights and privileges.

Although surveying history will be shown to be quite ancient, it does not overlook the fact that the United States is rich in many phases of the history of surveying and law that have widespread application within the individual states, territories, and the United States proper:

1. the English system that gave rise to English common law, which was used in the colonies and now forms many of the fundamental rules, as well as civil law, which evolved from Roman law and was subsequently referred to as the Napoleonic Code and is still found in vestiges in most of the remainder of the United States;
2. the Mexican and Spanish land grant systems;
3. the French system, used in the Louisiana Purchase area and elsewhere;
4. the sectionalized land system of the public domain;
5. land divisions under state laws, especially in Texas and some of the eastern seaboard states; and
6. the various other systems that developed in states such as Georgia and Maine.

The intent and meaning of the words in a deed should always be interpreted in the light of laws, words, and conditions existing at the date of the creation of the document. In New York, under Dutch rule, land dedicated for road purposes passed in fee title to the Crown. New York, on acquiring the streets, retained the fee title; hence vacated, former Dutch streets revert to the state. The same can be said of Texas roads dedicated during Spanish or Mexican control [2]. The ownership of stream beds or bodies of water is often dependent on which nation had jurisdiction at the time of the land's original alienation and on the effect of the laws in force at the time of the grant. An indispensable part of all boundary location is knowledge of the history of the development and settlement of the area. With this idea in mind, Chapter 8, on evidence of water boundaries, has been totally rewritten.

With a foundation of the historical development of real property and surveying, the surveyor or the student will have an opportunity to expand their knowledge by learning the procedures used in locating already described parcels and procedures used to create new parcels, including suggestions as to how to describe parcels in writings. If the reader feels that some of the aspects of property surveying appear to be treated too briefly or are not mentioned at all, it may be because these topics have been adequately discussed in other works previously published and available to the practitioner and student.

1.2 Definitions of Surveys and Surveyors

The terms *survey*, *surveying*, and *surveyor* have broad and possibly confusing connotations; the average citizen becomes confused when a survey is conducted as to public opinion but then sees a surveyor measuring the lot next door. We recognize that the term *survey*, without more exacting words, describes procedures as well as vague studies. Simply stated, the word *surveying* may have numerous implications.

First, being the verb: "I am surveying a description of a parcel of land." Then, going from verb to noun, a surveyor can state, "Here is my survey." Finally, the same word may be used as an adjective with the phrase "Here is my survey plat." When used with other terms, such as *land*, *property*, *boundary*, *geodetic*, or *cadastral*, the term then becomes more definite and can be discussed and is pertinent to the subject of this book. The word *cadastre* is defined as an official register of the quantity, value, and ownership of real estate used in apportioning taxes. *Cadastration* is the act or process of making a cadastre or cadastral survey. Cadastres and cadastral surveys are concerned with land [3], law, and people [4]. In popular use, land surveying is defined as the determination of boundaries and areas of a tract of land.

A *boundary survey* is understood by some as a survey that is conducted for the location and establishment of lines between legal estates, or it may be a physical feature erected to mark limits of a parcel or political units [4] and a *cadastral survey* is confined to the location and subdivision of the public domain [5]. To minimize confusion, the term *property surveying* may be used interchangeably with *boundary surveying* or *boundary line*. As such, the terms may not be considered as denoting property rights but may be employed throughout this book to denote the activity of locating, establishing, and delimiting boundaries of real property. The practice of property surveying is defined in many of the state registration and licensing laws. Such definitions usually include the measurements of area, length, and directions and the correct determinations of descriptions, especially when such property is to be conveyed or when the instrument of conveyance is to become a matter of public record.

Although not applicable here, we must mention that to the general public the term survey may address an opinion poll or political analysis, which may add additional confusion. The National Council of Examiners for Engineering and Surveying (NCEES) defined *land surveying* to mean the performance or practice of any professional service requiring education, training, and experience in the application of special knowledge in the mathematical, physical, and technical arts and sciences to such professional services as the establishment or relocation of land boundaries, the subdivision of land, the determination of land areas, the accurate and legal description of land areas, and the platting of land subdivisions for record [6].

In a model registration law approved by NCEES, the following statement appears [7]:

> The term Land Surveying used in this act shall mean and shall include assuming responsible charge for and/or executing: the surveying of areas for their correct determination and description and for conveyancing; the establishment of corners, lines, boundaries and monuments; the platting of land and subdivisions thereof including as required, the functions of topography, grading, street design, drainage and minor structures, and extensions of sewer and water lines; the defining and location of corners, lines, boundaries and monuments of land after they have been established; and preparing the maps and accurate records and descriptions thereof.

The American Congress on Surveying and Mapping (ACSM), now the National Society of Professional Surveyors (NSPS), jointly published *Terms & Definitions of Surveying and Associated Terms* in 1978 with the American Society of Civil Engineers (ASCE). The current edition, revised in 2005, provides the following definition of land surveying:

Land surveying is the art and science of

1. retracing cadastral surveys and land boundaries based on documents of record and historical evidence;
2. planning, designing, and establishing property, land, and boundaries; and
3. certifying surveys as required by statute or local ordinance such as subdivision plats, registered land surveys, judicial surveys, and space delineation.

Land surveying can include associated services such as mapping and related data accumulation; construction layout surveys; precision measurements of length, angle, elevation, area, and volume; horizontal and vertical control systems; and the analysis and utilization of survey data. In summary, the term *surveying* can be considered an ambiguous word.

1.3 Activities of Boundary Surveyor

Few other countries rely on the surveyor as people do in America. No other country identifies and applies property rights as we do. Property (boundary) surveyors are found in private practice, are employed by federal, state, county, and local government, and are associated with related business. In the past, those engaged in private practice and especially those in rural areas often had small organizations composed of the surveyor with one or more helpers. As a general rule, most surveyors were sole practitioners or employed by small firms. This still applies today. In larger cities and in densely populated areas it is not uncommon to find large surveying firms preparing subdivisions and locating parcels of land for sale. Many land surveyors have found that a small organization in a country setting is most enjoyable and offers many advantages. In many instances, small surveying firms or organizations have spanned several generations in the same family.

The federal government, through the Bureau of Land Management (BLM), is still engaged in the original subdivision, resurveying, and retracing the public domain; the U.S. Forest Service employs surveyors who are well versed in land surveying, as does the National Park Service, the Fish and Wildlife Service, and military organizations. One of the more important functions of State Departments of Transportation is the location of rights-of-way with respect to adjacent properties. Counties and cities often have similar problems, although confined to a more local area. In a few states, the county surveyor makes private property locations part of his or her official duties; but in other states the county surveyor's responsibility is narrowly confined to county government problems, as defined by the law.

Today many more governmental organizations employ surveyors or survey advisors for geographic information system (GIS), land information system (LIS), and global positioning system (GPS) work. The surveyor may not necessarily confine his or her work to property, boundary, surveys, often mapping topography and staking the outline of engineering projects such as buildings, sewer lines, water lines, curbs, sidewalks, and paving. Although these are important technical functions, they will not be treated in this book. This textbook will primarily be devoted to the location of boundaries of parcels of land. The modern surveyor will be intimately involved in creating maps or descriptions for GIS–LIS projects or acting as a consultant for the design or such projects. GIS should mean "get it surveyed."

1.4 The Surveyor in Society

History supports the fact that the practice of surveying, including property surveying, is as ancient as property ownership itself. In Babylon, over 3500 years ago, the name of a surveyor was inscribed on a boundary stone, giving testimony to his acts [8]. A visit to the National Museum in Cairo, Egypt, will expose the visitor to a tomb of a surveyor. In the National Museum in Athens, Greece, as well as in Rome, Italy, one can view the tombstone of a surveyor long dead.

It is recorded that for over 1000 years ancient Rome used surveyors to locate boundaries and survey roads and aqueducts. In fact, the Roman *agrimensores*, namely the surveyor, was required to pass an examination for competency. Because of the nature of surveying and the varied needs, Rome separated the "civil" surveyor from the "military" surveyor.

In early times, surveyors possessed special skills and talents that were regarded with almost reverent respect; they filled a necessary need in civilization, and they utilized the most advanced sciences known to the world. The same Roman surveyors were required to receive special training in the varied aspects of leveling and boundary law while in school. They were guided by a series of textbooks titled *The Corpus* [9].

There is little doubt that the practice of surveying today is a profession and should be performed by and through a professional upholding the standards of the higher meaning of the word. Yet as practiced today, surveying is very different from that practiced by Washington, Ellicott, Thoreau, and Lincoln and their contemporaries. The present rapid development of this country's lands and resources has created a need for professional surveyors, a demand much greater than graduates being supplied by the universities with surveying programs. Today, many surveyors have obtained professional status or recognition by apprenticeship, on-the-job training, and self-study programs, without the aid of much formal education. But as the demands of society become more complex, the surveyor of yesterday will be little recognized by the surveyor of the future. Such regeneration probably will not suffice in the future, and the surveyors will have to come from schools and colleges with students earning formal degrees in the various disciplines of surveying.

> Technology development has outpaced the development of the law.

Education was and will be the foundation of the surveyor of tomorrow, while experience was the foundation of the immediate past. In the formative years of the world and our nation, the surveyor was perhaps among the more educated individuals. The surveyor mastered mathematics, instrumentation, astronomy, and cartography, and he was the explorer and the mapmaker of new worlds. Today the new worlds the modern surveyors will conquer are the technical and the legal worlds.

Like all other professions, the demands on surveyors are more complex and have become greater and more differentiated, the practice of property surveying has to consider inexact laws in addition to exact engineering or measurement sciences. Few problems confronting the property surveyor can be solved completely by applying exact sciences only; answers will depend on law, an inexact science. The development of mature judgment, logic, and meaningful experience will be demanded of the student.

There are no known studies as to why young students wish to devote their lives to the profession of surveying. What type of person does this profession appeal to was best explained over nine decades ago by A.C. Mulford who finalized the book *Boundaries and Landmarks* with the following statement [10]:

> Yet it seems to me to a man of an active mind and high ideals the profession is singularly suited; for to the reasonable certainty of a modest income must be added the intellectual satisfaction of problems solved, a sense of knowledge and power increasing with the years, the respect of the community, the consciousness of responsibility met and work well done. It is a profession for men who believe that a man is measured by his work not by his purse, and to such I commend it.

Surveyors look with pride to the many famous men in history, including several presidents, who have been engaged in surveying as a profession during their lifetime. Property surveyors, perhaps more than their associates in engineering, are in constant contact with people. This gives the property surveyor an opportunity to present a professional image as a member of a profession with superior standing, integrity, and knowledge. Awareness of the many responsibilities and a conscientious fulfillment of professional obligations will enable the property surveyor to maintain a status in society.

1.5 Current Need for Surveyors

In the United States and now in other areas of the world there are millions of people with many basic needs that involve land. Food and minerals and fiber and forage are derived from the soil. All substances that go into essential goods are products of the land. Rain must be conserved and controlled as it falls on the earth. Lakes and rivers provide transportation, irrigation, and energy for those who live along their shores. Yet surveying is more than just the surveying of a lot or a subdivision. The various areas of surveying and what they bring to humanity is one of the basic needs of all peoples.

This century has been highlighted by many spectacular advances in technology. Structures, buildings, bridges, and tunnels, as well as dams and highways, all of which are located on land that has been here for millennia, require surveys or measurements. When these structures disappear, the land on which they stood will still be there. When miles of new highways are constructed, the location will depend on and require miles of new property lines to be located or surveyed and described!

An exploding population is creating demands for more and more residential lots. New land developments containing 50,000 home sites and small subdivisions with only a few lots are being created. For each parcel of land, there are owners' rights and privileges extending to the limits of their property, both skyward and to the depths of the Earth. These requirements are requiring that surveyors identify division lines on, below and above the Earth's surface. With expansion, a community requires additional lands for public and municipal purposes, such as waste disposal, service utilities, parks, schools, and rights-of-way.

In the early years of the country, real property sold for $1.25 per acre or less; today, portions of that same land may have values in excess of $2 million per acre, and some land is valued in terms of thousands of dollars per front-foot or square foot. Even "poor" land in rural areas may bring $1000 per acre or more.

Of the thousands of surveyors in the country today, many are not prepared to cope with these crucial problems, either technically, educationally, legally, or experience-wise, and, even if they were able, the numbers of trained and qualified surveyors are not sufficient to satisfy the needs of the growing population. This book is written with a twofold purpose in mind: first, to help those in practice to better understand the professional problems that are present and, second, to aid the surveying student in preparing for a profession that can be as rewarding as the engineering, medical, or legal profession.

1.6 Future Needs for Surveyors

No one can accurately predict the problems of the future, but some of the needs can be anticipated and prepared for accordingly. Certain needs are obvious. The population will increase, and thus more efficient use of land must be made. As the costs of land spiral upward, the delineation of property lines will become more critical. New divisions of land will continue to be made each day, and the ancient surveys will have to be identified and retraced. These divisions will not only be horizontal but also vertical. We will build upward as well as downward.

Even in the future, the property surveyor must not ignore the past, for a professional's problems go back as far as land ownership itself. If we will not learn from the mistakes of yesterday, we will provoke headaches for tomorrow, and new surveys must be performed more accurately and with greater precision. Rare are the areas where the value of real property has not risen considerably over the years. In addition, now that communism has failed, large areas of once communally held property will be divided and placed into private ownership.

The property or boundary surveyor of tomorrow will need to exercise far more technical skill and more in-depth knowledge of the law and use legal judgment than the professional surveyor of today, yet the modern surveyor must have an understanding and knowledge of the law. Just as the compass and chain are gone, the transit and tape are not sufficient to cope with future problems. Modern technology has developed many marvelous devices and techniques with which to make more precise surveys and aid in studying and solving property surveying problems. Such tools as photogrammetry, electronic computers, and microdistance devices are now commonplace. The surveyor will use new terms such as GPS, GIS, LIS, and perhaps some that have not yet been coined.

It is hoped the student surveyor will not lose sight of the fact that technology cannot replace hard work and experience that can only be gained from the exposure to field exercises and the necessity of having an understanding of the law. It is possible that the property surveyor of tomorrow will locate title boundaries many fathoms under the sea or meters above the ground. He or she may need to subdivide the Antarctic continent. Whatever are the problems, the property surveyor will need to be armed with all the tools, knowledge, and education of the profession and have the ability to exercise sound and mature reasoning.

One fact is certain: Education is the foundation of tomorrow's surveyors. Today, surveying programs are directed toward two general areas: one predicated on course-oriented subjects in modern aspects and a second directed toward basic education courses supplemented by extensive field training. There is no present answer as to which of the two will produce the better surveyor of tomorrow.

What the authors do see is that the surveyor of tomorrow will be more closely oriented toward the legal aspects. Surveying should be considered to be a combination of three separate and distinct areas: technical, legal, and administrative, professional, or business. Although each of these is separate, they have a common area of overlap in which all three are interrelated. Many textbooks adequately discuss the technical, and in most college courses, the technical aspect is thoroughly taught and examined, but the business or administrative and legal areas have few references to aid the student and surveyor. It is hoped that this book will meet this need for the student and the practicing surveyor (see Figure 1.1).

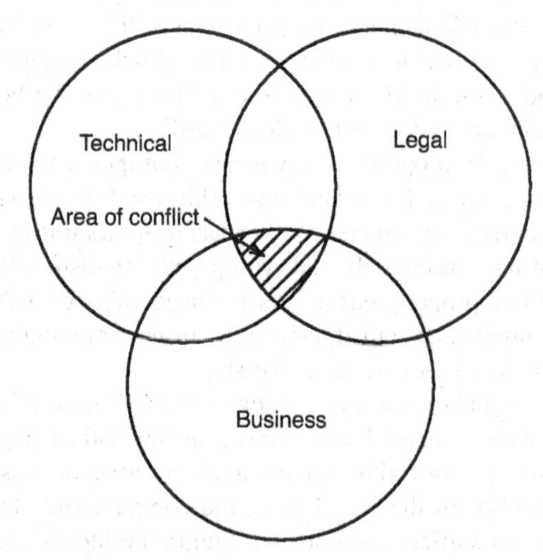

FIGURE 1-1. Three disciplines of surveying.

1.7 Land Data Systems

At this time, the need for restructuring and reorganizing the land data systems existing in most states is urgent. Most, if not all, states employ land systems that are antiquated and cumbersome. In some counties, land records are deplorable. The move toward modernization of land data systems includes pilot projects in various parts of the country, some of which have been in existence for a few years. Although needed, land data will vary from place to place; yet, the systems in general may be similar in many respects.

Geographic information systems are being introduced in many locales, yet few people realize that the basis of any system predicated on measurement requires, first, an adequate survey framework. Systems index parcels of land according to a unique number taken from a cadastral map, which is based on a rigid framework of horizontal and vertical control. Data are computerized and can include not only title information and deed descriptions but also information concerning natural resources, zoning, and the like. Most recording offices now use microfilm as a necessity for saving space.

Most countries are now using some type of computerized system. With advances being made in GISs, LISs, and GPSs, the coordination of land data of any nature is now faster, more accurate, and more readily available to the surveyor and the consuming public. In addition, the retrieval of such data is routine in many offices and governmental agencies and in some areas is available to the consuming public on a 24/7 basis. Sophisticated computer hardware and software are being developed and modified to meet any budget or need.

Surveying and engineering offices are moving rapidly toward computerization, microfilming, and retrieval of records to conserve time and storage space and to increase accuracy of the record. Database management programs are being designed to fit the needs of most offices, no matter how large or small.

The thrust of this book will be in the legal dimensions, along with some of the business aspects of the surveying profession. Surveyors should be an integral part of the development and maintenance of land data records; they should be the ones to see that the evidence of monuments and their position are included with all other information that has become a matter of public record.

The future in the land data systems, in GIS/LIS, and all other new systems will rest on a group of professionals. If surveyors are to lead, they must take the initiative and not let the system pass them by. In the world of tomorrow, there will be no place for the hesitant, for they will be lost. Those who do not look to the future will earn their "Doctorate of Plane Table Technology."

1.8 Global Positioning Systems

Surveyors have found a new useful tool with which to measure lines and locate positions. Like all tools, some believe a new innovation called *global positioning systems*, or GPS, will revolutionize their practice. However, caution is always advised, since all tools have their shortcomings and limitations. Although GPS will permit long-distance surveying far faster than ever before, and makes distance ties and surveys of six-mile lines practical for small-area surveys, there is still no substitute for ensuring the measurements between points or the occupation of points, or positions. It cannot be overemphasized that there is only one correct point for a corner, whether marked or not, and it is the position where it was originally placed that controls above all.

GPS, being a measurement tool, is subject to the same rules of law and technology as all other measurement tools of the past. For, after all, the GPS unit used today is nothing more than the historical extension of the Roman groma or the dioptra of Heron of Alexandria's era. Courts have always ruled that measurements are the least reliable of all items of evidence, and that they should only be used as a last resort. Measurements contain error; original corners (positions), by law [11], are without error.

REFERENCES

1. Robert J. Griffin, Retracement and Apportionment as Surveying Methods for Reestablishing Property Corners, *Marquette Law Review*, vol. 43, 1960, pp. 484–510.
2. *Mitchell v. Bass*, 33 Tex. 260 (1862).
3. *Webster's New International Dictionary* (Springfield, MA: Merriam-Webster, Inc., 1993).
4. P. F. Dale, *Cadastral Surveys within the Commonwealth* (New York: Wiley, 1978).
5. *Definitions of Surveying and Related Terms* (ACSM-ASCE, 1976).
6. Proceedings of the 34th Annual Meeting of the NCSBEE (NCEES), 1955.
7. NCEES Model Law, 1960.
8. L. W. King, ed. *Babylonian Boundary-Stones and Memorial Tablets in the British Museum* (London, 1912).
9. O. A. W. Dilke, *The Roman Land Surveyors* (New York: Barnes & Noble, 1971), pp. 37–45, 178–187, 227–230.
10. A. C. Mulford, *Boundaries and Landmarks*, 1912, Reprinted 2009, by W.G. Robillard.
11. *Cragin v. Powell*, 128 U.S. 691, La. 1888.

2

DEFINITION, SCOPE, AND NATURE OF EVIDENCE

2.1 Historical Concept of Evidence

In retracing prior surveys, regardless of their ages, evidence becomes paramount in being able to identify and locate both large and small parcels of land. Often the practicing surveyor, the attorney, and perhaps even some courts will find it difficult to make the distinction between facts and opinions. Evidence of land surveys takes many forms, including instrument measurements and physical objects. The interpretation, or the acceptance or rejection of that found evidence, is quite difficult. Evidence is a foundation for attorneys in that one of the first courses law students are required to master is on the Federal Rules of Evidence (FRE) and Federal Rules of Procedure (FRP). Surveyors may not be familiar with the FRE or FRP rules, but we often use rules within them, including the ancient documents rule. The search for and discovery of evidence are important. But the presentation of the found evidence is often just as important. Often, cases are either lost or won due to the presentation of the evidence through the teamwork of the attorney and the surveyor expert witness.

THERE IS NO TECHNICAL OR LEGAL SUBSTITUTE FOR FOUND ORIGINAL EVIDENCE.

"Evidence, its use, identification, and application are a team effort between the attorney and one who may be called his co-worker." This was paraphrased from a comment made nearly 100 years ago by A. C. Mulford.

The surveyor may create original evidence, identify it in a retracement, and catalog it for use by the client or in a judicial tribunal, but the attorney must have the ability to effectively use it and get it accepted by the court. Evidence is worthless without the unified harmonious teamwork of the attorney and the surveyor.

Historically, most early rules of evidence were judge-made, in that each jurisdiction and each judge in that jurisdiction were permitted to interpret and apply the rules for admitting evidence in their court rooms. In fact, there were major differences in the admissibility of evidence within the same jurisdictions, and as such, most rules of evidence accepted by the courts were closely tied to English common law.

After many years of chaos, an effort was made in 1942 to develop a codification of the *rules of evidence*. Unfortunately, they received little support from the various states. Then in 1975, the U.S. Supreme Court developed rules that were adopted by Congress that became the rules of evidence in federal courts. These *Rules of Evidence for U.S. District Courts and Magistrates* are in effect today and only apply to the federal courts.

Subsequently, individual states either adopted these rules in their entirety or adapted them to suit regional or state-specific needs. Currently, all states have rules of evidence as part of their civil and criminal court systems.

In essence, these rules, whether federal or state [1], for the most part are attempts to codify existing common law. As such, it becomes necessary that the students not only know the codes that identify evidence but also have knowledge of the case law of evidence from which these rules evolved. From these sources, the students will come to appreciate the importance of evidence in surveying or creating, as well as retracing boundaries.

All survey or boundary questions can be categorized into three areas:

1. Questions of fact, to be ascertained by the jury.
2. Questions or propositions of law, to be determined by the judge.
3. A combination of the first two.

In most litigation, questions of fact may exert greater influence on the ultimate outcome or decision of a boundary problem than will questions of law. *Evidence* is the material that is *offered* to persuade the jury, which is the trier of facts, that certain events are true. In a bench trial—that is, a trial without a jury, having only a judge—the judge becomes the trier of fact.

Although there are many definitions of evidence, which will be discussed later, perhaps a modern definition should be examined. To understand better what evidence is or is not, various historical legal experts should be consulted.

The student and the surveyor must first make a distinction between facts and evidence.

The actual corner point is a fact, along with all the information used to create, identify, describe, recover, or preserve that point that is evidence of that one unique point, the corner.

The discussion of evidence for the student may seem complex and difficult, but this is far from so. To make it more understandable, the topic of evidence has been separated into chapters. Once the student becomes a registered surveyor, the failure to understand the rules of evidence and to apply them in the retracement of boundaries and other aspects of surveying could possibly result in that surveyor being investigated by various licensing boards or in having a claim of malpractice brought against the individual [2].

First, in layman's words, *evidence* can consist of almost any object, action, or verbal statement, either oral or written. The *laws of evidence* consider the admissibility, effect, and relative importance of the evidence produced. A found monument is evidence; the control afforded the found monument is determined by the laws of evidence. Evidence and the laws of evidence are, of necessity, closely related, and the discussion of one invariably must include a discussion of the other.

In a broad sense, a discussion of evidence and the laws of evidence could include most phases of the boundary surveyor's practice. Initial land transfers must be in writing. Writings are evidence. Measurements are evidence to prove or to disprove where deed lines or corners are located. Evidence of possession may be proof of an unwritten conveyance, but it may also be used to show possession of unwritten rights. Scientific principles are evidence, as is judicial notice that is accepted by a court.

Although entire books have been written describing evidence [3], discussing it here would not be a logical approach for students or for surveyors. This chapter is devoted to the fundamental concepts of evidence and the laws of evidence. Later chapters will be devoted to discussions of the amount, kind, and sufficiency of evidence necessary to prove certain facts of prior surveys and of evidence that identifies prior boundaries. These volumes should be a part of any land surveyor's reference library.

One will find a distinction between the meaning of the word as used by surveyors and as used by legal scholars and the courts. Regardless of the definition, it can be said that there is no single good or positive definition of evidence. To have the student appreciate the problem, Bentham defines evidence as follows: "Evidence is any matter of fact, the effect, tendency or design of which is to produce in the mind a persuasion affirmative or disaffirmative of the existence of some matter of fact" [4].

This definition is important in that it is one of the very first definitions indicating what the witness's responsibility is as to evidence. The witness is a *storyteller*. He or she must present, through guidance by the attorney, a story linking the history of the evidence to the evidence found in order to produce a fact in the minds of the judge or jury. Unless the surveyor can effectively communicate his or her knowledge of the evidence to today's problems, facts may go unproven.

In more limited terms, evidence as used in legal proceedings and the courts has variously been defined as follows:

- Blackstone states, "Evidence signifies that which demonstrates, makes clear, or ascertains the truth of the fact or point in issue, either on one side or the other" [5].
- Greenleaf wrote, "Evidence is legal acceptance and includes all means by which any alleged matter of fact, the truth of which is being submitted to investigation, is established or disproved" [6].[1]

- Thayer wrote, "Evidence is any matter of fact which is furnished to a legal tribunal, otherwise by reasoning or a reference to what is noticed without proof, as the basis of inference in ascertaining some other matter of fact" [7].[2]
- The judicial scholar Wigmore writes: "Evidence represents any knowable fact or group of facts, not legal or a logical principle, considered with the view to its being offered before a legal tribunal for the purpose of producing persuasion, positive or negative, on the part of the tribunal, as to the truth of a proposition, not of law or of logic, on which the determination of the tribunal is to be asked" [8].
- The California Evidence Code defines evidence as "Testimony, writings, material objects, or other things presented to the senses that are offered to prove the existence or non-existence of a fact" [9].

Most of these definitions were written for attorneys and the courts, and these definitions lend themselves to the lawyer's and the court's definitions. However, to summarize the bulleted definitions, the surveyor should consider evidence as follows: any document, object, writing, action, thing, verbal statement, or other information that is identified to prove the fact that is in question.

It is *evidence* that perpetuates the location of corners and boundaries. Seldom does the investigator, *and the person who makes the subsequent evaluation of the evidence*, have clear facts—the older the boundary, the less clear it may be. Its evidence is of prime importance in finding its precise location. Whatever remaining evidence there is must be interpreted to determine what was established at some point in the past.

If one tried to summarize the definition of evidence in a neophyte's language or in the simplest words that can be used a simplistic definition could be as follows:

> Evidence is ANYTHING for which the effect, tendency, or design is to produce in the mind of a person a persuasion—either affirmative or negative—of the existence of some matter of fact.

2.2 Surveyor's Role in Evidence

A surveyor is intimately involved with evidence from the creation of the original survey to its subsequent recovery by a retracing surveyor. The span of time between the two may be a few days or many decades or possibly a century or more. Many things can happen from the time a surveyor creates the evidence until a retracing surveyor, acting as a witness, is asked to recover it, interpret it, and explain it to a client, an attorney, or a court of law.

The recovery of evidence and its use by a surveyor form but one function or stage in a complex chain. The chart in Figure 2.1 is an attempt to give readers insight and to provide an opinion as to what we see as the role of the surveyor in the area of evidence.

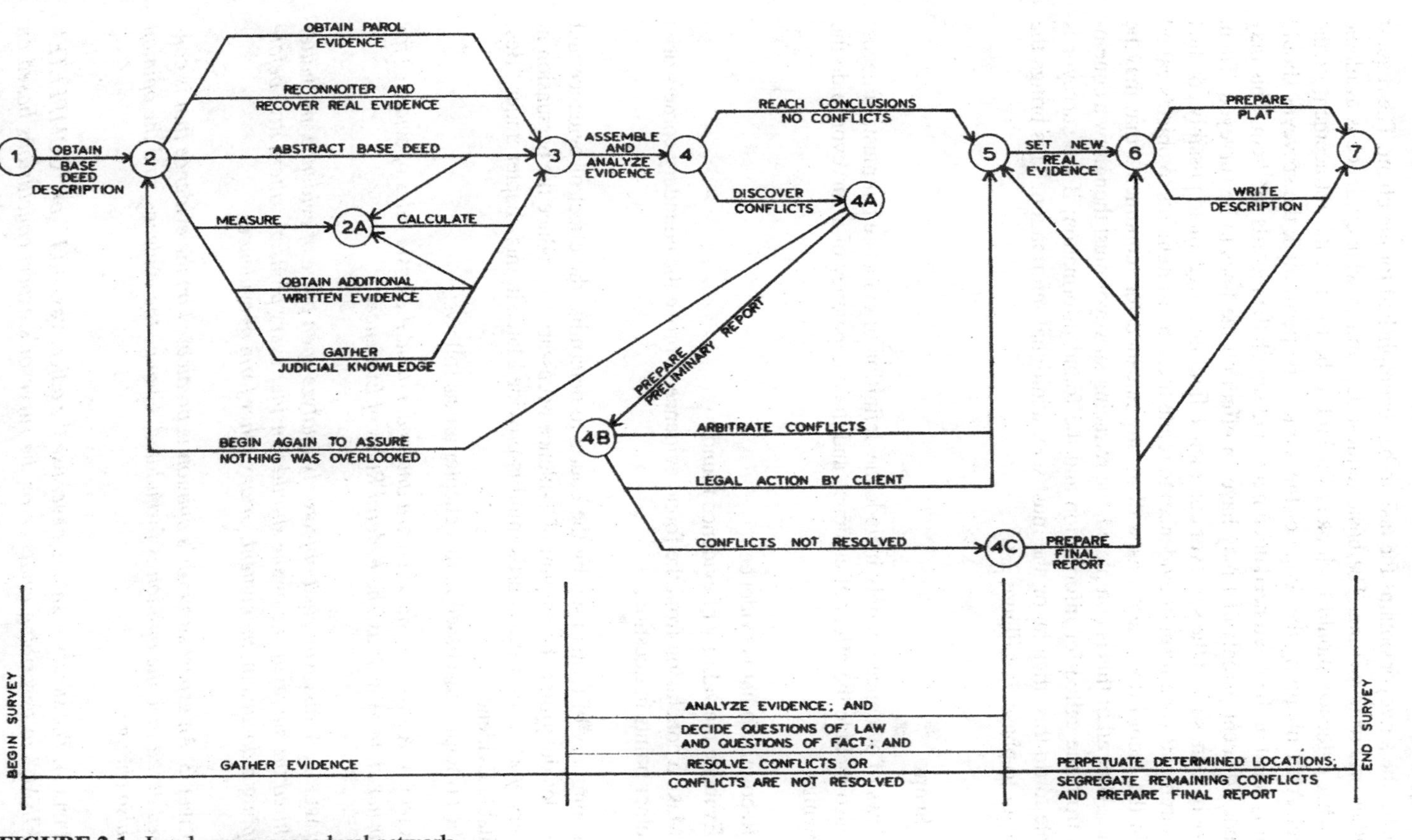

FIGURE 2-1. Land survey procedural network.
Source: Copyright ©1990 by George F. Butts. Reprinted with permission.

First, the surveyor obtains the task to be accomplished for the client. The request is either oral or in writing. The basic request determines what evidence is available and what evidence should be looked for and to be used in the retracement of the parcel. After obtaining the scope of the work to be performed, the surveyor should identify a list of the referenced documents and search the records for them. This may include adjacent parcels that may have an effect on the lines to be surveyed. Then, and only then, should the surveyor commence field investigation to possibly conduct measurements (one form of evidence) to relate these measurements to objects called for in the record documents, as well as any uncalled-for monuments that may be found. After all of this is completed, the retracing surveyor must then draw a conclusion from the called-for information and the found information; if the surveyor is unable to do this, then he or she must seek additional information. This brings the surveyor to phase 5 of Figure 2.1.

2.3 Scope

The laws of evidence not only include the definition of evidence but must, of necessity, also include the effect of evidence and the competency of a surveyor to do the following:

- Recognize what is evidence.
- Evaluate the validity of evidence found.
- Make a conclusion from the found evidence to prove the boundary corners or other points in question.

A summary of terms used by the courts to determine the effect of evidence and what rules determine the amount of evidence necessary to produce the persuasion in the surveyor's and/or the court's mind is provided later in this chapter under "Best Available Evidence."

The principles discussed in this chapter are as follows:

PRINCIPLE 1. *A surveyor should learn and understand the basic rules of evidence for his or her state as well as The Federal Rules of Evidence.*

PRINCIPLE 2. *A surveyor must objectively analyze each piece of evidence, no matter who hired the surveyor; he or she should consider every plausible alternative before accepting a document, monument, or any other form of evidence.*

PRINCIPLE 3. *An expert surveyor's opinion is predicated on the evidence that is considered at the time the opinion is formulated. Change the evidence, and the opinion may also change.*

PRINCIPLE 4. *The surveyor should never use the adjectives "ALL" or "COMPLETE" or "FINAL" in describing evidence or the surveyor's expert conclusions based on the evidence. Expert opinions are based on the available evidence. There may be other evidence that was not found by the surveyor or was withheld from the surveyor.*

PRINCIPLE 5. *Every survey of a conveyance must start from evidence that proves the position of at least two consecutive monuments somehow related to the written record.*

PRINCIPLE 6. *Evidence is not proof. Evidence leads to proof. A consideration of sufficient evidence and conclusions to be drawn from evidence, in accordance with the law of evidence, may produce proof.*

PRINCIPLE 7. *The affirmative party or plaintiff in a civil case has the duty of presenting sufficient evidence to convince either the judge or jury of the allegations. He or she has the burden of proof. A defendant has no obligation to present evidence.*

PRINCIPLE 8. *In civil cases that deal with boundary issues, it is not necessary to prove "beyond a reasonable doubt" as in criminal cases; it is only necessary to prove a "preponderance of evidence."*

PRINCIPLE 9. *A person claiming adverse possession, acquiescence, or loss of property rights usually must present clear and convincing evidence.*

PRINCIPLE 10. *A survey may be proved by any evidence of facts that are relevant and material, but this evidence may not be admissible.*

PRINCIPLE 11. *A surveyor is concerned about the relevancy of the evidence and not the admissibility of the evidence.*

PRINCIPLE 12. *A surveyor must be careful to make the distinction between a forensic survey and a utility survey when working with evidence. The evidence may be the same, but the manner in which it is presented and used may be different.*

PRINCIPLE 13. *Surveyors are presumed to know the laws of evidence pertaining to the location of land boundaries described by writings, and they are charged with the responsibility of knowing how to apply the laws of evidence when they locate deed boundaries.*

2.4 Importance and Necessity of Being Impartial

PRINCIPLE 1. *A surveyor should learn and understand the basic rules of evidence for his or her state as well as The Federal Rules of Evidence.*

Whenever one practices a profession, the individuals should be knowledgeable about and understand the "rules of engagement." For the retracing surveyor, an important topic is EVIDENCE. The surveyor creates the original evidence and then describes his or her monuments and accessories in field notes or legal descriptions. Then, many years later, expects some other surveyor to identify and recover this ancient evidence that has been hiding in plain sight for years or decades. It is imperative that the creating surveyor use standard practices in setting and describing the new monuments so that the retracing surveyor knows what to expect when they begin to search the property or the records office for evidence of the property boundary. Following the prescribed and accepted rules may eliminate individual or personal biases that may be present.

PRINCIPLE 2. *A surveyor must objectively analyze each piece of evidence, no matter who hired the surveyor; he or she should consider every plausible alternative before accepting a document, monument, or any other form of evidence.*

Impartiality is a key factor and an absolute for a professional surveying expert. Until the mid-1950s, there were no guidelines as to the uses and presentations of evidence to and in the courts. Congress initiated a study for the federal courts and the resultant document was *The Federal Rules of Evidence* that became applicable to all federal courts. The respective states started either adopting these rules *in toto* or modifying them to suit local desires and needs, so that today all of the federal and state courts have rules of evidence that judges, attorneys, and others must understand, recognize, and use. These rules cover all aspects, from defining evidence, to who can be a witness, to identifying the requirements for expert witnesses.

As a professional who conducts a retracement, the surveyor should remain independent of any disputes or problems. A surveyor who offers an opinion as to evidence becomes "biased" and may be proved as being "opinionated" as to that particular survey. To maintain the aura of being an independent third party and of respectability, the surveyor should leave no stone unturned. Several of the definitions given in Section 2.3 stressed the importance of having evidence on all sides of the question or issue.

The impartiality of the land surveyor in finding and evaluating evidence is key because the land surveyor will absolutely be scrutinized on cross-examination during deposition or trial. During the cross-examination, the opposing attorney's goal is show that the land surveyor (1) fell below the surveying standard of care and/or did not follow standard surveying practice or (2) ignored key evidence in order to purposefully help his or her client win the case. In either instance, the surveyor's testimony may be seriously impaired and cease to be effective if it is determined that both sides of the question have not been explored, considered, evaluated, and analyzed.

Thus, any surveyor who is asked to perform a survey for a boundary problem should always consider the evidence that supports the location of the boundary corner, the evidence that supports other possible corner locations, and then be ready to explain why the preponderance of the evidence supports why his or her opinion is correct.

No surveyor should become an advocate on behalf of any client. The surveyor should approach the evaluation and analysis of evidence in a wholly impartial manner, and, as such, the surveyor should consider all of the positive evidence for the position and all of the negative evidence. The surveyor should then form the expert's professional opinion based on the totality of the evidence collected and analyzed.

> Professional surveyors should maintain an impartial position in any boundary issue.

PRINCIPLE 3. *An expert surveyor's opinion is predicated on the evidence that is available at the time the opinion is formulated. Change the evidence, and the opinion may also change.*

A surveyor's opinion is based on the evidence that was originally created, then described and subsequently recovered and evaluated. Since evidence is a major factor in helping the surveyor to formulate the opinion, if and when new evidence is found, or the integrity of existing evidence becomes violated, then the expert's opinion may also be affected. Surveyors should be very careful in making premature opinions based on partial, incomplete, or faulty evidence before sufficient evidence is recovered and then evaluated.

A surveyor should never become 'boxed in" to any opinion. Before an opinion is rendered, sufficient evidence should be recovered and evaluated. Once the opinion is formulated, if additional evidence is recovered and presented, then the surveyor may render a new opinion based on the new evidence.

Before any field survey begins, the professional land surveyor (PLS) must conduct an exhaustive review of the publicly recorded documents that describe or convey the property in question. Then, the surveyor must search for physical evidence such as natural and artificial monuments during the field survey. The evidence collected in the county records combines with the field evidence to form an opinion of each property's boundary corner. If the surveyor meets the standards of care for the research and field survey, and new evidence is found by subsequent surveyors, then the PLS may alter his or her boundary opinions to reflect the newly-found evidence without censure.

PRINCIPLE 4. *The surveyor should never use the adjectives "ALL" or "COMPLETE" or "FINAL" in describing evidence or the surveyor's expert conclusions based on the evidence. Expert opinions are based on the available evidence. There may be other evidence that was not found by the surveyor or was withheld from the surveyor.*

When dealing with historic evidence, including documents and field information, or a multitude or complexity of evidence, there can be no absolute certainty that research and field recovery have uncovered all of the evidence. To write the word *all* or state at trial that "I considered all of the evidence" could make for an embarrassing situation when a document is found that was never examined or was overlooked while doing research, or when the opposing expert surveyor produces evidence that conflicts with your evidence.

Some clients, especially title companies, demand that survey plats contain certification statements that are absurd. Examples include statements such as "no encroachments exist" or "this property contains 100.0000 acres of land." These statements, which are impossible to prove, place added liability onto the surveyor and create a false sense of security for the landowners. Instead, the surveyor should only certify to what the evidence shows. An example of a valid statement would be "no *visible* encroachments exist at the time of the field survey" or "this property contains 100.00 acres of land, *more or less.*"

PRINCIPLE 5. *Every survey of a conveyance must start from evidence that proves the position of at least two consecutive monuments somehow related to the written record.*

An indispensable part of every survey that either creates the parcel or lines of the parcel and the subsequent retracement of an existing conveyance is the discovery and the evaluation of the original evidence.

Evidence is *not* proof! Evidence is the commencement of the recovery process from which one must make conclusions, and from these conclusions flows the proof. Proof is the establishment of the sufficient requisite evidence to instill in the mind of the trier of fact or the surveyor as to the facts at issue. It is the *totality of all of the evidence* that persuades the trier of fact (the surveyor or the jury) as to the truth of the location of the parcel being retraced.

The surveyor's use of evidence is the means or methodology that is used by the expert surveyor to convince the trier of fact or the jury that certain events took place that created the boundaries of the parcels being retraced.

Consider the question, "Is this an original corner?" Does the recovered evidence, as it relates to the documented evidence that describes that particular corner or parcel correlate to the found evidence on the ground? If so, then the evidence creates the following in the mind of the surveyor: "I have found the original point of the corner." The found evidence proves the original corner location.

In the process of performing a complete survey of any parcel of land, whether it is considered an original survey or a retracement, certain steps should be followed, usually in this order:

1. A request, preferably in writing, for the survey is obtained, usually from the client or preferably from the title company.
2. Title evidence is researched and obtained, either by an attorney or a surveyor. Usually this is written evidence of title in the form of deeds, abstracts, title policy, or plat.
3. The surveyor should rely on documents and information delivered by the client only in rare instances.
4. Evidence of maps, field notes, county and city records of surveys, state and other public agency records of surveys, and necessary written records that disclose evidence of monument positions, both pro and con, pertaining to the survey is obtained.
5. Adjoiner deeds for evidence of seniority or conflicts are examined or read.

6. Armed with evidence of existing monuments called for in the writings and witnessing evidence of possession and usage, the parcel is inspected.
7. Testimony (evidence) of the existence and location of old monuments and the history of possession is sought and evaluated.
8. Measurements are made from found monuments to determine search areas or locations to dig for missing monuments, and measurements (also evidence) to tie found monuments together are made.
9. Calculations (also a form of evidence) are made to confirm found monuments or to determine the validity of areas or the marking of lost corners.
10. From the evidence of monuments, measurements, testimony, and computations, conclusions in accordance with the laws of evidence are made.
11. Based on these conclusions, measurements are made to set new monuments in accordance with these conclusions.
12. Finally, a report should be prepared for the client (see Figure 2.1 and Appendix A) [10]. The report becomes new evidence.

Principle 5 must be understood by any practicing surveyor, in that a line is usually as terminating at a corner at each end, and may be described by a call for two corners, one at each end of the line, as well as by a bearing and a distance, and according to the dignity of calls, it can only be absolutely and legally fixed by the identification of both monuments. If only one monument is relied on, then the course (bearing and distance) has errors in position because both the bearing and the distance are subject to the normal errors one might expect in creating the line and then in retracing the same line.

If the evidence is changed, then the final conclusions may change.

Until a surveyor obtains sufficient knowledge of the available evidence, it is nearly impossible to make a correct boundary determination or location. Change the evidence, and the location of the line(s) may change, and the resulting opinion and the law may also change. The important aspects of any boundary survey are the ability to search, find, and discover all available evidence and to arrive at conclusions about where boundaries belong in accordance with the laws of evidence and the laws of boundaries. A surveyor may be able to compute, make drawings, use instruments, and stake precise engineering projects, but he or she is not qualified to make boundary locations without understanding the laws of property and boundaries and the laws of evidence.

2.5 Arrangement of Subject Matter

Because of its nature, the subject of evidence is comprehensively treated in the legal literature but is seldom found in technical surveying textbooks. Thus, we must first define terms and elementary rules of evidence that the student must know and understand.

Evidence in itself is not proof of facts; conclusions or inferences that can be drawn from evidence are proof. In defining what inferences or conclusions can be drawn from evidence, the law has evolved rules to aid in evaluating evidence. The law uses such terms as *presumption*, *rebuttable presumption*, *inference*, *burden of proof*, *extrinsic evidence*, *preponderance of evidence*, *clear and convincing evidence*, and evidence that is *beyond a reasonable doubt*. The surveyor should understand these terms as they are legally defined in order to be able to clearly understand the laws and rules of evidence as evidence affects original surveys and the resulting retracements thereof.

Evidence varies in significance, importance, weight, and application. Witnesses may give eyewitness testimony as to the location of a boundary (property) corner. But such evidence is incompetent and irrelevant to overcome the location of an original, undisturbed monument called for and proven by original evidence or even circumstantial evidence at a location other than that testified to. The evidence of a rock mound is of little importance where the original notes called for an oak tree. The evidence of a measurement is incompetent to prove an original monument to be in error. The retracing surveyor should not entirely discredit or eliminate measurements as they may be quite useful in supporting evidence of a lesser degree.

In reaching conclusions from evidence, the most important attribute a surveyor needs is the ability to recognize and know what is the best evidence of that available. Yet the term *best evidence* has an entirely different concept to an attorney or lawyer. After defining terms and after presenting legal concepts that determine the effect of evidence, the remainder of this chapter is arranged, in descending order of importance of evidence, that is, in the order of the best available evidence.

Many misconceptions as to the proper methods to locate boundary lines arise from the misconceptions of the rules of evidence concerning what is the best available evidence. Most surveyors understand that a monument called for in documents in either the survey chain of history or the chain of title of the parcel and then found and if undisturbed is given preference over calls for measurements; but it must be distinctly understood that, to have primary or controlling significance, the monument must be called for by the written evidence, either directly or by implication. There may be instances that an uncalled-for monument may control, but the control of this monument may be for other reasons.

Few surveyors realize that failure to know the law of evidence may result in disciplinary actions, claims of malpractice, and other claims of negligence.

The subject matter is arranged into four objectives:

1. Understanding the importance of presumptions when it comes to evidence.
2. Defining what is acceptable evidence.
3. Stressing the order of importance of evidence—that is, what is the best available evidence.
4. Stating the boundary surveyor's obligations with respect to evidence.

2.6 Kinds of Evidence

In Section 2.1, evidence was defined and identified by various legal scholars. Courts and legal scholars recognize at least five kinds of evidence which the surveyor should become familiar with:

1. Written evidence is evidence in the form of documents.
2. Real evidence consists of material objects addressed directly to the senses, including physical monuments.
3. Oral evidence or testimony is evidence given by witnesses.
4. Judicial notice is evidence in the form of knowledge. The courts may take judicial notice of certain facts such as (a) the true significance and meaning of all English words and phrases, (b) whatever is established by law, (c) the laws of nature, and (d) other well-known and commonly accepted facts.
5. Circumstantial evidence may be the most important of all of the forms or types of evidence when it relates to ancient boundaries.

2.7 Evidence, Conclusions, and Proof

PRINCIPLE 6. *Evidence is not proof. Evidence leads to proof. A consideration of sufficient evidence and conclusions to be drawn from evidence, in accordance with the law of evidence, may produce proof.*

Evidence is not proof of a fact; a conclusion or inference that may be drawn from evidence is the proof. A written deed is evidence of ownership; it is not proof of ownership. Land can be gained by unwritten means; hence, a paper title does not prove ownership. It is evidence only of the claim of ownership and the right of possession. A written deed may be void because of state statutes defining limitations or prior court decisions. If a person can prove by evidence that he or she has a written deed vesting title, that the person conveying the land was competent to do so, that no one adversely occupies the land described by the deed, that title has not passed by escheat, and that other items making up ownership do not operate against that person, he or she may have proof of ownership.

Proof may be considered the establishment of that necessary degree of belief in the *mind* of the trier of fact as to the facts at issue. It is the culmination or the *totality* of the evidence that persuades the trier of fact. With the question of proof comes the *burden of proof* that must be carried forth. In most litigation, that burden is placed on the plaintiff to prove his or her case. In this, the person on whom the burden is placed must rely on all forms of evidence, usually circumstantial and testimonial, to persuade the trier of fact that certain events occurred. The burden of proof is separated into two parts:

1. The obligation to produce.
2. Going forward with the evidence to prove the facts.

Raw recovered evidence is neutral; that is, until the surveyor draws conclusions from the evidence that is recovered, the evidence is neutral in weight and application. Each surveyor applies the evidence in light of his or her training, education, and experience, and usually no two surveyors see evidence in the same light.

Surveyors will use a legal description of property and mark it on the ground; they do not and should not, unless specifically requested by legal counsel, consider validity of signatures, possible insanity, whether by escheat, competency of the seller, or like considerations. They do consider senior rights and note possession not in agreement with the written deed.

It is in this area of evidence that clients, attorneys, and perhaps some judges cannot understand how (and why) two surveyors who use and analyze the same evidence can draw different conclusions, yet once in a while there are rays of hope, one of which was addressed in a federal court:

> It is a well-known fact that surveyors are apt to differ from each other, and surveyors employed by the United States government are not immune from the frailties of their profession . . . [11].

A second court wrote:

> It is a matter of common knowledge that surveys made by different surveyors seldom, if ever, completely agree. And that, more than likely, the greater number of surveys, the greater number of differences [12].

In a recent article in the *American Bar Association Journal*, Williams and Onsrud wrote:

> Surveyors occasionally disagree on the proper location of a boundary line—not necessarily because each surveyor measures better than the other, but more commonly because each surveyor has weighed the evidence differently and has formed different opinions. Just as two lawyers may draw different conclusions from the same line of cases, surveyors may disagree about the appropriate location for a boundary [13].

2.8 Classifications of Evidence

Evidence varies in weight and dignity. In general, evidence may be divided into 10 classifications:

1. **Indispensable evidence** is evidence that is necessary to prove a fact. Conveyance of property must be in writing; hence, a conveyance cannot be proved without proof that there was a written document.
2. **Conclusive evidence** is that which the law does not permit to be contradicted. As an example, the contents of conveyance writings (recital of a consideration excluded) are conclusive as between the parties, except for pleadings if illegality, fraud, mistake, or reformation. Also, the written document, or contract, cannot be altered by oral testimony, and, as is commonly and frequently stated, everyone is presumed to know the law.

3. **Prima facie** is that which suffices for proof of a fact until rebutted by other evidence. In the event that an original deed cannot be produced, a recorded deed is prima facie evidence of the contents of the original. In many areas, the results of the survey of certain official surveyors, such as the county surveyor, are prima facie evidence of the location of lines. Prima facie evidence may be disproved, but until it has been proved incorrect, it is assumed to be correct. The law specifies what is or is not prima facie evidence.
4. **Primary evidence** is that which is most certain. The contents of a written document are more certain than the oral testimony of what the document contained.
5. **Secondary evidence** is inferior to primary evidence. A copy of the original document is inferior to the original. Secondary evidence is used to prove the contents of lost or unavailable primary evidence.
6. **Direct evidence** proves a fact directly without resorting to presumptions or inference; for example, a witness testifies, "I saw the surveyor drive the stake into the ground."
7. **Circumstantial** or **indirect evidence** depends on inferences or presumptions that tend to prove a fact by proving another; for example, the witness testifies, "I saw the surveyor drive similar stakes at other corners."
8. **Partial evidence** establishes some detached fact. At times it is erroneously referred to as corroborative evidence.
9. **Extrinsic evidence** is derived from sources outside the writings.
10. **Corroborative evidence** is supplementary to evidence already given and tending to strengthen or confirm evidence already given. It is additional evidence of a different character. Corroborative evidence, if used, may also act in a negative aspect.

2.9 Types of Evidence Gathered and Considered by Surveyors

Evidence used by surveyors to prove boundary lines or deed line locations can be placed in the following categories. Before any evidence can be used or considered, it first must be tied to written document or documents in the chain of title. These can be any of the following six categories:

1. Written documents, maps, and historical facts directly traceable to the specific parcel.
2. Facts, laws, and documents of which the courts may take judicial notice.
3. Physical objects (real evidence) observed by the surveyor, for example, surveyors' stakes, trees, fences, rivers, and street improvements.
4. Parol evidence, which may be divided into:
   a. Witnesses who observed the former location of physical objects (a monument now destroyed).
   b. Witnesses who can explain a latent ambiguity.

c. Witnesses who can testify about commonly reported facts.
d. Witnesses who can describe the customs or conditions existing as of the date of the deed.

5. Measurements of distances, bearings, and angles that were conducted.
6. Mathematical calculations, including correlations and algorithms.

2.10 Scope

To understand the law of evidence, one must first answer the basic question, "What is evidence?" This first must include a definition of evidence, the effect of evidence, and the competency of evidence. A summary of terms used by the courts to determine the effect of evidence and a summary of those rules that determine the amount of evidence necessary to produce proof provide an introduction to the section on best available evidence.

2.11 The Law of Evidence

DEFINITION. *The law of evidence is a collection of general rules established either by statute law or by case law to accomplish the following:*

1. *Declare what is to be taken as true without proof.*
2. *Declare the presumptions of law, and identify those that may be rebuttable and those that may be irrebuttable as well as that may be defined as both those that are disputable and those that are conclusive.*
3. *Produce legal evidence.*
4. *Exclude whatever is not legal.*
5. *Determine in certain cases the value and effect of evidence.*

The surveyor is involved in evidence in two instances: first, as the individual who creates the original evidence of the original boundary or boundaries and, second, as the individual who is given the task or responsibility of recovering the evidence of the first surveyor. The success of the second surveyor is directly affected by the quality of the evidence returned by the first.

The two surveyors, the creating surveyor and the retracing surveyor, are seldom the same individual. In many instances, decades or centuries may separate the work of these two individuals.

The quality of each surveyor's tasks is directly dependent on his or her capabilities, training, and quality of work performed. On the one hand, if the creating surveyor was lax in creating the boundaries, that is, failed to do what he or she purported to do or failed to adequately describe what was done, then the task of the retracing surveyor is made more difficult because of the difficulty of recovering and identifying the evidence that was initially created and left for history.

On the other hand, if the retracing surveyor is unable to adequately recover the evidence because of lack of training or failure to recognize the original evidence, then the process is seriously flawed, and questionable results will be obtained. In this instance, the creating surveyor's evidence lacks effectiveness if the second surveyor cannot recover it.

The surveyor uses evidence to assist in locating and proving boundaries that identify property locations, and, unless correctly recovered, identified, and finally evaluated, evidence is worthless. The law of evidence—that is, the law that declares what evidence is admissible and what right is to be accorded to admissible evidence—is determined by codes, statutes, and common law. It is the law of evidence that the surveyor relies on to guide him or her in assigning proper values and weight to discovered and known evidence. Not all jurisdictions have the same rules of evidence; hence, variations of the law of evidence can be expected in different states as well as at the federal level.

It has been decided by the courts that the only true and correct location of a written deed is the position that a court of competent jurisdiction decrees to be the correct location. The court bases its decision on admissible evidence and the application of law to the evidence presented first by the plaintiff and in rebuttal by the defendant. Understanding the significance and value of a particular piece of evidence is just as important as understanding the statutory and common laws that pertain to boundary location. The courts sometimes disagree on the value and conclusiveness of evidence. A higher court has been known to reverse the lower court, and, on occasion, two judges with identical circumstances have rendered opposite opinions. In the U.S. Supreme Court, there have been numerous five-to-four decisions. If experts in interpreting the value of evidence are not in complete harmony, it can be expected that on occasion surveyors will not agree. Disagreement in itself is sometimes a desirable thing. Disagreement with existing theories of science has, without doubt, been the cause of new discoveries. Disagreements may bring advancements; but disagreement based on stupidity, lack of knowledge, or being unusually stubborn is undesirable and can be professionally devastating for the surveyors and financially devastating for the clients.

A Florida court addressed the question as who is the most qualified person to take a written description and place the words on the ground, and it adjudicated that a registered surveyor was the *only* qualified person to locate a written description on the ground [14]. (Note: At this time, the student should read this case in Chapter 19, Section 19.7.)

The decision of the Florida court has great insight as to the relationship of the original surveyor, who creates boundaries, and the retracing surveyor, who assumes the responsibility of retracing the originally created boundaries. The appellate court commenced its decision as follows: "Since time immemorial, parcels of land have been identified and described by reference to a series of lines or 'calls' or 'courses' that connect to completely encircle the perimeter or boundaries of a particular parcel" [15].

The court then went on to discuss the place of the original surveyor as well as the retracing surveyor. To the student, it will become obvious that evidence becomes a critical element in the survey of land and land descriptions.

Although it is admitted that sometimes the value of given evidence is debatable, usually evidence is not subject to alternate interpretations. Since the practicing surveyor can expect on occasion that his or her findings of boundary location will be tested in court, he or she should assign values to evidence in the same proportion that it will be accepted in court. For example, it is indeed a difficult task to get an unrecorded map accepted for court evidence unless it is accompanied by the surveyor's testimony or is declared ancient by law and is commonly reputed as being correct, in which case evidence may be admissible as exceptions to the "*hearsay rule*." Surveys based on unrecorded maps of unknown origin may be impossible to substantiate in court.

The value of evidence is relative and is subject to general, not definite, rules. In a borderline decision, what the law of evidence is will not be known until the conclusion of the case. The court then declares what the law is and further takes the attitude that this is the law as it always was, and this is the law as you should have known it to exist all the time. Fortunately, the majority of boundary surveys are not dependent on doubtful legal considerations.

> The burden of proving a boundary is placed on the person claiming the boundary and not on the neighbor.

2.12 Burden of Proof

The term *burden of proof* does not have a fixed and definite meaning and application. On the contrary, at times it is used indiscriminately to signify one or both of two distinct ideas or philosophies. There has been a concentrated effort to bring a clear and uniform understanding of its meaning.

In recent decisions, courts have held that the burden of proof signifies the duty or obligation of establishing a conviction in the minds of the jury or the judge of the ultimate issue or questions being tried. The interpretation of what the term *burden of proof* means has one meaning at law, but may have multiple meanings when applied by the various judges. To the surveyor, it should have only one meaning. In legal circles, the term is defined as follows [16]:

> BURDEN OF PROOF. Lat. *Onus probandi.*
>
> In the law of evidence. The necessity or duty of affirmatively proving a fact or facts in dispute on an issue raised between the parties in a cause.

In describing the term, an Illinois court wrote,

> [I]t is frequently said, however, to have two distinct meanings: (1) the duty of producing evidence as the case progresses; (2) the duty to establish the truth of the claim of the preponderance of the evidence and through the former may pass from party to party, the latter rests upon the party asserting the affirmative of the issue [17].

The term also signifies the duty that the party using the evidence has the burden of "going forward" with the evidence. This duty of going forward, in itself, is not evidence, but a rule of law, that the party who is carrying the burden, the plaintiff, must prove his or her case, and the defendant has no burden to prove anything. The party who has the burden is obligated to prove his or her case, and a failure to do so may be fatal as to the final adjudication.

The law points out two burdens:

1. The burden to produce enough evidence so that a reasonable conclusion can be reached.
2. The burden of persuasion or producing evidence so that the preponderance of the evidence will be in favor of the individual who raises the question.

PRINCIPLE 7. *The affirmative party or plaintiff in a civil case has the duty of presenting sufficient evidence to convince either the judge or jury of the allegations. He or she has the burden of proof. A defendant has no obligation to present evidence.*

The burden of proof, or the need to prove the case, lies with the party who wishes to prove the fact. The individual wishing to prove his or her title has that burden and cannot rely on the weakness of the other person's title. The surveyor seeking to prove a survey is correct cannot rely on the fact that the other party had no survey; however, this burden can shift under certain circumstances. The individual who seeks an affirmative defense or who seeks a counterclaim has the burden shifted.

Although more of a legal problem than a survey problem, the burden of proof has at least three separate legal concepts:

1. The plaintiff is obligated to produce the degree of evidence required to prove the facts that are being relied on.
2. The party is obligated to *introduce* or *go forward* with the evidence.
3. The burden of proof may shift during the course of a trial.

Thus, under concept 1, the moving party must persuade the jury that the facts being presented prove the facts they want to use in order to win the case. This is accomplished through the testimony of the expert surveyor witness.

Making a prima facie case means that the affirmative party must supply sufficient evidence to win the case. Since evidence is for consideration by a jury, it is the total accumulation of the evidence that establishes the degree of certainty in the mind of the surveyor to persuade that certain facts are true.

Simply put, if the moving party does not have sufficient evidence to prove the case, the defendant may move for a directed verdict; but if the defendant cannot produce sufficient evidence to counteract the plaintiff's evidence, the plaintiff may move for a directed verdict. This is a trial tactic that is usually conducted as a pretrial motion, before the start of the trial. Most attorneys use this pretrial motion routinely in the conduct of boundary issues.

At the close of the plaintiff's case in chief and at the request of the defendant, if the plaintiff fails to present sufficient evidence to prove his or her contentions, the court will leave the parties in the position in which it found them and hold for the defendant. In effect, the defendant wins, and the status quo is maintained.

This could result if it is determined that the measurements are incorrect or if the surveyor started from the wrong beginning point; thus, the survey and all testimony fail, and the respondent's case falls with it.

As an example, a boundary in dispute was located on the range lines created in a federal General Land Office (GLO) survey line. The land owned by the state was settled on and homesteaded by John F. Gates in about 1907. He constructed a fence on the east of his land, which was now claimed by the state as the true line.

The true boundaries of the lands could not be determined from the evidence. Gates employed no surveyor to locate his lines, and he testified, in substance, that the fence was built as a fence of convenience, not intended to represent the line to which he made absolute claim, and that he would have changed the fence to the true line if and when it should have become known. The point in dispute is whether the Robert survey and the U.S. government survey approved in 1927 or the Harvey survey is correct.

Harvey, who surveyed the same township, did not run a straight line through the township, only from the southeast corner of the township for a distance of 5 miles, then angled west, and then north, to connect with the northeast corner of the township. The witness stated this was an incorrect method. From the testimony it was clear to the court that the state failed to sustain the burden of showing that the true line between its land and the land of the defendant is that claimed by the state [18].

This situation happens many times when one party fails to properly conduct a retracement survey in accordance with the proper rules of survey. This usually occurs either by shortcutting the survey process or because of money or time restraints. When there is no question of title but only a question of boundary, in relying on a survey, the moving party should first show that its survey was properly conducted in accordance with proper rules or principles of retracements.

In one instance, the plaintiff sued in equity alleging ownership and possession of a small tract of land that was described as follows: "Bounded on the south by the lands of Wilk Witten; on the west by the lands of James LeMaster; on the north by the lands of G. Preston; on the east by the lands of W. W. Brown, and containing 15 acres, more or less." The defendants claimed this to be at the head of a hollow while appellants claimed it to be further down the same hollow. The true determination rested largely on the ascertainment of the boundary lines of an old patent. The court held that the disputed calls were "N 13° E, 10 poles to two maples and a small oak in a dogwood gap; N 23° E, 68 poles to a beech and maple; N 18° W, 80 poles to a beech on the bank of a branch." The court said these were the seventh, eighth, and ninth calls of the patent. A certified copy of the plat of the patent is the same except the call for 10 poles, in that the 10 poles are 100 poles. In addition, a number of witnesses testified as to the location of the dogwood gap. The court stated: "We are unable to determine which is correct, as the abutting lands are not described with any degree of certainty. The burden is on the plaintiff and the doubt should be resolved against him" [19].

Since no party was able to say with certainty the location of the lines in question, the court left them as they were. Ordinarily, the burden of proof is on the plaintiff, but a defendant has the burden of proof in matters of defense. Where the statute divides the burden of proof equally (as in Louisiana) [20], each side has an equal responsibility. The burden of proving surveys, monuments, agreements, acquiescence, or a change of a boundary is on the party asserting that fact.

In another situation, the court had to examine the "survey" of a registered surveyor and draw its own conclusions. Hageman, a civil engineer, presented a plat and stated that he examined the area in question and found the course of the stream was not as shown on the government plat but was as described on a plat that he produced that showed the stream flowing east of south instead of southwesterly into the Quillayute River and that, as a result, lot 3, if extended to the actual stream, would contain a little more than 82 acres of land instead of 39.50 acres. It was shown that Hageman made no actual survey using instruments and that the area indicated the stream was merely a visual survey without any actual measurements to delineate the area, without determining the exact course, sinuosities, and location of the actual stream. He had never even made any boundary surveys, although he was probably professionally competent to do so. The court held that what he did in this matter was merely that of a nonexpert, and it could not consider this evidence competent to prove the fact that it was offered to prove or sufficient to contradict or impeach the official government survey that fixed the character of the entry and determined whether the possession is adverse [21].

In a situation where there has been possession doubt in the location of a parcel (especially for long duration), location evidence is usually resolved to leave the status quo; proof of true lines must be positive.

As an example, three separate surveys were made by three surveyors, and each surveyor admitted he was unable to state whether or not the point from which he started his resurvey corresponded with the corner of the block as it was originally surveyed. The appeals court stated: "We think that the appellant has failed in this court to establish where the true boundary line lies. In case of disputed boundary lines, based on uncertain evidence the courts ought not to disturb boundary lines between lot owners which have been acquiesced in for years" [22].

2.13 Degree (Quantum) of Proof of Evidence

Courts consider the degrees of evidence—that is, they give legal ranking to it—yet seldom does a surveyor rank evidence as to its weight, that is, its dignity. To the surveyor, evidence is evidence. If it helps to prove a point, use it; if it cannot, do not use it. By contrast, courts and lawyers wish to give evidence a ranking as to the degree of proof needed to prove a fact or facts.

Any surveyor must do the amount of work, including research and surveying, needed to make certain that the burden required can be sustained. Anything less may lead to problems when it comes to "proving" that certain facts are true based on the evidence collected.

For routine boundary dispute cases, which are almost always civil, not criminal, the degree of proof required for the plaintiff to win their case is the preponderance of the evidence. In some rare causes, such as adverse possession or prescriptive easements, the degree of proof rises to clear and convincing.

Land tenure cases, including boundaries, easements, title, and water rights, are almost exclusively based on state law. Since these are not matters of federal statutes or federal case law, the plaintiffs must bring their cases to the court in which the property lies. For example, if the properties in dispute are in Sparta, North Carolina, then the proper venue is the District Court of Alleghany County, North Carolina.

If the cases are brought in state court, then the degree of proof required will be set by state statute or state case law. The Federal Rules will not apply. Therefore, it is important that the plaintiffs, attorneys, and land surveying expert witnesses appreciate the degree of proof required for each cause of action before bringing suit against the defendants.

In litigation that involves gaining or losing of interests in land and the modification of boundaries, which include adverse possession, acquiescence, and boundary agreements, the weight is then increased to the middle degree—that is, "clear and convincing" or "clear and positive." A definite and unambiguous definition of this category is not available, but it has been held to be "more than a preponderance but less than beyond a reasonable doubt."

2.14 Preponderance of Evidence

PRINCIPLE 8. *In civil cases that deal with boundary issues, it is not necessary to prove "beyond a reasonable doubt" as in criminal cases; it is only necessary to prove a "preponderance of evidence."*

Preponderance of evidence is the least demanding of the three categories or, according to some courts, "more creditable and convincing." Preponderance of the evidence is not considered by the number of the witnesses, but by the ability of one or more witnesses to "convince" the jury or the judge, and perhaps the opposing party; it is the greater weight of the evidence, not the "total weight" or totality of the witnesses. The surveyor cannot always prove conclusively "beyond a shadow of doubt" that found monuments are positively in their correct position or if they even are the original monuments. If they are to be upheld by the courts, the surveyor must be prepared to prove that the preponderance of evidence is in his or her favor. A second surveyor, in disagreement with the first, must be prepared to prove by a preponderance of evidence that the other is wrong. This *preponderance* criterion is the threshold from which all evidence is related. From this evidence base, the decision is the evidence to prove the facts, then it is sufficient. The more conclusive the evidence, the more certain are the facts.

Often, when the evidence is inconclusive, the jury leans toward the individual who can best communicate the story to them. The ability of the surveyor to "persuade" then becomes important.

Preponderance of evidence is not necessarily just more than 50 percent of the weight of the evidence, or even related or identified by a number, it should be one of convincing what is said to be solid evidence. It is not predicated on the number of witnesses, but what the witnesses have to say and what facts they will testify to. It has been variously defined as that which inclines "an impartial mind as to one side rather than the other" and that which removes "the cause from the realm of speculation." In a survey, the surveyor should be satisfied with the preponderance of evidence, giving due weight to presumptions, prima facie the evidence, and law, that the location is probably correct to the exclusion of other possible methods of monumenting the property.

The surveyor investigates all possibilities, excludes the unlikely or improbable, and monuments in accordance with the most certain.

The law does not require such a degree of proof as, excluding the possibility of error, produces absolute certainty because such proof is rarely possible. Moral certainty only is required or a degree of proof that produces conviction in an unprejudiced mind.

In a Louisiana case, the question to be determined was whether an oil well had been drilled on one side or the other of a line that divided two quarter sections of land in a section and township that had been surveyed and subdivided by the authority of the U.S. government. One might think that it was a simple question, but it was not. The court said:

> No one on earth can furnish the information necessary for its decision, save the gentlemen of the civil engineering and surveying profession, and those of them who have testified in that behalf in this case have *arrayed themselves on opposing sides.* Several surveys were made in order to re-establish the lines and corners originally established by Jones and Moore. Mr. W. E. Martin made his survey in which he located the line 22.2 feet east of the oil well. Mr. H. E. Barnes, on behalf of the defendants, found the line to be 15.7 feet west of the well. A private survey made by Mr. H. A. Jenkins, on behalf of the defendants, located the line 35 feet west of the well. Mr. A. D. Kidder, acting on behalf of the United States government, made a survey for purposes not connected with this litigation, located the line 1.74 feet east of the well. Mr. Welman Brandford, who was employed by the defendants, found the line to be 31.2 feet west of the well. Surveyors Martin and Williams, surveying under direction and instructions of the trial court, reported the line run under these instructions to be 14.7 feet west of the well [emphasis added] [23].

Many other surveyors testified in the case, and the diversity of opinions among them was most bewildering to the judges as well as to the lay minds of the jury and the witnesses. The gentlemen who were employed or consulted in the matter were all reputable members of the civil engineering and surveying professions; their work seems to have been done with care, and their opinions expressed only after deliberate consideration. It therefore seemed to be impossible to determine and to prove with mathematical and absolute certainty precisely where the corners designated on the sketch belong. This situation may likely have been the result of error in the original field notes, as in the resurvey made in connection with this litigation. The court said:

> We shall proceed to establish the limits of the property according to what we consider to be the preponderance of evidence in the case. Of the several surveys made, one stands out most as worthy of consideration by the court. It was not made on behalf of any of the parties in the interest of this litigation, but it was ordered by the United States Government, and it was executed under instructions from the General Land Office, whose stamp of approval has been placed upon it. The engineer under whose personal supervision was done was Mr. Arthur D. Kidder, supervisor of chief of surveys of the General Land Office [24].

It was found that in the field, an accurate north and south line was established to determine with absolute accuracy the variation of the needle. Neither he nor his assistants, though aware of a contest involving the ownership of an oil well, knew, at the time of making the survey, any of the litigants in the case, nor did they know on which side of the line in dispute the said litigants respectively claimed the oil well to be located. Under these circumstances, the recognized ability and competency of Mr. Kidder, the total absence of any possible bias on his part, the great care he exercised in the performance of his work, the most modern and scientific methods adopted by him, and the further fact that the results of his work bear the approval of the General Land Office were, in the court's opinion, sufficient to establish a preponderance of evidence in favor of the plaintiff to justify a decree based upon his findings under the law [25].

Preponderance of the evidence may not mean by weight; rather it usually refers to quality of the evidence, not quantity. Preponderance is the usual proof required in civil cases, in which questions of surveys and boundaries are litigated. As required by the court, this is simply the ability of one witness to induce a persuasion in the mind of the judge or the jury. The evidence may not be conclusive for either party. Preponderance means the weight, credit, and value of the totality of the evidence presented by either or both sides [26]. One could look at preponderance as being "most probable" or most convincing. Preponderance is not predicated on the number of witnesses who testify, but on the credibility of the witnesses.

2.15 Clear and Convincing

PRINCIPLE 9. *A person claiming adverse possession, acquiescence, or loss of property rights usually must present clear and convincing evidence.*

Modern courts require proof of property rights by unwritten means to be "clear and convincing," "clear and positive," "clear and satisfactory," or in some other way "clear."

One court attempted to give a definition, but in doing so only added confusion. It stated that "clear and convincing" was somewhere between preponderance and beyond a reasonable doubt [27].

To the boundary surveyor, this places an added burden or requirement for a solid survey, and the search for and recovery of more evidence of a positive nature than what ordinarily would be required to prove a simple survey. To the surveyor who testifies as to disputes in this area, a critical burden is added in that the surveyor whose testimony is being heard must have an ability to communicate the facts to the court and to the jury.

2.16 Beyond a Reasonable Doubt

Beyond a reasonable doubt is the proof usually required to prove guilt in criminal cases. Yet, this degree of evidence is also the degree of proof that is required by the Bureau of Land Management in differentiating between a lost and obliterated corner. *Reasonable doubt* is a term well understood by the courts but very difficult to define by the surveyor [28]. The difficulty is that each surveyor will apply a different standard to the terminology. What is reasonable to one surveyor may not be reasonable to another.

PRINCIPLE 10. *A survey may be proved by any evidence of facts that are relevant and material, but this evidence may not be admissible.*

The methods of proof can include many different forms of evidence. Judicial notice, judicial admissions, presumptions, and evidence, both direct and circumstantial, all may be considered methods (evidence) for proving facts.

2.17 Relevancy and Materiality

To be of value to the surveyor, all evidence must be relevant and material; but to be of value to the lawyer or to the judge, it must be relevant, material, *and admissible* in the court of law.

Many people consider relevant and material as being the same, but they are not. In trying to differentiate between the two words, relevancy refers to whether the evidence being offered has to deal with the fact(s) in issue. *The Federal Rules of Evidence* Rule 401 defines *relevant evidence* as "evidence having the tendency to make the existence of any fact that is of consequence to the determination of the action more probable or less probable than it would be without the evidence" [29].

Regardless of the definition used, the person seeking to determine if evidence is relevant must ask, "What issue is the evidence relevant to?" What is the attorney or surveyor trying to prove?

Then, considering *materiality*, the individual, when confronted with the second question, must ask, "Is the evidence offered upon a fact properly in issue?" If the surveyor or attorney is trying to prove an issue or a fact, then only those elements that tend to prove the issue are material (Figure 2.2).

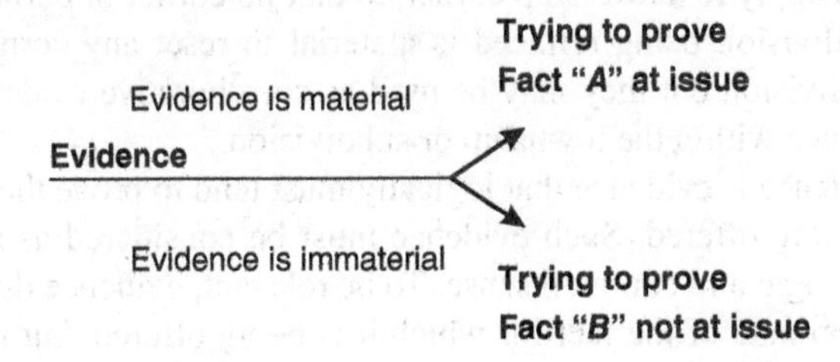

FIGURE 2-2. Material facts.

2.18 Admissibility, Relevancy, and Conclusiveness of Evidence

The purpose of any trial is to determine the truth regarding the issues that are presented by both sides and then make a decision based on the facts. In court, not all evidence is admissible to prove boundary location. Evidence, to be admissible, must be relevant to the issues, competent under established rules of law, and material in the sense of having some reasonable tendency to prove or dispute points in issue and then presented to the trier of fact for its consideration. The decision about what evidence will be admitted lies exclusively with the judge. Parol evidence, with some exceptions, must be based on direct observations and normally cannot be something the witness had heard someone else say. This is addressed in the rules of evidence in the area of hearsay. Lay witnesses generally do not express conclusions from evidence. Thus, there are rules about what evidence should be considered. The admissibility, relevancy, or conclusiveness of evidence is discussed in evaluating the various types of evidence used to ascertain boundaries.

DEFINITION. *Relevancy is a relationship between the evidence being used and the fact, theory, or proposition that the surveyor wishes to prove.*

The Federal Rules of Evidence Rule 401 defines relevancy as any "evidence having any tendency to make the existence of any fact that is in consequence to the determination of the action more probable or less probable than it would be without the evidence" [30]. This definition probably may be stated very simply as in "Does it tend to prove the fact with a great degree of probability?," that is, to a degree better than 75 percent. The surveyor must then ask if the evidence is relevant, and material and what is its probative value to the issues he or she wishes to prove.

Materiality addresses the question of whether the evidence that is offered relates to the question (issue) that is to be proved. If section corners of section 10 are in question, all section corners in that section are relevant, but the corners of the other sections are not material if all of the section corners of section 10 are recovered and proven. Since the law of surveying states that found original corners control and a surveyor cannot go beyond a found original corner, the other section corners only become material if one or more of the section corners of section 10 are considered as lost and the surveyor is required to proportion from the closest original corner. Once the attorneys determine or agree to the issues, the surveyor determines what must be proven; then the question of materiality is addressed.

This may also apply to a township corner, in that no corner or corners outside the township or subdivision being retraced is material to reset any corners within the township or subdivision but they may be used as corroborative evidence to support supporting evidence within the township or subdivision.

Probative evidence is evidence that logically must tend to prove the theory or fact for which it is being offered. Such evidence must be considered as one where the surveyor applies logic and common sense. To be relevant, evidence does not have to be actually determinate of the fact for which it is being offered, but it must tend to prove the fact or be probative [31].

At trial, collateral facts, even if not relevant, material, or probative, usually are admitted. The student must remember the admittance of any and all evidence is in the sole discretion of the trial judge. Such purposes as impeachment of a witness, rebuttal to material evidence introduced by the opposing side, and more important evidence may be used to establish a foundation for relevant evidence later in the trial.

Today lawyers, courts, and surveyors tend to consider both materiality and probative evidence as the same. When dealing with evidence, surveyors should not play lawyer. The surveyor should use or consider evidence for its ability to correlate documents to the evidence found at the time the retracement was conducted.

In an actual trial in which the expert for the defendant was being cross-examined, the plaintiff's attorney asked him a question relative to the evidence he used. The surveyor said, "I refuse to answer that." Then the attorney asked the judge to make the expert answer the question. The judge asked the surveyor why he would not answer the question? He responded, "Your honor, I cannot answer that question, because it is hearsay." After a long pause, the judge pushed his glasses to his forehead and stated, "I will worry about that; you worry about what your answer is, and I will worry about what the law is. So, answer the question." This is an excellent example where the surveyor was trying to play lawyer. Had the surveyor understood about his local rule of evidence, he would have known that the answer was an exception to the hearsay rule.

PRINCIPLE 11. *A surveyor is concerned about the relevancy of the evidence and not the admissibility of the evidence.*

2.19 Admissibility of Evidence

Admissibility is a legal question involving whether the court will permit relevant evidence to be admitted so as to be heard and considered by the jury. The basic rule is "Except as otherwise provided all relevant evidence is admissible" [32].

Whether evidence is admissible at a trial becomes quite important in that only admissible evidence should be considered by a jury in reaching its decision. There are many reasons why evidence may be inadmissible, which will not be discussed here. As evidence is used and then introduced for consideration, the attorney usually says, "Your honor, I would like to introduce this evidence into evidence." The judge then asks the opposing attorney, "Any objections?" Then if objections are made, the judge determines whether evidence is admitted or rejected. (In a recent trial, the day of the trial, the defendant's expert recovered a very important piece of evidence, which the plaintiff did not know of. After objections, the judge refused to admit the evidence, but the important element is, the jury did see it.)

Evidence may be admissible for one purpose but inadmissible for a second purpose. This will cause problems for attorneys because, regardless of limiting instructions, juries always feel that "evidence is evidence."

Some reasons that a court will not admit evidence may be the failure to authenticate its origin, hearsay, questions of authenticity, or its being totally unreliable. The question of admissibility is one for the judge to determine.

DEFINITION. *Presumptions of law are inferences that the law expressly directs to be deduced from certain known facts. Presumptions are either conclusive or rebuttable.*

Presumptions are for the purpose of expediting trials by reducing the amount of evidence necessary, and the time to present the evidence. Technically, presumptions are not evidence but are considered substitutes for evidence. A presumption affects only the burden of offering evidence; thus, it becomes a "procedural tool" based on considerations of (1) probability, (2) practical convenience, or (3) public policy.

In applying such a rule, proof can be offered that a properly stamped and addressed envelope was placed in the mailbox; this will give an inference that it was received; thus, the courts will "presume" it was received in the absence of evidence to the contrary.

- **Conclusive presumptions** are those that are irrebuttable, and this, by law, permits no contradiction. A commonly known conclusive presumption is "everyone is presumed to know the law." Whether a person does or does not in fact know the law is immaterial, since no evidence can upset this conclusive presumption. In the absence of pleadings of fraud or illegality, the truth of the facts recited in a written conveyance is a conclusive presumption as between the parties. But there is an exception in the recital of a consideration.
- **Disputable or rebuttable presumptions** are those that may be proved incorrect by other evidence; but, unless proved otherwise, the jury's findings must be according to the presumptions. Common-law presumptions usually have force in the various states, provided they are not in conflict with statutory presumptions or provisions. One who grants a thing is also presumed to grant whatever is essential to its use. Thus, a conveyance of land includes all existing easements necessary for the use of the land, whether said easements are recited or not. The burden of proof is with the person trying to prove that the easements did not pass with the property. Evidence willfully suppressed is presumed to be adverse if produced.

As applied to a surveying issue, presumptions could be as follows: In the GLO surveys, if the surveyor has a GLO plat accompanied by the field notes indicating courses, lines, corner monuments, and parcels were patented to that plat, then the conclusive presumption is that a survey was conducted according to the laws and instructions in effect at the time of the date of the plat. In conducting a retracement, after extensive field work no original evidence is found. Then the surveyor could say that there is a rebuttable presumption that the survey was never conducted. Then in a similar situation, if in conducting a retracement one finds scattered corners, but the field notes do not correlate with the topography on the ground, or that the field notes and the plat agree, but they do not fit the actual lines, the presumption is the survey is a "tent" survey or a "campfire survey."

A rebuttable presumption is one in which evidence can be presented to prove the presumption is in error.

If inferior evidence is produced, it is presumed that the higher evidence would be adverse. In many states a letter duly directed and mailed is presumed to be received in the regular course of mail. This is known as the *mailbox rule*. Each of these presumptions may be overcome by acceptable contrary evidence.

Many rules of surveyors are equivalent to presumptions. Normally, the order of importance of conflicting deed elements is senior rights, intentions of the parties, monuments, measurements, and area. Of course, this is not a hard-and-fast rule, but it is merely a good disputable presumption to be followed until contrary evidence is developed. The surveyor, in the quest for evidence, sometimes discovers or seeks facts for the purpose of proving a presumption wrong, but until he or she proves the contrary, the presumption governs.

Before a presumption can be assumed, the person who wants to seek the benefit must establish the "basic fact" that is the foundation to the "presumed fact." For the most part, courts hold that a presumption is not evidence but is a deduction that must be drawn from evidence.

In looking at evidence, courts have held that *presumptions are evidence or have the effect of evidence*. In using presumptions, the courts and even the parties may find themselves facing legal dilemmas when both parties invoke presumptions on the same issue.

2.20 Inferences as Substitute for Evidence

DEFINITION. *An inference is evidence in the form of a logical conclusion from a set of facts without express directions of the law to that effect.*

Surveyors frequently resort to inferences to prove a given property location. Inferences are not based on imagination or supposition; they are based on probabilities, and the drawing of inferences is a matter of discretion left up to the trier of fact. A failure to speak and explain when it is a duty to do so gives rise to an adverse inference. The negative testimony of a person in a position to see a monument in a particular location supports the inference that the monument was not there.

2.21 Extrinsic Evidence: When Used

Once an agreement or deed is reduced to writing, testimony cannot be used to overcome clear, unambiguous, written words. But if the words are not clear and need explanation, extrinsic evidence, that is, evidence other than the writing itself, may be sought to explain the words. A deed reading "beginning at the southeast corner of Jones' watermelon patch" may require a considerable amount of extrinsic evidence to explain where the southeast corner was located at the time the document was written.

Extrinsic evidence may also be taken to explain a local meaning to a particular word. In Texas and the West, the old Mexican *vara*, a unit of measurement, was used. Extrinsic evidence is sometimes needed to explain the length of a vara, since it was not the same in all localities.

At times, courts have generally taken extrinsic evidence to explain latent but not patent ambiguities in descriptions. A patent ambiguity is a defect appearing on the face of the conveyance itself. A deed reading "a house and lot on Main Street" describes nothing in particular and contains a patent ambiguity that cannot be remedied by other evidence. A latent defect—that is, a defect not apparent on the face of the instrument but apparent when the instrument is applied to matters outside the instrument, can be cured by extrinsic evidence. If an ambiguity has been raised by extrinsic evidence, it is only logical that the courts would allow the same kind of evidence to explain it.

The rigidity of the original rule defining these two types has been greatly relaxed. The current tendency is to regard a description as valid rather than void and to extend the word *latent* to its most liberal meaning. A patent ambiguity is construed to exist only when persons of competent skills are unable to interpret the deed. In general, if a competent surveyor can take the deed and locate the land on the ground from the description contained therein, with or without the aid of extrinsic evidence, the description will be held to be sufficient [33].

The general rule of law is that possession must start under a claim of color of title that purports to be valid. If the deed under which the claim is made is void or insufficiently formed to pass title, the possession is not adverse under our statutes. The description reads, "One tract of land lying and being in the county of the foresaid, adjoining the land of John J. Phillips and Pender, containing 20 acres more or less." This description fails to identify or furnish a means of identifying. It gives neither course nor distance of a single line, nor a single point, stake, or corner anywhere to begin at [34].

Where land is conveyed by a general description, extrinsic evidence is admissible to ascertain the location of adjoining tracts called for, so as to apply the conveyance to its proper subject matter. If with the aid of these the land granted can be sufficiently identified, that is all that is necessary [35].

2.22 Judging Effect or Value of Evidence

The jury judges the effect or value of evidence, but the judging must be in accordance with the laws of evidence. The jury, subject to the control of the court, is the "judges" of the effect or value of evidence addressed to them, except when it is declared to be conclusive. They are, however, to be instructed by the court on all proper occasions of the following:

1. Their power of judging of the effect of evidence is not arbitrary, but one exercised with legal discretion, and in subordination to the rules of evidence.
2. They are not bound to decide in conformity with the declarations of any number of witnesses, which do not produce conviction in their minds, against a lesser number or against a prescription or other evidence satisfying their minds.
3. A witness false in one part of his or her testimony is to be distrusted in other parts of testimony.

4. Testimony of an accomplice ought to be viewed with distrust and the evidence of the oral admissions of a party with caution.
5. In civil cases, the affirmative of the issue must be proved, and when the evidence is contradictory, the decision must be made according to the preponderance of evidence; that in criminal cases guilt must be established beyond reasonable doubt.
6. Evidence is to be estimated not only by its own intrinsic weight, but also according to the evidence that it is in the power of one side to produce and of the other to contradict.
7. If weaker and less satisfactory evidence is offered, when it appears that stronger and more satisfactory evidence was within the power of the party, then the evidence offered should be viewed with distrust.

2.23 Judicial Notice as a Substitute for Evidence

Although not evidence, courts and attorneys may accept judicial notice as a substitute for evidence. Judicial notice does not affect the surveyor directly, but it may affect the manner in which surveys are approached.

Judicial notice is usually used in situations for evidence of absolute certainty to the point that no trial evidence needs to be presented to confirm that it is certain. (On one occasion this author took judicial notice in an action for flooding damages that "water runs downhill.") Because of the nature of this form of evidence, the requirements and rules are somewhat vague and are left up to the discretion of the judge.

Usually, there are two requirements for the judge to take judicial notice of facts or laws, which are: (1) they are generally known within the territorial limits or jurisdiction of the court, or (2) they are capable of accurate and ready determination by sources whose accuracy cannot reasonably be questioned.

One of the very basic presumptions that any litigant assumes at trial is that the court and the jury are totally uninformed about the facts of the case [36]. The duty then falls on the attorney to present the case to the jury and the judge through the presentation of the evidence. Judicial notice states that there are certain facts that do not need to be proved, and thus, the court and the jury can rely on them. Although a substitute for evidence, judicial notice has equal force to evidence.

The entire presumption of judicial notice is that these are facts (evidence) that are so common that each person of average intelligence will (should) be knowledgeable about them.

The student should not get the terms *judicial notice* and *judicial knowledge* confused. They are entirely different. The source of judicial notice may be from textbooks, laws, dictionaries, and encyclopedias that are commonly seen as being accepted. The acceptance of judicial knowledge is not automatic, in that the court has the authority to determine what it will judicially notice.

All courts will take judicial notice of law, including the federal laws of surveying. Also rules and regulations may be judicially noticed, which for the surveyor may be the BLM Manuals and other publications.

Courts will also take judicial notice of scientific facts and mathematical calculations and facts. This could be extended to the measurements that are obtained from electronic distance-measuring equipment.

Judicial notice is a formal part of all trials and, as such, will not permit any judge or jury to substitute personal knowledge for their own. The basic rule that must be remembered is that judicial notice may be taken of any fact of "common, everyday knowledge" that is accepted as being "indisputable" by any person of average intelligence and knowledge [37], but interestingly enough, courts have permitted judicial notice being taken of both judicial knowledge and common knowledge possessed by the average individual that judges assume every informed individual possesses.

In a "whiskered" decision from Alabama, the early Appellate Court wrote, in respect to judicial notice; "Courts are authorized to take Judicial Notice of the United States [*sic*], in this State government surveys in this State, and of the location and the relative situation of the lands officially surveyed and mapped out under the authority and laws enacted by Congress" [38].

When a surveyor is employed as an expert, if certain information is of common knowledge to the surveying community in general, the expert should recommend to the attorney that the attorney seek to have that information accepted as judicial knowledge, and then he or she is relieved from having to plead the facts during the course of the trial.

2.24 Weight of New and Old Evidence

In a boundary dispute being settled in court, each party presents evidence that each hopes tends to prove his or her side of the case. After all evidence is presented, the judge decrees a property position as based on law, facts, and evidence. After a period of time, if the decree is not reversed by a higher court, the decree becomes final and the law then becomes controlling for that court.

The judge has the advantage: He or she draws a conclusion from the evidence and facts presented and from knowledge of law. After a decree is final, discovery of new evidence does not alter the property or boundary line location decreed. Because of this situation, it makes it imperative for the testifying surveyor to have sufficient and creditable evidence on which to make the expert's decision.

The surveyor, when making a boundary survey, gathers evidence that was created and left by the original surveyor. He or she then makes findings and draws conclusions and finally develops an opinion and the sets the survey or boundary markers. But the surveyor's findings do not have the finality of a court decree. Another surveyor, uncovering additional evidence, may come to a different conclusion and locate the lines in another position. The second position might be the correct location. Yet another surveyor might be more diligent and uncover further evidence that proves still another location to be proper. Thus, it is of the utmost importance that a surveyor seeks and finds all the evidence in the first instance, irrespective of costs. Many so-called lost monuments have been found on later surveys. In fact, *one of the major causes of disagreement between surveyors relates to the lack of discovery of sufficient and creditable evidence* at the time of the initial survey. If every surveyor uncovered all the evidence, differences would be reduced to a minimum, and the surveys would have a finality of location.

2.25 Scope

After boundaries are created and lines and corners described, and then a conveyance made, in a subsequent retracement the location of the land is determined by conclusions drawn from the best available discovered evidence, all in accordance with the law of evidence. The surveyor's area of necessary knowledge includes what is acceptable evidence, what conclusions can be drawn from evidence, and what evidence must be rejected. The law of evidence, in effect, assigns an order of importance to evidence. If two pieces of evidence are in conflict, conclusions are drawn in accordance with the evidence's importance as stressed in the law of evidence. The written title to land can be partially (rarely, fully) extinguished by the acts and behavior of adjoiners, usually with wrongful possession for a period of time, which must be in accordance with the statutes or common-law principles. This chapter pertains to written title rights. The transfer of title by acts of possession and behavior of adjoiners, for informational and not legal purposes, is treated in Chapter 3. Normally, surveyors locate land in accordance with a written description, but it is important that the surveyor understand the significance of legal as well as unauthorized possession rights (usually called *unwritten title*) for reasons of reporting as well as liability, and will be discussed in a later chapter. Occasionally, portions of land are sold to two parties; the first buyer (sometimes the first person to record) has the right of ownership. Consideration of this subject is included in this chapter.

PRINCIPLE 12. *A surveyor must be careful to make the distinction between a forensic survey and a utility survey when working with evidence. The evidence may be the same, but the manner in which it is presented may be different.*

Little is written as to the professional relationship of the surveyor with his and her client and the attorney. A surveyor is usually called on to perform his or her duties in one or more of the following areas:

1. To create surveys for the client.
2. To explain prior work performed for a client.
3. To defend prior work performed for a client.
4. To create new work for a client, after having conducted prior work.
5. To explain prior work performed by a second surveyor in contemplation of litigation.
6. To conduct an independent survey or surveys in anticipation of litigation, usually for a third party, who is often an attorney.

Items 1, 2, and 3 should be considered as utility surveys, in that these surveys were performed as routine work and only for the purpose specified to the client.

In many instances, the surveyor or the firm is called on usually several years after the survey was initially completed, and in many instances many, if not all, of the individuals who performed the original survey are dead or not available. This type of survey may be identified, for want of a better word, as a *utility survey*.

If, by contrast, litigation is anticipated relative to a prior survey, that is, a utility survey, where the complaining party wishes to determine the validity of a specific prior utility survey, and the attorney or the landowner engages the services of a surveyor who is an expert in that specific area, then that survey may be considered a *forensic survey*. For practical purposes, a forensic survey is conducted to either give validity to a prior utility survey or conduct an independent survey from which the expert will formulate opinions in anticipation of litigation.

In both instances, the evidence on which opinions are formulated may be the same, but to whom the opinion is directed and for what purpose the evidence and opinion are to be used differ.

In a utility survey, the surveyor may be ethically and contractually bound to the client for whom the work was originally performed, while in a forensic survey the work and the opinion are the sole product of the person who engaged the surveyor, in that the surveyor may be ethically and morally bound to the landowner but contractually bound to the attorney (such as when litigation is involved).

2.26 Understanding the Law as It Relates to Boundaries and Evidence

The surveyor's responsibility or job is not taking verbal evidence to explain the law but to apply the law. There are two areas about which the aspiring student surveyor as well as the experienced registered surveyor should be knowledgeable: the area of *boundaries* and the area of *evidence*. Each is separate and distinct.

The law is a matter of judicial notice and is gathered from statutes and writings of learned people. As already stated, it is an irrebuttable presumption that everyone knows the law, and this may not be overcome by contrary evidence. If the surveyor agrees to monument a certain written conveyance on the ground, he or she also agrees to locate the conveyance in accordance with the laws regulating the interpretations of written conveyances. The surveyor is a professional specialist who knows how to read and interpret deed words and how to set monuments in accordance with the words. Understanding and applying the correct law (including the laws of evidence) are unquestionably part of a surveyor's duties, and responsibilities. Failure to understand and properly apply these requirements may result in claims of negligence or malpractice.

Most law is written, but there also exists unwritten law. It is not the opinion of a lay witness. When performing a boundary survey, it is the prerogative and duty of a surveyor to interpret the meaning of the written words of a conveyance and then to apply his or her interpretation of these words to the evidence that was originally created and then recovered in a retracement. It is this fact that elevates a surveyor above the layman and promotes professional stature in the eyes of the public.

2.27 Duties of Surveyors in Finding Evidence

The parcel descriptions along with the descriptions of the boundaries found in deeds that are valid when formed are not made invalid because of a loss of evidence of location; the surveyor must find the *best available evidence* that determines the location of boundaries and elements for the parcel of land described in the deed on the ground. In those areas in which there has been widespread obliteration and loss of evidence, it may become necessary to accept evidence of an inferior type, such as hearsay and reputation, but whatever is accepted, it must be the best of that found after an extensive and complete search of the record, the ground, and adjoiners is completed.

The retracing surveyor should not confuse the term *best available evidence* with the meaning the attorney places on it. For the attorney, it has one meaning (legal) and, for the surveyor, it has a second meaning (survey and evidence).

The fact that a valid written deed has a location on the ground is not to be disputed; the surveyor is charged with finding the ground location in conformity with the law of evidence.

Ownership of land can be obtained by either of two methods:

1. A lawful written title.
2. A unwritten possession rights (called an unwritten title), as discussed in Chapter 13.

An unwritten right to land ownership can be transformed into a written right by either a court decree or a written agreement between adjoiners. Lawyers may classify written titles as marketable and unwritten titles as nonmarketable. These questions are not in the field of the surveyor. Lending agencies will not loan money on the basis of an unwritten title. In other words, land claimed by some unwritten right to title may have a lower market value than land claimed by a written title and possession. For this reason, surveyors may be asked to distinguish between each type of ownership right. These are legal questions, not survey questions.

It is not the surveyor's responsibility to determine admissibility. It is the surveyor's responsibility to determine relevancy and credibility.

Evidence is an essential part of every boundary line survey. For a property surveyor to gain prominence and remain free of liability for negligence, he or she must have an intimate acquaintance with the laws of boundaries and the laws of evidence to prove boundaries.

Surveyors rarely fail to come to an agreement about the distance existing between two known monuments. Disagreements more often result from differences of opinion over which of these two known monuments should be used if multiple ones are found to be present. Perhaps the worst disagreements arise from a failure of one surveyor to uncover all available evidence. Two surveyors having the same evidence, if equally educated and equally intelligent, should come to the same conclusions from the evidence collected and used. Unfortunately, all surveyors are not equally diligent in their search. The one with sufficient and credible evidence usually comes to the correct conclusion, whereas the one with partial evidence makes faulty locations.

PRINCIPLE 13. *Surveyors are presumed to know the laws of evidence pertaining to the location of land boundaries described by writings, and they are charged with the responsibility of knowing how to apply the laws of evidence when they locate deed boundaries.*

It is the responsibility of the surveyor to obtain all sufficient and credible evidence and then make a correct boundary line determination or location in accordance with controlling writings. Failure to find all the necessary evidence to make a correct location is not an excuse for an incorrect survey. The surveyor should correctly locate the written title lines and also report unwritten title rights while recovering, interpreting, and reporting this evidence.

Notes

1. This includes facts and arguments.
2. This is a very limited definition.

REFERENCES

1. *Rules of Evidence for U.S. District Courts and Magistrates*. (St. Paul, MN: West Publishing Co., 1979).
2. *Spainhour v. Huffman*, Supreme Court of Virginia 237 Va. 340; 377 S.E.2d 615; 1989 Va. LEXIS 49; 5 VLR 1884, March 3, 1989.
3. Burr W. *Jones, Jones on Evidence*, 6th ed. (Rochester, NY, Lawyers Co-operative Publishing Co., 1972).
4. Jeremy Bentham, *A Treatise on Judicial Evidence*, p. 17. See https://oll.libertyfund.org/title/bowring-the-works-of-jeremy-bentham-vol-6
5. Blackstone, 1970 (3, 367); *Lynch v. Rosenberger*, 249 P. 682 (1926, Ks).
6. Simon Greenleaf, A *Treatise on the Law of Evidence* (New York: Little, Brown, 1899), Sec. 1.
7. J. Thayer, "Presumptions and the Law of Evidence," *Harvard Law Review*, vol. 3 (1889), pp. 142, 147.
8. *Wigmore on Evidence*, 3rd ed. (Wolters Kluwer), Sec. 1.
9. California Evidence Code, sec. 140.
10. Some states, such as Indiana, require a written report as part of their minimum standards.
11. *U.S. v. State Investment Co.*, 246 U.S. 206, cited in *Hagerman et al. v. Thompson et al.*, 235 P. 2d 750 (Wy. 1951).
12. *Erickson v. Turnquiat*, 77 N.W. 2d 740 (Minn. 1956).
13. M. Williams and H. Onsrud, "What Every Lawyer Should Know About Title Surveys," reprinted in *Land Surveys: A Guide for Lawyers and Other Professionals*, 2nd ed. (Chicago, IL: American Bar Association, 2000).
14. *Rivers v. Lozeau*, 539 So. 2d 1147 (Fl. 1989).
15. Ibid.

16. *Spainhour v. Huffman.*
17. Burr W. Jones, *Jones on Evidence: Civil and Criminal*, 7th ed. See https://elibrary.law.psu.edu/fac_books/24/
18. *State v. Vanderpool*, 19 P. 2d 955 (Wy. 1933).
19. *Green v. Whitten*, 200 Ky. 725, 255 S.W. 519 (1923).
20. *Russell v. Producers Oil Co.*, 143 La. 217, 78 So. 473 (1912).
21. *Rue v. Oregon Washington Ry. Co.*, 109 Wash. 436, 108 P. 1074 (1920).
22. *Westgate v. Ohlmacher*, 251 ILL 538 (1911).
23. *Russell v. Producer's Oil*, 143 La. 317, 78 So. 473 (1912).
24. Ibid.
25. Ibid.
26. *Willcox v. Hines*, 100 Tenn. 534, 455 S.W. 781 (1897).
27. *Re. Estate of Fife, 164 Ohio St.* 449, 132 N.E. 2d 185 (1956).
28. *Hiller v. State*, 116 Neb. 582, 218 N.W. 386 (1928).
29. *Federal Rules of Evidence* (St. Paul, MN: West Publishing, 1987), p. 23.
30. Ibid., p. 23.
31. *U.S. v. Medera*, 574 F. 2d 1320 (1978).
32. *Federal Rules of Evidence*, Rule 402.
33. *Hume v. MacGregor*, 148 P. 2d 656 (Calif. 1944).
34. *Dickens v. Barnes*, 79 N.C. 490 (1878).
35. *The Sulphur Mines Co. of Va. v. Thompson Heirs*, 93 Va. 293, 25 S.E. 232 (1906).
36. *U.S. v. Wilson*, 7 Pet. 150, 8 La. Ed. 640. (1830).
37. *Varco v. Lee*, 180 Cal. 338 (1919).
38. *Knabe v. Burden*, 88 Ala. 436, 7 So. 92, (1889).

ADDITIONAL REFERENCES

American Jurisprudence, vols. 29, 30, Evidence. Rochester, NY: Lawyer's Co-Operative Publishing, 1989.

Delisle, R. J. Evidence*: Principles and Problems*, 5th ed. Toronto: Carswell, 1999.

Kaplin, J., and Waltz, J. *Evidence: Gilbert Law Summaries.* New York: Harcourt Brace Jovanovich, 1982; distributed by Law Distributers, Gardena, CA.

Rothstein, P. *Evidence in a Nutshell.* St. Paul, MN: West Publishing, 1970.

Siemer, D. *Tangible Evidence*. New York: Harcourt Brace Jovanovich, 1984.

Wright, E. T. Evidence*: How and When to Use the Rules to Win Cases*. Englewood Cliffs, NJ: Prentice Hall, 1990.

3

WORDS AS EVIDENCE

3.1 Value of Written and Spoken Words

This chapter is devoted to words as evidence [1], both written and spoken. Words are used in field notes, on plats and maps, in survey reports, and in legal documents. Words are used by surveyors, attorneys, judges, and clients to create, describe, expand, restrict, explain, and confuse. Words are translated from one language to another—and at times, the ideas are lost. To the surveyor, words are important, not only in creating boundaries but in retracing boundaries as well. It is the surveyor who usually first creates boundary lines, and then leaves evidence, both in the form of his or her "footsteps on the ground" as well as the documentary evidence in the form of field notes, reports, descriptions, and maps.

Many times, evidence is also created by attorneys or landowners in the form of descriptions contained in deeds, and the retracing surveyor may find that the words subsequently created by others may conflict with the words created by the surveyor. Words may be the only directions that the surveyor is asked to follow or the only means the original historical surveyor may have to communicate the locations of the boundary corners to a modern surveyor. Words are not just important, they are critical, both at the time they are written and at some later time when the boundaries are being retraced.

The professional land surveyor is no doubt an expert measurer. But that is not enough. The surveyor must understand how to use words, phrases, sentences, grammar, and syntax to convey complex ideas, descriptions, and concepts to future generations of surveyors. This is not as easy as it sounds because those deeds, descriptions, and other legal documents created by attorneys and laymen are often poorly written, consuming, and contradictory. Therefore, the surveyor must act more like an English professor and an expert measurer in the records research phase of his or her boundary survey.

The discussion of verbal evidence is unique in that both written and spoken evidence must be considered. This will include words in descriptions, in deeds, in reports, and especially on maps. This evidence may be in the form of parol evidence—written words, testimony, and words in every shape or form used by the surveyor or by other individuals, from the field survey crews to the draftsperson in the office.

Without written documents, deeds, descriptions, and/or field notes, found real tangible evidence is valueless or of little legal value to the surveyor, the attorney, and especially the courts. The key in any evidence evaluation is the ability to correlate the verbal (written, documentary, parol) evidence to the found evidence in the field, such as a rifle barrel or oak tree.

Certain principles apply when it comes to evaluating words. These may follow the path of first being part of the documentary evidence, then part of the survey field evidence. They show up in the reports and finally find their way into the legal documents used to convey land and ultimately to locate the same land years later. The following basic principles are discussed in this chapter:

PRINCIPLE 1. *A written conveyance is indispensable evidence in the claim and proof of ownership.*

PRINCIPLE 2. *All words in any written document should be interpreted as having the meaning at the time the document was created and not at the time today's surveyor attempts to retrace it.*

PRINCIPLE 3. *Descriptions do not identify themselves, and all that is required of a deed is that it furnishes evidence, or a key, of a means of identification. If the description in a deed is sufficient when the deed is made, no subsequent change in conditions or loss of evidence can render it insufficient.*

PRINCIPLE 4. *When the terms of an instrument, deed, or will have been reduced to writing, they are to be considered as containing all those terms, and there can be, between the parties or their representatives or successors in interest, no evidence of the terms of the instrument other than the contents of the writing, with several exceptions.*

PRINCIPLE 5. *Each deed should be construed as a whole; all parts are read in the light of other parts. Nothing stands alone and nothing is rejected unless it is impossible to give that part a meaning.*

PRINCIPLE 6. *Writings or plats, referred to in a conveyance for the purposes of identifying land, are to be regarded as a part of the conveyance itself as much as if incorporated into it.*

PRINCIPLE 7. *Where the law requires field notes as a part of a survey and the notes are stored in a public place, reference to the survey includes reference to all the field notes of that survey.*

PRINCIPLE 8. *Extrinsic evidence may be sought to explain (1) the meaning of words existing within a written conveyance and (2) the conditions existing as of the date of the deed.*

PRINCIPLE 9. *Evidence of practical location, written or parol, when appropriate, may be received to clarify ambiguity of writings.*

PRINCIPLE 10. *Ambiguous deeds or descriptions that cannot be cured by extrinsic evidence may be considered as void or voidable by the courts.*

PRINCIPLE 11. *An ancient survey recorded or accepted as a public document, made by a competent authority, and produced from proper custody is generally admissible as evidence to prove the location of a boundary line.*

PRINCIPLE 12. *The intentions of the parties to a conveyance, as expressed by the evidence of the writings, are the paramount considerations of the court in interpreting the meaning of a deed, and that intent is gathered exclusively from the written words of the deed, except where the written words have extrinsic ambiguities or where explanations of conditions existing as of the date of the deed are necessary.*

PRINCIPLE 13. *Depending on the evidence discovered, a conveyance is classified with respect to the adjoiner as being senior in rights, as being equal or simultaneous in rights, or as being junior in rights. An intent, gathered from the written evidence, cannot alter senior rights.*

PRINCIPLE 14. *The words of a written conveyance are conclusive evidence as between the buyer and seller, but they are not conclusive evidence as against a third party who has senior rights or a person with occupancy rights.*

PRINCIPLE 15. *Each deed is construed as a whole; parts are read in the light of other parts. Nothing stands alone and nothing is rejected if it is possible to give every part of the deed a meaning.*

PRINCIPLE 16. *Testimony of interested landowners is one form of evidence that must be considered by the surveyor as part of the totality of the evidence. Not to do so may be a serious flaw in the collection of any evidence.*

PRINCIPLE 17. *Words that are translated from one language to another have the meaning inferred in the original language and not in the second language.*

PRINCIPLE 18. *In ancient documents the words have the meaning that they had at the time they were written and not the meaning in today's definition.*

3.2 Writings as Indispensable Evidence

PRINCIPLE 1. *A written conveyance is indispensable evidence in the claim and proof of ownership.*

English common law and the *Statute of Frauds*, as accepted in the United States, require that any conveyance of land or interest in land must be in writing to be considered a valid conveyance. Because of this legal requirement in all fifty states, some form of written proof of title such as a contract for the sale of the property is indispensable evidence to show a claim of ownership. Furthermore, because of this legal requirement, surveyors rely on legal descriptions included in land conveyances such as deeds to begin the retracement of any parcel of land. After the description contained in the written conveyance (which is described with words) is located on the ground, the surveyor then records (using more words) his or her results in field books, then prepares a surveyor's report, in words, and usually prepares a map, a form of words and graphics. The surveyor's notes or reports on possession not in agreement with the writings and, at a minimum, informs the client of the possible significance of prolonged possession. Finally, the professional land surveyor is required in many states to author *another* legal description based on his or her own survey measurements and place that description on the face of the survey plat. Thus, creating another link of the proverbial chain between himself or herself and the last re-surveyor.

In addition to *creating* the legal description, map, or surveyor's report, a professional land surveyor must scrutinize every *found* document for errors, omissions, or typos. When using the *found* documents, he or she must be certain that the deed describes the correct property, the legal description is mathematically correct (it closes, more or less), and the document is legally complete, otherwise there is a possibility that the conveyance is not valid.

The causes of the error or omission contained within the conveyance are numerous. The document could be patently defective because the landowner wrote the deed himself. Or the drafting attorney could have simply copied and pasted a century-old legal description into modern-day closing documents. No matter, the surveyor must remember that there is always the possibility that the legal description contained in a conveyance document could have been changed or altered, either inadvertently or on purpose. Before using any deed, description, or alike, the surveyor must scrutinize each document.

In *Hunt v. Feese et al.* [2], three landowners were involved in a lawsuit over the riparian boundaries within a semi-circular cove. One surveyor was hired by Hunt and Pounds while another surveyor was retained by Feese. Since riparian boundaries are simply extensions of upland boundaries, the first surveyor conducted a full boundary survey of all properties within the cove and then retraced the riparian boundaries after locating all of the upland boundaries. The other surveyor was impeached on cross-examination by Hunt's attorney and admonished by the presiding judge because the surveyor (1) failed to follow the record conveyances of the properties; (2) failed to conduct an upland boundary survey of the properties before locating the riparian boundary lines; and (3) and placed an adjacent property's legal description on the face of the Feese's survey map. Hunt and Pounds ultimately prevailed. This case demonstrates the need for surveyors to follow the words as evidence of any property boundary and to always verify the correct information is contained in the survey plat. Otherwise, the surveyor opens himself or herself up to impeachment by fellow surveyors, attorneys, and the courts.

3.3 Identification of Property Descriptions

PRINCIPLE 2. *All words in any written document should be interpreted as having the meaning at the time the document was created and not at the time today's surveyor attempts to retrace a property.*

Any surveyor who accepts the responsibility of retracing or locating a parcel of land from a written description should ascertain the date that the description was written in order to give the true meaning to the words that must be retraced. The older the document in which the lines are described, the greater the necessity that the proper meaning be given to the words.

The lands that now make up the United States were acquired by European settlers from different sovereigns and at different time periods. An example of Principle 2 is boundary retracements in Texas. The citizens of current-day Texas have lived under six flags over the past four centuries: French, Spanish, Mexican, the Republic of Texas, the Confederate States of America, and the United States of America. Each sovereign created its own land tenure laws, adopted various forms of measurement, and used terminology that is not found in today's legal descriptions. Under Spanish and Mexican rule, leagues, labors, sitios, and haciendas were all parcels of land with a defined shape and size, based simply on their names. If John Smith was conveyed "a labor of land in Bexar county," he was allowed to go anywhere in the county, find an unclaimed property, and stake-out a 177-acre parcel. Modern surveyors without knowledge of such key terms would be ill equipped to retrace Mr. Smith's boundary today.

This applies not only to the meaning of the words but also to the names indicated in the documents.

One type of property description that was very popular in the past, and still is common today in some parts of the country, is the description by adjoinders [3]. In this type of description, the subject property is defined by nothing more than calls for contiguous parcels. This type of description is quick, cheap, and easy to create. But retracing the adjoinders description is difficult and costly because it requires the retracing surveyor to locate many senior parcels before the boundaries of the subject parcel can be located. Many old deeds will use such calls as "bounded on the North by the lands of Jones" or "thence to the center of Rabbittown Creek." The two controlling elements in the descriptions, the "lands of Jones" and "the center of Rabbittown Creek," may not be identifiable today. Through a series of successive transfers, the modern-day surveyor may determine the former "lands of Jones" is now in the ownership of Smith and Rabbittown Creek is nowhere to be found in the entire county. But after an examination and comparison of modern-day and ancient

U.S. Geological Survey (USGS) topographic maps that can be dated to within a few years of the date of the original description, a creek that is indicated as Rabun Creek in 1956 was originally called Rabbittown Creek on the 1898 edition of the same quadrangle. Then by being able to show that a clear chain of evidence exists from the time the original description was created to the present time, the surveyor made documents that may have been questionable, positive. The surveyor was able to bring the ancient description to today's meaning.

PRINCIPLE 3. *Descriptions do not identify themselves, and all that is required of a deed is that it furnishes evidence, or a key, of a means of identification. If the description in a deed is sufficient when the deed is made, no subsequent change in conditions or loss of evidence can render it insufficient.*

Although a description may have been sufficient at the time it was written, subsequent disappearance of evidence or changes in terminology may render the certainty of location exceedingly doubtful. But courts declare where property is located irrespective of the poor quality of evidence presented; the best available evidence is accepted even though the evidence may not be admissible in other types of litigation.

Courts have held that vagueness in a description may render a deed void or voidable. Written descriptions of property are to be interpreted in the light of surrounding facts and circumstances well known in the community; descriptions do not identify themselves, and all that is required of a deed is to furnish a means of identification. If the description in a deed is sufficient when the deed is made, no subsequent change in conditions can render it insufficient [4].

When it comes to placing a written description on the ground, the retracing surveyor is the final leg of a possibly extended line of surveyors. We assume that the original description was created from a survey that was completed by a surveyor, which was then placed in a description possibly written by an attorney, or even by the landowner, and then the description sat dormant for years or generations. The modern surveyor was asked to take that ancient description and place it "on the ground" with a great degree of certainty.

There has always been a question as to which of the professions, lawyers, surveyors, or engineers should write property descriptions. Any discussion would only show personal preference. Yet recently in Florida the question was raised in court as to which profession was most capable of interpreting a description. The Court of Appeals answered with a definitive response.

> Although title attorneys and others who regularly work with them develop an expertise as to land descriptions, the only professional authorized to locate land lines on the ground [from descriptions] is a registered land surveyor. In fact, the definition of a legally sufficient real property description is one that can be located on the ground by a surveyor [5].

One of the most difficult tasks of a land surveyor is that of having to take the written words in a description and place them on the ground with a degree of certainty that says, "Here is your land parcel. No ifs, ands, or buts!" Many times, the surveyor will find that the words contained in the description are related by the owner. Thus, we have a conflict between the written words of the owner and the spoken words of claim of ownership.

The older or more abbreviated the description, the more problems the surveyor will usually have in doing this. In those many instances where the surveyor cannot say with certainty where the land to be surveyed is located, the courts will permit the use of "keys" to locate the land. The key could be anything the court will permit.

In one state, Georgia, the Supreme Court was required to comment on three different problems and the descriptions that they contained.

One description was "his home place 20 acres of land immediately surrounding his land." The second was "ten acres of land which is hereafter to be surveyed, using the center of the school," and the third being "one and one-half acres out of a 21-acre parcel."

The court held that a key was present to locate the parcel and its elements, the boundaries, but no keys were present in the other two. In the first instance, the key was the word *surrounding*. They stated the description means a circle with a radius of 20 acres, using the house as its center [6]. The elements of the circle were not identified.

In the second instance, a description was held to be incapable of possessing a key. It held that "one and one-half acres" was not locatable even though the parent parcel of 21 acres was capable of being located [7].

Any surveyor who creates a boundary description must have a good command of the English language to be able to create, read, and interpret property descriptions.

3.4 Conclusiveness of Written Words of a Deed

PRINCIPLE 4. *When the terms of an instrument, deed, or will have been reduced to writing, they are to be considered as containing all those terms, and there can be, between the parties or their representatives or successors in interest, no evidence of the terms of the instrument other than the contents of the writing, with several exceptions.*

These exceptions are as follows:

1. Where a mistake of imperfection of the writing is put in issue by the pleadings.
2. Where the validity of the agreement, deed, or will is the fact in dispute.
3. To establish illegality or fraud.
4. To explain an extrinsic ambiguity.
5. To show the circumstances under which the instrument was made for the purpose of properly construing the instrument.

In the foregoing, the surveyor usually is not concerned with items 1–3. These are matters for title companies, abstractors, attorneys, and ultimately the courts. The surveyor does frequently seek evidence to explain extrinsic ambiguities and sometimes evidence to understand the circumstances under which the deed was written (was the basis of bearing magnetic or astronomic?). In addition to items 4 and 5, the surveyor is charged with "judicial notice" of the meaning of words used in land descriptions. This is especially so if the surveyor is ever qualified as an expert.

The survey of a parcel of land and the placing of corner monuments, with the view that they will be described in a subsequent conveyance, cannot be admitted to vary or control any deed that was made afterward, when that deed calls for none of the monuments that were placed in the survey and the description contained no reference to the survey.

In a 1926 decision, the court wrote: "Ancient deeds are to be construed as written in light of the then use of properties conveyed and adjacent land and cannot be cut down by vagueness in subsequent conveyances" [8].

When this happens, all prior negotiations must be taken, so far as the construction of the deed is concerned, to have been merged in that instrument; the conclusive presumption is that the whole engagement of the parties and the extent and manner of it were reduced to writings. The actual intention of the party, or the surveyor, is not admissible to affect the construction of the deed [9]. This is especially true when the court attempts to identify the "intent" of the parties to the conveyance.

The conclusiveness of the written words of a deed also applies with the same force to all documents called for in the deed. If a certain map is called for, all the writings and monuments called for on the map have as much conclusive control as though the writings and monuments were called for by the deed itself. Because of the conclusiveness of the written documents, the writings themselves are often called the best available evidence. This concept can be explained in the situation when the description of the property and its boundaries that is conveyed in a deed is definite, certain, and unambiguous. Extrinsic evidence cannot be introduced to show that it was the intention of the grantor to convey a different tract. Thus, where the land is described by course and distance and/or monuments, parol evidence is inadmissible to prove that the true boundary is a line of marked trees not mentioned in the deed, or that a deed stating a definite thing was intended to express another, or to alter or vary the legal import of monuments, the location of which has been ascertained. If the calls in the grant when applied to the land correspond with each other, parol evidence will not be admitted to vary them or show the fact they were not the calls of the survey actually made (but not called for) [4].

After a surveyor has completed a plat and has certified to the same that he or she has surveyed it, the testimony would be incompetent to prove that he or she did not in fact make a survey. The evidence on the plat, being writings, is not to be impeached by verbal testimony.

This was demonstrated when appellants contended that a surveyor whose certificate appeared on a plat admitted, while testifying as a witness, that he had not, in fact, made the survey. The plat was completed, certified by the surveyor, acknowledged by those claiming to own the land, and duly recorded, in compliance with the law. In this instance the law provided that plats "may be used in evidence to the same extent and with like effect in the case of deeds." The surveyor was not competent as a witness to verbally impeach his written certificate (writings) [10]. As with many rules of law, exceptions do occur in a few states under "uncalled-for monuments."

This situation has also occurred in instances when surveyors prepare multiple surveys and the resultant plats disagree with one another.

3.5 Rules for Interpretation of Writings

Often in a surveyor's professional life, interpretation of the meaning of various words and phrases contained in written evidence will be required. The following seven rules may help explain the significance of evidence and are generally applicable in most states:

1. The language of a writing is to be interpreted according to the meaning it bears in the place and at the time of its execution, unless the parties have reference to a different place or time.
2. In the construction of an instrument, the place of the judge is simply to ascertain and declare what is in terms or in substance contained therein, not to insert what has been omitted or to omit what has been inserted; where possible, to be adopted as will give effect to all.
3. In the construction of the instrument, the intentions of the parties as expressed by the writings are to be pursued, if possible; when a general and particular provision is inconsistent, the latter is paramount to the former. So, a particular intent will control a general one that is inconsistent with it.
4. For the proper construction of an instrument, the circumstances under which it was made, including the situation of the subject of the instrument, and of the parties to it, may also be shown, so that the judge is placed in the position of those whose language he or she is to interpret.
5. The words used in writings are presumed to have been used in their primary and general acceptance, but evidence is nevertheless admissible that they have a local, technical, or otherwise peculiar signification and were so used and understood in the particular instance, in which case the agreement must be construed accordingly.
6. When the characters in which an instrument is written are difficult to be deciphered, or the language of the instrument is not understood by the court, the evidence of persons skilled in deciphering the characters, or who understand the language, is admissible to declare the characters or the meaning of the language.
7. Words expressed in a deed or conveyance reflect the intent and meaning of words of the parties at the time of preparation, not necessarily the present meaning. Most frequently, the need for explanations of words arise from local customs such as a local name for trees, water bodies, and unique elements that may be specific to a local area. What does the term "spotted tree" mean in Maine? Is it a blazed, hacked, or marked tree? What is meant by "plug" (a 2-by-2-inch wooden stake)? Is a "branch" the same as a "run," "creek," or "bayou?" What is a "bar" that was called for in a deed? All of these terms have local meanings, which the boundary surveyor must know.

3.6 Deeds Construed as a Whole

PRINCIPLE 5. *Each deed should be construed as a whole; all parts are read in the light of other parts. Nothing stands alone and nothing is rejected unless it is impossible to give that part a meaning.*

Although surveyors and the courts may be required to reject certain phrases or words in a description if it is absolutely impossible to give them meaning or if there is an absolute conflict in several elements in the description, parts are rarely rejected. The following represents standard thinking throughout the United States.

A deed must be construed as a whole, and a meaning given to every part of it. This means that a simple decision to reject any portion will not be tolerated by the courts [11]. If possible, the surveyor should correlate all parts of the description to determine the true intent of the original parties. Courts seem to give great credence as to the "intent" of the parties to the original instrument, in that a court will attempt to duplicate the original intent. Without the original parties present, it would be virtually impossible to determine the true intent of the parties from the written words if the words were ambiguous. The formula used to determine the true intent is:

$$Intent\ of\ the\ grantor + Intent\ of\ the\ grantee = True\ intent$$

Thus, if there were any ambiguity in any of the words, then the true intent is elusive, and as such, courts will "step in" and determine what the parties intended. They may, of course, seek guidance from an expert surveyor.

Several rules are universally accepted:

- The main objective in construing a deed is to ascertain the intention of the parties from the language used and to effectuate such intention where not inconsistent with any rule of law [12].
- Grantor's intention controls unless it is in conflict with some positive rule of law [13].
- *Intention* as applied to the construction of a deed is a term of art and signifies a meaning of the writing [14].
- The cardinal rule for the interpretation of deeds is the intention of the parties as expressed in the instrument [15].

If the words in the description are ambiguous, and surveyors usually address the validity of the description and not the deed that carries the description within its "four corners," then two surveyors may determine the following: After reading the description, surveyor 1 states, "I can survey that description on the ground." Then surveyor 2, after reading that same description states, "That description defies being capable of being surveyed. There are ambiguous words and there is not enough information for me to locate it. IT IS VOID in its present form." Now this becomes a question of fact, and as such, the ultimate decision will be submitted to a jury for determination. Courts will usually permit expert surveyors to make or interpret certain words to make the meanings clear.

A particular description will control over a general description of boundaries. This application gives attorneys, courts, and surveyors problems. Before they can apply the principle, they must determine what is "particular" and what is "general" [16].

This principle may be applied in the situation where a description contains a positive identification of the land conveyed by a reference to recorded plats. In an effort to make a more "positive" identification of the parcels, an attorney includes a description of the dividing line between the lots conveyed (by the plat) and the parcels retained by the grantor that was identified by a described line. Where one part of a description in a deed is false and impossible but, by rejecting that, a perfect description remains, the false part should be rejected, and the deed held good [17]. Any particular of a description may be rejected if it is manifestly erroneous and enough remains to identify the land intended to be conveyed [18].

3.7 Reference Calls for Other Writings or Plats

PRINCIPLE 6. *Writings or plats, referred to in a conveyance for the purposes of identifying land, are to be regarded as a part of the conveyance itself as much as if incorporated into it [19].*

The evidence of writing (including the writings of plats) always includes in scope the document and all writings referred to by the document. Further, the document or plat referred to may call for additional writings that may also serve as evidence of the original intent.

This principle is important to a retracing surveyor. When he or she is given a description to retrace that includes reference to prior deeds, the question arises as to whether the surveyor is responsible for researching those prior descriptions. The answer could be yes or no, depending on the circumstances.

The answer is yes if there is no written contract in place identifying the exact scope of the duties expected. In this situation, the surveyor will be expected to retrace the original description that created the parcel to make certain there are no errors in the subsequent documents. If the surveyor has a written contract specifically stating the exact document that will be retraced, that is where the responsibility lays. Otherwise, the recited documents will control.

The no answer applies if the surveyor's employment contract specifically states which documents will be retraced. In the American Land Title Association and the American Congress on Surveying and Mapping (ALTA/ACSM) contracts, the surveyor is responsible only for those documents delivered by the client.

These remarks are equally applicable to all writings, including all words and symbols on plats.

When another deed is referred to for a description of the premises conveyed, the deed referred to is regarded as of the same effect as though it had been copied into the deed itself, and whatever is described in it will pass [20]. A very basic principle was adopted in Maine that is strengthened by a federal law and appellate decisions in many states. *This is where an actual survey is made and monuments marked or erected, and a plan afterward made intended to delineate such survey, and there is a variation between the plan and the survey, the survey must govern; but such is not the role where the survey was made subsequent to the plan [21].*

Many times, problems occur when the surveyor is unavailable to testify as to his or her work, and the documents that were created from the survey are entered as evidence. There are two possible scenarios: (1) the surveyor is unavailable to testify or (2) the surveyor is dead. In New Hampshire, the court has ruled that a deceased surveyor's ancient plans, minutes, or field books, clearly describing and identifying the lots and bounds in controversy, are admissible as evidence [22].

This decision is interesting in that it may be confusing. According to common law, if the surveyor is unavailable to authenticate the work, then basically it is inadmissible. Yet there are exceptions, the principal one being the ancient document exception that is prevalent in most states. This exception basically states that any document that is considered as ancient is admissible as an exception to the hearsay rule if the document is 30 years (20 years in federal courts and in many states) old, and free from alteration. Since the Morse decisions stated the document was ancient, perhaps that prevailed over the death and unavailability problem. However, if the document is not considered "ancient," it would not, in all probability, be admissible, for failure of authentication. In this area, surveyors should work very closely with attorneys.

In *Morse v. Emery* [22], wherein a deceased surveyor's written records were admissible evidence, the principle is not applicable in all states, especially where records are commonly filed. In the majority of states, only publicly stored field notes, as, for example, government-sectionalized land surveys, are admissible for the reason that a landowner should not be held to know what is contained in private files. In states where extensive obliteration of surveys has occurred and the location of the descriptions of monuments is unknown, the strict rules for boundary locations are often relaxed, allowing reputation and inferences to be of importance, especially in older states. The usual rule is that someone must testify as to the accuracy of field notes of private surveyors, and, even if the notes are admissible, they may not be used to prove that clearly stated words in a conveyance are in error.

3.8 Contrasting Plats and Writings as Evidence

Plats have certain advantages as evidence over written descriptions in that the symbols used on a plat convey definite meanings without using words. In Figure 3.1, the north arrow indicates direction, and the scale given indicates relative size. Without words the picture conveys the idea of direction of all lines (north and south or east and west), length of lines (100 feet multiplied by 200 feet as scaled), and a closed figure (rectangle). Some of the early maps contained no more information than this. Evidence of this type is graphic evidence, it is peculiar to maps or plats, and it is regulated in value by the law of evidence.

Surveyors, cartographers, and landowners have drawn maps on just about everything: in the sand, on animal skins, Mylar, linen, brown wrapping paper, chiseled on

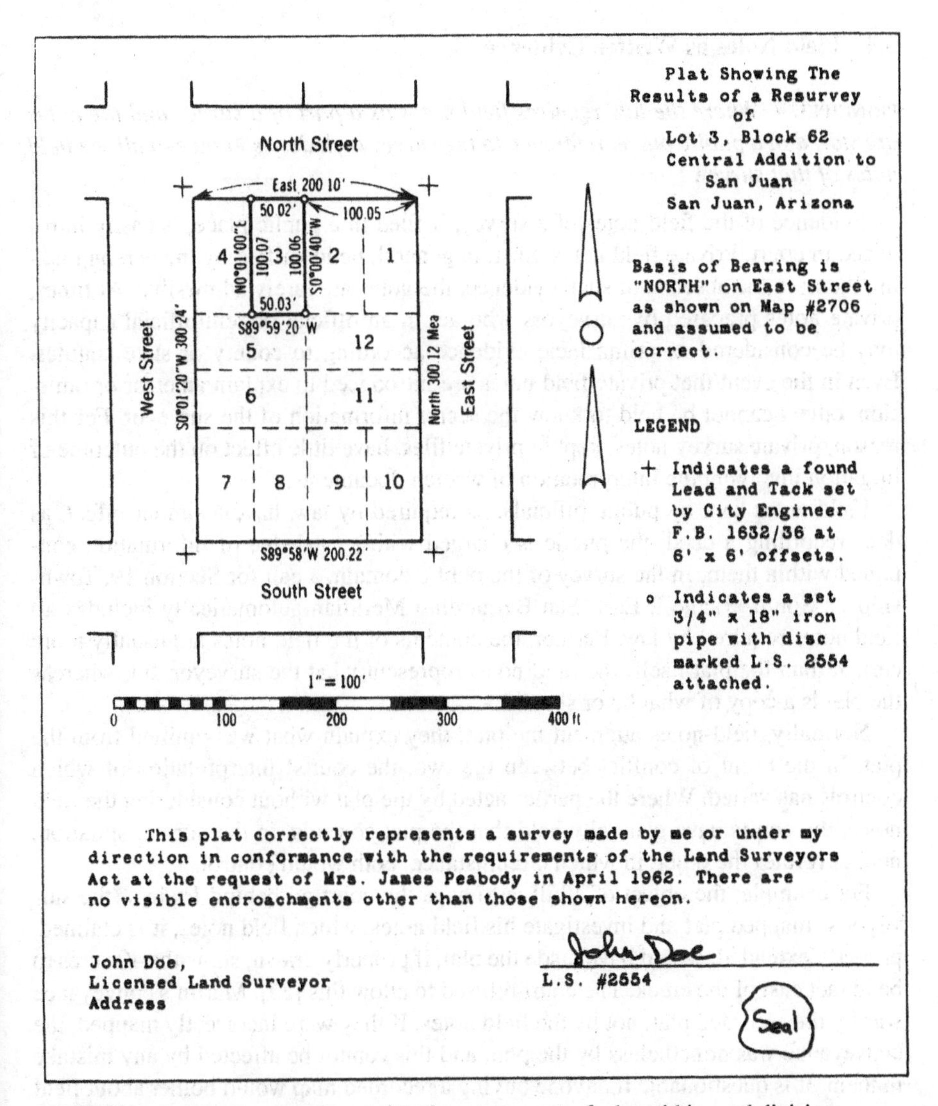

FIGURE 3-1. Plat showing the results of a retracement of a lot within a subdivision.

stone, on the bark of trees and metal-mounted paper. Maps have also come in every imaginable size, scale, and depiction. Round maps, square maps, upside-down maps, and maps in reverse have all taxed the creator's imaginations, as well as the technical capabilities of the people who had to read them.

The major objective of filing maps, plats, and field notes is to avoid the necessity of encumbering deeds or descriptions with lengthy wording.

> A wise person once said, "A picture is worth more than ten thousand words."

3.9 Field Notes as Written Evidence

PRINCIPLE 7. *Where the law requires field notes as a part of a survey and the notes are stored in a public place, reference to the survey includes reference to all the field notes of that survey.*

Evidence of the field notes of a survey, if filed in a public place, is easily introduced in court. Private field notes must, in general, be identified by the person making them; in the absence of such evidence, the notes are rarely admissible. At times, private notes prepared by surveyors who act in an official or semiofficial capacity may be considered as prima facie evidence according to county or state statutes. Even in the event that private field notes are introduced to explain an error or omission, others cannot be held to know the secret information of the surveyor. For this reason, private survey notes, kept in private files, have little effect on the outcome of litigation involving the interpretation of written documents.

Field notes kept by public officials, as required by law, have a similar effect, as does recording a deed; the public is charged with knowledge of information contained within them. In the survey of the public domain, a call for Section 19, Township 12 South, Range 2 East, San Bernardino Meridian automatically includes all field notes required by law. Further, the contents of the field notes are usually more certain than the plat itself; the field notes represent what the surveyor did, whereas the plat is a copy of what he or she did.

Normally, field notes augment the plat; they explain what was omitted from the plat. In the event of conflict between the two, the courts' interpretation of which controls has varied. Where the parties acted by the plat without considering the field notes, the courts have generally held that the plat controls. In the normal situation, neither refutes the other, in which circumstance, both would control.

For example, the object of a bill is to have the court go behind Irwin's (the surveyor's) mapped plat and investigate his field notes, which field notes, it is claimed, properly extended, would have made the plat, if properly drawn, show the 40 acres to be in fact east of the creek. The court refused to allow this [23]. Martin's conveyance was by the recorded plat, not by the field notes. If they were incorrectly mapped, the conveyance was nonetheless by the plat, and this cannot be affected by any mistake in them. It is questionable if anyone buying a recorded map would bother about field notes. Without the plat, Haley would have nothing. He is bound to claim under it, but wants to claim under it as incorrect, and to have the error corrected by the field notes. It is not possible that he can claim under the field notes, because the conveyance was not by them but by the plat.

Field notes are the highest and best evidence of what a surveyor did in the field, which become evidence of the utmost importance in following a surveyor's footsteps. When markers become lost, they must be replaced in the same position as originally placed, which is where the original surveyor placed them. Without a record of what the surveyor actually did, the retracing surveyor is handicapped and forced perhaps to rely on evidence of lesser importance. This evidence is likely to contain considerable error.

Field notes retained in private files do not have the same dignity as do government field notes. Whereas government field notes may have some evidentiary privileges, private field notes do not. Basically, private field notes are hearsay in nature, and they must be sponsored or their introduction in evidence should meet one of the hearsay exceptions.

Conflicts and ambiguities in field notes are resolved through regular rules of construction [24].

3.10 Extrinsic Evidence of Ambiguities in Writings

PRINCIPLE 8. *Extrinsic evidence may be sought to explain (1) the meaning of words existing within a written conveyance and (2) the conditions existing as of the date of the deed.*

Extrinsic ambiguities are ambiguities that must be explained from some source of evidence other than the writings. Once a conveyance is reduced to writing, words may not be added to or subtracted from a deed, but extrinsic evidence may be gathered to explain existing ambiguous or conflicting words of the description elements in the deed or description or to explain a latent ambiguity. Such an ambiguity, if serious enough, could invalidate the description for want of certainty. If a deed is exact and explicit, no amount of verbal testimony can change it. A deed reading "thence to a stone mound" goes to the stone mound regardless of subsequent verbal protests. But it must be proved by evidence that (1) the stone mound is the one referred to and (2) the stone mound has not been disturbed. Parol evidence may be taken to prove or explain these two points, but it may never be taken to refute the fact that the line goes to the spot occupied by the stone mound as of the date of the deed. Extrinsic evidence is often necessary to understand a deed. If a deed is written "my house and lot," it can be a perfectly valid description, provided verbal evidence or written evidence can prove that the seller had one and only one house and lot. If it can be shown that the seller had two houses and lots, verbal evidence cannot be taken to differentiate between the two houses. The deed would be invalid because of uncertainty. Words cannot be added to identify between two houses, but an explanation of where one house exists is permissible.

Most deeds may have words or conflicts that need explaining, but such explanations are proper only as long as the laws of evidence are not violated. "Beginning at Jim Brown's henhouse" certainly needs explanation about where the henhouse was located as of the date of the deed. The fact that the deed started at Brown's henhouse is indisputable; it is written in the deed. But the location of the henhouse may take real detective work. If the henhouse is now gone, old nails or a few fence posts may indicate where it used to be. Again, Farmer Brown may be called on to testify about the location of the old henhouse. One thing is certain, if the location of the henhouse as of the date of the deed can be established in accordance with rules of evidence, it will control the starting point. Frequently, extrinsic evidence must be used to identify the physical objects called for.

Parol evidence is admissible to locate and identify monuments and calls in a description of land [25].

When grants and conveyances of lands are made, the usual mode of describing the parcel conveyed is by reference to natural or artificial monuments as boundaries, and by this means the premises may be found and distinguished from other tracts or parcels of land. The same object is also attained by describing the premises conveyed by a specific name, but in either case the location is not always determined alone by the description in the conveyance, independent of extrinsic evidence. The deed description describes the objects bounding the premises, but parol evidence is usually resorted to for the purpose of identifying the objects themselves. It is well-accepted law that all monuments, and things referred to in a deed for the purpose of locating the land, may be established and identified by extrinsic evidence. In most instances, however perfect the description employed by the conveyance, the premise could not be located and identified without reference to extrinsic evidence, whether more approximate, and for the purpose of sustaining a grant, extrinsic evidence may always be used to identify, explain, or establish the objects of the call in a deed [26].

A trial judge may admit extrinsic evidence to show that by "due north" the parties to the boundary agreements meant that the call in the deeds should be surveyed only by a basis of bearings method. This type of evidence may also be used to explain such ambiguous items as *pace*, *step*, and *survey* or the meanings of little-known or used words.

In the case involving *Richfield Oil Co.* [27], Richfield contended that this evidence required reversal of the judgment, invoking another aspect of the parol evidence rule. The rule is that evidence of the circumstances surrounding execution of a written instrument is admissible to determine its meaning. Carpenter and Henderson and defendants, however, were not parties to the boundary agreements, and it was not shown that they had actual notice that *Richfield and Norris Oil Company* attached a particular meaning to the words *due north*. The conduct of Carpenter and Henderson indicated that they believed that the "due north" description in the March 15 quitclaim deed was interchangeable with the government section description in the 1947 sublease. Although evidence of the negotiations preceding execution of the boundary agreements may be admissible as between the parties thereto, a special interpretation of the quitclaim deed could not be enforced against a party relying on the words of the instrument and without knowledge of the meaning attached thereto by the other party to that deed. Thus, in the present case, the expert testimony of surveyors and engineers showing the proper method of surveying the calls in the deed was admissible as against both parties to the deed, since they must be deemed to know that "technical words are to be interpreted as usually understood by parties in the profession or business to which they relate," but evidence of the negotiations of the parties to the boundary agreements could not be admitted to the detriment of defendants, for it would be manifestly unjust to charge them with the secret interpretation of other parties after defendants had relied upon the ordinary meaning of the words of the instrument [28].

3.11 Practical Location and Descriptions

PRINCIPLE 9. *Evidence of practical location, written or parol, when appropriate, may be received to clarify ambiguity of writings.*

On occasion, deeds may not be definite in defining the limits of a particular thing. For example, a deed reading "all of Lot 1 and a 20-foot-wide road easement across Lot 2" lacks an exact location for the easement mentioned. But if the owner of Lot 1 uses a particular place for a period of time, that particular place becomes the correct easement line by practical location. The reasoning is simple; since that location was used, there must have been an agreement, either verbal or implied, between the parties. Seeming ambiguities may not be ambiguities at all; a simple inspection of the land for evidence of usage sometimes provides the answer.

Evidence of practical location may be received only to clarify an ambiguity; it may never be used to overcome clear, concise, unambiguous words in a deed or to create an ambiguity. If the easement just described were defined as to location in the deed, evidence of usage would be incompetent to overcome the written words unless such usage ripened into a possession right.

3.12 Patent Ambiguities

PRINCIPLE 10. *Ambiguous deeds or descriptions that cannot be cured by extrinsic evidence may be considered as void or voidable by the courts.*

"Ten acres in Section 21" is not locatable and is probably void. "My house and lot" is patently ambiguous if the party owns two houses and lots. A latent ambiguity—that is, an ambiguity that is not apparent until it is being located on the ground—can be explained by extrinsic evidence. A patent ambiguity—that is, an ambiguity that appears on the face of the document—cannot be explained by extrinsic evidence.

This area of case law is still ambiguous, and some cases seem to disregard this ancient theory. One such case reads:

> No deed or conveyance of land was ever made, however minute and specific the description, that did not require extrinsic evidence to ascertain its location; and this is so whether the description be by metes and bounds, reference to other deeds, to adjoining owners, watercourses, or other description of whatever character [29].

Courts will also look at the actual situs of the land and it is a common experience that descriptions of property in rural communities are somewhat indefinite, and even inaccurate, and courts deal leniently with such cases in seeking to ascertain the intention of the parties.

When discussing the four-corners rule, a New York court wrote, "the antiquated doctrine that a document must be construed solely within its four corners, no matter how puzzling the problem, is no longer the law of this state" [30].

3.13 Ancient Survey Plats and Documents

PRINCIPLE 11. *An ancient survey recorded or accepted as a public document, made by a competent authority, and produced from proper custody is generally admissible as evidence to prove the location of a boundary line.*

A private survey may be admissible on proof of its correctness by the party making it, whereas an ancient survey by proper authority and recorded as a public record may be accepted without further verification. Private surveys, in general, are not accorded the same standing as publicly recorded surveys or surveys and documents created by government agencies. Usually, a private survey requires authentication and the chain of custody usually requires proof.

An ancient map of the public roads of a county purporting to have been made by authority and coming from the proper custody is competent evidence to show the existence and location of the public roads of the county at the time it was made; in a contest between coterminous landowners, where a road delineated on the map is claimed to be a boundary, such a map is relevant to the question of the boundary. Many of the old maps are of such a small scale that they only depict a general depiction of the information.

Which ancient maps are received is determined by the theory that where the matter in controversy is ancient and not susceptible of better evidence, traditional reputation of matters of public general interest is competent evidence of the matters to which it relates. The admissibility in evidence of an ancient map of matters of a public general interest is not to be confounded with a map that a landowner causes to be made of his or her premises. It was doubted that the rule admitting a map 30 years old as an ancient document applied to private maps. An unofficial survey may be admissible in evidence when it has been proven to be correct [31].

The admissibility of any proffered evidence is at the sole discretion of the presiding judge.

Each state usually has rules of evidence declaring how old a document must be before it can be classified as ancient, and it is often 30 years, except in federal courts, where it must be 20 years old. Such items as deeds, wills, and other documents used in conveyances are most often applicable to surveying evidence. The following indicate a selection of cases found:

"Ancient deeds" more than 30 years old, are admissible to prove ancient possession [32]. By reason of statute, it would seem that in New York State a period of 20 years satisfies the age requirements to bring a map or survey within the ancient-document rule [33].

"A will executed more than 30 years before it was offered in evidence when produced from the proper custody and otherwise free from suspicion is entitled to the benefit of the same presumptions as an ancient deed" [34].

The ancient-document rule is 20 years in the federal courts and some state courts, but in a minority of state courts the recognized time frame is 30 years. Surveyors should become familiar with the rules of evidence for their particular state.

3.14 Interpreting Words in Conveyances

The meaning and intent of words may normally be the subject of judicial notice; that is, the meaning is sought from evidence found in books or authoritative sources but not from the opinions of the parties to the deed. It is only when words have a particular meaning for a particular locality or a particular profession that the courts resort to extrinsic evidence to determine the meaning of a term. Normally, in the matter of surveys, the surveyor is the expert who interprets the meaning of unusual words as applied to the profession. Under certain circumstances, it may be necessary to seek verbal explanations of a word from others. Problems are encountered when the verbal spoken evidence is in conflict with the verbal written evidence. A deed reading, "commencing at an oak tree" certainly needs verbal explanation about where the oak tree is located, but it needs no explanation of the fact that it is an oak tree. The conflict occurs when the subsequent spoken testimony identifies a pine tree.

Courts usually rely on *Black's Law Dictionary* and *Webster's Dictionary* for definitions of words. West Publishing Company publishes a set of books (100 volumes) entitled *Words and Phrases*.

The legal interpretation of the word *so-called* in a deed is not what the parties say they mean but what the public generally says about what they mean [35].

3.15 Evidence That Determines Intent of a Conveyance

PRINCIPLE 12. *The intentions of the parties to a conveyance, as expressed by the evidence of the writings, are the paramount considerations of the court in interpreting the meaning of a deed, and that intent is gathered exclusively from the written words of the deed, except where the written words have extrinsic ambiguities or where explanations of conditions existing as of the date of the deed are necessary.*

Surveyors do not ask a person what was intended; the surveyor reads the documents that were signed and takes into account explanations of extrinsic ambiguities and the legal and survey meaning of words and phrases; then the surveyor decides what the written intent was. The written deed is conclusive evidence; verbal testimony is of no avail except to explain an extrinsic (latent) ambiguity or to explain a condition existing as of the date of a deed.

Circumstances existing at the time the deed was executed may be inquired into to explain the situation or meaning of words as of that date. A description written in very old deeds may use terms that have become obsolete or may fail to make reference to elements such as the basis of bearing. Failure to specify the basis of bearings may be interpreted to be on a magnetic basis probably determined by prior case law decisions. Witness evidence and other documentary evidence revealed that this was the custom at the time. Allowing witness evidence to reverse the usual interpretation placed on bearings is not altering the meaning and intent of the words in a deed; it is merely explaining the meaning and intent of the words as of the date of the deed.

It cannot be overemphasized that it is the *intent* of a conveyance or document that is controlling, above all. The basic rule in the construction of documents is to ascertain and give effect to the real intention of the parties, particularly of the grantor, *as expressed by the language used*, when not contrary to settled principles of law and rules of property [36]. It is the intent of the grantor as spelled out in the deed itself that must be interpreted, not the grantor's (or parties') intent in general, or even what he or she or they may have intended [37].

Admitted mistakes after the fact will not change the wording in a document, nor will they, by themselves, void the document. If that were allowable, any document could be subject to change at a person's whim. A deed conveys the land actually described, regardless of the mistake of the parties. Land cannot be transferred by the intent of the parties alone [38]. The words used in a deed (description) become sealed in law and in time at the instant the deed becomes effective.

There seems to be some confusion or contradictory information as to what constitutes intent of the parties. Some rely on the "meaning and intending" clause in a document to define or explain the intent of the parties. References to prior conveyances are made for varying purposes. They are made sometimes to show the source of title; sometimes to show the identity of the land conveyed; sometimes, and generally by way of caution, to afford a more definite description. It is probably true that in the larger number of cases, the reference is made to show the source of title.

"Reference to prior deeds, unless expressly appearing otherwise, is only intended to help identify the premises conveyed, and not to determine the quality or quantity of the title" [39]. A clause in a deed, at the end of a particular description of premises by metes and bounds, "meaning and intending to convey the same premises conveyed to me," was held in a Maine case to be merely a help to trace the title, and did not enlarge (or limit) the grant [40].

Where an original survey is involved, the intention of the parties is considered to be essentially the same as that of the surveyor. The surveyor's intention is to be ascertained by scrutinizing what the surveyor actually did in making the survey as reflected by his or her field notes and the attending totality of circumstances of the survey [24].

3.16 Evidence of Intent of Senior, Equal, or Junior Rights

PRINCIPLE 13. *Depending on the evidence discovered, a conveyance is classified with respect to the adjoiner as being senior in rights, as being equal or simultaneous in rights, or as being junior in rights. An intent, gathered from the written evidence, cannot alter senior rights.*

To correctly survey a given parcel with its boundaries, the surveyor should identify the classification of priorities of prior conveyances, and such classification is sometimes difficult or impossible to derive from discovered documentary evidence.

A person who has conveyed part of his or her property to another cannot at a later date, legally, convey it to someone else, irrespective of the person's written intent in a new conveyance. The first deed (senior) has a right to all the land that is called for, and the seller (junior) owns the remainder. If a person owns a remainder, no excess or deficiency exists. A remainder does have a definite size, but it is "more or less" in character until measured. A person may have more or less land than he or she expects, but as long as he or she has a remainder, the unexpected quantity of land, be it large or small, all belongs to that person. It is not divided among several owners.

If parcels are created in sequence with a lapse of time between them, senior rights exist. Metes-and-bounds descriptions are usually created with a lapse of time between each creation and hence are created in sequence, whereas all lots on a subdivision map are normally created at the same moment of time (when the map is filed or accepted), even though the lots are sold in sequence. Lots created simultaneously will have equal rights.

Examples of simultaneous descriptions appear in the following:

1. Wills and gifts, wherein none of the heirs or benefactors are designated to receive a remainder.
2. Lots in subdivision, wherein a map is filed with a governing body and no lot is sold prior to filing the map.
3. Lots in any legal subdivision, wherein it is impossible to distinguish an intent to give senior rights to buyers in sequence.
4. Court proceedings in partition, wherein each litigant is given a proportionate share of the whole and no one is designated to receive the remainder.
5. Metes-and-bounds descriptions that are created simultaneously and no one is designated to receive a remainder.

PRINCIPLE 14. *The words of a written conveyance are conclusive evidence as between the buyer and seller, but they are not conclusive evidence as against a third party who has senior rights or a person with occupancy rights.*

If a person conveys part of the land, at a later date the seller cannot sell more than the remaining parcel. The first buyer has senior rights and the second buyer has junior, or remainder, rights. The senior buyer is entitled to all of the land that was conveyed in the initial deed, according to his or her description. The junior also is entitled to the entire description, provided none of the description interferes with the rights of the first. If it is determined that there is an interference, then the junior deed loses. In dealing with written evidence, the surveyor should first determine if junior and senior rights are important. If so, then they must be identified from the written evidence.

As an example, Jones sells the westerly 50 feet of his record 100-foot lot to James. Later, he sells the easterly 50 feet to Smith. A subsequent survey indicates the lot is only 98 feet in width. Thus, each of the two individuals have a conflict of title to 2 feet between the two portions. Each of the two have title to the same 2-foot portion. In this instance the written title of Smith does not prevail against James. James is senior to Smith to only the 2-foot portion. One of the problems lies in the fact that rarely does a seller call for the senior monuments, if they are in place, and rarely do they conduct an adequate survey when they have an identifiable lot with indicated data.

The problem usually occurs when descriptions are changed in the "middle of the stream" or title. When in the course of subsequent descriptions, the grantor goes from a description of, for example, a 10-acre parcel, E1/2 and west 1/2 to E 1/2 and the west 5 acres. The words 1/2 and *acres* do not have the same meaning. To paraphrase a Pennsylvania decision, the court held that in an ejectment action where the plaintiff's deed (senior deed) called for vacant land along the common disputed line, and the defendant's deed called for the plaintiff's survey, it was conclusively held that the plaintiff's survey was earlier (senior) in time than the defendant's and there was no question to submit to the jury [41].

The court held that the date of the conveyance and not the date of the recording determined the seniority of the deed. This may be true in Pennsylvania, but in many states any unrecorded deed usually will not prevail against an innocent third-party purchaser.

PRINCIPLE 15. *Each deed is construed as a whole; parts are read in the light of other parts. Nothing stands alone and nothing is rejected if it is possible to give every part of the deed a meaning.*

Although they may be rejected if it is absolutely impossible to give them meaning, clauses of deeds are rarely rejected. The following will give light to this reasoning.

In New Hampshire, as well as in the majority of the other states, the courts hold that a deed must be construed as a whole and a meaning given to each and every part [11]. The courts then go on to say that if any part of a description is false and impossible, but by rejecting that, a perfect description remains, the false part should be rejected, and the deed held good [17].

They also tried a little different approach: "Any particular portion of a description may be rejected, if it is manifestly erroneous, and enough remains to identify the land intended to be conveyed" [42].

3.17 Value of Spoken Evidence (Words)

The second type of verbal evidence is spoken words. Many individuals call this testimony and parol evidence. Usually, a witness believes that testimony is that given in a court during trial. This is far from true, but that is but one form of testimony. Testimony can be any spoken statements given to the surveyor or to the attorney.

In the course of a survey, interested individuals who are not parties or landowners affected by the survey relate statements or information to the surveyor, which become evidence. It is not up to the surveyor to discredit this testimony, and the surveyor must consider this testimony as part of the totality of the evidence considered.

PRINCIPLE 16. *Testimony of interested landowners is one form of evidence that must be considered by the surveyor as part of the totality of the evidence. Not to do so may be a serious flaw in the collection of any evidence.*

Failure of the surveyor to consider all testimony (verbal spoken evidence) could be a serious flaw in the survey. Testimony of interested individuals is critical in determining if a corner is lost or obliterated. This form or class of evidence is one of the major types of evidence that courts will permit in order to raise the dignity of a lost corner to an obliterated corner.

The strength the Bureau of Land Management (BLM) gives to testimony is quite compelling. The BLM defines an existent corner as follows (emphasis added):

> An existent corner is one whose position can be identified by verifying the evidence of the monument, or its accessories, by reference to the description that is contained in the field notes, or where the point can be located by an acceptable survey record, some physical evidence, or *testimony* [43].

The strength of testimony is that even though all of the physical evidence has disappeared, a corner will not be considered lost if its position can be recovered through testimony alone of one or more witnesses who have a dependable knowledge of the original position.

The problem is when two or more witnesses place the corner in different locations. Even if the surveyor may consider this a problem, it is still not a problem, because that is what juries are for. The jury is to listen, to observe, and to weigh the evidence (testimony) of all of the witnesses and then to pass judgment on what they wish to discredit and what they wish to believe.

In the same publication, the BLM goes on to discuss an obliterated corner. At law, each of these two classifications has equal standing:

> An obliterated corner is one at whose point there are no remaining traces of the monument. or its accessories, but whose location has been perpetuated, or the point for which may be recovered beyond reasonable doubt, by the acts and testimony of the interested landowners, competent surveyors, or other qualified local authorities, or witness, or by some acceptable record evidence [43].

This is interesting in that the position is based on collateral evidence, but this collateral evidence should be supported, generally through local acceptance and relationship to other corners and agreement with field notes (verbal evidence) or unquestionable testimony [43].

The surveyor has the terrible burden of trying to extract sufficient testimony that is unprejudiced and supported by a sound reasoning. If there is a shred of acceptable evidence of the original location, that position should be accepted.

PRINCIPLE 17. *Words that are translated from one language to another have the meaning inferred in the original language and not in the second language.*

In today's transient world, surveyors move from country to country—as do landowners as well as professional people, including attorneys. English is accepted as the official language in the United States, and all modern documents and court records are in that language. For surveyors whose second language is English, or English-speaking surveyors who must refer to documents written in second languages, it is imperative that any translation be compiled by a certified translator. Translation of survey and land documents are not like "tourist" translations. They require a translator who is knowledgeable as to nuances and words used specifically in land boundaries and surveying. As an example, in Hawaii, in order to become a licensed surveyor, the applicant must be able to translate a survey and legal description from the Native

Hawaiian into English. Those who are familiar with Puerto Rico find that the original documents are in Spanish, and all local courts are conducted in Spanish, but federal documents and the federal courts are all conducted in English. And, of course, many of the original documents and surveys in Texas and some in California are in Spanish as a result of their Spanish heritage.

PRINCIPLE 18. *In ancient documents the words have the same meaning they had at the time they were written and not the meaning in today's definition.*

In referencing old documents, one should use the meaning of the word used at the time the document was created and not the one in use today. Over the years, common words have changed meanings, and the original meaning should be applied because that will meet the intent if the original parties at the time the document was created.

3.18 Summary

Without written and spoken evidence or words, it would be difficult for the boundary surveyor to function effectively. There are those inexperienced surveyors who will attempt to conduct a retracement survey without obtaining an adequate description of the property or without doing the necessary research to obtain additional survey and deed information. There is no possibility that any retracing surveyor can attempt to retrace any original survey without first obtaining a copy of the field notes and plats that created the original parcel. This borders on negligence, and it has been held by several courts that the failure of the surveyor to do an adequate record search is justification to hold the surveyor liable for the resulting damages.

Although there may be no written standards, a surveyor, when conducting the retracement of a property boundary or description, must realize that any boundary has a minimum of two owners, one on each side, and it is an absolute necessity that both individuals be interviewed to obtain testimony as to beliefs and knowledge. Although the written word is paramount as to what was conveyed in a description and deed, the spoken word is just as important as to what the parties know and believe.

REFERENCES

1. *Black's Law Dictionary*, 4th ed. (St. Paul, MN. West Publishing, 1968).
2. *Don Hunt and Hunt family v. LLC v. Christopher Feese and Kathleen Feese v. Stephen H. Pounds* (in the circuit court of the ninth judicial circuit in and for Orange County, Florida, case no. 2014-ca-004736-0).
3. D. Wilson, *Interpreting Land Records* (Hoboken, NJ: Wiley, 2015).
4. *Blair v. Rorer's Administrators,* 135 Va. 1 (1923).
5. *Rivers v. Lozeau*, 359 So. 2d 1147 (Fla. 1989).
6. *Kauka Farms, Inc. v. Scott*, 256 Ga. 642, 352 S.E. 2d 373 (1987).
7. *Brasher v. Tanner*, 256 Ga. 812, 353 S.E. 2d 487 (1987).

8. *Harvey v. Inhabitants of Town of Sandwich*, 256 Mass. 379, 152 N.E. 625 (1926).
9. *Wells v. Jackson Iron Mfg. Co.*, 47 N.H. 235 (1866).
10. *Almendinger v. Mc Hie*, 189 Ill. 308, 59 N.E. 517 (1901).
11. *Lyford v. City of Laconia*, 75 N.H. 220, 72 A 1085 (1909).
12. 26 *C.J.S. Deeds*, sec. 83 (St. Paul, MN: West Publishing Co., 1956).
13. *Leeper v. Leeper*, 347 Mo. 442, 147 S.W. 2d 660 (1909).
14. *U.S. v. 15,883.5 Acres of Land*, 54 F. Supp. 849 (S.C.D.C. 1944).
15. *First Hartford Court v. Kennebec Water District*, 490 A. 2d 1209 (Me. 1985).
16. *Jerrick v. Hopkins*, 23 Me. 217 (1843).
17. *Wadleigh v. Cline*, 99 N.H. 202, 108 A 2d 38 (1954).
18. *Lane v. Thompson*, 43 N.H. 320 (1861).
19. *Ferris v. Coover*, 120 Cal. 590 (1858).
20. *Basso v. Veysey*, 118 Vt. 401, 110 A. 2d 706 (1954).
21. *Thomas v. Patan*, 13 Me. 329 (1836).
22. *Morse v. Emery*, 49 N.H. 239 (1890); *Smith v. Forrest*, 49 N.H. (1870).
23. *Haley v. Martin*, 85 Miss. 698, 38 So. 99 (1904).
24. *United States of America v. Champion Papers, Inc.*, 361 F. Supp. 481 (Texas, 1973).
25. *Proprietors of Claremont v. Carlton*, 2 N.H. 369 (1821).
26. *Williams v. Warren*, 211 Ill. 549 (1850).
27. *Richfield Oil Corp. v Crawford*, 39 Cal. 2d 729, 249 P. 2d 600 (1952).
28. Civil Code Proceedings. *Richfield Oil Corp. v. Crawford*, ibid., Cal. LEXIS 300, 1 Oil & Gas Rep. 1583.
29. *Preacher v. Strauss*, 47 Miss. 353 (1872).
30. *People v. McCall*, 129 Misc. 862, 223 N.Y. Supp. 257 (1927).
31. *Bunder v. Grimm*, 142 Ga. 448, 83 S.E. 299 (1914).
32. *Mentz v. Town of Greenwich*, 118 Conn. 137, 171 A 10 (1934).
33. *N.Y. v. Wilson*, 278 N.Y. 86, 15 N.E. 408 (1938).
34. Appeal of Jarbo, 91 Conn. 265, 99 A. 562 (1916).
35. *Madden v. Tucker*, 46 Me. *367* (1859).
36. *Gibson v. Pickett*, 512 S.W.2d 532 (Ark., 1974).
37. *Wilson v. DeGenaro*, 36 Conn. Sup. 200 (1979).
38. *Tilbury v. Osmundson*, 352 P.2d 102 (Colo., 1960).
39. *Perry v. Buswell*, 113 Me. 399 (1915).
40. *Brown v. Heard*, 27 A. 182 (Me., 1893).
41. *Collins v. Clough*, 222 Pa. 472, 71 A. 1077 (1909).
42. *Lane v. Thompson*, 43 N.H. 320 (1861).
43. U.S. Department of the Interior, Bureau of Land Management, *Restoration of Lost or Obliterated Corners and Subdivision of Sections* (Washington, DC: U.S. Government Printing Office, 1974), p. 9.

ADDITIONAL REFERENCES

Boundaries: Measurement in Horizontal Line or Along Surface or Contour, Geo. Gluck, ed. 80 ALR 2d. 1208 (Rochester, NY: Lawyers Co-Operative Publishing Co., 1961).

Chermside, Jr., H., *Boundaries: Description in Deed as Relating to Magnetic or True Meridian*, 70 ALR 3d. 1220 (Rochester, NY: Lawyers Co-Operative Publishing Co., 1976).

Donald A. Wilson., *Interpreting Land Records, 2nd Edition*. Hoboken: John Wiley & Sons, Inc. 2015).

4

EVIDENCE AND TECHNOLOGY

4.1 Introduction

Evidence has generally been described in very simple terms. To paraphrase the descriptions proffered in various legal publications and numerous legal decisions in the several states, the practicing surveyor has been cast into an entirely new world of evidence that he or she may not understand and may not be prepared to enter.

Evidence that ultimately identifies the positions of the original boundary lines and corners—other than the original evidence of boundary corners and lines utilized to locate or prove corners and the resulting boundary lines controlled by these corners—is a combination of the totality of all the evidence.

> In retracing boundaries, the *totality* of the evidence must be evaluated. Only the original evidence is absolute. All other evidence is secondary and is not controlling.
> All primary and secondary evidence must be correlated to writings that created the evidence of the original boundaries, which must be within the chain of title.
> Relocated lost corners and boundaries that are determined by modern precise means can be no more precise than the instruments that originally created them.
> When evidence fails to prove the original corner position, then and only then can the surveyor disregard the evidence and rely on measurements alone.
> Original corners and lines do not depend on technology for accuracy. Accuracy is determined by the law. Law has deemed all original corners accurate; regardless of their precision.

Lost corners that are created by methods of greater precision, other than the original method(s), can be no more precise than the method that created the original corners and boundaries. The application of modern technical methods of either greater precision or methodology than those that created the boundaries does nothing to increase the accuracy of the original corner positions and line locations. It does, however, aid in the subsequent retracement of those lines with more modern equipment.

Since the corners and the resulting lines are a product of both technical methodology and equipment as well as the application of laws, the two elements that created the corner positions are not technically compatible. Technology has changed geometrically, while the law has changed arithmetically.

In order to create a boundary line, the surveyor must first create two corners: one at each end of the line.

4.2 Principles to Be Applied to Modern Technology

The principles one should consider when applying technology to retracing original lines may be outlined as follows:

PRINCIPLE 1. *All original corner locations or positions are a product of two measurements: (a) angle (course) and (b) distance.*

PRINCIPLE 2. *When once created, all corners and lines created by an original survey are without error, in a legal sense.*

PRINCIPLE 3. *All corners and lines created by the original surveys should be equated in relationship to the standards of precision that were in effect at the time the lines and corners were created.*

PRINCIPLE 4. *Any error present in the original surveys will be in either direction (bearing) or distance, and these two error possibilities should not be considered to be in a direct relationship to each other. Errors may also be found in words.*

PRINCIPLE 5. *Each error should be independently considered in relationship to the standards that were in effect at the time the line measurements were created.*

PRINCIPLE 6. *The positive position of the original corner locations (positions) must be predicated on the recovery, identification, and interpretation of original evidence and not on applying modern measurements by the retracing surveyor.*

PRINCIPLE 7. *A surveyor cannot use more modern precise techniques and measurements of course (angle) and distance to set a lost corner, but only as supportive or combative evidence in support or in contradiction of lesser evidence.*

PRINCIPLE 8. *An original corner, once created, cannot be replaced or redefined by coordinates created by modern survey calculations and measurements using more precise modern methods.*

PRINCIPLE 9. *Corners cannot be proved from evidence of lines found. But lines are determined from the corners found.*

PRINCIPLE 10. *An original line cannot be ascertained from a single corner, but the line must be determined from proving two original corners, one on each terminus of the line.*

PRINCIPLE 11. *An original line cannot be ascertained with certainty with reference from or to a single corner.*

PRINCIPLE 12. *In order to conduct measurements for replacing a lost corner, the retracing surveyor should obtain the notes of the original survey.*

PRINCIPLE 13. *Any retracement of boundaries must be predicated on locating, with certainty, the evidence left by the creating surveyor. A retracement is not a matter of going into the field and "just looking for evidence," it is a task of knowing what to look for. One must have the original created information in order to have the information of what to look for. Failure to obtain and use this information may raise the question of negligence.*

PRINCIPLE 14. *A coordinate value computed from bearings and distances is less precise as were the original values from which it was calculated.*

PRINCIPLE 15. *With an understanding of the limitation of the original survey descriptions, modern technology can be used to support weak or questionable evidence, quickly and with reasonable cost. It cannot be used as a substitute for evidence.*

These principles set the tone for this chapter and will be discussed in the hope that present-day teachers, surveyors, attorneys, and courts will better appreciate and understand that the sanctity of the original corner positions and the resulting lines cannot and should not be replaced by modern measurements. Modern measurements should only be used to redefine the original lines.

The position of a corner located by modern measurements cannot be the final answer to the problems of retracements. It may only be the starting point for applying all of the various other legal and surveying principles that attorneys and courts need to know, and that surveyors must apply, to locate the positions of the corner or corners being replaced.

The specific rules of evidence that apply to retracements do not permit modern technology to substitute for evidence and legal specifics that are adopted by the courts. Technology cannot be a substitute for either the federal rules or state legal precedents.

Technology cannot be a substitute for legal principles or law.

When survey practitioners determine that new technological methods can be used as substitutes to replace accepted rules or statutes that are exclusively reserved for the courts, this will place a false sense of security on the surveyors who cast aside accepted rules to rely on such technology.

4.3 Applying Technology to Evidence

> Technology is not evidence, but evidence depends on technology in its creation and future retracement.

The foundation or basis of evidence is twofold: first, in its creation, and, second, in its retracement or its reidentification after it is once created.

Identifying the purpose of any retracement—that is, a survey that is being conducted for a specific purpose—no state has a law that requires a landowner to employ a surveyor to conduct a survey if the landowner wishes to prepare a description for a conveyance of a tract of land. If the person who is buying a parcel of land is satisfied, the landowner can prepare a land description based on anything—paces, a yardstick, a tape measure, a rope. The landowner may walk the parcel, pace the distances, and estimate the course angles; or the seller may estimate the distances, use a very crude angle determination and prepare the description. The seller may estimate the distances to determine the distances of the courses as well as angles and the area, and all of these practices would be legally acceptable for the conveyance of a land interest described and conveyed.

If any description is not sufficient, the instrument containing it is legally void in that no land interest was conveyed. The sole purpose of the description of land contained in a deed of conveyance is to identify subject matter of the grant [1]. If it does that, it is sufficient. In fact, descriptions really don't have to identify the land, but only furnish the means for identification—the "key" [2].

A deed will not be declared void for uncertainty of description if it is possible by any rules of construction to ascertain from the description, aided by extrinsic evidence, which property it is intended to convey, and it is sufficient if the description points out and indicates the premises so that by applying the description to the land involved it can be found and identified [1]. Judges will "bend over backward" to make a description valid.

4.4 Original Corners Control

PRINCIPLE 1. *All original corner locations or positions are a product of two measurements: (a) angle (course) and (b) distance.*

From the earliest textbooks known, the various elements of what constitutes a line have been described. John Love addressed the problem in 1688 when he wrote in his text, *Geodaesia*:

> There are but two material things (towards the measuring of a piece of land) to be done in the Field; the one is to measure the lines . . . and the other to take the quantity of an Angle included in these lines; for which there are almost as many Instruments as there are Surveyors . . . [3].

He went on to state that there are almost as many methods of measuring angles as there are surveyors.

Accepting the fact that a line is a component of an angle and a distance, which are products of measurements and errors, both elements need to be considered, not just one.

An original line has errors, some systematic and some random; some instrumental and some personal in nature. Therefore, legally speaking, to have a true retracement, the same instruments as well as the same methods should be utilized to retrace an original line. In addition, if possible, the same individual should be employed to do it, because errors are affected by whether the surveyor is right-handed, left-eyed, short, or tall. The original measurements can never be duplicated—including by the same person who originally created them. They can only be verified.

> The responsibility of the retracing surveyor is twofold:
>
> 1. Verify the evidence of the original surveyed lines and create new evidence for future surveyors.
> 2. Resurvey and describe the lines in accordance with present-day standards while making reference to yesterday's results.

4.5 Original Survey Controls

PRINCIPLE 2. *When once created, all corners and lines created by the original survey are without error, in a legal sense [4].*

Although the U.S. Public Land Survey System systematized the mode of surveying vast areas of the *public* domain, it also has had lingering and legal effects on the metes and bounds system of surveys that dominate nearly a third of the states—primarily the colonial states and Texas, as well as many surveys conducted in the public domain.

The residual effect is the adoption into our legal system of various survey principles that seem to cross legal boundaries. There were no legal requirements or laws precedence in the metes and bounds systems of surveys that the original corners are where the original surveyor created them by a function of measuring lines and angles and then setting monuments at the corner points. A far-reaching precedent whose decision has been repeated and adopted in the legal world in both the metes and bounds as well as the General Land Office states:

> The making of resurveys or corrective surveys . . . once proclaimed for sale is always at the hazard of interfering with private rights, and thereby introducing new complications. A resurvey, properly considered, is but a retracing, with a view to determine and

establish lines and boundaries of an original survey, . . . but the principle of retracing has been frequently departed from, where a resurvey (so called) has been made and new lines and boundaries have often been introduced, mischievously conflicting with the old, and thereby affecting the areas of tracts which . . . hav[e] previously [been] sold and otherwise disposed of.

It will be perceived that [the resurvey] not only disregards the old original survey making new lines and boundaries, but does so in contravention of the order from the Land Office that those resurveys should not be extended into this township.

That the power to make and correct surveys of the public lands belongs to the political department of the government and that, whilst the lands are subject to the supervision of the General Land Office, the decisions of that bureau in all such cases, like that of other special tribunals upon matters within their exclusive jurisdiction, are unassailable by the courts, except by a direct proceeding; and that the latter have no concurrent or original power to make similar corrections, if not an elementary principle of our land law, is settled by such a mass of decisions of this court that its mere statement is sufficient. and cases cited in that opinion; *United States v. San Jacinto Tin Co.*, 10 Sawyer, 639, affirmed in *United States v. Flint*, 4 Sawyer, 42, affirmed in *Ellicott v. Pearl*, 10 Pet. 412.

The reason of this rule, as stated by Justice Catron in the case of *Haydel v. Dufresne*, is that "great confusion and litigation would ensue if the judicial tribunals, state and federal, were permitted to interfere and overthrow the public surveys on no other ground than an opinion that they could have the work in the field better done and divisions more equitably made than the department of public lands could do" [5].

4.6 Standards of Creation Control Retracements

PRINCIPLE 3. *All corners and lines created by the original surveys should be equated in relationship to the standards of precision that were in effect at the time the lines and corners were created.*

Changes in the law and changes in technology are not related. Law changes at a more deliberate pace and more slowly than does technology.

Within a present-day's surveyor's lifetime, we have seen the technology change from the staff compass and link chain to global positioning units. By contrast, a line run by a staff compass could be expected to create an angle measurement precise to one-quarter of a degree of arc and a distance measurement was precise to one chain in a mile, with the corner being established at the end or termini of these survey-measured lines. Today's global positioning units create the end points first, and from the created points calculate the lines that were not run. This becomes a legal problem in that the law provides that the lines run on the ground indicate the intent of the boundaries of the creating surveyor as indicated by the end points (corners) and lines created by the original surveyor. The corners and lines created in the original survey provide the final question for the retracing surveyor to answer: "Am I following the lines that were originally created?"

This book does not advocate the return to the staff compass and chain, but the retracing surveyor should understand the relationship to distances and courses recited in the historic original descriptions. The following scenario is an example.

The retracing surveyor is given a description that was originally created in 1802 with the angles indicated to the nearest degree and the distances to the nearest full chain. Example: Thence North 56 degrees West a distance of 80 chains. All of the recited angles and distances are indicated the same. Applying the significant figure, the creating surveyor could have read the following in the survey to indicate those numbers.

The bearing could be North $55\frac{1}{2}$ degrees to $56\frac{1}{2}$ degrees. Remember, the compass could only be read to the nearest half degree. The ability to read the compass was unique to each individual. This would place an uncertainty of approximately 96 feet in a mile. Distance-wise, if the 80 chains are indicated to the nearest chain, the surveyor can call a distance of 79.5 chains to 80.5 chains, 80 chains. This would place the measured position for the placement of the corner in a sphere of $79\frac{1}{2}$ chains to $80\frac{1}{2}$ chains North and South and approximately 96 feet east and west. If this corner established were to be considered a lost corner and placed solely from measurements by a GPS unit, the spread would be only inches.

4.7 Errors in Original Surveys

PRINCIPLE 4. *Any error present in the original surveys will be in either direction (bearing) or distances, and these two error possibilities should not be considered to be in a direct relationship to each other. Errors may also be found in words.*

Although a course has two components, bearing and distance, there is no relationship between the two. The original measurements were made separately, by different instruments. In all probability, they were made by different individuals, in some instances at different times, with each method having separate and distinct errors.

Modern measurements made by modern equipment also have the same errors present, but a new error was introduced, that of errors of calculation, which in many instances cannot be detected. Thus, if a point were placed from an erroneous error, the retracing surveyor, not having the actual field notes available, will not be able to detect and correct for the error and thus find the original location.

Although many modern surveyors now discuss "positional tolerance" as a criterion for evaluating corners, one who understands the "what and how" of the original surveys realizes that applying the standards in effect at the time finds that the original corner positions also had a "positional tolerance."

All original corners have a positional tolerance location.

The retracing surveyor must look at the equipment and standards that were used at the time the corners and lines were created. As the sophistication of the equipment has developed, the original surveys remained static. The retracing surveyor must look at the equipment used by the creating surveyor. Since each line has two elements, a course, or angle from a meridian, and a distance, each of these two elements have numerous errors and limitations.

Using a practical example from the early 1800s, the majority of the equipment used was a surveyor's compass, usually reading to a quarter of a degree, and a two-pole chain of 50 links, or 33 feet. Realizing that the original surveyors were creating the boundaries and then leaving a field trail and a written record, the unwritten standards required of the original surveyor for the fieldwork was +/– 1/2 for the angle and +/– 1/2 chain. This probability of creation is the "positional tolerance" one can expect today. The distances and angles recited in the original notes and plats are not absolutes but are "uncertainties" of position.

4.8 Effect of Standards

PRINCIPLE 5. *Each error should be independently considered in relationship to the standards that were in effect at the time the line measurements were created.*

In analyzing found evidence, the relationship of the found evidence to the measurements should be to the standards in effect at the time the measurements were conducted and not to the modern measurements of today.

Applying this principle permits the surveyor to exercise latitude in accepting or rejecting evidence to prove the existence or nonexistence of a corner point, This also will permit the retracing surveyor to exercise personal initiative and experience in evaluating evidence and in accepting this evidence.

Legally, the courts frown on the total reliance on measurements to position corners. They would much prefer to accept a corner position proven by exhaustive field research than a corner set by a precise measurement. In evaluating found evidence, the retracing surveyor should use the standard "preponderance of the evidence" in the evaluation. Although the U.S. BLM identified the criteria as "beyond a reasonable doubt," this has been held by the various courts to be in error, and now have modified the standard to be "credible," exactly what the original was before the 1973 Manual was printed. With the lower criteria or degree of proof, for the retracing surveyor, it now becomes possible to permit a wider degree of latitude and possibilities to make "lost corners" obliterated.

4.9 Original Corners Based on Found Evidence and Not New Measurements

PRINCIPLE 6. *The positive position of the original corner locations (positions) must be predicated on the recovery, identification, and interpretation of original evidence and not on applying modern measurements by the retracing surveyor.*

Courts have been stung several times by the misapplication of technology data from experts. Although judges may have GPS units in their automobiles, this does not assure that the judge will not get lost. It will assure that if the judge does get lost, others may be able to tell him or her to one-quarter of a foot where they are [6].

This places a survey responsibility on both the creating surveyor to leave sufficient evidence behind and the retracing surveyor to know what to look for and then to understand what was found.

4.10 Original Evidence Superior to Modern Measurements

Modern precise measurements yield to original evidence of found and proven corners and lines.

There is no retracement principle that permits a retracing surveyor to disregard found original evidence or to change original measurements. If evidence of the original survey is recovered and identified at a position that modern precise measurements indicate that it theoretically cannot or should not be, the retracing surveyor is obligated to disregard that position and give the greater weight to recovered evidence.

It is easier for the untrained surveyor to rely on modern measurements that do not have to be proven than it is to place his or her professional reputation and career in jeopardy in having to argue the evidence recovered.

4.11 Technology and Setting Lost Corners

PRINCIPLE 7. *A surveyor cannot use more modern precise techniques and measurements of course (angle) and distance to set a lost corner, but only as supportive or combative evidence in support or in contradiction of lesser evidence.*

This principle places a degree of responsibility on the retracing surveyor, in that it first places a large responsibility on the retracing surveyor to understand the degree of proof for the evaluation of evidence and how this should be applied to the evidence found and how new technology should be used in replacing lost corners.

The beauty of new technology is twofold:

1. Coupled with an understanding of the precision of the original surveys, and then applying this understanding to the found evidence, a more precise modern measurement can be used to support weak and questionable evidence.
2. It can be used when a decision is made to accept the older evidence as the best remaining evidence of the original corner position. A more modern measurement can be used to identify the position, more precisely, that could be an aid for future recovery.
3. It can be used to "refine" the original measurements to modern relationships.

4.12 Coordinates and Corners

PRINCIPLE 8. *An original corner, once created, cannot be replaced or redefined by coordinates created by modern survey calculations and measurements using more precise modern methods.*

With the advent of computers and the software to accompany the rapid ability to complete computations, surveyors are relying more on the computed evidence than on the solid evidence found while retracing original surveys.

Recently in a trial, two surveyors were testifying, both as experts, as to the validity and reliability of their measurements and computations. The basic question was, is a measurement indicated on a plat as 231.14 feet better than one indicated as 231.136? Neither surveyor understood that the validity of the computed measurements is only as good as the field measurements. One distance was generated from a field measurement and the other from a GPS reading. The surveyor with the 231.14 feet measured it using an electronic distance measurement (EDM), measuring a random line between two corners, and then calculating a straight line between the two nonvisible distant points, while the second surveyor computed the coordinates from the two coordinate values from a GPS measurement between the two occupied corners.

Questions must be answered by the surveyor as to the validity of the corner points being measured between; the precision of the original field measurements; and the program used to compute the measurements. None of these were asked.

In retracing original measurements, most of the measurements indicated in the ancient descriptions are not balanced measurements of bearing and distance, but most likely are the original field measurements, as they should be.

4.13 Corners Control Over Lines

PRINCIPLE 9. *Corners cannot be proved from evidence of lines found. But lines are determined from the corners found.*

There are some surveyors, attorneys, and judges who, unknowingly, attempt to use lines to prove corners. These few individuals do not understand that without two proven corners, one on each end of the line, one cannot identify or prove an original line.

Once again, we can reach back to John Love, who defines both a point [corner] and a line as follows:

> A Point [corner] is that which hath neither Length nor breadth, the least thing which can be imagined, and which cannot be divided, commonly marked as a full Stop in Writings thus (.)
> A line has Length, but no Breadth nor thickness, and is made by many Points joined together in length, of which there are two sorts, *viz* Straight and Crooked [7].

This definition once again leads us to the realization that measurements create *points*, and it is the use of these points by placing monuments and referencing in descriptions that the points become corners. The corners, in turn, control future locations of the boundaries.

But in order to have a line created, the surveyor must have a minimum of two points, one at each end of the line. This is the minimum to create a line, and to absolutely locate the created line, the same two points must be recovered.

4.14 Two Points [Corners] Define a Line

PRINCIPLE 10. *An original line cannot be ascertained from a single corner, but the line must be determined from proving two original corners, one on each terminus of the line.*

As discussed, to have an absolute location of an original line, the two original corners that were established on each end of the line must be recovered and identified from reference to the original evidence. Without two original points or corners at each end of the line, the retracing surveyor must resort to evidence plus two independent and separate measurements, in two directions, from that original corner.

Since, by law, original corners' positions are without legal error, then the measurements from that original proven point is subject to four possible errors. For discussion, consider this scenario: We have four corners of a parcel: A, B, C, and D. Our research indicates only corner C. If we were to prove a second corner, we could resort to proportioning the lost corners, but that is not so. Thus, we use the found corner to locate our parcel on the ground, and the remaining three corners and four lines must be resurveyed for location.

Consider a description that was written in 1800. All angles were given to the full degree and lines were measured to the full pole. There are three ways to retrace this parcel:

1. We retrace it in 1965, using a 1-minute transit and a 100-foot steel tape marked in feet.
2. We retrace it in 1995, using an EDM that measures to .01 feet and a 1-second theodolite.
3. We survey the same description in 2005, using a three-unit GPS system that *does not* measure lines on the ground, but determines the position of three points and then ascertains, by computations, the distances between the points to any significant number you want.

Each method will give different results for the three lost points, and none will resemble the original four lines and corner positions.

Using the techniques in the first two examples, according to Love, we have errors in the measurements of the angle and the distances. For example, we have possible error in the measurement from each satellite, as well as measurements in the computations [1]. A great professional responsibility is placed on the original surveyor who created the corners and lines, in that the creating surveyor who created the original boundaries should create, leave, and identify sufficient evidence so that the retracing surveyor will not have a difficult time "*finding the footsteps*" that were originally created. This evidence should be in the form of marks in the field, notes, plats, and reports to aid the retracing surveyor of tomorrow.

4.15 Using Modern Technology Wisely

PRINCIPLE 11. *An original line cannot be ascertained with certainty with reference from or to a single corner.*

Since the definition of a line requires termini at each end of the line in order to have the line begin and end, corner points are required. With a single point the termination of the line is subject to the angle and the distances to the opposite end of the line. Yet the identification of the terminal point can positively be located by evidence to the degree of precision that the point was originally created.

PRINCIPLE 12. *In order to conduct measurements for replacing a lost corner, the retracing surveyor should obtain the notes of the original survey.*

PRINCIPLE 13. *Any retracement of boundaries must be predicated on locating, with certainty, the evidence left by the creating surveyor. A retracement is not a matter of going into the field and "just looking for evidence," it is a task of knowing what to look for. One must have the original created information in order to have the information of what to look for. Failure to obtain and use this information may raise the question of negligence.*

PRINCIPLE 14. *A coordinate value computed from bearings and distances is less precise as were the original values from which they were calculated.*

Every measurement has errors. Some instrumental, some personal, and some resulting from the tables used to obtain data for calculations. It is not a question that decimal points will make a measurement or a calculation any more precise. It is a result of what was the precision of the original measurement and calculations. As a result, two numbers are merged to compute a third number, which then is used to calculate coordinates, which then are placed back on the ground, with instruments that have errors of their own, which are different than the original instruments.

PRINCIPLE 15. *With an understanding of the limitation of the original survey descriptions, modern technology can be used to support weak or questionable evidence, quickly and with reasonable cost. It cannot be used as a substitute for evidence.*

Modern technology *is not* a final answer to retracements, but modern technology and its resulting instruments can be excellent tools to aid the modern-day surveyor, primarily in two areas:

1. Creating modern measurements for inclusion in modern descriptions that will provide the retracing surveyor of tomorrow with precise measurements, with little or no uncertainty, for retracing.
2. Giving credibility to weak or inconclusive evidence.

When conducting a survey, the surveyor is usually asked to either create new parcels of land or to retrace and locate parcels originally created. The use of new technology can certainly be an asset in either situation.

The more precise the original parcel location, the easier is the job of retracing for the retracing surveyor, because it takes the uncertainty out of the evidence recovered. The more precise the measurements made in the original surveys, the easier it is to retrace parcels based on descriptions created from these surveys. This is because the retraced measurements can easily be correlated and referenced to the retraced measurements, thus making the verification of suspect measurements much easier. The surveyor who is employed to perform this work needs to be technically trained.

The second point is to lend credibility to weak or inconclusive evidence. The retracing surveyor should have different qualifications from the creating surveyor. This does not mean that the surveyor needs to be technically versed in the modern methods and methodology to survey. In retracing old surveys, evidence becomes the criteria of proof. The experienced surveyor will distinguish between good evidence and bad evidence and know what weight to give to this evidence in order to reach sound conclusions. This is where an understanding of new technology can be beneficial, correlating and using other forms of evidence.

For example: A one-quarter corner ($\frac{1}{4}$) of a section is in question. Is it lost or obliterated? This is critical, in that to proportion it would place the resulting lines some 16 feet from an existing east–west fence line. The question is: In using the totality of all the evidence, testimony, monuments, and measurements, "Is the fence line the best remaining location of the original corner, or should it be proportioned for lack of positive evidence?"

Possible solution: The original field notes of the line indicated a total distance of 80 chains, or 40 chains and 40 chains. The retraced distance between the two proven section corners is 79.10 chains, between one existent corner and an obliterated corner, which would place the proportioned lost quarter corner at 39.55 chains between the two corners. Giving weight to the record from the corner from which the original line was run on the record bearing, the fence lies at 39.87, just 13 links shy of the record distance.

Additional information indicates the fence has been in place for over 65 years. It is a barbed-wire fence placed on the remains of a split-rail fence, and has been considered as the division line between two separate patents dated within 2 years of each other.

Applying the Cooley dictum, the surveyor accepts the fences as the best remaining evidence of the most probable location of the position of the original quarter corner over the resurvey proportioned location that is inconsistent with possession lines.

The surveyor cannot say this corner is based on the recovery of the original evidence, but the surveyor can say the following:

- I have two fences that are indication of possession based on title.
- I have a fence that measures within 13 links of record distance.
- I have an original line that was precise to within 50 links, and as such the remeasurement is entirely possible.
- The proportioned location places the proportioned line some 25 links off the north–south fence line, to the west.

This is the strength of modern technology—the ability to get quick and precise measurements and to make instant decisions, while within the field.

4.16 Conclusion

This chapter is not written to discredit technology. It is written in an attempt to indicate that new technology and its results should not be used to the exclusion of higher and more conclusive evidence. It should not be used to the exclusion of all other evidence, but it should be used along with other technological tools such as the transit, the compass, the chain, or any other of the tools as a supplement to help support better or contradictory evidence that causes the surveyor to ponder and to have problems evaluating the evidence recovered.

TECHNOLOGY BY ITSELF IS NOT EVIDENCE.

It is the resultant measurements from modern technology that can become evidence to corroborate lesser evidence that is conclusive by itself.

REFERENCES

1. *Kuklies v. Reinert*, 256 S.W.2d 435 (Tex. App., 1953).
2. *City of North Mankato v. Carlstrom*, 212 Minn. 32 (1942).
3. John Love, *Geodaesia, or the Art of Surveying and Measuring Land* (1688). Reprints available from Walt Robillard, robw@mindspring.com.
4. *Cragin v. Powell*, 128 U.S. 691 (1888).
5. *York and Md. Line RR. V. Winans*, USSC, 1884, 17 How. 30.
6. *Daubert v. Merrell Dow*, 507 U.S. 903, 113. S. Ct. 1245, 122: Ed.3d 645 (1993).
7. Love, *Geodaesia*, Chapter II.

5

OTHER TYPES OR SPECIES OF EVIDENCE

5.1 Evidence of a Survey

In previous chapters the general concepts of evidence, both verbal—evidence both spoken as written and demonstrated by words—and nonverbal have been discussed. Most surveyors and field personnel are more familiar with what we will call real or field evidence or, for want of a better description, other evidence. This type of evidence includes all other forms of evidence, except testimony and words, and measurements. This is the evidence of the field individual, the woodsman, the stalker, and the specialist other than the eminent measurer. When working with and evaluating evidence, the "basic Sherlock Holmes" that is within all of us must be drawn upon. Evidence should be looked at from both the technical as well as the forensic viewpoints. In this volume, we cannot teach this area; we can only point out guidelines that the student and the surveyor can learn.

For practical purposes, without an actual survey creating the parcel or parcels of land, and the monuments set, the chopped lines that were left in the woods, the bearing or witness trees that were scribed and referenced by course and distance, as well as the maps that were drawn by the light of a campfire or candle using quill pens, and the thousands of miles of roads and fences that were constructed in reliance of these original surveyed lines, there would be no other evidence of survey. We would of necessity be limited to the written words in the written descriptions from which surveyors conduct their work.

This is an area where only experience can be the teacher. No surveyor can appreciate the fact that an ancient fence line may be "aged" or that a blaze or scribe mark might be accurately, though perhaps not precisely, determined. One cannot appreciate determining the species of a stump hole from a few charcoal fragments that were recovered in the course of searching for an original corner or bearing tree. A surveyor may be a preeminent measurer of distance, able to use the most current and sophisticated surveying and electronic equipment, but until the art and science of the interpretation of other evidence are mastered, the surveyor will never be capable of conducting retracement surveys.

In this chapter, *field*, or other, evidence will be discussed. This term is used because no textbook or reference book in the field of surveying will be able to impart adequate knowledge to the surveyor. This area of surveying is a specialty area and, as such, can only be learned through a combination of study and practice in the field, with the latter probably the most difficult and most important and the most time-consuming.

In this chapter, the following principles will be discussed:

PRINCIPLE 1. *For a survey to be a consideration as part of a conveyance, it must be called for by the conveyance or it must be identified by law as a part of the conveyance proceedings. As an exception, in a few states, surveys conducted shortly after the conveyance is created are considered as contemporaneous with the conveyance.*

PRINCIPLE 2. *The evidence of intent of any survey called for in a conveyance is to be interpreted from the map of the survey, the field notes of the survey, and the acts of the surveyor, not from the unwritten, unexpressed intent of the surveyor.*

PRINCIPLE 3. *Generally, for a monument to be a controlling and considered as evidence, the monument must be called for in the written evidence proving the conveyance or it must have been required by law. In some states a call for a specific survey may also be considered a call for monuments set by that survey. The call for a monument is a call for the exact spot originally occupied by the monument as of the date of the written conveyance.*

PRINCIPLE 4. *No one corner or monument recited in a description has any greater dignity or legal weight than any other corner or monument recited in the same description.*

PRINCIPLE 5. *For an absolute location of a parcel, every survey of a conveyance must start from evidence that proves the original position of all of the corners called for in the description, and that must be positively related to the written record that created the parcel.*

PRINCIPLE 6. *The fewer the number of the original monuments of the parcel found and positively identified, the less certain the location of the original parcel.*

PRINCIPLE 7. *For an original monument to be of value and controlling, it must be located at the same spot occupied as of the date of the deed. The exception is waters that are subject to riparian doctrines.*

PRINCIPLE 8. *Witness objects called for by writings, if found undisturbed, are conclusive evidence proving original corner locations and have the same weight and dignity as the object that was set to identify the corner.*

PRINCIPLE 9. *Other than the original monument, the best evidence of a monument's original position is a continuous chain of history by acceptable records, usually written, back to the time of the original survey and its monumentation.*

PRINCIPLE 10. *In considering evidence of bearing trees and line trees, called for in an original survey, a surveyor should always have the belief that the evidence of the original survey will still be present and it can be recovered.*

PRINCIPLE 11. *In general, but not always, the county surveyor's records (or city engineer's) are prima facie evidence, whereas those of private surveyors are not.*

PRINCIPLE 12. *Witness evidence or testimony may be used to locate the former position of a corner monument. But witness evidence cannot refute a found, undisturbed monument called for by a written conveyance.*

PRINCIPLE 13. *Passing calls for objects do not have a high ranking in surveys. Passing calls are those calls for surveyed objects along a line. A passing call does not determine the start or terminus of a line and should not be given the dignity of location monuments.*

PRINCIPLE 14. *Evidence of possession representing the location of corners and original survey lines may be used as a form of evidence to prove original survey lines.*

PRINCIPLE 15. *By common report or by reputation, the location of the position of the original monuments or former monument positions may be proved by parol testimony (evidence). But title to land can never be proved by reputation or parol evidence, in that it can only be proved by documentary evidence.*

PRINCIPLE 16. *Fences that are not recited in a description are only one form of evidence to prove boundaries. Taken alone, they do not and cannot constitute proof of title or boundary lines. Fences are only one form of evidence of possession.*

PRINCIPLE 17. *All original corners have equal weight in location of the parcel. No single one is controlling, and they must be considered as evidence of that survey. This principle applies to both GLO and metes-and-bounds surveys.*

Some conveyances calling for a survey are of the map or plat type and other conveyances not calling for a survey are of the metes-and-bounds type, either of which may recite a call for a survey. In Texas, the state required a survey as a condition for obtaining a patent to state-owned land. East Texas lands, in general, although not always, were surveyed as the need arose—that is, land was not laid out in lots and blocks prior to needs. As a result, many Texas patents calling for a survey are sequence conveyances having junior and senior considerations. In the sectionalized land states, where surveys are called for, most patents are of the lot and block type (called sections, townships, and ranges). A call for a survey is not then limited to one type of conveyance; it may be inserted in either a sequence or simultaneous conveyance.

If the evidence of a survey is to have legal force, it must be called for in the writings or it must be presumed by law. A survey not called for by the written conveyance cannot be considered a part of the writings; words cannot be added to a conveyance. But where the law requires a survey to be made as a part of the conveyance, it is presumed that the law was obeyed. The United States required, by law, surveys of sectionalized lands; as such, a call for a GLO parcel includes, as adopted by case law, a call for all documents relating to that particular patent. Since 1879, Texas has required surveys of patents of land that were issued; hence, whether or not a survey was called for in the writings for Texas or U.S. patents, it is presumed that one was made. Contrary proof—that is, proof that no survey was ever made—will overcome this presumption, but the proof must be real, not merely surmised [1].

A Missouri decision addressed the problems when it stated as follows:

> [O]ur statute required that anyone laying out an addition should cause to be made out an accurate map or plat thereof, particularly setting forth and describing all lots for sale, by numbers, and their precise length and width. There are two strong reasons for finding that the lots and streets and their location were accurately surveyed and marked on the ground as a basis for the plat. The first reason is that the law presumes that such survey and markings were made. The plat bears on its face facts which, taken in connection with conceded extrinsic facts, show that the plat was based on a survey, not only of the boundaries of the addition but of the interior lot lines. The plat shows that the course of the railroad is not straight, but it curves to the eastward as it goes southward. The lots bordering on the right of way of the railroad are irregular in shape and the lengths of their boundaries are marked accordingly. It would have been difficult if not impossible to correctly indicate such distances without such survey. We have a right to take notice of the historical fact that on the invention of wire fences, hedges were no longer planted in this state, and that the hedge fence at or near the southwest corner of the addition was there when the addition was platted. We have a right to draw the inference that such fence was adopted by the maker of the plat as the west boundary; in other words as the center of the section. It was right on the line called for by the course and distance [1].

There are instances when a surveyor wishes to introduce or use the information or accept the evidence of a survey even when it was not mentioned in any document. An Oregon appellate decision stated:

> A deed which does not refer to a particular survey, and which is unambiguous in its terms, cannot be enlarged, or in any manner modified by the introduction of parol evidence tending to prove that one of the parties to it understood in descriptive words to have reference to a particular private survey [2].

5.2 Retracements and Evidence

PRINCIPLE 1. *For a survey to be a consideration as part of a conveyance, it must be called for by the conveyance or it must be identified by law as a part of the conveyance proceedings. As an exception, in a few states, surveys conducted shortly after the conveyance is created are considered as contemporaneous with the conveyance.*

The major portion of the modern-day surveyor's work is conducting retracements, that is taking the client's historic description and then using the evidence indicated in that description find the lines.

When considering the retracement of an original survey, one must consider special categories. A resurvey not shown to have been based on the original survey is not conclusive in determining boundaries and will ordinarily yield to a resurvey based on known monuments and boundaries of the original survey [3]. Like all other areas of surveying and law, there are exceptions. Exceptions to the stated general rule can be found in some of the original states, especially where widespread loss of evidence makes it necessary to rely on reputation and hearsay. Evidence of a survey that cannot be proved to have been made before or after a conveyance may be accepted merely because it has the reputation of being correct. Here are two cases involving surveys made after a conveyance that may help to explain this principle. In the first, the conveyance called for certain specific monuments that did not exist at the time of the conveyance but were set some months later by the same surveyor who created the original survey. The second case pertains to a survey made after the conveyance and approved by the acts and behavior of the adjoiners. The second case should be discussed under the subject of unwritten rights, but it is presented here to emphasize the differences.

Where land has been conveyed by deed and the description of the land in the deed makes reference to monuments not actually in existence at the time, but to be erected by the parties at a subsequent period:

> When the parties have once been upon the land and deliberately erected the monuments, they will be as much bound by them, as if they had been erected before the deed was made. In this case, there was a reference in the deed to monuments not actually existing at the time, but the parties soon after went upon the land with a surveyor, run it out, erected monuments and built their fences accordingly; and this is not all. They respectively occupied the land according to the line thus established, for nearly ten years. And there is now no evidence in this case of any mistake or misapprehension in establishing the line [4].

In the foregoing case, it should be noted that the monuments were called for by the deed but not set until some 18 months later. In New Hampshire, when such monuments are set by the parties, they become a controlling consideration. This should not be considered as the majority application of the law. In most jurisdictions, any monuments called for but not in place at the time of the conveyance are not controlling as long as it is proven that they were not in place at the time the conveyance was executed.

In the following case, the boundary or property line was run immediately after the conveyance, and the parties then built a fence according to that line. Several years later, it was discovered that the fence did not agree with the writings; the surveyor was wrong. The court was of the opinion that the acts of the parties following the deed indicated their intentions, and as such, were controlling. Whether this principle was applicable to surveys made by strangers to the original transaction or was applicable to surveys made many years after the original conveyance was not in issue. Because of the importance of this case in the northeast, it is quoted in its entirety. This decision represents a variation of *practical location* discussed elsewhere:

> Both parties claim title through the same grantor. Henry Smith, who, in the first instance, conveyed "parts of lots numbered 9 and 10, on the east side of Sandy river," to the defendant. After reciting the other boundaries, the description in the deed continues as follows, "thence easterly by a line parallel with the north line of lot No. 9 to the county road," the grantee taking the land north of the line now in dispute, and the grantor retaining the land south of it. The line was run and marked by a surveyor immediately after the conveyance, and the parties then built a fence on it, intending it for a division fence, Smith occupying to the fence on the south, and the defendant on the north side of the fence, for some six years, when Smith conveyed his remaining parcel to the plaintiff's grantor, describing the line in controversy as follows, "to land supposed to be owned by George Toothaker, thence easterly on said Toothaker's south line to the county road." About eight months afterwards, the grantee conveyed the last named premises to the plaintiff, describing it as "the same she purchased of Henry Smith." The plaintiff claims to hold to the line described as running "easterly by a line parallel with the north line of said lot No. 9 to the county road:" in Smith's deed to the defendant, which is several rods northerly of the fence, and the defendant claims to hold to the divisional line made by the fence; and the question is, which is the true line between the parties? The presiding judge ruled that the words, "on said Toothaker's south line," would limit the plaintiff's land to the line established by Toothaker and Smith, on which the division fence was built, and that she could not hold beyond this line, even if she could satisfy the jury that it did not conform to the original lot line; thereupon the parties agreed to submit the question to the law court, judgment to be rendered for the defendant if the ruling is correct; if not, the action is to stand for trial.
>
> But for the acts of the parties in interest, in running, marking, and locating the line, building a fence upon it immediately after the conveyance, and occupying up to it down to the commencement of this suit, the line on the course described in the deed, if it could be ascertained, would be the line between the two parcels. Did these acts fix and establish the divisional line as the true line?
>
> Early it was held that where a deed refers to a monument, not actually existing at the time, but which is subsequently placed there by the parties for the purpose of conforming to the deed, the monument so placed will govern the extent of the land, though it does not entirely coincide with the line described in the deed [5].

Again, it was held in Maine [6], that when parties agree on a boundary line and hold possession in accordance with it, so as to give title by disseisin, that is, by possessing the land, such boundary will not be disturbed, although it was found to have been erroneously established. In that situation the call in the deed was "a line extended west, so as to include" a certain number of acres, the boundaries on the other three sides having been accurately described. The parties to the deed agreed on and marked that line, erected a fence on it, and held possession according to it for 30 years.

The same doctrine was held by the Supreme Court of the United States, in giving construction to a line described in the deed as "running a due east course" from a given point [7]. So the court in Massachusetts, in giving effect to a deed, describing a line as "running a due west course" from a given point, held that the line located, laid out, assented to, and adopted by the parties was the true line, though it varied several degrees from "a due west course" [8]. Again in Maine the call in the deed was a line from a given point, "on such a course . . . as shall contain exactly one and a half acres" [9]. In addition:

> The lots to be conveyed were located upon the face of the earth by fixed monuments, erected by referees mutually agreed upon; and the parties to the several conveyances assented to and adopted the location before the deeds were given. Deeds intended to conform to the location thus made were then executed by the parties. The respective grantees entered under the deeds, built fences, and occupied in conformity with the location for fifteen years, when, it being found that more land was contained within the limits of the actual location upon the face of the earth than was embraced within the calls of the deed, a dispute arose. The court held that the monuments thus erected before the deed was given must control, thus extending the rule adopted in *Moody v. Nichols* to cases where the possession had not been long enough to give title by disseisin. That decision also makes the rule of construction the same, whether the location is first marked and established, and the deed is subsequently executed, intended to conform to such location, or whether monuments, not existing at the time, but referred to in the deed, are subsequently erected by the parties with like intention.
>
> In construing a deed, the first inquiry must be what was the intention of the parties? This is to be ascertained primarily from the language of the deed. If this description is so clear, unambiguous, and certain that it may be readily traced upon the face of the earth from the monuments mentioned, it must govern: but when, from the courses, distances, or quantity of land given in a deed, it is uncertain precisely where a particular line is located upon the face of the earth, the contemporaneous acts of the parties in anticipation of a deed to be made in conformity therewith, or in delineating and establishing a line given in a deed, are admissible to show what land was intended to be embraced in the deed. It is the tendency of recent decisions to give increased weight to such acts, both on the ground that they are the direct index of the intention of the parties in such cases, and, on the score of public policy, to quiet titles. The ordinary variation of the compass, local attraction, imperfection of the instruments used in surveying, or unskillful in their use and inequalities of surface, and various other causes, oftentimes render it impracticable to trace the course in a deed with entire accuracy. If to these considerations we add, what is too often apparent, the ignorance or carelessness of the scrivener in expressing the meaning of the parties, we shall find that the acts of the parties in running, marking, and locating a line, building a fence upon it, and occupying up to it, are more likely to disclose their intention as to where the line was intended to be, when the deed was given, than the course put down on paper, if there is a conflict between the two. Hence, the rule of law now is, that when, in a deed or grant, a line is described as running from a given point, and this line is afterwards run out and located, and marked

upon the face of the earth by the parties in interest and is afterwards recognized and acted on as the true line, the line thus actually marked out and acted on is conclusive, and must be adhered to, though it may be subsequently ascertained that it varies from the course given in the deed or grant. The acts of the defendant and Smith, through whom the plaintiff claims, in surveying and marking the line in dispute upon the face of the earth by stakes and stones and spotted trees, building a fence thereon, intending it to be the line between them, and occupying up to it, make and establish such line as the divisional line between the two lots.

The ruling of the presiding judge was in accordance with this construction of the deeds, and there was a judgment for the defendant [5].

A careful reading of this case indicates that, in Maine, where the parties to a deed, soon after it is consummated, make a practical location of the deed on the ground, by conducting a survey, the parties are bound by it whether it agrees with the writings or not. The theory is that the acts of the parties disclose their intentions of the deed. In a later case, unrelated to the Knowles case just cited, two adjoiners, not a part of their original deeds as of the time the land was originally divided, built a fence along a line that they believed to the true dividing line [10]. The court decreed that since the parties were trying to mark the true deed line, they were not bound by the fence. In other words, strangers to the *original transaction* of dividing the land, when mutually marking a line that they believe to be the true line, are not bound by the line unless it is in fact the true line. Proof of where the surveyor marked his or her lines is usually inferred from the evidence of monuments; hence, proof will be discussed later.

5.3 Historic Private Surveys

In most states, the records of private surveys of former surveyors, many of whom are now dead, usually are not admissible in evidence for the reason that landowners cannot be held to unavailable records. In a few states where evidence of surveys is not recorded and there is a widespread loss of both field and documentary evidence, courts have accepted ancient private survey information. As an exception to the hearsay rule, the court will permit, in limited instances, the testimony relative to surveys of dead surveyors. The discretion is left up to the judge. Unless a court decision exists in that state giving credence to approving old private survey records, the surveyor should assume that such records cannot be used in court at trial. This should not exclude the modern surveyor from examining these documents and then considering them as part of the totality of evidence. These may cause admissibility problems, not relevancy problems. The courts usually hold that such documents are hearsay, in that the party who completed the records is unavailable for cross-examination. Yet, such documents may be admissible as one of the many exceptions to the hearsay rule under ancient documents, ancient boundaries, or business records or possibly as a declaration against a proprietary interest. One such court decision occurred in New Hampshire where a deceased surveyor's ancient plans, minutes, or field books, clearly describing and identifying the lots and bounds in controversy, were admissible in evidence [11].

5.4 Intent of a Survey

PRINCIPLE 2. *The evidence of intent of any survey called for in a conveyance is to be interpreted from the map of the survey, the field notes of the survey, and the acts of the surveyor, not from the unwritten, unexpressed intent of the surveyor.*

A survey called for cannot be interpreted in light of any secret or hidden intentions of the creating surveyor. What the surveyor did and what he or she recorded in the evidence of writings are what count. The surveyor's testimony that he or she intended to include all the lands up to the adjoiner cannot enlarge a survey to include lands that were omitted by the original survey.

This was addressed in a Texas decision:

> The question seems to have been whether or not the calls for course and distance or those for lines of older surveys should prevail. Upon this question, we are of the opinion that the testimony of the surveyor stating his intention in making the survey was not admissible. In determining the location of the land in such cases, the courts seek to ascertain the true intention of the parties concerned in the survey; but the intention referred to is not that which exists only in the mind of the surveyor. When reference is made in the decisions to the intentions of the surveyor, the purpose deduced from what he did in making the survey and description of the land is meant, and not one which has not found expression in his acts. Hence, if the intention of the surveyor appears from his field notes and his acts done in making the survey, his evidence to prove his intention is superfluous, while if it does not so appear, it cannot control or affect the grant [12].

The surveyor, of course, can state what was intended about things he or she actually did, but not about intent. A statement by the surveyor that a certain found monument was the one that he or she set would have force, but a statement to refute writings would more than likely be rejected, since parol evidence is inferior to written evidence.

5.5 Value of Monuments as Evidence

From an evidence standpoint, field evidence of monuments can be considered to fall into two groups: the actual monuments and then the accessories to original monuments. A corner placed without a monument, has no certainty of position, if it has to be resurveyed in. Both are equally important, because a monument that is placed to sanctify the corner position has no error, by law; but the measurements have residual and human errors. The courts recognize monuments when by case law, and in some states by statute, natural monuments and artificial monuments rank high in the dignity of control.

5.6 Evidence of Monuments

PRINCIPLE 3. *Generally, for a monument to be a controlling and considered as evidence, the monument must be called for in the written evidence proving the conveyance or it must have been required by law. In some states a call for a specific survey may also be considered a call for monuments set by that survey. The call for a monument is a call for the exact spot originally occupied by the monument as of the date of the written conveyance.*

A monument to control the intent of a description must be called for either directly or indirectly by reference or be required by law. A deed may call for an oak tree in the writings, or the deed may call for a map, which, in turn, calls for an oak tree, or the deed may call for a survey. The surveyor's field notes may call for an oak tree. If the law requires a survey and set monuments, extrinsic evidence may be taken to explain what monuments and corners were set as required by law. One very important fact that is sometimes overlooked is that a call for a monument is in actuality a call for the particular spot occupied by the monument as of the date of the description. The monument itself is merely a symbol or object to mark the spot. A found monument that is uncalled for or is not referred to has no weight in substantiating that survey unless it can be shown by other evidence that it is occupying the spot of the original monument.

If there is a call for a monument, that monument, if discovered undisturbed and uncontradicted by the remainder of the writings, is conclusive. A deed that calls for bearing, without identifying the reference meridian, and distance but does not call for a monument directly, indirectly, or by reference and is not required by law cannot be altered by giving control to a monument found in the vicinity to the line described by the bearing and distance termination.

This is explained when a witness testified that he saw a rock pile in 1872, and for some years afterward, at the north end of the line run by a surveyor named Holman. It was the evident purpose to have the jury believe that this rock pile was made at place for the northeast corner of the Lampasas County School land. If it was made there for that purpose, it was wholly irrelevant and immaterial, unless it was placed there by the surveyor who made the original survey or by someone who knew that it was at the corner of the survey. There is no evidence how the rock pile got there. It was not called for by the original field notes [13].

In another example, three witnesses testified that they personally witnessed the scribed bearing trees of an original GLO survey present at a specific location. All three individuals were quite specific in describing the markings, the location, and the cutting of timber up to the corner. The jury did not believe the testimony and discredited the entire testimony. They strictly relied on measurements to locate the "lost" corner. The trial judge reversed the jury and gave credit to the testimony of the three witnesses. Then the appellate court reversed the trial judge and discredited the testimony of the three witnesses, in that the appellate court believed that three witnesses could "manufacture" a location by simply all agreeing to the same "facts." The court also stated that the jury had the opportunity to examine the personal demeanor of each witness when they testified as to the truth [14]. The appellate court little realized the judge also had the opportunity to observe the witnesses.

In surveying terminology, the term *original monument* is applied to the monument or monuments called for, either directly or indirectly, by the deed. Subsequently set monuments, so far as a particular deed is concerned, are not original monuments. With the possible exception of monuments called for in a senior deed, original monuments control to locate the parcel of land.

As noted, the spot occupied by the original monument, as of the date of the deed or as of the date of a survey called for by the deed, waters excepted, is the controlling consideration. All monument evidence seeks to identify where that particular spot exists on the ground for that one corner.

Discovery of the original monument itself is not a necessity, since many types of evidence can be resorted to that will suffice as proof of the original location. A disturbed monument may be of no value; the original spot occupied by the monument may not be identifiable. An *obliterated monument,* that is, one lost from view, may be restored to its former position by competent witness evidence. Evidence is to prove where it was as of the date of the deed, not where the measurements say it should have been set.

One of the most difficult decisions a surveyor may have to make is to ascertain whether the corner is lost or obliterated. That can only be a function of the evidence that is created and then recovered and considered, with the understanding that higher and more modern precise forms of evidence cannot be used to prove evidence created by less precise methods.

A *lost corner* is one whose position cannot be determined, beyond reasonable doubt [15]. To be lost, when applied to section or township corners, means more than that they have been merely obliterated, tampered with, or changed, but they must be so completely lost that they cannot be replaced by reference to any existing data or other sources of information [16]. Before courses and distances can determine a boundary, all means of ascertaining the location of the lost corner (or monument) must first be exhausted [17].

When the record distance from one interdependent corner to another varies substantially with the field measurement obtained in the retracement, the surveying method of apportionment may yield the most probable correct location of the lost corner [18]. Note that apportionment does not automatically provide the correct location of the corner, it provides the most probable location of the corner, all things being equal.

A corner should not be regarded as lost until all possibility of fixing its original location by the *totality* of the available evidence has been exhausted. It is so much more satisfactory to locate the corner than to regard it as "lost" and locate it by, or through, proportionate measurements [18]. Sometimes people hastily react regarding a missing monument and attempt to locate the corner, or the point where the monument once was, by apportionment. Courts have always taken a dim view of this practice, knowing that measurements contain error, and have stated that apportionment is only a rule of last resort: "Courses and distances are most unreliable in fixing the location of an old survey; distances being more uncertain than courses" [19]. To attempt to determine the position of a corner with measurements is almost a folly. The original (presented) measurements contain error, and the relocation measurements also contain error. If the errors are cumulative, the results could be disastrous and people will very much have a false sense of security.

One should remember that a corner and a monument are not one and the same, although in theory they should be in the same position. A monument should mark the position of the corner. In the absence of the monument, the position it once occupied controls.

The retracing surveyor should remember that courts have repeatedly held that in order to resort to proportioning, the corner most be "so lost" that none of the evidence recovered can support its original location.

Another Wisconsin decision stated that in retracing a tract of land according to a former plat or survey, the surveyor's only function or right is to relocate, on the best available evidence, the corners and lines at the same place originally located. Any departure from such purpose and effort is unprofessional and, so far as any effect is claimed for it, unlawful [20].

In a minority number of states, if soon after its formation the parties to a deed erect monuments to indicate their intent by a practical location, they are bound by them. This was the finding in a New Hampshire case where monuments were called for but not set until after the date of the deed [4]. Rephrasing, the court stated that where a monument does not exist at the time a deed is made and the parties afterward erect such a monument, with intent to conform to the deed, such monument will control. In most states the mutual designation of a property line by parties to the original conveyance is considered as an unwritten agreement and is called a practical location, which often controls.

It is pointed out that the practice of monumenting after the conveyance is in widespread practice today in the area of "mortgage" or "loan" surveys. Many times, in the survey for a lending institution that wishes to make a loan, the surveyor preparing the survey plat usually will indicate the monuments, but since the survey will not be paid for until the house closing, the monuments will not be set until after the deed is signed and delivered. This practice is dangerous and in some states unlawful.

5.7 Control of Original Monuments

PRINCIPLE 4. *No one corner or monument recited in a description has any greater dignity or legal weight than any other corner or monument recited in the same description.*

All original corners that are created or established for a parcel of land have equal weight in the location of the parcel. Each corner monument called for has equal control with every other corner or monument called for; all are to be given equal control, if possible, that is, if the monument or corner is properly identified, is undisturbed, and corresponds to the description called for. This applies only to written descriptions.

This principle is as equally applicable in the GLO states as it is in the metes-and-bounds states. Since the GLO states the acceptable method for locating a "lost" corner is by interrelated measurements to the closest original corners, always within the same township, then the lost-corner position is predicated on the positive identification and location of at least two of the original GLO corners.

Yet in metes-and-bounds states in the location of a metes-and-bounds parcel, a surveyor may commence at any corner recited in the deed to locate the parcel of land. No single corner or monument recited in the description has any greater control than any other.

5.8 Monuments as Indispensable Evidence

PRINCIPLE 5. *For an absolute location of a parcel, every survey of a conveyance must start from evidence that proves the original position of all of the corners called for in the description, and that must be positively related to the written record that created the parcel.*

When locating a parcel of land from a deed description, the locations of at least two monument positions are indispensable evidence. All measurements commence from a monument and go in a direction determined by another monument (a star, magnetic pole, the North Pole, or a physical object).

Normally, in the interpretation of the intent of a description in a deed, all monuments called for by the writings are given preference over conflicting calls of distance, directions, or area.

In locating written deed lines, not all found or discovered monuments are of value as evidence; some are accepted; others are rejected in accordance with the laws of evidence. In general, for a monument to be controlling as evidence, it must be called for by the writings, either directly or by reference. A call for an oak tree is a direct call for a monument. A call for a survey by a particular surveyor is a call for any monuments set by the surveyor. Many of the old maps have statements on them that read "surveyed by John Doe." Although on the face of the map there is no mention of monuments set, it is always necessary and proper to seek an explanation of what was meant by "surveyed by John Doe." If monuments were set by this person, they can possibly be accepted. This is a question of proof by using evidence originally called for and then evidence subsequently called for and found.

In some areas of the United States, early surveys were made, but it is unknown whether they were made as a consideration of a conveyance or because of widespread loss of original evidence. Where there is longstanding acquiescence in a boundary that could have been marked by an original surveyor and there is no evidence to the contrary, there is no reason to disturb the status quo. If this happens, the retracing surveyor must be certain that the original evidence will not be discovered at a later date. Most important, though, is that in conducting a modern retracement, the surveyor must be absolutely certain better evidence proving another location does not exist or has not been overlooked.

There can be no positive location of a land parcel from its description without the positive identification and location of at least two originally called-for monuments from which to initiate a retracement. Since a line is determined by both its beginning point and its endpoint, any line that is predicated on a single recovered corner position is then usually identified by a course, a bearing, and a distance to the other corner. The course is nothing more than a finger pointer as to where to look for the evidence of the second corner, and it is not controlling unless the corner and the monument have entirely disappeared. Every single course, consisting of a bearing and distance, has some error of position in it, whereas a found original corner has no error of position, so far as the law is concerned [21].

5.9 Sufficiency, Amount, and Kind of Evidence to Prove Original Corner Monuments

This is the area that separates the true retracing surveyor from the measuring surveyor or the surveyor who relies on evidence or the surveyor who relies exclusively on measurements.

The amount, kind, and quality of evidence necessary to prove the correctness of an original monument that marks a corner point are varied, depending on the circumstances. The laws of evidence are not exact laws but are relative and at times may be quite flexible. Stating general principles that may have flexibility depending on the circumstances leaves much to the surveyor to understand and apply. In questions of civil litigation, under which land disputes fall, the courts accept the premise of the "preponderance of the evidence," which class is the least demanding for the surveyor and the attorney, this standard is not the same as "beyond a reasonable doubt." Some courts have also referred to this standard as "creditable," "sufficient," and "satisfactory," indecisions. It is very difficult proving a monument, or rather, proving the position of the corner occupied by a monument, as of the date of the description in a deed, but may be accomplished with the following evidence:

- the physical characteristics of the monument itself;
- probability or possibility of being disturbed or moved;
- public and other records, maps, notes, and documents that prove historical sequence;
- witness evidence;
- hearsay evidence, as permitted by the rules of evidence;
- common rapport;
- measurements to prove proximity to record measurements;
- other evidence, such as witness monuments, old fences, and old lines of possession.

The problem of authenticity is not one for the surveyor in a dispute but is one for the jury. The surveyor's opinions are relative to evidence, while the jury's opinion may be more personal in nature. This was commented on by a Missouri court when it wrote:

> The monuments set by the original deputy United States surveyors for the west section corners must control as to the proper location of those corners. The question where they were located, if destroyed, is one of fact, and not of law, for the jury to determine under all the evidence argued and presented for its evaluation [22].

5.10 Physical Characteristics of Monuments

Monuments are classified by most modern courts as "natural" and "artificial." In all jurisdictions, the courts hold that natural monuments control over artificial monuments. This seems to be the restricted thinking of courts trying to place evidence into pigeonholes. Some monuments are conclusively identifiable by the evidence of their physical characteristics alone. No trouble should be encountered when identifying the Great Lakes or a particular river or a mountain, which are considered natural monuments. But the question that must be asked is, "Where on the Great Lakes or the river?" That is uncertain. To be certain, the exact spot that locates the average water mark on the shore or the exact location of the thread of the stream or its shore or bank may be difficult to determine, but to be absolutely certain, there should be no trouble identifying the monumented corner called for at the exact point. An oak tree with a particular type of blaze mark should not present identity troubles, but an oak tree without a blaze mark and located in an oak grove could give substantial trouble to the retracing surveyor. In the first case, if a blazed oak tree is found, witness evidence would be incompetent to overcome the location of the blazed tree. But if two blazed trees are found, witnesses may be able to differentiate between the two trees. In the case of no identifying marks, called-for measurements from other known monuments are the best evidence to distinguish between the various trees. However, if measurements are not certain and conclusive, witness evidence may be resorted to. Where the monumented corner is certain of identification from the physical characteristics of the monument itself, witness evidence is incompetent to prove any other monumented position. But if the wording of the writings is ambiguous so that alternate monument positions are possible, less conclusive witness evidence may be used to distinguish between the monuments.

The retracing surveyor should consider monuments as a single category, with no distinction being made as to category, being further divided into natural monuments and artificial monuments. A call for a marked tree on the shore of a lake is a call for both a natural monument, the lake, and an artificial monument, the tree—however, some courts have held these are two natural monuments. What if the surveyor discovers the tree, but it is not on the shore of the lake? When two natural monuments or one natural and one artificial monuments conflict, the question is, "What controls?" The complexity of natural monuments is displayed in Figures 5.1 and 5.2.

FIGURE 5-1. Copy of original GLO notes showing topographic call.
Source: Donald A.Wilson

FIGURE 5-2. Topographic call for a natural bridge. Line ran across bridge, or did it?
Source: Donald A. Wilson

In Figure 5.1, the call for the natural monument states, "42.00 cross natural bridge, 4 feet above water." As can be seen in Figure 5.2, the natural bridge is distinctive and would cause the retracing surveyor little problem to positively locate the crossing. Then in another natural call the call in the original notes is "49.73 cross stream, 60 lks (the abbreviation of links used by early surveyors) wide." The exact location of the stream is open for question.

Evidence concerning the physical characteristics of a monument, identifying a corner position, may or may not be described in the writings. If in a subdivision the original surveyor stated on his or her map that a stake was set at a particular corner in the *year 1900*, the stake must be identified as to species of wood with the proper markings. The fact that the surveyor set a stake is not disputable; the only question is, what kind of stake? Each surveyor has peculiarities and habits, and many old-time surveyors had certain specific and characteristic habits. In the San Diego vicinity, surveyors usually used redwood hubs or so-called plugs. In one area, 4 × 4

FIGURE 5-3. A scribed 3½ × 3½-inch redwood post set for the corner of quarter section 42 of Rancho de la Nación in 1868 and replaced in 1958. Paint protected the area below ground, and weathering deteriorated the portion above ground.

redwood posts were used; another area used 3 × 3 redwood posts (see Figures 5.3, 5.4, 5.5, and 5.6). In sectionalized land surveys, natural stone, chiseled stones, and rock mounds were common. Knowledge of what former surveyors did, what their habits were, and the customs of surveyors at the particular date are all evidence. In the case of a stake set in 1900 in San Diego, if anything other than a redwood stake and of the type commonly observed were found, the retracing surveyor should be put on actual notice to search further. A found Model T Ford axle being used as a

FIGURE 5-4. A scarfed redwood posted 2½ × 2½ × 18 inches set for the side line of Salinas Avenue, Del Mar Heights, Del Mar, California, in 1887. The post was stenciled with black paint, and weathering has embossed the letters so that they stand out. The paint is now gone.

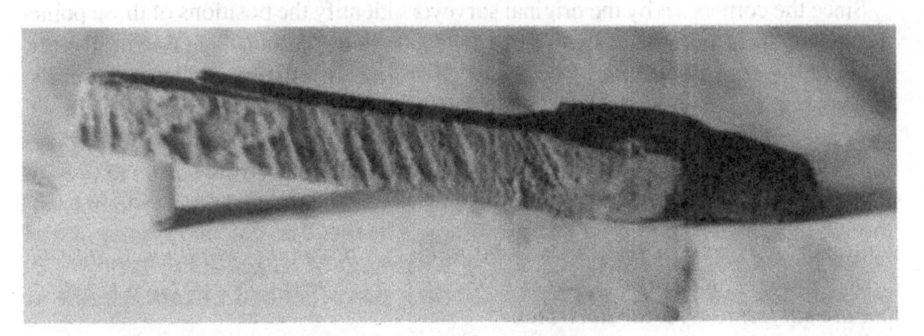

FIGURE 5-5. A scarfed redwood post set for lot 22. Weathering has removed much of the upper portion as compared to that below ground.

FIGURE 5-6. A witness stake to a concrete monument. White paint on the numbers protected the wood, whereas the unpainted wood was eroded away by the elements.

subdivision corner would put the surveyor on notice that Fords were not manufactured at that time. These minor, seemingly unimportant facts of evidence are often the difference between a successful retracement and an erroneous resurvey.

These principles can also be applied to fences that are called for in legal descriptions. If an 1804 deed calls for a fence, the question to be answered by the surveyor is, "Is the ancient wire fence the one identified?" Yet without knowledge that wire fences were not used until at least 1816 or even later, a wrong decision may be made. The call for a barbed wire fence in 1840 would be without meaning, because barbed wire was not manufactured until after 1850. One thing that may make the older, experienced surveyor more valuable than the new surveyor is the older surveyor's knowledge of local past history. The older surveyor may know what is correct; after all, some could have been there when the fence was constructed or repaired—or even placed on the ground. At least, that is what some younger surveyors believe who wish to rely totally on modern measurements.

5.11 Disturbed Evidence

PRINCIPLE 6. *The fewer the number of the original monuments of the parcel found and positively identified, the less certain the location of the original parcel.*

Since the corners set by the original surveyors identify the positions of those points and the law fixes those points with certainty and as being absolute and free from error, if one of the corners becomes lost, that is, if there is insufficient evidence to set the corner's point absolutely, if the retracing surveyor resorts to measurements, that is, an angle and a distance, to reposition that lost point, any errors in methodology to reposition that point are transferred to the lost position because the errors that were present in the original survey are different from the ones in the more modern measurements.

PRINCIPLE 7. *For an original monument to be of value and controlling, it must be located at the same spot occupied as of the date of the deed. The exception is waters that are subject to riparian doctrines.*

The rules of evidence permit both real and parol evidence to prove the location of original corner locations. Often the condition of the monument itself will prove whether it has been moved. A monument found in a cut bank is doubtful without further proof. A monument not in the measured proximity of where it is described in the written documents should cast doubt on its location and should suggest that further evidence must be sought. Or a monument that is reputed to have been in place for a long period of time but when examined by a surveyor is discovered to have tree leaves or grass in the hole should leave serious doubt in the surveyor's mind.

Water boundaries are an exception to the rule that the spot occupied by the original monument as of the date of the deed is the controlling consideration. Whenever a deed calls for naturally occurring waters (e.g., rivers, lakes, the ocean, but not artificial sources of water such as dams or canals), the location of the governing line of water at any particular moment is controlling. The subject of water boundaries is quite complex, and a more extensive discussion of this subject calls for a separate chapter, including evidence and location procedures.

This area is very important to the retracing surveyor, in that experience is needed to tell what is still undisturbed and what has been disturbed. The area of evidence of water boundaries is complex, confusing, and fraught with contradiction in the courts. Surveyors should approach any retracement of a water boundary with caution, and the young surveyor who attempts a retracement of a water boundary will be held to the same standard of expertise as the experienced surveyor.

5.12 Evidence of Witness Objects

PRINCIPLE 8. *Witness objects called for by writings, if found undisturbed, are conclusive evidence proving original corner locations and have the same weight and dignity as the object that was set to identify the corner.*

When alienating land, the creating power, either by law or by instructions, often required a survey, monumentation, and measurements to nearby identifiable witness objects. When found, these witness objects are considered equivalent in value to the original corner with its monument and may constitute proof of where the monument was placed. If a corner monument is easily movable, witness trees and other immovable objects may be a more certain means of identifying the original monument position than is the monument itself (Figures 5.7, 5.8, and 5.9).

FIGURE 5-7. Bearing tree with scribing exposed.

FIGURE 5-8. A pine tree with scribing exposed by cutting in with a chain saw; S25 is visible in the lower portion, R9E is visible above.

FIGURE 5-9. Backscribed beech tree in Alabama.

These objects that were originally measured in the original survey and referred to in the field notes or on the plats may consist of trees, marked stones, other natural features, and numerous other objects and as such become part the original survey. To be conclusive, each object must be referenced in the original survey, by writings, then identified with certainty and referenced by a bearing and a distance that is tied either to the original lines or corners.

At times, the descriptions may call for trees by species, and at times by diameter, and referenced by bearing and distance, usually from the corner to the trees. The more trees that are referenced in the notes, the greater the possibility for recovery and the greater the opportunity for having minor conflicts in the *absolute* location of the corner. The absolute location of one referenced bearing tree does not locate the original corner in its absolute position, because of the residual errors in the original bearings and distances.

5.13 Evidence Used to Identify Trees

Trees are intimately involved in the identification, location, and preservation of corners. The original federal laws of 1785, 1796, and 1805 directed federal surveyors to establish bearing trees as evidence to the original corners and then to refer to these trees in the field notes. In the metes-and-bounds states, surveyors constantly find references to tree corners and trees as "pointers," "bearings," and "witnesses."

In forested areas, marked trees were the most commonly used witness objects. Identification of these witness trees is a specialized science, and those who are expert in this area find their expertise valuable in litigation. As part of the basic knowledge that a surveyor should have to fully understand and attempt to conduct adequate retracements is a basic knowledge of tree and wood identification.

As a basic plateau of knowledge, when conducting retracements in forested areas, the retracing surveyor should be able to identify and relate today's tree names to historic tree names, and the retracing surveyor should have knowledge as to the characteristics of the indigenous tree species called for in the descriptions. In a recent retracement problem, a key corner was identified in 1799 as a hemlock. The evidence of very old fences indicated that a certain fence corner, by relating it to surveyed angles as well as natural features, could be a corner of the original survey. But the tree was not to be found. At the fence corner, a large decayed stump was found. On closer examination of the stump, the surveyor found a stump over 30 inches in diameter. The wood was badly decayed, to the point that it was nothing but "mush." Knowing that decayed hemlock is the color of milk chocolate and has the odor of cat urine, the surveyor was able to determine, by the preponderance of the evidence, that the stump was hemlock. Then applying the additional knowledge of fence age, supported by angles, and references to natural features, he was able to say "a hemlock" was "the hemlock." Then after conducting a total retracement of the entire 1,000-acre parcel, he was able to correlate 70 percent of the identified features in the 1799 survey to today's measurements. The evidence was sufficient to the point that the Federal District Court granted a summary judgment.

The first question that the retracing surveyor must address is: Was the original identification of the trees correct as to species and diameters? Names given to trees in one area or state were not necessarily the names given the same trees by other surveys in an adjacent state. During the initial days of the GLO surveys in the late 1700s and early 1800s, many surveyors who surveyed in Ohio and the southern United States commenced their careers in New England. They brought with them the local names of the trees they knew and if they did not know a tree, they made up a name. Probably one of the most recognized examples of this is the "Spanish oak" referred to in many of the original notes. The Spanish oak is in the red oak family, but the surveyors recognized it as a red oak, but one they were not familiar with. The leaf is quite elongated and, using one's imagination, it could take the form of a Spanish soldier's sword. Thus, it was named Spanish oak. In reality, the Spanish oak is a southern red oak, a recognized variation of the red oak and its northern red oak. The same applies to the "spruce pine" in New England and the Lake states. Many of the old notes recite this tree, but the species known as a spruce pine, *Pinus glabra*, grows mostly in southern Georgia, Alabama, and Mississippi and northern Florida and somewhat in Louisiana and South Carolina. There are three other trees called spruce pine, the most common being Eastern hemlock. In fact, the tree mentioned is neither a spruce nor a pine but the Eastern hemlock, *Tsuga canadensis*. The only positive identification for a tree is by its scientific name. Red oak is *Quercus rubra,* while Southern red oak is *Quercus falcata*. Local names of trees play a very important part in the recovery process. An excellent example is the sycamore, *Platanus occidentalis*. Some local names that settlers, surveyors, and woodsmen have given this tree are button ball tree (identifying its fruiting body), buttonwood tree (identifying its use for wooden buttons), and plane tree. In a 1798 survey for an Indian treaty, a critical corner was identified as a *buttonwood.* None of the local surveyors or attorneys understood what was called for. An expert was engaged who was also a forester. Upon identifying the true tree, the expert went out and found an area of sycamore sprouts, which helped in identifying the position of the corner.

In Figure 5.9, the first retracing surveyor discredited the original 1800 bearing tree because he did not see a blaze, and the markings were questionable. The second retracing surveyor accepted it as an authentic GLO bearing tree, because he knew that since this was a beech tree, *Fagus grandafolia,* and that at the time of the original survey in Alabama, Alabama was still a territory and not a state, the markings were authentic; The R = range 8; T = township 9; S = Section 6; and the UST stood for United States Territory.

There are over 50 species of oaks, with the two main categories being the red oaks and the white oaks. Distinction can be made by a visual examination of the leaves, and a microscopic examination can be made of the wood. Red oak leaves have "bristle-tip" pointed lobes, while white oaks have rounded lobes. The large, or early-wood pores in the wood of red oak are open, while those of white oak are clogged, or *occluded,* making separation of the two groups quite easy when magnified.

Appendix C illustrates how a field surveyor can make the distinction with a simple 10x hand lens.

Basswood is another example. Basswood has many names: American basswood, American linden, lynn, lime, lime-tree, linden, whitewood, bee tree, and white basswood. Regardless of what it is called, its scientific name is still *Tilia americiana*. To many people of German ancestry, the famous street in Berlin, Unter den Linden, brings back many memories. Available keys to tree identification permit the surveyor to have positive identification from leaves, twigs, fruit, or bark. Every retracing surveyor should have a field guide to leaves, twigs, and fruiting bodies to aid in the identification of trees; basswood being distinctive, for one.

The surveyor must have this knowledge to properly read the old descriptions and notes. If this knowledge is lacking, serious problems of misidentification could occur. All retracement surveyors should have a pocket guide of tree identification in their field packs. Many times the surveyor is asked to identify trees as part of a boundary dispute. Failure to be able to properly identify the species of tree could seriously limit the ability to be qualified as an expert in retracement. When the actual tree is present, surveyors use such evidence as hacks, notches, scribing, crosses, and blazes to single out a particular tree as a monument or line identifier. From such, a surveyor is able to date the time that the tree was hacked, and from the inference that this is the tree that was marked by the original surveyor, the line or tree can be dated with a reasonable degree of certainty, and the approximate age of the hack can be determined by counting the annual rings of the tree.

Many early instructions and laws of both federal surveys and some state surveys identified the manner in which trees would be marked during the course of a survey. Some early instructions as well as a state statute noted that identified trees should be marked by hacks. A hack is a single horizontal axe cut deep enough to penetrate beyond the bark into the cambium, the growing layer of the tree. In many of the metes-and-bounds states, the presence of three or four hack marks on each side of a tree was one of the usual methods of indicating a corner tree. When hack marks heal, scars are noticeable on the bark years later. In healing over the hack, wood grows as shown in Figure 5.10.

Notches or *hacks* are made by cutting in two directions so as to leave a horizontal notch extending into solid wood. Notches are larger than hacks and require longer periods of healing. Because of this, the danger of wood rot increases, especially if the notch extends through the sapwood into the heartwood. Notches are not recommended for line trees, since hacks will serve the purpose equally well. In most instances, the surveyors were instructed to notch or hack trees through which the boundary line passed. If one recovers a tree or trees that had been hacked, then the original line has been identified, and as such the retraced line should also pass through the same tree. Figure 5.10 shows the remains of two hacks that were made by chopping into a longleaf pine about 1815. The tree died, but the portion of the tree that had been hacked became impregnated with resin and was very resistant to decay. The portion of the tree with the original hack was recovered from a stump hole in the 1970s. This was "proof positive" of the original line. The original GLO field notes indicated "43.55 chs. Pine line tree."

Blazes are larger blocked-out areas extending below the bark and growth layer. The danger of wood rot, due to exposure to bacteria and virus, is greatly increased, and there is the possibility of tree death. Blazes should go just into the cambium layer, from the sapwood into the heartwood. Blazes have been and are being used for

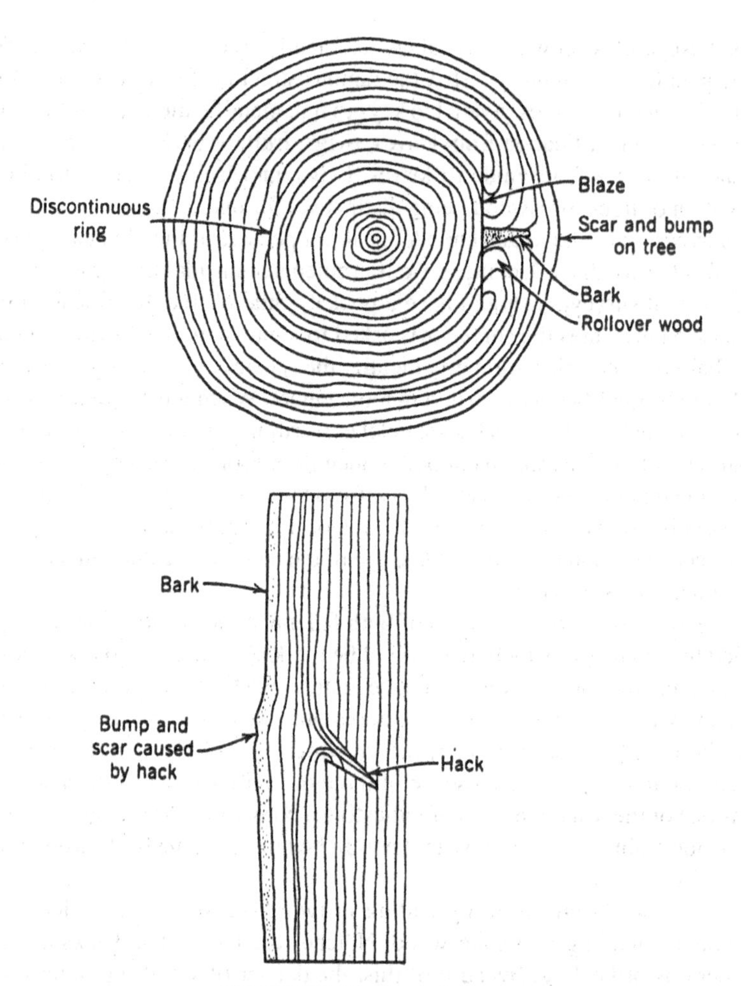

FIGURE 5-10. Annual rings.

marking (1) tree lines, (2) witness trees, (3) witness trees to a line, and (4) the monument itself. Marked line trees have a blaze on each side and usually have hack marks above and below each blaze. Trees are and have been blazed for many reasons, but the hack marks distinguished the surveyor's trademark. Trees directly on line were said to be fore-and-aft blazed. Side-line trees, that is, trees within about 3 feet of the line, were customarily marked with three hacks facing the line. Witness trees are blazed on the side facing the corner. If the corner itself happened to be a tree, the tree was blazed and often scribed to assist in its identification.

Recently while retracing a rancho line in California that called for a 60-inch redwood, one surveyor attested that he had found the tree and that a blaze was present in the wood. The second surveyor discredited that finding in that line in that live redwood trees usually were bark scribed because of the thick bark the trees had.

In another instance in an area that was covered with approximately 3 feet of water, a bearing tree for a section corner was identified as a black walnut. The stump of a badly decayed black walnut was identified, absent any markings. The surveyor "assuming" this could be the walnut "pulled off" the bearing and distance. About 2 feet from the reversed point, a badly decayed post was found for the corner.

When analyzing tree evidence of hacks and blazes, the surveyor must be very careful to make certain that some earlier surveyor did not disturb the original evidence on the tree. Figure 5.11 shows a cypress that was originally marked in 1849. Then, in 1952, it was "blocked out" to see the scribing. Between 1952 and 1979, the tree added new overgrowth. If the surveyor had not known of the original removal of the initial overgrowth, a false calculation of the growth rings would have resulted.

In Texas and some other areas, crosses rather than blazes were made on witness trees. This practice has merit since the danger of wood rot is lessened and a positive identifiable mark exists. In some areas of the country, many old surveyors had their own identifying marks, and to recover the mark was to recover the work of the surveyor.

FIGURE 5-11. Overgrowth on a cypress.

In other areas, numbers and/or letters that had been scribed into the wood of bearing trees, line trees, and station trees have been recovered as conclusive evidence of the original surveys. On a survey of a Spanish land grant in Florida, the 1823 survey of the Spanish grant called for trees that were scribed with an "H" for Hernandez. A third surveyor, following two earlier surveyors who had failed to find any of the original evidence, recovered one of the original trees that had been misidentified. The original field survey called for a "Cabbage marked with an H."

After conducting a complete retracement of the original line of the 10,000-acre grant, the third surveyor recovered a palm tree that had a scribed "H." The previous two surveyors had seen the tree but failed to recognize that "Cabbage" was a cabbage palm, and neither surveyor recognized the faint scribed "H" on the palm tree.

In identifying trees, the following six characteristics should be taken into consideration:

1. Species of tree and its characters (e.g., life expectancy, growth habits, susceptibility to disease).
2. Particular or peculiar original markings called for.
3. Stump particulars.
4. Ring count.
5. The site on which the tree is growing.
6. Particular name given to the tree in that area.

5.14 Trees and Their Characteristics

Identification of a particular tree, whether it is a witness tree, a bearing tree, a corner tree, or a reference point, may simply require identification by species, especially where only one tree of that species is found. But usually when one species is found, growth conditions are favorable and many other trees of the same species will exist. Particular markings, size, and age are then used to identify different trees. Sand pine rarely exceeds 100 years of age, and it is one of the pines that does not have a taproot; hence, any survey over 100 years old and calling for a young sand pine as a witness tree probably cannot be identified by the witness tree. A 50-year-old survey calling for a large loblolly pine may not be identifiable by the witness tree, since the tree may have been more than 50 years old when it was marked. Redwood or cedar trees may exist for more than a thousand years, and, regardless of the size of the tree called for, the chances of existence are probably for another few hundred years. Certain trees, such as redwood (*Sequoia sempervirens*), black oak, and sycamore, have the ability to regrow from the stumps or roots. Fire, disease, or humans may fell the main trunk, but regrowth from the roots will preserve the original tree's position, minus, of course, any particular original markings. Tree rot attacks some trees much more aggressively and readily than others. Trees susceptible to rot may be made open to attack through axe marks. Although the rot that enters may not destroy the entire tree, the original markings may become unidentifiable. When blazing trees, it is recommended that the blaze be painted to prevent rot. Certain trees, such as redwood and cedar, are resistant to rot and are ideal for blazing.

Many forested areas have been logged, burned, and altered both by man and nature, and as such, the original appearances may not resemble the original conditions. The combination of the two has destroyed many of the original marked trees. Where only one witness tree is called for, identification of a stump is very uncertain. If the tree happened to be unique to the area or the vicinity, an unlikely fact, then the identification of the wood stump by microscopic examination may be possible. If more than one witness tree was called for, say, three witness trees or, as instructed in the GLO surveys, four, then the relationship of bearing and distance between existing stumps may serve for identification, based on the theory of probability.

The size of a stump minus its estimated growth from the time of the original survey may be sufficient for identification. Figure 5.12 shows the discoloration of a pine bearing tree that was cut; subsequently the stump burned and only a discolored charcoal ring remains. The trained surveyor can determine the species of the tree that originally stood there by an examination of the charcoal.

In Figure 5.13, the surveyor was unable to recover the stump, but the remains of the pine roots were recovered showing where they attached to the stump. It was determined that the stump had been a pine stump by the resin smell of the stump and an existing tap root stump hole that was 4 feet deep.

Tree Growth Most ring porous trees grow annual rings from a single cell layer called the cambium layer. Bark growth is on the outside, and wood growth is on the inside. The inner part of a tree is composed of live sapwood and dead heartwood. Both sapwood and heartwood are incapable of reproducing cells, and any injury to the wood can be repaired only by fill-in grown from the cambium layer. Once an injury occurs to a tree, by an axe or other means, the cambium layer is stimulated

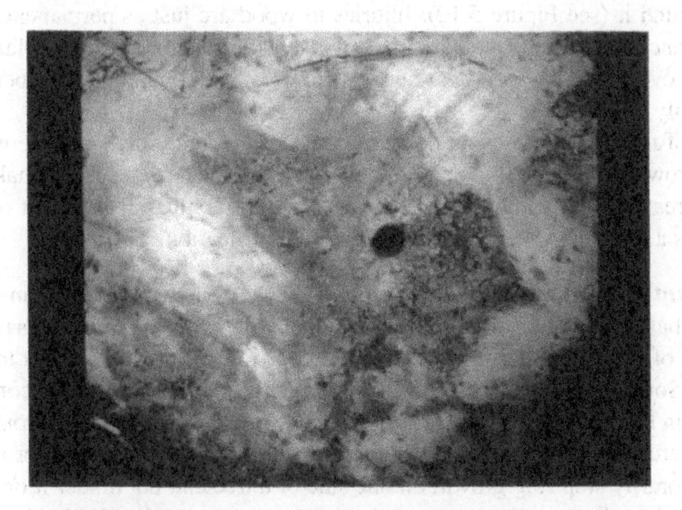

FIGURE 5-12. Charcoal discoloration of burned pine stump.

FIGURE 5-13. Remains of pine root leading from stump.

into more rapid growth at the point of injury. In some species a special enzyme is produced that stimulates growth and inhibits decay. Most pine trees will exude resin and impregnate the surrounding wood, making it resistant to decay from rot. The terms *light wood*, *lighter wood*, *fat pine*, and *heart pine* are all used to describe the corner posts and the wood that is usually found after the tree dies. In most species, the stimulated growth produces a hump or bump at the point of the injury. As the cambium layer grows outward to fill in the gap, the characteristic appearance has given rise to the term *rollover wood* (see Figures 5.10 and 5.11). Since sapwood cannot regrow, any injury caused by cutting into the tree leaves a permanent mark that is always identifiable. A nail driven into a tree or wire attached to a tree does not move up in height as the tree grows older; it stays in the same place and the wood grows around it (see Figure 5.10). Injuries to wood are just as permanent as, or in some instances more permanent than, the wood itself. Paint applied to a blaze before it became overgrown by wood may be exposed years later, and it will appear as if it were freshly painted.

Trees of the same species do not have the same growth rate. One tree may have a better growing site or soil than the other; one tree may have a genetic makeup that ensures greater size than another. The size of a tree comparative to that originally reported is a poor and uncertain means of age identification.

Ring Count Counting rings to determine the age of a tree is not an absolute indicator, but it may be accurate within a small margin of error. Differences in the rate of growth of wood for seasons of the year cause rings of greater density to deposit annually. Sometimes, as a result of various causes, annual rings are discontinuous, as shown in Figure 5.10. These "false" rings may have been caused by drought, fire, or other natural conditions. Loss of a tree crown, destruction of roots, or an injury may temporarily stop ring growth on one side of a tree and not hinder it on another. Ring count based on a segment or portion of a tree may not be absolutely accurate.

In addition, in the first few years of a tree the ring growth is probably below where the sample was taken. Ring count does not give the absolute age of a tree, but it is an indicator of within two to five years. In surveying, the ring count from a blaze outward is usually wanted, and it can usually be determined more accurately than the tree's absolute age. In some varieties of trees, the difference in the density of wood from different seasons is so slight that ring count is difficult or impossible to determine. Members of the palm family have no annual rings, and the age of palms cannot be determined.

Most ring counts are taken at diameter breast high (DBH) or 4.5 feet above ground level or at the level of the blaze. In these instances, the age that it took the tree to reach that height must be considered.

Even when a wound has completely overgrown, the surveyor still can determine, with a degree of certainty, the number of years from the time of retracement to the approximate time the blaze of hack was made. By using an increment borer, the metal bit is screwed into the wood and an extractor is placed inside the metal tube.

A core of wood can then be extracted. The blaze or hack can be ascertained from the discoloration and the separation of the wood. Figure 5.14 shows an increment borer in a tree, and Figure 5.15 shows the extracted core and a blaze. The surveyor can then count the number of rings and determine the approximate number of years since the tree was blazed by counting the number of rings outward. It should be cautioned that not all trees can be bored. Several of the hardwood species have wood that is so dense that the surveyor could break off the surgical steel shaft in the tree or the tree will close, or "freeze," the shaft in the tree, making extraction impossible.

Some trees, such as birch and maple, are called diffuse-porous trees—that is, they do not make easily identifiable annual rings like pine or oak. In these cases, the surveyor may experiment with using dyes and attempt to highlight the rings for counting.

FIGURE 5-14. Increment borer in tree.

FIGURE 5-15. Core sample from pine; note annual rings and blaze.

An interesting discussion of this is found in the literature:

The marking of trees and brush along lines was required by law as positively as the erection of monuments, by the act of 1796, which is still in force. The old rules therefore are unchanged. All lines on which are to be established the legal corner boundaries will be marked after this method, viz. Those trees that may be intersected by the line, will have two chops or notches cut on the sides facing the line, without any other marks whatever. These are called sight trees or line trees. A sufficient number of other trees standing within 50 links of the line, on either side of it, will be blazed on two sides diagonally or quarter towards the line, in order to render the line conspicuous, and readily to be traced in either direction, the blazes to be opposite each other, coinciding in direction with the line where the trees stand very near it, and to approach nearer each other toward the line, the farther the line passes from the blazed trees. In early surveys, an opposite practice prevailed. Due care will ever be taken to have the lines so well marked as to be readily followed, and to cut the blazes deep enough to leave recognizable scars as long as the trees stand. This can be attained only by blazing through the bark to the wood. Trees marked less thoroughly will not be considered sufficiently blazed. Where trees two inches or more in diameter occur along a line, the required blazes will not be omitted. Lines are also to be marked by clearing away enough of the undergrowth of bushes or other vegetation to facilitate correct sighting of instruments. Where lines cross deep wooded valleys, by sighting over the tops, the usual blazing of trees in the low ground when accessible will be performed, that settlers may find their proper limits land and timber without resurvey.

The practice of blazing a random line to a point some distance away from an objective corner, and leaving through timber a marked line which is not the true boundary, is unlawful, and no such surveys are acceptable. The decisions of some state courts make the marked trees valid evidence of the place of the legal boundary, even if such line is crooked, and has the quarter-section corner far off the blazed line.

On "trial" or random lines, therefore, the trees, will not be blazed, unless occasionally, from indispensable necessity, and then it will be done so guardedly as to prevent the possibility of confounding the marks of the trial line with the true, but bushes and

> limbs of trees may be lopped, and stakes set on the trial or random line, at every ten chains, to enable the surveyor on his return to follow and correct the trial line and establish the true line. To prevent confusion, the temporary stakes set on the trial or random line will be removed when the surveyor returns to establish the true line [22].

Surveyors must be careful about tree names in field notes and in land descriptions because colloquial names abound. Most trees have several different local names, some of which are very misleading, for example, larch or tamarack being called "juniper" or hemlock being called "spruce pine."

Sometimes trees or their blazes are scribed with corner information. Even when grown over, if the tree is in sound condition, this can be recovered by cutting into the tree and removing the overlaying layers of wood and bark. By taking out a section, the original marks can be uncovered and a mirror image exists on the removed section.

Remains of trees can often be identified by anatomy. Stumps, even rotten wood, and charcoal, all retain their internal structure, which can be identified by an expert. This, combined with other evidence, can aid in the identification of the true monument or corner.

A surveyor who desires to practice in the area of retracements must become knowledgeable about stumps and stump holes. Many species of trees produce either taproots or lateral roots. Knowledge of which species falls into the respective categories will greatly aid in the recovery of field information. Most pines are taproot trees, while most hardwoods are lateral-root trees. But there are always exceptions. A hickory may have a limited taproot while a sand pine may have only a lateral root. This information is important when it comes to identifying a stump hole as being a "hardwood" or a "softwood." An exception to the pine tap root is in a soil area in which a "hardpan" is present in the soil and the pine tap root cannot penetrate the hard soil layer.

Once a stump hole is identified, the surveyor can ascertain a taproot by "poking" the area with a probing bar. If a taproot is present, then the bar can easily be pushed deep into the remains of the hole. After it is ascertained that a possible stump hole is present, then the surveyor, using a systematic approach of uncovering the hole, can possibly obtain traces of wood or charcoal and the species can be positively identified. An individual trained in wood identification can identify the species of wood from remains of the wood or charcoal in the hole. Pinewood can be identified from the unique smell, as can hemlock, and oak wood can be separated into the "red" oak or the "white oak" groups from an examination of the wood. Using a 10x pocket hand lens, the surveyor can positively identify numerous species of wood. In most instances each species has a unique and distinct specialty about it that will permit the surveyor to be able to identify, with a fair degree of certainty, what species it is. Wood will retain its anatomical structure, chemical composition, and characteristics long after it has decayed. If a sample of either stump wood or wood that was found on the ground is too badly decayed to examine, some surveyors have taken wood samples, soaked them in water, and frozen them. After freezing, the samples can be cut and then examined with a hand lens for identification (see Appendix B).

If it is necessary to be able to identify pulpy masses of rotten wood and distinguish the sample as being a hardwood or a softwood, the surveyor can perform a chemical test on the pulp. The resulting tests that have been conducted by the U.S. Forest Service Forest Products Laboratory in Madison, Wisconsin, have proven to be 100 percent correct in control samples. This simple test can be found in Appendix C.

The "stump hole" surveyor is a specialist in the specific area, and as a well-trained individual will be a valuable asset to any retracement project. From the size and shape of the stump hole, from the remains uncovered and recovered, and from observations, the surveyor will be able to say with certainty the story the stump has to tell. A failure to search for stump evidence is a failure to do a complete recovery project.

All blaze marks on trees should be treated with a certain amount of skepticism. As Mulford stated, "It must be remembered that a small boy with a hatchet can mark up more trees in one Saturday afternoon than a dozen surveyors can in a year" [23].

Three examples of scribed trees are shown in Figures 5.7, 5.8, and 5.9.

5.15 Chain of History of Corners and Monuments

PRINCIPLE 9. *Other than the original monument, the best evidence of a monument's original position is a continuous chain of history by acceptable records, usually written, back to the time of the original survey and its monumentation.*

Deeds have a chain of title back to their inception, and the validity and correctness of a deed are based on this chain of title. Similarly, monuments should have a continuous chain of history. For example, the original surveyor set a stone mound for the original corner. Surveyor 2 finds the stone mound and sets a 2-inch iron pipe. Surveyor 3 finds the 2-inch pipe and sets reference points 30 feet on each side of a new proposed road. Surveyor 4 finds the reference monuments and resets the true original corner in the centerline of the new road. Surveyor 5 finds the new monument in the centerline and wants to prove its identity and the correctness of its position.

How can surveyor 5 accomplish his or her goal without a continuous record of what each previous surveyor did? It is because of the need for continuous records that many states have laws making it mandatory to file a record of survey under certain circumstances.

Many states have mandatory corner registration laws, and the products that are submitted are usually poorly completed, lack the necessary ties or references to the original information, and for the most part are worthless as records. This is due to the failure of the states to provide for adequate funding. A history or chain of record for the monument position is valuable evidence, but all too often, there is an interruption in the history and a continuous chain of records cannot be proved. In such an event, a different type of evidence must be resorted to. These records have little value unless the retracing surveyor can trace the present monument back to its original creation.

5.16 Surveyors' Records on Monument Location

PRINCIPLE 10. *In considering evidence of bearing trees and line trees, called for in an original survey, a surveyor should always have the belief that the evidence of the original survey will still be present and it can be recovered.*

As retracements become more remote from the date of the original surveys, the original evidence that was created during the original surveys becomes more difficult to recover because of the nature of the created evidence. Bearing trees have a limited life span, and the wooden posts that were set to bear witness to the corner points decay and are destroyed by humans and nature.

Regardless of the time lapse of the original surveys from the date of the retracement, the surveyor should remain optimistic that evidence of the original bearing trees and/or corner monument will be found or a record of the perpetuation of the original corner will be recovered.

This philosophy should be applied to surveys performed by governmental agencies as well as private surveys. The retracing surveyor should never have the attitude that since the survey is an ancient survey, there is no possibility of finding evidence.

In a GLO survey that was conducted in 1805 in Alabama, deputy surveyors established a section corner and marked four trees as witness trees. One of the trees, a 6-inch yellow poplar, was blazed and scribed. Applying all of the rules of longevity, the trees should have a life span of approximately 150 years. Thus, in a 1996 retracement, the surveyor would not have expected to find the tree still alive. But in the course of the 1996 retracement, a 60-inch-diameter yellow poplar was found with the proper blazing at the correct bearing and distance. Yet in the same township, no other trees from the original survey were found. One tree survived the odds.

PRINCIPLE 11. *In general, but not always, the county surveyor's records (or city engineer's) are prima facie evidence, whereas those of private surveyors are not.*

The proof of monument positions is frequently dependent on surveyors' records obtain, the quality of the fieldwork conducted, the surveyor's ability to communicate the findings to juries and judges, and then finally how it is received. It is accepted that record and testimony are critical, but not all records of surveyors can be admitted in evidence. With regard to admissibility of evidence, surveyors can be divided into two classes: (1) public and (2) private surveyors. A regularly written record of a public officeholder or public employee whose records were written as part of his or her job is usually acceptable in evidence; but the records of a private surveyor, without the testimony of the private surveyor, are difficult to use as evidence in court since they would be hearsay, unless they fall under a hearsay rule exception.

Public surveyors and deputy surveyors have an official duty to run lines, establish boundaries, and make and file reports of their results. When such reports are publicly filed, no question exists about their admissibility as evidence. Sectionalized land field notes are difficult or almost impossible to impeach despite the fact that numerous instances of definite questionable surveys are known.

Many individual surveyors have tried to question original GLO surveys, and the courts have responded as follows:

> The original surveys made by the United States are not to be taken as conclusive presumptions of law; they may be rebutted and impeached as to their correctness; but, prima facie, they are presumed to be correct until properly impeached [24].

Some states have made public survey returns prima facie evidence and consider them as presumptively correct; they have held that such surveys need not prove a monument set by public surveyors. Yet others may permit the surveyor to disprove the monuments set. Private surveyors do not enjoy a standing of this type. The value of a private surveyor's results must usually come from personal testimony, except in the case of death or common repute.

An Arkansas court instructed a jury that a survey made by the county surveyor for a Mrs. Coffman was prima facie evidence of the correct line so far as it appeared from the survey. This instruction was erroneous. A Missouri statute provides that the county surveyor shall keep a record of every survey made by him or her under the statutes and that a certified copy of this record under the hand of the surveyor shall be deemed prima facie evidence in any court of record.

In the case that was tried, the official record of the county surveyor was not placed in evidence, nor was it shown that notice that his survey would be made was given as provided by the statute. The statute was precise in prescribing that it is only a certified copy of the record of the county surveyor that shall be admitted as prima facie evidence. The oral evidence of the county surveyor gave his acts no more validity than the acts of any other surveyor [25].

Decisions as to the admissibility of evidence and their relevancy are looked on in terms of how relevant it is to the question being presented and usually are determined by the judge.

5.17 Witness Evidence to Prove Corners and Their Monuments

PRINCIPLE 12. *Witness evidence or testimony may be used to locate the former position of a corner monument. But witness evidence cannot refute a found, undisturbed monument called for by a written conveyance.*

Those who have personal knowledge may testify about where the monument was. If such evidence is found to be reliable and undisputed, it is conclusive.

In a South Dakota decision, the appellant contended, and rightfully so, that it is to be presumed that the government field notes are correct and that such notes placed the quarter corner on a direct line between the section corners. It will be presumed that the corner, as actually located, was on that line and that it will take clear and satisfactory proof that it was located elsewhere to justify a court in finding that it was located other than on such line. No witness swore that anyone had ever seen a government mound, or what was thought to be a government mound, on the direct line between the section corners; but there were four witnesses other than the plaintiff who were positive that there was formerly a mound and pits answering the description of government mounds and pits located at a point some distance south of the straight line running between the section corners. One of these witnesses had known of this mound ever since he was a boy. He testified positively to seeing the mound in 1909, that this mound was in a field that a person was then breaking, and that in breaking the field, the mound would be and was destroyed. He testified as to its location in reference to a road that formerly ran from section corner to section corner, which passed very close to the mound. The other witness also testified as to its location in relation to the old road. The court was satisfied that there was sufficient evidence from which to find that the original government corner had become obliterated and that such corner had been seen by those witnesses. It was therefore not a lost corner. It certainly cannot be contended, merely because a quarter corner has become obliterated and the exact location cannot be fixed, that the courts must treat it as though it were a lost corner and locate it on a direct line between the section corners [26].

Evidence has only as much value as will be accredited to it by the court. Obviously, the person giving the testimony must have had an opportunity to have observed the corner prior to its destruction; hence he or she must have personal knowledge and must have lived, or resided, or been in the area at the right time. If a person has too much financial interest in the corner's location—that is, if he or she stands to gain—the testimony should be viewed with suspicion. Hearsay evidence, that which someone hears someone else say, is generally not admissible evidence, except in the case of what a deceased person has said or to prove a contradiction. Parol evidence may not refute written evidence; hence, the undisturbed position of an original monument called for by the writings may not be altered by witnesses.

The lines of a grant must be established by the calls in the field notes. If these calls are inconsistent, then the courts might resort to certain rules of construction and even parol evidence to resolve any doubt and to establish the line that was actually run by the surveyor. It is but a case of a latent ambiguity in a written instrument. A writing, unambiguous on its face, may become doubtful when applied to the calls found in the field notes of a survey, there is no room for construction. The calls must speak for themselves. To permit the introduction of parol evidence to vary the calls would be to violate the familiar rule that extraneous evidence is not permissible to vary a written instrument [27].

5.18 Summary of Competent Parol Evidence

Verbal or evidence consisting of words is evidence that is not competent under rules of law and should be received but little consideration should be given to it by the surveyor. From the foregoing it is obvious that a lay witness cannot do any of the following:

- Testify as to the laws pertaining to boundaries.
- Testify in such a manner as to alter written words of deeds.
- Express an opinion.

The general rule is that testimony may not contradict, vary, or modify writings. Testimony in general is limited to these five items, although there may be others:

1. Testimony may be taken to explain a latent ambiguity that is not a question of law.
2. Testimony may be taken about the former location of a monument and the identity of a monument if the writings are not clear about whether the monument has been moved, but testimony may not be taken about the location of a monument identifiable from the clear and unambiguous written words of the conveyance.
3. Testimony may be taken about the usual customs and meaning of words as of the date of the deed.
4. Testimony, if needed, may be taken about the general reputation of a monument.
5. Testimony may be taken about the surrounding circumstances as of the date of the deed.

As to inferences, a lay witness may testify as to facts within his or her own perception but not about conclusions, opinions, or facts contrary to the writings.

A California decision found no merit in the contention that the parol evidence referred to was improperly admitted. Contrary to the appellant's claim, the evidence neither altered nor varied the terms of the written instrument. It was offered by the respondent and received by the court, not for the purpose of changing or adding to the deed, but for the sole purpose of explaining the language therein contained. That parol evidence was admissible for such a purpose cannot be denied. If the word *block,* as used by the common grantor in the subdivision map and several deeds executed pursuant thereto, was intended by him to have a particular and peculiar meaning, it was competent for the court to permit the introduction of parol evidence to establish that meaning [28].

In a Massachusetts case, parol evidence was admissible to show that at the date of a deed M Street was straight but its course was subsequently changed to a curve [29].

5.19 Passing Calls and/or Topo Calls

PRINCIPLE 13. *Passing calls for objects do not have a high ranking in surveys. Passing calls are those calls for surveyed objects along a line. A passing call does not determine the start or terminus of a line and should not be given the dignity of location monuments.*

Passing calls are more frequently found in surveyor's field notes, particularly in sectionalized land areas and in Texas surveys. Notes reading "thence north 3 chains cross a creek, 15 chains on a ridge, 40 chains set a stone" have two passing calls of a creek and a ridge and a locative call, "set a stone on a ridge." This stone is the controlling element. In the absence of the locative call, the passing calls may have probative force to indicate the surveyor's footsteps. *But they are not controlling* except in California. Courts might resort to passing calls for the purpose of ascertaining a located corner where the locative calls have disappeared or cannot be identified and there are no means other than the incidental calls of ascertaining the place where the monuments for the corner was placed by the surveyor [30].

While the sectionalized land area was being surveyed, notes all refer to numerous calls for passing objects and were written in the field notes. These calls were to aid in finding the corners and in preparing maps. Many of the state courts give strong probative value in the absence of the original monuments.

A minority have given control to these calls as identified in a California decision. Initially, these calls were used to show topographic features on maps. The California court adopted the topographic calls to determine the location of a line and rejected the proportional method. It should be noted that this is a minority decision. The court justified this with the following:

> On the west line of Section 6 Larson went north from the southwest corner thereof to the last recognizable call in the government field notes beyond the Garcia River, then there being no further ascertainable calls except that of "north," he continued north to the township line. On the east line of Section 6 Larson followed the calls from the southeast corner of said section north to the Garcia River. He could find no recognizable calls beyond that, so he followed the only call he could be sure of, and that was north to the township line. Cummins apparently made no attempt to follow the calls, even as far as the Garcia River. He started right out at the southeast corner of Section 6 and went on a straight line about 40 degrees off of a true north direction, ran many more chains than the notes called for, and in a different direction, and came out on a township line a mile westerly from where he would have had he gone north all the way from the corner he started from.
>
> *There can be no doubt, and I understand all parties to agree, that the government field notes beyond the Garcia River are grossly incorrect as to distances, and consequently as to courses* [emphasis added]. It does not seem possible that such errors could occur by one in the field. One can guess that parts of said lines were not actually run, but only estimated from afar (not good estimates either). It is the duty of a later surveyor to try and retrace steps of the original surveyor where such is possible. This, Larson attempted to do. Kenneth Cummins "found no relation between the line ran and the calls in the field notes." Larson followed the call north, Cummins did not. It is pointed out his closing point was between two streams and a spur, which description is found in some of the field notes with reference to other political subdivision. However, this is far from being conclusive when we ascertain that in order to get there we must go along a line which leads us to a point a mile away from where it would be if the line went straight north from the southeast corner to the northeast corner thereof said Section 6. And where the distance is so far off from the distance calls in the field notes and where it is not shown there were other places along the township line that could not qualify as coming within the same description, there is not much left to go on in that regard.
>
> All of the original government plats in evidence show the east and west lines of Section 6 to be straight lines north and south. The Larson line is a straight line north, the Cummins line is not. Such plats may be considered in the overall picture. Indeed, Russell Cummins testified the original surveyor may have intended to run on a true

> line north until it intersected the boundary of the township next north. If he made a correct survey that is what he did. The Court is forced to the conclusion that the great weight of the evidence shows that was what he did do. It would certainly be a peculiarly shaped section if the Cummins line was adopted as the east line thereof. Even so, if we were satisfied he had thereby traced the line of the original surveyor, that is the way it would have to be. But the evidence, including much of the testimony of Kenneth Cummins himself, shows that he did not do that. He ignored the calls for field notes from the southwest corner northerly altogether and started out on a line by the single proportionate method in a case where it was not authorized. This line cannot be adopted [31].

The question should be asked, did the court have to rely on topographic calls to reach its conclusions? Did it exceed its authority? Could this problem have been decided on using an agreed-upon boundary concept? Are not all of the requirements present to have an agreed-upon boundary?

Another interesting California case took 340 trial days, 36,302 transcript pages, 844 exhibits, and 1,072 pages of briefs to determine the same. One surveyor, William Wattles, was cross-examined for 46 days. The final decision was made on passing calls (topography noted in the field notes). The digested case is as follows: Obliteration of a monument does not justify adoption of the proportional method of locating a common corner as a lost corner where the surveyor's field notes refer to certain natural objects that can be found along the line mentioned so as to approximately locate it [32].

Some surveyors and courts place greater weight on a locative call then they should, but those who properly use these calls as they should add a great dimension to their surveying skills. In a Florida situation, the section corner was considered lost by the first retracing surveyor, who located the corner on a bluff, some 280 feet from the location of the second surveyor's corner. The field notes indicate "ties" from the west and from the south. The east-west line read, "79.45 cross stream, 5 lks.; wide; 80.20; Set pine post 6 in. by 6 in., 2 feet above ground, in a very swampy area." The line to the north read, thence, "Run North, 35 lks cross same stream, 6 lks. wide."

On a retracement, the first retracing surveyor considered the corner as lost and located it by proportioning, which placed the point on a bluff, some feet higher and out of the swamp. The second retracing surveyor, after determining the creek had remained relatively stable in location and the using the field measurements, to locate the general location of the corner, used the two distances indicated in the original field notes to position the corner.

Relying on the *topo calls* supported by the locative calls, "in a swampy area" and corroborated by a measurement, gave credence to surveyor number 2.

The task of the court, when confronted with an obliterated corner, uncertain boundary location, or the like, is to decide from the data appearing in evidence its approximate position when the exact spot cannot be found and fix the place at a point where it will best accord with the natural objects described in the field notes as being about it and found to exist on the ground, and where it will be least inconsistent with the distances mentioned in the notes and plat.

5.20 Possession Evidence

PRINCIPLE 14. *Evidence of possession representing the location of corners and original survey lines may be used as a form of evidence to prove original survey lines.*

Possession that represents the location of original monumented lines is distinctly different from unwritten title lines. In many instances, after all original monuments have disappeared, the best available evidence of the original lines is evidence of old fences built soon after the original stakes were set. In the late 1800s, Justice Thomas M. Cooley, chief justice of the Michigan Supreme Court, addressed this problem much better than any written discussion today. His paper deserves a position of authority and should be read by all students and lawyers and be read and applied by the courts. This paper is given in Appendix C and should be read at this time. Keeping in mind that the fences being addressed by Justice Cooley were of recent origin, having been constructed 15 to 25 years after the original surveys, most of the original monumentation should have still been in place. The rules presented in this document should become guiding principles for all future work.

A decision in Wisconsin indicates that the City of Racine conducted a resurvey and then located its road by proportionate measurements from distant points [33]. Whether a fence that had been maintained for more than 40 years was on the true line of the street or not was a question that was to be determined *according to the original plat made prior to the erection of the fence* and not according to a resurvey made by city authorities nearly 40 years later, by which after fixing the line of one street from one of the original monuments, the distances were apportioned between the several blocks and the streets changed accordingly.

Although according to the retracement, the fence in question was more than 2 feet within the street, evidence that it was built according to stakes set by the surveyor who made the original plat, that it is on a line with other fences, that old buildings were erected according to stakes set at the time the original plat was made, and that all the fences' no-win dispute had been maintained on substantially the same line for more than 40 years was held to show that the fence in question was built on the true line [33].

In Sacramento, California, a similar situation with the same conclusion was recorded [34].

For possession to represent evidence of the original lines created by the original surveyor, the following five facts should apply:

1. There was an early survey that, if located, is controlling the line between the adjoiners.
2. The lines of possession are along the lines surveyed or presumed to have been surveyed by the surveyor.
3. Usually, but not always, a series of possessions, in agreement with one another, substantiate one another.
4. Possession is an ancient matter of a former generation (if it is of a present generation, someone can testify about its origin).
5. Possession has the reputation of being on the correct survey lines.

The fact that the line between the parties, being the 1/16 line, was not at the time marked by original monuments deprives it of the right accorded ancient fences on the presumption that the fences were located on the original stakes then visible [35].

Not all possession represents controlling survey lines. Although some lines of occupancy may become title lines by the process of unwritten agreement, adverse rights, or other unwritten means, such lines should not be confused with original survey lines. The surveyor relates possession to his or her survey lines and tries to gather evidence explaining the origin of possession. If possession came about merely for the purpose of a cattle enclosure and the person erecting the fence had no idea of ownership lines or if possession follows a line that was not originally surveyed, obviously possession cannot represent original survey lines.

Where possession could be an original survey line, it is one of the duties of the surveyor to seek evidence to explain its origin. This does not permit the surveyor to use a line of possession (or fence line) from which to begin measurements. These lines of possession, fence lines, agreed-upon lines, and pointed-out lines are not controlling, but they can be used as points of reference for measurements that also are evidence. To use a line as a controlling element, it must be conclusively proven that the line is the original line called for, whether it is a section line, subdivision line, and so on.

Surveyors should use the evidence of measurements to help support lines of possession, coupled with testimony and other collateral evidence to establish the fact the lines of possession may be the best remaining evidence of the original survey (see Figure 5.15).

5.21 Common Report, Reputation, and Hearsay

PRINCIPLE 15. *By common report or by reputation, the location of the position of the original monuments or former monument positions may be proved by parol testimony (evidence). But title to land can never be proved by reputation or parol evidence, in that it can only be proved by documentary evidence.*

The importance that courts will place on the effort to get to the truth of the "original location of lines and corners" is exhibited in the *Rules of Evidence* that are recognized and accepted by federal and state courts. Hearsay testimony is excluded, but of the numerous exceptions that are recognized, this exception has been accepted for over 150 years in the various courts, and was legitimized in *The Federal Rules of Evidence*, especially in following exceptions [34]:

Rule 803 (a) (5); Recorded recollection(s).
Rule 803 (a) (6); Records of regularly conducted activity.
Rule 803 (a) (14 and 15); Records of documents affecting and interest in property.
Rule 803 (a) 16; Statements in ancient documents.
Rule 803 (a) 20; Reputation concerning boundaries.
Rule 804 (b) 1; Former testimony.

Although there is a general rule that hearsay evidence—that is, evidence of what you heard someone else say—is not admissible in court, several exceptions to this rule exist. In a trial over land boundaries, the court is charged with making a location of the boundaries based on the best available evidence, and if hearsay evidence is the best available, it may be used. The reputation of a monument as being correct is mere hearsay, but if better evidence of a monument's stature is not available, the reputation may be sufficient to establish its authenticity.

Although differences of opinion do exist about the application of the principle of reputation, none exists regarding the legal force of the principle. The rule is, of course, one of last resort.

Once a boundary or line is run and the survey is called for, the line is fixed in position, although it cannot be heard, seen, or felt. Actually, the true and correct boundary line is an invisible line that begins and ends at two corners, which may be monumented or not. Until some individual creates evidence along that invisible line, it has no positive actual location, although it is legally located. All living trees that were marked to identify the line eventually die, decay, and become forgotten. Stones, mounds, and other physical objects disappear. Finally, all original markings of lines may be gone. The original position of the ground remains the same, but how can it be monumented after the locative objects are gone? Can the certainty of title vanish with the objects? The answer is no! Then one should ask, "Can the land be located from the best available evidence?" The answer is yes.

In most states, boundaries over 30 years old are considered ancient boundaries. The federal courts will accept the time frame as 20 years. Reputation and hearsay are admissible to prove ancient private boundaries. Such testimony was admitted, and there is no escape from the conclusion that dating from the time when the original boundaries were freshly and plainly marked on the ground, the southern boundary of claim number 6 has been considered and established as coincident with the northern boundary of claim number 7 and was so recognized by those who were familiar with the stakes and other boundary markings [36].

A Kentucky decision held that the actual location of a deed may be satisfactorily established, not only by the natural objects found on the ground, but by the fact that all parties that knew the facts and were interested in the land located the deed in a certain way or acquiesced according to it. Time obscures all things, and facts that might be clearly shown 50 years ago may be incapable of proof now, when all the people of that generation have passed away. What remains [37]?

After all original monuments of a survey have been lost along with the suitable chain of records proving the new monuments to be replacements of the originals, no method exists, other than common report or reputation, to prove the status of some monuments. At times, all surveyors accept monuments and use these monuments, knowing full well that they cannot possibly prove them by direct evidence or a suitable chain of history dating back to the original. They will give authority to this evidence to be in the original positions. Reputation evidence is important to prove monuments that are not originals but are accepted as replacements of the originals.

The fact that a monument has been used by numerous surveyors does not make it an original corner without sufficient proof. The mere fact that all surveyors use a monument, without additional proof, does not and will not make it correct by continued use; the monument must initially be correct. Thus, in a superior court case in Alpine, California, it was shown that at an early date the state highway surveyors tied in a fence corner and for some unexplainable reason described it as a section corner [38]. A later surveyor in 1928 accepted the fence corner and set numerous corners from the accepted section corner. Up until 1950, some 10 or 15 surveyors filed maps and accepted the old fence corner as correct. When surveying an old holding dating back to 1900, another surveyor found that fences did not fit the proclaimed section corner. In a routine check it was discovered that the original government field notes stated, "Set a rock, mound 3 feet south of a 12-foot-high boulder." Not only was the 12-foot-high boulder found but also a witness testified that in 1898 he had seen a stone mound just south of the boulder. All the expert testimony, reputation, and recorded plats could not overcome the fact that the true corner was 70 feet east of the accepted fence corner. The best available evidence was the written government field notes, and these prevailed. Reputation evidence does not overcome contrary proof, but the contrary must be proved, not just surmised.

As a sidelight on the Alpine case, those with substantial enclosures were awarded title based on unwritten occupancy rights, and this occupancy was described from the old original location of the section corner.

Reputation evidence is pure and simple hearsay evidence, but it is an exception to the hearsay rule. Reputation is resorted to only when other means of proof are lost because of a long time lapse. The necessity of such evidence can arise only from the lack of better evidence. Certain safeguards have been set up by the courts so that the usage of reputation will not be abused. First, the reputation must be of ancient matters, such as an old fence of unknown origin and reputed to be a property line. Recent surveys are excluded. Second, the reputation is of a former generation. What has happened in the present generation is provable by other evidence. Third, the reputation must predate the boundary litigation; otherwise, the reputation will merely be a contention by one party. Fourth, if the reputation is based on the statements of an individual, such as a surveyor, the individual must be shown to be now dead and that he or she was disinterested at the time of the statements. Fifth, the statements of the individual are generally, although not always, required to be in reference to some monument or be supported by occupation. The reason is that a witness can remember a certain fence or monument but has no way of remembering isolated spots. It is also well settled that ancient boundaries may be proven by evidence of common reputation [39].

The same principle was applied in a metes-and-bounds state, Kentucky. Here the true location of the point where the gate post stood would settle the dividing line in dispute. The gate post had long since disappeared. It was shown in evidence that the adjoiners had a dispute as to the location of the gate post corner about 20 years ago. The appellant and her husband then applied to the county court for the appointment of processioners to establish and remark the obliterated corner. They were appointed and with the county surveyor met on the ground. They had called a number of old people living in the neighborhood, who were requested, and probably sworn, to locate the old gate post. Each of the witnesses so called stuck a stick down at the point where he remembered the post to have been. Each differed by some few yards. The surveyor placed his Jacob's staff in the center of the point selected and from there ran the line to the beech, which was then pointed out as the next corner. The county surveyor who was present at the time made a memorandum of what had been done, and signed it. But it was not signed by the processioners, nor was it returned to the county clerk, as was required by statute. On the trial of this case the surveyor who ran the line was called as a witness, testified to the facts, and produced the certificate that he had given at the time. His testimony was objected to. The old citizens who were called on by the processioners, and on whose statements the corners were established, were all dead.

This is an interesting case in that the surveyor started running from a questionable corner to a proven corner. He should have commenced at the good corner and run the course to help prove the questionable corner.

It is competent to prove the location of the corner or line of a public survey by reputation. In the nature of the thing, those who mark the original corner, and know personally of its location, will in time pass away, and so in some instances will the corners themselves disappear. Such matters of common knowledge are discussed in the neighborhood and are accepted and treated by those interested as being of a certain nature, so that their reputation becomes established and known of by all in the community. After the death of the original witnesses and the destruction by time of the monuments marking the corner, the only thing left by which its location might be identified is the reputation established and made notorious when both witnesses and corner were in existence. Therefore, the law receives the evidence of the reputation in proof of the fact to the location of such corners and lines as the best evidence obtainable in the nature of the case. The surveyor's certificate was not receivable because it did not conform to the statute, but what was said then to the surveyor by the persons who were then before him was evidence of reputation of the location of the original monument and corner. This evidence outweighed the conflicting and unsubstantiated statement of the appellee's witness that the gate post was north of the point located by Wood [40].

Where for more than 40 years the southeast corner of a block had been recognized as being at a certain place, and lots, blocks, and streets located, and buildings built with reference thereto, such universal usage and acquiescence outweigh indefinite notes of the surveyor who, many years before, replatted the block, and under which it is claimed, the corner is located 3 feet farther south [41].

In some areas, particularly along the eastern seaboard, where there has been an extensive loss of original records and destruction of monuments, the courts are inclined to accept ancient evidence of old boundaries in preference to modern measurements. This in no way means that a surveyor should accept all ancient boundaries: The surveyor must seek an explanation of the boundary. Many times old fences have been proved to be mere barriers of convenience. Accepting a location based on reputation is a rule of last resort.

5.22 Evidence of Fences to Prove Boundaries

PRINCIPLE 16. *Fences that are not recited in a description are only one form of evidence to prove boundaries. Taken alone, they do not and cannot constitute proof of title or boundary lines. Fences are only one form of evidence of possession.*

Some surveyors love using fences and at times may give them such definitions as "fence the section line," "fence the lot line," "fence the property line," "fence the aliquot section line," and even "fence is the boundary line." In reality, a fence is a fence, and unless called for in the conveyance or unless erected by the original surveyor on his or her survey line for the section or the lot, a fence is nothing more than one form of evidence, no better than any other evidence, but at times more uncertain than other forms of evidence. Of all the evidence the surveyor will use in determining the location of the original boundaries, fences give the greatest amount of concern and worry. The most difficult question to be answered in a retracement is, "What amount of weight should I give to the fences I find?" No person can answer that question without additional information.

A Virginia court found a surveyor negligent, as a matter of law, when he failed to follow the proper rules for surveying boundary lines described in deeds when he used a fence to locate a parcel of land. The court held that since the surveyor failed to understand and to apply the proper relationship of the "Dignity of Calls" that surveyors must understand and use, but rather used an old fence that "just happened" to be in the vicinity, he was negligent as "a matter of law" [42].

Since the first wire fence in 1816 and the barbed-wire fences of the late 1800s, fences have been made of all materials known to mankind: board fences, stump fences, stone walls, wattle fences, ditches, and many others. Surveyors may find a fence of several materials present along the survey line. Fences may have been erected many years after the wire was purchased, or the wire may have been used several times on several fences. However, knowing when a fence wire was manufactured or invented may help to date it and its lines of possession [43].

The law is: A landowner cannot erect a fence on a neighbor's land, but a fence can be erected anyplace on the parcel he or she owns. Many fences were nailed to trees without the benefit of a survey. The neighboring landowners would sight three poles through the woods keeping a straight line. As the line progressed, they would then erect the fence as they progressed at or near the line. In most instances, fences were nailed tresses or placed on posts that were set at or near the boundary line. Many times, one landowner would erect the fence, making certain to stay on his or her property. When parties did not know where the line was, they would erect a fence and agree that the fence would be the boundary between them.

In reading case law, all references are to "fences as evidence" of lines and not as the fence being the lines. Consider these three:

1. *Texas:* Ancient fences used by a surveyor in his attempt to reproduce an old survey are strong evidence of the location of the original lines and, if they

have been standing for many years, should be taken as indicating such lines as against the evidence of a survey that ignores such fences and is based on an assumed starting point [44].

2. *From the West:* Evidence of ancient fences and improvements is competent to prove boundary, where monuments and lines of original survey cannot be shown [45].
3. *Michigan:* A long-established fence is better evidence of actual boundaries settled by practical location than any modern survey made after the monument of the original survey has disappeared [46]. To determine the amount of strength to place on existing information that was revealed during the course of a survey, the surveyor should attempt to ascertain answers to the following questions:
   a. Who built the fence?
   b. When was the fence constructed? Have any subsequent repairs or location been done?
   c. Why was the fence built?
   d. Where was the fence built in relation to the boundary?
   e. How was the fence built? According to a surveyed line?

The surveyor should fully investigate and be able to answer these questions. This may include cutting into trees that have wire growing into them and counting the annual rings since the fence was built or scraping the ground to uncover a long-buried fence post hole of an ancient fence or searching the area with a metal locator to determine if any fence wire lies buried under the ground cover.

Many times evidence of ancient fences will be uncovered and long-forgotten fences, much older than the standing ones, are uncovered off the survey line that was determined to be the boundary line based on the acceptance of a recent fence. It will take all of the knowledge and experience of the surveyor in working with evidence to determine the acceptance or rejection of fences. This is an area in which the neophyte surveyor will find liability and serious legal problems.

5.23 Summary

The ability to relocate the lines of the original surveys is dependent on many factors: the knowledge, both formal and informal, of the surveyor; the ability to recognize evidence when it is seen or recovered; always asking if there is more evidence that was not recovered or if something was missed that some other surveyor will recover; the personal trait, either inborn or acquired, of never giving up, either in office research or in the field investigations; and most of all, asking people the important question, "Do you know anything that can help me?"

PRINCIPLE 17. *All original corners have equal weight in location of the parcel. No single one corner is controlling, and they must be considered as evidence of that survey. This principle applies to both GLO and metes-and-bounds surveys.*

A retracing surveyor should realize that all corners of a parcel that was created at one time usually by a single deed or survey have equal weight in retracing that parcel. It does not matter whether it is a GLO township containing thousands of acres or a simple lot survey containing a fraction of an acre. All corners recited have equal dignity in helping to retrace and relocate that parcel. The footsteps are the evidence the original surveyor created and described in his or her documents. This makes it convenient for the retracing surveyor, in that any original corner recovered has equal weight in the retracement.

The requirement of any retracement, either in metes-and-bounds states or in a GLO state, is basic to all: Find the footsteps of the original surveyor who created the original lines. Yet, it is impossible to follow the original surveyor, in that all he or she left is the evidence of the footsteps. Thus, the responsibility of the retracing surveyor is very simple, to find the evidence the creating surveyors left.

At times the surveyor will find that a thin thread of evidence is all that remains. The key factors are being persistent and asking questions. Figure 5.16 is a good example. The original GLO surveyor marked a 10-inch red oak in 1889 as one of four witness trees to a section corner. Most of that township had been farmed and had been granted to a tribe of Native Americans as a reservation. In a retracement of the township, little original evidence was recovered. A definite, positive identification of some evidence was needed. After surveying several miles of line to locate a "lost corner" and after having a local Native American watch these efforts for several days, the surveyor asked the Native American, "Do you know where the government corner was?" He responded, "Yes. I do." He promptly went into his barn and retrieved the stump of one of the original bearing trees with actual scribing still visible, showed it to the surveyor. Then he went over to a point approximately 15 feet from where the surveyors had been digging, a point that had *precision* of location, placed his foot on the ground at a point that had an *accurate* location, and said, "It came from here." The surveyor took his shovel, dug down, and found a piece of pottery that the person

FIGURE 5-16. Remains of original GLO bearing tree that has been "lost" for 40 years.

had placed in the hole from which the stump had been removed. On discussion, the Native American stated that when the road was constructed, the contractors pulled all four trees and burned three of them in a brush pile, but he could not let them destroy the tree that indicated where his land was located. He had kept that stump for over 40 years.

He often wondered why no other surveyor in those 40 years had not asked him that one important question, "*Do you know where the corner is?*"

REFERENCES

1. *Dolphin v. Klann*, 246 Mo. 477, 151 S.W. 956 (1912).
2. *Rowland v. McCown*, 20 Ore. 538 (1891).
3. *Pallis v. Daily*, 100 N.W. 2d 197 (Neb. 1960).
4. *Lerned v. Morrill*, 2 N.H. 197 (1820).
5. *Knowles v. Toothaker*, 58 Me. 172, 1870 Me. LEXIS 48.
6. *Modey v. Nichols*, 16 Me. 23 (1839).
7. *Missouri v. Iowa*, 6 How. 660. 93 U.S. 92 (1876).
8. *Kellogg v. Smith*, 7 Cush. 382 (Mass. 1857).
9. *Emery v. Fowler*, 38 Me. 102 (1854).
10. *Bemis v. Bradley*, 126 Me. 462, 139 A. 593 (1927).
11. *Morse v. Emery*, 49 N.H. 230 (1870); *Smith v. Forrest*, 49 N.H. 230 (1870).
12. *Blackwell v. Coleman County*, 94 Tex. 216, 59 S.W. 530 (1900).
13. *Runkle v. Smith*, 69 Tex. Civ. App. 549, 133 S.W. 745 (1911).
14. *Fehrman v. Bissell Lumber Co.*, 204 N.W. 582 (Wisc. 1925).
15. Bureau of Land Management, *Manual of Surveying Instructions* (Washington, DC: U.S.D.I. Bureau of Land Management, 1973).
16. *Mason et al. v. Braught*, 46 N.W. 687 (D.S., 1914).
17. *United States v. Doyle*, 468 F.2d 633 (C.A., Colo., 1972).
18. *Thomsen v. Keil*, 226 P. 309 (Nev., 1924).
19. *Barker v. Houssiere-Latreile Oil Co.*, 106 S. 672 (La., 1925).
20. *Pereles v. Gross*, 126 Wisc. 122, 105 N.W. 217 (1905).
21. *Riley v. Griffin*, 16 Ga. 141 (1854).
22. *Golterman v. Schiermeyer*, 111 Mo. 404, 19 S.W. 484 (1892), pp. 696–697.
23. A. C. Mulford, *Boundaries and Landmarks*, p. 28. (New York: D. Van Nostrand Co., 1912). (Reprinted by Carben Surveying Reprints, Columbus, OH, 1977).
24. *Brayton v. Merriman*, 6 Wisc. 14 (1857).
25. *Sherrin v. Coffman*, 143 Ark. 8, 219 S.W. 348 (1920).
26. *Kohlmorgan v. Roswell Township*, 41 S.D. 124, 169 N.W. 229 (1918).
27. *Thompson v. Langdon*, 87 Tex. 254, 28 S.W. 931 (1894).
28. *Ferris v. Emmons*, 6 P. 2d 950 (Calif. 1931).

29. *Abbott v. Frazier*, 240 Mass. 586, 134 N.E. 635 (1923).
30. *Davenport v. Bass*, 153 S.W. 471 (Tex. 1941).
31. *Hanes v. Hollow Tree Lumber Co.*, 12 Cal. Rptr. 713, 19 Cal. App. 2d 658 (1961).
32. *Chandler v. Hibbard*, 332 P. 2d 133 (Calif. 1958).
33. *City of Racine v. Emerson*, 85 Wisc. 80, 55 N.W. 177 (1893).
34. *Perich v. Maurer*, 29 Cal. App. 293, 155 P 471 (1915).
35. *Wollman v. Ruehle*, 104 Wisc. 606, 80 N.W. 919 (1899).
36. *Rickert v. Thompson*, 8 Alaska 398, 72 F. 2d 897 (1934).
37. *Wilson v. Commonwealth*, 243 Ky. 333, 48 S.W. 2d 3 (1932).
38. *County of Alpine v. County of Tuolumne*, 49 Cal. 2nd 757 (1958) 332 P. 2d 449.
39. *Cockrell v. Works*, 94 S.W. 2d 784 (Tex. Civ. App., 1936).
40. *Phillips v. Stewart*, 133 Ky. 134, 97 S.W. 6 (1996).
41. *Crandall v. Mary*, 67 Ore. 18, 135 P. 188 (1913).
42. *Spainhour v. Huffman*, 377 S.E. 2d 615 (Va. 1989).
43. R. T. Clifton, *Barbs, Prongs, Prickers, and Stickers* (Norman, OK: University of Oklahoma Press, 1973).
44. *James v. Hitchcock*, 309 S.W. 2d 909 (Tex. Civ. App. San Antonio, 1958).
45. *Day v. Stenger*, 45 Idaho 253, 274 P. 253 (1929).
46. *Diehl v. Zanger*, 39 Mich. 601 (1878).

6

CALCULATIONS AND MEASUREMENTS AS EVIDENCE

6.1 Introduction

As one of the more important forms of evidence pertinent to boundary location, measurements are the specialty of the boundary surveyor. Today, most modern measurements are calculations that should be correlated with modern technology as well as historical measurements. Like other forms of evidence, measurements are necessary first to determine quantity or area and the location of title lines described in descriptions contained in deeds and then to relocate or replace lost corners or to corroborate obliterated corners. When all other forms of evidence fail to identify evidence of the originally created corners and lines, modern measurements may become the controlling element, either by necessity or by law [1]. The surveyor who specializes in property or boundary surveys or one who conducts retracements must be skilled in making measurement observations and also should be competent in the evaluation of measurements, both historical and current.

In court, the few attorneys who ask questions on measurements usually talk about numbers as they pertain to accuracy or procedures—probably because they lack sufficient background or knowledge on how to form questions, either in direct or cross-examination. These attorneys will usually ask about error of closure but seldom ask about monuments or found original corners. However, this does not mean that the surveyor need not be prepared to answer extensive questions in this area.

The surveyor should realize that there is always that exceptional attorney who is extra diligent, or who has acquired adequate survey knowledge and can quickly bring out shortcomings of the less-qualified surveyor. A knowledgeable attorney can and will always use measurements to either support or discredit the expert surveyor. Because of this, the practicing surveyor must be knowledgeable about measurements. Several states have adopted, by statute or through case law, the practice of having the surveyor follow certain procedures for retracements of metes-and-bounds surveys, as well as surveys in the GLO states. These procedures are given in the *Manual of Instructions for the Survey of the Public Lands of the United States* [2]. Asked by an attorney in court, the question, "Did you use a two-pole chain in your survey?" can result in problems, since the several federal statutes requiring the use of a two-pole chain have never been repealed. The answer to the question is no, even though the *Manual of Instructions* authorizes the use of other measuring devices. A background of the knowledge of past and present measurement methods is necessary.

The purpose of this chapter is threefold:

1. Discuss the dependability and accuracy of existing measuring devices, including the most modern and the historical.
2. Analyze errors and uncertainties inherent in original measurements, the instruments used to create them, and procedures that relate these to today's retracements.
3. Define the accuracy expected of a professional surveyor when locating boundary lines—that is, to make a statement of allowable uncertainty of measurements.

Techniques of measurements (e.g., how to operate a transit, how to hold a plumb bob, and how to drive a pipe with a sledgehammer) are not treated in this book. It is assumed that the professional surveyor has developed sufficient experience and skills in such areas. The discussion here is more from the point of view of how measurements are used to prove boundary locations as supported by other evidence, rather than how to make measurements.

The principles that will be discussed in this chapter are as follows:

PRINCIPLE 1. *By law, either by statute or by case law, there is no error in an original measurement that created the original corners and connecting lines and the recited bearings.*

PRINCIPLE 2. *When modern measurements are related to original measurements, the comparison must be in terms of the original creating units of measurement and not in terms of the more modern units of measurements.*

PRINCIPLE 3. *For any conveyance or description of real property, the length of the unit of measurement is that measurement that was used and recited as of the date of the deed or survey.*

PRINCIPLE 4. *Every measurement of distance or angle is subject to errors, either known or unknown. There is no perfect measurement.*

PRINCIPLE 5. *Measurements may be used to prove the validity of corners and monuments. Such monuments, to be acceptable, should be within reasonable proximity of the record measurements, and supported by other evidence.*

PRINCIPLE 6. *In the GLO states, unless proved otherwise, measurements of distances are presumed to be horizontal, while in the early surveys of the metes-and-bounds states, measurements may have been made with the "lay of the land," according to the statute in effect at the time.*

PRINCIPLE 7. *In the public land states, the legal presumption is that bearings are relative to astronomic, or true, north, unless otherwise specified, but in the metes-and-bounds states, the presumption is that bearings are in reference to magnetic meridian, unless otherwise specified.*

PRINCIPLE 8. *If there is a conflict within a deed and a choice must be made between bearing and distance regarding which controls, no uniform rule has been laid down by the courts. Variations between states occur, as well as between the GLO states and the metes-and-bounds states.*

PRINCIPLE 9. *When modern measurements are related to original measurements, the analysis must be in terms of the original creating units of measurement and not in terms of the more modern units of measurements.*

Many excellent books and articles have been written on the geometric theory and practice of surveying. This chapter does not duplicate this material, but, rather, it considers the evidence element, which is less understood and, at times, improperly applied.

Units of length and direction are only means of expression. The surveyor should be aware of the uncertainties that exist in any given measurement, how these may be reduced, and by what means the most probable values may be obtained. To have a created described line, there must be a reference to an angle as well as a distance. As early as 1687, John Love mentioned the two elements of a line:

> There are but two material things (towards the measuring of a piece of Land) to be done in the Field; the one is to measure the Lines (which I have shewed you haw to do by the Chain) and the other is to take the quantity of the Angle included by these Lines; . . . [3]

The problem of measurements is probably more serious with the boundary surveyor than it is with engineering surveyors. Those working on topographic maps, construction layout, and engineering quantities fill a need: The property surveyor produces measurements not only for an immediate purpose but also for measurement calls that will be on record for many years and in specific instances should relate these to the original measurements. These future measurements will probably be reproduced by a method that was unknown at the time the original measurements were created and as such these may be in conflict in relationships as well as in testimony.

When they were surveying with their two-pole chain or 33-foot link chain of 50 links in the mid-1800s, the original GLO surveyors did not realize that today's surveyor would be using electronic distance-measuring (EDM) equipment or global positioning system (GPS) units to retrace the original measurements and to create new boundaries as well as to retrace ancient boundaries.

6.2 Types of Measurements

There are two basic types of measurements: horizontal or angle and distance. The two of these are combined to make a line or a course, which has a point of beginning and hopefully a termination. In reality, a line cannot exist as a single unit except only in theory and that only when the distance is at infinity. The line must have an end point or a corner.

To analyze courses (bearings) and distances, the surveyor must first determine what errors exist, and this is dependent on whether the courses were a result of the creating or the retracing surveyor. Thus, a creating surveyor may have actual error in the angles and their resultant courses and lines identified by recited distances, but so far as the law of property retracement is concerned there are no legal errors in that line.

This philosophy was cemented law first by the Land Act of February 11, 1805, and then into case in 1885 in the U.S. Supreme Court decision *Cragin v. Powell*, in which the justices held that the original approved surveys have no error.

Original approved surveys have no error.

PRINCIPLE 1. *By law, either by statute or by case law, there is no error in an original measurement that created the original corners and connecting lines and the recited bearings.*

As early as 1805, in the Public Land Survey System it was legally determined, by statute, that the original units of measurement were legally and presumptively correct and without error. The Land Act of February 11, 1805, decreed that all of the original lines and distances were legally, if not technically, correct and that no error existed in the original survey. This legislation was enacted so as to prohibit all future courts and surveyors from correcting the original surveys when errors were found. Congress realized that surveying, especially measurements, defied exactness. So, to prohibit future surveyors and the courts from correcting any discovered "errors," they made the lines actually run in the field and "returned in the field notes" to be legally without errors.

PRINCIPLE 2. *When modern measurements are related to original measurements, the comparison must be in terms of the original creating units of measurement and not in terms of the more modern units of measurements.*

Once again, an original measurement has no error. If in a retracement, the retracing surveyor determines his or her retraced distances/angles are not what was reported, then the difference should be reported, and the differences noted from the record.

All original lines have both a unit of measurement and the manner in which it was laid out or created—that is, whether it was created as a slope measurement (surface) or horizontal. In the public land surveys, the presumption is that all distances were laid out or created on horizontal planes, although there are occasions when slope or surface distance must also be considered. In the mountainous areas of the colonial states (metes and bounds), the presumption is that the distances were usually measured along the slope, except for more recent surveys.

Today, most modern surveyors measure their surveyed lines in feet. They are comfortable in reciting their distance in feet, and their chains or electronic instruments indicate feet; thus, they usually indicate their survey returns in feet, even though the original deeds or surveys may indicate a different unit of measurement (e.g., chains, poles, rods, or *varas*).

Even though a surveyor may choose not to measure a retracement in the same units, today the proper approach is to indicate in the deed or the field notes and plat any analysis between resurveyed measurements, and retraced measurements should be made in the same units as the original measurement. If the original measurements were in chains, then the modern analysis should be made in chains; if the original deed description were in poles, then the analysis should be made in poles, not feet. By referencing the original units to the retraced units, any anomalies between the two will become readily apparent, can be analyzed and then explained. When correlating retraced measurements to the original measurements, the relationship should be made in the same units of reference as they were created (i.e., poles = poles; *varas* = *varas*; feet = feet; steps = steps). The relationship should not be poles to feet or yards to feet.

6.3 Distance

When making references to distances, the retracing surveyor should know the historical facts that the reference is now to horizontal distance relative to some actually defined unit. In early U.S. history, surface measurement usually was the standard practice in some states—usually, the metes-and-bounds states. Boundary surveying in the United States has been subject to many different units of length, usually of historical significance or application and are of particular importance in retracing historical descriptions.

In the General Land Office (GLO) states, the equipment used to measure distance was indicated by statute. The Land Act of 1796 directed that "a chain of two-poles, 50 links, shall be used." Thus, we know what the surveyors were directed to use, and as such we can presume they held to the letter of the law. However, although four poles were made and were available, they were much too heavy for surveyors to use. This is not so in several of the New England states in which the authors were told four pole chains were used. Each pole probably weighed in excess of 10 pounds. In the metes-and-bounds states, early surveys did not have the benefit of information directing surveyors to use one chain or another, and we direct the reader to Appendix D, an excerpt from one of the leading survey textbooks of the time, *Geodaesia, or the Art of Surveying,* written and compiled by the colonial surveyor, John Love. The book states, "But that which is most in use among Surveyors (as being indeed the best) is Mr. Gunter's, which is 4 Pole long, containing 100 Links, each Link being 7.92 Inches." Then on the next page he writes, "If you find the Chain too long for your use, as for some Lands it is, especially in America, you may then take the half of the Chain, and measure as before" [3].

In terms of distance, the most common unit in use today in the United States is the foot, while the remainder of the world embraces the meter as its standard. Courses or bearings and perhaps early references to azimuth angles in the United States are usually referenced to degrees, minutes, and seconds in relation to a circle that consists of 360 degrees; the remainder of the world uses this but also includes a circle composed of 400 grads.

As the standard unit, the foot is defined as one-third of a yard, standardized by the U.S. Bureau of Standards, as being equal to 0.9144 meter, or 1 foot = 1200/3937 meter. In 1959, a new definition of comparison, the foot–meter relationship, was agreed on: 1 inch = 2.54 centimeters exactly. At the time, it was agreed that the U.S. survey foot would remain the same. Although the difference between the two units of measurement for the foot amounts to two parts in a million, the difference does exist. This minute difference is insignificant for boundary surveying, where most distances are very short, but it is becomes quite significant for geodetic surveying, where distances may extend for miles.

The surveyor must not assume his or her own value for the foot, nor must the manufacturer's word that the length of the tape is consistent with the standard value be accepted without further checking.

By statute, the Land Act of 1796 states that all section lines be "measured with a chain." The Land Act of 1805 then states the chain "shall be 2 poles long" but distances will be "indicated in 4 pole chains" [4] The standard unit of measurement for the Public Land Survey system and the one adopted by the BLM is the chain, consisting of four poles of 161/2 feet for a total of 66 feet, or 100 links. Thus, by law, a section line is 80 chains long, or 320 poles, not 5,280 feet, or 80 chains converted to feet. This becomes quite significant in retracement work and in presenting testimony. A major problem is encountered when different units are being related to each other.

6.4 Standardization of Measuring Devices

PRINCIPLE 3. *For any conveyance or description of real property, the length of the unit of measurement is that measurement that was used and recited as of the date of the deed or survey.*

The surveyor may have the length of his or her tape determined by sending it to the U.S. Bureau of Standards, Washington, DC. Under laboratory conditions, the bureau compares the length with a bench measure and issues a certificate or a report stating the values of the length for certain conditions of temperature, support, and pull. Tapes conforming to the specifications for standard steel tapes will be certified by the bureau, and a precision seal showing the year of standardization will be placed on the tape. A tape not conforming to the specifications will be tested by the bureau and a report will be issued.

The practicing surveyor can also self-calibrate a field tape to a calibrated invar or lovar steel tape manufactured by one of several firms in the world. In a controlled laboratory calibration, the standard tension is 10 pounds for tapes 25–100 feet, or 10–30 meters in length, and 20 pounds for tapes longer than 100 feet, or 30 meters. Realistically, no surveyor measures actual tension of pulls in the field.

More realistically, "self-calibration" can be accomplished by the surveyor making the comparison between two monuments set with a specific distance inscribed on the tablets. It is advisable for the surveyor to keep historic and accurate records of these calibrations and measurements.

The surveyor who has an extensive practice and employs several tapes or other types of surveying equipment would benefit by having one tape reserved solely for comparison. This tape may be certified by the Bureau of Standards and can be kept on a reel for use only to check the length of the tapes used in daily work. The standard tape, or other types of measuring instruments, should be recertified periodically. During the manufacturing process, internal strain develops, and while sitting on a shelf, the relaxation of the strain will cause the tape to change in length. In one recorded case, a lovar tape expanded almost 1.14 inches in a four-year period.

Today, in the field, few surveyors rely on steel tapes for measurements and very few, if any, modern surveyors rely on the chain. For the most part, the majority of the surveyors use EDM equipment. And currently, more and more surveyors are relying on GPS to conduct retracement surveys, little knowing that, once the lines are created, this may give future surveyors serious problems for conducting retracement surveys, as well as original surveys.

When this book was first written, it was stated that if a surveyor is going to court and is to testify based on a distance measured with an electronic measuring device, the surveyor should be prepared to prove that he or she has frequently standardized the equipment. Most important is to prove that the instrument is consistent within given limits under variable conditions of temperature and atmospheric humidity. When that was written, few surveyors used electronic equipment in the everyday course of surveying; but today the opposite is true. Few surveyors now use tapes, even for relatively short distances, and electronic equipment is the standard with the majority of surveying firms and their employees.

This places a greater burden on the surveyor, for full and total reliance is placed on electronic equipment that measures light waves, electronically reads the distance in light, and then converts this distance first to meters and then to feet. This requires the surveyor to check two areas: the original internal measurements and the reliability of the everyday measurements. The first may require calibration by the factory or checking to a standardized baseline, and the second daily or weekly checking before and after surveying to make certain that the instrument is functioning properly.

6.5 Units of Length

The official standard of length in common use in the United States is the foot, which is used in everyday surveying, including boundary surveys. However, the original surveys were conducted using various units of length, depending on the origin of the settlers. Historically, with the origin of most of titles, the English mile formed the basis for property granted by the British Crown for many original major grants. In the areas that were once under Spanish or Mexican sovereignty, the *vara* was the prevalent unit, and in Texas it is still the unit used. In many French areas, the *arpent*, the *league*, and the *toise* were the basis of area and distance. In an effort to aid the surveyor, tables were derived, and these customary units of length, according to the Mendenhall Order of April 5, 1893, are to be derived from the U.S. prototype meter. Table 6.1 lists some of these units and their equivalents.

TABLE 6-1. Units of lengths.

Units	Inches	Links	Feet	Yards	Rods	Chains	Miles	Meters
1 inch	1	0.126263	0.0833333	0.037778	0.005050	0.001262	0.000015783	0.02540005
1 link	7.92	1	0.66	0.22	0.04	0.01	0.000125	0.2011684
1 foot	12	1.515152	1	0.333333	0.060606	0.015151	0.000189394	0.3048006
1 yard	36	4.54545	3	1	0.181818	0.045454	0.000568182	0.9144018
1 rod	198	25	16.5	5.5	1	0.25	0.003125	5.029210
1 chain	792	100	66	22	4	1	0.0125	20.11684
1 mile	63,360	8000	5280	1760	320	80	1	1609.3472
1 meter	39.37	4.970960	3.280833	1.093611	0.198838	0.049710	0.000621370	1

Source: Modified From Units of Weight and Measure: (U.S. Customary and Metric) Definitions and Tables of Equivalents, United States. National Bureau of Standards, 1955, U.S. Government Printing Office.

PRINCIPLE 3. *For any conveyance or description of real property, the length of the unit of measurement is that measurement that was used and recited as of the date of the deed or survey.*

An example of this is the Texas *vara*. In the history of Texas surveys, one can find three different values of the *vara*: the Spanish value, the Mexican value, and the Texas value. In order to give guidance, the Texas legislature "standardized" the *vara*. Now there is only one value, regardless what the original value was.

Values of measurement for the same unit sometimes varied from locality to locality. In the year 1919, the legislature of Texas standardized the *vara* to be 33$^1/_3$ inches. In no way is this retroactive, and conveyances made prior to 1919 in Texas are not bound by the legislative act. When retracing original survey lines, unless the surveyor uses the same chain length as well as the same instrument that was used originally, the retracing surveyor will not reach the same point, or its location, as did the original surveyor and he or she will not be following the original surveyor's footsteps. Possibly it is for this reason that the courts hold that once you have found and proven the corner points, you have the line.

PRINCIPLE 4. *Every measurement of distance or angle is subject to errors, either known or unknown. There is no perfect measurement.*

Perfect measurements of land boundaries do not exist. Distances—whether made by a tape, electronic distance equipment, or the application of trigonometry or estimated—are subject to numerous errors. The surveyor should strive to make each measurement as close to the standard as possible. Many applied corrections to tapes and other equipment are dependent on a formula developed under laboratory conditions and are theoretical and will differ when applied in the field. Every applied correction has some error.

When one sees distances recited (i.e., 89 chains, 100.00 feet, or 32.34 meters), it is immediately assumed that the distance is a directly measured distance. In many instances, this is far from the truth. From the early days of surveying to today's modern world of the most sophisticated instruments, seldom do we have actually measured distances. The distances cited are more often than not indirect distances that are computed in some manner using mathematical principles in applying computed distances from their relationships with field-measured distances. These indirect methods require calculations that are based on formula. Today with the indirect methods of measurement, seldom are the actual lines measured; they are a result of calculations, usually between non–intervisible points.

Unfortunately, some surveyors, engineers, and courts accept a distance (e.g., that of 100.00 feet) as being absolute, when in reality it is a distance measured under conditions that were unique at that time only and cannot be repeated.

6.6 Historical Determination of Distance

Length has always fascinated humans. How long is a unit of length? Throughout history, mankind has used units of measurement to record the intended length of a line, a distance between two points, or a part of the body, but unfortunately, the definition of the length of the unit used was not always preserved in history. The mille, cubit, stadia, and rod are all historical measurements. Far back in the Egyptian culture, the problem of distance was solved by comparison. The rope stretchers (see Figure 9.1 on p. 246) compared the length of rope with a distance measured on the ground. Aristophanes, in determining the circumference of the Earth, had the computations correct, but the control distance troubled him. He apparently "calibrated" a camel's pace to determine the distance of 500 stadia. After comparison to modern precise calculations, his computations were found to be in error due to the miscalibration of the camel's pace.

In early Roman times, measuring bars were used for a standard, and these bars and rods have also been used for the same comparison as recently as 1900, when similar bars were used for baseline measurements by the United States Coast and Geodetic Survey (USC&GS) and USGS, while land surveys were measured with poles. In the 1800s, the Survey of India created base lines using calibrated measuring poles. The invention of the surveyor's chain in 1620 by Edmund Gunter, an English astronomer, was an improvement over the ropes and poles used at that time. The English rod or perch was once defined as the total length of the left feet of 16 men, tall and short, selected on a random basis as they came from church on a certain Sunday. The chain has been mentioned in many early textbooks, but it was not until the 1620s that John Love, an Englishman, fully described what the chain was and how it could be used to lay off angles and measure lines. This surveying instrument (Figure 9.4) became the most common in use and was the means for laying off most of the public domain in this country, as well as surveys in the metes-and-bounds states. In fact, in the early land laws, Congress, by statute, made the chain the only legal unit. This also applied to several of the metes-and-bounds states, Georgia being one that legislatively adopted the two-pole chain as the "official" unit of measurement in 1803. The chain was manufactured in various lengths, such as 66, 33, and 100 feet, and some chains were made too long intentionally to compensate for errors. All were bulky to handle, and usually the chain would be dragged on the ground, and the length would be in error because of slope, kinks, bad alignment, and other conditions. The acre is a unit of area that can be related only to the chain. Research indicates that no state has identified the value of the acre. It is only described in history by legislation passed by the English Parliament as being a "parcel of land comprising 160 square poles." It is an area that is enclosed by a rectangle 10 chains on one side and 1 chain on the second, or 160 square poles. To calculate acreage, one simply divides the area in square chains by 10.

Steel or ribbon tapes came into popular use at about the beginning of the twentieth century. The use of these tapes reduced the errors in measurement and permitted the surveying team to measure distances in a shorter time with greater precision.

Steel tapes are available in almost any conceivable length, from a few feet to 1,000 feet. They are graduated in various units such as feet, meters, yards, *varas*, and chains. A common tape once used by boundary surveyors was the 100-foot band chain. Today, the EDM has replaced the chain or tape as the primary equipment used in measuring distances.

One of the newer methods of measurement is the use of GPS (Global Positioning Survey) system. GPS does not measure lines, but does measure individual points, usually in latitude and longitude, from which coordinates are determined. Then, from two or more points, through calculations, lines are determined. GPS calculations do not measure lines; they measure position of end points of lines.

Today, where surveyors commonly use electronic distance-measuring equipment, it must be remembered that the distances established for most conveyances were based on old measurement methods. Precision attainable in former times has no relationship to precision attainable today. For those situations where taping was not practical, tachometric methods were sometimes used. In the United States, the stadia was one of the best-known tachometric methods, used more than any other, and the precision of this method, if carefully done, was not more than 1 part in 500. The subtense bar had limited use after the development of optical angle measuring instruments capable of 1-second measurements. Tachometric methods found popularity in Europe, with some instruments giving precision closures of 1 part in 5,000 under controlled conditions.

More details on the tachometric surveys may be found in many popular textbooks on surveying, as well as in journals published by the technical societies.

Many surveys were made in early times with a wheel or odometer giving surface distance. Such devices are useful for special-purposes today. Good results depend on a smooth rolling surface, which is not often obtainable.

In recent years, electromagnetic and electro-optical devices have come into use for measuring distances. Today, whereas these devices are of great value in making remeasurements, they were not available in former times. Sometimes in retracing earlier surveys, it is better to use the equipment available at the time, especially when retracing a survey originally made with a compass. Local magnetic variations are not detected in modern directional devices, and errors of slope measurement do not show up. The object in a retracement survey is to *retrace the footsteps* of the original surveyor, but in some instances this cannot be done by modern equipment that skips over part of the ground using traversing methods or one that uses the radial method of surveying. Marked trees and corners can be easily overlooked. After a survey is accurately retraced with older equipment, the surveyor should return and then precisely determine the position of found or set monuments; he or she is merely accomplishing a refinement of the earlier survey.

6.7 Early Determination of Bearing References

Since a line is composed of both a direction and a length, the next discussion must be orientation of the line. The usual, both historical and modern, reference is that of north. Although some areas have made reference to others, north still is the major reference for surveys and deeds in the United States proper. True north, geodetic north, magnetic north, and assumed north all use north as the point of reference. The major problem of the surveyor is to determine what the original surveyor used as the point of meridian reference for boundary lines. Some references to "north" defy identification. In Hawaii, references to bearing in the early Hawaiian deeds use the center of the island or the high point of the mountain as reference. Many of the surveyed lines also use south as their point of reference in deeds. Any line of direction must be relative to two fixed points, and from the facts known about the Earth today, it is most logical to use the direction formed by the two poles as a reference. In early history there was much debate about whether the Earth was round or flat. Earlier people, especially seafaring men, did observe that certain stars remained relatively fixed and others seemed to rotate around them.

Polaris, from which many surveyors now determine north, was not always the North Star. At the time of the pyramids, Thuban (Alpha Draconis) was within 4° of the celestial pole, and Polaris was 25° away. From about 1800 BC to AD 1000, no bright star existed near the North Pole. Today, Polaris is less than 1° from the pole, and it will reach within 00.35′ before it travels away from the pole. Some 10,000 to 12,000 years from now, Vega will be in a position to be considered the North Star.

Substantial proof exists that the North Star of 2900 BC was used to construct the Great Pyramid's entrance passage, because it was so built that Thuban at lower culmination (3°42′ below the pole when pointing due north) could be seen from the bottom of the passage. The base of the entrance is only 3′06″ west of north! Other evidence indicates the pole star, Thuban, was probably in use in 5000 BC.

An early reference of direction was the magnetic needle. Although it is generally recognized that the Chinese developed the "magnetic chariot" that constantly pointed to the north and the first compass was probably of Chinese origin, little is actually known about the compass and its development. The Greeks and Romans had developed a sundial that incorporated the magnetic needle, and the Norse used lodestones to guide ships, but they were not in general practice for surveying. For a period of over 100 years in Medieval England, anyone who used compasses was considered a heretic and was put in prison or put to death for using witchcraft. It was not until the sixteenth century that magnetism was finally understood. It is generally accepted in surveying circles that the magnetic meridian is probably the least reliable reference for meridian reference, but many states still permit the practicing surveyor to reference surveys to the elusive reference, magnetic north. Unfortunately, in today's records, one can examine hundreds of plats on which the elusive phrase "according to the magnetic meridian" can be found.

6.8 Direction

Distance alone does not define the terminus of a boundary line. To adequately describe a line, a minimum of two elements and a maximum of four are required.

First, a line is described by a bearing and a distance or an angle from two known points. A line so described can be positioned any place on the face of the Earth. Then add a bearing, a distance, and one point at one terminus of the line, and the line is fixed on the face of the Earth and the resulting line is fixed in bearing and distance from the initial point. The end of the line that is the resultant of the bearing and the distance is not fixed positively, in that its location is a product of the bearing and the distance, with the resultant errors in these two elements. Now add a second endpoint at the opposite end of the line, and we now have a line that is definitively fixed at its endpoints, which are now identified by two elements: a bearing and a distance. This, now, is where the law enters into the equation. The law dictates that if these two endpoints, which we will call *corners*, are first called for in a written conveyance and then are recovered as a part of the survey, then they will control, regardless of the bearing and distance called for, but with direction and its location identified by the actual measurements. Direction is determined by angular or circular measure in a horizontal plane. In the United States, the sexagesimal system of angular measure, which divides the circle into 360 degrees, each degree into 60 minutes, and each minute into 60 seconds, is in common use. The angular measure affects the position in direct proportion to the distance. As an example, 1 minute subtends an arc of about three-tenths of a foot at 1,000 feet. The accuracy of directions is dependent on the instrument used, the procedure employed, and the length of sight plus the eyesight of the observer. Understanding this principle is important when it comes to retracing historically described lines.

The discovery of the Earth's magnetic field and the invention of the compass came before Columbus's time. The use of the magnetic meridian for determining north was known in the twelfth century, and the Chinese knew of the properties of lodestone about 1000 BC. Early navigators in Western Europe were blessed with an almost zero declination, an advantage that no longer exists. A belief arose that the compass always pointed to the true north, and during Columbus's voyage, his men were frightened by the discovery that the compass deviated from true north. In 1635, Henry Gillibrand discovered changes in the magnetic needle, and it was in 1722 that daily compass variations were discovered. The first isogonic chart appeared in the nineteenth century.

The Earth's magnetic field is not static; it is subject to annual changes, daily variations, and periodic, unpredictable variations, such as magnetic storms as well as local attraction from unknown sources. At present, no certain or absolute method exists to prove what the magnetic declination was on a particular date in the past without positive observations on that date, and it is even more difficult to predict what future declination will be. Directional calls found in old deeds and usually based on magnetic observations are difficult to interpret and then locate. This one phase causes inexperienced surveyors agony in attempting to determine the correlation between old bearings and the present reading they must use to identify and then retrace ancient lines. Many surveyors who conduct retracements are woefully ignorant of magnetic declination and its application to retracements. If a surveyor plans to retrace descriptions contained in old deeds, a complete understanding of magnetic declination is a prerequisite.

6.9 Reckoning

Directions must be referred to a meridian. Often, the meridian is assumed along one line purely for the purpose of providing relative bearings for each of the sides. The U.S. Rectangular System was intended to be oriented to astronomic north, although later surveys reveal considerable deviation from this belief.

"True" north may be defined by one of a multiplicity of meridians: astronomic (north as defined by the celestial poles); geodetic north based on a reference spheroid (generally Clarke's); or geographic north. Although these differ very little (only a few seconds), true north in property surveying is generally considered as that determined from star observations.

Once the direction is reckoned with respect to a meridian, all other lines may be related by angle measurements. Metes-and-bounds descriptions are sometimes oriented by carefully reading the compass needle on one of the sides and then relating the other sides by the deflected angles measured or by reading each line. When the survey is retraced years later, the magnetic direction of any line can be determined within a limited accuracy, by applying the change in magnetic declination for the elapsed period of years. By using this method, the retracing surveyor may be able to find the original position of the line today.

Originally, most of the early surveys of the public lands were surveyed with the compass. First, it was the staff compass and later it was the solar compass. Today, such methods are not generally satisfactory, although many surveyors still use the compass needle for retracing old lines and for a basis of orientation. Some states prohibit the use of a compass in land surveying.

In those areas where grid systems are employed, grid north or grid south may be the reference meridian. This system has many advantages, as well as disadvantages, since it is plane rectangular; that is, the meridian lines are parallel and can easily be corrected to astronomic north.

In a retracement, a factor that gives the surveyor, the attorneys, and ultimately the courts difficulties is having to "find" or determine the original bearing reference. This then will permit the court to make its own determination what the basis of bearing is or was. For the most part, the early surveys in the metes-and-bounds states were run by the magnetic meridian. But then in the GLO states, the Land Act of 1785 directed that "all lines will be run by the true meridian."

Some of the early survey lines were created by *dead reckoning*. This term is nothing more than *deduced reckoning*—a term for "we guessed what it was."

6.10 Methods of Observing Directions

The early method of observing magnetic bearings for boundary surveying has passed into obsolescence and, in fact, is prohibited by law in some of the states. Besides the annular and secular changes and the danger of local attraction on a compass needle, readings can, at best, and based on the precision of the early instruments used, be read to the nearest quarter degree. Little discussion is ever seen concerning the instrument that was primarily and exclusively developed in the United States and then ultimately required by law to be used in the surveys of the public domain or the GLO surveys—namely, the solar compass. This instrument was developed in 1836 by a U.S. deputy surveyor, William A. Burt, and it remained the workhorse of the GLO surveyors for over 100 years. This instrument revolutionized the survey of the public domain in that accuracies never before known to original surveys were obtained independent of the magnetic meridian.

Although plane (plain) tables and alidades were used by colonial boundary surveyors, the modern transit is relatively new for boundary surveying. Roman *agrimensores* used a similar instrument called the dioptra but this instrument never found use in America. It permitted running lines using right angles only. The terms *transit* and *theodolite* are not well defined by popular usage. The instrument's telescope must turn about the vertical axis to be properly termed a transit. The theodolite is generally associated with more precise instruments, although it is possible to have a theodolite that will transit and a transit that is a theodolite. Today, most surveyors use a form of theodolite that is commonly referred to as a *total station*—that is, the angular measurement capabilities and the distance-measuring capabilities are integrated into one instrument. This instrument was not perfected until the early 1980s. The newer instruments do not read a vernier but are optical reading and usually now have a measuring system built into the angular instrument, giving what some now call *total stations*. In reality, the instrument should be called an integrated system, since it combines a distance-measuring system with an angular measuring system. This instrument usually incorporates an "on-board" computer that takes the slope distance measured by the EDM, combines it with the vertical angle measured by the instrument, and creates the horizontal distance and the difference in elevation automatically, which is then stored into an electronic notebook.

Early instruments and methods of observation are classed as either *repeating* or *directional*. By alternating the use of the upper and lower motions, the value of an angle may be repeated or "run up" several times so that small fractions can be measured even when smaller than the smallest division on the instrument could not be detected. By repeating an angle, more precision can be obtained than is possible by a single reading.

Direction instruments have no lower motion and are usually of higher precision than repeating instruments. The angles may be derived from the difference in the observed directions of two lines. A number of readings with a direction instrument will improve the accuracy but not the precision.

The quality of direction should be directly related with the quality of the distance. It would be ridiculous to use a magnetic compass with a carefully taped distance and absurd to read angles to a second when the distances were determined by stadia. As in all survey operations, the observer should provide for a check against blunders when reading angles and distances. This may be realized easily by "doubling" the deflection or angle to the right, or "closing the horizon," even though a single reading will satisfy the precision requirements.

Systematic readings will minimize blunders and increase efficiency. Many systems for turning angles are described in the various textbooks on surveying and are not discussed here.

The surveyor should decide on the procedure that best suits his or her needs, and then these methods should be followed in an orderly fashion. The vernier transit, which has been the workhorse for boundary surveyors for many years, is now generally replaced by optical reading instruments for many purposes.

Historically, the general history of boundary survey methods has been to have less precise instruments to create the boundary surveys and then the descriptions are created from these less precise surveys, but then, at a later date, using more precise instruments to retrace these original lines used in the recovery of the earlier lines. This practice has led surveyors down erroneous and at times perilous paths when attempting retracements.

When one creates corners and boundaries by original surveys, the greater the precision of the instruments used, the easier the job of retracing those original lines.

6.11 Historical Application of Measurements

With the advent of easy calculations through the use of computers and handheld calculators and as a result of the education and training, surveyors feel comfortable in converting between totally unrelated measurements. One will see a section line of an original survey as being identified in chains (e.g., 80 chains), yet retracing surveyors will abandon the original chain measurement and show their distances in feet. This practice defeats the responsibility of the retracing surveyor, whose responsibility is to "follow the footsteps" and makes it virtually impossible to relate retracement distances to the original distances. Even though the method of retracement is modern, the modern measurement should be converted to the original to determine correlations; the original measurements should not be converted into modern measurements. When thinking and speaking about measurements, the retracing surveyor should think and speak in the original units that were identified by the creating surveyor. This should be done for the correlation of distances only and not for computation of closure.

6.12 Survey Computations

Within the last few years, remarkable developments in the science of computations have occurred in ease and accuracy. Electronic calculators take raw data, compute the error of closure, adjust the traverse, and calculate the area in a matter of seconds, whereas it formerly took hours, while redacting from printed tables and then check and rechecking the numbers. Many of the early tables required the surveyor to interpret numbers. Most of the drudgery and errors of office work have been eliminated.

In the early days of surveying, textbooks had tables of logarithms in the appendix. At one time, computation sheets were stored with every job. It is now quicker to recompute than it is to try to understand old handwritten computation sheets with sine, cosine, bearing, and distance. Because of the difficulty and tedious nature of computations in early times, the cause of errors can be understood; proof by computations was often omitted.

Mechanical calculators came into common use by 1930, electric calculators by 1950, and electronic calculators capable of solving for the sine, cosine, and tangent by the 1970s. Most computations prior to 1920, if done, were generally by logarithms or sine tables. Few surveyors' offices were without a book of sine, cosine, and tangent tables, usually identified to 1 second. Today, instruments or total stations can carry thousands of points and complete calculations and adjustments in the field.

The main purposes for calculating an error of closure is to determine the precision of the survey—that is, the relationship of angular measurements and distance measurements to each other—and then to calculate area. Area cannot be calculated from field measurements. The error of closure does not indicate the accuracy of property corners. The law holds that the original corners are accurate, regardless of the precision. Assumptions are made when a traverse is balanced that any errors are distributed uniformly in each course, yet this is not necessarily true. No one is precluded from placing any corrections in those lines in which the error most probably occurred. The selection of the proper lines requires experience and a knowledge of the theory of errors and adequate field experience. Balancing surveying errors permits one to isolate errors and then finally to compute area, for area cannot be computed from field measurements.

Since most surveyors now rely on computer software calculations to balance field data and then to adjust and compute the final area within the lines of the survey, they have little idea or information as to how the field data are balanced or the method of area determination. In all probability, different software programs use different methods for balancing field data and for area determination. Two different programs using the same field data may compute different balancing adjustments and then compute a different area from the same field data.

Once adjusted lines are computed, area can be determined. The calculated area of a balanced, closed traverse is determined by the method used. No two methods of area computation will give the same results and may vary in the first or second decimal place. When the surveyor indicates either on the plat or in writing the area of a parcel, the method of error calculation as well as the method used to compute the area should be indicated.

6.13 Consistency of Significant Figures

Since measurements are but one form of evidence, all measurements should not be considered as equal. A measurement must be evaluated in terms of what it tells the individual. Since it has been discussed that measurements should be evaluated in the units of the original measurement, the key to their evaluation is one of significant figures. The number of significant figures in a numerical expression is the number of digits in the expression minus any zeros. Retracing surveyors now consider their main purpose being to indicate the position of the retraced corners points to the decimal point.

In applying significant numbers, surveyors must look at their evaluation as part of the overall evaluation of the evidence.

Built into the expression of any numerical value is an error of half the magnitude of the unit of that last significant figure (e.g., 1,432 chains or feet numerically stands for those values between 1,431.5 and 1,432.5; therefore, the error in expressing the value is ±0.5 chains or feet).

Measurement evidence is nothing more than a matter of numbers and how one applies those numbers. Table 6.2 relates significant figures to other numbers.

The error in the expression of a measured quantity should be consistent between linear and angular measurements. There is a generalization using the following particular example (see Figure 6.1). The position of a point *B* from point *A* may be defined in terms of a distance *D* and an angle from some reference line.

TABLE 6-2. Significant figures related to other numbers.

Value	Significant figures	Error of expression	Relative error
0.023	2	±0.0005	1/46
172	3	±0.5	1/344
84.15	4	±0.005	1/16,830
3,956	4	±0.5	1/7,912
1,320	4	±0.0005	1/2640
32,100	3–5	±50 to ±0.5	1/642 to 1/64,200

Significant figures	Relative error
1	1/2 to 1/18
2	1/20 to 1/198
3	1/200 to 1/1,998
4	1,200 to 1/19,998
5	1,20,000 to 1/199,998

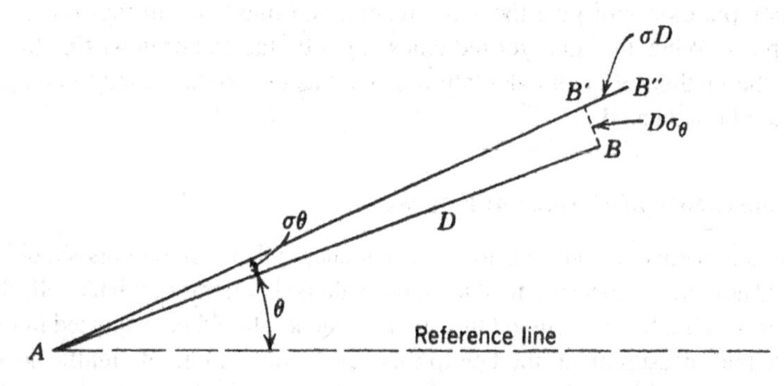

FIGURE 6-1. Relationship between angles and distances.

An error in the angle would produce a shift in the position of the point to B'. An error in the distance (D) would produce a shift in the position of the point from B' to B''.

The relationship between an error of closure or relative error and angular error may be expressed as follows:

$$BB' = B'B''$$

To be consistent, therefore,

$$\sigma D = D\sigma\theta \text{ or } \frac{\sigma D}{D} = \sigma\theta$$

Relative error in distance = Angular error in radians

Relative error	Angular error
1/57	1°
1/690	5′
1/3,440	1′
1/10,300	20′
1/206,300	1′

The error in the expression of a measured quantity should be consistent with the error in the measurement itself. Thus, if a distance was measured and recorded as 319.42 feet but the error in the measurement was 0.5 foot, it should be recorded as 319 feet. The last two figures are meaningless, and the expression merits only three significant figures instead of the five.

This equates to taking a series of measurements with a tape or chain between two corners by dropping a stone or kicking the ground and then at the last chain length using plumb bobs to refine the final measurement to a distance of 0.01 foot.

Surveyors readily work between units with little consideration of what they are doing. If one value is used to compute another value, the rounding-off error will be introduced as a systematic error into the computation and may become significant. Thus, rounding off should not be done until the final result is obtained or should not be done in the computed value. For most computations, this statement can be modified by saying that one significant figure more than is necessary should be maintained until the final result is obtained.

Using the same philosophy, when an original unit of measurement is recorded as 80 chains, in applying significant numbers, the actual distance may be between 79.5 and 80.5 chains. This would give a latitude of distance of 1 chain, or, if converted to feet, of 66 feet. But if one were to initially convert the original measurement of 80 chains to 5,280 feet and then proceed to work in the foot unit, the impression would be that the 5,280 in reality is a distance between either 5,275 and 5,285 or 5,279.5 and 5,280.5. Neither of these two measurements correlates with the original measurement of 80 chains.

6.14 Significant Numbers in Angular Measurements

In evaluating evidence, the surveyor must not only consider the angular and distance measurements but also understand the significant numbers of the measurement.

In many early field notes, distances were recorded in degrees and in fractions rather than degrees and minutes. Yet today, when some surveyors attempt to analyze these early distances, they fail to fully understand this significance.

An angle of 5½ degrees is readily translated by surveyors, attorneys, and the courts to 5 degrees and 30 minutes. In evaluating and applying evidence, this is not so. Applying significant numbers 5½ degrees ranges from 5⅜ to 5⅝ degrees; while 5 degrees and 30 minutes ranges from 5 degrees 29.5 to 5 degrees 30.5 minutes, or a much smaller spread. An early seventeenth-century surveying textbook, *Geodaesia,* by John Love, had several chapters explaining how a surveyor could determine angles from horizontal measurements and their relationships (see Appendix D).

This also applies to original field notes. A surveyor may find that the distances indicated in the field notes of many original surveys were recorded to the nearest full chain, such as "3.00 chains, to a creek or 4.00 chains to a beech." In surveying, it is virtually impossible that such a situation exists. Thus, one can presume that the distances were actually measured to a more refined distance but were recorded to the nearest full chain; that is, a field measurement of 7.88 chains was recorded in the field book as 8.00 chains.

6.15 Meridian Determination from Celestial Observations

For many centuries, the stars have been the guide for people to locate direction. The fact that the Earth spins on its axis causes the stars to appear to rotate about a point in the sky (the north celestial pole), which point has become known as true north. One star, Polaris, is in close proximity to the North Pole and is a convenient star for observers in the middle latitudes of the Northern Hemisphere. All stars cross the observer's meridian each day. If the bearing of the star is computed for the time the observation was made, the meridian can be located.

Many sections of the international boundary of the United States, most of the state boundaries, many of the county boundaries, and most of the U.S. Rectangular System are defined in terms of "true" bearings. Although the monuments set to perpetuate their location are controlling, the intent was to orient many of these lines with the cardinal direction of astronomic north. In most states, surveys that are described by metes and bounds will be oriented with astronomic north if another system is not named or implied. For the surveys that are based on other meridians, it will often be more efficient to initiate, terminate, or check directions with a star observation.

With present-day surveying equipment and time pieces, all star observations are relatively simple to perform in the field, and with programmable computers of the electronic type, computations are no problem. Yet, seldom do surveyors spend nights taking Polaris observations or "sun shots." Reliance is made on measurements from technology. In performing subdivision surveys and any survey to create a new parcel, directions should be based either on a sun or Polaris observation or on a plane coordinate system with sufficient monument density if such a system has been established by the state, so as to make retracements easier.

Today, GPSs are being recommended by surveyors as possible substitutes for azimuth use. Since all computations of position of the survey points are predicated on computations from satellites, and since the azimuth determination is itself predicated on computed numbers of the survey stations, we have the same problem—our reliance is totally on a computed and not a directly observed value.

6.16 Photographs as Evidence

Photogrammetry is almost as old as photography. In the broadest sense, it is the science and art of making measurements from both aerial and terrestrial photographs. The intent of this brief discussion is not to present the theory of the science or the technique of the art but to point out some of the proven and accepted applications that may be made in property surveys by knowledgeable surveyors In this chapter, the use of photographs is basically dependent on the "whim" of each judge. Attorneys and surveyors go into court armed with both aerial and "candid camera" photographs, hoping to introduce them into evidence and to influence the jury and the judge with the "facts" indicated on them and with their abilities to interpret the images. A surveyor does not become an expert simply by obtaining copies for court use. Before a surveyor ever attempts to use aerial photos as evidence in a court setting, he or she should have substantial and creditable knowledge about various aspects of photographs. Such elements as photo scale, types, and scale and image portrayal are necessary to become one who understands and is able to testify. Simply being a registered surveyor does not make one versed in the interpretation of aerial photographs.

While in court, many judges will ask the surveyor to show their map (the aerial photograph).

AERIAL PHOTOGRAPHS ARE NOT MAPS.

An aerial photo does not have a constant scale, visual images depict the data, all lines are depicted as surface lines and do not have a common plane of reference.

The ability to visualize images depends on the scale of the aerial photograph as well as the emulsion of the photograph. Today, the basic types of images are digital, panchromatic, infrared, and color. Each of these provides different imagery.

The practical beginning of aerial photogrammetry in the United States was in the mid-1920s, nearly a generation after the first start was made in Europe. It is not known when these measurements were first made for cadastral surveys, but as early as 1938, property damages caused by forest fires were surveyed with the aid of aerial photographs, and boundary lines were located for reservoirs by the Tennessee Valley Authority. By the close of World War II, the use of aerial photographs for boundary line location was common.

Many experiments and practical applications of modern photogrammetric principles to boundary location were performed in Europe. The earliest known work occurred in Italy in 1931. Reallotment surveys in Germany made use of coordinates of property corners obtained by photogrammetric methods. The accuracy was quite satisfactory, and over a large-scale operation with high-density ground control monuments, it was successful for property registration purposes. The project required the identification of all points on the photographs using small targets. Similar projects were done in the Netherlands, Sweden, and France. A recent land registry program in Austria made use of the teleprinter and electronic computer, undoubtedly the most automatic property survey performed up to that time.

In recent years, there have been several noteworthy experiments in the United States. The U.S. Forest Service conducted a survey to locate sections of forest lands for timber management in Missouri and the Lake Tahoe area. Together with ground control, the measurements were made photogrammetrically from high-quality aerial photographs. In a test area in Utah, the BLM subdivided several townships with the use of photogrammetric methods. Picture points served to locate the position of the corner, with plane tables used in making the actual establishment. This test was well within the required accuracy stated in the *Manual of Instructions* [2], but the costs were 50 percent higher than those of the usual conventional methods. In Minnesota, the center of sections and 1/16 corners were set by photogrammetric methods.

Currently, photogrammetry in the United States is used in many ways, such as pipeline location, "paper" surveys for highways, and subdivision planning. In all cases, final location and measurements must be made on the ground.

The student, surveyor, attorney, and court should make a distinction between measurements from photographs and photogrammetric measurements (see Section 6.17). They are entirely different, distinct, and independent.

When considering measurements from photographs we envision working with the actual contact photographic prints (usually 9 × 9 inches) and making measurements using precise stereoscopic methods. They are different. Personal measurements made from the contact print and stereoscopic measurements give different results.

Aerial photographs, as well as terrestrial photographs, and contact prints from "regular" cameras can all be used to make measurements. Each of these measurements will have a large degree of measurement uncertainty due to errors. Absolute measurements cannot be made from photographs without applying and rectifying the inherent and residual errors in photographs due to lens distortion, changes in elevation, misidentification, scale, and many other errors, both actual and theoretical.

Measurements from aerial photographs can be used to locate search areas for corner location to a degree of certainty, depending on the scale of the photograph, but without precise photographic measurements, the same methods cannot be used to set lost corners. Aerial photographs can be used to place overlays with deed plottings to position areas or parcels, but they cannot be used to locate the actual deed lines, and measurements from photographs can be used to help identify lines of possession and use. With the introduction of computer-aided design (CAD) systems and scanning capabilities by computers, new dimensions are being opened up for surveyors to use aerial photographs. Correlating ancient surveys to images that can be observed permits the surveyor to locate ancient parcels.

6.17 Photogrammetric Measurements

Before distance, direction, or position can be determined from photographs, identification of points on the pictures is essential. The corners may be either marked before the photography is flown over or identified afterward by field inspection. When the corners are marked with a target, there is danger of it being moved, destroyed, or confused with some other object. The practice of premarking points is quite prevalent in Europe, where small, colored cards are centered over the monuments, or other devices are used that will cause them to contrast well on the photograph. Lime or plastic strips spread on the ground are sometimes used for premarking points. When the identification is left until after the photographs have been taken, some additional cost may result.

Lines and corner locations can be ascertained by directly reading from known points on the photographs. Basically, there are two forms of measurements: (1) direct from the photographs and (2) using stereoscopic mapping instruments. Direct measurements are difficult to obtain and have little degree of accuracy.

The discovery of accepted boundary corners is usually the biggest task for the property surveyor, but after the discovery of the corners the geometric location of these points may not present a great problem. It is possible to determine these positions photogrammetrically if extremely precise methods are used. The photography needed is of the highest order, at a very large scale, and taken under ideal conditions. A considerable amount of ground control is necessary to orient the photogrammetric models. With high-precision plotters, measurements have been made from photographs with a certainty of less than 1 inch that would satisfy the requirements of many property surveys. Whether the courts will accept such measurements as proper evidence of property title has not been tested.

Such a modern method of surveying can be applied to the layout of an urban subdivision, but as yet it has proved too costly for other than topographic plotting. It is anticipated that in the near future, photogrammetric methods will be used more extensively. The future of photogrammetry as a tool for surveyors is unlimited.

The practicing surveyor should have confidence in using all forms of evidence that may be available to assist in the location of ancient parcels and boundaries.

6.18 Availability of Photographs

The United States has been completely covered by aerial photographs of various scales and film emulsion types. In some areas, especially around large cities, the coverage has been repeated numerous times. Most of these photographs are available to the surveyor at a nominal cost. Governmental agencies such as the U.S. Geological Survey, Production and Marketing Administration of the Tennessee Valley Authority, Corps of Engineers, Agricultural Stabilization and Conservation Service, U.S. Forest Service, and others keep on file many photographs of various ages and scales and have the means of supplying prints to individuals. Many states, counties, cities, and private companies have photographs that have been taken in connection with some special purpose, and these can be purchased for a small charge.

Some practicing surveyors take their own photographs. It is not always essential to use an aerial camera; press cameras with a large negative size have proven to be satisfactory for many uses. Using a small chartered airplane and a conventional press camera, an entire township can be photographed. The information obtained may be worth several thousand dollars. These will have little technical value and should be used only for interpretation.

One of the largest repositories of aerial photographs is the various State Departments of Transportation. The surveyor should also contact the various aerial photogrammetric mapping firms listed in the Yellow Pages.

One should not overlook those photographs taken by the surveyor or the client or survey crew. One must realize each of these photographs must be sponsored and accounted for.

Errors in Measurements Every measurement, either horizontal or angular, has errors. *There are no perfect and absolute measurements.* Whether the measurement errors are caused by improper methods of surveying or are inherent in the instruments or caused by nature, errors are present. The responsibility of the surveyor is to be able to understand and ascertain the types and magnitude of the known and unknown errors present in every measurement.

6.19 Precision and Accuracy

In retracements or other types of surveys, attorneys, judges, clients, and some surveyors equate numbers to quality of the final product. Many assume that a survey that has a closure of 1/23,000 is better, legally, than a survey that closes 1/300 or even 1/500. This problem was addressed in 1912 by A. C. Mulford, who stated, "It is better to have a faulty (inaccurate) measurement where the line truly exists, than it is to have an extremely accurate (precise) measurement of the place where the line does not exist at all" [5]. In this simple statement Mulford described the problem and differences between precision and accuracy. Precision refers to the degree of refinement by which a quantity is determined. *Accuracy* is a measure of how close to the true or exact value the determination is. In some circumstances, the number of decimal places is a measure of the precision. It is not to be construed that higher precision will ensure more accurate results. Consider the example that follows, in which three angles of the same triangle were measured by a 1-minute transit and a 1-second theodolite.

Transit	Theodolite
64° 40′	64° 41′ 22′
42° 25′	42° 24′ 52′
72° 55′	72° 53′ 40′
72° 55′	72° 53′ 40′
180° 00′	179° 59′ 54′

The three angles measured by the transit total to the required 180° 00′, whereas the angle measured with the theodolite fail to close by 6″. Obviously, the quantities measured by the theodolite are more precise, but those determined by the transit are accurate within their limit of expression.

Precision without accuracy is meaningless, and better accuracies can be obtained only through increased skill and more refined equipment. Precision is achieved by employing equipment of high quality with skilled techniques under the most favorable conditions. Each must be compatible with the others or the effort is for naught. Thus, a precise survey that does not properly identify the original corners or fails to provide an area of search for the original corners is meaningless. The question that must be addressed is: What good is a precise survey from a corner that is misidentified or even lost? All of the precision available cannot properly locate the parcel if it is misidentified. In boundary surveying, the surveyor must exhibit both accuracy and precision. If one is more important than the other, then accuracy must be the paramount element.

In most cases, the true accuracy cannot be ascertained except by careful analysis of the errors consistent with the precision used and an estimate of the accuracy that should be achieved.

6.20 Classification of Errors

An error may be defined as the difference between the true value and the observed quantity. Errors must not be confused with blunders, which also cause differences in the derived values but are illegitimate and can and should be removed.

In general, all errors may be classed into two groups: systematic and accidental. *Systematic errors* are those that have predictable quantities and signs and, in turn, can be compensated for and possibly eliminated. With sufficient collateral evidence, the effect of these errors may be removed by correction. A typical example of a systematic error is the thermal expansion of a steel tape due to the change in temperature. Another is the correction for humidity and temperature in an electronic measurement or a prism offset error. If the temperature coefficient of expansion for the tape and the length are known, the correction can be applied mathematically or made physically. These corrections can be made either at the time the survey is conducted or later, when the calculations are made. It should be noted that systematic errors behave in a predictable way; the magnitude and direction of these errors are functions of determinable variables.

Recently, a survey was computed for a second-order geodetic control net that was going to be used for a GIS/LIS project. A large variation, or error, was observed between two known, first-order geodetic control stations, and the closure was far beyond the contract requirements. The surveyor assumed that there was an error in one of the stations. After hiring a GPS firm to check the validity of the stations, the GPS data reaffirmed the coordinate positions of the two primary stations.

The firm then engaged a private consultant to check all of the data. Starting from the initial data, the equipment was functioning properly; the question was then asked: What were the meteorological corrections applied? It was revealed that the survey crew did not apply any meteorological corrections to the instruments, but they did record temperature, humidity, and barometric pressure in their field books. After applying these meteorological corrections to the raw observed data, the expert was able to show that the final computations of closure were well within the contract requirements. Had the survey crew not tied between the two separate geodetic stations, they never would have known the problem existed.

An excellent example of this is the failure to properly set the prism offset, or a tape that has an error. At one time a manufacturer sold an invar tape to an individual.

The tape was used for several months on traverses that were independent of any other work. Then the surveyor tried to check between two first-order triangulation stations. The traverse tie line would not close. He ran it again. The same results were obtained. Then he checked the tape with a standardized tape and found that the new tape was 1 foot short. Due to a manufacturing error, the tape was 99 feet long and not the 100 feet stated on the reel.

Accidental errors may be divided into two types: random and constant. *Random errors* have equal chances of being plus or minus. An example of this is setting a chaining pin at the end of the tape or setting up a tripod for a prism. Say that the chainman can set the pin within +0.005 foot. Therefore, each time the pin is set, the chances are equal of its being advanced or set back by the 0.005 foot.

A *constant* accidental error is one whose quantity is unknown but that will have its effect in the same direction throughout the operation. Alignment error is one such example. It is impossible to determine how much error is introduced by reason of the tape not being in a straight line, but the tape will always have the effect of producing too large an observation; therefore, the effect is constant in direction. The calibration error of the tape is another fine example of constant error. If a certain tape is certified to be 100.00 long, its length is uncertain by 0.005 foot at least. We do not know if it is too long or too short, but, whichever, it will be the same every time it is used, and the error will accumulate in direct proportion to the number of times it is used.

Even though systematic errors can be corrected, there will always be some accidental errors accompanying each systematic correction. In the preceding example, the temperature correction is not complete because of uncertainties in reading the thermometer. Many of these "residual" errors are small or even insignificant, but often they must be considered when writing specifications. These errors become significant when they accumulate. Today, a major source of this type of error is instruments that use tribrachs for mounting the instrument to the tripod. In reality, the tribrach is a miniature surveying instrument in its own right. If this instrument is not properly adjusted, systematic errors will be introduced at each setup.

Blunders and mistakes have to be avoided or the work must be redone. Such acts as misreading the tape, transposing numbers, and dropping a tape length cannot be corrected for and have no place in surveying of a professional caliber. In analysis the student and practicing surveyor can see that many of these errors are individual in nature to the individual's instruments, yet some surveyors realize that there is one minor piece of equipment that is never mentioned—the tribach, that minute piece on which the EDM or the total station sits. This instrument holds the tripod to the measuring instrument. This small unit is as important and as necessary as the Total Station or EDM, yet seldom does anyone check it for errors.

6.21 Theoretical Uncertainty Analysis (TU)

All surveyors realize that no quantity can be measured to the absolute true value and that each stated measurement has some inherent doubt. It is essential that each measurement be accompanied with a statement of its reliability. In control surveys, such as are conducted by the governmental agencies, the customary method of expressing the quality of the measurement is by stating the *relative error of closure*. These ratios are computed by dividing the indicated closure discrepancy by the total distances. For this, familiar expressions such as 1 in 5,000 and 1 in 10,000 are used.

A survey that has been conducted to locate the title ownership of real property must be certain to a degree that any conflicts arising over encroachments and the like cannot be attributed to the poor quality of the survey. Closures of traverses must be good, certainly, but it is possible for a traverse, displaying a small closure discrepancy, to contain excessive uncertainties. The concept of relative error of closure is not very meaningful in this circumstance, for it would permit a larger uncertainty in a survey of a large lot than in one of half the size. It is, perhaps, better for the surveyor to examine his or her methods, procedure, and equipment to estimate the uncertainty in derived positions of the points through which the survey has passed. This expression of quality stated with the quantity will be of great help to the client, the court, retracing surveyors, and all who have interest in the property in question. As an example, a property line may be expressed as "659.74 feet, to .08." This would indicate that the true length will probably lie between 659.66 and 659.82 feet and the most likely value is 659.74. If a suit between the coterminous owners were to result over an encroachment of 1 inch, it would be obvious that this much uncertainty exists in the survey itself.

The value ±0.08 cannot be determined by casual guess. It must be calculated from sound theoretical analysis. The surveyor has, of course, corrected the distances for all known systematic errors and has stated the uncertainty in the accumulated results of accidental errors only. For many accidental errors, it is quite difficult to ascertain the magnitude, whereas for others, the effect of the error is quite obvious. A number of the errors will be so insignificant that their effect will be overshadowed by the others.

It is likely that the random errors will accumulate as the square root of the opportunities, expressed by the equation

$$E = \pm e\sqrt{n}$$

where E is the total effect of a particular error (e) and n is the number of opportunities. Constant errors accumulate in direct proportion to the number of opportunities, or

$$E = \pm e \cdot n$$

When the total error has been calculated for each of the contributing factors, the resultant is found by the expression

$$E_T = \pm\sqrt{E_1^2 + E_2^2 + E_3^2 + \ldots}$$

where E_T is the final resulting error in the observation and E_1, E_2, . . . , are the full effect of each of the contributing errors.

The value E_T may be expressed in linear units, rotation, or whatever is appropriate to the situation. This value is termed the *theoretical uncertainty (TU)*. This value has very little to do with the closure, except that the experienced discrepancy should fall within this expected range. *It is just as likely for a closure to be zero as for it to be any other number within the expected range.*

The relative error of closure is dependent on the length of the survey as well as the quality with which it was executed. If need be, the surveyor can express the theoretical uncertainty as a ratio to the perimeter and have a *relative uncertainty (TU)*.

Today TU is referenced in applying today's ALTA-ACSM minimum standards.

6.22 Errors in Taped Distances

Today few surveyors rely solely on taped distances, and for the most part the only time most young surveyors have any exposure to taping errors is when they take their examination for licensing or while in school. Because of residual inherent errors present in electronic measuring equipment, short distances of less than 250 feet in length are not recommended. The surveyor in practice must be familiar with the various corrections necessary to remove the effect of systematic errors in taped distances. In general, these involve only two parameters: length and alignment. Length (L) is affected by temperature, tension, and the standard dimension of the tape. Alignment error can be in both the vertical and horizontal planes. Vertical alignment or slope can be corrected mathematically or by leveling the ends of the tape. Alignment errors, although accidental in character, have a systematic effect; and if proper values of the amount of mal-alignment can be determined, a correction can be applied to the observed value:

$$Ca = -\left(n / 2L\right)$$

where n is the number of sides, L is the length of tape, and a is the standard deviation of the alignment error. When the tape is allowed to sag, there is a shortening that is a form of vertical misalignment.

Corrections to the length of tape for temperature, elastic stretch, and sag can be taken from the family of curves that can be found in most technical surveying textbooks. More complete information on corrections to taped distances will be found in any standard textbook on surveying.

Accidental errors in taped distances occur also in alignment and length. For each correction applied to remove a systematic error, there will be a *residual accidental error*. Uncertainties in determining the temperature, inherent error in the standardized length, and the like all have accidental errors of either the random or constant class. Human limitations account for some of the accidental errors in distance measurements such as estimating the fractions, setting the pins or stakes, and other elements with which surveyors are involved. To calculate the uncertainty of the distance, a careful analysis must be made of all the contributing factors. These elements will include the following:

1. Calibrated length
2. Variation in temperature
3. Uncertainty in tension
4. Error in determining slope
5. Error in setting the pins
6. Fraction and end reading
7. Horizontal alignment
8. The ability to interpret between divisions on the tape

Other errors may be present, but for the purpose of the following example, it is considered that those listed are the only ones that have a significant effect. After estimating the magnitude of each of these contributors, the resulting uncertainty can be computed as follows: tape used was carbon steel, calibrated at 68° F, and found to be 100.02 feet long when fully supported on a level surface with a tension of 10 pounds. The tape weighed 1.8 pounds. For a distance of 360 feet, the tape was fully supported on the ground, and the tension of 10 pounds was exerted. For the remaining distance, the tape was suspended from the endpoints; 25 pounds of tension was applied; and the plumb bobs were used to mark the endpoints. The ground was sloping 4 1/4° generally. The average temperature was observed to be 92° F. The computation of uncertainty is shown in Table 6.3.

6.23 Errors in Observed Directions

If directions are derived from angle measurements at each of the sides, the new side will inherit the uncertainty of the previous one. When directions are measured directly, such as with the magnetic compass, each side is independent of the others

TABLE 6-3. Computation of uncertainty.

Error	Magnitude	Frequency	Total	Total squared
1	0.005	6.6	0.03	0.0009
2	0.02	1	0.02	0.0004
3	0.003	16.6	0.001	0.0001
4	0.02	7	0.05	0.0025
5	0.005	6	0.01	0.0001
6	0.005	8	0.01	0.0001
7	0.016	7	0.04	0.0016
				0.0016
				0.006

ET = 0.006 = ±0.08 foot for 659.74 feet.

and the carryover is not in effect. Unfortunately, the errors in reading the compass are so large that this method is not worthy of consideration here.

Accidental errors associated with the measurement of angles involve such considerations as pointing, reading the instrument, centering over the station, centering the signal over the sighted station, instrumental errors, and others. It is seldom that systematic errors are involved in this operation, and residuals need not be considered. Of the accidental errors, those that are random are most prevalent and lend themselves conveniently to error analysis.

Careful and skilled operation can minimize the effect of pointing and centering over the station so that their contribution is insignificant. Under ordinary situations, the instrumental error can be minimized by an appropriate program of observations. It must be emphasized that these factors cannot always be ignored, but a typical situation would require consideration of only the pointing error, the inherent error, and the reading error. The following example is for an angle measurement made on a backsight 300 feet away and a foresight 659.74 feet distant. The instrument is a half-minute vernier transit with insignificant instrumental errors. The angles were turned once and doubled. A plumb bob string was used on the short backsight and a range pole on the foresight. In this example, the errors of instrument centering, instrument adjustment, pointing, and the like are considered insignificant. The point occupied is the fourth station from the initial corner.

Reading Uncertainty As the instrument must be set on zero, the single and double angles observed, three readings are involved, but only the initial and the final are used for deriving the angles. The limitation of the instrument is ±15″. The full effect of the reading error therefore is

$$\frac{\pm 15''\sqrt{2}}{2} = \pm 11''$$

Signal Error On the short sight, the plumb bob can probably be held to within ±0.005 foot of the station, and this error contributes 0.005/300 = 0.000017 radian, or about ±4″. On the foresight, the range pole is centered within ±0.05 foot and introduces 0.05/660 = 0.000076 radian, or ±17″ error.

The result of these three uncertainties is the square root of the sum of the squares, or $\sqrt{11^2 + 4^2 + 17^2} = \pm 21'$, the uncertainty in the angle measurement. If this had been the uncertainty for each of the preceding four sides, the inherited uncertainty is $\pm 21\sqrt{4} = \pm 42'$. The uncertainty of the direction of the foresight will then be $\sqrt{42^2 + 21^2} = \pm 48'$.

6.24 Reliability of Meridian Observations

Observations taken on the stars for azimuth determination are subject to errors, and the results will be uncertain to some extent. The contributing errors will appear in the instrumentation, ambient conditions, and attending observations.

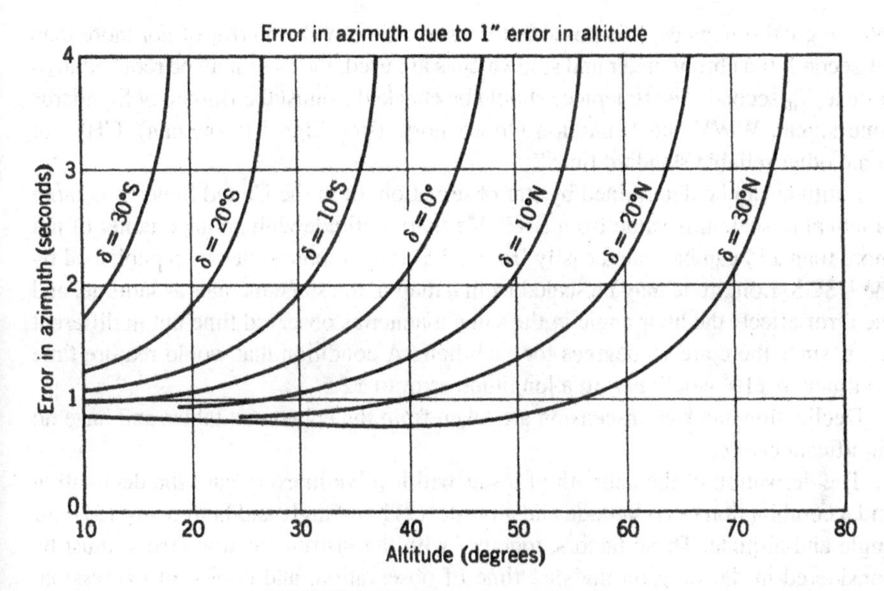

FIGURE 6-2. Errors due to altitude.

Instrumentation When extreme vertical angles are observed, some of the instrument adjustments become quite critical. If observations are being taken at 20° or more above the horizon, the manufacturer's manual and a good textbook on surveying should be consulted. Maladjustments of the line of sight, horizontal axis, and plate bubbles contribute to the uncertainty in the horizontal angle reading. A typical reading taken on Polaris (altitude of 400) may be uncertain by ±0.3′. This—combined with the pointing, centering, targeting, and reading errors—may result in an uncertainty of ±0.5′ for a typical Polaris observation.

When altitude is observed, it is dependent on the horizon as determined by spirit bubbles or a pendulum. On vernier transits, the vertical angle can be read to the nearest minute only, and there is no practical way of doubling the angle. The uncertainty in the observed vertical angle of a star with a vernier transit is about ±: 1′ and with the optical transits about ±:20″ (see Figure 6.2).

Ambient Conditions Temperature and barometric pressure observations are necessary to correct for atmospheric refraction. The uncertainties in temperature and pressure will not be significant, but it must be realized that the corrections are empirical only and will not completely remove the error because of refraction. It is for this reason that low-altitude observations are uncertain by several minutes when below 10°.

Attending Observations Time, latitude, longitude, declination, and right ascension are not ordinarily a result of field measurements but are derived from other observations and sources. When the azimuth is derived from an expression involving the hour angle, time must be considered. Under some conditions, standard time to the nearest minute will be sufficient, and in others a fraction of a second is necessary.

With a good timepiece, it is possible to observe time with an error of not more than ±1 second. If a chronometer and stopwatches are used, the error may be reduced to as little as $^{1}/_{10}$ second. The timepiece should be checked against the Bureau of Standards time signal, WWV, the Dominion Observatory Time Signal (Canadian), CHU, or some other reliable standard time.

Latitude can be determined by star observation, or in the United States it is more practical to scale this value from a reliable map. Latitude with an uncertainty of not more than ±1″ can be scaled easily from a 7.5′ topographic series map published by the USGS. Longitude may be scaled from a map in the same manner as latitude, and the error affects the hour angle in the same manner as observed time but in different units, since there are 15 degrees for each hour. A condition that would require time accuracy to $\pm 10^{S}$ would permit a longitude error to ±2.5′.

Declination and right ascension are taken from the ephemeral tables and have no significant errors.

The derivation of the azimuth of a star will involve in every case the declination and a combination of (1) latitude and altitude; (2) hour angle and latitude; or (3) hour angle and altitude. These factors, together with the instrumentation errors, must be considered in deciding on the star, time of observation, and choice of expression. There is no reason why the surveyor cannot obtain an azimuth reliable to ±0.3′ with ordinary equipment by observing either the sun or Polaris.

Today, many computational problems are eliminated by the availability of read-only memory (ROM) programs that can be used with handheld calculators. It is recommended that any surveyor who anticipates routinely using astronomical observations purchase one of the programs.

6.25 Measurement Evidence to Prove the Proximity of Corners and Their Monuments

PRINCIPLE 5. *Measurements may be used to prove the validity of corners and monuments. Such monuments, to be acceptable, should be within reasonable proximity of the record measurements and corroborated by other evidence.*

Measurements may be used in conjunction with other forms or classes of evidence to prove and then accept or reject found monuments. Although survey law does permit the total reliance on measurements in the event that a surveyor determines that a corner is a lost corner, surveyors should also realize that measurements are but one form of evidence and should be used in conjunction with all other found evidence in its totality.

In many circumstances, the meaning of the word *reasonable* can vary. Unfortunately, many courts use this term in such discussions as "reasonable reliance" or the "reasonable and prudent surveyor," but they have not yet been able to adequately define just what "reasonable" means. Some deeds are written without the benefit of a survey; hence, measurements may be nothing more than estimates. Surveyors have been known to make gross errors. But no definite rule exists that identifies exactly how close a monument to the retraced distance should come to the record measurement in order to accept it. If an iron pipe shows signs of having been disturbed, it is only right that it should be bent back a fraction of a foot to its measured position. However, if an oak tree is identified as the one called for, a substantial measurement error probably would not disprove it. An error of 100 feet in going to the ocean may not cause the surveyor anguish, but an error of 100 feet measured to an easily movable stone monument would present serious doubts and should cause concern to the retracing surveyor. The amount of measurement error allowed is dependent on how permanent the monument is and the certainty of identification of the spot occupied by the monument as of the date of the description that is contained in the deed when it was created.

When there is a discrepancy between the noted distance or angle and the remeasured distance or angle, the courts will allow a reasonable variance. The definition of "reasonable" is left to the discretion of the court.

A measurement between a questionable section corner and a proven existing quarter corner that has a plat distance of 40 chains is found to be 39.55 chains. The section corner cannot be proved from the original GLO notes, but it is near a fence corner, and testimony of landowners indicates that the fence is in excess of 90 years old, that it has always been recognized as the section corner, and that to rely on the distance would cause serious problems between neighbors. In this instance, the measurement of 39.55 chains, knowing that the original survey was precise to approximately 1/80, could be used to add additional evidence to the acceptance of the corner as an obliterated corner, rather than a lost corner (see Figure 6.3).

FIGURE 6-3. Surveyors converted 80 chains to 5,280 feet. Monument set at the distance. Fence has been accepted as section line for 90+ years.

6.26 Evidence of Measurement

In the order of importance of items that may prove the intent of a deed, evidence of measurement ranks below senior rights of adjoiners and monuments. A surveyor who needs his or her ego deflated need only reflect on the fact that in the matter of interpreting deeds, the courts have placed measurements very low on the evidence scale.

PRINCIPLE 6. *In the GLO states, unless proved otherwise, measurements of distances are presumed to be horizontal, while in the early surveys of the metes-and-bounds states, measurements may have been made with the "lay of the land," according to the statute in effect at the time.*

In the colonies early measurements were often surface distances, and when proved as such, surface measurements may control. In some areas of the United States, especially along the eastern seaboard, early linear measurements were made on the ground or "with the lay of the land" without corrections for slope. When it can be shown that the original surveyor did use surface measurements as a local custom, the presumption of horizontal distance is overcome. The following cases will help illustrate this.

In Kentucky, two streets differed in elevation by about 40 feet and were about 100 feet apart. McCoy's lot called for 60 feet "up the hill." If 60 feet was measured horizontally, the measurement would be about 70 feet on the surface and would include parts of the houses of the upper lot. The plaintiffs contended that there was an absolute rule requiring the use of horizontal measurements; the court's opinion was that surface measurements were proper when such was the custom of the locality or was dictated by the circumstances of the case [6].

In some areas, the courts will recognize that while the surface and not the level or horizontal mode of measurement is generally adopted in surveys, and the general presumption is that a survey of the surface was contemplated by the parties to a deed, that presumption prevails only where it appears feasible and reasonable to have pursued that course.

As an actual example, a surveyor is retained to retrace an 1820 description in which all distances are recited in poles. Using the historical conversion of 16½ feet to a pole, he went into the field looking for evidence of the corners, He found nothing. Then casting precision aside, he used an actual two-pole chain to conduct his retraced measurements. He finally completed the retracement by finding evidence of over 40 percent of the original corners.

One court stated:

> Where a line of survey crossed a perpendicular cliff at a place where it could not be climbed, and to give the quantity of land called for by the survey and to take the line to a boundary shown to have been marked in an old survey, it was necessary to exclude the distance up the face of the cliff, it was not error to instruct the jury to exclude it in determining the boundary [7].

Some courts have exhibited an unusual degree of understanding. In a Pennsylvania decision, the justices wrote:

> Surface measurement is the only kind in practice by the district surveyors of this state. But in making that it is usual for the surveyor in chaining over an uneven surface to

> make allowance by elevating the chain. There was nothing to show that this was not the practice when the original survey was executed; whatever it was, would regulate the measurement now; and we see no error, therefore, in the answer of the learned judge to the defendant's point that they were entitled to the measurement adopted by the surveyor who located the warrants. In other words, to the ordinary measurement of official surveyors [8].

As can be observed from these decisions, it is a matter of proof by evidence about whether the original surveyor made his measurements on the ground surface or horizontally. That is, it is a matter of proof in those states that permit it. In most states west of the Appalachian Mountains, the standard of measuring horizontally cannot be refuted. In Tennessee, in 1860, a U.S. court rejected surface measurements even though the original surveyor testified that he made the survey on which grant no. 22,261 was founded. At trial, he stated that "that no actual survey was made in 1838 of said land, except the first line from A to H. That the other three lines of the grant were not run, but merely platted" [9].

He then testified that the proper way of making surveys was by horizontal measurement but that he had not been in the habit of making them in that way; in making the line in the field from A to H, in this survey, he had measured the surface of the ground. He also testified that the custom of the country was to adopt surface measure and that he had made the survey in accordance with such custom.

The court held that the grantee was bound to abide by the marked line from A to H, but the other lines must be governed by a legal rule, which a local custom cannot change. In ascertaining the southwest corner of the tract at 894 poles from the poplar corner, the mode of measuring will be to level the chain, as is usual with chain carriers when measuring up and down mountainsides [9].

All private surveys today are, of course, relative to the level standard foot and in the near future will be relative to the meter. The federal government used the chain as the official unit of measurement as prescribed by federal law, and new descriptions are written in chains.

The foregoing cases are cited only to explain some of the unusual customs of the past, because the retracing surveyor must understand and follow what was done in the past and that evidence controls. In the sectionalized land system commencing in 1785, horizontal measurement was specified and has always been used. In most states, the assumption that horizontal measurements were made cannot be refuted. Certain cities, such as Miami, Florida, are imposing metric system by virtue of an influx of foreign-trained surveyors. And in early 1993, all government agencies required metric measurements for government contracts. No matter what distance is used, the presumption of slope or horizontal will have to be addressed.

PRINCIPLE 7. *In the public land states, the legal presumption is that bearings are relative to astronomic, or true, north, unless otherwise specified, but in the metes-and-bounds states, the presumption is that bearings are in reference to magnetic meridian, unless otherwise specified.*

In a few states, especially along the eastern seaboard, the presumption for older surveys is that the magnetic meridian as of the date of the measurement was used as the basis of bearings for angles.

Basic presumptions can be overcome by positive contrary evidence. In the western mining areas, as well as Spanish, English, and French land grants, where it was shown that the custom was to survey on the basis of magnetic north, the courts have upheld such basis of bearings. In the original 13 colonies, until about 1800, and in certain places after that, it was the custom to survey by the compass without correcting for declination. Which method was used, magnetic north or true north, is a question that must be proved by evidence, including historical information, but if contrary evidence is not presented, the presumption will prevail.

In an 1815 U.S. Supreme Court decision, Chief Justice John Marshall noted that in North Carolina it was undoubtedly the practice of surveyors to express on their plats and certificates of survey the courses that were designated by the needle, and if nothing exists to control the course, the land must be bounded by the courses according to the magnetic meridian [10]. In an 1854 case, the Supreme Court of Georgia held that if nothing exists to control the call for course and distance, the land must be bounded by courses and distance of the grant, according to the magnetic meridian [11].

A similar result was found in 1813 in Kentucky, wherein the judge observed that calls for course were usually and generally understood to be according to the magnetic meridian. Furthermore, instruments were not then available to determine the true meridian. To this, the judge noted that the magnetic meridian to be used was as of the date of the original survey, not as of date of the resurvey [12]. In New Hampshire in 1866, a judge found that the term *due north* did not mean by the true meridian; as a matter of common knowledge, boundaries had almost uniformly been run out according to the magnetic meridian. It was common law in the state of New Hampshire that the course in deeds of private lands are to be run by the magnetic meridian when no other is specifically designated [13].

In 1953, in North Carolina [14], the court ordered a new trial because the judge had failed to instruct the jury to take into consideration the change in magnetic north since 1898.

In 1858, in Ohio [15], the court ruled that the custom in private boundaries was to use the magnetic meridian, and this was done even though sectionalized lands were done on the true meridian.

In the Act of Congress of May 18, 1796, all sectionalized lands were to be surveyed as based on true north (astronomical north). Although the compass was to be used, the direction of magnetic north was to be corrected to true north. From the foregoing, it is readily apparent that the burden of proving the basis of bearing to be used in retracing a deed is the responsibility of the surveyor. After it has been decided that the original deed was based on a magnetic bearing comes the task of determining the magnetic variation as of the date of the surveyor deed. The fact that magnetic north and the direction the needle points vary from time to time and from hour to hour in the day is well established. Basically, three natural causes of magnetic change are present: (1) daily or diurnal changes, (2) annual changes, and (3) magnetic storm changes or local attraction caused by anomalies of magnetic ore deposits. The magnitude of these daily changes can be as much as 7 minutes and is brought about by the effect of the sun on the Earth. The annual change is gradual and can vary from 1 minute up to several minutes per year.

The magnetic needle on a compass can be affected not only by natural causes but also by man-made causes, such as metal eye glass frames, a knife, or a firearm.

Although the change for one day or one year may not be significant, the change over several years can become very significant when it comes to determining a relationship of an ancient bearing into a modern bearing reference. The change may be in one direction for several years and then change to the opposite direction. Although charts and books are available predicting variation and annual rates of change, these are nothing more than predictions and will not reflect the actual changes that occur. Surveyors cannot predict the magnetic changes with certainty, but they can tell, within limits, what the changes were. Magnetic storms that occur at infrequent irregular intervals may cause as much as 7° variation of the needle during the time they are in progress; thus, there always will be a shadow of doubt about what occurred in the past.

Local attraction can affect the compass needle. If there is a problem in relocation of an original compass-surveyed line, the best solution is to find two monuments that existed as of the time of the survey and were referred to and then make an actual determination of the present magnetic direction on that same line. The differences between the two readings will give the change in magnetic deflection or declination. In the absence of monuments to make a direct determination, tables published by the U.S. government can be resorted to, for example, by the U.S. Coast and Geodetic Survey (USC&GS), now the National Geodetic Survey (NGS), or, more recently, National Oceanic & Atmospheric Administration (NOAA). Tables and maps published by the NGS show the magnetic declination as it existed in the United States at various times. Today, this information is easily and readily available on the Internet.

PRINCIPLE 8. *If there is a conflict within a deed and a choice must be made between bearing and distance regarding which controls, no uniform rule has been laid down by the courts. Variations between states occur, as well as between the GLO states and the metes-and-bounds states.*

In the sectionalized land system created by the federal government, in replacing lost corners in accordance with BLM rules, distances are always kept in their original proportions and bearings yield to proven corners.

In the metes-and-bounds states, including Texas, Kentucky, North and South Carolina, and the New England states, the bearing is given definite preference to distance. Since it takes both a bearing and a distance to determine a line, it would be a rare instance to give one control over the other. In the usual situation, the surveyor gives control to as many items in the description as is possible and rejects as few elements as possible. In the situation N 20° 10′ E 100 feet to the south line of Palm Avenue, the bearing would be given force and the 100 feet would yield. In the situation N 80° 10′ E along Palm Avenue, the distance would hold and the 100 feet would go along Palm Avenue regardless of the bearing. The custom is to reject as few terms in a deed as possible to ascertain the best fit of all the elements in the description.

Figure 6.4 shows a block with regular lots of 50 × 115 feet each. A resurvey discloses that the block frontage measures 300.60 feet instead of 300.00, as originally recorded. The rear of the lots measures 301.20 feet instead of 300.00. Assuming that

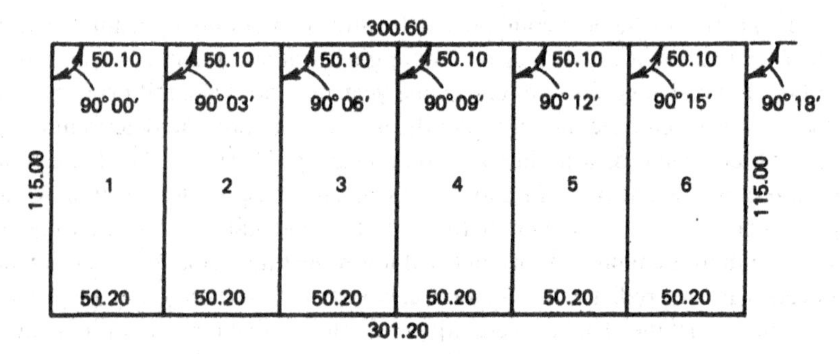

FIGURE 6-4. In some states, direction is considered more important than distance. In this subdivision, excess exists. If the excess is apportioned by distance, the results are almost identical to apportionment by direction.

only block corners were originally set, in most states the excess would be evenly divided between all the lots (frontage is 50.10 feet; the rear, 50.20 feet). In a state where direction is more important than distance, the angle would be prorated as shown in Figure 6.4. Although the two methods theoretically yield differences in lot dimensions, the differences are too small to measure; both methods are comparable.

In some states, direction is considered more important than distance. In the just-discussed subdivision, excess exists. If the excess is apportioned by distance, the results are almost identical to apportionment by direction.

In Kentucky, the rule is definite. Give direction (course) control over distance as stated in an 1801 case:

> When a departure from either course or distance becomes necessary, reason as well as law seems to suggest that the distances, taken in our mode of measurement, ought to yield, as being much the most uncertain of the two. Indeed, it never has been held that in laying off vacant land purchased from government, that a line established by a surveyor can be altered on account of its being longer or shorter than the distance specified on the plat and certificate of return. These considerations seem to dictate an answer to the question that has been stated: From one of the adjacent corners which remain, the courses and distances of the lost lines should be run, as called for in the plat and certificate of survey, and if they close with the other adjacent corner that remains, the true situations of the lost corners, and consequently the true situations of the lost lines, will be satisfactorily ascertained. But, if the courses and distances thus run do not close the survey, it must be accomplished by running the same courses, and either lengthening or shortening the distances, as each case may require, and in proportion to the length of each line, as called for in the plat and certificate of each survey. And if the survey cannot be made to close by this means, then, and not otherwise, a deviation from the courses called for must also aid in accomplishing the purpose. For example, where all the corners of a survey are lost but one, lost corners ought to be ascertained by running comfortably to both the courses and distances specified in the plat and certificate of survey. Where there is but one corner of a survey lost, the courses called for in the plat and certificate of survey ought only to be regarded: and nothing more is necessary than to extend those courses from the adjacent corners that remain, until they intersect each other, and the place of intersection will be the situation of the corner to be ascertained [16].

Once again, in a federal decision applying state law it was held that "compass courses called for in a deed prevail over measurements" [17].

New Jersey reaffirmed this: "When a departure from either course or distance becomes necessary, the distance should yield" [18].

In conducting a retracement of a 1798 original survey on Long Island, the field notes indicated the direction of the lines to be N 72° W. Of course, the usual practice was a staff compass and reading the magnetic meridian. The first retracing surveyor assumed this bearing was true and placed the line on the ground. The line "cut" into the neighbor's house by 0.03 feet. The second retracing surveyor, after determining the change in declination from 1798 to 2009, indicated by his retracement that the line missed the house by 0.5 feet.

6.27 Errors in Position of Corners and Lines

All corners have measurement errors in actual position.

Error in position is the combined effect of all the uncertainties in distance and in direction. When rectangular coordinates are employed, it is most convenient to express the uncertainty in terms of latitudes and departures.

Although accidental errors may tend to compensate, the uncertainty will continue to increase as the errors accumulate. As will be seen, this uncertainty is not the same as the error of closure, for the closure may fall anywhere from zero to the full amount of the uncertainty. The closure merely serves to provide assurance that there have been no excessive errors or blunders. One of the useful purposes to which this method of analysis can be put is in preparing specifications. If it is required that the position of a point be certain within specific limits, this type of calculation can aid the surveyor in planning the method of taping and other measurements and to provide for a program of observations, how refined the temperature measurements and other corrections should be, and the like.

6.28 Errors in Traversing

The effect of errors in both distance and direction is reflected in the resulting uncertainty of position. This is a product of both the angle and the distance. Instrument errors and errors in measurement all apply, whether an open traverse or a closed polygon. Figures 6.5 and 6.6 show possible errors from two sources, and Figure 6.7 shows the combination of these errors.

If the older methods are employed using nonelectronic equipment or equipment of a lower precision, to meet the degree of precision that is now built into the newer equipment, the surveyor must consider the various errors that can be calculated and compensated for. These include but are not limited to trunnion error (Figure 6.5), altitude error (Figure 6.6), and other errors associated with measuring, including tribrach adjustment.

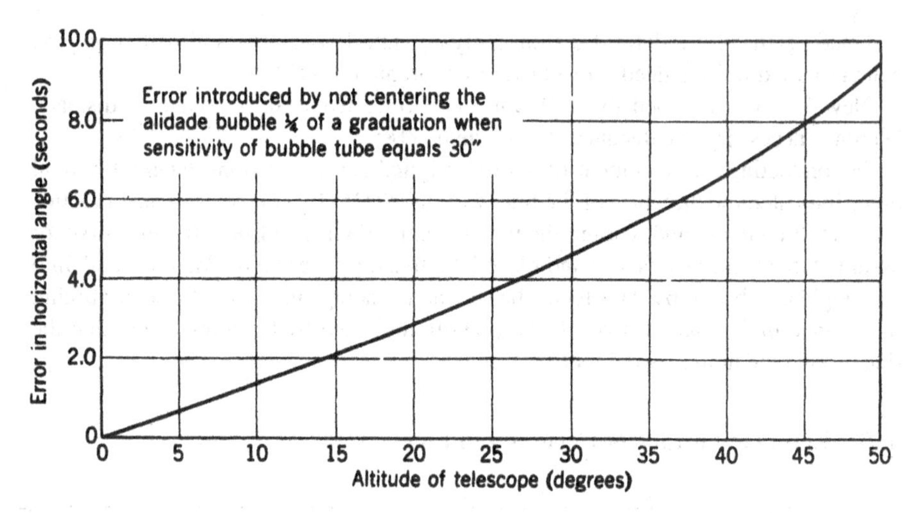

FIGURE 6-5. Trunnion error.

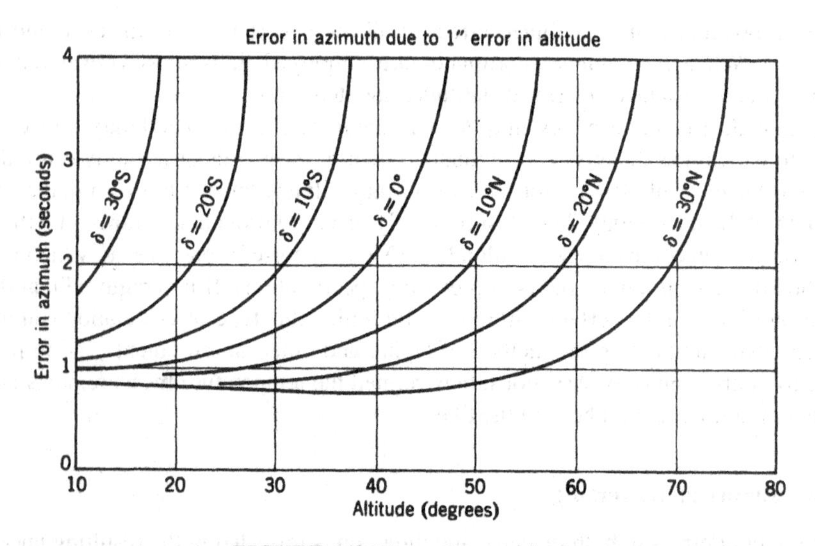

FIGURE 6-6. Errors due to altitude.

Today, the older methods of traversing are seldom used. No longer does a survey crew run miles of traverse by transit, measuring each segment or leg with a tape. Nor do survey crews worry about the chain, plumb bobs, and chaining pins to conduct their work today.

The educators and theorists should commence studying the effects of traverses measured with EDM or the computation of line and positions by GPS. Many older theories of traversing are now ancient history, and in a few years they will be relegated to historical surveying textbooks.

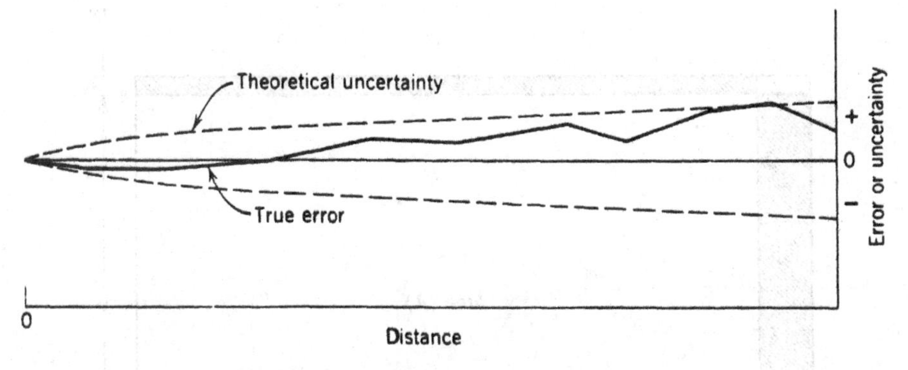

FIGURE 6-7. Traverse errors and uncertainties.

6.29 Uncertainties in Area

The recitation of area in a deed or description is the only element that the courts will not say is without error. In the original calculation of area, the two least reliable elements, which are the two lowest elements in the Priority List, bearing and distance, are the elements that are used to determine area.

There will always be individuals who will want to use traversing and the older methods, and this will apply to them.

When areas are calculated, there is often the temptation to extend the numbers beyond a reasonable significance commensurate with the field survey methods. As area is computed by the product of two numbers or a sum of several such products, it will have uncertainties accordingly. The area in Figure 6.8 is determined by measuring x and y and applying the expression $A = xy$, where A equals 600 feet^2.

If x (30 feet) were uncertain by ±0.5 foot and y (20 feet) had an uncertainty of ±0.2 foot, the theoretical uncertainty could be computed as follows:

$$\sigma A = \pm\sqrt{(x\sigma y)^2 + (y\sigma x)^2}$$
$$= \pm\sqrt{(30\times 0.2)^2 + (20\times 0.5)^2}$$
$$= \pm 12\, feet^2$$

A more convenient and generally satisfactory means of predicting the uncertainty in the area quantity is provided when the significant figures are respected (see Section 6.13).

Many surveyors want to determine the certainty of the area without the necessity of computations. As a rule of thumb, a surveyor can determine the certainty of the area by determining the error of closure of the closed figure and then dividing by 2. That is a "rule of thumb" calculation of the relative area accuracy. As an example, the area of a parcel is computed as 7.99 acres, and the error of closure is 1 in 12,000.

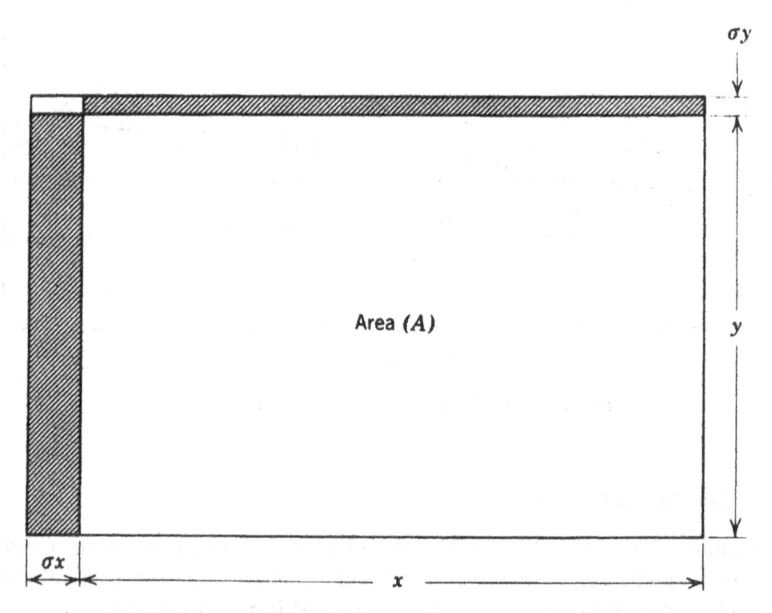

FIGURE 6-8. Area uncertainty due to measurements.

The 1 in 12,000 divided by 2 gives 1 in 6,000. The expected accuracy of area is then 1 in 6,000, or 0.001 acre in 7.99 acres as a close approximation.

Surveyors should realize the area is a product of two elements: (1) measurements or calculations to obtain the lengths and widths; and (2) the actual measurements themselves. Since a closed figure must be balanced (closed) before area can be calculated, the final area is subject to variations. This can be because of the methods by which the various sides were balanced and the method of calculating the area. Two surveyors balancing a closed figure by two different methods and each calculating the final area, one by double meridian distance (DMD) and the other by coordinate multiplication, will arrive at different areas for the same figure. Each will be correct. Surveyors should *never* "guarantee" area, for to do so may lead to litigation and possible damages.

The value of area depends on many factors: How the original traverse was balanced, the method used to compute the area from the adjusted numbers, and how precise the original survey was. The answers to these questions will change the result of the final area computation.

Whenever an area is recited in a description or on a plat, the surveyor should indicate the method used to balance the field angles and the method used to compute the area on the final plat or in the final description.

6.30 Purpose of Survey Specifications

Many organizations such as the Massachusetts Land Court, various title insurance associations, and the ALTA and ACSM have drafted standards and specifications governing boundary surveys. Each of these has served well to bring about some uniformity in the property surveys and to improve the quality of the work. With advances in technology and the ever-increasing value of real property, it may be necessary to reexamine the question of standards and present criteria that are workable and realistic and ensure a minimum of title and boundary location difficulties.

The purpose of survey standards is to specify the quality of work to be performed. Only the accepted theories on probability and effort can be of help in estimating the uncertainties in expressed positions.

The specifications suggested in the following pages are a guide for performance standards as well as academic discussion. Requiring field crews to follow rigid specifications for performance is better assurance than is error and omission insurance.

6.31 Adaptability of Existing Standards

Many standards of accuracy are in existence. They have been developed by and for the governmental agencies engaged in surveying activity, state surveying societies, state legislatures, title companies, and many others. Unfortunately, there is no one national standard that can be applied to all surveyors. The accepted standards that have been adopted by many local organizations are those developed by the NGS for horizontal and vertical control. Nongovernmental organizations have written contracts using these national standards or a modified version. The new standards are in metric units and have additional classes of survey. See Table 6.4.

It must be borne in mind that these standards were devised to suit the needs of federal agencies engaged in control surveys, which usually involved large areas. The closure discrepancy (not really an error) is but an indication of the quality of the

TABLE 6-4. Table of traverse closures.

	First order	Second order	Third order
Number of azimuth courses between azimuth checks not to exceed	15	25	50
Astronomical azimuth, probable error of result	0.5″	2.0″	5.0″
Azimuth closure at azimuth, checkpoints not to exceed[a]	2 second N or 1.0 second per station[b]	10 second N or 3.0 second per station	30 second or 8.0 N second per station
Distance measurements, accurate within	1 in 35,000	1 in 15,000	1 in 7500
After azimuth adjustment, closing error in position not to exceed[b]	0.66 foot M or 1 in 25,000[c]	1.67 feet M or 1 in 10,000	3.34 feet M or 1 in 5000

[a] The expressions for closing errors in traverse surveys are given in two forms. The expression containing the square root is designed for longer lines where higher proportional accuracy is required. The formula giving the smaller permissible closure should be used.

[b] *N* is the number of stations for carrying azimuth.

[c] *M* is the distance in miles.

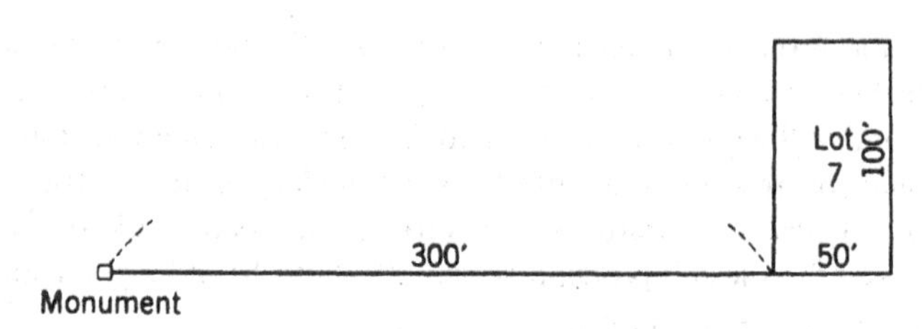

FIGURE 6-9. Example of a survey problem.

observations and does not necessarily reflect the amount of the uncertainty in the quantities [19]. Property surveys, although better than none at all, are not suited for that purpose. As an example, consider a city lot that is 50 × 100 feet in size and 300 feet from the nearest acceptable monument (see Figure 6.9).

Suppose the boundary surveyor is required to conform with second-order traverse standards as defined in the standards. The "closing error" in position must not exceed 1.67 meters, or 1 in 10,000. The traverse that encloses the perimeter of this sample city lot would be 50 + 100 + 50 + 100, or 300 feet. The traverse needed to connect the lot with the known monument would be 2 × 300, or 600 feet long. Considering the perimeter traverse first, a closure of 0.22 or 0.03 foot is called for, and since it is required to take the smaller of the two, the 0.03 would be controlling.

If distance and direction contribute equally to this discrepancy, the share for distance would be 0.03 *M*, or ±0.02 foot. This would limit the error to no more than +0.01 per side or 1 in 10,000 for the long sides and 1 in 5,000 for the shorter ones. If 1 in 8,000 is used for each side, the ±0.02 can be satisfied just as well.

The azimuth closure is required to be 0° 00′ 20″ or 0° 00′ 12″; and since it is required to take the smaller, 0° 00′ 12′ will govern. Three seconds per station will introduce an error of ±0.0015 foot on each of the long sides and ±0.0008 foot for each of the short sides. The resulting effect of this restriction is about ~*i* ± 0.0024 foot, or only about one-tenth of the amount that will still be small enough to satisfy the traverse closure of 0.0103 foot. If the direction error is to carry the same share as distance (±0.02 foot), an uncertainty of about 0° 00′ 25″ per angle will be adequate to realize the required closure. Permissible angle and distance errors are not in balance.

The traverse needed to locate the lot within the block is somewhat longer but would require only two sides. The limiting closure, according to these standards, would be somewhat relaxed because of the much longer peripheral distance.

From the foregoing, it is obvious that to comply with the closure specifications, it is not necessary to comply also with the distance and direction requirements. In fact, it is practically impossible to determine a distance of 50 feet on the ground within 1 to 15,000. It is equally impossible to obtain directions to ±3″ for such short sights, even if this were necessary. Certainly, the location of the lot in the block is more critical for title purposes than is the relative size of the lot itself. The use of the government standards reverses this order.

6.32 Uncertainty Expression

A statement that Mr. Jones's property corners were located within 1/10,000 is meaningless. If the surveyor had commenced at a point located 2 miles away, the error could be approximately 2 feet, but if the nearest original monument position was only 10 feet away, the error could only be 0.001 foot. A person erecting a building, a fence, or any improvement would only be bewildered by an accuracy ratio; but if the person were told the error could be ±0.5 inch, he or she would have concrete information on which they could rely.

A statement that a building extends over the surveyed deed line by 2 inches is positive in terms of recognized units; it is never expressed as an accuracy ratio. The surveyor should state facts and not legal conclusions or opinions. The statement "building encroaches 2 inches" is either a legal conclusion or an opinion. Surveyors are not required to pass on the validity of possession. They are only paid to survey the description and then indicate variances to the description. The statement should probably read, "The building is placed 2 inches over the deed line."

Every client should be entitled to a quantitative statement of uncertainty of position. Survey standards of accuracy should likewise be expressed relative to positive units rather than as an accuracy ratio.

Error ratios are helpful to evaluate errors due to a particular element of measurements. An error of 15° in temperature contributes a nearly 0.01-foot error for 100 feet measured, or, as more conveniently expressed, 1/10,000. This is a proper usage of error ratio: The error is in direct proportion to the distance.

When making a given set of surveying measurements, the sum total of all errors can probably be said to be some function of the distance traversed, but that function is not in direct proportion to distance, as would be indicated by standards based on error ratio. Survey standards based solely on a given permissible error ratio are arbitrary.

6.33 Theoretical Uncertainty

The theoretical uncertainty (TU) is a value from the theory of probability and the propagation of accidental errors. If all features are taken into proper account, it will provide an indication of the quality of the position. The statement "TU = 0.10 foot" indicates that the actual quantity, as based on probability, has a 70 percent chance of being a value of anything from 0.10 foot less than to 0.10 foot more than the stated number. In the statement 327.62 ± 0.10, the 327.62 is the best quantity from the given evidence and observation. It is impossible to ascertain the true value, but it is within a tenth either way.

It should be pointed out that these figures are those corner points that must be located by survey. There is no measurement substitute for doing adequate research and investigation to determine what measurements should be present and what ones were actually found. This could also apply in those instances when a closure is made of the original notes in order to determine search areas for lost corners. Using the closure information, one could determine the certainty of the theoretical positions of lost corners.

6.34 Value of Property

There is much merit to the viewpoint that the value, cost, or potential of a property should not be a consideration of the care with which the survey is made. Who can predict the value of a tract of land a few years or a hundred years from now? Also, at times, expensive structures are erected on relatively inexpensive land parcels. The fact that present value is low should not preclude a more precise and accurate survey.

The value of the parcel should not affect the quality of the work.

6.35 Specifications for Location of Property Boundaries

The uncertainty of boundary location is the theoretical uncertainty determined by proper analysis. The estimate of this quantity can be determined by two different methods. If a number of redundant observations are taken (a closed-figure traverse usually has two redundants), the standard deviation may be computed by applying statistical theory. If no redundants or only a few are provided, the uncertainty can be determined by compounding the various contributing uncertainties according to the best theories of probability. The results should be about the same in the absence of blunders. Since it is impractical to take a large number of observations or to measure the traverse many times, analysis of the contributing uncertainties is generally the best approach for surveying locations. Such a theoretical uncertainty has a likelihood of being 70 percent in error. To allow for the 100 percent uncertainty, an infinite value would result, but twice the standard deviation will be approximately 95 percent in error. Figure 6.10 shows the probability curve and where these theoretical values lie.

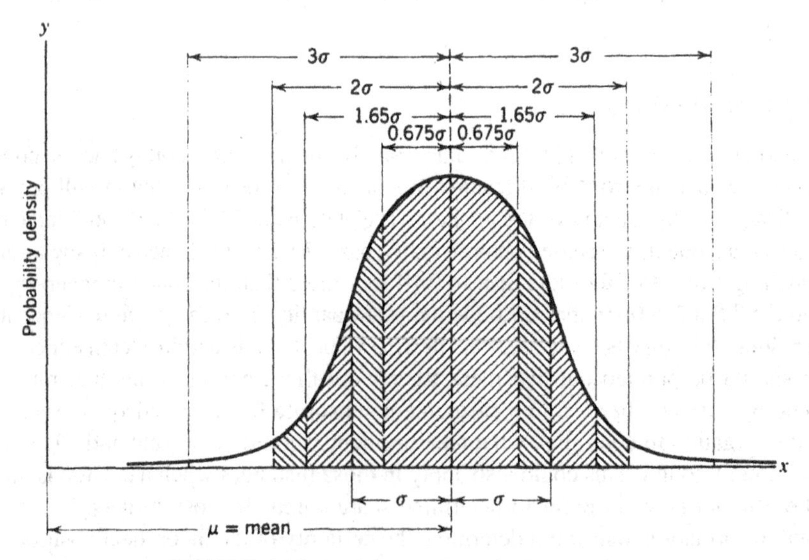

FIGURE 6-10. Bell curve, theoretical uncertainty.

6.36 Size of Properties for Each Class

Assuming that each property will be located with equal diligence and care irrespective of the value of the land, the listed classes of surveys may fall into the following:

Class A Small areas wherein dense monument controls exist, as in a downtown commercial area, where lots are 50 × 100 feet.

Class B Longest side is under 250 feet.

Class C Longest side is from 250 to 2,500 feet, and no side is under 100 feet, unless the periphery exceeds 500 feet.

Class D All sides are 1,000 feet or larger; lots have a periphery of 5,000 feet or more.

A farm survey of 40 acres would probably be included in class D. It would also be possible for a small rural farm tract of 1 acre to fall in class B and a large city property to be under class C or D.

As it is in actual practice, most states as well as federal agencies and organizations have their own classifications of this nature, and adjoining states may disagree. Some states classify the standard to be used by the terms *urban*, *rural*, *farm*, and so on, and the classification of what class the property to be surveyed falls in is left to the surveyor to determine. This is unfortunate, because no two individuals will classify the same parcel the same.

Whatever method of measurement is used (e.g., EDM, total stations, tape, transit, or theodolite), the surveyor must make certain that the procedures used for distance and angles are such that the resulting uncertainty in the position of the endpoint of the line does not exceed 65 percent of the resulting TU.

6.37 Closures

Table 6.4 gives theoretical values (see Section 6.31) and will be in effect only if the proper procedures have been followed and if there have been no blunders. This is effected by providing at least one redundant observation or check to compare the closure with the theoretical. Realizing that the closure has as good a chance or being zero as any other number and that it can fall anywhere in the range, plus or minus, its limiting factor is a function of the TU and the distance to the closure point. In most cases, the closure tolerance can be more than that of the TU. The following expression will serve as a guide:

$$LC = tu\sqrt{D_c D_p}$$

where LC is the linear closure, D_c is the total distance to the closure point, and D_p is the distance to the point.

6.38 Adjustment

All positions must be adjusted to balance out the closure discrepancy and provide a more probable value for the position of the points. The adjustment method must be chosen with reason and logic to suit the surveying conditions. Generally, a "least-squares" adjustment may be satisfactory, although it is not justified in many cases. If the uncertainty has been caused by distance and direction in equal proportions and is a function of the distance, the "compass rule" will be satisfactory and will probably produce the best adjustment. But if the uncertainty is in the angles, then the transit rule may produce the best results. There is no requirement that any one method of adjustment is preferred over the other, and as such the surveyor is free to select any method for adjustment, including the PTEWTEO (put the error where the error occurred). In any case, the adjustment must not shift the position of the point more than the limiting TU. The method used in adjusting the field data should be indicated on the plat or in the surveyor's report, as well as the method of area calculation.

6.39 Monuments

The refinement of measurements cannot be a substitute for adequate monumentation. When original surveys are made, the monuments must be chosen for their durability and to ensure adequate future identification. Wood should not be used in those areas where soil conditions will cause early decay. If iron pipe or pins are used, they should be long enough to hold their position against frost, durable enough to resist rusting for a reasonable length of time, and identifiable at a future date. Concrete monuments and stones are most satisfactory. All monuments should be set low enough so that they will not be disturbed by traffic. Referencing is essential for all monuments.

The law still gives preference to adequate and identifiable monumentation over measurements, in that monuments called for in the documents and descriptions are legally superior over any subsequent measurements that are made to locate parcels, and as such the responsibility is placed on the retracing surveyor to "find" the parcel.

In retracements the retracing surveyor must make certain that adequate recovery records are kept and then made a matter of public record for future generations of surveyors. All of the precision of measurements and calculations cannot be a substitute for adequate monumentation of the original surveys and for the adequate recovery records of the retracing surveyor.

6.40 Computations

Fieldwork should be conducted so that adequate measurements are made to ensure that sufficient internal checks can be made in the office calculations so as to keep error to a minimum.

The method of computations must not be such as will detract from the value of the fieldwork. All calculations should be independently checked, and sufficient significant figures must be used so that the computations will be adequate for the quantities of the survey. The surveyor should understand the principles that are used in the calculations, including adjustments and area calculation. The surveyor and the attorney should not let precise calculations give a false sense of security in a boundary survey. After all, a property survey includes not only the technical measurements but also the accuracy of the location of the land itself.

Good hunting.

References

1. *Riley v. Griffin*, 16 Ga. 141 (1854).
2. Bureau of Land Management, *Manual of Instructions for the Survey of the Public Lands of the United States, Technical Bulletin 6* (Washington, DC: U.S. Government Printing Office, 1973).
3. John Love, *Geodaesia*, 1687, reprint by Walter Robillard.
4. Land Act, February 11, 1805, Title 43 U.S.C.A., Sec. 572.
5. A. C. Mulford, *Boundaries and Landmarks* (New York: Van Nostrand Co, 1912, reprinted by Carben Surveying Reprints, Columbus, OH, 1977), p. 3.
6. *Justice v. McCoy*, 332 S.W. 2d 846 (Ky. App. 1960).
7. *Stack v. Pepper*, 119 N.C. 434, 25 S.E. 961 (1896).
8. *Boynton v. Urian*, 55 Pa. 142 (1897).
9. *McEwen v. Den*, 24 Howard (U.S.) 242 (1860).
10. *M'Iver's Lessee v. Walker*, 13 U.S. 173 (1815).
11. *Riley v. Griffin*, 16 Ga. 141 (1854).
12. *Vance v. Marshall*, 6 Ky. 148 (1813).
13. *Wells v. Jackson, Iron Mfg. Co.*, 47 N.H. 255 (1866).
14. *Goodwin v. Greene*, 237 N.C. 244, 74 S.E. 2d 630 (1953).
15. *McKinney v. McKinney*, 8 Ohio St. 423 (1858).
16. *Beckley v. Bryan*, 2 Ky. 91 (1801), 16 Ky. 91 (1809).
17. *U.S. v. Murray*, 41 F. 862 (Me. 1890).
18. *Curtis v. Aaronsen*, 49 N.J. Law 68, 7 A. 886 (1887).
19. *Specifications to Support Classifications: Standards of Accuracy and General Specifications of Geodetic Control Surveys* (Rockville, MD: U.S. Department of Commerce, July 1975).

ADDITIONAL REFERENCE

Judson, L. V. (1960). *Calibration of Line Standards of Length and Measuring Tapes at the National Bureau of Standards*, NBS Monograph 15. Washington, DC: National Bureau of Standards.

7

PLATS AS EVIDENCE

7.1 Introduction

The use of drawings and or plats as evidence to identify land predates written descriptions. Humans have in the past and will continue in the future to rely on drawings to supplement words and phrases to locate land parcels and their respective boundaries, and many conveyances depend on a plat to depict the intent of contracting parties. In this chapter, the features of survey plats, as well as the effect they may have on boundary location, are discussed, but drafting techniques and cartographic expression are not included.

7.2 Definition of a Survey Plat

The terms *maps*, *plats*, *plans*, and *plots* are often used interchangeably. Legally, there may be distinctions. A map is a representation of the Earth's surface, or some portion of it, showing the relative parts represented, usually on a flat surface. It has also been held that a map is but a transcript of the region that it portrays [1].

Plat and plot have been defined as a map or representation on paper of a portion of land subdivided into lots with, for example, alleys and streets, usually drawn to scale [2]. A plan is defined as a map, chart, or design, being a delineation of a projection on a plane surface of ground lines of a house, farm, street, city, and so on. It has also been held to represent an ocular view of the result of a survey, constituting a visual demonstration of the work done.

The three terms are used casually without a specific meaning. State statutes are equally confusing and refuse to use a single, all-encompassing term. *Plat* or *map* is used in 10 states. In several states, the term *plan* is used. Four states use the term *map* exclusively. Three states use the terms *plan* and *map*. One uses *plat* but also includes *diagrams*, *drawing*, and *map*. Twenty-two states use the term *plat* by itself.

As it was stated: "All plats are maps, all plats are plans, all plats may be plots, all plats are diagrams, all plats are charts and . . . all plats are illustrations. All maps are not plats, all plans are not plats, all diagrams are not plats, and all illustrations are not plats" [3]. This can be reduced to a plat as a special kind of map and a particular kind of plan. It is a certain type of diagram and an unusual illustration.

To properly apply the terms, a map can be considered as the overall representation of the general area depicting its various parts, and a plat is that representation depicting the minute parts of the total. A plat is more detailed and descriptive. Some people consider all drawings, regardless of detail and quality, as plats.

It can be seen that a map may be prepared from an actual survey, from collected information, from one's imagination, or from hearsay. The plat or plot, to prepare the minute parts of the total, requires a survey. Graphic representations, including photographs, topographic maps and other types of maps, and plats are admissible if they will assist the jury in understanding testimony and if they fairly represent what they purport to represent [4].

A survey plat is a surveyor's diagram showing land boundaries and/or a subdivision of land. A map, as contrasted with a plat, graphically represents, at a scale or to scale, the physical features of an area and may show some general land boundaries, especially political boundaries.

A plat is more restrictive in scope than a map, and it has added features as well as possibly identifying a smaller area. As used by surveyors, a plat is a plan showing property lines and the interrelationship of property lines with dimensional data on lines; it does not normally express relief. A map rarely has dimensional data; the quantities have been determined by scaling. On a plat, dimensional data may eliminate the necessity of scaling, and the value of the plat is not dependent on the accuracy with which points are plotted [5].

7.3 Types of Plats

Unfortunately, many surveyors and courts place more credence on compilation plats than they deserve. In most instances, plats are nothing more than opinions of the individuals who created them, and may or may not have been based on field recorded information.

When using plats, the surveyor will find that nearly everyone who is encountered will have used maps or plats at one time or the other and will automatically consider himself or herself capable of reading and interpreting any plat that is used or created by the surveyor. In this chapter, the following principles will be discussed:

PRINCIPLE 1. *The essential purposes to be accomplished by a survey plat are: (1) to represent the correct size and shape of a property to a specific scale, (2) to define by dimensions the correct size and shape of a single and unique parcel of land, (3) to specify locative elements (physical monuments, including cultural features), and (4) to show title identity (record monuments).*

PRINCIPLE 2. *Unless otherwise proved by scaling, distances given on a plat are from the nearest point on each side of the dimension as written.*

PRINCIPLE 3. *A distance that is not indicated by writing or numbers on a map cannot be scaled to prove its true distance.*

PRINCIPLE 4. *Reference to an "official plat" will include all of the field notes and all of the instructions that created the plat.*

PRINCIPLE 5. *A map prepared by a private surveyor must be authenticated before it can be proffered in evidence, unless it is a permitted hearsay exception.*

PRINCIPLE 6. *The courts have held that tax maps should only be used by the surveyor and the attorney to indicate to whom the taxes are billed and not for boundary location.*

PRINCIPLE 7. *The courts may or may not make the distinction between a plat and a map.*

7.4 Purpose of Survey Plats

PRINCIPLE 1. *The essential purposes to be accomplished by a survey plat are: (1) to represent the correct size and shape of a property to a specific scale, (2) to define by dimensions the correct size and shape of a single and unique parcel of land, (3) to specify locative elements (physical monuments, including cultural features), and (4) to show title identity (record monuments).*

In addition to the purposes listed, plats also show data that lend authority to the plat, such as date, the surveyor's name, the certificate of accuracy, and the client's name.

Compilation plats generally lack expression of locative points and correct dimensional data, as would be revealed by a recent survey.

Some states have specific requirements for plats whose data or information are identified or created by survey.

7.5 Features of Plats

Cartographic expressions—such as symbols, north arrow, dimension arrows, and the like—are generally identical for all types of plats, but cadastral features, such as legal descriptions, monuments found, and setbacks, are variable within each parcel of land surveyed and identified. Since the purpose is not to discuss the techniques of plat making, much of the cartographic expression is only briefly mentioned.

Although land surveyors in private practice are not required to do so, it is recommended that the surveyors adopt and use standard symbols identified in national and state-specific map standards.

7.6 Drawings

The size, shape, and kind of drawing needed to record the results of a survey are so variable, depending on the state, need, and purpose, that no standard exists. In some areas the law requires recordation of plats, such as a record of survey in California, where it specifies the details.

When a surveyor or the attorney wishes to use a map as evidence in the course of a trial, several considerations must be addressed. The first is whether the court would keep the map for its records. When the court retains the map, if the original is used, then that original may be lost or never returned to the attorney or the surveyor. To keep this from happening, the original should be introduced in evidence and a copy substituted. In court trials the resultant print needs to be large enough that a juror or judge may see it from considerable distance. For ordinary lot surveys a small drawing is usually adequate (see Figure 7.1). At times it is necessary to make one large-scale plat or several detailed plats.

Drawings should be reproducible. Many types of tracing paper, cloth, and film are available, although no standard exists. Today, surveyors may prepare multiple plats for routine purposes or during the course of litigation from the same CAD file.

A plat should tell a complete story; it should show sufficient information to allow any subsequent surveyor to understand how the survey was made and why the survey is correct. It should also show complete information on alleged encroachments to enable any attorney or others to evaluate properly the effect of continued possession by both parties.

A surveyor should be extremely careful about signing and affixing a seal, in that any original map that is sealed and signed and delivered to the client is capable of being altered. It is suggested that only copies be given to the client and that the original be retained in the records of the creating surveyor and all distributed copies numbered with each number identified on a retained file.

No set rule exists about how much area needs to be shown. If a lot within a block is being surveyed, often it is necessary to show only the dimensions and size of that particular block. But if a parcel is abutting a land-grant line, it may then be necessary to show the land-grant boundary monuments found several miles apart. If a portion of a section is being measured, complete proof of the survey, as a minimum, calls for four quarter-corner locations and at least one section corner. Controlling corners should be indicated for all boundary lines indicated on the plat.

7.7 Title of Plat

One should immediately be able to identify the survey without having to look over the entire map. The purpose of the survey should stand out. One's eyes should instantly see a title that identifies the survey and its purpose. This means the title should be prominent, short, and clear. Ordinarily, the words "Plat of a Survey" precede the general description to follow. Appropriate titles are "Plat of a Survey of Rancho Santa Fe;" "Plat of a Survey of Lot A Block 12, Horton's Addition;" or "Plat of a Surveyor's Portion of S.12, T.13 S., R.2 W, as described in Book 1213, page 21,

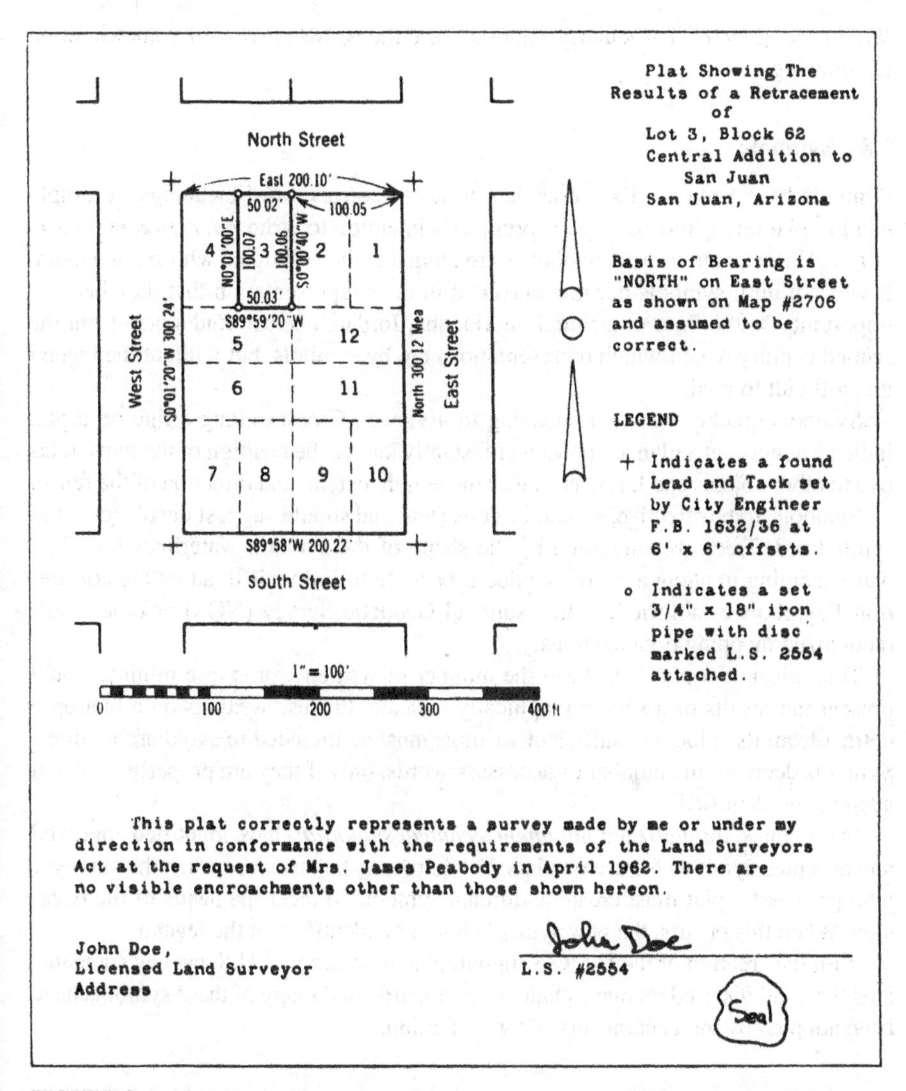

FIGURE 7-1. Plat showing the results of a retracement survey of a lot within a subdivision.

Official Records." If the survey is a retracement, a topographic survey, or a boundary survey it should be identified. The surveyor, simply as a matter of protecting any possible future claims or problems, should in the initial contacts with the client identify the type of survey work that will be conducted for the client. The word *survey* really does not connote any specific purpose.

There may be instances when a plat is prepared for someone other than the record owner, such as when an attorney or a potential purchaser of a parcel calls and requests a survey of a tract of land whose title is in the name of a third person. The title of the survey, or plan, should reflect this with wording, for example, *as requested by* or *title in the name of.* Particularly important are the words *survey for* somewhere on the drawing.

7.8 Symbols

Symbols have been used since ancient times to convey word meanings. A simple hand-drawn line could convey the preparer's intention to write *river*, *stream*, *house*, or *tree*. Unfortunately, most symbols were unique to the individual who created them. It was not until symbols became universal in their representation that they became important. On the floor of a church in Madaba, Jordan, one can find a map from the second century AD, in which representations are by symbols, but without the legend it is difficult to read.

Symbols quickly convey a meaning to a viewer. Crosses along a line on a plat indicate a fence; at a glance, the viewer instantly knows the position of the fence relative to other objects and lines, as well as the length, extent, and direction of the fence.

Symbols, to be effective, should be consistent and should suggest the object being symbolized. Trees are suggested by the shape of the symbol; waterlines look like waves coming in along a shore. A good procedure to follow is to adopt the conventional symbols established by the National Geodetic Survey (NGS) or other well-recognized mapping organizations.

The object of a plat is to keep the number of written words to a minimum and present the results of a survey graphically to scale. Profuse writings on a plat only detract from its value, but sufficient writings must be included to avoid ambiguities. Symbols decrease the number of necessary words, only if they are properly used and adequately identified.

Many states, through their *minimum technical standards*, have identified approved and accepted symbols for use on plats. Yet there may be instances when the surveyor who prepared a plat must create additional symbols to meet the needs of the occasion. When this occurs, the new symbol should be identified in the legend.

With the creation of the U.S.G.S topographic map series, a U.S. agency standardized the symbols used on maps made for public use, and many of these symbols have been adopted by the general surveying profession.

7.9 Scale

The relation between the plat dimensions and those on the ground can be expressed by (1) a bar scale, (2) a ratio, or (3) a representative fraction. The bar scale is most useful because its value is not changed when the plat is reproduced by either an enlargement or reduction process. Along with the bar scale, a ratio scale is indicated as 1 inch = 100 feet. Architects' scales ($\frac{1}{16}$ inch = 1 foot) are not convenient to use on survey plats and should be avoided.

A representative fraction, such as 1:100 (meaning 1 inch = 100 inches), is commonly used on government topographic maps but is not used on private survey plats.

It is suggested that bar scales be used in the event the original map is enlarged or reduced.

7.10 Indicating Direction

Every map, plat, or sketch should bear an orientation arrow, and if a grid system is used, grid north is indicated. Ordinarily, the north arrow refers to astronomic north, but under some circumstances the magnetic meridian is indicated. If the magnetic meridian is used, the plat should bear a date, have a statement that the magnetic meridian is used, and show the relationship between magnetic and astronomic meridian (declination). Often, alongside the north arrow, the scale is indicated.

One should not assume that the top of the map is north, or that the reference meridian is north. In selected areas, south is the major reference meridian for maps. With some early religious maps, east is the reference meridian.

When convenient, the drawing should be oriented with north at the top of the sheet. If this is not practical, a larger-than-usual north arrow should be included on the plat. The north arrow need not be elaborate. The ancients were sometimes carried away with embellishments on their compass rose, but plats made today can best be served with a simple, efficient arrow (see Figure 7.1). If other than astronomic north is used, it should be noted at the bottom of the arrow or as a special note indicating the basis of bearings.

Once again, as required by the minimum technical standards, a north arrow is essential, but failure to include one on a plat may not be critical. Like all other professions, surveying has some unwritten standards. In their formative years, the authors were taught that, whenever possible, a map should be placed on a sheet of paper so that north would always be at the top of the paper.

7.11 Basis of Bearings

All plats that show the results of a survey should carry on their face a note explaining the basis of bearings. Without doubt, the most frequent cause of ambiguity and litigation is the result of failure to specify the meridian reference. For every survey and resulting survey plat, the surveyor must take one of four actions:

1. Assume a meridian, if one is not indicated.
2. Establish a meridian by astronomical observation.
3. Establish a meridian from geodetic control.
4. Establish a meridian by a magnetic observation.

By some type of note on the plat, the surveyor must explain what he has done. Following are examples of notes found on plats:

> Basis of bearings is a Polaris observation as made Jan. 12, 1962, at corner No. 1.
>
> Basis of bearings is a sun observation made July 2, 1961, at corner No. 3. Basis of bearings is N 0° 12′00″ E along the centerline of 6th Street as assumed from the city engineer's records.
>
> Basis of bearings is N 7° 21′30″ as shown on Map 2132 for the easterly line of Lot 13 between monuments indicated.

A magnetic observation was made to determine the bearing of the east line of the property platted; all other bearings were determined by calculation from deflected angles as measured by a 1-minute transit.

Basis of bearings is grid north as based on . . . State Plane Coordinate System (the drawing will show the stations occupied).

Finally, and perhaps the most difficult to duplicate for a retracement, is "basis of bearings is assumed."

All platting laws should require the surveyor to specify his basis of bearings, and, where a suitable grid control exists, the basis of bearings should be specified as grid north. If a magnetic basis of bearings is permitted, it is particularly important that the date of magnetic observation be given. Magnetic north changes with time. When using magnetic bearings, it is helpful to place a diagram on the north arrow indicating the direction of the declination. By doing this, future problems may be eliminated. There are many instances when the field surveyor and the office draftsperson will apply the magnetic correction in the wrong quadrant.

7.12 Elevation Datum

Elevation disputes rarely are the subject of boundary line measurements, and resulting disputes, but occasionally a deed is defined to run along the mean high-tide line or along some other water contour line (for dams or lakes). For original plats, mean sea-level datum is preferred and should be defined on the plat. The use of colloquial expressions such as "U.S.G.S. datum" or "government datum" is not definite and should be avoided.

For any survey based on the record, the record usually defines the datum, and, of course, that datum must be used to determine boundary location. On the plat, the datum used should be explained, and it may be advisable to relate the datum to other data. When using "mean sea level," the surveyors should remember that this description, by definition, is an ambulatory boundary that is referenced to a computed line, and it is always changing.

This is especially important when a contour line is the boundary line. Many times, the boundary will state "640 foot contour," with no reference as to what the basis datum is. With any contour reference, the basis of the reference elevation should be indicated. It should be noted that when referencing a contour elevation, as the land may build up by accretion or be lost by erosion, the boundary usually does not change.

7.13 Dimensional Data

One of the essential features of a survey plat is the dimensioning of the size and shape of the land represented by a drawing. Too many dimensions detract from a clear picture, yet too few dimensions can create ambiguity and possibly make the conveyance void for uncertainty.

Dimensions, as symbolized on plats, are reduced to their simplest expression; dimension lines are omitted when not needed for clarity.

PRINCIPLE 2. *Unless otherwise proved by scaling, distances given on a plat are from the nearest point on each side of the dimension as written.*

Often, dimension lines are unnecessary and do not contribute to a plat; in fact, too many lines will detract from the clarity of a plat. Plats are drawn to a scale; most distance data, uncertain in the mind of the viewer, can be made certain by scaling. Scaling should be used as one form of evidence in order to add certainty to distances and not for absolute values. This should apply to the scaling of distances of lines as well as angles between lines. Occasionally, dimension lines are necessary to prevent ambiguity in indicating limits of measurement, in which event, dimension lines, as shown in Figure 7.1, are used. Arrows of the type used on house plans are not used.

Most subdivisions, developed today, are rarely rectangular with a gridiron pattern. Curved street lines and nonuniform bearings require dimensions on every lot line, and so do radial bearings (see Figure 7.2). Good platting laws require complete dimensional data on every line.

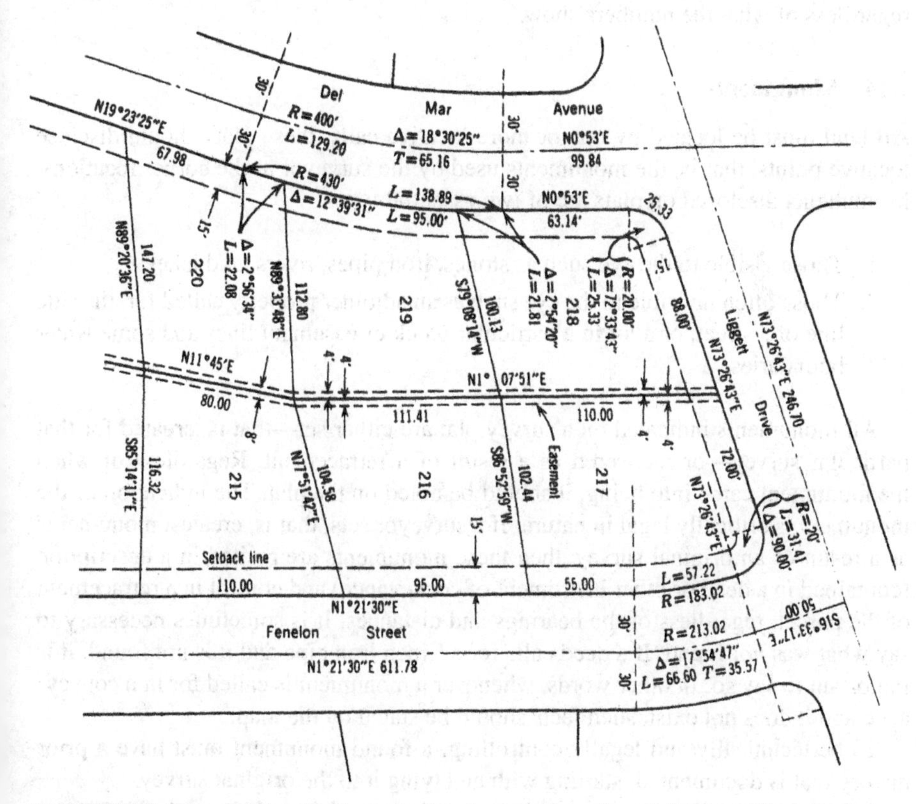

FIGURE 7-2. Plat showing lot dimensioning for a portion of a subdivision. Often platting acts require an additional curve such as the chord and tangent. If the title sheet does not specify what the easements are for, it must be stated on the face of the plat.

Distances are expressed consistently, usually to the nearest 0.01 foot, except where absolute values are implied. A street shown as "50 feet" will be construed consistent with other units as 50.00 feet.

Most subdivision acts require each lot to mathematically close; for this reason, angles are often shown to an accuracy greater than that which can be measured. Angles shown to the nearest second for lines 50 feet long cannot be measured that precisely, but they comply with the law. Sectionalized land subdivisions (U.S. government) have all dimensions written on the map as measured; seldom does a section close mathematically. As long as closures come within permitted errors, the survey is accepted, and dimensions are written on the map as shown in the field book.

Formerly, the use of ditto marks on plats was permitted to allow expression of the same dimension; presently, because of ambiguities caused by the ditto marks, most laws prohibit their use, and they are not recommended.

When reading and referencing descriptions, the reader should realize that there are two types of closures: mathematical closure and legal closure. While the surveyor is concerned about having his or her balanced survey close 0.00 N and 0.00 E, the lawyer wants it to close with such words as "and thence to the point of beginning," regardless of what the numbers show.

7.14 Monuments

All land must be located by one or more locative calls; these plats should disclose locative points, that is, the monuments used by the surveyor at the corner locations. Monuments disclosed on plats are of two essential types:

1. Those visible to the eye such as stones, iron pipes, rivers, and lakes.
2. Those often invisible to the eye such as an adjoiner property called for, the side line of a street, or a lot in a particular block or easement lines and some water boundaries.

All monuments indicated on a survey plat are either set—that is, created for that particular survey—or recovered as a result of a retracement. Regardless of when the monument came into being, it should be noted on the plat. The indication of the monument is critically legal in nature. If a surveyor sets, that is, creates, monuments as a result of an original survey, then these monuments are recited in a description (contained in a deed or other instrument of conveyance) and control in a retracement of the parcel, regardless of the bearings and distances. It is sometimes necessary to say what was *not* found. If a deed calls for a l-inch iron pipe and it is not found, it is important to say so. In other words, whenever a monument is called for in a conveyance and it does not exist, such facts should be stated on the map.

To be technically and legally controlling, a found monument must have a prior history that is documented, starting with and tying it to the original survey.

A found monument has a past history, and as much past history that is known should be included. The following all convey that a found existing monument was present when the retracement survey was conducted: "Fd. orig. mon. as per

Map 2102," "found city engineer's 2″ × 2″ stake per field book 56, pg. 32," "found old barbed-wire fence known to have existed more than 20 years," "gradient line of Red River," "mean high-tide line," "centerline of existing concrete paving," "Eucalyptus tree," "hole in ice caused by warm spring," or "found stone marked with three notches south and two notches west." If the person who originally set the found monument can be identified, this information should also be mentioned.

All monuments set should be identifiable in the future by some unique means such as the surveyor's license number firmly attached or stamped on all set monuments. For practical, as well as legal requirements, not only should those monuments called for in prior descriptions be indicated, but also those monuments that fall into that broad category identified as "uncalled for" should be noted.

7.15 Cultural Improvements

Cultural improvements such as cultivation, barns, houses, streets, paving, sidewalks, storm drains, sewers, power lines, telephone lines, and any possession over the boundary lines, either permissive or without permission, may have a controlling consideration on property ownership and are usually shown on survey plats. In litigation involving unwritten rights, these features are essential data for the court. Ordinarily, the position of any of these improvements is shown relative to some boundary line as defined by a written conveyance. Conventional symbols, as used on topographic maps, are used to represent cultural features.

7.16 Title Identity

Title identity is established by identifying the position of "record monuments" relative to the property being surveyed. A call of "lot 10, block 3, Barn's Addition" is a call for a record monument. It can happen that a single physical monument may not mark that lot 10, yet lot 10 is a record monument. A call to "the east line of section 12" is a call for a record monument.

Locative points (physical monuments) or measurements from locative points identify record monuments on the ground. Both record and locative monuments are essential to identify property, and both are necessary features of a survey plat.

In the current trend, irrespective of whether a plat is for the purpose of showing the results of an original survey, a resurvey, or a survey of a property based on the record, it is always necessary to show title identity. In former years it was possible to survey and describe a parcel of land that was not near any other parcel of land, but that condition no longer exists. For the purpose of obtaining a clear title, it is now necessary to show the relationship of the parcel being surveyed to that of other parcels of record.

This does not mean that title identity must be established for junior landholders. Ordinarily, only those with senior or equal rights (e.g., senior parcels, lot lines, section lines, and original lines marked and surveyed) are identified on plats. At times junior title holders are shown, but it should be made clear that they are junior in standing.

PLATS OF SURVEY RESULTS

7.17 Effect of a Plat Showing Survey Results

A survey plat properly certified, as provided by law, will serve as evidence of two things:

1. Limits of title rights of the client
2. The surveyor–client contract

A survey plat, showing the results of a survey, is merely an opinion of the creating surveyor and evidence for the benefit of both the client and the surveyor. The client can use the plat to show to neighbors or buyers to prove area for taxation, as evidence in court, and as a basis for an American Title Association policy. The surveyor can use the plat for ready reference when performing other surveys in the area, and he or she can clearly state the limits of the surveyor's responsibility. The plat serves as a conclusion to a contract and should clearly define what has been done and what is certified in exchange for the fee paid. The retracing or creating surveyor should be hesitant to indicate any information that is beyond the boundaries of the parcel being surveyed. Any information indicated should be that information that lies physically within the parcel boundaries. If information outside the parcel is indicated and it is found to be erroneous, the surveyor may be found negligent if it is found to be wrong.

7.18 Plats with Omitted Information

Many plats fail to indicate sufficient information of courses and distances on indicated lines. Surveyors and attorneys often scale the omitted distances and angles from the map and attempt to pass them off as being the true distance. When individuals resort to scaled distances to locate definitive distances or angles from maps to "prove" a line distance or to set a boundary line or corner, they have little substantive information. A Texas court disregarded scaled distances from a map whose scale was 1 inch = 800 feet by stating, "We attach no significance whatsoever to the mere scaling of the maps" and the trial court evidently took the same view [6].

PRINCIPLE 3. *A distance that is not indicated by writing or numbers on a map cannot be scaled to prove its true distance.*

A line on a map or plat that has no distance or angle indicated cannot be scaled to determine the true or correct distance. Any scaled distance is a result of five elements:

1. The scale of the map
2. The refinement of the measurement
3. The medium on which it was drafted
4. The ability of the individual to measure the line
5. The thickness of the pen point that competed the draft

In most instances, a person can scale to approximately 1/50 of an inch. Thus, the final distance calculated is only as close as the ability to measure. One must also consider the base material on which the plat is drawn. Many materials are not stable and change with the moisture conditions, either shrinking or expanding.

7.19 Plats of Title Surveys for Title Associations

Various title associations have established minimum standards for acceptable title surveys wherein location as well as title matters may be insured. The proposed minimum standards for land surveys made for title insurance purposes in Florida are given as follows.

> A land survey to be acceptable for title insurance purposes must be a full and complete survey showing every detail affecting title and should be certified to the insuring agency as meeting the following minimum standard requirements approved by the Florida Surveying and Mapping Society (FSMS) and the Florida Land Title Association (FLTA). The certificate should read as follows: "I hereby certify that the survey represented hereon meets the minimum standard requirements approved and adopted by the FSMS and the FLTA."

All measurements made in the field must be in accordance with the U.S. standards and made with a transit and steel tape or other modern devices proven equal or superior. All measurements should refer to the horizontal plane. All computed distances and bearings must be supported by careful and accurate preliminary measurements made as required. Wherever possible, the accuracy of the fieldwork thus performed should be substantiated by the computations of a closed traverse. The relative error of closure usually permissible should be no less than the following:

Locality	Maximum error of closure
Commercial and high-risk areas	1 foot in 10,000 feet
Suburban	1 foot in 7,500 feet
Rural areas	1 foot in 5,000 feet

The plat drawn of that survey must bear the name, address, certificate number, and signature of the land surveyor in charge. The date of the survey must be shown.

A reference to all bearings shown must be clearly stated: for example, "Bearings shown refer to True North," "Bearings shown refer to Grid North as established for the Eastern Peninsular portion of Florida by the USC & GS," "Bearings shown refer to Assumed North based on a bearing of S 5° 30′ 00″ W used for the centerline of County Road No. 100," or "Bearings shown refer to the Deed Call of N 22° E for the easterly line of Lot No. 4." References to magnetic north should be avoided except in those cases where a comparison is necessitated by a deed call. Where bearings are recited in the deed description or on an original plat of the lands being surveyed, a comparison of the deed or plat bearings with the bearings used must be shown on all courses. In all cases, the bearings used should be referenced to some well-established line.

A "north arrow" should be prominently shown and, whenever possible, placed in the upper-right-hand corner of the plat.

The caption of the prepared plat must be in complete agreement with the record title. Whenever possible, the correct caption will be furnished by the title company concerned along with a record description to which the survey must refer. In all cases, the survey must make reference to and identify the source of information used in making that survey, such as the recorded deed description or other conveyance, a recorded or unrecorded plat, or other claim of rights.

Where inconsistencies are found, such as overlapping descriptions, hiatuses, excess or deficiency, erroneously located boundary lines and monuments, or where any doubt as to the location on the ground of survey lines or property rights exists, the nature of the difficulty should be shown and the range of possible differences indicated on the plat.

All outer boundaries and any interior subdivision lines lying within the lands surveyed must be shown on the plat, with all the data necessary to determine the correctness of the survey by both mathematical computations and plotting. All angles must either be given directly or indicated by the bearings shown. Where lines are curved, the significant elements of the curve, such as the radius, the arc length, and the bearing and distance of the chord, must be shown. Centerline data for an adjoining curved right-of-way should also show sufficient information to permit the computing or plotting of the curve in its entirety, such as the central angle, the tangent (T), and the radius (R). The degree of curve should only be shown in those cases where it is recited in the deed description or shown upon a controlling plat.

In all areas where lots and blocks are established, the distance to the nearest intersecting street or right-of-way must be shown. Distances to intersecting streets or rights-of-way in both directions must be shown if either of the distances varies from the original plat. Surveys of parcels described by metes and bounds within a large tract of land should show the relationship of that parcel to at least one of the exterior lines of that tract, preferably by the bearing and distance along an established right-of-way.

If the survey is of all or any part of a lot or lots are part of a recorded subdivision, all lot numbers, including those of adjoining lots, and the block number must be shown on the plat.

All legal public and private rights-of-way shown on record plats adjoining or crossing the lands surveyed should be located and shown on the plat. Where ties are made to an intersecting street or other right-of-way, pertinent data concerning the right of way should also be given. Changes in rights of way should be noted and the date of and the authority for the change shown. Where a proposed change in a right of way is known, such as the knowledge that appropriations have been made by the authorities to acquire additional right of way for the widening of a public thoroughfare, that proposed change should be shown and noted as such on the plat. The land surveyor, however, is not to be held responsible for the accumulation of such information. If streets abutting the lands surveyed are not physically opened, a note to this effect should be shown.

The character of all evidence of possession along the boundary lines, whether fence, wall, buildings, monuments, or otherwise, should be shown and the location identified carefully, given in relation to reference or record description lines.

Encroachments such as eaves, cornices, doors, blinds, fire escapes, bay windows, windows that open out, flue pipes, stoops, steps, and trim—by or on adjoining properties or on abutting streets—must be indicated with the extent of such encroachment clearly shown. Openings, such as windows and doors, in walls of premises or adjoining premises adjacent to the boundary lines, must be noted. Whenever possible, foundations near the boundary lines should be located and shown on the plat. In all cases, where the location of foundations is prevented by pavements or other obstructions, the failure to determine the location should be noted on the plat. If a building on the premises has no independent wall but uses any wall of adjoining premises, this condition must be shown and explained. This also applies when conditions are reversed. Encroachments are legal in nature. The surveyor should not determine its validity, but the boundary line and any infraction should be noted.

All physical evidence of easements or rights of way created by roads, surface drains, telephone or telegraph lines, and electric or other utility lines on or across the lands surveyed must be located and noted on the plat. Surface indications or markers of underground easements should also be shown. If location of easements of record, other than those on record plats, is required, this information must be furnished to the land surveyor.

The character, construction, and location of all buildings on the lands surveyed must be shown and referenced to the boundaries. On large tracts, buildings and structures remote to the boundary lines may be shown in approximate scaled positions and thus noted on the plat. If interior improvements are not located, a note to this effect such as "interior improvements, if any, not located" must be shown. The same rules apply with respect to interior cross fences. Joint or common driveways must be indicated. Independent driveways must be shown completely to the property line.

Cemeteries and burial grounds located within the premises must be properly located and shown on the plat.

Monuments, stakes, and marks found or placed must be adequately described and data given to show the actual location on the ground in relation to boundary lines. Redates of "foundation" or "under construction" surveys must take into consideration all of the requirements contained herein. Special attention should be given to any indication of changes and to any encroachments that may have arisen since the date of the original survey, including, but not limited to, possible encroachments of driveways or aprons at the corners of irregular lots on curved streets.

If the survey order indicates FHA (Federal Housing Administration) or VA (Veterans Administration) insurance of the title for which the survey is to be used, compliance with the appropriate requirements must be made. City, county, or state requirements must be complied with where applicable.

Most of these standards, as in the foregoing, fail to use an effective method of specifying accuracy. The error of closure, as noted in Chapter 9, is never an indication of accuracy; it merely indicates a possible absence of blunders.

7.20 Surveyor's Certificates

On every survey plat the surveyor should certify about what was done. Careful consideration should be given as to reporting what was asked, what was examined, what was found, and what decisions were made. The form used varies, depending on the purpose. In populated areas, usually the client's purpose of having a property surveyed is to determine the status of encroachments or uses over the identified boundaries, as in the following certificate:

Certificate of Survey

This is to certify that the here-in-before plat correctly portrays a survey made under my direction of Lot 12, Block 3, of Grantville, and that the monuments shown thereon are to the best of my knowledge and belief wholly in accordance with Map thereof No. 43W filed in the office of the County Recorder of . . . County, State of . . . and that there are encroachments thereon as shown above, but no others.

John Doe
Licensed Land Surveyor
No. 2554 Sealed

In connection with title insurance surveys, the Pennsylvania Title Association requires the following:

To Whom It May Concern:

This is to certify that this plat and the survey on which it is based were made in accordance with Minimum Standards for Title Surveys adopted by the Pennsylvania Title Association on May 20, 2001, and that this plat correctly sets forth all data obtained in the said survey.

Signed by the surveyor.
License No . . .

The title company usually limits its policy to persons named on the face of the policy, yet it asks surveyors to guarantee location for present as well as future owners by using "to whom it may concern." The better procedure for the surveyor is to state specifically for whom the survey was made as John Doe and ABC Title Company.

A surveyor should not make any warranties or guarantees as to area or measurements. A surveyor can only certify that "according to the survey and the calculations, this area contains 23.34 acres" (or whatever is calculated) according to his or her survey.

These standards are now being replaced with positional tolerance criteria that have been identified in the newer ALTA standards.

Certification is relative to the plat and the standards in effect at the time the work was undertaken and not at a later date, if the standards change.

7.21 Contents of Plats

A plat should be complete in itself and the information and calculations indicated should present sufficient evidence of monuments (record and locative) and measurements so that any other surveyor can clearly, without ambiguity, find the locative points and follow the reasoning of the surveyor. The information should be sufficient in nature so as to provide the means for the client to locate the parcel on the ground, without any difficulty. A plat does not show the client's land alone; it shows all ties necessary to prove the correctness of location. If it is necessary to measure from a mile away to correctly locate a property, that tie, as measured, is shown.

On most plats it is necessary to show the following 14 items:

1. Record of monuments called for (title identity), including abutting streets and easements.
2. Found physical monuments that locate the record monuments and a complete description of these monuments whether called for or not.
3. Proof of correctness of the found monuments (history).
4. Notation of monuments called for but not found.
5. Notation of monuments found but not called for.
6. Basis of bearings used on the plat.
7. Expression of measurements on those lines surveyed, by reference to
   a. Direction
   b. Distance
   c. Coordinates (not always given) and only if relevant and certain
   d. Curve data (e.g., central angle, radius, length, chord, tangent).
8. Monuments set and their descriptions (including a description of monuments replaced).
9. Oaths or witness evidence.
10. Date of survey.
11. Client's name.
12. Surveyor's certificate (signature, seal, statement of accuracy, and guarantee of location in accordance with a particular description furnished).
13. Easements located in accordance with descriptions furnished.
14. Unauthorized possession (actual and title) and possession on the title lines as described in the title furnished.

ORIGINAL PLATS

7.22 Purpose

Original surveys have two parts: (1) a boundary survey that is a survey for the creation of the boundaries of a land parcel, from which an original plat is prepared along with the field notes and other necessary information; and (2) a resurvey that is preceded by a retracement that was conducted to determine the status of the remains of the earlier original survey. A dependent resurvey is based on giving full faith and credence to the recovered corners of the original survey and then relocating the lost corners in accordance with the correct rules of survey. Then there is an independent resurvey, which disregards all of the original interior corners and lines and recreates new corners and lines within the original exterior boundaries.

7.23 Laws Regulating Platting

The creation of new parcels of land is usually, though not always, regulated by law. When the law does regulate how land must be divided, the surveyor must obey the law; in many areas, how plats should be made and their scale and size, the amount of information disclosed, and numerous other items are specified by platting acts. One state, California, writes:

> Regulations (California Subdivision Map Act) controlling the contents of plats, often include that the final plat should conform to the following provisions:
>
> a. It must be a map legibly drawn, printed, or reproduced by a process guaranteeing a permanent record in black on tracing cloth or polyester base film, including affidavits, certificates, and acknowledgments, except that such certificates, affidavits, and acknowledgments may be legibly stamped or printed on the map with opaque ink when recommended by the county recorder and authorized by the local governing body by ordinance. If ink is used on polyester base film, the ink surface should be coated with a suitable substance to assure permanent legibility.
> b. The size of each sheet should be 18 × 26 inches. A marginal line should be drawn completely around each sheet, leaving an entirely blank margin of 1 inch. The scale of the map should be large enough to show all details clearly and enough sheets used to accomplish this end. The number of the sheet and the total number of sheets composing the map should be stated on each sheet, and its relation to each adjoining sheet should be clearly shown.
> c. It should show all survey and mathematical information and data necessary to locate all monuments and to locate and retrace any and all interior and exterior boundary lines appearing thereon, including bearings and distances of straight lines and radii and arc lengths of chord bearings and lengths for all curves, and such information as may be necessary to determine the location of the centers of each curve.
> d. Each lot should be numbered and each block may be numbered or lettered. Each street should be named.
> e. The exterior boundary of the land included within the subdivision should be indicated by a colored border. The map should show the definite location of the subdivision, particularly its relation to surrounding surveys.
> f. It should satisfy any additional survey and map requirements of the local ordinance.

7.24 Compilation Plats

To the surveyor, the most useful compilation plat is the arbitrary plat prepared by an abstractor or title-insuring agency. It is a compilation of those land descriptions within an area showing controlling measurements and other data. Not all deed dimensions are shown; those that are a controlling consideration are always emphasized to show their importance. In a deed reading "N 20° 10′ E, 128.02′ to the westerly line of that land described in . . . ," the notation on the platted line would be "N 20° 10′E, 128.02 feet." The abstractor has shown that the call of 128.02′ is more or less in character, since the distance must go to the record monument called for.

Abstractor plats vary from office to office, and the contents of the plats depend on local customs or local abbreviations. Each state has its own recording system or method of numbering plats, and the method of disclosing recording data varies. The following notations and their meaning are sometimes used:

TK 10,921 (Tract Map 10,921)
B1721/231 OR (Book 1721, page 231, Official Records)
B122/31 D (Book 122, page 31, Deeds)
SC52127 (Superior Court Case 52,127)

Using these notations, an abstractor's plat could be as shown in Figure 7.3. In Figure 7.3, each parcel is given an arbitrary number; the plat is called an *arbitrary plat*, or just an *arb plat*. On the plat, arbitrary lot 1 is clearly a 200 × 200-foot parallelogram, as measured along each street. Arbitrary lot 2 is 200 × 200 feet (westerly

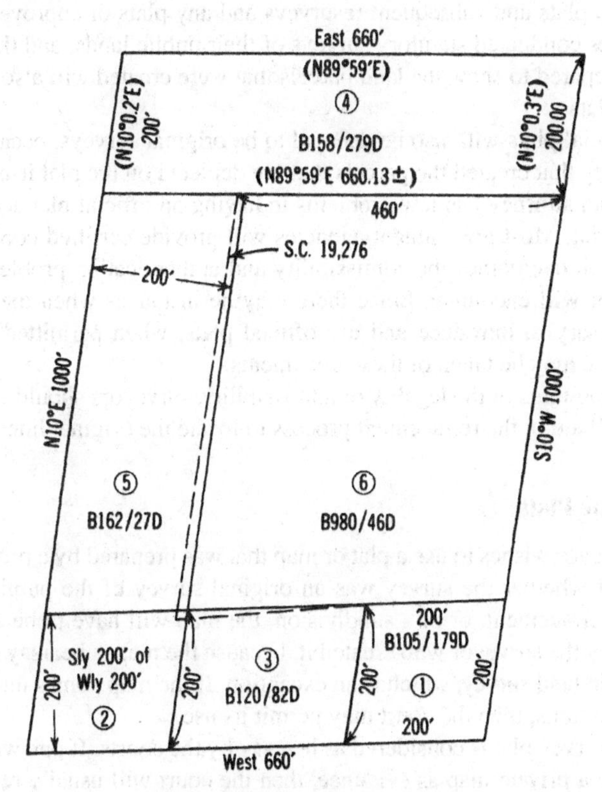

FIGURE 7-3. Abstractor's plat.

200 feet of the southerly 200 feet of lot 10) as measured at right angles to each street. Arbitrary lot 3 is a remainder between land 2 and is formed by a straight line connecting the corners of land 2, as shown. Lot 4 was created next and presents no difficulties. Lot 5 was sold as the westerly 200 feet, minus 2, 3, and 4. When lot 6 was sold, it overlapped 5 and 3. The deed author failed to recognize the 200 feet as being measured at 90° to the street. The overlap on 5 was settled by Superior Court Case No. 19276. The overlap on 3 still exists, although 3 is senior and is entitled to the disputed area. This arbitrary plat clearly shows the conditions existing and who has senior rights as defined by written titles. Today, title abstracts and abstractors' plats are being replaced with title insurance policies based on the "probability" of having to pay a claim. Title abstracts are becoming historical documents.

7.25 Official Plats

PRINCIPLE 4. *Reference to an "official plat" will include all of the field notes and all of the instructions that created the plat.*

The term *official plat* is reserved for those plats, usually GLO plats, that were prepared under a governmental authority. In the GLO states this will include the original township plats and subsequent resurveys and any plats of approved land grants. Several states conducted statutory surveys of their public lands, and thus those plats that were prepared to show the land parcels that were created will also be deemed to be official plats.

Most official plats will also be deemed to be original surveys, because it was the official survey that created the parcels that are depicted on the plat itself.

Usually, an attorney has few problems in having an official plat accepted as evidence in a trial. Most government agencies will provide certified copies from their records, which overcomes the admissibility and authentication problems that a private surveyor will encounter. Since there may be instances when the attorney will find it necessary to introduce and use official plats, when permitted by the court, judicial notice may be taken of these documents.

Without thoughts of the legality or admissibility, surveyors should use any plat or plats that will aid in the retracement process to locate the original lines in question.

7.26 Private Plats

When a surveyor wishes to use a plat or map that was prepared by a private surveyor, regardless of whether the survey was an original survey of the public lands or its subsequent retracement, or of a subdivision, the map will have to be authenticated, if possible, by the surveyor who created it, because the map is hearsay unless it is an official public land survey, which is an exception. If the map comes under one of the hearsay exceptions, then the court may permit its use.

A recent survey plat is considered as hearsay by the courts. If one wishes to introduce and use a private map as evidence, then the court will usually require the surveyor who conducted the fieldwork and who created the map to authenticate it. This rule will not apply if it can be shown that the map falls under one of the hearsay exceptions.

Understanding these rules of evidence becomes critical for the attorney and the surveyor, since the hearsay exceptions clause provides several exceptions for hearsay evidence.

PRINCIPLE 5. *A map prepared by a private surveyor must be authenticated before it can be proffered in evidence, unless it is a permitted hearsay exception.*

This area may become important, in that in today's business world, a field crew usually conducts the fieldwork, then turns the field data over to an office CAD operator to create the plat and then at trial, the principal of the firm usually testifies as to the facts on the plat, without ever having seen the lines on the ground. A good defense attorney is able to keep any of these maps from being used or introduced as evidence for several reasons as being hearsay.

Litigants in a court action believe that if a map depicts any line of the dispute, or the line has been in possession of one of the parties, it can be proffered in evidence. This is not so. Maps are considered as hearsay, and as such the party must be given the right to cross-examine the information indicated on them. This requires that the surveyor who prepared the map must be available to substantiate the information and also to be cross-examined. There are many maps that are permitted to be used because they fall under one of the hearsay exceptions. Most notable is the ancient-document exception that permits documents to be introduced if they meet the age requirements, usually 20 years old in the federal courts [Federal Rules of Evidence (FRE), 803 (15)] and 30 years in some state courts. Rule 803 (20) also permits any evidence concerning boundaries that originated before the litigation commenced as well as Rule 803 (5 or 6), recorded recollections or business records.

7.27 Assessor's and Tax Plats

PRINCIPLE 6. *The courts have held that tax maps should only be used by the surveyor and the attorney to indicate to whom the taxes are billed and not for boundary location.*

Courts have repeatedly held that tax maps and assessors' maps are hearsay and as such do not meet any exception to excluding hearsay evidence, and as such are inadmissible. Yet the prudent surveyor usually consults them for the information they contain.

Many surveyors, when confronted with a survey problem, usually run to the local tax office to obtain copies of the tax maps to aid in their work. Then when it comes to trial, they try to introduce these tax maps to prove their boundaries [7].

The courts have been fairly uniform in excluding these maps as a source of information for the determination of boundary problems. For the most part they have held that so far as a question of boundary matters, they are hearsay, but they can be used to ascertain either who paid the taxes of the land parcel or against whom the taxes were assessed.

Thus, practicing surveyors should be very hesitant in placing undue reliance and confidence on tax maps or abstractor's compilations that claim to depict boundaries (see Figure 7.3). Many of these documents are not prepared by professional surveyors and many of these tax maps are put together without reference to reading the references deeds.

7.28 Plats as Evidence

PRINCIPLE 7. *The courts may or may not make the distinction between a plat and a map.*

When attorneys or surveyors anticipate using maps or plats as evidence at a trial or for purposes of discovery, they should realize that simply because it is a map or plat of the area in question will not make it automatically admissible. The attorney must consider two elements: relevancy and admissibility. Although the surveyors' main concern is relevancy, it is useful to help prove the facts it wasn't used to prove.

A map is relevant if it depicts the area in question, either historically or showing information as it currently exists. But the question of its admissibility must be addressed to make it usable. There are two basic sources of maps, and the courts will treat each differently. For lack of a better terminology, one will be identified as *official* and the second as *private*.

The more abbreviated the map, the less useful it is. The problem becomes important when the attorney and the surveyor attempt to use a map or plat for purposes other than what it was designed for. At times, surveyors are asked to scale distances from maps or plats where the distances are not indicated on the map itself. Many surveyors fall into this trap.

7.29 Conflicts between Plat and Field Notes

At times, a surveyor or an attorney will encounter a problem when it is discovered that there is a conflict between the information indicated on a plat and that indicated in the field notes of a survey.

There is no set rule to solve the conflict except that usually the field notes will control, in that the plat is usually drafted from the field note record, usually by a third person. In all probability, the solution will be different for "official" GLO plats and those plats that were prepared by private surveyors. Since under federal decisions land is patented according to the plat, the plat becomes the reference for the conveyance and thus should control. Actually, the "field notes" that are prepared are copies of the running notes of the actual fieldwork and, for practical purposes, are not the originals. In the PLS states in many instances, the original notes created in the field were destroyed when the notes were copied into their final form.

Private plats that differ from the field notes will probably be treated in a different manner, depending on whether the map is made reference to in the conveyance—that is, the parcel was conveyed "according to a plat of survey." In this case, the aggrieved party will probably be permitted to introduce the field notes to show the discrepancy and the true intent of the parties. But if the map is not referred to in the conveyance, then the notes will probably control, because the plat is a product of the notes and thus is junior to the notes.

7.30 Summary

When discussing plats, a distinction must be made as to the plat itself and the information that is depicted on the plat. There are plats of original surveys and original plats—that is, the original plats, which are the first plat of the parcel. Plats of original surveys are usually created after the fieldwork has been completed, and as such, they have been held by case law as being a part of the survey itself. An original plat is the first plat that was created to depict the work.

There may be an original plat of a retracement survey or an original plat of a resurvey. Although the surveyor usually does not make a distinction between the two, courts may give different weights to them.

Original survey plats, resurvey plats, and retracement plats are similar in cartographic expression. Original survey plats are called for by a conveyance, and the contents of the plat are considered as part of the writings. A resurvey plat serves as evidence of what the surveyor did in following and relating evidence to the ground.

On an original plat, the object is to explain how to identify and locate a newly created parcel of land. On a resurvey plat, the object is to present sufficient evidence to prove that the surveyor correctly identified and located a previously described parcel or parcels.

An original plat usually shows the results of a resurvey of a boundary. The object of all survey plats is to show sufficient information to enable any competent surveyor to identify with certainty a given parcel of land by the locative points (physical and record monuments), by the measurements, and by the scaled drawing shown. Retracement plats are nothing more than an expression of an opinion of where the retracing surveyors believe the original lines and monuments the creating surveyor placed on the ground and identified from the evidence that was left for future surveyors to find.

REFERENCES

1. *Banker v. Caldwell*, 3 Minn. 46 (1959).
2. *McDaniel v. Mace*, 47 Iowa 510 (1872).
3. *Bartlett v. Bangor*, 67 Me. 460 (1878).
4. *Rokes v. City of Bridgeport*, 191 Conn. 62, 462 A. 2d 252 (1983).
5. The preceding paragraphs are modified from W. G. Robillard and L. J. Bouman, *Clark on Surveying and Boundaries*, 6th ed. (Charlottesville, VA: Michie Co., 1993).
6. *Blaffer v. State*, 315 W. 2d 172 (Tex. Civ. App. 1930).
7. *Resseau v. Bland*, 491 S.E. 2d 809 (Ga. 1997).

8

EVIDENCE OF WATER BOUNDARIES

8.1 Introduction

Water boundaries are perhaps the oldest and, at times, the most frustrating, elusive, and least understood boundaries. Finding the original rifle barrel or a lighter wood post surrounded by rusted barbed-wire provides a "eureka" moment for even the most novice surveyor. But there is no such thing as a "found IP" in the middle of a lake. If there is one area of boundary retracement that the courts have not mastered, it is the adjudication and determination of water boundaries, both tidal and non-tidal.

Yet, despite the elusive qualities of ambulatory retracements, water boundary surveying is a key aspect of the profession because the most expensive properties in any state are often along a beautiful stretch of meandering river or contiguous with the white, sandy beach that we all love. Given the high cost of real estate, the love of the property by its owners, and the lack of physical monuments to delineate the property corners, water boundaries in today's society are some of the most frequently contested and bitterly fought disputes within the civil court system.

The edge of water forms an excellent natural boundary because everyone can see the topology: water meets land, and that's where the boundary must be. It is easily defended and easily recognized. However, the shore's boundary is deceptive because the location of the land/water boundary is ambulatory; the land–water interface is dynamic because the water surfaces of most water bodies are constantly changing due to the tides, bathymetric topography, winds, or meteorological conditions. In addition, the shoreline in many areas is constantly changing due to the natural processes of erosion and accretion as well as man-made structures such as dams, dikes, and the like. Consequently, a unique set of laws based on statutes and civil court cases, as well as advanced techniques pioneered by federal and state agencies has been created to define and locate water boundaries.

8.2 Determination of Public Title Interest

A unique characteristic of many water boundaries is that they represent the line between privately owned upland and submerged land held in the public trust. This public interest is by virtue of the *public trust doctrine*, which, although chiefly a product of English common law, has roots as far back as the ancient Roman Civil Code of Emperor Justinian I, written about AD 500. Under that code, the sea as well as rivers were considered to be *res communes* or commonly owned by all mankind. This philosophy is expressed in one translation of a portion of that code as: "By the Law of Nature, these things are common to all mankind—the air, running water, the sea, and consequently the shores of the sea" [1].

During the rise of the great maritime nations of the Middle Ages, some nations laid exclusive claims to entire seas or oceans, as opposed to the early Roman concept of the ocean as being common to all people. Gradually, however, such claims were found to be unrealistic and impossible to defend. Therefore, a new doctrine appeared with claims by various nations limited to a marginal sea of a width that could be defended and with the high seas considered to be common to all mankind.

Perhaps the earliest mention of this theory in English literature was by Thomas Digges, an engineer, surveyor, and lawyer during the reign of Queen Elizabeth I [2]. His treatise formed the basis for the Crown's claim to the submerged lands along the coastline of the kingdom.

In the United States, this doctrine began to take its present shape with a series of U.S. Supreme Court cases beginning with *Martin v. Waddell* [3] in 1842. According to the court in that case:

> When the revolution took place the people of each state became themselves sovereign and in that character hold the absolute right to all their navigable waters in the soils under them for their own common use.

In 1845, in *Pollard's Lessee v. Hagan* [4], one of the very earliest of American water interest decisions, the court ruled that states admitted to the Union after the original 13 colonies had the same rights as did the original 13 colonies. Thus, it may be generally stated that in the United States the several states, in their sovereign capacity, hold title to the beds under navigable waters abutting the coastlines of their respective states. The public trust doctrine also extends to navigable nontidal waters in many areas.

This extension is not necessarily a modern innovation. As an illustration of this, the Roman Institutes of Justinian expressly declared rivers to be owned in common by the people in whose territory they lie [1]. In contrast, early English common law considered only tidal waters to be sovereign. This is illustrated by the following: "That rivers not navigable (that is, freshwaters rivers of what kind so ever) do, of common right belong to the owners of the soil adjacent. But that rivers, where the tide ebbs and flows, belong to the State or public" [5]. This distinction is perhaps due to the fact that as a small island kingdom, England has few inland waters with a capacity for public navigation.

American case law has differed from English common law in this regard and has not limited the public trust to tidal waters. In 1876, the case of *Barney v. Keokuk* ruled that state title in navigable waters extended to inland waters as well as tidal waters with the following words:

> The confusion of navigable with tide water, found in the monuments of the common law, long prevailed in this country, notwithstanding the broad differences existing between the extent and topography of the British island and that of the American continent . . . And since this court . . . has declared that the Great Lakes and other navigable waters of the country, are, in the strictest sense, entitled to the denomination of navigable waters, and amenable to the admiralty jurisdiction, there seems to be no sound reason for adhering to the old rule as to the proprietorship of the beds and shores of such waters. It properly belongs to the States by their inherent sovereignty. [6]

Thus, it may be generally stated that in the United States, with few exceptions, the several states, in their sovereign capacity, hold title to the beds under all navigable waters, regardless of whether the waters are tidally affected. The conditions for such title are that the water body must be natural; it must have been considered navigable and in federal ownership at the time of statehood of the state in which it is located and therefore conveyed to that state by virtue of statehood; and it must not have been subsequently conveyed by the state to other parties.

Often the most controversial issue in determining whether a water body is within the public trust is the question of whether it is navigable. This is due to the fact that there are many definitions of navigability. In addition, there are some water bodies that are navigable-in-law although not necessarily navigable-in-fact, while others are navigable-in-fact although not necessarily considered navigable for public trust purposes.

In nontidal waters, navigability for title purposes generally is a question of navigability-in-fact. In Florida, for example, case law [7] offers specific clarification to that state's definition in nontidal waters. In that case, the court held that "Florida's test for navigability is similar, if not identical, to the Federal Title Test." The Federal Title Test was defined as being "based on the body's potential for commercial use in its ordinary and natural condition."

Further understanding of the Federal Title Test may be obtained from the following early statement of this test:

> And they are navigable in fact when they are used or susceptible of being used in their ordinary condition, as highways for commerce, over which trade and travel are or may be conducted in the customary modes of trade and travel over water [8].

It appears that, based on these and subsequent cases, the Federal Test does not allow consideration of artificial improvements to the water body. It must be navigable in its ordinary and natural condition. The test has been clarified to indicate that waters are considered navigable even if the waters are not navigable during the entire year [9, 10]. However, the implication has been that the lack of navigability should be an exceptional circumstance as opposed to a normal or predominant condition.

One possible exception to navigability-in-fact being the criterion for public trust nontidal waters may be found in the New England states. An early Massachusetts

Colonial law, which appears to define all lakes greater than 10 acres in area as public waters, states:

> . . . no town shall appropriate to any particular person or persons, any great pond, containing more than ten acres of land. . . . And for great ponds lying in common, though within the bounds of some town, it shall be free for any man to fish and fowl there, and may pass and repass on foot through any man's propriety for that end, so they trespass not upon any man's corn or meadow [11].

In tidal waters, navigability for title purposes appears to be not always based on navigability-in-fact. In some states, public ownership appears to include submerged lands subject to the ebb and flow of the tide, regardless of actual navigability. One authority indicates that, in Louisiana, Maryland, Mississippi, New Jersey, New York, and Texas, state ownership extends to all waters subject to tidal ebb and flow; while, in California, Connecticut, Florida, North Carolina, and Washington, public ownership is based on navigability-in-fact. The same article states that "Alabama, Oregon and South Carolina find tidal watercourses prima facie navigable and thus presume the land beneath the watercourses to be sovereign land, but this presumption of state ownership may be rebutted by a finding of non-navigability" [12].

It is instructive to review a challenge to the concept of ebb and flow versus navigability-in-fact in the state of Mississippi. As previously indicated, that state has generally been considered to be an ebb and flow state. The case (*Cinque Bambini Partnership v. State of Mississippi*), at both the chancery court and in the Mississippi Supreme Court, was decided in favor of ebb and flow. Salient excerpts from the state Supreme Court opinion are as follows:

> The early federal cases refer to the trust as including all lands within the ebb and flow of the tide . . . it is our view that as a matter of federal law, the United States granted to this State in 1817 all lands subject to the ebb and flow of the tide and up to the mean high water level, without regard to navigability. Yet so long as by unbroken water course—when the level of the waters is at mean high water mark—one may hoist a sail upon a toothpick and without interruption navigate from the navigable channel/area to land, always afloat, the waters traversed and the lands beneath them are within the inland boundaries we consider the United States set for the properties granted the state in trust [13].

The case was appealed to the U.S. Supreme Court, which concurred with the state court in ruling that all coastal states received all lands over which tidal waters flow, and that Mississippi still does. The opinion noted that this ruling "will not upset titles in all coastal states" since it "does nothing to change ownership rights in states which previously relinquished a public trust claim to tidelands such as those at issue here" [14]. Therefore, it is presumed that the ruling has no net effect in those states that have established navigability-in-fact case law.

8.3 Boundaries of Public Trust Waters

Tidal Waters under the Law In areas where land title has its roots in grants from a sovereign power where civil law, as opposed to Anglo/American common law, prevailed, the boundaries of public trust waters are typically defined in terms of seasonal water-level changes. As an example, a translation of a portion of the Roman

Institutes of Justinian reads: "The sea-shore, that is, the shore as far as the waves go at furthest, was considered to belong to all men . . . The sea shore extends as far as the greatest winter floods runs up" [15].

The difference between the common law and civil law boundary definitions was clearly made in *Borax Consolidated Ltd. v. City of Los Angeles* as: "By the civil law, the shore extends as far as the highest waves reach in winter. But by the common law, the shore "is confined to the flux and reflux of the sea at ordinary tides" [16].

As an example of the use of the civil law definition, Louisiana has adopted the civil law boundary of the line of the "highest tide during the winter season" [17]. However, in that state, the boundary of tidally affected lakes is considered to be the mean high water line [18]. Another such example is Texas, which has recognized the civil law definition in a number of cases in areas of the state with the origins of land title in Spanish or Mexican land grants. For such grants, it has been held that the limit of ownership is controlled by Spanish law contained in *Las Siete Partidas* [19], written in the thirteenth century, which closely tracks the Roman *Institutes of Justinian*.

In 1958, the Supreme Court of Texas attempted to apply the use of daily tides to civil law definitions. In that trial, testimony provided various interpretive translations of *Las Siete Partidas*, which suggest that the proper meaning is alternatively the highest swell of the year, the highest tide of the year, or an average high tide. The court concluded that the language of *Las Siete Partidas* implied an average tide and that the applicable rule is that of the "average of highest daily water computed over or corrected to the regular tidal cycle of 18.6 years." In a response to a motion for a rehearing, the court clarified its intent as follows:

> It was our intention to hold, and we do hold, that the line under the Spanish (Mexican) law is that of mean higher high tide, as distinguished from the mean high tide of the Anglo-American law [20].

Regarding this decision, it is noted that the tides along the Texas coast are predominately diurnal in nature. Therefore, there is an almost insignificant difference between mean higher high water and mean high water in that region. In addition, some coastal areas in Texas have a relatively small average range of daily tide in relation to the annual cycle related to the sun's declination and seasonal winds. In some such areas, the relatively large seasonal variation combined with an extremely flat coastal slope results in wide tidal flats which may be either exposed or inundated depending on the season of the year and wind conditions. In such areas, the line of mean higher high water and mean high water may be found considerably waterward of that Spanish grant shoreline indicated by the grant confirmation surveys. Therefore, mean higher high water, although offering mathematical certainty, may not represent the true intent of such grants, which were typically placed by a meander line at the upland limit of such flats, nor of *Las Siete Partidas* [19]. Recently, a trial court decision in Texas [21] held that the methodology suggested in the *Luttes* decision would not apply in such areas. In that case, the court accepted a boundary determined by physical shoreline features located several miles landward of the mean higher high water line using daily tidal heights.

Tidal Waters under Anglo/American Common Law: Boundary Definitions As with all other areas of law and surveying, water boundary/riparian law has its own terminology. There are certain words and phrases that students and professional surveyors must understand in order to effectively work in the area of water boundaries. Following the previously discussed claims of submerged lands on behalf of the Crown made by Thomas Digges circa 1568 [2], there apparently was not immediate judicial acceptance of the claim. In the century following, however, the doctrine became generally accepted as evidenced by the writings of Lord Matthew Hale [22], a jurist who was to become the British Chief Justice. He espoused the public trust doctrine in his treatise *De Jure Maris*, written about 1666. In that writing, Hale concluded that the foreshore, which is overflowed by "ordinary tides or neap tides, which happen between the full and change of the moon," belonged to the Crown.

With our knowledge of the tides today, it is obvious that Lord Hale was incorrect in equating "neap tides" with "ordinary tides." At the very least, his definition was ambiguous. In 1854, this definition was clarified in English common law by the case of *Attorney General v. Chambers*. The *Chambers* case, reflecting tidal theory developed after Hale's writings, ruled that the ordinary high water mark was to be found by "the average of the medium tides in each quarter of a lunar evolution during the year (which line) gives the limit, in the absence of all usage, to the rights of the Crown on the seashore" [23].

In the United States, there apparently was no clarification to the boundary until 1935 and the U.S. Supreme Court's landmark decision in *Borax Consolidated Ltd. v. City of Los Angeles* [16]. In essence, this decision called for the application of modern scientific techniques for precisely defining the boundary. This decision has been contested in the State Court of Hawaii in the decision *In Re Ashford. Borax* went on to say:

> In view of the definition of the mean high tide, as given by the United States Coast and Geodetic Survey that "mean high water at any place is the average height of all the high waters at that place over a considerable period of time," and the further observation that "from theoretical considerations of an astronomic character" there should be "a periodic variation in the rise of water above sea level having a period of 18.6 years," the Court of Appeals directed that in order to ascertain the mean high tide line with requisite certainty in fixing the boundary of valuable tidelands, such as those here in question appear to be, "an average of 18.6 years should be determined as near as possible." We find no error in that instruction [24].

As this language demonstrates, the *Borax* decision applied modern technical knowledge and set forth a workable technique for precisely locating the boundary in question that still prevails in U.S. common law. As may be seen from the previous definition, the mean high water line represents an attempt to define the upper reach of the daily tide as the boundary between publicly owned submerged lands and uplands, subject to private ownership. Since the upper reach of the tide varies from day to day, the use of an average value attempts to split the difference for a compromise line. This results in a line that is exceeded by the high tide on approximately one-half of the tidal cycles.

Case law in the various coastal states has, in the main, followed English common law and its updated definition as put forth in the *Borax* decision. Sixteen states (Alabama, Alaska, California, Connecticut, Florida, Georgia, Maryland, Mississippi, New

Jersey, New York, North Carolina, Oregon, Rhode Island, South Carolina, Texas, and Washington) have followed this course [25]. Some states have codified their common law on this subject. For example, in Florida, the Coastal Mapping Act of 1974 [26] declares that "mean high water line along the shores of lands immediately bordering on navigable waters is recognized and declared to be the boundary between the foreshore owned by the State in its sovereign capacity and upland subject to private ownership." The statute also defines the mean high water line using the *Borax* definition.

8.4 Evidence and Procedures for Tidal Boundaries

The tide is the alternating rise and fall in sea level produced by the gravitational force of the moon and the sun. Other nonastronomical factors, such as meteorological forces, ocean floor topography, and coastline configuration, also play an important role in shaping the tide. To understand the mechanics of the tide-producing forces, one may visualize a moon orbiting around an Earth covered only with a layer of water. Due to the attractive force of the moon, there would be a bulge in the water on both sides of the Earth in line with the moon and a low water zone in between. The high water on the side of the Earth closest to the moon is caused by the moon's pull on the fluid water. On the side opposite the moon, the lesser gravitational force and the centrifugal force as the Earth and moon spin cause the high water. These waves of water follow the moon in its revolution about the Earth. Similarly, there are waves following the apparent rotation of the sun about the Earth.

The sea-level rises caused by many other relationships between the sun, Earth, and moon may also be considered as waves of specific periods. For example, the elliptical orbit of the moon about the Earth results in a constituent wave with a period of 27.55 days with highest water at the time of perigee (closest point to the Earth) and lowest water when the moon is at apogee (greatest distance from Earth). In addition to the constituents of different periods, another wavelike cycle that is associated with the regression of the moon's nodes with a period of 18.6 years affects the amplitudes of the various constituents. Regression of the moon's nodes refers to the movement of the intersection of the moon's orbital plane and the plane of the Earth's equator, which completes a 360° circuit in 18.6 years. The resultant tide is the composite, or algebraic sum, of all the previously mentioned constituent cycles as modified by the 18.6-year nodal cycle.

A *tidal datum* is a plane of reference for elevations that is based on average tidal height [27]. Considering the preceding discussion, it is obvious that to be statistically significant, a tidal datum should include all periodic variations in tidal height. Therefore, a tidal datum is usually considered to be the average of all occurrences of a certain tidal extreme for a period of 19 years (18.6 years, rounded to the nearest whole year to include a multiple of the annual cycle associated with the declination of the sun). Such a period is called a *tidal epoch*.

As examples of tidal datum planes, *mean high water* (MHW) is defined as the average height of all the high waters occurring over a period of 19 years. Likewise, *mean low water* (MLW) is defined as the average height of all the low tides over a 19-year tidal epoch. *Mean tide level* (MTL), or *half-tide level*, is the plane halfway between

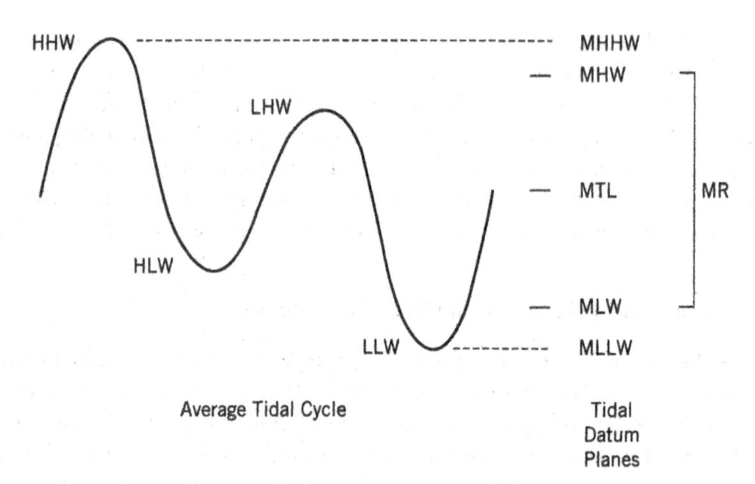

FIGURE 8-1. Common tidal datum planes.

mean high water and mean low water. This should not be confused with mean sea level, which is defined as the average level of the sea as measured from hourly heights over a tidal epoch. The relationship between mean sea level and mean tide level varies from location to location, depending upon the phase and amplitude relationships of the various tidal constituents at each location. These data are illustrated in Figure 8.1.

In addition to the above, there are two other datum planes illustrated in Figure 8.1 that may be of significance to water boundaries. *Mean higher high water* (MHHW) is the average of the higher of the high tides occurring each day. *Mean lower low water* (MLLW) is the average of the lower of the low tides occurring each day. Both averages are calculated over a tidal epoch.

From the foregoing, it may be seen that the primary determination of a tidal datum involves the relatively simple determination of the arithmetic mean, or average, of all the occurrences of a certain tidal extreme over a 19-year tidal epoch. In practice, this is usually accomplished by computing mean values of the various tidal extremes for each calendar month and then annual mean values by averaging the 12 monthly means for each extreme for each calendar year. Finally, the mean values for the tidal epoch used are determined by averaging the annual mean values for the 19 years in the epoch. It should be noted that the tidal extremes used in the calculation of tidal data are not necessarily the highest and lowest water that occur in a tidal cycle. In most areas, especially on the open coast, wind wave action causes a constant and frequent variation in water level. For consistency, since wind wave action is essentially unpredictable, tidal gauges measure the height of stilled water, which is approximately one halfway between the crest and trough of the wind waves. In areas with sizable wind waves, there can be a significant vertical distance between the crest (or trough) of the wind wave and the stilled water level of high and low water.

A tidal datum is a local phenomenon because of numerous local topographic forces shaping the tide. There are a number of different topographic factors that shape the tidal wave resulting in significant differences in elevation of a tidal datum from point to point in even the same general vicinity [28]. Therefore, a tidal datum should be determined in the immediate area of its intended use.

As previously mentioned, a tidal datum is defined as an average over a 19-year period known as a *tidal epoch*. For consistency, most published datum values are referred to a specific epoch called the *National Tidal Datum Epoch*. A specific 19-year period is used since apparent nonperiodic variation in mean sea level has been noted from one 19-year period to another. It is not known if such variation is truly nonperiodic or a part of some long-term oscillation. These apparently nonperiodic changes are possibly due to "glacial-eustacy, thermal volumetric changes, vertical land movements, and both climatological and oceanographic trends" [28].

During the last 100 years or so since systematic sea-level monitoring has been available, a worldwide trend of continual rise in sea level has been noted. To correct for this rise in the United States, a new epoch has historically been adopted every two or three decades when significant change has occurred. At such times, adjustments are made to all datum elevations. In effect, a quantum jump occurs in the elevations of all tidal datum planes for stations published by the National Oceanic and Atmospheric Administration (NOAA) at those times. An example of this occurred in 1981 when a change in epoch was adopted. Previously, the epoch of 1941–1957 was used. The National Tidal Datum Epoch adopted in 1981 is 1960–1978.

In recent years, sea level has been rising at an average rate of 0.0066 feet/year in the United States [29]. However, some sections of the coast have a much higher rate, especially along the western portion of the Gulf Coast.

Obviously, in areas with above-average rates of sea-level rise and even in areas with average rates of sea-level rise, a significant difference can at times exist between an elevation computed on the National Tidal Datum Epoch and a datum computed on the most recent 19 years. Therefore, it may sometimes be more desirable to recompute data on a more current epoch than use published data computed on the National Tidal Datum Epoch.

In many cases where a tidal datum is required in an area with an existing well-planned network of tidal datum points, a precise datum may be established by interpolation, as opposed to supplemental tidal observations at desired locations. On long stretches of regular and unbroken open ocean coastlines, tidal range variation generally is relatively linear. In such areas, linear interpolation is generally an acceptable method of determining a local tidal datum elevation. This may be accomplished by determining the elevation, in relation to a common datum, of the nearest existing tide station on either side of the desired point. Such an approach is not recommended in inlets or estuaries, where there are intervening inlets, where either existing tide station is near an inlet, where the coastline is irregular, or where the distance between the two stations is excessive.

Where interpolation between known tidal datum points will not yield satisfactory results, a new tidal datum must be determined. The primary determination of a tidal datum involves a relatively simple calculation of the arithmetic mean, or average, of all the occurrences of a tidal extreme over a 19-year tidal epoch. Most tidal datum elevations, however, are determined from observations of less than 19 years. Methods have been developed for correcting such short-term observations at the desired point and at a control station at which 19-year mean values are known. The average of the observed tidal extremes may then be reduced to a value equivalent to a 19-year mean by a correlation process using a ratio of tide ranges observed at the two stations. Methodology for this follows.

Standard Method The classic method [30] for accomplishing this process, also known as the range ratio method, is based on the following equations:

Notation
MHW 19-year mean high water
MTL 19-year mean tide level
*MLW*19-year mean low water
MR 19-year mean range
TL mean tide level for observation period
R mean range for observation period
s subscript used to denote subordinate station
c subscript used to denote control station

The equivalent 19-year mean range at the subordinate station may be calculated as

$$\frac{MR_s}{R_s} = \frac{MR}{R_c} \tag{8.1a}$$

Equation (8.1a) may be restated as

$$MR_s = \frac{MR_c R_s}{R_c} \tag{8.1b}$$

The subordinate 19-year mean tide level may be calculated as

$$TL_c - MTL_c = TL_s - MTL_s \tag{8.2a}$$

Equation (8.2a) may be restated as

$$MTL_s = TL_s + MTL_c - TL_c \tag{8.2b}$$

The 19-year mean high and mean low water levels may then be calculated by applying half of the mean range to the mean tide level as:

$$MHW_s = MTL_s + \frac{1}{2} MR_s \tag{8.3}$$

$$MLW_s = MTL_s - \frac{1}{2} MR_s \tag{8.4}$$

Regression analysis may be used as a substitute for the standard equations provided [31]. Such an approach provides a more statistically valid correlation as well as providing a means of evaluating the precision of the results.

Amplitude Ratio Method The standard method requires observation of both the high and low tidal extremes at the control and subordinate stations during the period of observation. In recent years, a demand has developed for determination of mean high water elevations in areas where only the upper portion of most tidal cycles are observable. These are areas such as marshes, mud flats, and tidal creeks with wide, flat intertidal zones. The recommended process for a short-term determination under such conditions is the amplitude ratio method [32]. This method was derived to mathematically duplicate the results of the standard method when only the top portion of the tidal cycle is available at the unknown or subordinate station. This method recognizes that for two waves of similar wavelength, the differences in height between the peaks of the cycles and the points at which the curves "fit" time periods of equal length are proportional to the ratio of the two amplitudes. It is noted that this method requires a computation on each observed cycle rather than using mean data for the entire observational period. An average of the results from the individual computations for the period of observation is usually used for the final value.

This method may be defined as follows (see Figure 8.2). Equation 8.5 calculates what the observed range at the subordinate station would have been if the entire cycle could have been observed:

$$R_s = \frac{R_c A_s}{A_c} \tag{8.5}$$

where A is the observed interval between peak high water and the elevation on the tide curve at the selected time interval.

Equation 8.6 calculates the equivalent 19-year mean range at the subordinate station:

$$MR_s = \frac{MR_c A_s}{A_c} \tag{8.6}$$

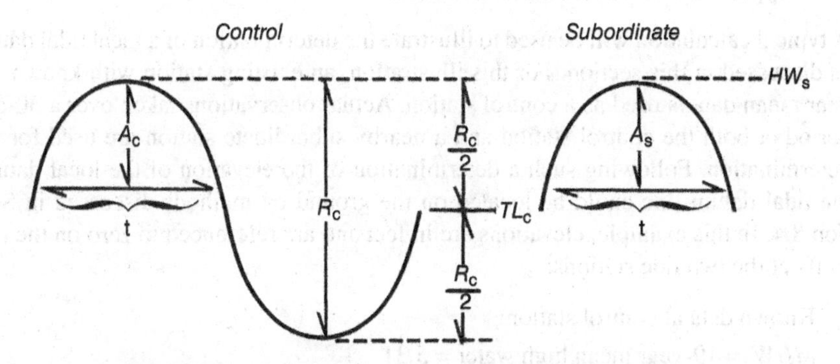

FIGURE 8-2. Computational procedure for amplitude ratio method.

Mean tide level and mean high and mean low water may then be calculated by Equations (8.2a), (8.3), and (8.4) of the standard method.

The distinction between the datum or elevation of a tidal boundary and the *tidal datum line* is an important concept in water boundaries. A tidal datum will remain relatively constant, for practical purposes, at a given location over the years. By contrast, a tidal datum line, which is the intersection of the datum with the rising land, may vary considerably as the land erodes or accretes under the same elevation of water. Therefore, the line may be ambulatory and its location should be related to a specific point in time.

There are two commonly used methods for locating the corresponding tidal datum line in the area around the datum determination. These are the observation of the leading edge of the water and leveling.

Often, the most practical of these, especially in areas without significant wave action, is the observation of the leading edge of the water. This method allows the water itself to define the line. As an illustration of this method, assume that the correct reading on a staff for a tidal datum has been previously determined. On a tide that is predicted to reach or exceed that value, the staff is observed. When the water level reaches the predetermined staff reading for the desired tidal datum, a signal is given. At the signal, personnel in the area around the staff place a series of stakes at frequent intervals along the incoming edge of the water. These stakes, defining the tidal datum line, may then be mapped by any of various horizontal surveying procedures. One obvious advantage of this method is that the surveyor actually sees the line on the ground and can identify all inflection points.

The leveling method consists of assuming that the local datum line is a topographic contour in the immediate area around the datum determination. This contour line is then located by leveling and is mapped by horizontal surveying procedures. Caution is necessary with this method because only points on the line are being located as contrasted with a continuous line as in the staking method. Therefore, significant breaks and inflections in the line may be overlooked unless care is taken.

Various ground-controlled photogrammetric means have also been used under certain conditions for locating tidal datum lines in areas with known tidal datum elevations.

8.5 Typical Tidal Datum Determination

A typical calculation will be used to illustrate the determination of a local tidal datum as discussed in this section. For this illustration, an existing station with known 19-year mean data is used as a control station. Actual observations taken over a 30-day period at both the control station and a nearby subordinate station are used for the determination. Following such a determination of the elevation of the local datum, the tidal datum line could be located on the ground by methods discussed in Section 8.4. In this example, elevations are in feet and are referenced to zero on the tide staffs at the two tide stations:

Known data at control station:

MHW_c = 19-year mean high water = 5.31

MLW_c = 19-year mean low water = 4.34

MR_c = 19-year mean range = 5.31 − 4.34 = 0.97
MTL_c = 19-year mean tide level = 1/2 (5.31 + 4.34) = 4.82

Observed data at control station:
HW_c = mean observed high water = 5.42
LW_c = mean observed low water = 4.32
R_c = mean observed range = 5.42 − 4.32 = 1.10
TL_c = mean observed tide level = 1/2 (5.42 + 4.32) = 4.87

Observed data at subordinate station:
HW_s = mean observed high water = 1.92
LW_s = mean observed low water = 0.63
R_s = mean observed range = 1.92 − 0.63 = 1.29
TL_s = mean observed tide level = 1/2 (1.92 + 0.63) = 1.28

Calculation of subordinate data [using Equations (8.1a), (8.2a) and (8.3)]:

$$MR_s = \text{19-year mean range} = \frac{MR_c R_s}{R_c} = \frac{0.97 \times 1.29}{1.10} = 1.14$$

$$MTL_s = \text{19-year mean tide level} = TL_s - (TL_c - MTL_c) = 1.28 - 4.87 + 4.82 = 1.23$$

$$MHW_s = MTL_s + 1/2\, MR_s = 1.23 + 1/2\,(1.14) = 1.80$$

8.6 Tidal Waters under Law Departing from Civil and Anglo/American Common Law

Six northeastern states (Delaware, Maine, Massachusetts, New Hampshire, Pennsylvania, and Virginia) recognize the mean *low* water line as the sovereign/upland boundary. For many of these low water states, that boundary is based on an early Massachusetts colonial ordinance of 1649 that provided as follows:

> it is declared that in all creeks, coves and other places, about and upon salt water where the Sea ebbs and flows, the Proprietor of the land adjoining shall have properties to the low water mark where the Sea doth not ebb above a hundred rods, and not more wheresoever it ebbs farther [33].

Therefore, based on recent interpretations of this law, the boundary between the uplands, subject to private ownership, and the public trust submerged lands is the mean low water line or a line 100 rods (1,650 feet) seaward of the mean high water line when the mean low water line lies further seaward. For such jurisdictions, procedures would be used which are similar to those covered in Section 8.5 for areas subject to Anglo/American common law. A local tidal datum for both mean high and mean low water would first be determined. Then it would be necessary to locate both the mean high and mean low water (and possibly the line lying 100 rods seaward of the mean high water line) by the methods described in Section 8.5.

Another state with a unique boundary for public trust tidally affected waters is Hawaii. The Supreme Court of that state found that a boundary described as the intersection of the shore with the horizontal plane of mean high water was erroneous and ruled that the boundary "is along the upper reaches of the wash of the waves, usually evidenced by the edge of vegetation or by the line of debris left by the wash of waves." In that case, the court decided that the mean high water line was not consistent with Hawaii's ancient traditions [34].

Assuming that the "wash of waves" referred to is the "the wash of waves during ordinary high tide" as contended by the State of Hawaii in this case, the difference between such a line and the mean high water line may be accounted for by two factors: the amplitude of waves and the run-up of the waves. Thus, the line defined in this case may be considered as being at the elevation of mean high water plus one-half of the average amplitude of the waves plus the average run-up caused by such waves at the coastline being mapped.

8.7 Boundaries of Nontidal Waters: Definitions

With the lack of a predictable rising and falling of the water level found in tidal waters, obviously different definitions must apply for sovereign/upland boundaries in waters not affected by tides. To distinguish from the mathematically derived boundary of tidal waters (mean high water), the boundary of nontidal waters is generally called the *ordinary high water line* or *ordinary high water mark.* Some states do differ. For example, Louisiana considers the ordinary low water to be the boundary for rivers and streams while adhering to the high water mark for nontidal lakes [17].

English common law offers little guidance regarding nontidal water boundaries. As has been previously discussed, during the period when tidal boundaries were being defined in England, only tidal waters were considered public domain. As has also been indicated previously, American case law recognized the topographic differences between the British Isles and the American continent early on and declared the Great Lakes and other navigable nontidal waters of this country to be sovereign. This presumably created somewhat of a dilemma for water boundary determination due to the lack of repeating tidal cycles, which was the accepted basis for tidal water boundaries. To resolve this, American case law adopted the physical fact test to determine the equivalent of mean high water in nontidal waters.

The leading definition [35] in federal case law gives the following instructions for determining the boundary of such waters:

> This line is to be found by examining the bed and banks and ascertaining where the presence and action of waters are so common and usual and so long continued in all ordinary years, as to mark upon the soil of the bed a character distinct from that of the banks, in respect to vegetation, as well as in respect to the nature of the soil itself [36].

Most state case law conforms substantially with federal law on this subject. A Florida case illustrates this:

> High-water mark, as a line between a riparian owner and the public, is to be determined by examining the bed and banks, and ascertaining where the presence and action of the water as so common and usual, and so long continued in all ordinary years as to mark

> upon the soil of the bed a character distinct from that of the banks, in respect to vegetation as well as respects the nature of the soil itself. High-water mark means what its language imports—a water mark [37].

Traditionally, in nontidal waters, the courts have allowed the use of botanical and geological evidence, as evidenced by the above decisions, and disallowed the use of mathematical averaging of water levels. This is typified by the court's decision in *Kelley's Creek and Northwestern Railroad v. United States*: "The high water mark is not to be determined by arithmetical calculation; it is a physical fact to be determined by inspection of the river bank" [38].

Recently, however, there has been an apparent trend to place more reliability on water level records, possibly due to the growing need for the precision, repeatability, and lack of ambiguity, which results from a mathematical solution. Typical of this are two Florida cases: *U.S. v. Parker* [39] and *U.S. v. Joder Cameron* [40]. The court in the *Cameron* case found as follows:

> There is no logical reason why a fourth approach to determining the line or ordinary high water may not consist of comparing reliable water stage and elevation data. Indeed, for a body of water whose levels fluctuate considerably with changes in climate, accurate water stage and elevation data may provide the most suitable method for determining the ordinary high-water mark [40].

8.8 Types of Evidence

Since significantly different definitions of sovereign/upland boundaries exist in nontidal waters from those used in tidal waters, significantly different techniques are used for their location. The emphasis of the location of the ordinary high water line in nontidal waters has been, traditionally, on the use of physical features rather than on mathematical averaging of water-level data.

When considering nontidal water boundaries, it is helpful to understand the general characteristics of a shoreline. The margin of most water bodies tends to form a similar profile consisting of a relatively flat floodplain between the high ground and the high water mark, a more steeply sloped foreshore between the high and low water marks, and a less steep near-shore slope lying below the low water mark. Obviously, the widths and slopes of the floodplain, foreshore, and near-shore slope will vary considerably with different water bodies. In addition, in riverine systems, there may be considerable variation due to the meandering of the river, including the existence of a natural levee at the waterward edge of the floodplain. However, the general characteristics should be similar for most water bodies.

Based on various case law, the floodplain is generally not considered to be within the bed of the water body. Therefore, the objective of an ordinary high water determination is to locate the dividing line between the floodplain and the foreshore. In addressing this problem, the Bureau of Land Management states:

> The most reliable indicator of mean (ordinary) high water elevation is the evidence made by the water's action at its various stages, which are generally well marked in the soil. In timbered localities, a very certain indication of the locus of the various important water levels is found in the belting of the native forest species [41].

The following sections describe the various types of evidence, including that just prescribed, which might be used for such a boundary determination [42].

8.9 Geomorphological Features

Some of the most graphic indicators of the location of the ordinary high water line are various geomorphological features such as natural levees, scarps, and beach ridges (berms). *Natural levees* are low ridges that parallel a river course. They are the highest near the river and slope gradually away from it into a low-lying area or swale. In larger rivers, they may be several feet high and a mile or more in width. In other rivers, however, they may be only a few inches high and a few feet wide. Levees owe their greater height near the stream channel to the cumulative effect of deposition associated with a sudden loss in transporting power when a river overflows its banks [43]. Therefore, the ordinary high water level is usually on the steep or riverside and below the crest of such features. Erosional features, such as escarpments, may often be found upon the riverside of levees.

In Texas, the use of natural levees has been refined to very specific techniques for determining boundaries of streams [44]. The boundaries resulting from such techniques have been endorsed by both federal and state courts in that state [45]. The first step in the Texas process is to select the "lowest qualified bank" in the area of interest. Such a bank should be an "accretion bank" (levee) as opposed to an erosion bank (escarpment), should have a well-defined top and bottom, should have a depression or swale on its landward side, and should be the lowest of such banks in the area. The second step is to determine the "basic point" which is the elevation halfway between the top and bottom of the lowest qualified bank. The third step is to measure the difference of elevation between the basic point and the current surface of the water at the lowest qualified bank. The boundary may then be determined by applying the same difference of elevation to the water level up and down the stream. By this method, the boundary at any point is the same difference above or below the water surface at that point as at the lowest qualified bank.

An *escarpment* or scarp is a miniature cliff cut into the shore by wave action. A *beach ridge* is a depositional feature or the wave-cut slope. Beach ridges usually have a convex shape and are systematic with the apex offset landward [46]. Ridges often form at various levels in a lake, but only the highest ridge is significant in boundary determination.

An interesting discussion on the use of scarps for the location of the ordinary high water mark is found in the *Manual of Instructions*:

> Mean (ordinary) high water elevation is found at the margin of the area occupied by the water for the greater portion of each average year. At this level a definite escarpment in the soil is generally traceable, at the top of which is the true position for the meander line. A pronounced escarpment, the result of the action of storm and flood waters is often found above the principal water level, and separated from the latter by the storm or flood beach. Another, less evident, escarpment is often found at the average low water level, especially of lakes, the lower escarpment being separated from the principal escarpment by the normal beach or shore. While these principles properly belong in the realm of geology, they should not be overlooked in the survey of a meander line [47].

As mentioned in the *Manual of Instructions*, scarps are also found, especially in river systems, at the extremes of the floodplain. In rivers, this may be some distance from the ordinary high-water mark. The more significant scarp would be found in the form of undercut slopes and cut banks near the meander channel.

As may be seen, geomorphological features are useful in locating the elevation of ordinary high water. They should be used with caution, however, since they can take a relatively long time to develop. If a water body is in the process of reliction or rising in elevation, there could be several sets of these features. At such time, other types of evidence are useful in resolving the ambiguity.

8.10 Changes in the Composition of the Soil

For lakes, another very repeatable indicator of the location of the ordinary high water line is a change in the composition of the soil. This may be evidence such as a change in the organic content of the soil or the landward termination of stratified beach deposits. Stratified beach deposits occur more graphically on lake shorelines subject to beach erosion, often at the base of beach scarps. These deposits are the result of wave erosion that tends to transport the resulting detritus away from the uplands. Generally, this transportation results in a systematic decrease in average grain size and a tendency for the particles to become more equal in size [48]. Therefore, a graphic difference can often be seen between the upland or parent material and the eroded material.

To make a determination of the elevation of the changes in character of the soil, it has been found that digging a narrow trench at approximately right angles to the shoreline allows a good cross-sectional view of the sedimentary and erosion features. A topographic profile of the shore along the trench should be made. Soil samples should then be taken along the profile at a few centimeters below the surface. Even when a change in soil character is not obvious, it may be advisable to take samples since laboratory analysis will sometimes indicate differences that are not readily visible to the eye.

The primary information desired from these samples is a sediment particle size analysis. There are various means of making this determination, including sieving, observation of settling velocity, and microscopic examination. However, sieving is probably the most practical for sediment of the size normally found along lakes and rivers. This is a method of passing the dried sediment through a series of standard size screens. From statistical analysis of the results of this process, two factors may be determined: the average grain diameter of the sediment and the degree of sorting (the extent to which the grain sizes spread on either side of the average diameter). From the previous discussion, the features for which one would look at the ordinary high-water mark would be a sudden improvement in the sorting together with the occurrence of the largest average grain size.

It should be noted that since rivers often form a wide floodplain due to seasonal flooding and the meandering process, this evidence probably has limited application for such water bodies.

8.11 Botanical Evidence

Another type of evidence cited in various case law is the lower limit of "terrestrial plant life" [49]. The *Manual of Instructions* gives the following directions regarding the use of such evidence:

> Where native forest trees are found in abundance bordering bodies of water, those trees showing evidence of having grown under favorable site conditions will be found belted along contour lines. Certain mixed varieties common to a particular region are found only on the lands seldom, if ever, overflowed. Another group is found on the lands that are inundated only a small portion of the growing season each year, and indicate the area that should be included in the classification of the uplands. Other varieties of the native forest trees are found only within the zone of swamp and overflowed lands. All timber growth normally ceases at the margin of permanent water [50].

The rationale for the use of such evidence is quite evident. It has been observed and well established that many forms of plant life are distinctly related to the amount and duration of water to which they are subjected. Some of these plants have distinct preferences for water over and around themselves, or over their roots and lower parts. There are others that do not tolerate water over the soil except for short periods of time [51]. Therefore, it seems reasonable that with knowledge of the water tolerances of the plant life for a particular geographical area, patterns of such growth may offer significant assistance in the location of the ordinary high water line.

In addition to upland vegetation, species more tolerant of the presence of water may also be helpful, especially in water bodies with gently sloping shorelines. In such areas, distinct zones of vegetation may be seen, based on the water tolerance characteristics of the various species present. Correlated with other evidence, these zones may help define the location of the boundary.

Another type of botanical evidence is the lack of vegetative growth that may often be found in a narrow zone at and slightly below the ordinary high-water line. This situation is often evident on exposed coasts in lakes where there is abundant shore vegetation above and aquatic vegetation below the line. Wave action in the erosion zone tends to prevent the growth of either the upland or aquatic species. Often, one sees an unvegetated sand bottom at such places.

Unfortunately, botanical evidence can often lead to ambiguity and misinterpretation, primarily because many species of vegetation can be very adaptive and unpredictable. The following quote may add insight to this problem:

> It is important to note that it would be impractical and unrealistic to strictly apply the ordinary high water definition where the situation calls for some departure. . . . Certainly the presence or absence of vegetation is not always conclusive. The Iowa Supreme Court stated in State v. Sorenson, for example, that large trees may sometimes continue to grow although covered with water at their bottoms for some period . . . This and other cases imply the converse as well. That is, even where aquatic vegetation is found some distance inland, in marshland or other poorly drained areas, for example, the finding of a realistic ordinary high water line should not be upset [52].

Therefore, botanical evidence should be used cautiously and with benefit of collaborative evidence. Generalized patterns of vegetation should be used rather than reliance on a specific species.

8.12 Hydrological Evidence

Water-level records are also a potentially valuable class of evidence for determination of nontidal water boundaries as is the case for tidal water boundaries. Resolution of the best method for interpretation of such data, however, is still an unsettled question. The *Manual of Instructions* provides only very cryptic instruction on this topic: "Practically all inland water bodies pass through an annual cycle of changes, between the extremes of which will be found mean (ordinary) high water" [53].

Likewise, some case law hints at possible approaches such as the following:

> This word (ordinary high water) . . . does not mean the abnormally low level of a lake during one of a series of excessively dry years, or the abnormally high level of a lake during an exceptional wet year or series of wet years, but the average or mean level obtaining [*sic*] under fairly normal or average weather conditions, allowing the proper range between high and low water mark in average years [37].

One approach for the utilization of hydrological data [54] is to use an average high water level, similar to that used for tidal waters. Instead of averaging the peaks of daily tidal cycles, however, this method averages the peaks of short-term variations of water level caused by rainfall. For an averaging period, this method uses the high waters from the entire interval over which observations are available. Instead of a simple average, however, a weighted mean is determined by regression analysis to correct for any long-term changes in water level. This is accomplished by determining an average high water for each of the years in the observational period and then calculating a least-squares line through the annual high water points. The point on the least-squares line that corresponds with the date of determination may then be used as the ordinary high water for that date. Tests of this method have indicated good agreement with traditional physical feature evidence. In addition, this method has the advantage of yielding the same elevation as that determined by tidal analysis in transitional areas such as at the upper limit of tidal effect in rivers. Presumedly, the same approach could be used to determine the ordinary low water in jurisdictions using that as a boundary for nontidal waters.

8.13 Typical Determination of Ordinary High Water Mark

A typical determination of an ordinary high water mark of a navigable river will be used to illustrate the application of the above principles. In the selected river, an abundance of both geomorphological and botanical evidence may be seen. Figure 8.3 provides a photograph (Figure 8.3a) as well as a generalized cross-section of the riverbank (Figure 8.3b). At the time of the field study, the river was at a relatively low stage. As may be seen from Figure 8.3b, three escarpments were observed.

(a)

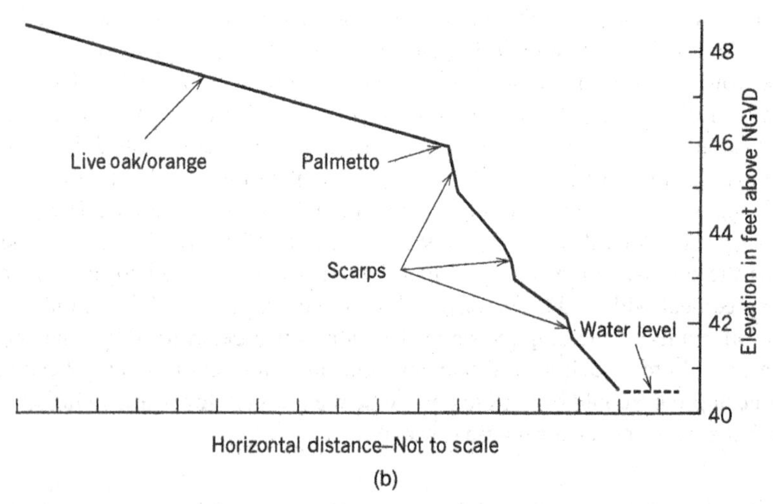

(b)

FIGURE 8-3. **(a)** Photograph and (b) topographic profile of case study river.
Source: Donald A. Wilson

The lowest was slight and located just above the water level at the time of the study. The middle escarpment (which is being pointed to in Figure 8.3a) was somewhat more prominent and contained numerous matted roots and marked the lower limit of transitional vegetation. The highest escarpment was covered, above and slightly below, by various upland vegetation including saw palmetto. Upland trees such as live oak and citrus grow down to slightly above (vertically) the upper escarpment. During times of seasonal flooding, the water level has gone well above all of the escarpments. Based on these indicators, it appears that the ordinary high water line should be located along the middle escarpment, the base of which is at an elevation of approximately 43.0 feet.

In addition to studying the geomorphological and botanical evidence, an analysis was performed of existing hydrological evidence. Approximately 26 years of water-level gauging by the U.S. Geological Survey was available for a site near the study location. Using the previously described method for hydrological data, an average of all the high waters occurring in each year was computed, and a least-squares trend line was determined for the annual mean high waters. The result of this is illustrated in Figure 8.4. Using the hydrological data in this manner resulted in a computed average high water of 42.8 feet, which provides excellent correlation with the middle escarpment indicated by the geomorphological and botanical evidence.

8.14 Boundaries of Nonpublic Trust Waters

Boundaries in Streams Perhaps the most elementary case of a nonpublic trust water boundary is that involving a nonnavigable stream as the boundary between two parcels of land. Where the deeds call "to the stream," the center of the stream would be the boundary. There is general agreement on the part of most courts on this issue.

However, there is a complicating issue relating to stream boundaries in which there appears to be a divergence among court opinions. This is the question of what constitutes the "center" of the stream. One school of thought considers the boundary

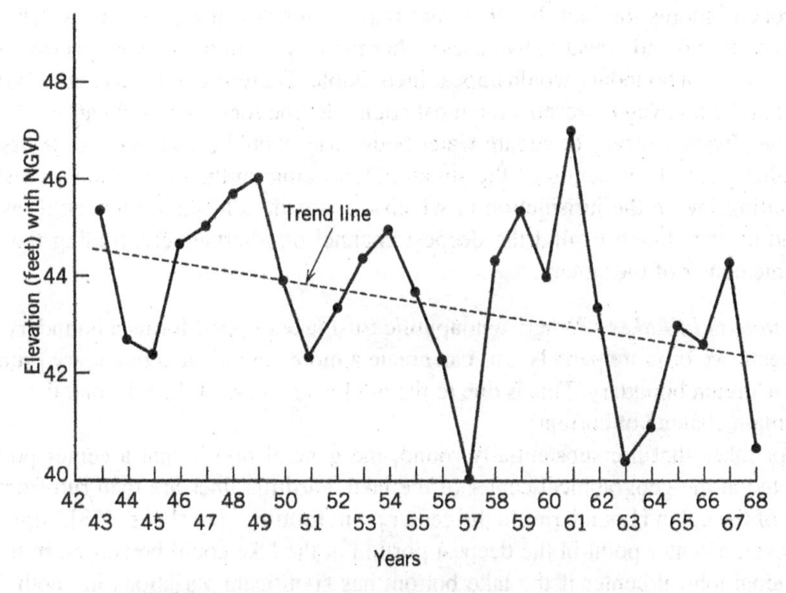

FIGURE 8-4. Plot of hydrological data for case study river.

to be the "thread of the stream" which is defined as the line lying equidistant between the banks. For example, one source states:

> The term 'thread of the stream' means the geographic center of the stream at ordinary or medium stage of the water, disregarding slight and exceptional irregularities in the banks. It is fixed without regard to the main channel of the stream . . . If the stream is made a boundary in a private conveyance, then the thread of the stream will be the stream boundary [55].

A number of court cases in several states support this position. An example quotes *Black's Law Dictionary* and defines the thread as "*a middle line; a line running through the middle of a stream*" [56].

The second approach to the center of the stream question holds the boundary to be the *thalweg*, or deepest part, of the channel. As with the former approach, a number of court cases also appear to support this opinion. An example illustrating this position states:

> Upon principle, therefore, it would appear that the thread of a nonnavigable river is the line of water at its lowest stage. The thread or center of a channel, as the term is above employed, must be the line which would give to the landowners on either side access to the water, whatever its stage might be and particularly at its lowest stage [57].

Upon analysis, it may be seen that the former approach has the advantage of equally dividing the submerged lands beneath the stream while the latter allows equal access to the water, even if the stream drops to an extremely low stage. Some authorities suggest that while the deepest channel is the correct boundary for navigable, nonsovereign streams, the geographical center of the steam is the correct boundary for nonnavigable streams. On the surface, this may appear logical. However, this approach ignores the fact that there are reasons for desiring access to water (such as recreational and consumptive uses) other than navigation. Denying access to one party to such a boundary would appear inequitable. Therefore, it is this author's opinion that the thalweg represents the most equitable line for either application.

The physical survey of stream water boundaries would therefore involve first the establishment of the center of the stream. Depending on the circumstances and the prevailing law in the jurisdiction in which the stream is located, this could involve sounding the stream to find the deepest channel or, alternatively, finding the geographic center of the stream.

Boundaries in Lakes Where a nonpublic trust lake or pond forms a boundary line between two or more parcels, this can create a more complicated boundary problem than a stream boundary. This is due to the nonlinear shape of the lake and the lack of a main channel or current.

For lakes that are substantially round, the general rule is that a center point is selected at the geographical center of the lake. Partition lines are then run from the ends of the upland boundaries to the center point forming "pie slices" [58]. Arguably, however, a center point in the deepest portion in the lake could be chosen in lieu of the geographical center if the lake bottom has significant variations in depth. This would give equal access to the water if the lake drops to a low stage.

For long or irregular lakes, the stream formula is applied to the body of the lake and the round-lake formula at the ends. This involves running a center line the length of the lake to points that are the center of arcs at the ends of the lake. At the ends, lateral lines would converge at the endpoints. Along the length of the lake, lateral lines would run to the center line at right angles to that line [59]. As with streams, the center could be a median line or could be considered to be along the deepest portion of the lake. Another case concisely states this approach as follows:

> Where a lake is very long in comparison with its width, the method applied to rivers and streams would probably be the most suitable for adjusting riparian rights in the lake bottom along its sides and the use of converging lines would only be required at its two ends [60].

8.15 Methods of Riparian Allocation

Land surveyors—as the name implies—are most often concerned with the size, shape, and topography of the Earth, but surveyors are also tasked with locating and allocating riparian rights among contiguous property owners. Many surveyors produce maps of riparian rights, but few professionals understand the methodologies of the five primary methods of riparian allocations and this fundamental lack of understanding has led to neighbors spending substantial sums of money to litigate their water boundaries.

Property owners pay a substantial sum of money for waterfront property—and for good reason—being able to fish, boat, float, and just enjoy the water has been described by many as something that soothes the soul. But when neighbors engage in a protracted dispute concerning the size, shape, and sinuosity of their waterfront boundaries, Shangri-La becomes a nightmare.

The two most common disputes involve either two neighbors fighting over the lateral boundaries of their common property line or an individual fighting with the city/county over the extent of the owner's "private" area where docks, wharfs, and other structures can be built. No matter the parties, the same rules apply when surveyors perform a field survey to locate riparian lines of ownership.

The first question to ask is: "Did the owner of the parcel already subdivide his or her riparian rights?" In some cases, the developers of a waterfront subdivision subdivide the riparian rights of each parcel before selling the lots in order to ensure an equitably distribute of riparian rights among all the owners. But the developers rarely have the foresight to do such a thing. Therefore, the common law riparian rights along a waterway must be equitably distributed by land surveyors in order to find the riparian boundaries for a client.

There are six common law methods of allocating riparian rights among waterfront owners:

1. the round lake method
2. the long lake method
3. the proportionate medial line method
4. the colonial method
5. the property line extension method
6. the proportionate acreage method

In this chapter, we will discuss four of them. The purpose of any riparian survey is to equitably subdivide the riparian rights within a lake, river, or ocean, among the contiguous property owners [61]. "Equity" is most often defined as "equitable or ratable allocation of the waterfront area." This translates to one's ability to enjoy their specific rights, including the rights of wharf-out, ingress/egress, build a dock, and most often, "direct access from their entire shoreline frontage to their equitable share of the line of navigability" [62]. These rights will be discussed in subsequent chapters.

The Round Lake Method The round lake method divides the lake, or portion of a lake, into pieces of a pie. The pie/round method should be used in two circumstances: (1) when the surveyor encounters a (substantially) round lake; and (2) at the ends of a long lake. Both circumstances are graphically described in Figures 8.5a and 8.5b.

The Long Lake Method (Proportionate Thread of the Stream Method) The long lake method creates equitably-sized polygons between the center of the waterbody and the boundaries of the waterfront properties. The size of the polygon is dependent on the size of each property's waterfrontage. The long lake method is most applicable for elongated bodies of water (lakes than are longer than they are wide). Examples are shown in Figure 8.6. The landward boundaries of each riparian allocation are set by the corners of each property, but the surveyor may either use the medial line or the thalweg of the waterbody as the dividing line that runs through the center of the lake based on which of the two creates the more equitable division of riparian rights within the lake, river, etc. As the name implies, the "long lake" method applies only to lakes, ponds, and similarly shaped waterbodies; it is not used in rivers and oceans.

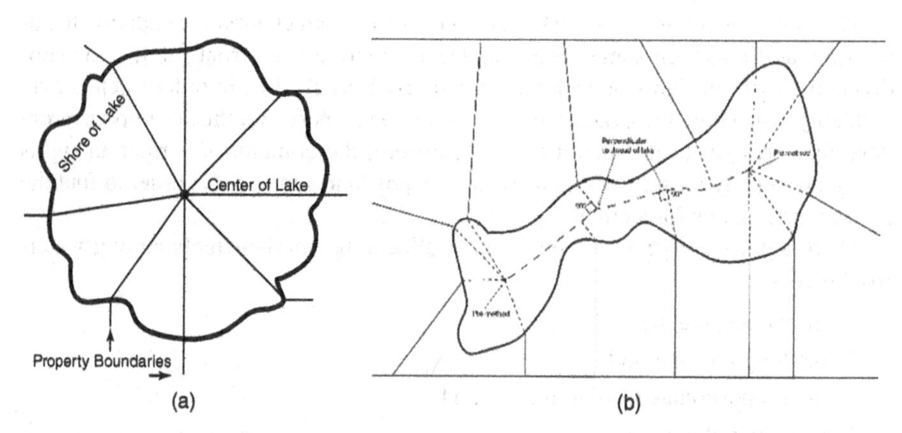

FIGURE 8-5. Example sketches of (a) under the round lake; and (b) under the long lake.

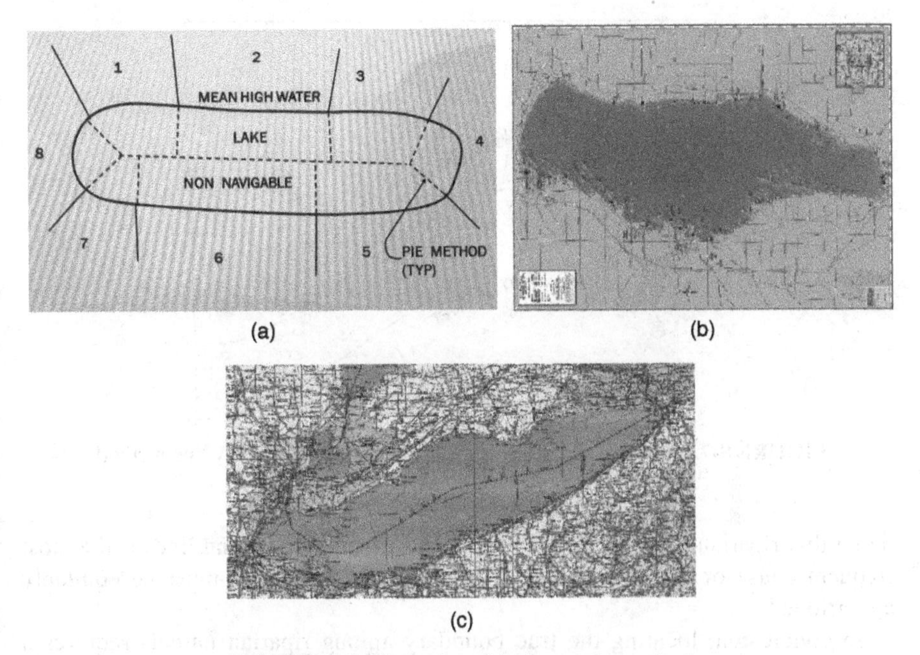

FIGURE 8-6. Examples of the long lake method.

The Proportionate Medial Line Method The proportionate medial line method is similar to the long lake method. But instead of turning right angles from the medial line to property corners, the proportionate medial line method creates a ratio between the medial line length and the waterfront frontage. For example, in Figure 8.6a, the medial line of the lake runs 4859 feet, the total length of the lake's left side is 8574', (adding the water frontage of all property owners on the left side of the lake), therefore each parcel will receive 0.5667 feet of medial-line for every 1 foot of water frontage (4859/8574 = 0.5667 or 56.67%). This is a very fair and equitable method because the size of the riparian parcel each property owner receives is not based on the geometry of their property lines, the geometry of the lake, or any other factor except the proportion of medial line vs. property line.

The Prolongation of Property Lines Method The prolongation of property line method is by far the easiest method of apportioning riparian rights on a lake, river, ocean, or any other body of water. The surveyor simply measures the property lines that extend to the water and then projects the property lines at the same angle into the water body.

This is the only method discussed in this section that takes no account of equity; only the angle of the property lines (Figure 8.7). Therefore, this method has been summarily rejected by courts throughout the United States. The Florida Department of Environmental Protection [63] stated that: "The direction of upland boundaries is largely ignored when apportioning riparian rights. The public's mistaken

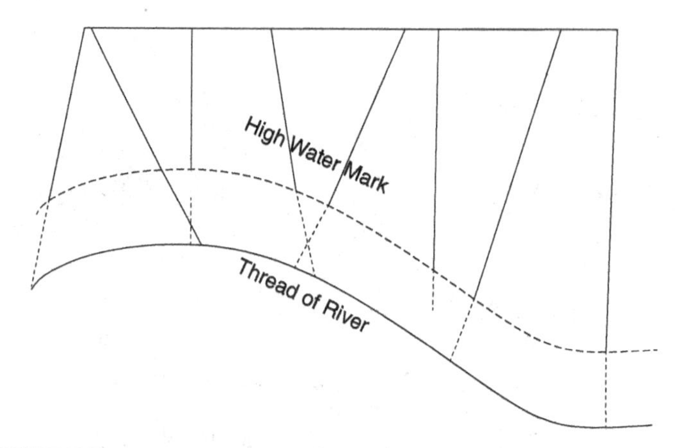

FIGURE 8-7. An example sketch of the prolongation of property line method.

belief that riparian lines are on the extension of their side upland lines is the most frequent cause of riparian disputes. Instead, the water body must be equitably apportioned . . ."

In conclusion, locating the true boundary among riparian parcels requires a specialized skill set and knowledge base of which many surveyors have little experience. Research, patience, and a deep search of state case law are required before conducting any riparian survey. Even then, it may be difficult to choose the "best" method for locating the boundary lines between docks, piers, and other structures.

8.16 Changes in Water Boundaries

Shorelines are dynamic in nature, and their locations may change considerably with time. It is helpful to consider such changes as being in two categories. These categories consist of changes associated with the permanent lowering (reliction) or rise of water level and those associated with changes in landform. The latter category would include erosion (gradual wearing away of the land due to water and wind action) and accretion (gradual building up of land). In addition, the latter category would include more sudden or avulsive changes in landform such as those caused by a specific storm or by dredging or filling.

In general, water boundaries change with shoreline changes. However, sudden changes due to avulsive activities generally do not result in a boundary shift. Also, in most states, a riparian owner may not benefit from shifts due to artificially caused accretion or avulsion resulting from his or her actions. Therefore, a knowledge of the cause of the shoreline change is desirable for a correct interpretation of the boundary location. Whenever the shoreline change has been the result of avulsion or human activity, it may be necessary to use the location of the shoreline at the last natural (or pre-alteration) position.

REFERENCES

1. T. C. Sanders, *The Institutes of Justinian* (Chicago: Callaghan & Co., 1876), p. 158.
2. T. Digges, "Proofs of the Queen's Interest in Land Left by the Sea and the Salt Shores Thereof. As reported in Fraser, 'Title to the soil under public waters," *Minnesota Law Review*, 1918.
3. *Martin v. Waddell*, 41 U.S. (16) 367 (1842).
4. *Pollards Lessee v. Hagen*, 3 How. 212, 44 U.S. (1845).
5. M. Haler, 1666. As printed in F. Hargrave, *The First Part of the Institutes of the Laws of England* (printed for J. Moore, Dublin, 1791).
6. *Barney v. Keokuk*, 94 U.S. 324 (1876).
7. *Odom v. Deltona*, 341 So. 2d 977, (1976).
8. *The Daniel Ball*, 77 U.S. 557 (1870).
9. *U.S. v. 2,899 Acres of Land*, 269 F Supp. 903 (MD. 1967).
10. *Buckie v. Cone*, 6 So. 160 (1889).
11. Public laws of 1649 of the Mass. Bay Colony as printed in M. M. Frankel, *Law of Seashore Waters and Water Courses Maine and Mass.* (Forge Valley, MA: Murray Printing, 1969), p. 3.
12. F. Maloney and R. C. Ausness, "The Use and Legal Significance of the Mean High/Water Line in Coastal Boundary Mapping," *North Carolina Law Review* (December 1974), p. 126.
13. *Cinque Bambini Partnership v. State of Miss.*, 491 So. 2d 508 (1986).
14. *Phillips Pet. v. Mississippi*, 484 U.S. 469 (1988).
15. Sanders, *The Institutes of Justinian*, p. 159.
16. *Borax Consolidated v. City of Los Angeles*, 296 U.S. 10 (1935).
17. LA Civil Code Annotated, Article 451. In *West's Louisiana Statutes* (St. Paul, MN: West Publishing Co. 1980).
18. *New Orleans Land Co. v. Board of Levee Comm.*, 132 So. 121 (1930).
19. *Las Siete Partidas* [The Seven-Part Code], see en.wikipedia.org/wiki/Siete_Partidas
20. *Luttes v. State*, 324 S.W. 2d 167 (Tex. 1958).
21. *Bright et al. v. Mauro*, Case No. 93-05265, Travis Co. Dist. Ct. 200 Jud. Dist. Texas (1996).
22. 78 Am. Jur. 2d, Waters Section 380. M. Hale, *De Jure Maris*, 1787; reprinted in S.A. Moore, *A History of the Foreshore and the Law Relating Thereto* (Rochester, NY: Lawyers Coop. Pub. Co., 1975).
23. *Attorney General v. Chambers*, 17 Eng. Rule. Cas. 555 (1854).
24. *Application of Ashford*, 440 P. 2d 76 (Hawaii 1968).
25. Maloney and Ausness; and G. M. Cole, "Tidal Boundary Surveying," *Proceedings of the American Congress on Surveying and Mapping*, 1977, pp. 235–249.
26. Florida Coastal Mapping Act 1974, in *Florida Statutes*, Chapter 77, Part II, pp. 1444–1446.
27. George M. Cole, "Tidal Boundary Surveying," pp. 235–249.
28. S. D. Hicks, et al., *Sea Level Variations for the U.S., 1855-1980* (Washington, DC: U.S. Department of Commerce, NOAA, NOS, 1983), p. 2.

29. S. D. Hicks and L. E. Hickman Jr., "United States Sea Level Variations Through 1986," *Shore and Beach*, American Shore and Beach Preservation quarterly journal (July 1988).

30. H. A. Marmer, "Tidal Datum Planes," Special Publication No. 135 (Washington, DC: U.S. Coast and Geodetic Survey, 1951).

31. G. M. Cole, F. M. Speed, and J. C. Fugate, "The Use of Regression Analysis for Estimating Tidal Data," in *Proceedings of DPH Conference* (Washington, DC: Marine Technology Society, 1989).

32. G. M. Cole, "Proposed New Method for Determining Mean High Water Elevations in Intertidal Zones," Technical Papers, in *Proceedings of the American Congress on Surveying and Mapping*, 1981, pp. 356–366.

33. Public laws of 1649 of the Mass. Bay Colony, p. 2.

34. *Application of Ashford*, p. 2.

35. F. E. Maloney, "The Ordinary High Water Mark: Attempts at Settling an Unsettled Boundary Line," *University of Wyoming Land and Water Law Review*, vol. 13, no. 2 (1978), pp. 289–298.

36. *Howard v. Ingersoll*, 54 U.S. 381, 427 (1881).

37. *Tilden v. Smith*, 113 So. 708 (Fla. 1927).

38. *Kelley's Creek and Northwestern Railroad v. United States*, 100 Ct. Cl. 396 (W. Va. 1943).

39. *U.S. v. Parker*, No. 75-34, N.D. Fla. (1976).

40. *U.S. v. Joder Cameron*, 446 Fed. Supp. 1099 (1979).

41. U.S.D.I., Bureau of Land Management, *Manual of Instructions for the Survey of the Public Lands of the United States* (Washington, DC: U.S. Government Printing Office, 1973), pp. 94–95.

42. G. M. Cole, "Non-Tidal Water Boundaries," *Florida Bar Real Property, Probate and Trust Law Newsletter* (January 1979).

43. W. D. Thornbury, *Principles of Geomorphology* (New York: John Wiley & Sons, 1954).

44. A. A. Stiles, "The Gradient Boundary," *Texas Law Review*, vol. 30, no. 3 (1952), pp. 305–322.

45. *Oklahoma v. Texas*, 260 U.S. 606, 261 U.S. 340 (1923); *Motl v. Boyd*, 286 W.W. 458, 116 Tex. 82, 286 S.W. 458 (1926); *Heard v. State*, 204 S.W. 2d 344 (Tex. 1947).

46. D. D. Knockenmus, "Shoreline Features as Indicators of High Lake Levels," U.S. Geological Professional Paper No. 575-C, 1967.

47. U.S.D.I., Bureau of Land Management, *Manual of Instructions*, p. 95.

48. W. C. Krumbein and L. L. Sloss, *Stratigraphy and Sedimentation* (San Francisco: Freeman & Co., 1963).

49. *Goose Creek Hunting Club v. U.S.*, 518 F. 2d 579 (La. 1975).

50. U.S.D.I., Bureau of Land Management, *Manual of Instructions*, pp. 95–96.

51. J. H. Davis, "Establishment of Mean High Water Lines in Florida's Lakes," Water Resources Research Center, University of Florida Publication No. 24 (Gainesville, FL: University of Florida, 1972).

52. *Tilden v. Smith*, p. 475.

53. U.S.D.I., Bureau of Land Management, *Manual of Instructions*, p. 94.

54. G. M. Cole, "Use of Hydrology for Determining Ordinary High Water in Non-Tidal Waters," Technical Papers, in *Proceedings of the American Congress on Surveying and Mapping* (Spring 1988).

55. E. S. Bade, "Title, Points and Lines in Lakes and Streams," *Minnesota Law Review*, vol. 24 (1940), pp. 305–307.

56. *Carlton v. Cent. & So. Fla. Flood Control Dist.*, 118 So. 2d 656 (Fla. 1996).

57. *Stubblefield v. Osborn*, 31 N.W. 547 (Neb. 1948).

58. *Markuson v. Mortonson*, 116 N.W. 1021 (Minn. 1908).

59. *Rooney v. Sterns County Board*, 153 N.W. 858 (Minn. 1915).

60. *Hardin v. Jordan*, 140 US 371, 1891.

61. *Bay Marina, Inc. v Grover* [149 A.D.2d 660,] 662, Gibson DEP 2013.

62. *Muraca v. Meyerowitz*, 13 Misc. 3d 348, 818 N.Y.S.2d 450, 2006 N.Y. Misc. LEXIS 1857, 2006 NY Slip Op 26279.

63. Florida Department of Environmental Protection, "Survey Requirements Package for Leases and Private Easements" (2013), p. 7. See https://apps.sfwmd.gov/entsb/docdownload?object_id=0900eeea8a95bcd3

9

USING HISTORICAL KNOWLEDGE AS EVIDENCE

9.1 Purpose and Scope

This chapter may cause the reader or student to wish they had stayed awake in eighth-grade history. For the retracing surveyor, knowledge of both local history and national history and possibly international history and even history of one's profession is a necessity. In addition, the surveyor must know how to apply the history to retracement endeavors. Knowledge on both a local and national basis of what has happened in the past is an indispensable tool of the retracement surveyor. At times, the use of this knowledge, and its introduction and use as evidence, may become critical in determining the location of ancient boundary lines and the interpretation of ancient documents as well as being able to convince a judge and possibly the jury as to pertinent facts. This chapter is the most comprehensive and the longest of this book simply because the knowledge of history is power to the retracement surveyor and the lack thereof may put the retracing surveyor at a great disadvantage. History is the foundation of all of the surveyor's responsibilities. For example, a surveyor must know the basis of measurements, how the measurements were made, the accuracy that could be attained with available instruments at the time of the creation of the boundaries and corners, the customary materials used for monuments, the laws in force at a given time, and the practical and learned knowledge that surveyors had. These are necessary bits of background information to relocate former surveyed lines intelligently. The retracement surveyor's duty, often stated as, "Follow in the footsteps of the original surveyor," is a feat that cannot be done without a comprehensive knowledge of past events. A history of surveying and knowledge of the early customs and practices surrounding the use and ownership of land helps to clarify complexities of land laws and to explain why surveyors must do as they do.

Prior to the discovery of the New World, Native Americans understood and respected property rights. Their comprehension of property rights would soon come into conflict with the European concept of property ownership and would remain in conflict into the 1990s, when the U.S. Supreme Court would be asked to identify these rights and the rights of ownership and to restore some that had been unlawfully taken.

After Columbus's discovery, the North and South American continents were claimed by various foreign sovereign powers who made conflicting land grants in the newly discovered lands, many of which were overlapping in nature and as such have become a fruitful source for today's litigation. With minor exceptions, settlement was from east to west. The development of the survey systems in this chapter is treated in much of a geochronological order, as is possible, beginning with the colonial era followed by the French, Spanish, Mexican, and other sovereign influences on this country's boundaries. Following this are the divisions of the public domain and the various economic and geographic events that have influenced the shape of this country.

Underlying and, in many sections of the United States, superior to the U.S. Public Land Survey System, was the oldest system known to the world—namely, the metes-and-bounds system of land descriptions and surveys, which is predicated on land grants from sovereign nations. Most U.S. systems of surveys are based on English measurements and methods. The establishment of the mile and the origin of the chain both play key roles in the definition of metes used to describe real property.

The history of surveying can be categorized into four epochs: pre-colonial, colonial, post-colonial, and the modern era. Each epoch is characterized by specific contributions, people, equipment, methods, and laws.

In this chapter, the technology and the legal aspects are discussed. Events of history help explain some of the characteristics of boundary lines found today. In the southwest, the Army was sent to protect surveyors from the Apache Indians. To keep the soldiers busy, they were instructed to erect large stone mounds for section and quarter corners. This is now one of the best monumented areas in the United States; the mounds can be seen more than half a mile away. In other areas, the Indians, including Florida in 1799 with Andrew Ellicott, secretly followed the deputy surveyors and destroyed the newly set monuments. In the extreme northern section of California, the subdivision of townships was being carried on during the Modic Indian War (1873–1874). The original notes in Illinois, Iowa, and Michigan related how surveying crews were being set upon and killed. These all explain the many interruptions because of hostile action from the Indians. Deputy surveyors were forced at times to carry weapons. This might explain variances in the points of the magnetic compass.

Even earlier, in 1799–1803, Andrew Ellicott ran the 31st parallel with a little help from the Spanish authorities.

9.2 Planned and Indiscriminate Land Conveyancing

In some states along the eastern seaboard, land was sold in accordance with a plan made prior to a sale, especially in New England. In other eastern seaboard states, land was conveyed in an indiscriminate manner. The purchaser was given a warrant for a given quantity of land, and he hunted for and located whatever suited him.

After locating a parcel of vacant land, the purchaser arranged for a survey, and the warrant became effective on recording the prepared deed. In many instances, the survey and description encroached on previous surveys, either inadvertently or deliberately. In 24 counties of Georgia, containing about 8,718,000 acres, 29,090,000 acres of land were granted. In North Carolina, with a land area of about 55,000 square miles, the state granted over 100,000 square miles of land both in early English grants, as well as state-sanctioned grants! Thomas Lincoln, father of Abraham Lincoln, bought a farm in Hardin County, Kentucky, and when he sold it, he lost 38 acres because of overlapping boundary claims. In a second purchase he lost the down payment, plus the cost of litigation, when a better title was presented. At Knob Creek, he lost a third farm in a court case. This was enough; Thomas Lincoln moved to Illinois, where Abraham Lincoln, age 8 at the time of the move, was reared.

Daniel Boone was constantly plagued by adverse claims on his homesteads and finally died landless. The adverse results from indiscriminate locations and the cost of litigation that followed were among the reasons why the Continental Congress developed a presurvey system for the public domain.

Principal areas where indiscriminate locations or settlement prior to survey took place are as follows: eastern New York, eastern Pennsylvania, the colonies south of Pennsylvania, Tennessee, Kentucky, southeastern Georgia, West Virginia, parts of Ohio, eastern Texas, and the Mexican or Spanish grants, particularly in California and Texas. Although the Mexican or Spanish grants located within the public domain were indiscriminate in original site selection, each claimant was required to present proof of ownership to the federal court. As part of the court procedure, a survey was required. In effect, both location and title were established by the court.

9.3 Civilization and Land Ownership: Precolonial Surveys

The era of precolonial surveys covers that period when time began until the early or mid-1600s. Probably it is best known for the foundations of the laws for land ownership and some of the basic instruments for measurements of distances and angles.

The earliest recorded accounts of property surveys are Egyptian, but it must be assumed that the Babylonians practiced such an art even as early as 2500 BC. A Babylonian boundary stone set in 1200 BC. was inscribed with the translated meaning "Itti-Marduk-balatu, the king's officer, measured that field."

The civilization that developed along the fertile banks of the Nile greatly depended on the fruits of this rich bottomland. The annual spring floods posed a serious problem when the markers set to define the little tracts of cultivated land were washed away. It is believed by some that the Egyptians devised the science of geometry to restore lost monuments. According to Herodotus, a certain king named Sesostris should be credited with the invention of geometry:

> This king divided the land among all Egyptians so as to give each one a quadrangle of equal size and to draw from each his revenues, by imposing a tax to be levied yearly. But everyone from whose part the river tore anything away, had to go to him to notify what had happened. He then sent overseers who measured out how much the land had diminished in order that the owner might pay on what was left, in proportion to the entire tax imposed. In this way, it appears that geometry originated, and passed thence to Hellas [1].

It is also believed that a system of coordinates, referenced to points on high ground was used to perpetuate monuments.

Shown on a wall of the tomb of Thebes and carved on a stone coffin in the Cairo Museum in Egypt are drawings of "rope stretchers" measuring in a field of grain. Witnessing the act are three officials with writing materials (see Figure 9.1). The Great Pyramid of Gizeh (2900 BC) has an error of about 8 inches in its 750-foot base, or a relative error of 1 part in 1,000 on each side. The Egyptian measurements in land surveying were made by means of a rope, and such linear accuracy is entirely possible. Is it not logical to assume that the same techniques and accuracy could have been used to relocate the lost corners along the Nile? Some of the boundary monuments set as early as 1300 BC are in existence today, and distance measurements between them agree with the ancient records.

Scholars of ancient Egyptian culture believe that these peoples respected land ownership, and that land was a measure of wealth. The Egyptians have left evidence of tax registration. In 2000 BC, in ancient Mesopotamia, clay tablets recorded land surveys wherein areas were divided into rectangles and triangles. The British Museum has preserved a clay cuneiform tablet dating back to 2400 BC that contains the measurements and statistics for 11 fields from the reign of Bur-sin, king of Ur.

According to Herodotus, the science of surveying originated in Egypt. The first extant treatise on surveying is that of Hero of Alexandria (about 130 BC).

About the same time as the Egyptian beginnings, the Chinese recognized land ownership and protected rights to land. Under a type of feudal system, there was little or no transfer of land ownership; property belonged to the family or clan.

The Greek and Roman cultures contributed greatly to the development of cadastral science. It is recorded that Mark Anthony employed "geometers" to divide land. *De Agrorum Qualitate* [2] by Frontinus, written 2000 years ago, is the basis of property law and defined things "movable" and "immovable." The Roman Empire was supported in part by the collection of taxes, mostly on land.

Today, even U.S. courts are having to examine early Native American philosophy on land ownership and rights.

FIGURE 9-1. The rope stretchers.
Source: Courtesy of Dean James Kip Finch.

The human body was the basic reference for the Egyptians.
The following can be documented from history:

Egyptian	Former Equivalent	Present
Cubit c 6000 BC		8–19 inches
Cubit c 4000 BC		18.24 inches
Digit	1/24 of a cubit	¾ inch
Thumb Breadth	1/12 foot Roman inch	
Palm	1/6 of a cubit	3 inches
Span	½ of a cubit	9 inches
Foot	⅔ of a cubit = 4 palms	
Fathom	4 cubits	6 feet
Reed	8 cubits	12 feet

FIGURE 9-2. Egyptian measurements.

9.4 Measurement

The surveyor must not take measurements for granted. The history of measurements is as important as the history of land rights. Since the beginning of time, humans have needed to measure. The original units of measure were referenced to body parts: the foot, the arm, the finger, the pace—even the hair on a caveman's head. This has been accepted today in that we still reference the height of a horse by the number of hands.

Initially, measurements were two separate and independent units; angle and distance. Later on, they became merged into angle and lines, a single unit.

Historically, such units as the reed and cubit all became part of the surveying history. Most areas developed their own systems of measurement, many of which are history and in little use today. Throughout history, certain terms became universal. Today, the world recognizes the Egyptian unit, the *cubit.* Figure 9.2 shows the equivalents for Egyptian measurements.

Probably the earliest known mile was the one used by the Persians, being 1,000 fathoms in length. The name comes from the Roman *mille,* which means one thousand, and was established in Rome by 1,000 paces, the pace being equal to 5 Roman feet. Of all of the Roman and Egyptian measurements, only the mile, foot, and fathom survive today. The units of measurement can be traced to early England, where the barleycorn became the standard unit. Figure 9.3 shows the relation of early English measurements. Most early land measurements were conducted in poles or rods, which, in turn, resulted in chains.

9.5 Biblical References to Land Surveying

The Bible is a fruitful guide for references relating to the ownership of land. Joshua directed the subdivision of the unoccupied lands to the Israelites. The land, according to the Old Testament, Book of Joshua, Chapter 18, was divided among the seven tribes in seven parts.

The land on which Abraham and Lot were living was too poor to support them, so they separated. In Genesis 13:14–18, there is the account of how the Lord spoke to Abraham giving the land to him and his heirs.

12 inches = 1 foot
9 inches = 1 span
5 spans = 1 ell
3 feet = 1 yard
3 feet = 1 pace
5 feet = 1 pace
125 feet = 1 furlong
5½ yards = 1 rod
40 rods = 1 furlong
8 furlongs = 1 mile
3 miles = 1 league
Later to have a greater refinement into smaller units
12 human hairs = 1 poppy seed
4 poppy seeds = 1 barley corn
At this then we calculate that the following can be determined:
9,123,840 ± human hairs = 1 mile.

FIGURE 9-3. Historical English measurements.

One of the earliest recorded transfers of land is found in Genesis, Chapter 23. Abraham wanted a suitable burial ground for Sarah, who had died. He negotiated with Ephron for a field and a cave to bury his dead (Genesis 23:16–18):

> And Abraham hearkened unto Ephron; and Abraham weighed to Ephron the silver, which he had named in the audience of the sons of Heth, four hundred shekels of silver, current money with the merchant. And the field of Ephron which was in Machpelah, which was before Mamre, the field, and the cave which was therein, and all the trees that were in the field, that were in all the borders round about, were made sure unto Abraham for a possession in the presence of the children of Heth, before all that went in at the gate of his city.

The basis of all real property title in the Christian world may be found in Genesis 1:1: "In the beginning, God created the heaven and the earth." Kings were given the sovereign right to land by the Pope, and they, in turn, gave ownership to members of the court. The account of Jacob's purchase of land is in Genesis 33:18–20:

> Jacob came to Shalem, a city of Shechem, which is in the land of Canaan, when he came from Padanaram: and pitched his tent before the city. And he bought a parcel of a field, where he had spread his tent, at the hand of the children of Hamor, Shechem's father, for a hundred pieces of money. And he erected there an altar and called it Elelohe-Israel.

Ahab, king of Samaria, wanted to acquire a vineyard that was near his palace, belonging to Naboth. Ahab offered to purchase the vineyard, but Naboth refused, "The Lord forbid it me that I should give the inheritance of my father unto thee." Jezebel, Ahab's wife, perpetrated a false charge of blasphemy against Naboth, and he was stoned to death. Ahab now possessed the land, but not for long, since the wrath of God overtook him and punished him for his wrongdoing (I Kings, Chapter 21).

Many accounts appear in the Old Testament relating to boundary monuments:

- Deuteronomy 19:14:
  - Thou shalt not remove thy neighbor's landmark, which they of old time have set in thine inheritance, which thou shalt inherit in the land that the Lord thy God giveth thee to possess it.
- Deuteronomy 27:17:
  - Cursed be he that removeth his neighbor's landmark. And all the people shall say, Amen.
- Job 24:2:
  - Some remove the landmarks: they violently take away flock and feed thereof.
- Proverbs 22:28:
  - Remove not the ancient landmark which thy fathers have set.

9.6 From the Greeks

The *Iliad* of Homer, Book 21, lines 499–503 (Morris translation) contains the following text: "She only stepped backward a space, and with her powerful hand lifted a stone that lay upon the plain, black, huge, and jagged, which the men of old had placed there for a landmark."

Even the gods were not to be left out. There is the account of the goddess Pallas in conflict with the god Mars in Book 12, lines 503–508 (Morris translation): "As two men upon a field with measuring-rods in the hands, disputing stand over the common boundary, in small space, each one contending for the right he claims, so, kept asunder by the breastworks, fought the warriors over it."

9.7 Sovereignty and Ownership

In the early Judeo-Christian era, the early belief was that possession and ownership came directly from God. From the beginning of the Christian era, the Pope vested sovereign rights in rulers. Even in non-Christian cultures and pagan societies, the ruler has been recognized as having eminent rights over the land of the nation.

Rulers conveyed parcels to members of their court and expected, in turn, tribute in the form of taxes, rents, services, or royalties. The term *estate*, derived from the word *status*, represents the interests one has in property—that is, the extent of the rights and privileges one can enjoy therein.

One of the inherent and necessary attributes of sovereignty is the right of eminent domain, which is for the king to claim private property for his own use. In the broadest sense, a nation has the right or power to take private property for public use: It is superior to all private property rights. The common-law principle of eminent domain has been greatly modified and amplified by the constitutions and statutes of nations and their legal states.

9.8 The Feudal Land System

After the Battle of Hastings in 1066, William the Conqueror took all of Great Britain and immediately claimed all the land as his own. Half of this he gave in parcels to his favorite Norman lords. A survey, the first written basis for taxes, was ordered in 1086 to compile a list of every farm and owner, known as the Domesday Book. The survey was executed by groups of officers called *legati,* who visited each county and conducted a public inquiry. By listing all feudal estates, both lay and ecclesiastical, the Domesday Book enabled William to further strengthen his authority by exacting oaths of allegiance from all tenants on the land as well as from the nobles and churchmen on whose land the tenants lived.

The feudal landlord conducted his life quite independent of the king and his neighbors. He offered protection to those who lived within his domain, and, in return, the tenants were required to pay rent, share the crops of their harvests, and fight for their lord when ordered to do so. The lord of the manor often leased land, termed *fiefs* or *feuds*, to other noblemen or vassals.

The landlord would enclose his private estate for hunting and other personal pleasures: The area not enclosed was termed *commons*. Peasants were permitted to farm, pasture their livestock, and try to satisfy other basic needs on the commons. The Royal estates were "off limits" to the common man.

More and more land was enclosed, until many peasants could not find enough common land on which to subsist. Stealing and other crime increased during the Middle Ages, and many peasants were executed for petty land trespasses. Some landlords obtained large tracts by conquest of neighbors, and others enlarged their estates by bartering, murder, and chicanery.

At the close of the fifteenth century, England was not very densely populated, but people were starving for want of land. It was nearly impossible for anyone other than those of noble birth to acquire title to property.

In the 1600s, many changes occurred in the relationship of interests in land as well as developments in surveying equipment (technology) and surveying methodology. The theodolite was developed to aid in the measurement of angles. Later in the century an instrument that could "transit" its telescope was developed. Then the plane table was developed to "make" the lay of the land.

This period also saw surveyors who heretofore had no textbooks on which to rely create the first "how-to" books for the practical surveyor. One of the more famous was a book first published in 1687/1688 by John Love entitled *Geodaesia.* The introductory pages to that volume can be found in Appendix D.

Along with the development of instrumentation came development in the law. Such terms as *contract*, *tort*, and *trespass* found their way into the English language and law. One of the first cases on the ownership of a tree that sat on a common dividing boundary line was decided in 1620, in the decision *Masters v. Pollie*, which determined that a tree that grew on the boundary line between two estates was owned as "tenants in common."

Beginning in the early 1600s, both the technical and the legal aspects of surveying were sprouting new wings that would be felt and have an influence today.

9.9 Livery of Seisin

Written conveyances of real property may have been executed at the beginning of civilization. Much land in England and her American colonies was conveyed by a ritual known as the *livery of seisin* (delivery of possession). The parties to the transfer would meet on or in sight of the land and, through a series of solemn acts such as the handing over of a twig, a handful of soil, or a signet ring, would memorialize the contract. Other demonstrations, such as throwing stones, driving stakes into the ground, and shouting and uttering such words as "I give," were witnessed to bind the conveyance. If the ritual was performed on the land, it was termed a livery in deed, or if within sight of the land, a livery in law. The grantor was required to practice abjuration—that is, to vacate the land within six months. Written contracts may have accompanied the ritual, but these were only evidence of the conveyance, not the conveyance instrument. Recently, this concept was argued in Connecticut, where the Plaintiff bought a parcel of land whose boundaries were created in the late 1600s. The parcel was described by four adjoining landowners. A new title search did not reveal a "modern" description. Using the general description and a search of the adjoining parcels, the parcel was generally located from the general description, but since it lacked a specific description calling for monuments and lines, it was virtually impossible to locate on the ground, except for the "key," which was a reference to "an oak of gigantic proportions." On a detailed field examination of the general location, the surveyor found stone walls that had been in place possibly for a century or more. At the intersection of three walls he recovered a white oak stump, probably 3 feet in diameter, which he assumed to be the called-for oak. In discussions with the trial attorney, the surveyor suggested that since no specific description was found, it was entirely possible that this parcel and its boundaries were the vestigial remains of a conveyance of a parcel by "livery of seisin" and that the stone walls were the "best remaining" evidence of the intent of the conveyance by the livery. The attorney refused to raise that process at the trial. The client lost since he could not make the nexus (connection) back to the original deed that lacked the specific calls for lines and monuments.

After the adverse decision, the attorney commented that he "really did not understand the process, and anyway it was old hat." It was the surveyor's opinion that the people did not need a description since the historical parties in interest all considered the stone walls as the definitive boundaries.

Although livery of seisin is replaced by delivery of a written conveyance today, certain parts of the ritual still remain in parts of the country, and the ritual was practiced in England until 1845. An example of the livery of seisin is found in a document dated July 1824, to Robert Millicam from the Mexican government:

> We put the said Robert Millicam in possession of said tract of land, taking him by the hand and causing him to walk around it and telling him in loud and audible words, that by virtue of the commission and powers vested in name of the Government of the Mexican Nation we put him in possession of said tract of land with all the uses, customs, rights, and services thereof, unto him, his heirs and assigns, and the said Robert Millicam, in evidence of being in real and personal possession of said tract of land, without any contradiction, cried out, pulled twigs, threw stones, drove stakes, and performed the other necessary solemn acts [3].

9.10 Statute of Frauds

In this dynamic period of the 1600s in which the technology of surveying and the law or real property were developing, Parliament enacted legislation that has come to be known as and recited by most as the *Statute of Frauds*. In reality, the true and correct nomenclature for the Act is "The Statute to Prevent Fraud." The Parliamentary Act of 1677 required certain contracts to be in writing in order to have a cause of action at law.

Section 4 of the *Statute of Frauds* states:

> No action shall be brought unless the agreement upon which such action shall be brought, or some memorandum or note thereof, shall be in writing, and signed by the party to be charged therewith or some other person thereunto by him lawfully authorized.

Of the five classes of contracts, one was: "Any contract or sale of lands, tenements, or hereditaments or any interest in or concerning them." Much of the *Statute of Frauds* has been overruled, but the requirements for land areas conveyances to be in writing to be enforceable and the need for the grantor to sign such an instrument are the basis of conveyance laws in this country [3].

The Statute of Enrollments rendered a conveyance void unless recorded within six months.

The basic unit of English measurement was the rod. In England, the rod varied from 12 to 22 feet, depending on the area to which it was used. A surveyor can find different definitions for the rod in the United States of 16.5 and 18 feet. The latter is usual in areas settled by the Scotch-Irish. This unit was referred to in several textbooks of the 1600s.

Without this history, the modern surveyor may be retracing with a 16.5-foot rod when the original measurements were made with an 18-foot rod. As late as 1964, the Supreme Court of Vermont was asked to determine the value of the "pace." The court determined that the pace was a unit of measurement that was employed, but it was not precise. Even so, it did not render the deed void. It did consider *pace* as "approximately three feet."

The authority for this is a Vermont case that stated that "pace," the unit of measurement employed by deed, was not precise, but did not render the deed so ambiguous as to justify resort to extraneous evidence and could be treated as approximately 3 feet [4].

As early as 1633, John Love recommended a survey chain of two poles for use in the Americas (see Appendix D).

The basis of all real property title in the Christian world may be found in Genesis 1:1: "In the beginning God created the heaven and the earth." Kings were given the sovereign right to land by the Pope, and they in turn gave ownership to members of the court. The account of Jacob's purchase of land is also in Genesis.

9.11 Early Property Surveys in the New World

The methodology of surveying remained fairly constant in the New World from the time of exploration until 1776, when a unique metamorphosis took place. The earliest surveys in the New World were to map the country rather than to delineate properties and their accompanying boundaries; information for suitable homestead land was quite meager, and surveyors were needed to explore and map rivers and other waterways that were the natural means of travel or the highways. Among the surveyors sent to Virginia to locate land allotments was William Claiborne, who, in 1621, at age 38, was paid 20 pounds per annum and furnished with a house. In 1629, for his part in defeating the Indians, he was granted a tract of land on the Pemunkey River.

Claiborne's trading post, founded on Kent Island in 1631, was included in the grant to George Calvert; a dispute continued over the title until 1776.

The men who conducted surveys during this time came from a variety of backgrounds. Some were astronomers and mathematicians, such as Mason and Dixon, and some were frontiersmen like Thomas Hutchins. Then as now there was little formal schooling in surveying, and surveyors with scientific background were rare. The colonial surveyors usually held high social positions and were generally able to advance themselves financially. One basic fiber surveyors usually possessed that placed them in the class of better-educated citizens was a solid foundation in mathematics and astronomy. Surveyors were often selected from the teaching or ministry professions, and surveying often was a second or part-time profession.

From 1621 to 1624, the surveyor general for Virginia was appointed by the governor and from 1624 to 1693 by the king. On February 8, 1693, the College of William and Mary was charged with the responsibility of surveyor general appointments as well as the examination and licensing of surveyors. Among the famous men appointed as surveyors, but not necessarily county surveyors, were George Washington (in 1749 at age 17) and Thomas Jefferson. Peter Jefferson, Thomas's father, was a surveyor, as was Peter Jefferson's grandfather. Thomas Jefferson was appointed surveyor of Albemarle on October 14, 1773, the same county in which his father was assistant surveyor. Joshua Fry and Peter Jefferson were commissioned in 1749 to run a part of the Virginia–Carolina boundary line [5].

George Washington was appointed official county surveyor for Culpepper County, Virginia, but was not a registered county surveyor. Culpepper was a new county, and George was its first surveyor. His duties mainly consisted of marking out small farms for settlers, as well as land for some wealthy Virginia landowners. George's grandfather, Colonel John Washington, came to Virginia in 1657 with wealth and influence. Although he purchased large tracts of land, the wealth of the Washington family soon dissipated. George was brought up at Mount Vernon, where it is supposed that he learned the "art" of land surveying from his half-brother, Lawrence Washington. George's surveying practice, as such, lasted only about two years. During this time, he used his position to further his land speculation and to locate lands for some of the wealthy Virginia plantation owners, the principal one being Thomas Fairfax in the Shenandoah Valley. In 1751, George and Lawrence joined the British army against the insurgent French. While in the army, George was able to obtain title to 32,373 acres in West Virginia from other officers.. With land speculation in full swing, he sent an agent to England to promote buyers for his tracts. Being a very shrewd businessman and having the advantage of being a surveyor, Washington built a large fortune in land; by the time he died in 1799, his holdings included 49,000 acres in the rich Kanawha River Valley of West Virginia, as well as vast tracts in Virginia, New York, Ohio, and other places.

Many exploratory surveys were conducted by the Army for fortifications and other uses. Thomas Hutchins and Henry Bouquet explored numerous rivers and the frontier country; others conducted surveys for homesteads. The Virginia assembly in 1783 created a board to survey the 150,000 acres given to George Rogers Clark and his men for services rendered.

> Methods and instruments used in colonial surveys were little better than those used in ancient times. The determination of latitude was usually derived from solar observations with a sextant or circumferenter that, at best, was within a half-minute of arc. Mason and Dixon apparently made astrolabe observations on several stars to attain their excellent consistency.

The determination of longitude was not practical in the early days because of the lack of accurate time propagation. Early grant descriptions called for differences in longitude that had to be measured on the ground. Considering the lack of geodetic knowledge, these distances were computed with a fair degree of accuracy.

In colonial days direction was usually determined by the magnetic compass. Star observations (Polaris at elongation) were made to check the variation in the needle. The solar compass did not come into use until about 1835. Straight lines were run with pickets by a succession of compass readings or by the circumferenter (Figure 9.4), a French instrument in general use until about 1800.

An English mathematician, Edmund Gunter (1581–1626), contributed to the science of surveying by such discoveries as magnetic variation, a portable quadrant for star observations, the introduction of the words cosine and cotangent to trigonometry, and a practical measuring device called Gunter's chain (see Figure 9.5). The chain was four rods long, or 66 feet. There are 80 chains to the mile. The 1 chain tape was composed of 100 wire links. The acre, today's measure of area, was defined by him as 10 square chains. The length of a chain was said to be 4 poles (later known as rods or perches) long. Even though the chain was a great improvement over ropes

FIGURE 9-4. A circumferenter.

FIGURE 9-5. A Gunter chain.

and bars used before, there was still considerable error in its use caused by the lack of standardization, frequent blundering, and surface measurement. Steel ribbon tapes were not in general use until early in the twentieth century.

A practical telescope, short enough for ordinary surveying instruments, was not available in colonial times and did not appear among the popular instruments until the early nineteenth century. The circumferenter had sight vanes, a small compass, and a spirit level and was mounted on either a staff or a tripod. The sextant was usually hand-held and resembled the instrument used by mariners of that day. On land, when making latitude observations, it had to be used with an artificial horizon, with an expected accuracy of location of 1 foot to 1 mile of the correct latitude. Perhaps the best device used was the early surveyor's own ingenuity.

9.12 The Concept of Area

When a person obtains a deed to a parcel of land, little regard is given to the boundary elements. Usually the question is, "How many acres are there?" The usual grantee cares little as to the validity of the boundaries. They seem more interested in whether they get all of the area they were supposed to get in the deed. Legally, the courts care little about the area, unless the area was the major element of the contract. This is exhibited in the *Priority of Calls* in which area is listed as the very last element of control. This is reinforced by the fact that the law will permit a valid conveyance (deed) to be delivered without any reference to area, as long as the boundaries are legal, discernible, and locatable.

The historical reference to area in the English Common historical area is the acre. If you, the surveyor were to stop any person on the street and ask them "How big is an acre?" the average American has always heard of the word but they have no idea how big it is, nor could they "step it off" with any degree of certainty. The fact is, no state in the United States has defined an acre by statute. The term has been one of common-law reference and acceptance, that had its beginning in the agrarian society as being the amount of land a man with a yoke of oxen can plough in one day. (They did not identify the number of hours in the day.) This has been held, by common practice to be a strip of land 1 chain wide and 10 chains long, or 10 square chains in area. This has become the common-law acceptance of the area.

In the only historical record uncovered in research, John Love in his historical survey text *Geodaesia* wrote that Parliament had deemed that for legal purposes an acre shall be 160 square poles. This is reaffirmed in many descriptions in the colonial states, where deeds will refer to a parcel of land as containing 4 acres 2 roods, and 12 poles.

Once again history enters. Love was the first person to identify the rood as equivalent to a quarter of an acre. Many surveyors and attorneys mistake the rood as a unit of horizontal measurement.

Today many surveyors, attorneys, and judges will state that an acre contains 43,560 square feet. This is far from the historical truth.

As a practical note, people have used "short-cut methods" for many elements in describing land and its boundaries. Some people have stepped off distances, used a yardstick to make measurements, estimated angles by "flapping arms" and North by looking at moss on trees, to the point that they have taken short cuts to determine area or acres. An old story is that a person bought a tract of land supposedly to have contained 10 acres, and after a survey, found it was short of area. He tried to blame the surveyor. The surveyor told him, "You have all within your boundaries." The surveyor asked the farmer, "How big is an acre?" The farmer replied, "About a stone's throw." Dumbfounded, the surveyor asked, "How big a stone?" The farmer replied, "That depends on whether you are buying land or selling land!"

9.13 The Concept of Title

Modern concepts of land ownership have developed from Christianity and feudalism. Fee tail ownership evolved from absentee ownership and feudalism. Fee simple ownership at first was reserved for the favored few. *Title* may be defined as the legal basis or grounds for land ownership and may be transferred by descent through heirs or by purchase. According to law, conveyances include that acquired not only by buying but also by devise, will, grant, adverse possession, escheat, condemnation, and the numerous other ways land title can be transferred.

Real property is not a creation of humankind; therefore, original ownership can come only from conquest or discovery. Private ownership of land has many injustices, but perhaps it is the least of several evils. An alternate (although entirely undesirable) plan would be nationalization of land, as prevails in some noncapitalistic countries. As civilization progressed, strong-armed men assumed ownership by force. *All titles today can be traced back to and are maintained by force*, either actual or legal.

Real property in Europe has been divided so many times that title is composed of a multitude of slivers. In many instances, reapportioning is necessary to establish equity in the chaos.

Today, people accept title and title transfer of real property. With the present-day high value of property, any delineation must be exacting and interpretation made with extreme caution. The boundary (retracing) surveyor must have an adequate understanding of the background of land titles to approach boundary problems intelligently.

Land titles today are held by many people and are protected by more legislation than any other singular right or possession. The property surveyor, in locating a title, is charged with a tremendous responsibility that cannot be discharged lightly.

SURVEY SYSTEMS OF THE EAST AND SOUTH

9.14 Origin of Title

For practical purposes, the colonial era of surveying in the United States probably commenced when the settlers arrived. Whether they were English, French, or Spanish, they brought with them their instruments, methods, and laws pertaining to surveys of parcels for settlement.

The English settlers' thrust was for title, that is, land was the combined bundle of rights. These rights could be not only possessed but also bought and sold. The law the English brought with them was English common law, one predicated on sovereignty of a king, passing edicts and making decisions.

The Spanish and French settlers brought with them a legal system patterned after the Roman system of law. Today in America, most states follow English common law, while Louisiana recognizes the Napoleonic code, which is patterned after Roman law. One of the major differences that the two systems address is the concept of adverse possession—that is, a method of acquiring title for long, continued, adverse possession of another person's land. English common law recognizes superior title, or as the English say, "A man's home is his castle," as being the criterion for possession, while the Napoleonic code accepts the concept that possession is nine-tenths of the law.

In English settled areas, colonial governments, both under the English Crown and later, and then state governments, adopted, either by proclamation or legislative acts, *English common law as practiced in England.*

For instruments they had the magnetic compass and the circumferenter, as well as the "playne" or "plain" table (early English spellings); measurements were made with a link chain either in poles or rods, probably some in varas, whereas perhaps the French brought their chains, which probably measured in toises. Angle plottings were made with a brass protractor approximately 6 inches in diameter calibrated to one degree and distances were plotted with a basswood scale, divided into tenths.

Double meridian distance (DMD) and coordinate computations were unknown, so area was determined by adding up all of the rectangles and triangles in a plot of land. It is with this basic knowledge that the colonial surveyor brought his expertise to the New World. These are the early footsteps the retracing surveyor must ascertain today.

An ideal situation could have existed if all this nation's land had been acquired at one time and from one source and included no prior grants. This nation is indeed fortunate to have so vast an area covered by the Rectangular System that, in spite of its problems, has been a tremendous help to title retracements or resurveys and other functions necessary to land ownership. Twenty states have not been subdivided under the National Land System (Figure 9.6), and parts of the rest of the states have exceptions and prior grants.

English, Dutch, French, Spanish, Mexican, and even Russian grants have been upheld in America, and the title going back to these sovereign nations reflects the system (or lack of one) imposed by custom or royal decree. It is a dangerous practice to generalize in regard to any one system or area, and even a conservative treatment of the various survey systems in the United States would fill several volumes. It is the authors' intent to now draw a few examples of these complex survey systems and to point out that such extremely local problems must be studied thoroughly on the local level. A successful Illinois surveyor on reaching retirement age migrated to Florida so that he could continue his practice in the comfort of a warm climate. In spite of the fact that both states were public land states, the differences in the local systems were so great that the new business soon failed, and the surveyor was forced to return to Illinois.

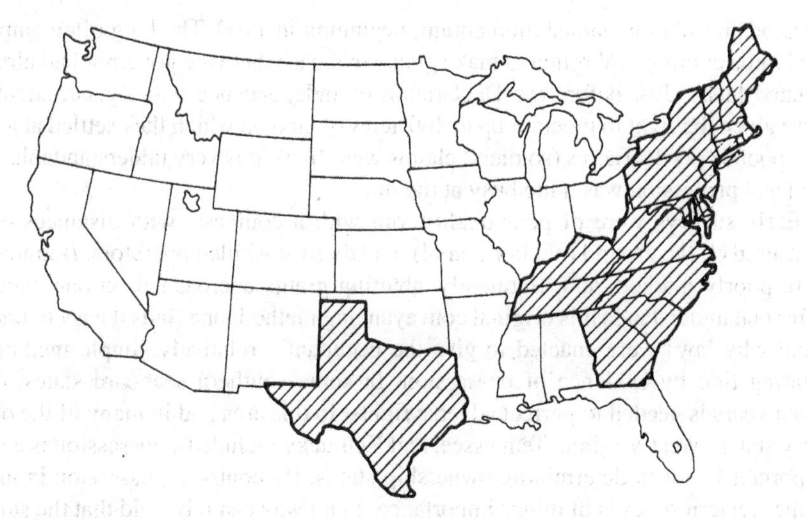

FIGURE 9-6. Shaded states are without public domain, Rectangular System (Hawaii not shown).

9.15 Surveys in the Early States

Virginia The original charter for the colony of Virginia was granted in 1606 and reissued again in 1609. The description was quite indefinite, calling for "West and northwest to the South Sea." The description in the 1609 charter reads, in part:

> All the territory two hundred miles northward and two hundred miles southward of Old Point Comfort and all that space and circuit of land lying from sea coast of the precinct aforesaid, up into the land throughout from sea to sea, west and northwest [6].

In 1776, Virginia gave up claims to area now in Maryland, Pennsylvania, North Carolina, and South Carolina. In 1792, Kentucky was allowed to become a state, and West Virginia was separated in 1866. At the time of the Revolutionary War, Virginia claimed lands in Ohio, Indiana, and Illinois. The grant to the "South Sea" was to the Pacific Ocean, and it was later reduced to the Mississippi River by a treaty with France.

The London Company, which held the original charter, planned to have lands surveyed and assigned to shareholders at the rate of 50 acres per share: later, the rate was raised to 100 acres per share. Also introduced by the London Company was the *head right concept:* The head of the family could claim 50 acres per head for all servants and persons with paid transportation to the colony. After the dissolution of the London Company because of a lack of funds, the Virginia colony claimed the charter, but later kings gave grants of the Virginia area to others such as Maryland to the Calverts in 1632, Pennsylvania to William Penn in 1681, and the Carolinas to the Carolina proprietors in 1663 and 1665.

Land speculation gained momentum, beginning in 1700. The king often empowered lords coming to Virginia to make grants in areas where the governor had already granted lands. Just before the Declaration of Independence was signed, squatters were given the right to preempt up to 400 acres of land on which they settled at a cost of 3 pounds per 100 acres (soldiers' claims were free). It is very understandable why the legal profession was quite busy at the time.

Early surveys were of poor quality, run with a compass, with distances often measured on the ground (not horizontal), mostly by unskilled operators. Boundaries were poorly marked, and, frequently, abutting grants encroached on one another. After one understands this original conveyancing method, one finds it easy to understand why laws were enacted to give the occupant a relatively simple method of clearing title by evidence of possession. In some southern seaboard states, only seven years is needed to perfect adverse rights. In this area and in many of the other early states (West Virginia, Tennessee, and Kentucky included), possession is a very important factor in determining ownership status. By contrast, possession in many of the western states is of minor importance. In no way can it be said that the survey problems in the older states are comparable to those of the newer states. Although the laws for interpreting the meanings of deed wordings are similar all over the country, the applications of the laws, because of differences in original conveyance methods, vary considerably.

New England The Plymouth Company was organized in England in 1606 and in 1607 received a charter from the king of England at the same time as the Virginia charter. The operation of the Plymouth Company was not a great success, and in 1620 it was reorganized as the Council of New England. Grievances among the colony inhabitants finally led to the House of Commons declaring the charter forfeited in 1635. New England was not developed as a unit; many different grants were given to different areas. The Massachusetts Bay Company received a charter in 1628, and the charter contained, in part, the following description:

> All that part of new England in America which lies and extends between a great River ther [*sic*] commonly called Monomack River alias Merrimack River and a certain other River then, called Charles River, being in the Bottom of a certain Bay ther, commonly called Massachusetts, alias Mattachusetts, alias Massatusetts Bay; and also all and singular those lands and hereditaments whatsoever, lying within the Space of Three English miles on the South part of the said River, called Charles River, or of any, or every part thereof; and also all and singular the lands and hereditaments whatsoever, lying and being within the space of three English miles to the southward of the Southern most part of the said Bay, called Massachusetts, alias Mattachusetts, alias Massatusetts Bay; and also all these lands and hereditaments whatsoever, which lie and be within the Space of Three English miles in the Northward of the said River called Monomack, alias Merrimack, or to the Northward of any and every part thereof, and all lands and hereditaments whatsoever, lying with the limits aforesaid, North and South, in latitude and breadth, and in length and longitude, of and within all the Breadth aforesaid, throughout the main lands ther, from the Atlantic and Western Sea and Ocean on the East Part, to the South Sea on the West part [7].

This charter was surrendered and a new charter was granted by William and Mary in 1691. Connecticut received a charter in 1662 and Narragansett Bay, Rhode Island, in 1663. New Hampshire (then including Vermont) received recognition as a province in 1680.

The current northern boundary of Massachusetts was first surveyed and marked in 1741. The east–west line with Rhode Island was agreed on by commissioners appointed from the two colonies in 1711, but it was not surveyed until 1719. Because the line was not very accurately located by this survey, almost ever since there have been suits in the U.S. Supreme Court contesting the position of the boundary.

The earliest grants of land in New England were of irregular size and generally without a plan. Soon it became apparent that it was advisable to have uniform regulations for the disposal of land. As early as 1634, a committee was formed to locate the boundaries of unsurveyed towns and to settle boundary disputes. One result of the committee's work was a law that required the walking of town boundaries every three years. Since this law was never repealed, many areas still practice walking of town boundaries, although it may be every five or seven years. Figure 9.7 shows a town line marker set by surveyor W in 1934 as part of perambulation.

When the need arose for additional land for settlement, commoners (proprietors) petitioned the court for permission to establish a town in accordance with a plan presented. The plan usually had some house lots, land areas of variable sizes (often 40, 60, 80, or 100 acres each), reservations for the school, church, and minister, and areas to be held in common. Rangeways (land set aside for roads) were allowed for between ranges and tiers of lots. Since the roads were not granted, they, in many cases, remain in the town. Since these "future roads" were laid out on paper with no regard for topography, many could not be used. When attempts were made to use them, considerable variation was necessary to avoid steep hills, ledges, ponds, and other obstacles.

After lotting plans were prepared, the custom was to "draw" at random and often in accordance with the number of shares held by the settler or proprietor. The settler could be entitled to more than one lot, and most settlers were entitled to a home lot. If the lots drawn were deficient in either quality or quantity, the settler could petition for another parcel. In some places "pitches" were made, meaning that the settler knew how much land he was entitled to and asked to enter the area of yet ungranted land and stake out his area so that it would not encroach on another.

Kingston, New Hampshire, had lots of 20, 40, 60, and 200 acres, home lots, meadow grants, and common areas for the public. As an illustration of a town plan, Candia, New Hampshire, is shown in Figure 9.8. The lots were generally 80 or 100 acres in size with a few of variable sizes and all were dated 1719. The following is example of a Candia deed: in 1764, Caleb Brown purchased the west half of Lot 57 by the following description:

> A certain piece or parcel of land lying and being in Candia, containing forty acres be it more or less and is the westerly half part of the 80 acre Lott of land be it more or less which I bought of John Taylor of North Hampton in ye provence aforesaid, which lott was laid out to the right of Joseph Bacheldor and said lott is bounded as follows: First

FIGURE 9-7. An early New England ordinance required town lines to be perambulated at regular intervals. In the process, town line posts were often erected as pictured in this figure. The "T" and the "L" stand for town line and the "W" for the surveyor who set the post; 1934 is barely visible. "Sapling" is scribed on the left side of the post, and "Misery Gore" is on the right side, each facing the towns of those names.
Source: Donald A.Wilson

> at the Northwest corner at a Beech tree which is the Northeast bounds of the 56th lott, then East South East 72 rods to a Hemlock No. 57, then South 29° West 180 rods to a beech tree No. 57; then west northwest by the highway 72 rods to a beech tree which is the South East Bound of the 56th Lott, then in straight to the first mentioned bounds [8].

From the foregoing a survey must have been made; tree monuments were called for. It is possible that 260 years later the trees are gone and possession may be the best indicator of where the original surveyor traveled. Several of the roads shown on the Candia plan (Figure 9.8) were never built because of difficult terrain, and those that were built were not straight.

In New Hampshire, a surveyor of highways and a fence viewer were elected to locate new roads, settle fence arguments, and perambulate town boundaries.

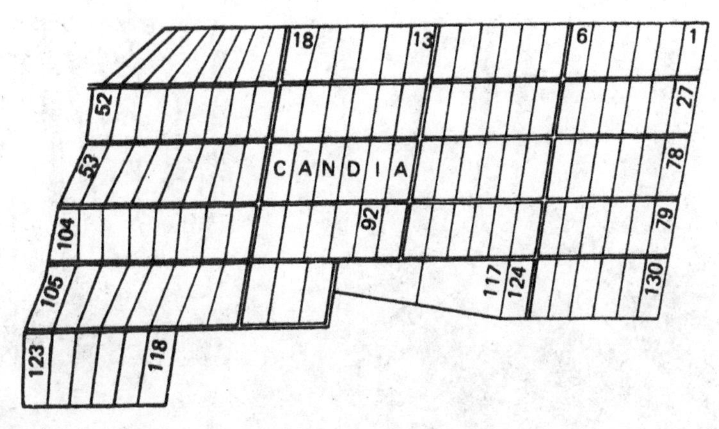

FIGURE 9-8. Plan for the town of Candia prepared prior to 1719. The lots of 80–100 acres, numbered in the sequence indicated, were granted to proprietors according to a drawing held in 1719. None of the original proprietors settled on the lots, but by the Revolutionary War, many of the lots were occupied. In New England, the word *town* is not to be confused with city; a *town* is a political unit, not a city.

Caleb Brown served in both posts and settled on lot 92 of Figure 9.8 about the time of the Revolutionary War. In addition to locating the "lotts" in New England, the surveyor usually judged the quality of the land; poor lots usually have an overrun of area. Ten percent overrun is not uncommon. Surveyors and assistants, as well as chainmen, were under oath and returned a report about what was done. The plan and description of the addition to Alexandria, New Hampshire, contained this note:

> June ye 1: 1773. This Plan Sheweth ye Number of Lotts in the addition of Alaxandrea Joyning on the South westerly Sid of said Alaxandr and the Easton side of the Patten Line to Grate Sunipe Pond then by Said Pond to fisherfield then Easterly on fisherfield to Parryton North Line then down Parrytown Line Easterly to a Beach Tree which is the Comer of Said Alaxandra: it is Divided in to 137 Lotts Each Lott Containing 150 acrs numbered as is Set down in the Plan Layd Down by a Scale of one Mild to one Inch and as Convenant as the Land wold alow I Laid Tew Senter Squars of Ten acrs Each for Publick Uses as marked in the Plan the Ponds are as Near as Posable Laid in their Shape and Bignes and the Streems Drawd as thay Run throw Said Land Said Streems Run Easterly from the Patten Line measured by me.
>
> **Jeremiah Page, Syayar of Land**
>
> John Tolford and Robart Mcmrphey, Junr Cheen man June ye 23: 1773 Then Jeremiah Page appeared and made Solomn oath that this Plan of the addition of Alaxandrea by him Drafted is Just and Trew acording to the Best of his Skill and Judgment . . . Benja Day Just Peace [9].

In 1727, new towns along the Connecticut and Merrimack rivers were instituted. Two townships were established in Maine and five in Massachusetts. In Vermont, in 1760, Governor John Wentworth planned a subdivision with 6-mile squares; for this,

FIGURE 9-9. In New England, rock walls commonly mark property lines. This is a picture of a lane with rocks on each side.
Source: Donald A.Wilson

Joseph Blanchard surveyed along the Connecticut River, marking township corner trees every 6 miles for about 60 miles. Many later town charters referred to this plan, but purchasers had to locate the bounds at their own expense. Herein lies the problem found in most of New England; early survey records are not available. Whereas it is probable that surveys were made by approximate methods—that is, approximate by today's standards—the task of the surveyor today is to try to prove what was done at an earlier date. In many recent surveys, the weakest forms of evidence must be resorted to, such as reputation. As compared to the Virginia indiscriminate method of disposal of land, New England town developments by a plan were far superior; only one line existed between lots, and no overlap between lots existed, as frequently occurred in Virginia.

Much of New England was cleared of trees, and the large rocks in the fields were moved to boundary lines. Although not all rock walls are boundary lines, many are (Figure 9.9). During the Great Depression, many of the lands lay idle, and second growth of trees filled in where pastures formerly existed.

Connecticut In 1630, the Plymouth council made a grant to the Earl of Warwick. A later charter, granted by Charles II, was described as follows:

> [A]ll that part of our dominions in New England in America, bounded on the east by Narraganset River commonly called Narraganset Bay, where the said River falleth into the sea; and on the North by the line of the Massachusetts plantation on the South by

> the sea; and in longitude as the line of the Massachusetts colony, running from east to west, that is to say, from the said Narraganset Bay on east, to the South Sea on the West part with the islands thereunto adjoining [10].

Later, Connecticut made a trade of equal parts with New York and, at the time of the Revolutionary War, still held western lands in Ohio, Michigan, Indiana, and Illinois. Because of the many disputes over the state boundaries, resurveys were conducted as recently as 1925 to confirm positions.

Rhode Island Settled by Roger Williams and followers in 1636, this small area was granted a charter by King Charles II in 1663. The description calls for natural objects, recites metes, and bounds, and was originally included in the Charles I grant to Connecticut. Although some dispute developed over location of the boundaries, the state never had claim to western lands. Today, several of these original boundaries are in question.

New Hampshire The territory of New Hampshire was included with the original charter to Virginia in 1606 as well as in two other grants of 1620 and 1622. The area, described by metes and bounds, was granted by the president of New England to John Mason in 1629. Controversies over the borders (including the one with Canada) flourished for many years. New Hampshire had no western lands.

New York The area of New York, claimed by Henry IV of France as part of the territory of Acadia in 1603, was included in the grant to the Virginia colony by James I of England in 1606. The French and Indians held territory west of the Hudson River and were the cause of many years of dispute. In 1613, the Dutch established trading posts along the Hudson and claimed jurisdiction for the territory between the Connecticut and Delaware rivers. The United New Netherlands Company (chartered in 1616) and later the Dutch West India Company (chartered in 1621) were vested with the government of this Dutch settlement. In 1664, King Charles II of England granted to his brother, the Duke of York, a large territory in America that included, with other lands, the area between the Connecticut and Delaware rivers. In 1664, with an armed fleet, the Duke of York took New Amsterdam and renamed it New York. In 1673, the Dutch recaptured New York and held it until 1674 when it was restored to the British by treaty.

The original area as described in the Duke of York's charter was reduced by many sales, cessions, and trades. With the exception of the relatively small triangle on Lake Eric, the colony relinquished no western lands.

The border between New York and Canada, intended to be along the 45th parallel, was found by the Collins and Southern survey of 1773 to vary as much as half a mile from this location. By the treaty of 1842, this survey was confirmed as part of the northern boundary of the United States. The monumented line between New York and New Jersey also varied from the intended straight line, giving New York about 10 square miles more than stated in the grant from the Duke of York to Lord John

Berkeley and Carteret. In 1834, each state ratified an agreement accepting the line as it was marked on the ground.

For the most part, the wilderness of New York was subdivided by privately owned land companies, and these companies could divide and sell their purchases as they saw fit. Little uniformity resulted between the various parts of the state. Most of the land companies patterned their names after the U.S. Rectangular System, using the terms *range* and *township*. Ranges varied in widths from 4 to 8 miles, were numbered from either the east or west sides, and were oriented as much as 45° off north. *Townships*, like ranges, were reasonably rectangular but varied in dimensions and directions, containing 40–144 lots. The lots, so called and found inside townships, were not usually marked on the ground prior to passing title. This is not so in New Hampshire.

The Military Tract for Revolutionary War veterans was an instance of the state's making a prior survey of a large area. Originally the area was to include 640,000 acres, but it was later expanded to include more than 750,000 acres. Two ranges of six townships apiece had each township divided into approximately 93 lots.

One of the outstanding private land subdivisions was the *Holland Land Purchase* located in western New York. The tract was referred to a *baseline* and *transit lines* previously surveyed. Pennsylvania's boundary line formed the baseline and lines run due north formed the transit lines. Created under the direction of Joseph Ellicott in 1797, the system contained townships nearly 6 miles on a side, with lot corners every three-quarters of a mile. With 64 lots in each township, numbered 1–64, beginning at the southeast corner and progressing north along each tier, a lot contained 360 acres nominally. Occasionally, some townships were divided into large lots, 1 to 5½ miles on a side. Townships were numbered toward the north from the baseline and westward from the east transit line. Joseph Ellicott, the brother of Andrew Ellicott, was intimately involved with this endeavor.

Tracts in the Phelps and Gorham purchases, located further west in the state, were subdivided by a similar plan.

In 1883, the state legislature ordered New York public lands to be resurveyed because of confused and undeterminable lines. During the years following, the survey resulted in the first accurate location of public lands. Triangulation, using 12- and 20-inch theodolites across the Adirondack Mountains, was the primary control. To complete this survey, it was necessary to locate almost all the boundaries of the large land purchases. The original notes of some of the earlier land company surveys were still in existence in 1884 when Verplanck Colvin, superintendent of surveys, conducted the state land surveys. Careful study of these old notes, along with Colvin's notes, would be necessary before a modern surveyor would attempt to resurvey the property lines in much of rural New York state.

Most of the New York state lands were titled under English common-law principles, but in the New York City area some titles were based on Dutch law; and, in these instances, the title to the streets and highways may rest with the state. A few titles go back to Swedish grants.

As surveyor general of the state of New York, Simeon Dewitt supervised subdivision of most of the lands in the northern and western parts of the state. Together with James Clinton and Andrew Ellicott, he surveyed the Pennsylvania–New York boundary from March 31, 1785, until the summer of 1786 at a cost of $4,750.27 for 90 miles. Every 20–32 miles they took sights on six or more stars with a 5-foot sector and accepted only those observations that agreed with the mean within 4 seconds. It was thought that their position error was within ½ second or 50.5 feet.

Simeon De Witt's brother, Moses, was a compass man on the Pennsylvania–New York boundary line survey in 1786. In 1826, De Witt made a detailed atlas of the state.

New Jersey The Virginia grant of 1606 included New Jersey. The first specific grant for this territory is the grant that all New Jersey title chains go back to, that of 1664 from the Duke of York to Lord John Berkeley and Sir George Carteret. In 1676, the lands of New Jersey were divided into Carteret's East New Jersey and William Penn's and other Quakers' West New Jersey by a line running from the extreme northwest corner of the state on the Delaware River to the southern tip at Little Egg Harbor. In 1702, the two parts were joined into the province of New Jersey and later the colony of New Jersey. The division line was surveyed by John Lawrence, George Washington's half-brother, in 1743, and it cut through several of the present-day counties.

New Jersey was the last state in which title to vacant lands resided wholly in proprietors. A court decision took away title to tide lands in favor of the state. The bounds of the grant were the Delaware Bay or River, the Atlantic Ocean, and a straight line connecting a point on the upper Delaware River with a point on the Hudson River. Considerable dispute has arisen over the jurisdiction of the islands in the Delaware River and over the location of the line between New Jersey and New York. New Jersey held no western lands.

In the early history of New Jersey, feudal tenures were established; later they were abolished by the following laws (still on the books):

> The feudal tenure estates, and the incidents thereof, taken away, discharged and abolished from and after March twelfth, one thousand six hundred and sixty-four, by section two of an act entitled "An act concerning tenures," passing February eighteenth, one thousand seven hundred and ninety-five, shall so continue to be taken away, discharged and abolished; and no such estate, or any incident thereof, shall, at any time, be created in any manner whatsoever. The tenures of honors, manors, lands, tenements, or hereditaments, or of estates or inheritance at the common law, held either of the king of England, or of any other person or body politic or corporate, at any time before July fourth, one thousand seven hundred and seventy-six, and declared, by section three of an act entitled "An act concerning tenures" passed February eighteenth, one thousand seven hundred and ninety-five, to be turned into holdings by free and common socage from the time of their creation and forever thereafter, shall continue to be held in free and common socage, discharged of all the tenures, charges and incidents enumerated in said section three [10].

Pennsylvania As a result of the careless royal grants, many areas were ambiguously described, and William Penn was in dispute with all his neighbors over boundaries. The original grant defined the boundary as being 12 miles distant from New Castle and along the 40th latitude north. It should be noted that the early geographers counted the parallels in a different manner, such that this would be the 39th parallel. Because of inadequate knowledge of geography and poor descriptions, the line between Pennsylvania, Maryland, and Delaware was in dispute for nearly 100 years. In 1760, the line was finally decided on and surveyed (1763–1767) by two English astronomers, Charles Mason and Jeremiah Dixon. Besides the "New Castle Arc," they were to run the south line on a parallel for a distance of 5 miles west of the Delaware, or 244 miles. Mason and Dixon determined the northeast corner of Maryland to be at latitude 39° 43′ 17.212, and later surveys have found the position to be 39° 43′ 19.91″, or about 180 feet in error. Huge limestone posts, cut in England with the respective coats of arms on the faces, served as monuments. The work of the survey was halted by Indian action short of the western border of Pennsylvania. The accuracy of the line is still a source of amazement; recent surveys reveal a variation of only 1″ or 2″ of latitude along this ancient line. A report by the legislature of Maryland (1909) lists a 2000-entry bibliography of manuscripts and documents relating to the line. The eastern part of the territory was originally held and occupied by the Swedish (Swedish/West India Company) in 1625. In 1655, the land was surrendered to the Dutch, who yielded to the British in 1664 during the conquest by the Duke of York.

A dispute resulted over the southern border, the location of which was not completely settled until 1921. The large area known as Westmoreland County was under dispute with Connecticut until it was granted to Pennsylvania in 1782. The Erie triangle, comprising 324 square miles, was originally part of New York and was ceded to the United States in 1781, which, in turn, deeded it to Pennsylvania in 1792 so that the state could have more access to Lake Erie. The conveyance was signed by George Washington and carried a consideration of $151,640.25, or about 75 cents per acre. Pennsylvania held no western lands.

In parts of Pennsylvania, the usual procedure for obtaining land was to buy a warrant for a given quantity of acreage in a given county near a developed area. After the selected site was surveyed and recorded at the expense of the purchaser, a patent became valid. In places, squatter settlements arose, and the settlers formed a mutual compact to settle all disputes over boundaries. Some titles, even today, are based on "squatter's rights" dating back to about 1718. With the diverse origin of titles, there have been many boundary disputes.

Maryland Although the territory was included in the earlier charters of Virginia, Lord Baltimore, in 1632, received a royal grant of the Province of Maryland. The governor of Virginia, also Lord Baltimore, proclaimed a recognition of the Province of Maryland in 1638. The treaty of 1638, which was reaffirmed in the constitution of 1776, formally gave the lands to Maryland. Controversy over the boundary between Maryland and Virginia continued through the years and was not finally resolved until 1929.

Another lengthy dispute continued between Pennsylvania and Maryland and was settled by the Mason–Dixon line survey. Lord Baltimore's grant was described, for the most part, by natural objects and called for no western lands.

Delaware Like Pennsylvania, the territory of Delaware was originally settled by the Swedish and in 1655 was surrendered to the Dutch—who, in turn, lost it to the Duke of York. William Penn obtained a 10,000-year lease of the area, although the Duke of York had an uncertain title. Later, Penn obtained a better title by royal grant. Under a proprietary government of Penn, the people of New Castle, Kent, and Sussex counties petitioned for separate statehood. Delaware, which never had claims on western lands, is still very nearly the original size of the three counties.

Carolina The charter to the Carolina territory (1663–1665) was described as running to the "South Sea." In 1729, the territory was separated into North Carolina and South Carolina provinces because of their widely separated settlements. The present state of Tennessee was ceded to the United States in 1789 and became a state in 1796. In 1787, South Carolina ceded to the United States a small strip of land about 12 miles wide lying south of the 35th parallel. Other western land claims were later ceded to the United States.

Carolina, like so many of the colonies, had poorly defined metes-and-bounds surveys that did not clarify the relationship to adjoiners. The lack of systematic procedures resulted in controversy and confusion among landowners and resulted in litigation.

Georgia In 1752, King George II separated the territory of Georgia from Carolina by a description calling for land to the South Sea. The exchange of cessions between Georgia and the United States netted the state $6,200,000, or about 11 cents per acre. The larger parts of Mississippi and Alabama were once in the original Georgia grant.

The separation of Georgia from the Carolinas began in 1717, when Sir Robert Montgomery obtained a grant from the Carolina proprietors for the land between the Savannah and Altamaha rivers and called it Azilia: "the most delightful country in the universe!" This land was to have square districts, 20 miles square, subdivided into various smaller squares, the least being 640 acres. The plan was never adopted. Also in South Carolina, by order of the king in 1729, 11 square townships of 20,000 acres were to be created, but only 9 were actually laid out. Not all were square, and not all contained the required 20,000 acres.

Georgia, the youngest of the 13 original colonies, was chartered by King George II as a charitable trust. In 1730, when the early Georgia settlers arrived with James Oglethorpe (Georgia's founder), Noble Jones, a surveyor, was second in command. The basic title to all lands emanated from King George II, King George III, or the state of Georgia. Oglethorpe had authority to grant lands in the new colony, and all grants of record in the surveyor general's office in Atlanta were made between the years 1755 and 1909.

The original boundary of Georgia extended to the South Seas but for all practical purposes stopped at the Mississippi River. Under the grants of land, the grantee was free to seek out only the most desirable lands. Large Revolutionary

War bounties were made in the Creek Cession to the Oconee River. These early grants were based on a specific number of acres for each person or head of a family, including slaves. Additional grants were made for sawmills, grist mills, and other purposes.

Starting in 1805, a series of six land lotteries of land parcels were surveyed for the purpose of granting the lands to private individuals by lottery. Each eligible person's name was registered in his county of residence, sent to the capital, and placed in a container. The presurveyed lot numbers were placed in another container, and the drawing of one paper from each container matched a person with a lot. After payment of a fee, the person was granted title. Survey instructions for the lotting were minutely detailed, and each county was divided into numbered land districts, with each district subdivided into square lots. Lot sizes varied from 490, 250, 2021/2, and 160 acres and 40-acre gold lots. Except for gold lots, lots were predicated on the value of the land or the producing capability of the soil (Figure 9.10).

This system was an abrupt departure from the metes-and-bounds surveys of the southern colonies and provided for surveys prior to disposal. Today, the office of surveyor general in the Georgia archives building in Atlanta maintains all grant information, original survey notes, and survey plats, plus all lottery information. Unfortunately, like so many of the early subdivided lands, few if any original corners are now recoverable.

Trans-Appalachian States Development of trans-Appalachian states was slow; Indians and a mountain barrier discouraged settlement. In addition, the French claimed the entire Mississippi Valley. The British sought to restrain settlement beyond the Appalachians in 1763 by establishing a series of forts and by proclamation in 1763 refused to issue title to any lands west of the Appalachians. After the end of the French and Indian wars, settlement was rapid. Kentucky and West Virginia were largely controlled by Virginia's liberal granting of warrants for land; settlement was indiscriminate, with numerous overlapping claims.

Tennessee Rights to land titles transcend English, state, and federal distribution of lands. The first settlement was made in the Watauga River Valley in 1769 by people who took possession without title or authority to settle. In 1772, these settlers negotiated specific leases for lands along the river.

The area of Tennessee was included in the original charter of North Carolina. After the Revolutionary War, North Carolina attempted to maintain the sovereignty of the Indians, but the settlers claimed the lands by conquest. The state granted tracts of 640 acres to settlers on a preemptive basis and also granted over 3 million acres as a military reservation. In all, over 8 million acres of Tennessee lands were granted prior to 1789, when the remaining lands were ceded to the federal government. The early entries were strictly selected on an indiscriminate basis; the best lands went first.

In the six years that Tennessee was a territory under the jurisdiction of the federal government, no specific intent was made to introduce the Rectangular System of surveys. Also during this period, North Carolina continued to grant lands in the area

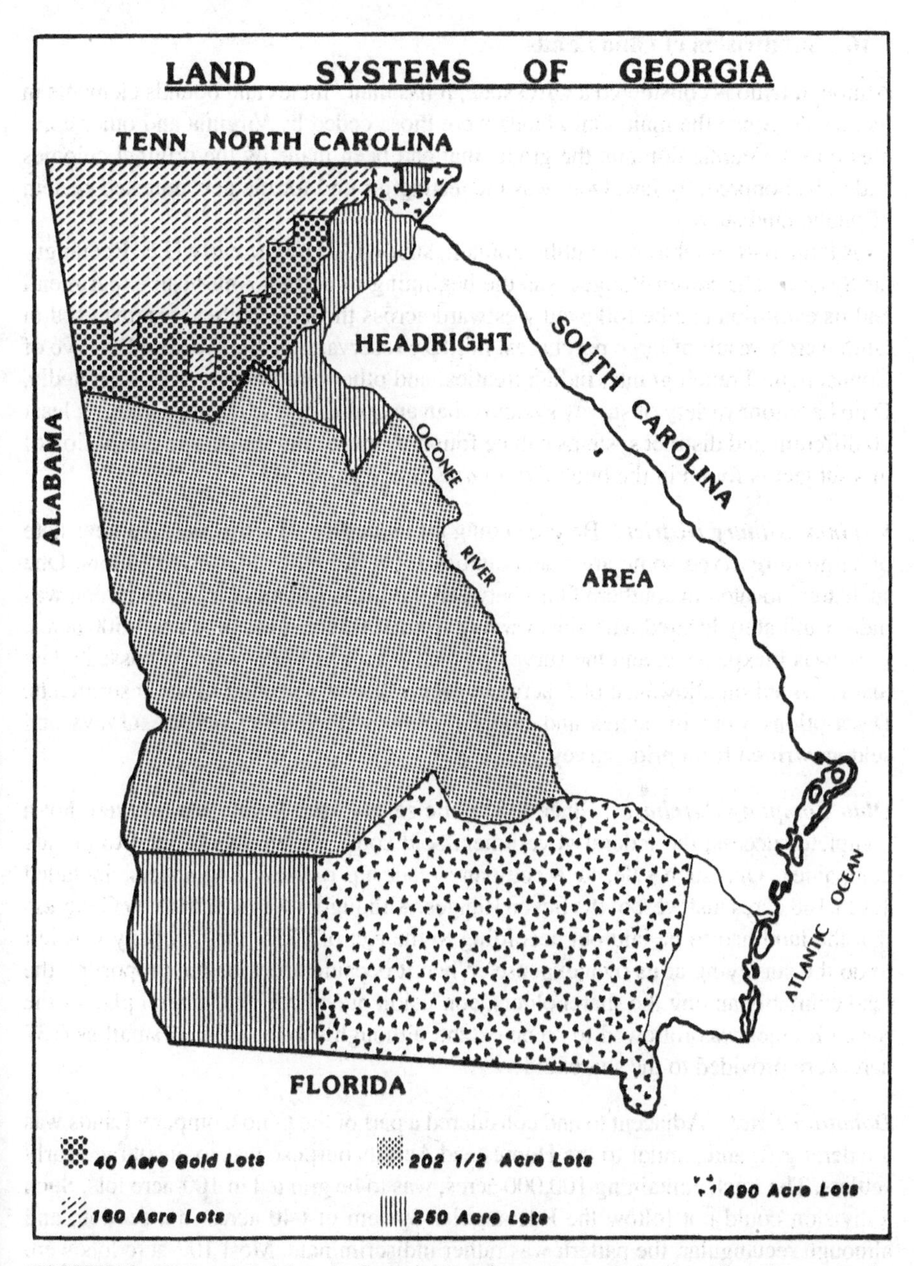

FIGURE 9-10. Land divisions in Georgia. Lot sizes were supposed to reflect the productivity of the soil. The 40-acre lots occur in the area of the first gold find in the United States.

ceded to the United States. A system predicated on townships, ranges, and sections is still in existence in middle Tennessee adjacent to the Georgia state line, with most land descriptions referred to the aliquot parts. Long ago, the field books and notes of the original survey disappeared.

9.16 Subdivision of Ohio Lands

Although Ohio is considered a GLO state, it has many metes and bounds elements in its history. Since the main Ohio lands were those ceded by Virginia and other colonies into the public domain, the grants that had been made by the original colonies had to be honored, by law. Ohio was the testing ground for the new untested system of public land surveys.

A large part of Ohio was public domain, subdivided by the early U.S. Rectangular System. The Seven Ranges was the beginning of a great American experiment, and its evolution can be followed westward across the state. Other systems used in Ohio were a result of large purchases, military reservations, the Western Reserve of Connecticut, French grants, Indian treaties, and other circumstances. Undoubtedly, Ohio has more variety of survey systems than any other public domain state. At least 20 different and distinct systems can be found. A more comprehensive discussion of this subject is found in the book *Ohio Land Subdivisions* [11].

Virginia Military District Before ceding all its northwest territories, the new state of Virginia reserved some areas as bounties for returning soldiers and sailors. One such area, located in southern Ohio between the Little Miami and Scioto rivers, was indiscriminately located with sites varying from 100 acres to as much as 1,500 acres. Land was inexpensive, and the surveying was crude. In 1779, Virginia passed a law that provided on allowance of 5 acres in every 100 for variation of the instruments. Descriptions were by metes and bounds, often written from partial surveys and seldom written from prior surveys.

Ohio Company Purchase Since the public sales of the Seven Ranges were not a complete success, the federal government sold some large tracts of land to private companies. One such sale, to a company made up of New Englanders, included 1,128,168 acres just west of the Seven Ranges. Although it was specified by Congress that the land had to be laid out according to the Act of 1785, the company was left to do the surveying at its own expense. When it became necessary to proportion the land equitably among the shareholders, confusion prevailed. The general plan of the Seven Ranges was followed to some extent, although subdivisions as small as 0.37 acre were provided to the shareholders.

Donation Tract Adjacent to and considered a part of the Ohio Company Lands was a federal gift; and similar to the Homestead Act, its purpose was to encourage early settlers. The tract, containing 100,000 acres, was to be granted in 100-acre lots. Such a division could not follow the Rectangular System of 640 acres in a section, and although rectangular, the pattern was rather indiscriminate. Most 100-acre lots were further divided into numbered lots.

French Grants To provide relief for some unfortunate French immigrants, several small tracts were granted along the Ohio River. Unlike the French grants along the Mississippi, these were not settled prior to the formation of the union but were made for people who were land-swindle victims. The tracts totaled to a little more than 25,000 acres and were subdivided into Ohio River front splinters. Ohio had surprisingly few grants prior to the Revolutionary War.

Symmes Purchase Between the Miami and Little Miami rivers are several tracts not included in the U.S. Rectangular System. One was the Symmes Purchase, which included 248,540 acres north of and including Cincinnati, all purchased at 67 cents per acre. The surveys within the Symmes Purchase were quite irregular, although the section numbering was the same as that prescribed under the Act of 1785.

Western Reserve The colony of Connecticut claimed title to lands between the 41st and 42nd parallels and from Pennsylvania 120 miles westward. On the basis of this claim, the state of Connecticut reserved a strip 120 miles long in the Cleveland area, and all titles to this land go back to Connecticut, although the jurisdiction over the area is and was under the state of Ohio. Using the Pennsylvania border as a meridian line, the land was subdivided into townships of 5 miles. In the western part of the reserve, 500,000 acres were set aside for relief of those who had lost their property in the Revolutionary War. Some of these 5-mile townships were divided into quarters and numbered clockwise; others were divided into sections of 1 mile square or laid out into tracts containing lots.

United States Military District The land set aside to pay the officers of the Revolutionary army was subdivided in a similar manner as that of the western reserve. Many of the townships (5-mile) were subdivided into quarter townships or 100-acre tracts.

Northwestern Ohio The last part of the state to be subdivided contained many different situations. Indian reservations, military posts, and lands donated for roads and canals under the "swampland acts" created irregularities. In general, when applicable, these lands were subdivided according to the U.S. Rectangular System, but many grants and reservations resulted in exceptions.

9.17 Prior Land Grants in Louisiana Territory

Prior to the purchase of Louisiana in 1803, the French and Spanish grants were made extensively along the tributary rivers to the lower Mississippi. Many French grants called for a certain number of arpents (1 arpent equals 191.8 feet, approximately) fronting on a river or bayou and extending back 40 to 80 arpents. Although the frontage on the water was usually definite, the direction and depth were often vague and indefinite. Land was plentiful, and the back portion was often swampy and unimportant.

Moses Austin, whose son Stephen Austin is known as the Father of the Texas Republic, and his associates were granted land in Missouri in 1797; but the depression following the Napoleonic Wars caused bankruptcy. Before his death in 1821, Moses received permission to settle 300 families in Texas. His son, Stephen, completed the grant.

Shortly after 1803, the United States appointed a commission charged with reviewing title claims from France or Spain. Proof of title right consisted of legal documents, surveys, or witness proof that land had been occupied for more than 10 years. After these claims were certified to, U.S. deputy surveyors segregated the land from the public domain by survey. In each case, the deputies were instructed to favor the claimant as much as possible and to make every effort to establish lines between claimants to the satisfaction of each.

HISTORY OF PROPERTY OWNERSHIP IN THE SOUTHWEST

9.18 Lands of Spanish Origin

The Spanish era began with the discovery of America and ended with the Mexican independence in 1821. The Mexican era in the southwest (encompassing California, Arizona, New Mexico, Texas, Utah, and parts of Colorado, Wyoming, Kansas, and Oklahoma) ended with the Treaty of Guadalupe-Hidalgo in 1848 and the Gadsden Purchase in 1853. After the Mexican rights were extinguished, Texas retained title to its vacant lands. Patents to lands in the southwest, after the Mexican era, were obtained from either Texas or the United States.

At the time of independence Mexico agreed to recognize Spanish titles, and at the Treaty of Guadalupe-Hidalgo both Texas and the United States agreed to recognize acceptable Spanish and Mexican titles. In the southwest there are four sovereign sources of original titles: (1) Spain, (2) Mexico, (3) Texas, and (4) the United States. Most of the patents issued by the United States were based on the Rectangular System, whereas the others were dependent on the system in force at the time and were often of the metes-and-bounds type.

Although many of our western states, including Texas, trace their historical titles and surveys to Spanish origin, it should be pointed out that only Texas requires a special license when it comes to retracing the boundaries of certain historical descriptions.

9.19 Seniority of Titles

Land grants and patents from Spain, Mexico, and Texas were, in general, but not always, made in sequence, with senior rights attaching to the older grant or patent. Because of this, the order of title alienation from the sovereign is important information needed to establish overlapping title rights.

Patents issued by the United States were, in general, surveyed prior to being offered for sale or for homestead. Descriptions of U.S. lands were usually of the lot-and-block type (called "sections" or "townships") simultaneously created without senior rights attaching to older patents. This fundamental difference (senior rights vs. lack of senior rights) has created a substantial difference in resurvey procedures; lands patented from the United States usually have proportionate rights to surplus or deficiency, whereas lands with senior rights do not share excess or deficiency (one person or the sovereign has title to excess, and another takes the loss).

9.20 Rights Included with Spanish and Mexican Grants

The foundation of all land titles is conquest followed by grants from the sovereign. At the time the Mexican rights were extinguished, the new sovereigns, Texas and the United States, could have refused to recognize existing land ownership but chose to recognize all valid titles as a matter of principle. Although Article 10 of the Treaty of Guadalupe-Hidalgo did stipulate recognition of all Spanish or Mexican titles, it was never ratified by the U.S. Senate on two grounds. First, the public lands of Texas belonged to that state, and the United States had no control over them. Second, with respect to lands in other areas, Article 10 was unnecessary, since valid titles to land in such areas were unaffected by the change in sovereignty.

9.21 Minerals

Land grants from either the Spanish Crown or Mexico did not pass title to minerals, not because of the language in the grants, but because of the general laws in force applying to all grants. The United States and Texas, each acquiring the Mexican and Spanish lands, became the owners of all minerals.

The Republic of Texas, by quitclaim, released all mines and minerals as of the date of the Constitution, 1866, to the owners of the soil.

9.22 Spanish Water Laws (1958)

The derivation of Spanish, Mexican, and English laws pertaining to waters and riverbeds stems from the Roman civil war [12]. In interpreting Roman law, the English adopted the construction that public interest attached only to the water and that the bed belonged to the adjoiner. In England, the sovereign owns the bed only insofar as the tide ebbs and flows.

By Spanish law, particularly in the United States, the general principle adopted was that the sovereign attached rights not only to the water but also to the bed. A distinction was generally made between navigable and nonnavigable streams.

In Spanish areas, the term *perennial stream* was used to describe the limit of sovereign rights to the bed. All nonperennial stream beds, sometimes called "torrential stream beds," belong to adjacent owners, whereas the bed of perennial streams was retained by the sovereign. In general, a perennial stream is considered to be one that flows all or most of the year, except in times of drought. A torrent or nonperennial stream is caused by an abundant rain or abnormal melting of snow and is dry a greater part of the year.

Until 1837, lands within the boundaries of Texas followed these rules. After that date, "the 30-foot-rule," by statute law, modified riparian ownership:

> All lands surveyed for individuals, lying on navigable water course shall form one-half of the square on the water course and the line running at right angles with the general course of the stream, if circumstances of the lines previously surveyed under the laws will permit. All streams so far as they retain an average width of thirty feet from the mouth up shall be considered navigable streams within the meaning hereof, and they shall not be crossed by the lines of any survey. All surveys not made upon navigable water courses shall be in a square, so far as lines previously surveyed will permit [13].

Prior to 1837, in Texas and the southwest in general, if the stream bed was perennial and lay wholly within a grant, title to the bed was reserved in the crown. If the grant bordered on a nonperennial stream, title of the bed, up to its center, passed with the grant.

After 1837, in Texas, where the bed of the stream was more than 30 feet, the sovereign retained title to the bed. If the bed was less than 30 feet, the title passed with the grant, and if the grant bordered the bed, up to the middle of the stream was granted.

This so-called 30-foot rule complicated ownership, since the law was and is ambiguous. Does "so far as they retain an average width of 30 feet from the mouth up" mean average width in front of the grant or average width of all measurements to such point that the average is 30 feet? A stream could be 100 feet wide at its mouth and narrow to 10 feet and yet have an average width of 30 feet. To date, the court has not ruled on this point.

The 30-foot statute also contained a clause that Texas grants "shall be square, so far as lines previously surveyed will permit." Most Texas land holdings are square or as nearly square as may be.

According to Spanish water laws, water could be appropriated and used. Unlike the remainder of the United States, water rights could be established by usage, and after the rights were established, they could be sold. California and the southwest in general recognized this principle. At the time Boulder Dam was built on the Colorado River, a treaty was made between the lower basin states (California, Arizona, and New Mexico) and the upper basin states, apportioning water between them. California became the biggest user of lower-basin water and claimed it on the basis of use. The U.S. Supreme Court decided that the treaty at the time of Boulder Dam's construction was binding; usage was not the deciding issue.

9.23 Road Beds

Because roads were reserved for the crown of Mexico in fee, abandoned roads in Mexican and Spanish grants, existing prior to 1848, revert to the state of Texas or the United States.

9.24 Early Settlements

Despite the rule that no settlement or town could be legally established without the approval of the king of Spain, many were. The Crown soon realized that towns or settlements could not exist without means of livelihood; hence, certain amounts of land were set aside and granted to the people to form *pueblos* or villages. Land was given to those who made conquests, and other land was sold; the remaining land could be used by the king's subjects to reap the benefits of natural resources. Although this land could be used, it could not be sold, tilled, or enclosed without permission. These common lands were known as *valdios* or *realengos*.

9.25 *Ordenanza de Intendentes*

Although there were rights in land granted prior to 1754, the *Ordenanza de Intendentes* of October 15, 1754, was the beginning for most land grants in the United States (and Texas). Its more important provisions were: (1) the appointment of officials who could legally approve titles in the name of the king; and (2) the granting of the right of perfecting a title within a specified time to those holding public lands (*valdios* or *realengos*) since 1700.

As a result of the Act of 1754, a commission granted some 400 *porciones* of land (1,000–1,500 *varas* by approximately 25,000 *varas* deep, or about 4,500–6,500 acres) along the Rio Grande River, and about 140 of these are in the state of Texas.

9.26 Mexican Land Grants

Although Mexico did not become independent until 1821, many events that started in 1808 led to the final declaration of independence. Following Mexican independence, land grants and colonizations were regulated by the central government. Later, colonizations were delegated to the states, and in what is now Texas, three different states (Coahuila and Texas, Chihuahua, and Tamaulipas) passed their own liberal land laws. As a result of Santa Ana's declaring a dictatorship on March 2, 1836, Texas colonizers proclaimed a republic independent of Mexico, and, from that date on, no further Mexican land grants were made in Texas. Texas joined the United States in 1845. In 1848, the Treaty of Guadalupe-Hidalgo ceded the southwest to the United States.

9.27 The Empresario System

Within the state of Coahuila and Texas the empresario system of colonization was established. The empresario was given land in exchange for bringing in families, and each empresario was given a specific area in which to operate.

Until 1820 the U.S. Congress sold land at $2 per acre, whereas in Texas it was much less. Texas grew rapidly, and about one-seventh of Texas was alienated by Spain and Mexico up until the time of independence.

In some Texas surveys, after some county boundary lines were created by running in the field, there was an attempt to try to adopt the GLO system of surveys by creating sections, townships, and ranges on a limited basis. Apparently, these were done under the auspices of the Texas General Land Office, but few land descriptions reference these surveys in modern land descriptions.

9.28 Suits Against the Sovereign

The king or sovereign cannot be brought to court without his consent; that is, the king can do no wrong. To straighten out many of the Texas title problems, the Texas legislature passed enabling acts allowing the citizens to bring title suits against the state. At various times the legislature of Texas passed laws voiding titles not recorded as of a specific date, but such acts were declared a violation of the Treaty of Guadalupe-Hidalgo.

9.29 Instructions to Surveyors

While Texas was under the government of Mexico, the following instructions were sent to the colonial surveyors by J. A. Nixon, commissioner for the eastern part of Texas:

> Sir: In consequence of your known honor, integrity, and ability, you are hereby appointed surveyor for the colonies granted to the following Empressarios by Supreme Governor of the State, to wit, to Lorenzo de Zavala on the 12th of March, 1829, to David G. Burnett on the 22nd December 1826, and to Joseph Vehlein and company on the 21st December 1826, and also another grant to Joseph Vehlein and company dated on the 11th of October 1828. And in the exercise of the duties of your office, you will be governed by the following instructions and such other as from time to time may be forwarded to you. Article 1st. Each and every surveyor shall provide himself with a compass after Rittenhouse construction. Article 2nd. All the lines shall be run in conformity with the TRUE MERIDIAN, and the greatest care shall be taken to have the horizontal measurement obtained by the chain carriers. Article 3rd In surveying, chains of iron or brass 10 varas long shall be used and the length of each link shall be 63 2/3 inches and the pins used for surveying shall not exceed 12 inches. Article 4th Field books must be provided to keep the notes. Article 5th The initials of the Grantee's name must be cut on the bearing trees at each corner with a marking iron, and a mound raised three feet high, and three feet in diameter at the base around the stake and the timber shall be so blazed near the line that it may be followed with ease. Line trees shall be blazed, and a notch cut above and below the blaze. Article 6th Rivers, large streams, and lakes must be considered natural boundaries, and no survey shall cross them, but their courses must be correctly taken and the contents of all surveys must be calculated by latitude and departure. Article 7th All surveys that do not close by 50 varas must be corrected and make each league contain 25,000,000 square varas as near as practicable. Article 8th On all natural boundaries, one half league front shall be allowed to each league of land so on in proportion to the whole quantity that may be surveyed. Article 9th The field notes must be forwarded to this office so soon as the surveys are completed. Article 10th No surveys will be acknowledged unless expressly ordered in writing by the Commissioner. Article 11th Special care shall be taken that no vacant land be left between the

possessions. Article 12th A report must be made qualifying the lands and giving as near as practicable the quantity of arable and grazing land contained in each survey. Article 13th All chain carriers shall be sworn by the Surveyor before commencing the survey, to perform their duties truly and faithfully according to the best of their ability and no person akin to the parties interested nearer than the fourth degree, shall be appointed to that survey [14].

John P. Borden, who was the first commissioner of the General Land Office under the Republic of Texas, issued these instructions:

The measure to be used will be varas and tenths of varas, or Spanish yards; each surveyor will therefore regulate his chain to the length of 10 varas, or, what is the same, 27 feet 9 1/3 inches, the vara being 33 1/3 inches; 5,099.01 varas on each side of a square will constitute a league and labor of land, or 4,605.53 acres; a labor of land will be equal to a square of 1,000 varas on each side, or 177.136 acres; and one acre, 5,645.75 square varas [15].

9.30 Survey of Spanish and Mexican Grants

A summary of the procedure for obtaining a Mexican land grant and the methods of surveying was written by Mario Lozano, a public land surveyor:

After complying with all necessary Judicial Proceedings, every one of which was a properly notarized instrument, the entire retinue composed of all the parties joining in the proceedings, would go to the PLACE OF BEGINNING where the first official act in connection with the actual survey was performed. The steps followed from this point on through the actual survey were as follows: The Judge of Measurements would command the surveyor to properly wax a rope, usually a hemp rope, and afterwards to measure 50 varas over said rope while held taut. The vara measure used in this ceremony was supposed to be a standard and properly certified vara. After this was done in the presence of the entire party, it was made part of the record after attestation. The next step, of course, was to start the actual survey. The instruments dealing with this phase of the proceedings are the actual Field Notes. In these instruments they recorded Direction or Bearing, passing calls for ravines, hills or any other physical characteristics of the terrain over which they measured and they often indicated the watershed. At the end of each day the work done was attested and made part of the Expediente. In connection with these proceedings it is interesting to note that what we often deduce as corners in an Expediente often were what they called "Parajes." Now a Paraje is a place. To be exact, a stopping place.

Often times they would get to a corner or at least to what in reading the Expedient we would expect to be a corner and this was called a Paraje also. This, even though it was a corner of the grant, was not necessarily an exact point in the sense we think of the corners we set today. A paraje can be said to be a site of small proportions and often this particular site constituted a corner. Furthermore, almost all Parajes designated as corners were given a name. The name usually being connected to some physical characteristic of the site, with some incident or with the Saint of the particular day when it was reached.

Not all Expedientes record the actual measurement of angles. But in instances when angles were measured, mention is made of the Graphometer: this is the case on some Mexican Grants. In other instances mention is made of the "Agujon" which was the

> name given the Compass. In all surveys the Judge of Measurements would cause the cordel[1] to be checked every fifty times it was used. After the survey was completed the Judge would command the Surveyor compute the areas and make his map. After this was done and made part of the Expediente, the Appraisers were appointed. Their acceptance of appointment was made part of the record and also their findings. The Judge would then draw an instrument of transmittal and send the Expediente to the Treasury official or Department, which in turn would add the necessary instruments to show that the appraisal value had been paid. After this, the Expediente was sent to the proper authorities for approval and confirmation. When this was done, the Expediente was returned to the Judge or Notary that acted through all these proceedings with instructions to place the Grantee in possession of his land. At this point the Judge would take the Grantee to his newly acquired land, admonish him to comply with the law by properly marking his corners with Monuments, and place him in actual possession by taking him by the hand, walking him over the land where he would dig, touch a piece of timber, pull some grass, take some water to irrigate the land (by sprinkling), and go through various other acts in keeping with the ritual or ceremony connected with giving him actual possession. After the act of putting the man in possession of his land in the execution of the proper instrument, the Expediente remained in the custody of the Judge or Notary that officiated and became part of the Protocol of said Notary. When a copy of the entire proceedings was requested and issued, it, of course, was a certified copy or "Testimonio" [15].

The procedure used in acquiring a grant when Mexico was New Spain and later when Mexico became independent are quite similar, since the procedure required to be followed for Mexican grants was in substance the same as that used in acquiring Spanish grants.

9.31 Gradient Boundary

In the Red River case of *Oklahoma v. Texas* in 1923 (260 U.S. 606, 261 U.S. 340, 265 U.S. 493) [16], the Supreme Court of the United States declared in an unprecedented action that the boundary between Oklahoma and Texas was to be determined by the "medial line" between gradient lines on each bank. Prior authority for such rule did not exist:

> One of the questions involved in the riparian claims relates to what was intended by the terms "middle of the main channel" and "mid-channel:" as used in defining the southerly boundary of the treaty reservation and of the Big Pasture. When applied to navigable streams, such terms usually refer to the thread of the navigable current, and, if there be several, to the thread of the one best suited and ordinarily used for navigation. But this section of Red River obviously is not navigable. It is without a continuous or dependable flow, has a relatively level bed of loose sand over which the water is well distributed when there is a substantial volume, and has no channel of any permanence other than that of which this sand bed is the bottom. The mere ribbons of shallow water which in relatively dry seasons find their way over the sand bed, readily and frequently shifting from one side to the other, cannot be regarded as channels in the sense intended. Evidently something less transient and better suited to mark a boundary

[1] A *cordel* is 50 varas (about 138.9 feet) and, as used, is a rope 50 varas long.

> was in mind. We think it was the channel extending from one cut-bank to the other, which carries the water in times of a substantial flow. That was the only real channel and therefore the main channel. So its medial line must be what was designated as the Indian boundary [16].

If the river had been navigable, the rule of "thread of the navigable current" would have applied. Determination of the "gradient boundary," a line on the bank, is difficult and requires great judgment. The gradient boundary cannot be determined without surveying long stretches of a river; hence it is costly to establish. After establishment, changing river conditions will change its location. Colonel Stiles discusses in detail the gradient boundary in the *Texas Law Review* [17], and surveyors attempting to establish a gradient boundary should certainly read this report.

Since the Red River case, Texas courts have made it clear that the gradient boundary principle is applicable to all boundaries between the state and private owners, even for grants made prior to 1837.

The student and the registered surveyor should realize the Gradient Boundary determination is unique only to Texas and cannot be applied to any other state. It is mentioned here because of the uniqueness of a single entity departing from the general path of surveying principles as to boundaries.

9.32 Effect of Native Americans

The accuracy of early surveys was sometimes in inverse proportion to the density of the native population. Indians were well aware that settlers soon followed the surveyor; by exterminating the surveyors and the results of their surveys, the Indians believed they could prevent the stealing of their land. Field notes and letters expressed concern of the troubles the surveyors were encountering with the various tribes from Florida to New Mexico. Native Americans tried their best to keep surveyors from doing their job.

Standard survey equipment included chains, axes, compasses, rifles, and pistols. Constant vigilance to avoid surprise attacks caused anxiety. No wonder occasional chain lengths were forgotten and corners falling in forbidding thickets were not set. Captain Barlett Sims' survey party in 1846 was surprised by Indians, and three of the four were killed. A party of ten surveying along the Guadalupe River, Texas, was attacked and all were killed. Captain John Ervey's survey party of ten had three killed in 1839. In 1838, Henderson's survey party of about 25 was attacked by Kickapoo Indians; in Texas history, this is called the "Surveyors' Fight." Eighteen were probably killed, and seven escaped. In Indian country, bobtail surveys were not uncommon; the surveyor set one or two corners and calculated the remainder [18].

9.33 Early California History

The coast of California was the object of a search for a shortcut to the Orient. Balboa, in 1513, claimed the Pacific shores in the name of the king of Spain. Cabrillo stopped at San Diego Bay in 1542. The Spanish government discontinued exploration of the California area after Viacaino's voyage in 1603.

By 1768, the Spanish government realized the need for colonization if the threat of English and Russians occupying the territory were to be prevented. The Jesuits were ousted from control of the missions in Lower California, and in 1767, the Franciscans assumed control. Father Serra established the first mission in upper California (San Diego) in 1769. Numerous missions followed, and by 1823 they numbered 21.

The soldiers and priests soon clashed over ownership of land. The Franciscans maintained, and at first successfully so, that all land was held in trust for the Indians until such time as they became civilized. The soldiers contended that the land belonged to the Crown, and, in accordance with the Spanish laws of 1773 and 1786, land could be granted to both the Indians and soldiers. By the time Mexico attained independence in 1821, only 20 private ranches existed in California. Soon Mexican law called for the expulsion of all missionaries, and a stringent secularization act in 1833 and 1834 ended the hold of the missions in California.

Under Mexican rule, land grants became frequent to those held in favor, and, during the early era, no records other than brief notes were kept. Who owned what was simply a matter of public knowledge.

During the Mexican era, many U.S. citizens moved into California, especially seagoing men. By claiming Mexican citizenship and proclaiming themselves Catholics, some obtained land grants. After war with Mexico started, the presence of Americans in California eased the problem of conquest by the United States.

By the end of the Mexican era (26 years), in 1847, the number of land grants in California had increased from 20 to 800. And many of these grants were made by Governor Pio Pico in the last days of his reign. The size of the grants varied from 28.39 acres (Rancho de la Canada de los Coches, which means "glen of the hogs") to one-fourth the size of Rhode Island (Rancho Santa Margarita y Las Flores in San Diego, Orange, and Riverside County contained, at one time, 226,000 acres, or 354 square miles) [19].

The historical knowledge goes beyond who did the surveys, it could also reach into local practices. In a recent trial concerning a ranchero in the redwood region of California, a young surveyor testified as to how he had found a blaze on a redwood stump that he considered as an original blaze from an 1866 survey. The second expert, a forester, testified that redwood bark was much thicker than normal trees and the usual trees and as such the majority of the blazes were "bark" blazes, and as such, the original blaze did not cut into the cambium of the tree, which was needed to create the blaze effect. Since the second expert was not from the local community, the judge disregarded his testimony and determined the blaze found was an original blaze.

9.34 Resurvey of Land Grants of the Public Domain

After the Treaty of Guadalupe-Hidalgo, all Mexican and Spanish land grants in the public domain had to be processed through the federal courts. As part of the procedure, a survey by a U.S. deputy surveyor had to be made, and all grants were made in conformance with the final approved survey. All indiscriminate parcels were described relative to the Rectangular System. Overlaps, except for one grant on another, were avoided.

PRINCIPLE 1. *In the retracement or resurvey of a Spanish or Mexican land grant, the surveyor cannot go behind the court decree that established the grant; the field notes and survey are a part of the grant, and the grant is senior to all sectionalized lands.*

9.35 Units of Measurements or Length

A surveyor working in the various sections of the United States or even the world will encounter references in deeds and descriptions that may cause problems or questions in determining if the reference is to area or length. Some of these are as shown in Table 9.1.

The arpent applies in the French settled areas as a unit of measurement as well as area. Like the rod, different values apply in different places. It is necessary when working with old descriptions with foreign units of measurement to determine what was applied in that area.

As stated earlier, the rod varied from 12 to 22 feet, depending on how it was used. Frequently, it was considered to be 16½ or 17½ to 18 feet; it was also a unit of area with an acre containing 160 square rods or a closed figure 1 chain (4 rods) by 10 chains (40 rods) = 160 square rods.

Of lesser importance, unless one practices in Texas, is the vara. There were numerous measurements for the vara, but Texas legislatively solved the problem by fixing the vara at 33⅓ inches.

Experience in retracing historic land grants seems to indicate that the surveys of these prior land claims were by the magnetic meridian, with little reliance on the existing federal statutes.

9.36 Summary

From the foregoing, it is apparent that the conveyance systems and methods of surveys within nonsectionalized states varied significantly from state to state. In older areas where anyone could perform a survey or prepare a map, many land location problems exist to this day. Today's surveyors should not be blamed for the unregulated systems existing in the past; the lack of regulations and the resultant chaos point

TABLE 9.1. Examples of variation in measurement definitions of the arpent. Similar variations occur with other units of measurement.

4 acres and 40 poles	=	4 acres and 40/160 square poles
1 arpent	=	0.8507 acres; Arkansas, Missouri
1 arpent	=	0.0845 acres; Louisiana
1 arpent	=	0.84725 acres; Louisiana, Alabama, Mississippi, northern Florida
1 arpent	=	32,400 square pieds (feet); Paris, Canada 36,800.67 square feet, 0.844,827.1 acre, 0.341,889.4 hectare
1 league square	=	1056 arpents = 6002.5 acres

to the necessity of good licensing laws qualifying who may practice land surveying. The system of allowing individuals to declare themselves as surveyors has to go the way of other obsolete ideas. As one can see, the responsibilities of the creating surveyor and all subsequent retracing surveyors as to evidence are quite different. Where the creating surveyor creates the boundaries with the underlying responsibility of leaving sufficient evidence for the following surveyors to find and to identify; the retracing surveyors must look for the remains of the original evidence and then utilize this evidence to retrace the original lines.

9.37 The Public Domain or GLO

As one can see, the metes-and-bounds system is an ancient system with no basic rules as to how surveys would be conducted by the field surveyor, and as such, the laws under which land disputes are litigated are common law that was dictated and molded as the need arose. The second major system in the United States is the Public Domain or Public Land Survey System (PLSS) or GLO system, which is a statutory system of surveys that was created under federal laws and is usually retraced under state laws.

> The major legal distinction between the metes-and-bounds system and the GLO system of surveys is that the GLO system is predicated on statute law, while the metes-and-bounds system is predicated on common (case) law.

As the colonial period of exploration and settlement came to a close, a new philosophy evolved, dictated and directed by a need of the "common man" to own his own land and be responsible to no person. One of the main catalysts of the American Revolution was not the price of tea, but it was the land policy adhered to by the king of England. Land grants made to favored settlers were restricted to an area east of the Appalachian Mountains. As settlers moved ever westward, the British Crown would permit no settlement of people wishing to homestead the "wild lands." British troops made regular thrusts into the "wilderness" and would destroy any homes and settlement that had been made, illegally. Most of the major land holdings were made to royalty in large grants, some in excess of 10,000 acres.

All of that changed in 1776 with the signing of the Declaration of Independence, and as a result of a change of land policy, more legislation was conducted by the Continental Congress concerning preemptors (illegal settlers, or squatters) than about financing the war. Immediately after the signing of the Treaty of Peace, Congress enacted the Land Act of 1785, an act that knew no precedent in any other country, then or now.

With large areas of wild land available for settlement, Congress had to find some manner to finance the late war, and having large areas of surrendered lands, they wished to use these lands as a source of financing the new nation.

Knowing that several of the former allies (France and Spain) and the former enemy (Great Britain) had previously granted millions of acres to individuals, Congress decided to recognize any and all legitimate land claims. That still left millions of additional acres available for sale to raise money.

To make certain the land would be conveyed in an orderly manner and to permit absentee owners to buy parcels without having to survey the boundaries, as in the metes-and-bounds states, a system was devised to permit a conveyance of a parcel with an absolute description, without the buyer ever having to set foot on the land or ever seeing it. Thus, the Public Land Survey System was devised and enacted into law. This system had three basic premises or requirements:

1. All adverse titles of land granted by the prior sovereigns would be recognized.
2. All title of the Indians would first be extinguished.
3. There would be no settlement prior to the land being surveyed.

This was quite a change from the metes-and-bounds system when the settler would go on the land, clear the land, and then survey what he either wanted or had under cultivation.

The philosophy and legal dictates of the PLSS (later to be called the General Land Office (GLO) system, after the agency which was given the responsibility of conducting the fieldwork and preparing patents), was developed within the first 20 years with three important acts, given in Appendix E. The student should read these laws and commit them to memory:

1785: The Northwest Ordinance created the system. It was very basic and left considerable latitude to the surveyors in the field. It provided for the sale of large areas of land. No instructions were provided as to equipment to be used; it only addressed methodology.

1796: The Act of May 18, 1796, refined the system and provided for the survey and sale of smaller parcels. It identified the compass and designated that lands were to be surveyed with a two-pole chain.

1805: The Act of February 11, 1805, is more important to today's surveyor for its legal ramifications than its survey ramifications. This Act provided for the legal fact that original surveys of the public lands, when conducted in accordance with the law, are unalterable and unassailable, after a patent had been issued. The basic legal premise the retracing surveyor must understand is that the original surveys of public lands are legally without error, and no court or subsequent surveyor can attack an approved survey.

The story of public domain is a story of force and fraud that reached from the local politician into the halls of Congress. Many volumes have been written as to how lands were acquired (some say stolen) from France (some say from Spain) and disposed of (some say legally and some say illegally). To say the least, there never was and never will be a land survey of that magnitude. The surveys have been in constant motion for over 200 years, creating new boundaries and reidentifying old ones.

Throughout this nation, about three-quarters of the area of real property may be traced back to the public domain. There are, at present, about 720,004,000 acres of land area still in the public domain.

Public domain is the vacant land held in trust by the government for the people. This area does not include federal reservations or other government lands set aside or purchased for specific purposes. It was acquired by purchases and cessions at a total cost of $85,179,222, or less than 5 cents per acre. The largest-cost item has been Indian claims, some of which are as high as $2.50 per acre plus interest. Table 9.2 shows the contributions to the public domain, and Figure 9.11 shows the land's location. The public domain system was one created by and conducted under federal law. The laws dictated how it would be surveyed, how it would be disposed of, and how it would be resurveyed. Public domain (PD) is used in the GLO (General Land Office) system, which conducted many of the surveys, and the Public Land Survey System (PLSS). The GLO and PLS systems probably embrace one-third to two-thirds of the land area of the United States, yet a surveyor will find metes-and-bounds surveys within surveyed townships and what many surveyors consider as metes-and-bounds states will also have Rectangular System surveys. Some of the better-known states that contain both are Maine, Vermont, Texas, South Carolina, Tennessee, Georgia, New York, and Pennsylvania.

As one becomes more versed in the various systems of surveys, a very interesting fact will become apparent: The majority of land and boundary litigation occurs in the metes-and-bounds states rather than the PLS states. Several states have adopted, either by statute or by case law, the laws that affect and that created the PLS System.

9.38 The Ordinance of 1785

In 1784, a task committee, chaired by Thomas Jefferson, was created to draft "an ordinance establishing a land office for the United States." As a law graduate from William and Mary College, an attorney, and a man experienced in surveying (his father, Peter Jefferson, was a surveyor—actually, a cartographer), Jefferson was well qualified to suggest new and better methods. Although much credit is due Jefferson for introducing the Rectangular System to America, it was not entirely original with him. It is interesting to note that Jefferson's initial proposal would be a rectangular system that would be 10 "reformed" miles square, having 100 square sections of 100 reformed acres each. Jefferson's mile would be approximately equal to a nautical mile. This was never enacted into legislation.

Egyptians were known to use a rectangular system in the Nile Valley surveys. Land divisions by meridians and parallels had been of religious and mystical origin. The words *decumanus* (east and west) and *cardo* (north and south, the cardinal directions) were in use with an instrument called the groma. One-hundredth of a nautical mile was the ancient division of the shire in England. This was adopted in America by Delaware, Maryland, and Virginia, but most of the land in the colonies was being sold with indiscriminate location, particularly in the South. North Carolina had a law requiring orientation with true north, although there were many permitted exceptions. After much debate, a quadrangle with seven English miles on

TABLE 9.2. PRINCIPAL SOURCES OF LAND TO THE PUBLIC DOMAIN.

Date	Acquisition	Area (acres)	Total cost (dollars)	Cost per acre (cents)	States included
1781–1787	Cessions from the original states	236,825,600	6,200,000[1]	11	Wisconsin, Michigan, Ohio, Illinois, Indiana, Alabama, Mississippi
1803	Louisiana Purchase	529,911,680	23,213,568	4	Louisiana, Mississippi, Arkansas, Missouri, Oklahoma, Nebraska, Kansas, Iowa, Minnesota, North Dakota, South Dakota, Wyoming, Idaho, Montana, Colorado
1819	Spanish cession of Florida	46,144,640	6,674,057	14	Florida, Louisiana
1825	Oregon compromise	183,386,240			Washington, Oregon, Idaho, Montana, Wyoming
1848	Mexican cession	338,680,960	16,295,149	5	California, Nevada, Utah, Wyoming, Colorado, Arizona, New Mexico
1850	Texas purchase	78,926,720	15,496,448	20	New Mexico, Colorado, Wyoming
1853	Gadsden Purchase	18,988,800	10,000,000	53	Arizona, New Mexico
1967	Alaska purchase from Russia	375,296,000	7,200,000	2	Alaska
Total		1,808,160,640	85,179,222	(5¢ average)	

[1] Part of the cession from Georgia.

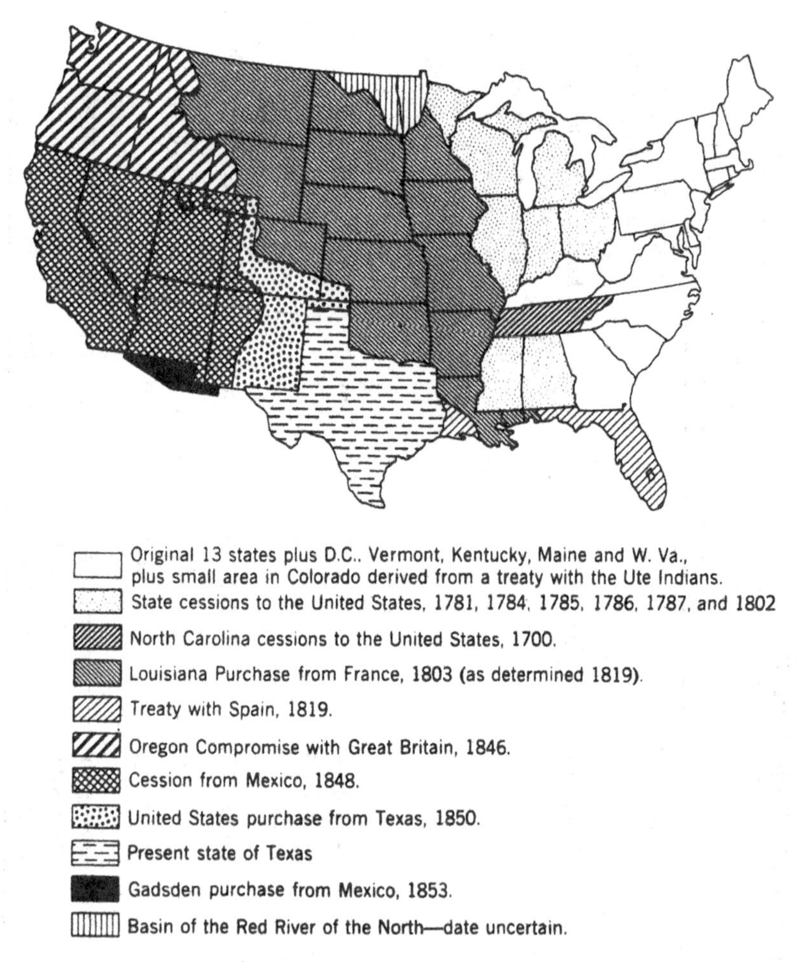

FIGURE 9-11. Acquisition of territories.

a side, containing 49 parts or square mile lots, was called for. Seven soon gave way to 6, but the total of 49 was overlooked. Later, the law provided for a numbering system: "numbered 1 to 36, always beginning the succeeding range of the lots with the number next to that with which the preceding one concluded." The final plan began numbering in the southeast corner, progressing north, then progressing south on the next tier, and so on. About half of the plats prepared for the Seven Ranges carried this scheme on the face, but it is felt that the surveyors did not follow such a pattern in the field.

9.39 Prior Surveys

The survey of the Seven Ranges introduced the system of prior surveys—that is, surveys prior to selling lands.

(a) (b)

FIGURE 9-12. Original GLO bearing tree (a) serving as modern fence brace; and (b) showing overgrowth over original blaze.

Townships and sections were a compromise between the North and the South: the North wanted smaller tracts that would be available to the common man, whereas the Southern plantation owners thought more in terms of thousands of acres. The 640-acre unit was thought to be within the reach of homesteaders and small farmers.

The new law specified lines to be run in the cardinal direction, after correcting for the variation of the needle. Certification was required by the surveyor: The surveyor was to mark trees and to erect corner monuments and set "bearing trees" (Figure 9.12).

The plats and notes were to include information on the topographical and cultural features on the land so that purchasers could judge the value of the lands.

9.40 Geographer of United States

The organizational structure of the Rectangular System called for the geographer of the United States to direct surveys and transmit plats of completed surveys to the Board of Treasury. Until 1784, Simeon De Witt occupied this post. Upon resignation, he was succeeded by Thomas Hutchins.

9.41 Beginning Survey of Seven Ranges

The nearest land available, that adjoining Pennsylvania, which was part of the original 13 colonies, was selected for the start of public land surveys, but before the land could be divided, the western line of Pennsylvania had to be established. In 1784, a survey was necessary to establish a beginning point. In 1784, a group of commissioners from Pennsylvania and Virginia began extending the Mason–Dixon line the final 22 miles needed to reach Pennsylvania's southwest corner. More recent surveys have shown that the mean error in this line was less than 1 second of arc. From the southwest corner of Pennsylvania, four commissioners, among whom were David Rittenhouse representing Pennsylvania and Andrew Ellicott from Virginia, ran a line northward for 63 miles until the Ohio River was reached. It is believed that the commissioners used a transit instrument equipped with a telescope and that points were set on the tops of hills at ½- to 2-mile intervals.

Polaris as well as other stars were observed to maintain true direction. At the end of this line an error of about 50 feet exists, an accuracy that was not equaled for many years to come. In recent years the starting point on the north bank of the Ohio River was perpetuated by a large granite monument erected in memory of the momentous occasion.

David Rittenhouse, born in Germantown, Pennsylvania, in 1732, became an astronomer, calculator, clock builder, and instrument maker. He had run the circular portion or the boundary between Pennsylvania, Delaware, and Maryland (a circle of 12-mile radius with Newcastle as its center). He also established the point from which the dividing line between New York and Pennsylvania was to be traced westward (42° of latitude) [20].

The western border of Pennsylvania has become known as "Ellicott's line." This is not to be confused with Ellicott's line of the 31stt parallel. Ellicott later designed the streets in Washington, D.C., and surveyed the boundary between the United States and Florida. Thomas Hutchins began running the "geographer's line" west from the point on the north bank of the Ohio River in 1785. Hutchins' sextant determination of latitude of the beginning point was in error by about half a mile. Because of an Indian attack in October, only 4 miles of this baseline was run in the first year. With verbosity that has never been equaled, his report to Congress, written closely spaced on eight foolscap pages, reads in part:

> For the distance of Forty six Chains and Eighty six links West . . . the Land is remarkably rich, with a deep black Mould. free from Stone, excepting a rising piece of ground on which there is an improvement of about 3½ Acres, where there are a few Grey and Sand Stones thinly scattered. The whole of the above distance is shaded with black and white Walnut trees, also with Black, Red and an abundance of white Oaks, some Cherry Tree, Elm Hoop-Ash, and great quantities of Hickory, Sassarfrax, Dogwood, and innumerable and uncommonly large Grape Vines producing well tasted Grapes of which Wine may be made. All the Hills in this part of the Country seem to be properly disposed for the growth of the Vine. Near the termination of the above mentioned measurement is a thicket of Shoemack, Hazel and Spice bushes, through which a passage was cut for the Chain-carriers. The first of these Bushes produces an Acid berry well answering the purposes of sowering for Punch, the Hazel yield an abundance of Nuts, and the Spice bushes bear a berry, red when ripe of an aromatic smell, as is also the Shrub on which it growes: the berry is about the size of a large Pea, of an Oval shape possessing some Medicinal virtues, and has often been used as a substitute for Tea by sick and indisposed persons. The Dogwood, the bark is used by the inhabitants and is said to be little inferior

to Jesuits bark in the cure of Agues: the Tree produces a berry about the size of a large Cranberry when ripe, but something longer and smaller toward the Ends, excellent for bitters: and decoctions made of the budds or blossoms have proved very salutary in several disorders, particularly in Bilious complaints. The whole of the above described Land is too rich to produce Wheat, the aforementioned rising ground excepted, but it is well adapted for Indian Corn, Tobacco, Hemp, Flax, Oats &c. and every species of Garden Vegetables, it abounds with great quantities of Pea Vine, Grass, and nutritious Weeds of which Cattle are very fond, and on which they soon grow fat [21].

The greatest influence Hutchins had on the beginning of the Rectangular System was the precedent established by him for meticulous notes and descriptive plats. The crude locations can be explained in part by the threat of Indians, untrained help, and the lack of suitable instruments.

At the completion of each six miles of baseline established by Hutchins, one group of surveyors was to take off to the south. The selection of whom was to be first was determined by lot, but, as it turned out, Hutchins did not make six miles the first year. The "secant method" was employed in running the parallel of latitude, and the direction was determined by observing a star or stars with a sextant. The beginning of the baseline had an error of 25.2 seconds of latitude and the western end, when completed, was 1,500 feet south (15 seconds) of its intended latitude. It is interesting to compare this accuracy for 42 miles of line with that of the Mason–Dixon line, where 244 miles of line were established with only 2.3 seconds of error.

9.42 The Survey of Seven Ranges

Each state was to send a surveyor, but only eight surveyors showed up for duty, and none of these performed work the first year. When actually surveying, none turned in good results with their circumferenters (see Figure 9.4). On May 12, 1786, the requirement of running lines "by the true meridian" was repealed, and township lines were projected by magnetic compass.

The length of the Gunter chains used for distances was checked by Hutchins, and a horizontal measure was advocated. Closures on townships were quite poor and nonuniform in character. The numbering of townships progressed south to north, and ranges were numbered from east to west, for the most part, there were no common section corners.

Surveyors failed to show in their notes what had transpired and contributed little, except for good marking of trees, fair record of points set, and recorded distances measured.

9.43 The Ordinance of 1796

The Act of May 18, 1796, amending and superseding the Act of 1785, was a victory for advocates of prior surveys and the Rectangular System. Small buyers were provided for in alternate townships wherein section lines were marked at two-mile intervals. Although all interior section corners were not marked, the new plan would provide the establishment of three corners for each mile square. Large buyers were sold land in other alternate townships wherein only township exterior mile posts were preset.

For some unknown reason, the numbering system of sections in townships was changed to the following:

> The sections shall be numbered, respectively, beginning with the number one in the northeast section and proceeding west and east alternately, through the township, with progressive numbers till the thirty-sixth be completed [22].

The previous described system was more logical, since it followed more closely the procedure used to divide townships, but the numbering system adopted in 1796 is still in force today. The pay of surveyors was raised from $2 per mile of line to a maximum of $3 for all expenses except occasional military escort along Indian boundaries.

More specifications appeared in the Act of 1796. The marking of trees, the notation of cultural and topographic features, the standard length of the chain, and the office of surveyor general in Marietta, Ohio, were now written into the law. Provision was made for special surveys and deviations from the plan when the lands were adjacent to prior grants, bodies of water, and other irregular boundaries. Townships of 5 miles square, marked at 2½-mile intervals, were called for in the military district and the missionary tracts. In all, the Act of 1796 was clearer and more explicit than the Act of 1785.

After working under the Statute of 1785 for 10 years, and with the changes permitted as to the size of the federal patents to smaller parcels, this Act of 1796 permitted the placement of more exterior section corners in the interior townships. It also corrected the omission of the reference to the length of the chain that would be used, identifying the two-pole chain of 50 links.

9.44 The Act of May 10, 1800

The Act of Congress, approved May 10, 1800, required

> [the] townships west of the Muskingum, which . . . are directed to be sold in quarter townships to be subdivided into half sections or three hundred and twenty acres each, as nearly as may be, by running parallel lines through the same from east to west, and from south to north, at the distance or one mile from each other, and marking corners, at the distance of each half mile, on the lines running from east to west, and at the distance of each mile on those running from south to north . . . And in all cases where the exterior lines of the townships thus to be subdivided into sections or half sections shall exceed, or shall not extend, six miles, the excess or deficiency shall be specially noted, and added to or deducted from the western and northern ranges of sections or half sections in such townships, according as the error may be in running the lines from east to west, or from south to north [23].

By the year 1800, the placing of excess or deficiency in the north and west tiers of sections was established. But the subdivision into quarter sections was not recognized.

9.45 The Act of February 11, 1805

Of all the statutes enacted by Congress, this Act has had the most far-reaching implications for the retracing surveyor. It is the foundation of the legal requirements of the retracing surveyor, in that the retracing surveyor is mandated to look for the evidence the original surveyors created.

The importance of this Act, as it relates to the metes-and-bounds states, is that several of the courts in the metes and bounds, in writing decisions, have used the exact wording in rendering a decision.

The law also provided for the subdivision of sections into quarter sections:

> Corners of half and quarter sections not marked shall be placed, as nearly as possible, equidistant from those two corners which stand on the same line. Boundary lines (center lines of Sections) which shall not have been actually run and marked as aforesaid shall be ascertained by running straight lines from the established corners to the opposite corresponding corners: but in those portions of the fractional townships, where no such opposite or corresponding corners have been or can be fixed, the said boundary lines shall be ascertained by running from the established corners due north and south or east and west lines, as the case may be, to the external boundary of such fractional township [24].

The direction due north and south or east and west has been interpreted by the courts to mean due north and south or east and west in the average direction run by the original surveyor as determined and tested by nearby lines. It is not generally interpreted to mean astronomic north.

Perhaps the five most important elements of this Act are as follows:

1. The original surveys have no error.
2. No person can correct an original survey, except for the agency that created it.
3. Area recited in the patents is the least important; boundaries are paramount.
4. The least size of a regular aliquot parcel is 40 acres, a quarter-quarter of a section.
5. Corners not set, but called for in the law, are legal, even though not monumented.

9.46 Principal Meridians and Baselines

No principal meridian was established in the Seven Ranges, although the point of beginning was on the western boundary of Pennsylvania. The geographer's line served as Hutchins's baseline and was in keeping with Jefferson's original plan of 1794 to base the surveys on the states' boundaries. What is now known as the First Principal Meridian and is also the boundary between the states of Ohio and Indiana was run north from the Great Miami River. The 41st parallel of latitude served as the baseline. To avoid rough terrain, the Second Principal Meridian was run through the eastern end of the Vincennes Tract in Indiana. The baseline, run by Ebenezer Buckingham Jr., in 1804, was started from a point in Illinois 67.5 miles away and was located in the southern part of the state, since the Indian claim to lands had not been settled in the north.

Beginning at the confluence of the Ohio and Mississippi rivers, the Third Principal Meridian ran the length of Illinois. The baseline for this territory was one extended westward from the Second Principal Meridian. To define the lands contained in a 2,000,000-acre military tract in northwest Illinois, the Fourth Principal Meridian was extended from a point north of the Illinois River on the meridian of the mouth of that river. The baseline was established westward from this point to the Mississippi. Another baseline, coincident with the Wisconsin border (latitude 42° N), serves the area northward in Wisconsin and parts of Minnesota.

Of the 40-plus principal meridians in the United States (Figure 9.13), the first six have numbers, and the remainder are named. Most of the locations were arbitrary or according to some natural features instead of coinciding with the state boundaries as Jefferson had planned.

Very little was stated in the early instructions about the establishment of the principal meridians and baselines. It is presumed that the surveyors followed customs laid down by the men who surveyed in the Seven Ranges. Even though the surveyors were relieved from running according to "true" north, the directions were taken with considerably more care than in the other surveys, but the distances were no better than that observed elsewhere.

9.47 The Surveyor General

As provided in the Act of 1796, the post of surveyor general, replacing the geographer, at a pay rate of $2,000 per year, was to be appointed to a man who would engage skilled deputy surveyors, administer oaths, frame regulations, and issue instructions. General Rufus Putnam, the first surveyor general and Washington's aide during the Revolutionary War, had long experience as a surveyor, both in the colonies and in the states, and was an advocate for the rectangular surveys. During the seven years that he served (1796–1803), he faced many problems in Ohio. He established the First Principal Meridian and baseline in accordance with Jefferson's plan; it is suspected that he was the author of the current system of numbering of sections, and he was the first to place excess and deficiency into the north and west tiers.

Jared Mansfield, a man of great learning and scientific background, succeeded Putnam in 1803, becoming the second surveyor general of the United States. He held this office until 1814. Mansfield's most noteworthy contributions to the Rectangular System were his proposal and execution of a general framework for the division of land into townships and his resolution of the conflict between rectangularity and the convergence of principal meridians.

Although the idea was not completely new, it was Mansfield who clarified the necessity of baselines and principal meridians. His first application was the laying out of the Second Principal Meridian through the present state of Indiana and the baseline from Clark's Grant to the Wabash River in 1804. He later extended this baseline to the Mississippi River, and he established the Third Principal Meridian in Illinois. From this framework, he ran the township lines, devising a new numbering system, whereby the townships were counted from the principal meridians and baselines rather than from natural boundaries [25].

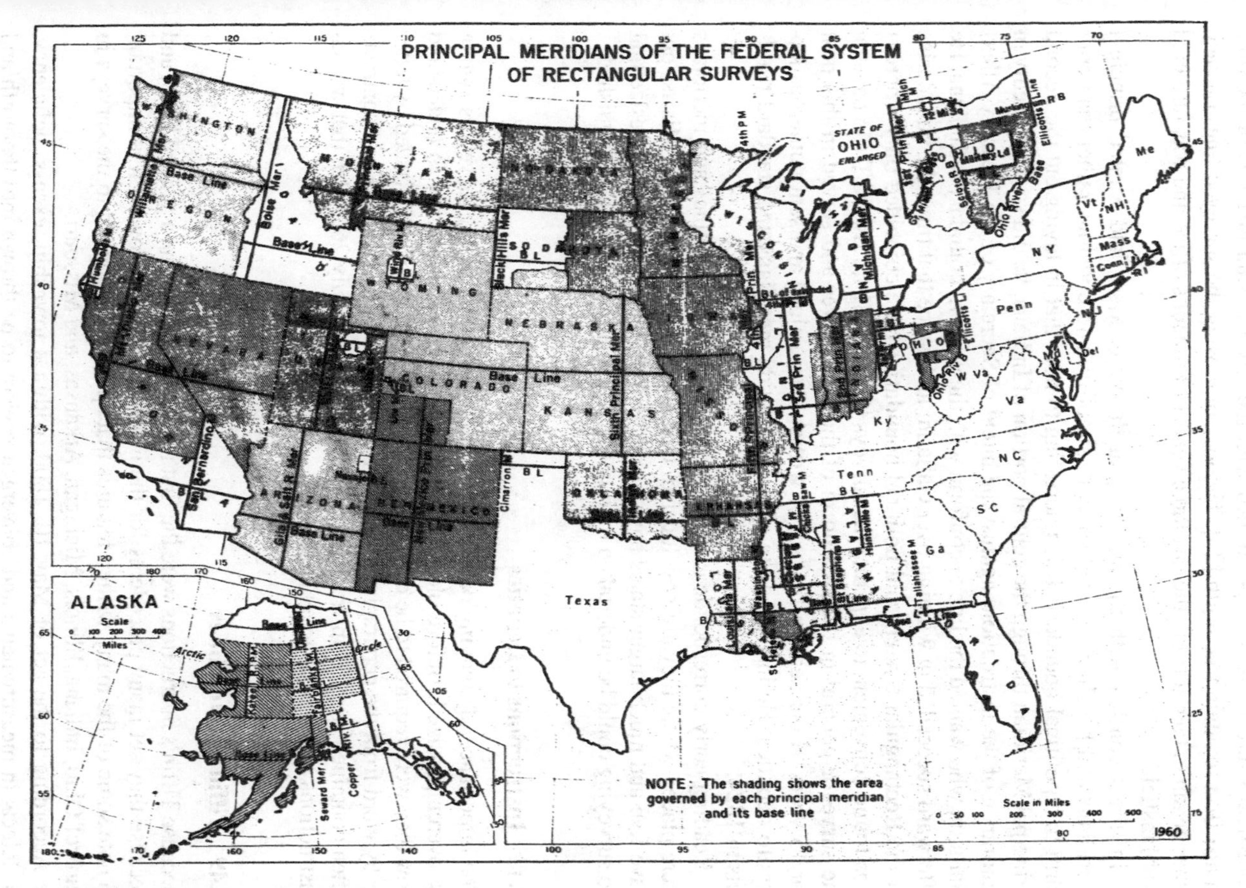

FIGURE 9-13. Principal meridians and baselines.

In his instructions, Putnam did not require true meridian lines and allowed meridians to run by the compass with slight regard for deviations. The lines run under Mansfield's directions were performed with the best celestial methods of the time in an effort to attain true north [26]. The idea of closing the surveyed lines on each other began to take form in the last three years of Putnam's office and was furthered by Mansfield.

In direct conflict with the authority at the time, Mansfield wanted principal meridians and baselines located so as to simplify land divisions rather than conform to a political boundary. It is interesting to note that in laying out the Second Principal Meridian through the central portion of Indiana, he did so (supposedly in ignorance of the instruction) against the intent of Secretary of the Treasury Albert Gallatin, who wanted the line to conform with the western boundary of Indiana. He afterward succeeded in winning over Gallatin; and, since that time, the boundaries of the Rectangular System and political boundaries have coincided only where it was found convenient to expedite the surveys [27]. In 1814, Edward Tiffin became the surveyor general in the area north of the Ohio River and served in this post for 15 years. He issued the first set of written instructions and sent numerous letters to deputies. He instituted the scheme of guide meridians and standard parallels in 1824.

Reading early correspondence gives no clues why the baselines and meridians were placed where they are now located. One would ask, "Why would Mississippi have five and much larger states have one or two?" One can only hypothesize that no surveying could be undertaken until native land claims had been extinguished.

9.48 Instructions to Deputies

The importance of studying original instructions issued to deputy surveyors cannot be overemphasized. Before old lines can be retraced, it is necessary to understand how they were required to be established. The first instructions were issued by letter or by word from the geographer and later from the surveyor general. Early contracts often contained specific instructions, but it was not until July 26, 1815, that written instructions were issued to all deputies.

9.49 Tiffin's Instructions

Because Tiffin's instructions were the first issued instructions on a general basis, and because they set many precedents for the instructions that followed, it is important to note some of the many written features that specifically applied to the surveys in parts of Ohio, Indiana, Illinois, Michigan, Arkansas, and Missouri.

According to the act of May 10, 1800, within a township, excess, deficiency, or defects in measurements and convergence were to be thrown into the north and west tiers of sections. Quarter section corners were to be established on a straight line between section corners and, except for closing lines, were to be at half-mile intervals. These instructions are similar to that used today, but in closing lines, quoting from Tiffin's 1815 instructions, there was a significant difference:

> Also in running to the western (township) boundary, unless your sectional lines fall in with the posts established there for the corners of sections in the adjacent townships, you must see posts and mark bearing trees at the points of intersection of your line with the town boundaries, and take the distances of your corners from the corners of the sections of the adjacent townships, and note that and the side on which it varies in chains, or links or both. The sections must be made to close by running a random line from one corner to another except on the north and west ranges of sections [28].

Double corners were permitted on the north and west sides of a township; in effect, this meant double corners were permitted to exist on any side of townships. The practice of setting double corners was partially abolished in 1843 and almost completely abolished in 1846 (double corners are used occasionally today). The procedure in subdividing townships was to begin at the southeast corner of the township, where the variation of the needle was checked with the township line, then retracing along the south line of the township, restoring the quarter corner if not in place. At 1 mile (south corner of sections 35 and 36), a meridian line was run northward, possibly parallel with the adjacent meridian according to Tiffin's suggestion, and a quarter corner was erected at 40 chains. At 80 chains, the corner common to sections 25, 26, 35, and 36 was erected. From here a random line was run to the east until it intersected the township line. The amount of falling to the north or south of the standard corner was noted, as well as the chainage of the line. The true line connecting section corners was marked by offsets, and the quarter corner was set on line and equidistant. Next, the meridian line between sections 25 and 26 was established, and so on, until the last mile was reached. On the meridian line between sections 1 and 2, a quarter corner was placed at 40 chains, and the line was extended until it intersected the north line of the township. Here a "closing corner" was erected, and the falling east or west of the standard corner was noted, as well as the total chainage of the line. This procedure was continued for all the tiers of sections, progressing from east to west, until the last tier was reached. When the corner common to sections 29, 30, 31, and 32 was established, and a random line to the east was run, and a true line returned, a line, at right angles to the meridian, was run westward and a quarter corner erected at 40 chains. The line was continued westward; at the intersection with the township line, a closing corner was set, and the falling north or south of the standard corner was noted, as well as the distance between the previously set quarter corner and the township line (Figure 9.14).

This procedure was, in many respects, the same as is followed today. The order of running lines and setting corners was and is the same. Of course, the instruments used were not as good or as accurate as those used today; the direction north, as determined by the compass, could not have been very accurate.

Today, the instructions are definite that all section lines shall be run parallel with the township line. Parallel lines running northward cannot have the same bearings;

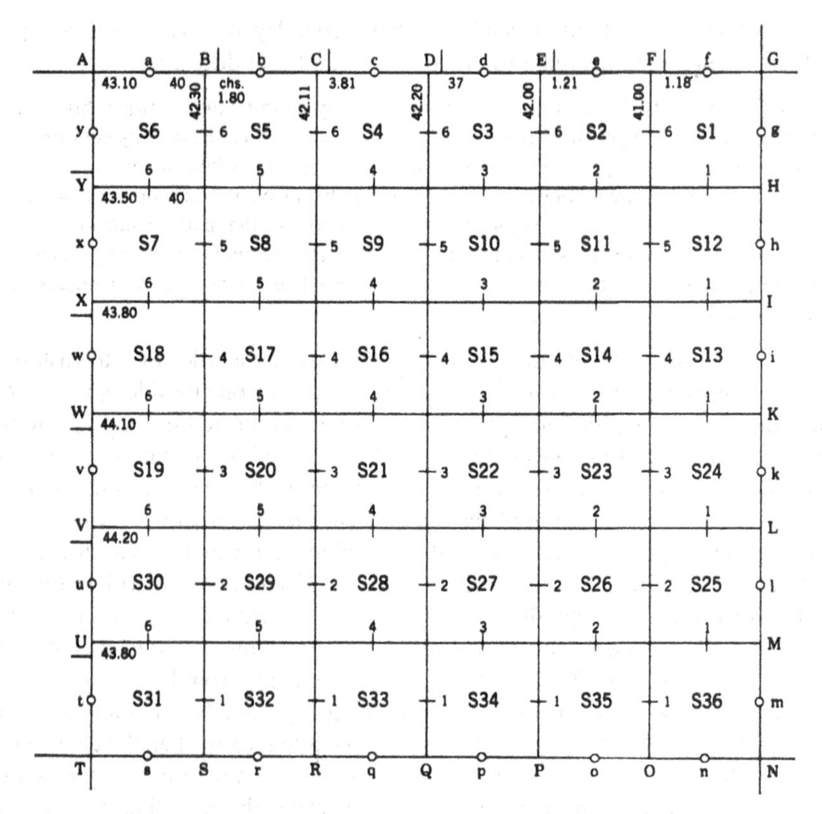

FIGURE 9-14. Tiffin's method of running original lines.

meridians converge toward the pole. There is a difference between running parallel with a township line and running north; lines running parallel will have different bearings for each line run. But because the difference was and is small and the error amounted to less than that introduced by the inaccuracy of compasses, in Tiffin's time, this instruction would have been without much significance; instruments used could not detect a change of a few minutes in bearings.

Double corners, permitted on the north and east sides of a township, were the most significant difference from today's instructions. Today, under certain circumstances, double corners will be permitted but not on every north and west township line. If a new area is being surveyed and it is adjoining a defective former survey, the errors of the former survey will not be incorporated into the new survey. This will necessitate double corners on the defective line.

Instructions to place errors in the north and west tiers of sections in the township have remained the same since Tiffin's time. Tiffin warned the deputy surveyors about errors. He stated that if the closure is too far north or south, they should mistrust the chain, and when there was excessive overrun or deficiency, they should mistrust the compass. The instructions point out that the principal source of error is the measurement by chain; the two-pole chain (33 feet) should be used horizontally with tally pins. Tiffin describes the method of breaking the chain and refers to a standard of length. When meandering rivers, bearings should be taken of the river course, and the contents of the fraction of the submerged lands should be made by latitudes and departures.

All courses were taken with compass sights and with the needle corrected. The Rittenhouse compass (Figure 9.15) was specified, and the deputy was to allow no one else to do the work of running the lines. It was not until some time after 1836 that the compass was abandoned.

Posts for the corners were to be driven well into the ground, and one or two sight trees were to be marked on each line. Bearing trees, located in each section, were to witness the section corner.

Corner posts that could be a tree or a post at least 3 inches thick were to rise at 3 feet above the ground and were to have notches on opposite corners for the number of miles to township lines (Figure 9.16). Bearing trees were to be referenced by distance and bearing, and the sizes and species of the trees were to be noted. Except for the corner post material (iron pipes and brass caps are now used), these instructions are similar to those given today.

Specifications for the notes included the manner of recording distances along the left margin as well as comments on timber, undergrowth, swamps, ponds, stone quarries, the quality of the soil, rivers, streams (navigable or not), and any improvements found. The entries were to be made at the close of each day and "written with a fair hand." (Today, final notes are accepted after being typewritten.)

FIGURE 9-15. Surveying compass.

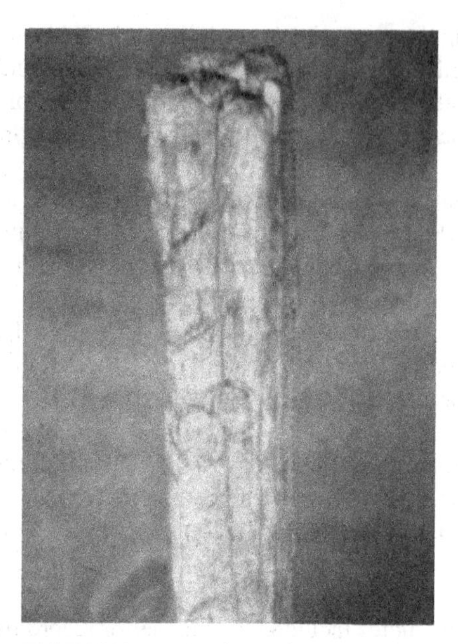

FIGURE 9-16. Section post with notches.

The plat was to be prepared on durable paper at a scale of 2 inches to a mile or 40 chains to an inch. A diagram was included to show magnetic north. The surveyor was required to certify the work as follows:

> Pursuant to a contract with, and instructions from . . . Surveyor General of the United States, bearing date the . . . day of . . . I have admeasured, laid out and surveyed the above described township (or fractional part) and do hereby certify that it had such marks and bound, both natural and artificial as are represented on said plat and described in the field notes made thereof, and returned with the plat in the Surveyor General's office. Certified this day of. . . . [29].

As a result of these first instructions issued in written form, the surveyors were required to check their chains and compasses and submit uniform and complete notes. The precedent for plats and notes was not established.

Some modern historians state the Tiffin Instructions did not originate from Tiffin, but were the work product of Mansfield, because several years ago a manuscript of the same instruction were discovered written in Mansfield's handwriting. We may never know the truth.

9.50 Later Instructions

Numerous instructions followed those of Tiffin, although some have become lost. The General Land Office was reorganized in 1836 and published the first *Manual of Instructions* in 1855 [30].

The *General Instructions* in 1843, applicable to Arkansas and elsewhere, made it mandatory to run to the existing section corner on the western side of the township and correct back on a true line. But in closing on the north township line, double corners were still permitted [31]. In the General Instructions of 1846, applicable to parts of Wisconsin and Iowa, section lines closing on both the north and west township lines were to close on the existing section corners. Since that date, no instructions, other than to correct defective survey conditions, have called for double corners.

In the instructions of 1846 is this note:

> Base, meridian, correction and township lines are to be run with an instrument that operates independently of the magnetic needle, which is to be employed only to show the true magnetic variation. Section, meander and all other lines interior of a township, may be run either with the same instrument, or with the Plain Compass, provided it is of approved construction and furnished with a vernier or nonius [32].

The *Oregon Manual* of 1851 should probably be considered as the first manual of instructions for the survey of the public domain; however, in the haste to put it in the hands of Oregon and California surveyors, several figures were omitted (later put in the 1855 *Manual*). There were minor changes in 1855.

In the instructions of 1855 is noted:

> Where uniformity in the variation of the needle is not found, the public surveys must be made with an instrument operating independently of the magnetic needle. Burt's improved solar compass, or other instrument of equal utility, must be used of necessity in such cases. In the instruction of 1850 are provisions for a telescope with stadia (two parallel lines in the principal focus) and permission to measure meander lines with stadia [33].

The *Manual* of 1855 was prepared under the direction of the commissioner of the General Land Office by John M. Moore and was approved by an Act of Congress, May 30, 1862 (Sec. 2399, RS).

> That the printed manual of instructions relating to the public surveys, prepared at the General Land Office, and bearing the date February twenty-second, eighteen hundred and fifty-five, the instructions of the Commissioner of the General Land Office, and the special instructions of the surveyor general, when not in conflict with said printed manual or the instructions of said Commissioner, shall be taken and deemed to be a part of every contract for surveying the public lands of the United States [34].

Although before this time Congress had not officially approved by law the instructions issued by those in charge, the courts have recognized instructions as being evidence of what was done.

Until 1855, little attention was paid to the problem of convergence of meridians. The *Manual* of 1855 provided for standard parallels, and among other interesting notes are the following:

> Standard Parallels (usually called correction lines) are established at stated intervals to provide for or counteract the error that otherwise would result from the convergency of meridians, and also to arrest error arising from inaccuracies in measurements on meridian lines, which, however, must ever be studiously avoided. On the north of the

> principal base line it is proposed to have these standards run at distances of every four townships, or twenty-four miles, and on the south of the principal base, at distances of every five townships, or thirty miles. [Note: this provision of five townships does not appear in later manuals.] Where uniformity of the variations of the needle is not found, the public surveys must be made with an instrument operating independently of the magnetic needle. Burt's improved solar compass, or other instrument of equal utility, must be used of necessity in such cases. The township lines, and the subdivision lines, will usually be measured by a two-pole chain of thirty-three feet in length, consisting of fifty links, and each link being seven inches and ninety-two hundredths of an inch long. On uniform and level ground, however, the four-pole chain may be used. You will use eleven tally pins made of steel, not exceeding fourteen inches in length, weighty enough towards the point to make them drop perpendicularly. In measuring lines with a two-pole chain, every five chains are called "a tally," because at that distance the last of the ten tally pins with which the forward chainman set out will have been stuck. He then cries "tally," which cry is repeated by the other chainman, and each registers the distance by slipping a thimble, button, or ring of leather, or something of the kind, on a belt worn for that purpose or by some other convenient method [35].

These instructions point out that in chaining, errors of counting could occur in any multiple of half chains (33 feet) or could occur by dropping a tally of 10 half chains (330 feet). Numerous discrepancies of this nature have been noted by surveyors. It is also interesting to note that where a four-pole chain (66 feet) was permitted, it was generally recognized, as of that time, to be 66.06 feet long. This ensured "full" mile measurements [36].

Burt's "improved solar compass," mentioned in the foregoing quotation, was not invented until 1836; few lines before that date were run independent of the needle.

The *Manual* of 1881 states that the initial point was to be located astronomically. The baselines, to be tested in direction every 12 miles, were to be made by a solar instrument. If a transit were used, the tangent method of baseline was to be followed.

The *Manual* of 1890 prohibited the use of the magnetic needle in major lines and permitted its use in subdividing and meandering when local attraction did not exist. The *Manual* of 1894 required all surveys to be independent of the needle.

The *Manual* of 1902 required that the initial points be in conspicuous locations and perpetuated by indestructible monuments. Among the many instructions of that time is the note:

> The surveys of the public lands of the United States, embracing the establishment of base lines, principal meridians, standard parallels, meander lines, and the subdivisions of townships, will be made with instruments provided with the accessories necessary to determine a direction with reference to the true meridian, independently of the magnetic needle [37].

In the *Manual* of 1930, the error of closure requirement was introduced. In terms of latitudes and departures, it was not to exceed 1/640 in either, with no more than 1/452 as a result. Distance was to close within 12.5 links per mile, or 50 links per section, 175 for a tier, and 300 for a township.

Comprehensive discussions on the instructions from the General Land Office will be found in references listed in Dodds [38].

A new edition of the *Manual edition of 2009–2010,* has recently been published. One of the major differences is the degree of proof required for the distinction and the definition of *lost and obliterated corners,* as well as discussions of riparian boundaries.

The new *Manual* is designed for internal use by employees of the BLM in surveying the public lands, and not for use in the metes and bounds states or by private surveyors in private surveys.

9.51 Reorganization of the Land Office

The General Land Office, established in 1812, was reorganized on July 4, 1836. The contract system of subdividing lands began with the Act of 1796 and was abolished in 1910 when a form of civil service was adopted. In 1946, the General Land Office and the Grazing Service were combined to form the Bureau of Land Management (BLM) under the Department of Interior. This agency employs professional cadastral surveyors who use the latest methods and equipment to locate and subdivide the public domain. Much of their present activity is in the state of Alaska, although there is a considerable amount of unsurveyed land in the western states. In 1973, the BLM published a manual of instruction that is presently in effect. A new manual is now being written.

9.52 The 1947 Surveying Instructions

According to the 1947 *Manual of Instructions*, the following points are appropriate to the division of lands in the public domain.

Initial Point New points will be established in Alaska by the director, and the position (latitude and longitude) will be accurately determined.

Principal Meridian These shall conform to the true meridian and will be extended north or south from the initial point. Corners are to be set each 40 chains and at intersection with bodies of water. Two independent sets of measurements are to be employed and must agree by 14 links per 80 chains. Also, a test must be made to ensure alignment within 0° 3′ 00″ of true direction.

Baseline The baseline is extended east and west from the initial point and will have corners set at 40 chains distant and at points of meander. The same accuracy standards are required as given for principal meridians. The line is to be run on a true latitude employing the solar method, or, if not, the tangent or secant methods are to be used.

Standard Parallels Standard parallels, called *correction lines,* are run east and west from the principal meridian at intervals of 24 miles north and south of the baseline. They are to be run in the same manner as the baselines, with corners set at each 40 chains and at meander points.

Guide Meridians At intervals of 24 miles east and west from the principal meridian and extending north from the baseline or standard parallels, guide meridians are projected on a true meridian. Excess or deficiency is thrown in the last half mile on the north, and a closing corner is placed at the intersection on the next northerly baseline or the standard parallel (see Figure 9.17).

Township Exteriors The south and east boundaries of the 24-mile tracts are the governing lines for the township exteriors. The random lines are run (ideally) from east to west with temporary points set, then corrected on the true line. The true line is to be blazed through timber, with notes taken on the character of the land. The field notes will omit the random lines except where necessary to explain triangulation and any other unusual procedures. The limit of closure for the township exteriors is set at 1/640.

Subdivision of Townships The boundaries of the townships previously run are to control. The meridian section lines are to begin on the south boundary line and will be run north parallel to the governing east boundary. Corners are to be set at each half mile, with fractions thrown in the last half mile. Lines are to be corrected back on

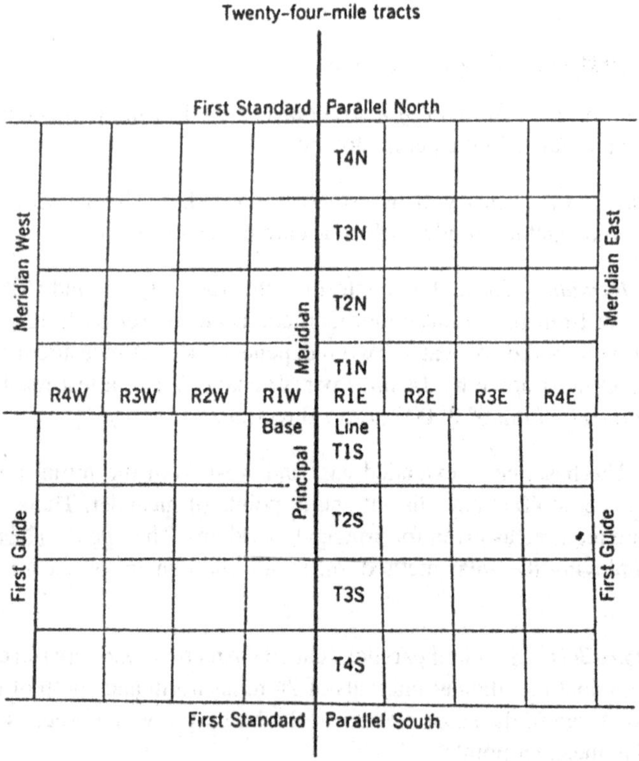

FIGURE 9-17. Twenty-four-mile tract.

true course. Meander corners are to be set where the meander line leaves the section line. Traverse angles of meander lines should be adjusted to 1/4°. Distances in whole chains or multiples of ten links are to be used for the meander traverse, with the fraction only in the last course.

A river that is three chains wide or wider is termed *navigable*. Lakes with an area of 25 acres or more will be meandered. If the lake is entirely within the section, a traverse is run from one of the quarter lines, or if entirely within the quarter, an auxiliary meander line is run. Artificial lakes and reservoirs are not to be segregated from the public domain. Islands may be determined by triangulation, and the title, name, or identification is to be noted.

Limits of Closure A relative error of 1/452 is permitted, provided neither latitude nor departure exceeds 1/640. A distance error not to exceed 2.5 links per mile in either latitude or departure is required. The latitudes and departures of a normal section will each close within 50 links with a maximum of 175 links for a tier and 300 links for a normal township. Fractional sections, including irregular lines, must close within 1/640.

Marking Lines between Corners Blazes and hacks would be made so that they will not be mistaken for other marks. Sight trees or line trees, as well as others, are to be marked within 50 links of the line.

Corner Monuments No longer is the type of monument that will be set at each corner left to the decision of the surveyor. Today, the standard iron post is 30 inches long, 2 inches inside diameter, split on the bottom, and set with a brass tablet riveted to the top (Figure 9.18). The standard tablet may be set in a boulder of, if a sound living tree falls in the corner location, it may be used to mark the corner. Cairns are to be used to protect and draw attention to the corner monument.

Running Section Lines The latitudinal section lines will be run random, parallel to the south line of the township. Temporary quarter corners are set at the half mile, and after the line is corrected back to true line, the permanent west random lines are run to the township line, then corrected back on the true line. The quarter corners are left at 40 chains. The westward line is run parallel with the controlling south line. The meridianal section line will have precedence in the order of execution. The order of progression is the same as outlined in Tiffin's instructions of 1815.

Subdivision of Sections The basic unit of subdivision is the quarter–quarter section. Sections are not subdivided in the field by the BLM; this is a function of the local surveyor. Subdivisions are made by protraction and are shown by broken straight lines on the plat. Fractional sections are protracted into "government lots," usually numbered in a counterclockwise direction. The fractional sections along the north and west tiers will have lots 1 to 4, except for section 2, which will have lots 1 to 7. Some of these lots will be shown on the plat by their dimensions, and some will be stated by area. Fractional lots will also be caused by meanderable rivers, meanderable lakes, mineral claims, defective boundaries, prior title, and so on.

FIGURE 9-18. Pipe and brass cap.

Meandering All navigable bodies of water and other important rivers and lakes that are to be segregated from the public domain will be traversed to determine the sinuosities of the shore. The chief purpose of this survey will be to determine the quantity of the land remaining. Normally, the mean high water is to be taken. In no case is the meander line intended to be a boundary line.

Corner Accessories Four bearing trees, one in each section, are to be marked with deep scribing indicating the township, range, and section in which the tree stands. If trees are not close enough to the corner, mounds of stone are to be erected so that the location of the corner can be restored by intersecting lines. Memorials, such as glassware, stoneware, a stone marked with an "X," a charred stake, pieces of metal, and the like, are to be buried at the corner. In desert country, pits are dug, and mounds are to be erected at the site of the corner.

Instruments A solar transit that complies with exacting standards is the usual instrument for the running of section lines. Steel tapes of two to eight chains in length are in common use. The lengths must be reduced to horizontal and to mean sea level. In some circumstances, distances by stadia are permitted. The instruments must have a stadia coefficient of 1/132 and a focal constant of 1.2 links. Stadia readings, when made on a targeted rod, may have a relative error as small as 1/1000.

9.53 The *Manual* of 1973

The most recent *Manual of Instructions* (1973), although important to present government surveyors, is not of great importance to the retracement surveyor; the manual in force at the time of the original survey should be referred to. It tells what was supposed to be done at the time. The latest manual permits new measuring devices (electronic, photography, etc.) and requires greater accuracy. The "limit of closure" set for the public land surveys may now be expressed by the fraction 1/905, provided that the limit of closure in neither latitude nor departure exceeds 1/1280. When a survey qualifies under the latter limit, the former is bound to be satisfied. Of considerable interest to the surveyor are the sections on apportionment of accretions and lakes (included elsewhere in this book). Also, the section on mineral surveys is new.

The retracement surveyor must again be cautioned; in a resurvey the procedure to be followed is that laid down by the courts. For example, in the relocation of a lost quarter corner in a regular section, the courts say it is to be located exactly halfway between section corners, not halfway with a permissible error of 1/905 feet. Original instructions are a guide about what was supposed to be done; the field notes say what was done; the courts guide one in the procedure to use in a resurvey, and that procedure may or may not agree with original instructions of the manual.

Many GLO states have adopted the *Manual of Instructions* either by statute or by case law. The surveyor must know whether the *Manual* is fixed by law or only suggestive as to methods and procedures. In most GLO states, when a surveyor or a lawyer reads case law, references to the *Manual* or to the pamphlet, *The Restoration of Lost or Obliterated Corners* [39] are often seen in the decisions, and in many instances references to the specific statutes are recited. As this edition goes to press, a new edition of the *Manual* is in progress.

Since the authors of this book have not had the opportunity to examine a copy of the 2010 *Manual,* we can only comment as to what the proposed changes were, as identified in Section 9.50. It should be remembered that once the new *Manual* is published, it can only be applied from the date of issue and not to prior work or litigation. It should also be remembered that this manual is only a guideline for the private surveyor and is not mandatory, unless provided for by state regulations or statute.

9.54 The Rectangular System Protraction Program

The outstanding feature of the U.S. Rectangular System has been the precedent of prior surveys; yet it may be that it is evolving into a master protraction scheme. In the vast unsurveyed lands of Alaska, expedient action has become necessary to identify lands for oil and gas leases. In the Umiat Principal Meridian, established December 31, 1956, a USCGS monument was chosen as the initial point. This meridian controls about 180,000 square miles of Alaskan lands where very little, if any, survey data exist.

Protraction diagrams are prepared on a Mylar base and show an area of one township per sheet. All township and section lines are indicated, with delineated lease areas of four sections (nine per township). All dimensions of the sections other than a full mile are shown, as well as the boundaries of fractional sections adjacent to water bodies. The geographic coordinates of the township corners are stated and used as a basis for description and location (presumably relative to USCGS monuments) of any of the smaller parts within the whole.

Alaska contains about 20,000 whole or fractional townships, and most of the titles will be located on this protraction system. Once the diagram has been approved and entered into the record, the geographic coordinates of the section corners become controlling, and it is then necessary for the BLM to locate the tract according to these values. This system has the advantage of subdividing lands at a much lower cost than the conventional procedure and will provide a more realistic method of location and restoration. It is quite possible that this method of protraction will revolutionize the process and philosophy of cadastral subdivision.

9.55 Summary of Public Land Surveys

Instructions on how to subdivide the public domain have varied from time to time. In resurvey procedures, the instructions issued at the time of the original survey are to be used as the guide for resurveys. The most recent *Manual of Instructions* is not the manual to be used for resurveying most areas within the United States; the manual in effect or the instructions issued as of the date of the original survey should be used as a guide. It would indeed be foolish to try and apply the 1973 *Manual* to surveys in Ohio.

9.56 The Timber Culture Act

In 1873, Congress passed an Act to encourage the growth of timber on western prairies. This law provided that any person who was the head of a family or who had arrived at the age of 22 and was a citizen of the United States or who had filed a declaration of intention to become such could secure a title, at the end of 10 years, to not more than a quarter section of land by planting and keeping in a healthy growing condition 40 acres of trees or a corresponding less amount if less than 160 acres were entered. Not less than 40 acres could be entered. Later, this law was amended to allow the settler to prove the claim at the end of 8 years and required the planting of only 10 acres of trees on each quarter section. Only one-quarter in each section could be entered for timber culture. This law was repealed in 1891.

9.57 Classification of National Forest Lands

Some of the lands of the United States were set aside and reserved as national forest land (mostly from 1891 to 1911). Since these lands were never under private ownership, the federal government has conclusive jurisdiction and rights of survey or resurvey. Under authority of the Weeks Law, passed by Congress on March 1, 1911, the federal government was authorized to purchase forest land privately owned.

All of the 20,000,000 acres so acquired by purchase or exchange under the Weeks Law was at one time patented and under state jurisdiction. After purchase by the federal government, the lands remain under state jurisdiction. As a result, all legal matters pertaining to the boundaries of such lands are tried by the states in accordance with state laws. Under proper circumstances, long-continued possession of Weeks lands can result in an unwritten title right against the federal government. Against lands continuously under U.S. ownership, possession cannot ripen into a fee title against the U.S. government. The private surveyor, as licensed by state laws, has the rights of resurvey for Weeks lands.

9.58 Land Acquired by Homesteading

Initially, there were no provisions for homesteading. It took the Civil War, when the Southern states left the Union, to enable legislation to be enacted.

The idea of developing a new area by granting lands to people who would occupy, grow crops, and otherwise improve the area originated several centuries ago. Even in the colonies there were "headright" grants that permitted people, without payment of land values, to gain title to lands after a certain period of occupation. A homestead act was passed by the Republic of Texas in 1839, and the first Federal Homestead Act in the United States was passed May 20, 1862. This act permitted the head of a family who was at least 21 years of age and a citizen or who had served in the army or navy to enter and take possession of 160 acres or less. Persons who were already in the territory could take smaller adjacent parcels if the aggregate did not exceed 160 acres. Under this first Act, the entryman was required to reside on the land for a period of 5 years to get an absolute title. Several other homestead acts have been passed since this first one, and in general the steps were as follows:

1. The person must file application stating his or her intent and qualifications.
2. A land office fee must be paid. This is usually a nominal $1.25 per acre and in some cases less.
3. The entryman must reside in or occupy the tract for a given period of time (3–5 years) and in some cases must make certain improvements.
4. Taxes are applied as soon as entry is made, but title is not perfected until the residence time and other stipulations are fulfilled.

Most of the United States, including Alaska, is no longer open to homesteading. A large part of the public domain was titled by homesteads. Homesteading differs from the colonial headright grants in that the homesteader had to be of sound financial status.

Under state laws, a homestead right, wherein a limited amount of land and improvements thereon are exempt from liability for certain debts, is established by filing a declaration with proper officials. This is not to be confused with the process of acquiring land by homesteading.

9.59 Obtaining Small Tracts within Public Domain

Under an Act of Congress on June 1, 1938, qualified individuals and certain groups are able to obtain small tracts (normally not more than 5 acres), unreserved public lands for residence, recreation, and other purposes. These tracts may be leased, purchased, or leased with an option to purchase. The sale price will be at least as much as the appraised price, and the sales are conducted at a public auction.

A suitable tract of land may be a regular subdivision of a section such as the N 1/2 of SW 1/4 of NE 1/4 of NW 1/4, Section 12, or an irregular parcel described by metes and bounds. Under unusual conditions, tracts as large as 7.5 acres may be acquired. Residence and improvements are not required, except as may be imposed by local zoning and planning ordinances. The tract holder has no rights to the timber or minerals on the land, since these are reserved in the public domain.

9.60 Land Ownership in Hawaii

Captain James Cook discovered the Hawaiian islands in 1778 and named them for the Earl of Sandwich. These eight inhabited islands were well populated by people who lived under a strict feudal system. The land belonged to the king, who doled it out to his chiefs. The "Great Mahele" of 1846 was a homestead reform that gave much of the public lands to the people. Title entries were made in the Mahele book and titles of property today refer to this record.

The United States annexed Hawaii on July 7, 1898. The Act stated "to cede and transfer to the United States the absolute fee and ownership of all public, Government or Crown lands." It was further provided that "the existing laws of the United States relative to public lands shall not apply to such lands in the Hawaiian Islands but the Congress of the United States shall enact special laws for their management and disposition" [40]. At the present time, public land is about 40 percent of the land area of the eight inhabited islands.

In 1920, the Hawaii Homes Commission was created to grant 99-year land leases to persons who were not less than one-half Hawaiian by blood.

Two systems of land registration are practiced in Hawaii: a recorder system and the land court. About 40 percent of the island of Oahu (containing Honolulu) and about 8 percent of the remaining lands are registered in the land court. A lot-and-block system tied to a rectangular grid system is used in the land court; the recorder system has many of its tracts described by metes and bounds.

As a matter of historical interest, one should read the decision *Midcall v. Hawaiian Housing Authority* in order to become acquainted with the history of Hawaiian lands [1].

9.61 Lands in Alaska

Alaska was discovered by Vitus Bering on July 16, 1741, while he was on a Russian expedition in search of fur. A Russian–American company followed with the establishment of small coastal settlements. Russia sold the entire claim to the United States in 1867 when virtually all the territory was vacant public land.

The federal acts pertaining to the disposition and management of public lands apply to Alaska, and at the present time slightly more than half of the public domain is situated in this state. Under the statehood land grant, Alaska is to acquire 103 million acres in a period of 25 years.

Alaska was an active area of homesteading, but this no longer applies. Many oil, gas, and mineral leases have been located in Alaska, and title surveying is active.

9.62 Court Reports

One of the more important sources of written historical records of survey procedures is found in case reports in law libraries. What the law was and how it was interpreted at a given date are usually summarized at the beginning (syllabus) of each case. The opinion of the judge, as recorded, gives an insight to the customs and thoughts of the period.

An example illustrates this point. In the year 1844, in Missouri, two neighbors, Gamble and Clark, got into a squabble over the location of a quarter corner. It seems that the section corners were located nearly a mile apart; yet the quarter corner was 20 poles (330 feet) nearer one corner than the other. (Note: This error corresponds to a tally mistake using a two-pole chain, as was customary at that time.) The fact that the quarter corner found was the original was amply proved. Today, this would not present trouble; but then, it must be remembered, many of the sections surveyed east of the Mississippi did not have quarter corners set, and it was common knowledge, in such cases, that the quarter corner was set halfway. In the summary of the report a historical sequence of public domain land laws pertaining to setting section corners was given. The judge's opinion, in accepting the corner as found, was as follows:

> From this review of the laws of the United States relating to the survey of the Public Domain, it will be seen, that at one time no subdivision of sections was made, it being divided into townships and sections only; consequently, no half or quarter section corners were established on the public survey. The act of 1800 authorized the sale of half sections, and prescribed a mode of ascertaining the contents thereof; the act of 26 March, 1804, authorized the sale of the public lands in quarter sections, which were required to be surveyed and the contents thereof ascertained at the expense of the purchaser. Then came the act of 1805, above cited, which prescribed the mode of ascertaining the corners of sections and quarter sections at the expense of the United States. No lands were surveyed in this state until long after the act of 1805. The Act of March 3, 1811, first provided for the survey and sale of land west of the Mississippi. At that time the quarter section corners were required to be marked on the survey and these corners were, by the act of 1805, to be taken as the true corners; and the provisions of the act, which required the corners of half and quarter sections not marked on the survey to be placed equidistant from the two corners of the sections, were intended to be applied to surveys made prior to the time when the sale of half sections and quarter section was authorized. Mr. Gallatin, then at the head of the treasury department, and to whom was entrusted the superintendence of the survey and disposal of the public lands, shortly after the passage of the act of 1805, in a letter of instructions to the Surveyor General, thus speaks of it—"You will perceive, from the enclosed act, that the principal object which Congress has in view is, that the corners and boundaries of the section

> and subdivision of sections should be definitely fixed, and that the ascertainment of the precise contents of each is not considered as equally important." Indeed, it is not so material, either for the United States or for individuals, that purchasers should actually hold a few acres more or less than their survey may call for, as they should know with precision, and so as to avoid any litigation, what are the certain boundaries of their tract. Now, that section may be divided among so many holders, and as valuable improvements are frequently made near lines dividing sections, it would produce great confusion and litigation if the established corners were to be abandoned in subdividing sections, and lines were to be run equidistant from the corners of sections, in order to subdivide it into quarter sections [41].

The historical sequence and timing of laws, in the foregoing case, are important information for resurvey procedures in the areas mentioned.

Many of the principles have been and will be outlined in this book on evidence as it is traceable back to the beginnings of the United States or to the original surveys. History in itself is often evidence used by the courts. In an early Massachusetts case, dated October 1804, in a dispute between Paul Revere (a mill owner) and Jonathan Lenord (an adjoiner), the question of the grantor's right or absence of right to testify about his intent was at issue.

In the course of the trial, it became important to ascertain the place the former dam mentioned in the deed was located. The parties did not agree as to its location. The counsel for the defendants moved that Mr. Robbins might be sworn as a witness to prove what was understood and intended by the parties to the deed, as to the forge-dam therein mentioned. He was objected to as not being a competent witness on the grounds, first, that he was interested and, second, that it is against the rule of law to permit a grantor to establish his own grant; that the intent is to be collected from the deed itself, except in the case of a latent ambiguity, which may, indeed, be explained by parol evidence but not by the testimony of the grantor himself.

As to the first grounds for objection, the court held that as Mr. Robbins was not interested in the event of the suit, the objection could not avail. But as to the second, the grantor was never permitted to explain his own grant, even in the case of latent ambiguity; that in such cases, although it was competent and necessary to explain the same by the testimony of witnesses, such testimony could not come from the grantor.

Sedgwick, the judge, charged the jury that it was against every legal principle to go out of the deed itself, by an inquiry into existing facts, to ascertain the meaning of those and such like words, there being in them no ambiguity of any kind; that what their meaning is, was merely a question of law, of which the courts are to judge by the words themselves [42].

Maine has had numerous court decisions over land boundaries, no doubt because of poor early conveyancing and surveying methods. Among the many questions that arose was the question of measurement index. If an original surveyor consistently makes long measurements, cannot the resurveyor, when resetting a lost monument, also make allowances and use long measurements [43]? In a dispute between Joseph Otis and Moody Moulton, this was the point in issue, and the report, to paraphrase the judge's opinion follows: At the trial to prove his title, the plaintiff read a deed from the Commonwealth of Massachusetts to Jarvis, dated February 16, 1794, conveying "all the unappropriated land lying between Penobscot River and the Lottery Township in the county of Hancock, Nos. 7 and 8, surveyed by John Peters in 1786; also the gore of land lying north of said township No. 8"; also a deed from said Jarvis to the Union Bank dated December 26, 1800, of seven-eighths of the land in No. 8; also a deed from the Union Bank to Sarra Russel dated November 10, 1816, in the same premise, and from Sarra Russel to Otis, dated November 21, 1816. All of the deeds were duly acknowledged, recorded, and legal.

He also introduced the grant from the Province of Massachusetts Bay to James Duncan and others of six townships, each township being conveyed by a separate description, and being numbered from one to six inclusive.

The opinion of the court, by Judge Shepley, was as follows:

> On the report in motion for a new trial two questions are presented for consideration: One arises out of the testimony relating to the boundaries of the town of Bucksport; the other out of and relating to the occupation and the title of the tenant. The bounds of the first six townships, being Bucksport, are described as beginning on the east side of the Penobscot River, "at a hemlock tree marked and running into the land in a course N 70° E, 5 miles and 184 rods to a stone monument, from thence along a line (which forms a boundary of the first and second of the said townships on the northeast and runs on a course S 76° E, 9 miles and 40 poles in the whole) upon a stone monument set up thereon which marks the east corner of the said first township; and from thence by a line S 53° W, 5 miles and 232 poles to a monument on the northwest side of the east branch of the Penobscot River and down the said branch one mile and 56 poles unto another monument on said branch, and from thence S 56° W one mile and one hundred and thirty-two poles to a monument on the east side of the river Penobscot," and from thence along the river to the first bounds. The place of starting from the Penobscot River at the first bound is not disputed, and the course of the line, making allowance for the variation, is found to be correct and not disputed. The stone monument named as the second bound is not found; and there is no proof of the original survey of line upon the earth between the two monuments, by which its length can be ascertained. The length should be ascertained by admeasurement upon the earth. It is contended, however, that in measuring it, the proprietors should not be limited to the exact measure named, but should be allowed a larger measure, corresponding to the measure found on other parts of the land, compared with that stated on the plan . . . the rule has been too well established to be now disturbed, that the admeasurement should be made upon the earth, and not by the scale upon the plan. And this case illustrates the propriety and necessity of the rule; for although it has been stated by a witness that in ten different admeasurements there was a larger measure in each case upon the earth, than that stated on the plan, yet there was no uniformity in the excess [44].

This early case points out some interesting facts: (1) townships in Massachusetts existed prior to 1794; (2) the term *gore* is an old one; and (3) a measurement index was not tolerated in Maine.

Georgia was one of the older states that granted land patents. Seemingly, from the facts stated in the following case, land had to be surveyed as part of the patent proceedings. In a difference of opinion between Margaret Riley and George W. Griffin (16 Ga. 141), dated August 1854, a controversy arose over measurements, monuments, and possession. The syllabus quoted the law at that time, and it is interesting to note the similarity to present laws:

> A possession which is the result of ignorance, inadvertence, misapprehension or mistake, will not work a disseisin. Marked trees as actually run, must control the line, which course and distance would indicate. If nothing exists to control the call for course and distance, the land must be bound by the course and distance of the grant, according to the magnetic meridian; but course and distance must yield to natural objects. All lands are supposed to be actually surveyed (note: that is in Georgia); and the intention of the grant is to convey the land according to the actual survey. If marked trees and marked corners are found, distance must be lengthened or shortened, and course carried so as to conform to these objects. Where the calls of the deed or other instrument, are for natural, as well as known artificial objects, both course and distance, when inconsistent, must be disregarded. And this rule is supposed to prevail in most of the states of the Union. Whenever a natural boundary is called for in a grant or deed the object is to determine at it; however wide of the course called for, it may be, or however short, or beyond the distance specified. Whenever it can be proved that there was a line actually run by the surveyor, or was marked, and a corner made, the party claiming under the grant or deed, should hold accordingly, notwithstanding a mistake in description of the land in the grant or deed. When the lines or courses of an adjoining tract are called for in a deed or grant, the line shall be extended to them, without regard to the distances provided these lines and courses by sufficiently established. When there are no natural boundaries called for, or no marked courses or trees to be found, nor the place where they once stood, ascertained and identified by evidence, or where no lines or courses of an adjacent tract are called for, courts are of necessity confined to the courses and distances described in the grant or deed. Courses and distances occupy the lowest rather than the highest grade, in the scale of evidence, as to the identification or land. Any natural object, and the more prominent and permanent the object, the more controlling as a locator, when distinctly called for and satisfactorily proved, becomes a landmark not to be rejected, because the certainty which it affords excludes probability or mistake. Courses and distances depend for their correctness on a great variety of circumstances and are constantly liable to be incorrect: A difference in the instrument used and in the care of surveyors and their assistants leads to different results. In ascertaining boundaries, the location (here meaning monuments and lines) or the original surveyor, so far as they can be found, are to be resorted to, and where they vary from the proprietor's land, the location actually made will control the plan. Whenever, in the conveyance, the deed refers to monuments actually erected as the boundaries of the land, it is well settled that these monuments must prevail, whatever mistakes the deed may contain, as to course and distance. Course and distance are pointers and guides to help ascertain the natural objects or boundary.
>
> Where a given line is exceeded (long) in the grant, according to the course and distance, evidence may be given of long occupation under it to prove the boundaries. Boundaries and courses may be proved by hearsay from the actual necessity of the case. See Chapter 19 for this case [45].

9.63 Necessity of Legal History

It is impossible for a surveyor to be able to conduct any retracement in an area without first having knowledge of not only local history but history in general as to the methods and equipment used in the creation of the original boundaries of land parcels.

History is the necessary foundation of the retracing surveyor.

Many legal decisions have extensive historical background to support the legal decision. The authors believe it necessary to acquaint the reader as to the necessity of knowing the history so that they could understand the legal decision. The individuals who write these decisions have access to extensive libraries and as such write extensive historical accounts of land, property, and surveying in all of the states. One of the better written decisions is the decision from the Virgin Islands in which the historical Danish surveys were discussed to some extent [1].

This decision is made available at the end of Chapter 19.

In this chapter, we have attempted to show that the practicing surveyor must have a better than average knowledge of federal, state, local, and legal history to be able to conduct adequate retracements.

The student of surveying should have an appreciation that the manipulation of equipment and the ability to determine coordinates to within fractions of a meter are just a part of a retracement. Knowing how the original surveyors created the original boundaries is an absolute necessity.

Recently a conversation was overheard between two attorneys who were trying a boundary dispute. One attorney stated to the second attorney, "I never realized how much history a surveyor had to know just to put the line where it should be."

Not all significant history has been discussed. In the area of legal history there are many solid decisions relative to surveys that are quite relevant and still are viable today.

A few selected decisions that surveyors should be familiar with are as follows:

- *Cherry v. Slade,* 7 N.C. (1805), 3 Murph. (N.C.) 82 (1819).
- *Cragin v. Powell*, 128 U.S. 691 (1888).
- *Riley v. Griffin,* 16 Ga. 141 (1854).
- *U.S. v. Doyle,* 468 F.2d 633 (1972).
- *Rivers v. Lozeau,* 539 So. 2d 1147 (1989).
- *Hawaiian Housing Authority v. Midkiff,* 467 U.S. 229, 81 L ed. 184, 104 S. ct. 1231 (1984)
- *Newfound Mgmt. Corp.*, 885 F. Supp. 727 (1995)

REFERENCES

1. David Thomas, ed., *Thompson on Real Property,* Second Thomas edition, Chapter 1 (Charlottesville, VA: Michie Co., 1994).
2. O. A. W. Dilke, *The Roman Land Surveyors* (New York: Barnes & Noble, 1971), pp. 40, 64, 86, 95, 96, 105–108.
3. T. R. Newton, "Land Is a Precious Thing," *Surveying and Mapping*, vol. 13, no. 3 (1935), p. 365.
4. *Haklits v. Oldenburg*, 201 A. 2d 690, 124 Vt. 199 (1964).
5. "The History of Surveying in the United States," *Surveying and Mapping*, vol. 18, no. 2 (1958).
6. Thomas, ed., *Thompson on Real Property*, vol. 1, p. 279.
7. Ibid., pp. 303–306.
8. Ibid.
9. Ibid., pp. 352–355.
10. Ibid., pp. 418–426.
11. C. E. Sherman, *Ohio Land Subdivisions* (Columbus, OH: The Ohio State University, 1925).
12. Section 9.22 is taken from K. Roberts, "Problems Relating to Riverbeds," Seventh Annual Surveyor's Short Course, Texas Surveyor's Association, Austin, 1958.
13. Texas Civil Statutes, art. 5302.
14. R. J. McMahon, "Perpetuating Land Corners in Texas," *First Texas Surveyors' Course, December* 16–17, 1940.
15. M. Lozano, Eighth Annual Texas Surveyor's Association Short Course.
16. *Oklahoma v. Texas*, 258 U.S. 574 (1923).
17. A. Stiles, "The Gradient Boundary: The Line Between Texas and Oklahoma Along the Red River," *Texas Law Review*, vol. 30 (1952), p. 305.
18. F. Daniel, "The Thing That Steals the Land," *Surveying and Mapping*, vol. 15, no. 4 (1955), pp. 461–467.
19. *A History of San Diego County Ranches* (San Diego: Union Title Insurance Co., 1960). See also Curtis M. Brown and Michael J. Pallamary, *History of San Diego Land Surveying Experiences* (privately published, 1988).
20. W. D. Pattison, *Beginnings of the American Rectangular Land Survey System, 1784–1800* (Chicago: University of Chicago Press, 1957), pp. 50, 68–72.
21. T. Hutchins, "Account of Soil and Timber in the Seven Ranges (1785)," December 27, 1785, Instructions as printed in *A History of the Rectangular Survey System.*
22. C. A. White, *A History of the Rectangular Survey System* (Washington, DC: U.S.D.I. Bureau of Land Management, 1926).
23. Ibid., pp. 38–42.
24. Ibid., pp. 54–57.
25. Pattison, *Beginnings of the American Rectangular Land Survey System*, p. 210.
26. Ibid, p. 215.

27. Ibid, p. 214.
28. J. S. Dodds, *Original Instructions Governing Public Land Surveys, 1815–1855* (Ames, IA, privately published, 1944), pp. 5–10.
29. Ibid., p. 9.
30. Bureau of Land Management, *Manual of Instructions for the Survey of the Public Lands of the United States* (Washington, DC: U.S. Government Printing Office, 1973, 1947, 1930, 1902, 1855).
31. Dodds, *Original Instructions Governing Public Land Surveys, 1815–1855.*
32. Ibid., pp. 53–63.
33. Ibid., p. 173.
34. Bureau of Land Management, *Manual of Instructions for the Survey of the Public Lands of the United States*, p. 101.
35. Ibid., pp. 101, 171, 771.
36. L. D. Stewart, *Public Land Surveys* (Ames, IA: Collegiate Press, Inc., 1935), p. 173.
37. Dodds, *Original Instructions Governing Public Land Surveys, 1815–1855*, p. 28.
38. Ibid.; see Bureau of Land Management, *Manual of Instructions for the Survey of the Public Lands of the United States.*
39. Bureau of Land Management, *The Restoration of Lost or Obliterated Corners* (Washington, DC: U.S. Government Printing Office, 1974).
40. U.S. Statutes at Large, 30, 750–751.
41. *Campbell v. Clark*, 8 Mo. 553 (1844).
42. *Revere v. Lenord*, 1 Mass. 91 (1804).
43. C. M. Brown, *Boundary Control and Legal Principles* (New York: John Wiley & Sons Inc., 1969), p. 172.
44. *Otis v. Moulton*, 20 Me. 205 (July 1841).
45. *Riley v. Griffin*, 16 Ga. 141 (1854).

ADDITIONAL REFERENCES

Akagi, R. H. 1963. *The Town Proprietors of the New England Colonies*. Gloucester, MA: Peter Smith.

Allcorn, B. 1959. *History of Texas Lands*. Austin, TX: General Land Office.

Bureau of Land Management. 1973, [1947, 1930, 1902, 1855]. *Manual of Instructions for the Survey of the Public Lands of the United States*. Washington, DC: Department of the Interior, U.S. Government Printing Office.

Chandler, A. N. 1945. *Land Title Origins*. New York: Robert Schalkenback Foundation.

Donaldson, T. 1950. *The Public Domain. New York: Johnson (reprint).*

Hilliard, S. B. H. 1973. "An Introduction to Land Survey Systems in the Southwest." *Studies in the Social Sciences*, vol. 12.

Marschner, F. J. 1960. *Boundaries and Records in the Territory of Early Settlement from Canada to Florida*. Washington, DC: Agricultural Research Service, U.S. Department of Agriculture.

McEntyre, J. G. 1978. *Land Survey Systems*. New York: John Wiley & Sons.

McLendon, S. G. 1924. *History of the Public Domain of Georgia*. Atlanta, GA: Foot & Davies.

Price, J. K. 1976. *Tennessee: History of Survey and Land Law*. Kingsport, TN: Kingsport Press.

Stewart, L. O. 1935. *Public Land Survey: History, Instruction, Methods*. Ames, IA: Collegiate Press.

Uzes, F. D. 1977. *Chaining the Land*. Rancho Cordova, CA: Landmark Enterprises.

White, C. A. 1990. *A History of the Rectangular Survey System*. Washington, D.C.: U.S. Department of the Interior, Government Printing Office. U.S. Statutes at Large 30,: 750–751.

10

RECORDING AND PRESERVING EVIDENCE

10.1 Scope and Purpose

The scope of this book has been directed primarily toward presenting students and surveyors with an appreciation of what is expected of professional surveyors and attorneys when locating, relocating, and litigating identified parcels of land, together with possible insight into the role of attorneys in this process. Although this chapter may not be the longest, it may well be the most pertinent to the surveyor.

In the location or creation of original parcels, the emphasis should be on adequate precise surveys and the placement of substantial monuments that will permit their future identification. In retracing land descriptions, the emphasis is on accurate surveys that identify the original monuments and lines. This requires that all conveyances be related to known monument positions that are usually located in written descriptions. If monuments are destroyed, certainty of location of the parcel being located may be compromised. The object of this chapter is to discuss suggested methods for preserving evidence called for in former conveyances and subsequently recovered by the retracing surveyor. The chapter will also discuss the professional duty of the surveyor in preserving this evidence, as the authors see this duty.

More and more state laws place a legal duty on surveyors to preserve evidence of the original surveys, both in the field as well as in reports, and to identify any subsequent evidence they create. This is accomplished through requiring the recording of public documents disclosing evidence discovered in the course of surveys, disclosing witness evidence taken under oath, and disclosing new monuments and measurements perpetuating original monument positions that were established.

Measurements to preserve any monumented positions are of two types:

1. Those made prior to conveyancing and called for by the conveyance identifying monumented corners in the reference documents and maps.
2. Those conducted after completion of the conveyance that were intended to identify the original position of the original nonmonumented corner and measurements to any subsequent monuments set in the course of a retracement.

Measurements taken after a conveyance is made, if they are to serve to reidentify monument positions as acceptable evidence, must be made by an expert and must be recorded so that the evidence is available for reference in the event of monument destruction. In addition, they must be related to the original monuments that were established at the time of the original survey.

There are many other means of preserving evidence that are given little attention or are so modern that few realize their potential uses as a means of preserving survey evidence or evidence of boundaries. These include aerial photographs, state plane coordinates generated from triangulation control stations or through the use of global positioning methods, and the storage of survey data in what are called *data collectors*.

If all surveyors always looked at evidence the same way, then the boundary lines being retraced would be located in the same positions by all who retrace them, without controversy, and the prestige and standing of the profession would be greatly enhanced in the eyes of the public. It is not uncommon that surveyors disagree about the length of a line or the angular direction between lines. When closely examined, surveyors' differences usually arise from varying opinions about which monument they should commence measurements from and the dignity that should be given to other, often uncalled-for, monuments.

This also holds true if all surveyors had equal knowledge and understanding of evidence and its application to monument position, and the certainty of all surveyors locating lines on the same position should be increased. For example, in Figure 10.1, different individuals will see the image in the frame differently. Try it. What do you see? What does your colleague see? Now turn it upside down.

The basic principles that will be discussed in this chapter are as follows:

PRINCIPLE 1. *A second-generation corner monument for which there is a complete chronological written history from the original monument and for which this history is reliable has legal dignity equal to the original monument.*

PRINCIPLE 2. *A found monument without a complete authenticated chronological history from the original corner monument is of little value as evidence of the original corner; other corroborative evidence may be used to provide for the acceptance of that monument, even though the monument may not be original.*

PRINCIPLE 3. *Evidence of original surveys collected by private surveyors and private surveys should be made available to other surveyors who wish to examine that evidence.*

FIGURE 10-1. Image used to illustrate how different individuals view the same image.

PRINCIPLE 4. *The surveyor and the attorney should consider the totality of the evidence recovered and used in the determination of boundary issues in retracements.*

10.2 Vanishing Evidence

Although the early surveyors, both GLO and colonial, usually set stakes as monuments, they also blazed trees and recorded the facts in field notes. However, precious little, if any, of their original evidence is left today. This statement can also be said for surveys conducted by various governmental agencies. The construction of new roads, cultivation, timber cutting, improvements, and the lack of general durability of the monuments mean that evidence of the original surveys has disappeared. The inventions of the bulldozer, the plow, and the ditch digger have contributed substantially to the loss of once-permanent markers that were set in the original surveys in hopes of being permanent in nature.

Time eradicates every living thing, including much of the evidence of earlier surveys. The nature of the physical evidence determines the durability or length of time the evidence will remain visible and recoverable. Trees eventually die, and with their decay, corners and witness monuments to corners become unrecoverable. People die, and with them vanishes knowledge or evidence in the form of testimony, and the resultant testimony of old corner locations is lost forever. Erosion and weathering will eventually erase mounds, pits, and even stones. The durability of the monument is dependent on the nature of the soil or the stone and the type of monument established in the creating survey. Iron will rust and all traces of iron pipes will eventually fade away. The worst scenario is that necessary documents are lost by fire, age, or destruction through the negligence of people.

In early times, trees were readily available and became the most common witness objects to the corners. Today, it has become quite obvious that trees fall far short of being completely satisfactory. In some states, witness trees have almost completely disappeared. Even in states that are still wooded, fire and old age have caused the original trees to vanish.

Evidence of any physical object should not be expected to last forever. Learning from past experience, the surveyor today, with a better selection of monument material and better methods of referencing monuments and the ability to duplicate and replicate measurements of distances and angles, can preserve the certainty of a monument's position.

As depicted in Figure 10.2, the percentage of original tree evidence that one can expect to be recovered is plotted over the time curve from the creation of the evidence of the original survey. This prediction is based on using wooden monuments and trees for accessories to the original corners.

An examination of Figure 10.2 will lead one to assume that the more remote the evidence is from the time of creation, the less you would expect to find. The deterioration is rather gradual until about the 75th to 100th year; then it becomes quite rapid. There will always be the isolated pine post or bearing tree that will defy all of the ravages of time and the odds of statistics and will stand to give testimony 200 or more years after it was placed by the creating surveyor.

This chapter is devoted to understanding the probability of recovering evidence and the means of perpetuating monument positions when they are recovered.

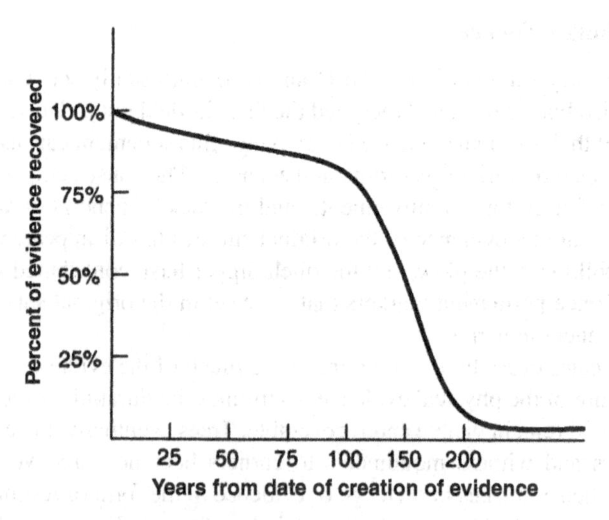

FIGURE 10-2. Theoretical percentage of evidence that one should expect to recover plotted over time from date of original survey.

PRINCIPLE 1. *A second-generation corner monument for which there is a complete chronological written history from the original monument and for which this history is reliable has legal dignity equal to the original monument.*

A retracing surveyor should never give up hope of recovering original evidence.

As previously pointed out, the spot occupied by the original monument at the time it was set is the controlling consideration. Hence, if its former position is known or can be determined with a degree of certainty, the corner is merely obliterated, not lost, even though the original corner monument may be destroyed. After an original surveyor sets a monument at the corner, the problem is one of perpetuating the corner's position. Even though the physical evidence of the monument may be destroyed, the corner always continues to exist.

The question of whether a corner is lost or obliterated may, at times, be unclear, and it is the surveyor, not the attorneys or courts, who must determine the answer.

The determination of whether a corner is obliterated or is lost is a matter of the quality of the evidence, not the quantity coupled with the ability of the retracing surveyor to recover and then to evaluate the original evidence. The law seldom accepts that a corner is totally destroyed. If the physical evidence that identifies the corner is insufficient to place the corner with a degree of certainty, then it permits the use or measurements (one form of evidence) as the controlling form of evidence over all other. After a corner located by measurements is set and at a later time adequate evidence of the original corner and monument, including accessories, is discovered or recovered, the position that was established by reliance on the measurements alone becomes invalid.

10.3 Degree of Proof Needed

Proof is not evidence. It only is the effect of evidence. It is the establishment of a fact by the evidence presented by a party. Proof is any evidence that is presented to make a fact more probable than not. Proof is the effect of evidence, while evidence is the medium or nexus to get to proof. Some legal scholars say that proof is the perfection of evidence, for without evidence, there is no proof. The law recognizes three degrees of proof in order of evidence needed to prove the fact:

1. Positive proof; also called preponderance, creditable or substantial.
2. Clear and convincing; reserved for property rights.
3. Beyond a reasonable doubt; applied to criminal complaints.

10.4 Perpetuation of Evidence

It is the duty of the original surveyor to set monuments, indicating the footsteps that were created. Although it is not the creating surveyor's duty to see that those monuments are preserved forever, it is the surveyor's duty to see that all efforts are in that direction. The original surveyor can do much to establish the permanency of monumentation by selecting better monument materials, setting them in positions unlikely to be disturbed, selecting adequate reference monuments, and making adequate written records to be used by future surveyors in conducting retracements. The creation of accessories to important monuments, both found and set, should be considered in many cases. If that is done, when the monument itself is disturbed, the accessory becomes the controlling evidence with as much weight or dignity as the monument itself. Obviously, the surveyor cannot prevent future destruction by others but may be able to minimize the destruction or provide alternative means of recovery of position. Once an owner enters on land, the surveyor's as well as the government's control over the monuments that were set is limited or completely ceases.

By law, the surveyor is often given the exclusive right to prepare and survey subdivision plats and to retrace original boundaries of lands. As part of these exclusive rights, legal obligations are imposed. Those obligations may include a requirement of devising some means of perpetuation of monument positions and recording information that was recovered and established. This also applies to retracing original surveys.

The U.S. government devised one of the best survey systems; one that is based on statute law. The system that is known today is the PLS (Public Land Survey) System or the GLO (General Land Office) System and it has not been duplicated in the world. As required by statute, the PLS (GLO) System required that numerous monuments be set prior to occupation, and all patents were made to those monuments that were set. Unfortunately, perpetuation of this system was not provided for after the patents were issued. The perpetuation of the PLS System was delegated to the states as their responsibility. Until recently, no state had accepted this responsibility, but now a few states have enacted laws that assist perpetuation. However, most are doing nothing. It is the disappearance of monument evidence and the lack of remonumentation, along with the lack of creating public records, that cause difficulty in retracing the footsteps of the original surveyors today.

10.5 Authority to Perpetuate

The authority to perpetuate monument positions and the responsibility of doing this are not the same. Many states delegate by law the responsibility of allowing licensed surveyors to remonument and replace original section corners, but they do not realize that these remonumented positions have no legal authority for permanency. The usual provision of the law is that surveyors may, among other duties, locate, relocate, establish, reestablish, or retrace any deed or described line or boundary of any parcel of land or of any road, right-of-way, easement, or alignment; perform any survey for the subdivision or resubdivision of any tract of land; and set, reset, or replace monuments called for in written conveyances, including monuments or reference points.

In most instances if a surveyor does remonumentation and if these acts of remonumentation as well as the evidence discovered are maintained as a personal secret, all is lost after the surveyor dies. Normally, what a surveyor did or found is not allowed as evidence unless supported by the personal testimony of the surveyor. Even though the surveyor may leave evidence in the form of writings, it is usually impossible to introduce and use these writings as evidence.

In the usual course of a survey, the surveyor discovers a section corner post that is deteriorated or badly rotted and may qualify as an obliterated corner. Then the post is replaced with a 2-inch iron pipe. Since the post is now gone, how can any other surveyor or the court prove the correctness of the pipe? How can it be proven that it was not used to stake out a cow or that a pipe was set where the corner was believed to be? The only link in the chain of evidence is the surveyor, who does not live or remember forever. Only the surveyor's properly recorded information may be everlasting. Delegating authority to remonument old original corners without the responsibility of perpetuating the chain of evidence is futile.

10.6 Responsibility for Perpetuating Evidence

If the surveyor is delegated the privilege of remonumentation of deteriorated corners, he or she should also be delegated the responsibility of perpetuating the evidence. In a few states, the land surveyor's law has such provisions as these: After making a survey in conformity with the practice of land surveying, the surveyor may file with the county recorder in the county in which the survey was made a record of such survey. One requires the following:

> Within 90 days after the establishment of points or lines the licensed land surveyor shall file with the County Recorder, in the county in which the survey was made, a record of such survey relating to land boundaries or property lines, which discloses
>
> 1. Material evidence, which in whole or in part does not appear on any map or record previously recorded or filed in the office of the County Recorder, County Clerk, Municipal or County Surveying Department, or in the records of the Bureau of Land Management of the United States.
> 2. A material discrepancy with such record.
> 3. Evidence that, by reasonable analysis, might result in alternate positions of lines or points.

Several states have similar acts that require the surveyor to perpetuate corner positions. The laws place the responsibility of perpetuating discovered evidence on the private practitioner. If evidence of monument positions is preserved by public records and if new monuments are set with a continuous chain of evidence from the time of the original monuments, the problems of future location of land parcels are greatly diminished.

The filing of a public record of survey is insufficient unless there is a mandatory provision to require the disclosure of evidence found and monuments set. In conjunction with the foregoing law in California, it is mandatory to show four qualities:

1. All monuments found, set, reset, replaced, or removed; a description of those monuments; the composition, size, location, and other data related to them.
2. Bearing or witness monuments, basis of bearing, bearing and length of lines, and scale of map.

3. Name and legal designation of tract or grant in which the survey is located and ties to adjoining tracts.
4. A memorandum of oaths.

Within the last 30 years, surveying organizations have been active in enacting laws requiring land surveyors to preserve evidence in public offices. One such law, enacted in 1963 in Colorado, illustrates the modern trend:

> 38-53-101. Legislative declaration. It is hereby declared to be a public policy of this state to encourage the establishment and preservation of accurate land boundaries, including durable monuments and complete public records, and to minimize the occurrence of land boundary disputes and discrepancies.

The next section of the code gives definitions that are to be applied by the surveyor:

> 38-53-102. Definitions. As used in this article, unless the context otherwise requires:
>
> 1. "Accessory" means any physical evidence in the vicinity of a survey monument, the relative location of which is of public record and which is used to help perpetuate the location of the monument. Accessories shall be construed to include the accessories recorded in the original survey notes and additional reference points and dimensions furnished by subsequent land surveyors or attested to in writing by persons having personal knowledge of the original location of the monument.
> 2. "Aliquot corner" means any corner in the public land survey system created by subdividing a section of land according to the rules of procedure set forth in the "Manual of Instructions for the Survey of the Public Lands of the United States."
> 3. "Bench mark" means any relatively immovable point on the earth whose elevation above or below an adopted datum is known.
> 4. "Board" means the state board of registration for professional engineers and land surveyors.
> 5. "Control corner" means any land survey monument whose position controls the location of the boundaries of a tract or parcel of land. The control corner may or may not be included within the perimeter of said tract or parcel.
> 6. "Monument record" means a written and illustrated document describing the physical appearance of a survey monument and its accessories or of a bench mark.
> 7. "Public land survey monument" means any land boundary monument established on the ground by a cadastral survey of the United States government and any mineral survey monument established by a United States mineral surveyor and made a part of the United States public land records.

> 38-53-103. Filing of a monument record required. Whenever a land surveyor conducts a survey which uses as a control corner any public land survey monument or any United States geological surveyor United States Coast and Geodetic Survey monument, he shall file with the board a monument record describing such monument, but said

monument record need not be filed if the monument and its accessories are substantially as described in an existing monument record previously filed pursuant to this section. A surveyor shall also file a monument record whenever he establishes, reestablishes, restores, or rehabilitates any public land survey monument. Each monument record shall describe at least two accessories or reference points. Monument records must be filed within six months of the date on which the monument was used as control or was established, reestablished, restored, or rehabilitated.

38-53-104. Surveyor must rehabilitate monuments. Whenever a monument record of a public land survey corner is required to be filed under the provisions of this article, the surveyor shall, if field conditions require it, restore or rehabilitate the corner monument so as to leave it readily identifiable and reasonably durable.

38-53-105. Forms to be prescribed by board. The board shall establish and revise, whenever necessary, the forms to be used for monument records and shall prescribe the information to be presented on said forms. These forms and necessary instructions shall be furnished to all Colorado registered land surveyors without charge.

38-53-106. Monument records to be certified. No monument record shall be accepted for filing unless it is signed and sealed by the land surveyor who was in responsible charge of the work.

38-53-107. Filing permitted on any survey monument. A land surveyor may file with the board a monument record describing any land survey monument, accessory, or bench mark.

38-51-104. Violations. Any person, including the responsible official of any agency of state, county, or local government, who willfully and knowingly violates any of the provisions of this article is guilty of a misdemeanor and, upon conviction there of, shall be punished by a fine of not less than twenty-five dollars nor more than two hundred fifty dollars. It is the responsibility of all district attorneys of this state in all cases of suspected willful and knowing violation of any of the provisions of this article to prosecute the person committing such violation. The state board of registration for professional engineers and land surveyors may revoke the registration of any land surveyor convicted under the provisions of this article. Such a person is entitled to a hearing on the revocation, pursuant to article 4 of title 24, C.RS. 1973.

On many occasions, in the conduct of surveys, surveyors recover evidence of original bearing trees and monuments. At times surveyors are required to cut into trees to verify markings, yet to do this they may have to enter private land not directly related to the immediate project at hand. Unfortunately, in the recovery process much original evidence is sometimes overlooked or inadvertently destroyed because of surveyors who are untrained in this important phase of surveying.

Initially, all monuments and their accessories belonged to the government, and surveyors could do whatever was necessary to prove their validity. Today, there seems to be some confusion about responsibility and liability. The courts have repeatedly held that the surveyor is bound to recover all of the best available evidence, yet the land no longer is in the public domain and trespass charges are possible.

Recently, a surveyor was sued for defacing an original GLO bearing tree. The court held that the surveyor was not liable because of the necessity to prove the authenticity of the corner, yet the following question was not addressed: What if the surveyor had cut into the tree and it proved not to be the tree he or she was looking for? Several states have enacted "right of trespass" laws that permit the surveyor to trespass on private land in the conduct of legal duties. The laws are very weak when it comes to enforcement should a landowner refuse permission.

In visiting surveying offices, one may witness examples of bearing trees, monuments, and other "spoils" exhibited in display cases, shelves, and as doorstops. Although the question has never been addressed, there is some question about the authority to remove these items and make them personal property. If this practice of altering and removing evidence is conducted and no public recording is made of the found evidence, the authenticity of the original corner could be seriously jeopardized.

In those states where it is not mandatory by law for the surveyor to file and record public records of found and altered evidence of monument positions, it should be the duty of the state surveyor's organization to back the enactment of legislation to help preserve evidence.

Except in a few very rare cases, positions marked on a plat as "t.b.s." (to be set) are not adequate and do not meet the requirement of the law. Too often they do not become set and then are either misleading to the public or the next surveyor or require further work (often by another surveyor), resulting in increased costs to unsuspecting landowners. At the least, this should be considered unprofessional conduct on the part of the original surveyor.

Retracements may be conducted by the individual who completed the original survey or by some other person who is remote from the original survey. Any retracements conducted by a third person are secondary only and do not carry with them the sanctity of authority. The best description of a retracement can be found in the decision *Rivers v. Lozeau* [1], in which the court held that a retracement was nothing more than an exercise to discover evidence and as such a retracement could not change any of the original measurements, lines, or locations of corners. The original records that described the original lines are an absolute necessity to conduct a retracement.

10.7 Oaths and Witness Evidence

In the event that an original corner monument is obliterated and there are reliable witnesses to testify about the former location of the monument, the surveyor should be authorized to swear in witnesses and record their testimony under oath. Witnesses do not live forever; it is highly desirable to perpetuate the evidence of witnesses while it is available.

In the event that the state does not permit the surveyor to administer an oath, then the surveyor could become a Notary Public and have the witness sign a notarized statement or affidavit. Although affidavits relative to land and surveys in and of themselves cannot be used as evidence, the law of Georgia provides that an affidavit concerning real property can be submitted as evidence if it is filed on record pursuant to the requirements of the law.

Although some governmental offices still require that officials administer oaths to surveyors before they undertake a survey, it has been held that a failure of a surveyor to take such an oath is not fatal to the work performed.

Some states, as well as government surveys, require the assistants to take an oath "that they will faithfully carry out their duties." This has been held unnecessary today.

10.8 Identifying Marks on Lines and Monuments

In the conduct of original surveys, the original surveyors usually created identifying marks along the lines as well as on the monuments that were set at the corner positions. The marks along the lines were usually cut into trees in the form of blazes or hacks—required of all U.S. surveyors but also the practice of private surveyors. This information was usually recorded in field books that became part of the permanent record, and many of the historic books are still available today for reference. The old field books are a necessity in order to conduct a retracement, but the information available is only as good as the original records that were created.

In the GLO surveys, monuments that were set at corner positions were required to be marked so they could be specifically identified, separate and distinct. In one Idaho decision, the markings cut into the stone monument came under scrutiny as to a single number, on a 1/4 corner [2]. The corner was marked with a chiseled 1/4, but the evidentiary question was the 4. The mark was an open four, and an expert testified that the original government surveyors never made open 4s on their monuments.

A set monument is worthless if it is unidentifiable in the future. The property owner may at any time set markers along his or her claimed line and can use iron pipes or any other material similar to surveyor's monuments. A found monument without a background history of who set it and how it got there is of little value as evidence.

Without doubt, the most certain method of identification of monuments set by surveyors is a legal requirement that their license number be permanently attached or marked on each monument. The marking not only lends authority to the monument but also serves as a means of checking its past history.

10.9 Recording Documents

The most certain means of perpetuating evidence is to record copies of original documents in a public place. This has three advantages:

1. The document itself is not apt to be destroyed; two copies exist, one in the possession of the owner and the other in the public records.
2. The general public is charged with knowledge of the contents of the document.
3. The recorded document is admissible as evidence in court actions.

Deeds, easements, quitclaims, and similar documents related to land are those most commonly recorded. Maps and plats are the second most important recordings.

All land must be located with reference to monuments, and one of the most important documents disclosing evidence of monuments and monument positions is the recorded plat and description.

Plats admissible as evidence include all publicly stored maps that a public official is charged with preparing. Road surveys, maps prepared for taxing purposes, and maps showing monuments or property or boundary line locations that are prepared by the county surveyor as part of official duties are similar to recorded plats.

The law often provides for the recording of private surveys. Recording has many advantages. Evidence disclosed in the way of monuments is preserved. If a client adversely occupies lands not his or her own, a recorded document showing adverse possession is public notice. Recorded surveys tend to decrease arguments between surveyors; both act on the same evidence. After a surveyor's death, publicly recorded documents are admissible evidence.

The surveyor can perpetuate evidence by including descriptions of all monuments found within the writings describing newly created parcels of land such as this: "Beginning at the southwest corner of Section 10, said corner being marked by a stone 10 inches in diameter and 18 inches long; thence . . ." But the opportunity for including such evidence is limited, since not all resurveys are accompanied by a new parcel description.

The failure of a landowner or a surveyor to record does not invalidate the document, but it may be important in the event title to the property is ever questioned. The courts will then be required to look at and adjudicate the doctrines of legal notice and actual notice.

10.10 Private Survey Records

Historically, the practice of the surveyor was to keep survey records of original surveys, as well as retracements in field books, that contained the records of their deeds and in many instances accompanied by sketches. Today these records are invaluable to the retracing surveyor to complete his or her "chain of history" as far back to the original work as possible.

After the death of the surveyor, private survey records are rarely of value as evidence, unless they are available to future surveyors and attorneys. Many times, a surveyor's records are purchased by other surveyors for the information that is contained in them, and they may or may not be made available for others to use and reference. Generally speaking, unless properly identified by witnesses to the act of surveying, unrecorded notes, since they are hearsay, are not admissible as evidence. If the surveyor is alive and testifies about the correctness of the notes or plats, the facts are evidence.

Private notes are not entirely useless after the death of the surveyor; the notes enable another surveyor to duplicate measurements and to discover places to seek evidence. If the former private survey can be restored by a current surveyor, the new surveyor can testify about what he or she did.

A problem with many of these historic field books and maps is using them at trial. There is no question that they may be relevant to the retracing surveyor to find evidence of the original lines, but the next question is, "Are they admissible as evidence at trial?" An attorney who understands evidence will realize that these field books are hearsay, and as such are inadmissible, but the rules of evidence provided for exceptions, under which these field books may fall. Several categories are as follows: ancient documents, office records, to refresh recollections, statements about ancient boundaries, and statements about real property.

10.11 Use of Aerial and Terrestrial Photographs to Preserve Evidence

An ancient Chinese philosopher once said, "A picture is worth ten thousand words." We can paraphrase this, saying, "A picture (photograph), in the hands of a qualified expert, is worth ten thousand words. But, like all other evidence, if it is misused, it is worthless, but in the proper hands, it is valuable as evidence."

The photograph in Figure 10.3 is of more value to the property surveyor than for the measurements made from it. Following are some of the more important uses to which aerial (and terrestrial) photographs may be put to aid, implement, and enhance land surveys.

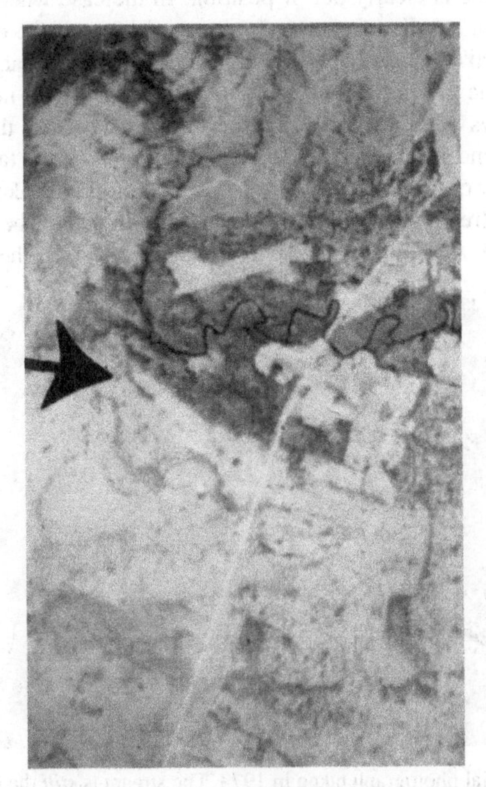

FIGURE 10-3. Aerial photograph taken in 1935. The stream is a boundary line.

Identification Many organizations photo-identify corner points as soon as they are found in the field. The U.S. Forest Service, the Bureau of Land Management, the Geological Survey, and other such agencies will, by photograph identification or other positive identification means, tie the location of a found corner or other boundary evidence to other monuments on the ground. Reference ties to three or more points that are easily identified on the photograph will, in effect, reference the corner to all the images on the photograph. This identification forms a permanent record for location (on the ground) after all the references have disappeared.

Use of Old Photographs Old photographs show evidence of ancient lines and conditions at the exact time the photograph was taken. After a road has been obliterated, a fence removed, or a dam built on a stream by one of the property owners, in which the stream was the property line, the traces of these lines may still be seen on the photograph, even though no evidence appears on the ground.

Comparing old photographs with new ones will indicate some of the changes that have taken place. Thus, in a 1938 aerial photograph, a farm fence that was accepted as a property line between two tracts was discernible. A recent picture of the same area after a new subdivision was improved shows that a road purporting to be along the old property line is clearly out of position. In the case where the center of the stream was the property line for several owners and one of the owners constructed a dam, causing a lake to cover not only his lands but also the lands of several other riparian owners, the original boundary was obliterated. It will not be lost, because the dam can always be destroyed to find the stream or corners that were inundated (see Figures 10.3 and 10.4). Figure 10.3 is an aerial photograph taken in 1935 showing the natural flow of the stream. Figure 10.4 is a photograph taken in 1973 showing the status of the stream at that time. Now the stream cannot be positively located in the flooded area. Additional aerial photographs pinpointed the approximate time

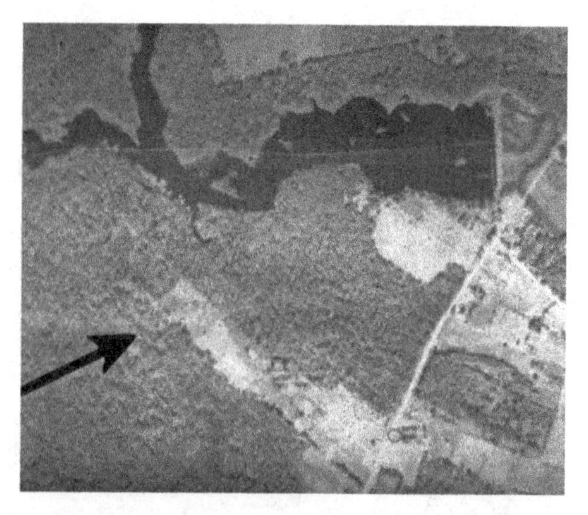

FIGURE 10-4. Aerial photograph taken in 1973. The stream is *still* the boundary. The questions are, where is it, and will prescriptive rights be acquired?

the dam was built. In this case, it was not constructed in 1935, but it was present in 1941; thus, it was built sometime between 1935 and 1941. These photographs can also be used to help claim prescriptive rights to the use of the dammed water by the riparian owners.

Unfortunately, most older aerial photographs of the United States go back to only about the mid-1930s. It would have been a great advantage to have photographic coverage of this country dating back to the time of the original subdivisions, especially for riparian lines.

In another situation, the surveyor was asked to retrace a 1798 treaty that was described both by a state statute as well as a survey that was executed in 1798, apparently using a staff compass and a two-pole (33-foot) link chain. In the conduct of the original survey, the surveyor read his angles to the closest 30 minutes and his distances for the most part were recorded in chains. A number of topographic calls, road and stream crossings, were indicated in the field survey notes. The exact location of the parcel was in doubt, but on examining aerial photographs that were identified from aerial photograph indices on record with the National Archives in Washington, the aerial photographs for the general area were obtained, indicating the earliest photographs on record were taken in 1938. As part of the project, new aerial photographs were taken of the general area in 1999. After plotting the field survey using CAD (see Figure 10.5), two successive angles were indicated as being unique; in fact, one angle in particular indicated a field angle of 56 degrees and 30 minutes. While examining the 1938 aerial photographs of the general area (see Figure 10.6), an image of a fence line was observed on the aerial photographs that had a similar configuration to the surveyed field angle. After digitizing the aerial photographs and superimposing the aerial photographs with the field survey notes with the plotting of the creek

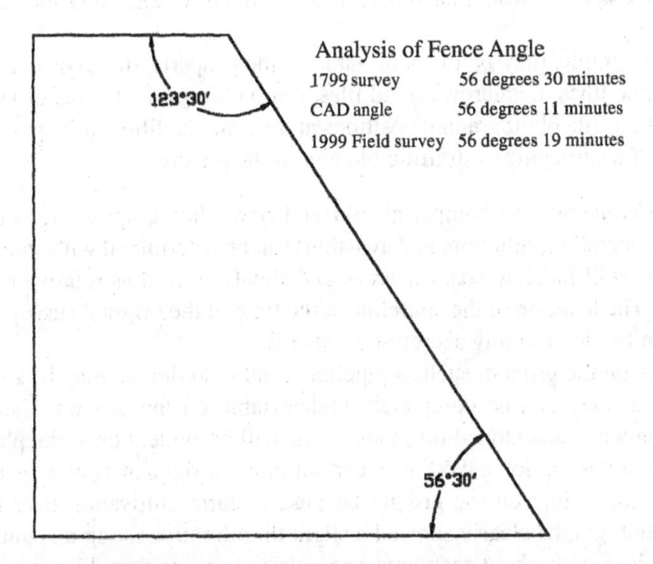

FIGURE 10-5. Plot of 1798 survey from field notes.

FIGURE 10-6. Fence angle depicted on (*a*) 1938 and (*b*) 1999 aerial photographs.

crossings and other topographic details, the retracement commenced at the point where the fence line and the survey angle were similar.

A complete retracement of the original field notes of the 1798 survey was conducted, and this survey was correlated to the imagery on both the 1938 and the 1999 aerial photographs. An analysis of the angles of the fence line intersection was quite favorable, as indicated in Figure 10.5. Without the imagery obtained from the 1938 aerial photographs, it would have been most difficult to positively locate the parcel with certainty.

Old photographs may be of more value to the property surveyor than old maps. Ancient fence lines, hedgerows, field tiles, old buildings, and ruins of buildings all appear in their true photographic positions at the time the film was exposed. There is no danger of a surveying or drafting blunder in the picture.

Riparian Evidence By comparing old and new photographs, the action of the water (e.g., accretion, reliction, and avulsion) can be determined with some certainty. Photographs will indicate shallow areas and shoals as well as relative beaches and shorelines. The location of the shoreline at the time of the original survey is essential information for determining accretion rights [3].

Evidence on the ground, such as pipelines, and field drains, may be valuable title evidence, but they can be completely undetectable on the ground. Usually, these effects, even when abandoned for many years, will be evident on aerial photographs. In one case, a description called for a certain railroad right-of-way long since abandoned and not visible on the ground because of farm cultivation over the former location. Photographs clearly showed a slight discoloration along a county road and through a field with about the same geometric shape as the old railroad. Careful measurements from the pictures enabled the surveyor to retrace the old alignment until an old stone culvert was recovered. From there on, more physical evidence was uncovered until the old right-of-way was positively located.

When analyzing water information, it may pay to have the client obtain infrared photographs. This type of photography depicts water in a different "light" or shades of gray.

Special Photography Infrared photography will reveal even the most subtle changes in the character of the land. Law enforcement officers were perplexed by the operation of an illegal still in the remote pine area of one of the Gulf Coast states. Infrared photographs were taken from an airplane. Because such film is sensitive to heat, a strange, light spot among the forest trees disclosed the position of the still, well hidden under the pines.

Today major advances have been made in false-color or color infrared (CIR) photography, which will show changes by the coloration of trees. In fact, the instant a live tree dies, it can be immediately detected on aerial photographs of this type.

Identification of Conflicts A building wall or corner may appear to be over the boundary line on a photograph. The extent of overhang or the obstruction of natural drainage is often clearly illustrated on the pictures. Public usage can be determined from examining aerial photographs, and such evidence has been admitted in court.

Identification of Lost Tracts Sometimes tracts described by metes and bounds have insufficient title identity. If these parcels are platted to the same scale as an aerial photograph, and if this shape is tried like a piece of a jigsaw puzzle until a similar pattern on the aerial photograph is discovered, often title identity can be determined. In a township tax map in an East Coast state, many record descriptions were identified by this method. Overlaps, gores, and gaps were revealed.

This method can also be used to locate large parcels identified by certain natural boundaries, or indicating the remains of stone walls or fence lines. Old fence lines that have been destroyed can be identified by the discoloration of the vegetation caused by bird droppings as they sat on the fence.

Locations of Monuments The search for ancient cornerstones, landmarks, and section corners can be greatly aided by a thorough study of the aerial photograph. Faint field lines can be projected, and their intersection will localize the area to search. In one trial in South Carolina, the 1929 deed called for a 50-foot white oak as the corner. No evidence could be found at the time of the survey in 1979. Recent aerial photographs revealed no lines, no usage, and so on. The surveyor was able to examine aerial photographs of the area that had been taken in 1934, just five years after the deed. While examining these ancient photographs under a stereoscope, he noticed a large tree in the vicinity of the corner that "stood out" above all of the other trees. Using the newer photographs, he went to the general area and found a gigantic stump, which he identified as white oak. Without these early photographs, this corner would probably have remained lost.

Although most surveyors use single aerial photographs, they are used most advantageously when viewed in stereo viewing. It takes little training to become proficient in the use of aerial photographs.

The ability to interpret aerial photographs depends on the scale of the photographs, which is a function of the focal length of the camera (f) and the flying height of the airplane above the ground (H).

Enlarging aerial photographs by photographic means does not increase the ability to "see" more objects. All it does is takes the objects present on the original print and enlarges the initial pixels.

10.12 Daily Use of Photographs

As a daily working tool, each office should maintain a permanent file of the most recent aerial photographs in the area of practice, and a print of an aerial photograph, when available, should be filed with each survey. All corners, whether located, recovered, or restored, when photo-identified, become permanent evidence, and surveyors should avail themselves of this means of preserving evidence. Geodetic control points, if available, should also be photo-identified whether used in the survey or not. In the future, these photographs can be rectified into more recent photographs from which precise measurements can be made. The photograph on file will serve as a permanent record of the land at the time of the survey.

Present-day photographs and old pictures should be preserved for cadastral purposes. These may do more to prevent costly boundary litigation than anything else. The process of making precise measurements from photographs, although in the infant stage, is improving at a tremendous rate.

Such measurements have considerably more freedom from blunders than do usual ground surveys, and the error is of more uniform character, not a function of the length of the line as in ground surveys.

10.13 Preservation of Evidence

As provided by both case law and the *Federal Rules of Evidence*, original corner positions, if lost, can be restored to their former position, provided (1) some acceptable witness remembered its former position or (2) measurements were known from other monuments or even from other evidence that can be related to the original corner position.

The National Geodetic Survey (NGS), managed by the NOAA, has numerous monuments set and interconnected with one another with precise measurements. The surveyor, by tying his or her discovered original monuments into this network, can greatly ensure the certainty of perpetuation of a monument's position.

Arrangements should be made to preserve all evidence, whether it is used at trial or not. If information from original historical field books is needed, the surveyor should make copies of the information that is needed in the field books and give the books and the copies to the attorney. Then the attorney should substitute the copies for the original field books. Once the books go into evidence, they, in all probability, may become lost, and the courts are not responsible for lost evidence.

10.14 Video Presentations

With the advent of relatively compact handheld video cameras, new means may be expanding to record field evidence and information. Surveyors can videotape evidence as it is being recovered in the field. Corner monuments, line trees, blazes, and possession data can all be preserved on videotape for future information. This also applies to digital cameras.

With digital cameras, surveyors can now take photographs and "download" them at the end of each workday onto their computers. Pictures can then be printed and placed in reports and used for court purposes.

10.15 Preservation by State Plane Coordinates

Retracement survey work and the location of boundaries and land parcels based on the record are interpretive in nature; once the lines called for are located, tying the monuments into a plane-coordinate network is an extra burden of cost to which the property owner will object. Where it is not a legal necessity, it is seldom done. The density of control monuments, with known coordinate positions, is a factor in voluntary usage of plane-coordinate nets. If control monuments are 10 miles apart and a surveyor has to run 3 miles in one direction and 7 in another to obtain tie-out information, the likelihood of it being done is small. But if control monuments are found every half-mile, the problem of usage is reduced.

The retracing surveyor must realize that with the advent of more precise ability to determine coordinates (latitude and longitude) from using GPS methods, there has yet to be developed a GPS with the capability of analyzing data and making determinations of the application of evidence to the facts. With new methods of GPS surveying, surveyors no longer attempt to "follow the footsteps" of the creating surveyors in a literal sense.

Assigning coordinate values to discover monument evidence does not in itself mean certainty of future location; the coordinates must be correctly determined by tying into monuments not too far away. A surveyor measuring 3 miles to tie into a monument has an uncertainty of measurement of 1½ *feet,* more or less. In reestablishing the position after it is lost, remeasuring the 3 miles will introduce uncertainty of another 1½ feet. Thus, nearby monuments might be far more certain than coordinates. Coordinates have no greater value than other known measurements; it is a question of what the best available evidence is. Coordinates carefully determined under the right circumstances could be the most certain evidence.

Establishing a net of state plane-coordinate positions is a cost impossible to bear by private surveyors, but after the net is established, it is not unreasonable to expect that the surveyors should use it. In some areas, such as the City of Los Angeles, the state plane-coordinate net is sufficiently complete that all surveys ought to be related to the system. But in other states the enabling legislation has yet to be passed.

Surveyors who utilize state plane coordinates must understand basic control surveying. Each basic control monument is subject to some error of measurement and adjustment. The land surveyor must identify the monuments from which his or her coordinates were calculated so that a subsequent surveyor who wants to "follow the footsteps" may duplicate not only the work but the inherent errors as well.

Without question, if a state plane-coordinate system were accurately established with a high density of monument locations and all surveys (including resurveys) were tied to it, an ideal system to perpetuate evidence would exist. But until such time as the public appropriates sufficient funds to provide for the density of monuments needed in the net, the property surveyor will continue with local monument control.

In the matter of new land divisions, the public can regulate how new land divisions will be surveyed, and it is within the public's power to require state plane-coordinate data.

REFERENCES

1. *Rivers v. Lozeau*, 539 So. 2d 1147 (Fla. 1989).
2. *Pointner v. Johnson*, 695 P. 2d 399 (Id. 1985).
3. Curtis M. Brown, *Boundary Control and Legal Principles*, 4th ed. (New York: John Wiley & Sons, Inc., 1995), Chapter 9.

11

PROCEDURES FOR LOCATING BOUNDARIES DESCRIBED BY WORDS

11.1 Introduction

The procedures used to locate boundaries of land parcels described by words, either written or used on plats and maps, are the subject of this chapter. Retracements of both public and private lands, location of tracts described by metes and bounds, location of lots and blocks, or the location of parcels described for any property right or interest are included. Original surveys made to create new land divisions are a separate subject and are discussed later.

The purpose of this chapter is twofold: to discuss procedures on how to locate boundaries already described and to discuss the duties, liabilities, and obligations of the surveyor who chooses to locate previously described boundaries.

When presenting a problem to a surveyor, most clients feel the surveyor should immediately rush into the field and commence surveying by conducting measurements—at times with no basis or understanding what it is he or she is looking for. Such a procedure is unrealistic because documentary evidence and map evidence must first be researched, examined, and verified, going back, if possible, to the original documents that created the original boundaries. A complete analysis of the problem is essential. In some instances, as much as two-thirds of the actual time, as well as a major portion of the actual cost of the survey, may be spent in the *preparation* for fieldwork and office completion with only one-third devoted to the actual and most visible fieldwork. If the area that is being surveyed is void of monuments due to the usual causes of obliteration, considerable time must be spent questioning old residents and others acquainted with the historical background of the area. The determination of a proper point from which to begin measurements and measuring may require days, whereas the act of measuring after the reference monuments have been located may take only a few hours. The principles that should be followed in recognizing and then in solving these problems are discussed in this chapter and are as follows:

PRINCIPLE 1. *Original surveys create boundaries. They must be considered in any conveyance made for the purpose of identifying land on the ground prior to or as a consideration of conveyancing.*

PRINCIPLE 2. *The retracement of original boundaries requires the retracing surveyor to go to the field armed with the original documents that are the evidence created by the creating surveyor.*

PRINCIPLE 3. *Resurveys of original surveys are for the purpose of relocating the original surveyor's lines in the same position as they were originally marked, and as such they can only be conducted by the entity that created the original boundaries.*

PRINCIPLE 4. *A subsequent surveyor who follows after the original surveyor, except one who may be in privity with the original surveyor, only conducts a retracement, and as such the work is open to collateral attack.*

PRINCIPLE 5. *Original surveys that divide land are or may be regulated by statute or other legislative action; but once a conveyance is made according to the land division, the location of the land parcels described is to be interpreted by the courts.*

PRINCIPLE 6. *The boundary surveyor has no judicial authority when resurveying or retracing boundaries for clients. The force of the property surveyor's authority is derived from reputation and respect. Judicial authority can be granted only by and through the courts.*

PRINCIPLE 7. *The surveyor must uncover sufficient evidence and facts about the boundaries being retraced; in this sense, the surveyor is a fact finder. The surveyor must then reach conclusions from the facts; it is the quality of these conclusions that is the mark of a professional.*

PRINCIPLE 8. *As a minimum, a boundary surveyor who decides to make a survey or a retracement from a written description assumes the responsibility for obtaining copies of (1) necessary adjoining conveyances called for in the legal description furnished, (2) all maps called for, (3) pertinent recorded adjoining surveys, (4) public agency maps that are kept in such a form that they are available, and (5) in the GLO states, original government township plats and field notes.*

PRINCIPLE 9. *The final decision of which documents should be used to locate a survey should be made by the surveyor.*

PRINCIPLE 10. *The boundary surveyor does not decide who owns land or rights in land. The surveyor's responsibility is only to locate land boundaries and, except for special agreements with respect to unwritten rights, only to locate land in accordance with written documents.*

PRINCIPLE 11. *Surveyors should never agree to locate all existing easements relating to or affecting property; the surveyor should merely agree to locate those easements in accordance with furnished descriptions and those that are visible or of public record.*

PRINCIPLE 12. *In a description, no one corner, whether monumented or not, is superior to any other corner. Each has equal survey and legal weight in retracing a description.*

PRINCIPLE 13. *Except where a senior right is interfered with, record or legal monuments, if called for in a conveyance and if found undisturbed, indicate the true intent of the original parties and, as such, control. If called-for monuments cannot be found or if they are found disturbed, their former position may be identified by competent witness testimony or acceptable physical indicators of boundaries.*

PRINCIPLE 14. *The surveyor should hunt and search in the field until the best available evidence is found on which to base the boundary retracement survey. Time should not be a consideration.*

PRINCIPLE 15. *Possession may memorialize original survey lines, and, as such, may be the best or the worst evidence of original lines.*

PRINCIPLE 16. *An original monument found undisturbed usually expresses the intent of the parties of the conveyance, fixes the point as between the parties, and, as such, has no error in position. All restored monuments established by measurements have some error in position.*

PRINCIPLE 17. *The magnitude of permissible uncertainty of measurements is always determined by a court's interpretation.*

PRINCIPLE 18. *The error of measurement originally permitted when tying original monuments together is independent of the accuracy required to reestablish an original lost-monument position.*

PRINCIPLE 19. *In the absence of the owner's specifying an unusually high precision coupled with an accurate survey, it is presumed that the surveyor will work to that precision consistent with the purpose for which it will be used or the standards of the profession in that locality.*

PRINCIPLE 20. *Every property survey should result in the preparation and delivery of a report or plat, whether or not it is to be recorded.*

PRINCIPLE 21. *The surveyor should conduct each survey as if it will ultimately be contested before a court.*

11.2 Boundaries Defined by Written Documents

Words become the basis of retracing boundaries, because not only did the creating surveyor leave the evidence in the field but he or she should have left a "paper trail" of words in descriptions, field notes, maps, and reports. All of this evidence has one element in common. They all require words to communicate what was done and what the evidence was created.

Land descriptions must be written in documents to be enforceable under the Statute of Frauds. These writings are generally used to describe an ownership (deed, will, etc.) or an interest in land and are indispensable evidence. Under English common law, transfer of real estate title or real property interests must be in writing to be enforceable. The legal description of a parcel—that is, the writings competent to convey title to the parcel—can be considered as an indispensable guide giving the surveyor necessary information about the location of the client's property. Some individuals use the terms *deed* and *description* interchangeably. This is not correct. A deed is the document that contains the description. As part of a legal document—the deed—a description is seldom complete by itself; it will generally refer to other documents such as maps, surveys, adjoiners, monuments, and prior conveyances in the chain of title. With as much available information as possible, the legal description and all reference documents, either directly called for or implied, together with adjoiner surveys, the surveyor goes to the site and locates the parcel in accordance with the best available evidence. As previously stated, deeds valid when originally created are not made invalid because of loss of evidence. Something must be used to locate the deed. Provided that better evidence does not exist or is not recovered, reputation of ancient existing *occupancy* may be used as the best evidence of where the *original boundaries* may be located.

Locating land described in a written conveyance is a great challenge to the surveyor; it is detective work; it is the process of discovering what transpired in the past; it is the understanding of past events, including the meaning of words, then and now, the evidence left by them as predicated by evidence and history; it is the final conclusions from the facts discovered; it is a professional service to the client.

Many of the surveyor's acts are based on judgment, reasoning, and conclusions; these attributes are not subject to standardized rules or applications. But if all surveyors followed systematic procedures of starting with the original documents that created the lines and corners and uncovered sufficient evidence of this original evidence, each surveyor should arrive at the same conclusions as would another surveyor using equal care.

11.3 Arrangement of Subject Matter and Systematic Procedures in Conducting the Survey

The functions of boundary surveyors are described by loosely used terms rather than precise ones. Because of this, clarification of these terms, along with an explanation of the surveyor's authority to perform boundary surveys, is necessary. Although it can be said that no two surveying problems are the same, it is possible to group the majority of situations into a few categories and perhaps systematize the necessary procedures in such a way that nothing is overlooked, and the client is then assured of the professional service deserved. Basically, all questions of boundary and land problems fall into one or more of three categories:

1. Questions of title (the quality of title or who owns the land or interest in the parcel).
2. Questions of where title lines or boundary lines are located.
3. A combination of the two previous questions.

Concerning questions regarding boundary lines, "What boundaries are [is] a question of law and where boundaries are [is] a question of fact" [1].

The steps outlined in this chapter will not fit all situations, nor will they perfectly fit any specific survey problem, but they will provide guidelines for general situations.

Even though survey questions may fall into separate categories, the surveyor should consider a systematic approach to all survey problems. Evidence may be grouped into five phases: contact, research, fieldwork, compilation of evidence, and presentation. Intermingled in the discussion of these steps are the responsibilities and obligations of the surveyor. The importance of each step will shift with different types of surveys.

NATURE OF LOCATION SURVEYS

11.4 Definition of Terms and Surveyor's Functions

When locating the boundary lines of any parcel that are called for in an existing conveyance, the surveyor may be required to do some or all of the following four steps:

1. Retrace lines run in an original survey.
2. Conduct an original survey in accordance with instructions in a written description.
3. Resurvey for the purpose of correcting or repositioning a former survey.
4. Survey the possession lines.

PRINCIPLE 1. *Original surveys create boundaries. They must be considered in any conveyance made for the purpose of identifying land on the ground prior to or as a consideration of conveyancing.*

As such, locating the original lines run is the highest element in the priority of calls. When the term *original survey* is properly used, it means a survey called for or presumed to have been made at the time a parcel or parcels were created. The essential character of an original survey is that, except where a senior right or a lawful fee right obtained by possession is interfered with, the location of the survey as set on the ground is legally correct, regardless of any errors that were made in the original measurements or in the original calculations. An original survey *creates* boundaries; it does not ascertain them.

A resurvey, when properly conducted, consists of two phases: a retracement to recover evidence of an original survey and a resurvey to replace lost corners, according to the proper rules of surveying. To have a valid resurvey, it should have been conducted by the entity that conducted the original survey or who created the boundaries that are being resurveyed. The purpose of a resurvey is to determine where the footsteps of the original surveyor were located [2]. The concept of "footsteps" is one of determining where the evidence of the original survey is located.

In some areas of the United States, especially in the northeast, the determination of whether an original survey was or was not made is difficult; early records of land surveys have been lost or destroyed or are deficient in explanations. In such areas, where land was sold by a plan, the basic presumption is that an original survey was made even though direct proof may be lacking. In other areas, especially the Public Land Survey states, direct proof that an original survey was made or was not made is almost always available; a need for such presumption is remote. The only presumption that must be made is that the survey was made in accordance with the published instructions and regulations and laws that were in effect at the time the survey was conducted.

The third class, after an original survey, is the retracement. The majority of what we erroneously call resurveys or surveys are actually retracements. A retracement, when properly conducted, is but an identification of the remaining evidence created by the original survey and the placing of lost corners in accordance with the proper rules of survey. The major distinction between a resurvey and a retracement is the dignity of the lost corners. In a resurvey, the lost corners positioned by the new survey stand in equal dignity of the original corners, whereas in a retracement any lost corners that are set are open to collateral attack by other surveyors and the courts. The terms *resurvey* and *retracement* are often used interchangeably, but in reality they are distinct and should not be confused.

For sectionalized lands, the last General Land Office (GLO) survey prior to the government's disposing of the land is the original survey. Some townships may have several original surveys within their boundaries. At present, when a subdivision is created, most states require a survey prior to the sale of any lot; that survey is the original survey. Boundaries can be created in two ways: by a survey from which a description is prepared or by a written description without benefit of survey. If a survey is made for a metes-and-bounds description and the survey is called for in the description, the survey called for is an original survey. Metes-and-bounds surveys made after the formation of a conveyance, in most areas, are not considered original surveys. In some states, such as Maine and New Hampshire, monuments set soon after a conveyance have been elevated to the status of an original monument. Where a monument does not exist at the time a deed is made and the parties afterward erect such a monument with intent to conform to the deed, such monument will control [3].

If a conveyance of land refers its boundaries to monuments not actually existing at the time the deed was signed and the parties afterward deliberately erect the monuments as and for those described, they will be bound by them in the same manner as if those monuments had been placed before the conveyance [4].

The monuments, whether they embrace more or less land than the deed, are controlling whether the deed is given to conform to them or they are erected to conform to the deed [5].

These cases raise questions which must be answered by the surveyor:

1. Are the monuments described a result of a survey?
2. Is the survey a result of the monuments that were created in a conveyance?

In each of these areas where monuments were set *after* the execution of the conveyance, the setting of monuments control only when done by the parties to the deed. The U.S. Supreme Court's theory was that where the parties to the deed set the monuments, the deed showed their original intent. Since later purchasers were not a party to the original deed, their acts do not disclose the original intent of the parties to the deed and thus are inferior in the chain of evidence.

The archaic term *free survey* is seldom used to designate a survey of property that is arbitrarily selected by the owner and is "free" in the sense that the lines of the survey are not fixed by a former conveyance. This term and *original survey* are, at times, used synonymously, but original survey is preferred. Although in the Public Land Survey states the land surveyor rarely performs original surveys of townships, local surveyors often create, within subdivisions of the townships, lots that are independent of any former surveyor conveyance and are free or original surveys. If a surveyor creates a new lot that does not touch the boundary of a new subdivision and the survey is a consideration of the subdivision or if a new parcel of land, wholly within an ownership, describes the parcel by using a metes-and-bounds description calling for that particular survey that was performed, an original or free survey may be created. Most original surveys are dependent on a resurvey to establish a record point of beginning. Original surveys today are more often called by the generic name "subdivision surveys."

The true concept of an original survey usually applies to Public Land Surveys. The deputy surveyors, in running the lines, setting monuments, making field notes, and preparing plats, created the original land net;. All these elements, when taken as a combined collection, constituted the original survey, yet this could also apply to a metes-and-bounds survey.

The term *resurvey*, in a sense, is reserved to mean first the identification of the remnants of the original survey and then the relocation of lines marked during an original survey. In a broad sense, a resurvey can mean the relocation of any former survey, irrespective of whether the former survey was an original survey. For example, a surveyor locates the called-for mean high tide line on a certain day, and at a later date a new survey could be conducted for the purpose of (l) locating the present mean high-tide line (not likely to be in the same place as in the first survey because of erosion and accretion, and even a difference if the tidal epoch used) or (2) surveying the former surveyor's lines. Neither would be a resurvey of an original survey as already defined.

At times, the surveyor must determine whether he or she is retracing an "original survey" or a "first survey." (An original survey was already described.) Initially, the surveyor must determine whether the creating surveyor actually ran the creating line and then reduced the survey to notes or the description was created on paper and then a surveyor subsequently placed that description on the ground. When a parcel or parcels are created on paper, without a survey being conducted, and the surveyor is later requested to place one of these paper-described parcels on the ground, this survey should be considered the "first" survey, in that it is the first survey to be placed on the ground after the description. The difference is that whereas the original survey controls, the first survey is nothing more than an opinion by the surveyor of where the written description should be placed. As such, it is always open to collateral attack.

To have a better understanding, the student should read the Florida decision *Rivers v. Lozeau*. This decision examines the various types of surveys, and as such the student will have a better understanding as to how an appellate court analyzed the various surveys [6].

PRINCIPLE 2. *The retracement of original boundaries requires the retracing surveyor to go to the field armed with the original documents that are the evidence created by the creating surveyor.*

A retracement, in the strictest sense is not a resurvey, but the surveyor uses survey methods to do the investigative work. For practical purposes, it is work that is conducted to find evidence using surveying methods. To know where to begin and what to look for, the retracing surveyor must go back to the original documents that created the line and the corners of the original tract.

As generally used, *resurvey*, without qualifications, has this connotation. Location of land described by a lot number within a recorded subdivision is generally a resurvey, since practically all laws regulating land divisions by plats require resurveying. There are exceptions where the land platted was not originally surveyed.

A description of land to be surveyed that reads, "Beginning at an oak tree located 33 feet northerly of the intersection of Foss and South Grade Roads; thence N 22°01′ E, 207.01 feet; thence N 89°07′ E, 242.01 feet; thence . . ." is presumed to have been surveyed, even though no survey has been called for. The surveyor, in monumenting these lines, is not performing a resurvey (no original survey was called for) but is performing a survey as based on the written record. Currently, no suitable term exists that adequately describes this function; herein, it will be called surveys based on the record.

The term *metes-and-bounds survey*, when used by surveyors and attorneys, is a broad term that can mean any of three survey functions performed in conjunction with metes-and-bounds descriptions. It includes (1) a survey of the previously described land, (2) original surveys being made for the purpose of preparing a metes-and-bounds description, and (3) if a metes-and-bounds description calls for an original survey, a later location of the land described would be a *retracement* as well as a metes-and-bounds survey.

A metes-and-bounds description can be written without the benefit of a survey, and any survey made to locate this already described land would be a survey based on the record.

Possession surveys are performed to locate land that may be gained or lost by acts of possession or to show the relationship between title lines that are described in land descriptions and possession lines. When conducting this type of survey, the surveyor should not discuss the validity of possession that is inconsistent with the deed lines. These are legal questions that can be validated only by the courts.

Overlaps and gaps may be found when surveying or locating defects found in titles. The terms *hiatus*, *compound hiatus*, *confusion*, *point of confusion*, *area of confusion*, and *gore* are used to express overlaps, gaps, and indefinite ownership areas between adjoiners.

A GLO survey is a survey that describes a system and an agency. Although the GLO has been superseded by the Bureau of Land Management, the type of surveys that are predicated on sections, townships, and ranges still are used in the majority of the United States. A GLO survey is distinguished from a metes-and-bounds survey.

Hiatus, according to *Webster's New International Dictionary*, means "an opening; a gap, a space where something is wanting." In title problems the meaning should be confined to a gap rather than overlap (see Figure 11.1).

Compound hiatus means multiple gaps on the same adjoiner. Confusion, at law and with respect to land, is the intermixture of the land (paper titles alone or paper titles and possession) of two or more owners so that their respective portions can no longer be distinguished. This term includes the concept of *uncertainty*, which may be either a gap or an overlap.

An area of confusion is defined by the limits (width and length) where confusion exists.

Compound confusion exists where one title has confusion with several adjoiners.

A point of confusion is that place where confusion commences or, as often used, the point of change from a possible gap to a possible overlap (see Figure 11.1).

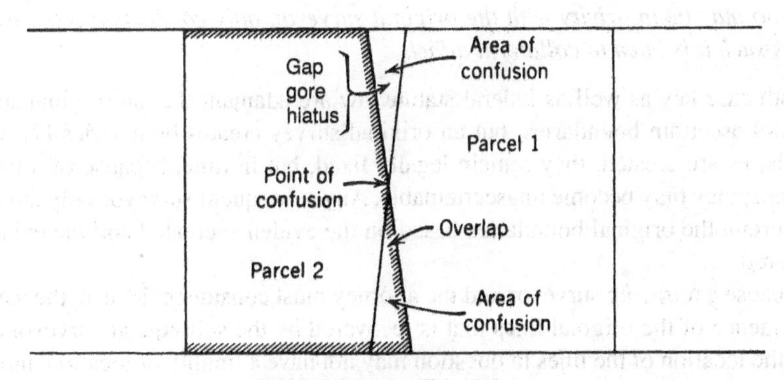

FIGURE 11-1. Title defects.

A gore is a small, triangular piece of land. The term, originally applied in New England, usually refers to a narrow, sharp-pointed, triangular parcel of land, more often occurring as a gap than an overlap.

An *overlap* is an extension of a written title over and beyond another written title. On the ground, there may be only one line of possession, but the writings may be in conflict so as to create a "paper" overlap. Ordinarily, the overlap belongs to the senior title holder, although possession may dictate otherwise.

Encroachment is a legal term and not a surveying term, and as such, surveyors should not use it on maps or in reports. The terms *encroachment* and trespass have no place in a surveyor's vocabulary, because if a boundary line is in dispute, until the correct location of the boundary is determined, it may be discovered that the use across the line was lawful, and as such there was no trespass or encroachment.

Even at law the terms *encroachment* and *trespass* defy an exactness of meaning. Thus, a surveyor can attest to such terms as overlap, gore, and hiatus, because these are factual questions, but the legal questions of encroachment and trespass should not be addressed by surveyors. As a professional, the surveyor should understand the doctrines and be able to give the attorney insight into these doctrines. The surveyor should not give any client legal advice as to whether the client has perfected the elements of any of the legal doctrines. If the surveyor does give such advice and the client relies on the advice and it is then proven to be erroneous, then in all probability the surveyor may be found negligent.

PRINCIPLE 3. *Resurveys of original surveys are for the purpose of relocating the original surveyor's lines in the same position as they were originally marked, and as such they can only be conducted by the entity that created the original boundaries.*

Perhaps the term resurvey is a misnomer, in that there are preliminary requirements that must be met before a resurvey, either a dependent or an independent, can be conducted. The retracement, which is a "fact-finding" endeavor prior to a resurvey, can be conducted by any surveyor. But a resurvey in which boundaries are involved can be conducted only by the person or agency who conducted the original survey.

PRINCIPLE 4. *A subsequent surveyor who follows after the original surveyor, except one who may be in privity with the original surveyor, only conducts a retracement, and as such it is open to collateral attack.*

Both case law as well as federal statute law are adamant that an original survey does not ascertain boundaries, but an original survey creates boundaries [7]. Once boundaries are created, they remain legally fixed, but in time, because of a loss of evidence, they may become unascertainable. Any subsequent surveyor only attempts to ascertain the original boundaries based on the evidence created and the evidence recovered.

Because a retracing surveyor and the attorney must consider evidence, the weaker the evidence of the original lines that is recovered by the subsequent surveyors, the more the location of the lines in question may not have a finality of location until the evidence is presented to, considered by, and then adjudicated by a judge/and or a jury.

11.5 Legal Authority for Regulating Boundary Surveys

Original surveys of property are for the purpose of creating the boundaries of new parcels of land. They are mainly regulated by legislative action, either federal or state.

How land may be subdivided, how it must be monumented, what measurement accuracies must be maintained, and what use may be made of the land are all questions within the power of the jurisdiction and may be covered by regulations. This philosophy dates from the Land Act of 1785 with the federal surveys.

The purpose of a resurvey is to relocate the lines described in an existing conveyance and then to replace lost corners in accordance with surveying principles. How it should be done in any given circumstance depends upon the interpretation of the meaning of a conveyance. Courts interpret documents. Although the legislature may regulate how a conveyance shall be formed, once it is formed, the courts interpret, in light of the laws existing as of the date of the deed, where the location shall be. All surveys based on the record are of this nature; the courts interpret their meaning and location. Legislatures may properly regulate who may perform resurveys, what kind of monuments are to be set on resurveys, and how survey plats must be created and delivered, but it is not within their authority to legislate how any court is to interpret conveyances or what evidence the court must use. Subsequently, any surveyor who conducts a survey of a parcel of land actually performs a retracement that is always open to collateral attack.

PRINCIPLE 5. *Original surveys that divide land are or may be regulated by statute or other legislative action; but once a conveyance is made according to the land division, the location of the land parcels described is to be interpreted by the courts.*

When subdividing or platting new parcels of land, the surveyor complies with statute law to create the parcels, but when the surveyor is locating the land described in a conveyance, court rules and past decisions determine how to interpret the writings and where the boundaries should be located.

11.6 The Boundary Surveyor's Authority

PRINCIPLE 6. *The boundary surveyor has no judicial authority when resurveying or retracing boundaries for clients. The force of the property surveyor's authority is derived from reputation and respect. Judicial authority can be granted only by and through the courts.*

This is most difficult for young surveyors to understand; in fact, the practicing surveyor may encounter situations in a retracement in which a third party, including the client, either does not accept or is forced to accept the surveyor's findings. The surveyor's monuments are given force only by the consent of landowners or the court.

After a surveyor has monumented the land boundaries, the landowner will view the results and accept or reject them. In many instances. possession will vary from marked lines, and the owner will refuse to accept the results. Sometimes the adjoiner objects and withholds consent. If a trial ensues, testimony usually becomes the "battle of the surveyors," but the court will declare whether the survey results are final. But in no instance can the property surveyor force his or her opinion on others but must convince the owners or the court that the survey is correct.

Without trust, a boundary surveyor cannot function effectively. The boundary surveyor is engaged by people because they believe him or her capable of delivering a product or service and that the product, usually a map, will be correct. If people believe that a surveyor is dishonest, the work will be questioned, and the findings will not be approved. All the surveyor's efforts will be for naught. Boundary surveyors depend on public acceptance, or the courts, for final approval of their work. They collect evidence, make decisions, draw conclusions, and may even act as arbitrators. Above all, they serve the public. The boundary or property surveyor is not an advocate; yet when the surveyor has reached conclusions based on the evidence recovered, he or she becomes fixed to the opinion(s) expressed.

A boundary surveyor is not an officer of any court, and as such, the surveyor's findings and conclusions have no binding effect on any person whose property touches the boundary lines. The surveyor has no authority to determine legal boundaries based on agreements, estoppels, or any other method of boundary creation or alteration. The surveyor is a collector of evidence and interprets it and, if needed, presents it in court.

Some countries do permit the registered surveyor to make legal conclusions and to negotiate disputes for the court, but this is not so in the United States.

11.7 Fact Finding and Conclusions

PRINCIPLE 7. *The surveyor must uncover sufficient evidence and facts about the boundaries being retraced; in this sense, the surveyor is a fact finder. The surveyor must then reach conclusions from the facts; it is the quality of these conclusions that is the mark of a professional.*

Frequently, the surveyor is referred to as a fact finder of information pertaining to boundary line locations. However, this is not the limit of responsibility and duty. After the surveyor has gathered sufficient evidence on both sides of the question, the surveyor should then draw conclusions from the evidence gathered, being careful to make certain he or she is not giving legal advice. The surveyor must come to a conclusion about where deed lines exist and also whether ownership lines might have changed because of prolonged possession. These are strictly legal conclusions and have no binding effect on any of the parties. This is the area where the surveyor should tread very softly in advising clients, and if the surveyor lacks experience in this area, no decision should be made.

ERRONEOUS DECISIONS MAY LEAD TO CLAIMS OF NEGLIGENCE OR MALPRACTICE.

Finding the facts about any particular line of property is often delegated to a subordinate or to office help, after which they may reach conclusions. Deriving a conclusion from the facts is a professional responsibility. Unfinished or uncompleted work does not relieve one of the burden of responsibility and liability.

The collection of maps, adjoining deeds, and measurements represent a necessary preliminary step prior to the final conclusions, but this is not what elevates a good surveyor above a mediocre surveyor. The surveyor's final correct conclusion and the ability to say what is right and what is wrong separate the professional from others. Although research is essential and no surveyor can come to a correct conclusion without adequate facts, surveyors should not be thought of solely as fact finders. Finding the necessary facts, and then evaluating these facts and drawing proper conclusions, completes the responsibility.

Fact finding can be divided into four parts:

1. Facts furnished by the client.
2. Searching pertinent written records (e.g., deeds, adjoining descriptions, maps, and old field books) and, above all, public records.
3. Fieldwork, searching for monuments, locating possession, and making measurements.
4. Seeking testimony and information from old residents and other surveyors.

CONTACT WITH CLIENTS

11.8 Initial Contact with Client

As a matter of professional standing, professionals should not actively solicit clients in the manner of business people; they should not advertise or seek recognition in a self-laudatory fashion. Standing in the surveying community, knowledge, prestige, and personal integrity are the foundations for success. Obtaining clients results from referral, personal contact, and discrete notices of services rendered. The client may become aware of the need for a survey from various outside contacts. He or she may have had similar experiences in the past and realize that this is the proper course of action or an attorney may advise that a survey is essential to settle a title problem. Occasionally, the client learns of surveying services from descriptive literature found in abstractor's offices, real estate offices, title insurance offices, and the like. Much public education is needed to inform property owners when they need a survey of line described in their deed descriptions. They usually request a surveyor to conduct one; this can be likened to purchasing a life insurance policy on one's property.

Referrals by previous clients are the most common means of client contact. A property owner will preferably engage a surveyor who has worked in the community rather than a complete stranger.

A surveyor who specializes in certain areas may find it difficult to act as an expert, because at times the client base is not local, but may be outside of the geographical area of practice, perhaps in an adjoining state. The practicing surveyor should be familiar with the local laws and standards when he or she is asked to consult outside his or her local area, even if within the same state, because the surveyor expert will find the local courts are very parochial in nature and at times distrust "non-area" surveyors.

The request may fall into two categories of surveys, which we call a *utility survey* and a *forensic survey*. The distinction may be made as to who requested the survey and for what purpose the survey will be used.

A utility survey is so-called because it is the type most often performed by surveyors. This is the individual lot survey or the subdivision survey that was performed for a client in the everyday practice of conducting a business operation. In most instances, these surveys are accepted and never questioned, but there are those utility surveys that are questioned, and either the surveyor or the landowner is called to task to either explain or to defend the lines located, the monuments set, and the plats prepared.

The attorney or the landowner may retain an expert to "prove" or demonstrate the first surveyor was in error. This survey is then used to either discredit or substantiate the earlier utility survey. This type of survey is entirely different from the utility survey, for it is used for litigation purposes and should be identified as a forensic survey.

Whereas the utility surveyor's responsibility is to the client, either by contract or in tort, the forensic surveyor's responsibility should be to the attorney who is litigating the dispute. This relationship gives both the client (the attorney) and the expert (the surveyor) certain legal privileges that will not attach to the utility surveyor.

11.9 Initial Conference with Client

The initial conference is generally an inquiry by the client about what services the surveyor can render. For such inquiry, a fee usually is not appropriate, but if a fee is charged, it should be minimal, unless the conference is detailed and consumes a considerable amount of the surveyor's time, since this should be considered an opportunity to educate the public in the need for expert help.

It is indeed the rare client who knows precisely what is needed, although they may think they do. The property owner sometimes contacts the surveyor when legal counsel would be more appropriate. The usual statement of the prospective client is, "I need my property surveyed." After explaining the services offered, the surveyor follows these steps:

1. Explains what services he or she can offer.
2. Tries to reach an agreement with the client relative to these services and the type of work to be accomplished.
3. Presents an estimate of cost and, if appropriate, an estimate of time for completion.
4. Tries to arrive at an agreement to do the work, preferably in writing.
5. Explains to the client if the surveyor lacks expertise in this area.
6. Explains the need to work under a written contract.

11.10 Contracts with Clients

Apart from the professional responsibility and pride essential to a good surveyor, from a business standpoint, the most important rule in this book is as follows:

> Never undertake any survey project without first having a written contract in place and a retainer from the client.

An agreement between a client and the surveyor arising from an "offer" by the client to provide "consideration" for the surveyor to locate a particular parcel of land and the "acceptance" by the surveyor to do so for the consideration is a contract, provided the agreement is possible and legal. The law will enforce contracts, both written and implied. Not only will the law enforce performance of a duty required by a contract but also a certain quality of work (called implied warranty) will be upheld. Most liability suits result from work of inferior quality or disputes over payment for the surveyor's services.

Historically, surveyors perform their work without a written contract. The client usually contacts the surveyor and requests that a map be made. The surveyor then performs the survey and delivers the product, expecting to be paid for the services rendered. Problems may develop when clients feel that they did not receive what they thought they had ordered. Surveyors have encountered problems by completing a project, without a contract and retainer, then do the work and, upon completion of the work, then "hope" to be paid, only to be sued for some minor reason. Historically, courts discourage professionals from suing their clients.

With a written contract, most or all of the essential elements are identified in writing, and problems are usually kept to the minimum. The essential elements of a contract are usually the identification of the parties, the work or service to be performed, the method and amount of payment, and the time of commencement and completion. A simple sample professional contract can be found in Hermansen, *Handbook of Annotated Forms for the Surveying Practice* [8].

11.11 Agreements and Disagreements with Clients

One of the most difficult areas of the surveying profession can be disagreements with clients, for example, over the work performed, the cost of the services, or the product delivered.

> Clients can sue in tort or in contract.

With a written contract in place, the client knows what to expect and what the cost will be and the surveyor knows to what standard he or she is to perform. In any dispute, a client may sue either in tort or in contract.

Liability usually results from the apparent failure to fulfill an agreement; hence, the surveyor should agree to do only those things that can be performed through the firm. At this point, the surveyor should make it clear to a client that the surveyor, without a court order, has authority to locate only written deed lines and that deed lines are not necessarily *ownership lines*. Sometimes, a client thinks a surveyor is capable of determining ownership lines; thus, when a client requests a survey of his or her property, the land surveyor, to avoid future embarrassment, should condition his or her acceptance on the premise that the survey is to be made in accordance with a deed to be furnished by the client.

When a client requests a property, boundary, or any other type of survey, it is logical to assume that the client intends to ask, "Do my possession lines agree with my written description contained in my deed?" The client already knows where his or her fences are, so why would he or she want them located? Unless it can be shown by reasonable evidence that fences represent monuments of original survey lines or are referenced in deeds, surveyors should not survey from fences. If, by some circumstances, the client does want fences located, especially for unwritten title action, the surveyor must be certain that the client understands that the survey is not a location of written conveyance lines. If this is not made clear, the surveyor will surely be embarrassed in the event of a later boundary dispute.

If there is any dispute as to final performance, having a written contract in place will aid the surveyor, should litigation be employed by either the landowner or the client. In a dispute over fees, a landowner may owe the surveyor for work performed, without a written contract in place, it becomes a question of, "Who is telling the truth?" As a businessman, the surveyor will find that a higher burden will be placed on the professional, and usually the surveyor will lose.

The second most important statement in this textbook relative to the aspects of the surveying business is as follows:

> Without a written executed contract, do not sue to collect a fee. In most instances, you will lose.

Sound business practice requires that the request for the survey and a specific offer be put in writing to reduce or eliminate future problems or misunderstandings.

RESEARCH OF RECORDS AND DOCUMENTS

11.12 Research Responsibilities

PRINCIPLE 8. *As a minimum, a boundary surveyor who decides to make a survey or a retracement from a written description assumes the responsibility for obtaining copies of (1) necessary adjoining conveyances called for in the legal description furnished, (2) all maps called for, (3) pertinent recorded adjoining surveys, (4) public agency maps that are kept in such a form that they are available, and (5) in the GLO states, original government township plats and field notes.*

In some instances, the responsibility for research goes beyond those items just named, especially in searching for adjoining deeds with their descriptions. In some areas abutting deeds are readily available from a title company or an abstractor.

The initial question that a surveyor should ask is, "How much or how little research should be done?" There is no definitive answer to that question. It has been said that a surveyor can make a precise survey but not an accurate or correct survey without research. Lines and positions can be accurately located with respect to other lines and monuments, but if the lines and monuments used to measure from are incorrect, the precisely located lines are without meaning or effect. It is far better to have an imprecise measurement of where the actual boundary lines are located than it is to have an extremely precise measurement of where the lines do not exist.

Title examination, or searching, and boundary, or deed, research are not the same thing even though many of the same documents are viewed in some of the same locations. Consequently, neither can substitute for the other. In addition, each is governed by its own set of minimum standards.

Title examination in some states is defined as searching back a given number of years, or to a resting document. In other jurisdictions, examinations are traced back to the sovereign, or to the original patent. Survey research should trace the subject parcel back to its origin, or creation, whenever possible in order to understand the complete history of the parcel. To do anything less may be inviting misleading or incomplete information.

A recent Maryland case addressed this concern, whereby the two litigants were seeking a decision as to which had the better title to a particular parcel of land. The court found that after the chains of title were traced back to the sovereign, neither chain of title contained the parcel in question. The court noted as follows:

> A requisite for valid title to real property is an original conveyance of public land from the State. Absent such a conveyance, one purporting to transfer an ownership interest in such property transfers nothing, and no quantity of successive transfers by deed nor the mere passage of time will metamorphose good title from void title [9].

It would seem then, from this case (as well as from common sense) that traditional title searches based on a number of years may not be sufficient. In the *Ski Roundtop* case [9], the court found that despite a series of deeds in the chain(s) of title, the title was void because an original land patent was lacking.

Most states have minimum standards for research that indicate to what extent research should be taken. It is vitally important for the surveyor to research and to trace description documents to their origin. This is not to ensure that title but to ensure the description is the one that created the original boundary lines. Some of the reasons are as follows:

1. Since many descriptions are copied from one deed to the other, errors in transcribing and copying occur, and, worst of all, some elements of the description are omitted.

2. It is necessary to know with certainty the date of the description (conveyance) so that the intent of the parties can be determined as to the words and phrases included in the description. At law, all words have the intent and meaning at the time of the writing. When it comes to the interpretation of deeds and descriptions, the cardinal rule for construction of deeds is that the intention of the parties is to be ascertained and given effect, as gathered from the entire instrument or as some courts say, "within the four corners," together with the surrounding circumstances, unless such intention is in conflict with some unbending canon of construction or settled rule of property or is repugnant to the terms of the grant.
3. To determine junior–senior rights and descriptions, the surveyor must know the sequence of conveyances from the description. In fact, one cannot know whether it is a sequential or a simultaneous conveying situation without examining the creating documents.
4. To ascertain the origins of the courses identified in the original description. Since many bearings in early descriptions are magnetic bearings, the date of the course creation must be known so that the proper declination can be applied to the retracement bearings.
5. To find early references to surveys, agreements, and easements that may not have been carried forth into recent descriptions. This is especially critical relative to those indexes to surveys by name of the owner. Without the names of the early owners to look up in the index, important survey information, even original surveys, may be missed.
6. To determine whether any agreements concerning boundary location have been filed since the creation of the title and boundaries.
7. To determine whether any easements have been created that either benefit or burden the parcel being searched. Often, easements are created early on, or in the beginning document and are not mentioned in subsequent documents. In addition, implied easements may only be detected by determining from whence came the creation of the original tract being investigated.

A person attempting to survey a parcel without a chain of title and history back to the creating conveyance cannot know the items in the foregoing list and therefore assumes great risk in undertaking the work. The work, according to the definition of surveying, may end up not being a survey at all, if it fails to locate the parcel on the surface of the Earth at the precise position(s) where it was originally placed.

In a recent California decision, that is now on appeal, the trial judge determined the retracing surveyor could do a "resurvey" without referring to or obtaining copies of the original rancho GLO plats and field notes. The expert testified he found "original blazes" on stumps some 150 feet remote from the straight line between the corners. The judge ruled the two rancho corners on each end of the line were lost, but the surveyor "found" the line from the blazes.

The same situation happened in Mississippi several years ago, when the "expert" did not find any corners, but found recent blazes along the line. The trial judge determined the "expert" had found the section line.

The objective of record research is to gather any written information (including plats) pertaining to evidence of title and evidence of monument locations. Although fieldwork follows the search in the records, evidence discovered in the field frequently suggests additional sources one could search for pertinent information. Each state, county, or local government has its own manner of storing and filing property location data. A local plat library can be very useful to the retracement surveyor.

The surveyor, as an expert, will be expected to have examined maps and writings from commonly known sources of written information that might disclose evidence of boundary location. Most survey offices have large investments in a plat library, and often most of the important information will be found within private files.

Where and how a surveyor conducts the research are so variable that it would be impossible to describe. The number of sources of information is one reason why the experienced surveyor in a community has the advantage. The surveyor should know where to find data. Also, the surveyor will usually have a large amount of historical reference materials and previous surveys, plats in personal files, and an extensive personal research library.

When a new surveyor enters the profession, consideration should be given to establishing a reference library for research purposes. Many personal files are prime sources of information.

In some sections of the country, legal descriptions have deteriorated to the point where it becomes routine and necessary to read former deeds in the chain of title to determine the description of what was originally conveyed. Although the first conveyance may give clear instructions about how to locate a line, later conveyances performed by the unqualified or uninformed may show that important wording was altered, either intentionally or unintentionally. The altered wording will not indicate how to locate the boundaries, because confusion is present in the "instructions." Without adequate record research, the surveyor does not know what monumentation controls, what should be looked for, or what is superior. It becomes a matter of stumbling in the dark. Generally, the question is not *should* research be undertaken, but *how much* research should be done. Without research, a correct location cannot be made.

Generally speaking, public records are the most fruitful source of boundary location information, although not always. Road survey plats, field books with reference ties, sewer surveys, water line surveys, railroad surveys, and numerous others often show locations of former monuments.

The courthouse and the recorder's office have the reports of court decrees, public records, and probate information. The state highway division, public utilities, tax assessor's office, railroads, special commissions, abstract offices, and title companies, to mention a few, are places in which to search for information.

Original government survey notes concerning sections are readily available from either national or state sources. Most original notes are now filed in state offices, and copies may be found in many counties. In many counties, the surveyor may find that the county surveyor's records have historical information of early work on the recovery of PLSS surveys.

Neighbors of the client may be a fruitful source of information, such as previous surveys. Adjoining deed and plats and writings can disclose valuable evidence. Not all survey plats are recorded, and, generally, they can be discovered only through local inquiry.

The mark of a true surveying professional in any community is whether the individual shares survey information with others. The key to the use of any information is the ability to retrieve it. Today, the computer, with its ability to store and retrieve data, has made the surveyor's task of storage and retrieval easier and more efficient.

It is also suggested that the young surveyor review court calendars to see when boundary trials are being heard. This will give the new surveyor an opportunity to see how the three people who are important to boundaries—the surveyor, the attorney, and the judge—work.

11.13 Documents Used for Boundary Surveys

PRINCIPLE 9. *The final decision of which documents should be used to locate a survey should be made by the surveyor.*

The basic documents needed for any retracement are the original information of the surveys that created the original boundaries.

The client usually informs the surveyor what he or she wants surveyed; the surveyor then accepts or rejects or negotiates what will be done. Ordinarily, the surveyor places on the client the burden of furnishing a description on which the survey is to be based. In some areas of the United States, the surveyor must search the records and personally determine the contents of the client's deed.

Not all documents on which a survey can be based are of equal dignity or value for surveying the parcel. A deed to property rarely recites all easements of record; a survey based on a deed can result in property monuments being located in the streets. Tax deeds are notorious for their errors, yet courts, attorneys, and surveyors may still place great reliance on them. It should be remembered that these are notices for taxation purposes only, not descriptions of property.

As the surveyor gains experience, he or she will become more confident when it comes to evaluating documents on which reliance can be placed and may inform the client of which documents were used or relied on. But the surveyor who is new to the area of retracement surveys should be very careful in advising a client as to the validity of documents. The validity of a document is primarily a legal question, not a survey question.

Boundary surveys are performed to either describe a client's boundaries and/or protect the client's interests. If the surveyor agrees to survey from a document that is potentially faulty, without first notifying the client about the possible dangers that are contained in that document, a truly professional service is not being provided, and such an act could result in malpractice claim.

Ordinarily, the documents preferred are in no specific order of importance. Such documents or information include abstractors' opinion or a title insurance policy, deeds, and mortgage or loan descriptions. Tax deeds should be used only as a last resort.

11.14 Ownership and Location of Land

PRINCIPLE 10. *The boundary surveyor does not decide who owns land or rights in land. The surveyor's responsibility is only to locate land boundaries and, except for special agreements with respect to unwritten rights, only to locate land in accordance with written documents.*

Ownership of land is very complex and is determined by legal principles, some of which are centuries old. To determine true ownership, investigations must be made into mortgages, trust deeds, validity of signatures, fraud, competency of witnesses, payment of taxes and special assessments, possible heirs, validity of possession, probate records, in addition to many other things, including prior court decisions, that may affect the property to be surveyed. Not all of these are survey problems. In the United States, the boundary surveyor has no authority to decide the ownership of land, and in all probability, until mandatory education becomes a requirement for registration, coupled with apprenticeship, this authority will never be granted. It may be tempting to the young, newly registered surveyor to advise the client that the "fence line is your property line" or "you own the land by adverse possession." These are questions of law, not survey. All the surveyor can state is, "The fence is 3 feet from your deed line," or, "The fence has been in place for 25 years." His or her statements can be based only on the evidence recovered and how it was interpreted. These are factual statements the surveyor must consider and, if necessary, be prepared to say in court.

The legality of a document is a matter to be referred to an attorney or title company, with the final decision resting with the courts. If a client has paid for an abstractor's opinion or obtained title insurance, the description in the insurance policy or in the abstractor's opinion is guaranteed by others. Unless there is an obvious error, the surveyor usually accepts the description at its face value.

There may be situations when the attorney or even the judge may seek the opinion of the surveyor. Under these circumstances, the surveyor may express an opinion, realizing it is only an advisory opinion and not a final opinion.

11.15 Location of Easements

PRINCIPLE 11. *Surveyors should never agree to locate all existing easements relating to or affecting property; the surveyor should merely agree to locate those easements in accordance with furnished descriptions and those that are visible or of public record.*

Easements necessary for the enjoyment of a property may automatically be transferred, whether mentioned in a conveyance or not. Title insurance policies guarantee to define every "recorded" easement as of the date of the policy, but easements may be created after the title policy is issued. The property surveyor should not lead a client to believe that no easements exist by any verbal or written statement. Only those that visibly exist or that are a matter of a written record deed should be certified.

Being an interest in real property, an easement should have definite, distinct, and definable boundaries. In fact, any easement requires three descriptions: the description of the easement itself, a description of the dominant parcel (the one that is enjoying the easement), and a description of the servient parcel (the land that is being burdened). Unless all three descriptions can be defined and located, the description may legally fail.

When an easement is appurtenant to a parcel, a conveyance of the parcel will carry with it the easement, whether mentioned in the deed or not, even though it may not be necessary to the enjoyment of the land by the grantee [10].

A court wrote:

> Where an easement has become appurtenant to an estate, a conveyance carries with it the easement belonging to it, whether mentioned in the deed or not, though it is not necessary to the enjoyment of the estate of the grantee [11].

One must consider that if an appurtenant easement makes the dominant parcel more valuable, it may make the servient parcel less valuable.

FIELDWORK

11.16 Planning for Fieldwork

The second phase of research that may be considered as fieldwork is the fieldwork proper, which is equally important to the office work. The ultimate quality of the fieldwork is totally dependent on the quality of the initial office research and the information that is taken to the field.

It is unfortunate that the licensed surveyor does not usually conduct his or her own fieldwork. One usually finds that the surveyor is necessarily too occupied with computations, platting, contacting new clients, and attending to general business activities to be active in the field process. As any surveying practice becomes larger, this procedure becomes more commonplace.

The surveyor must ensure that field subordinates are given adequate written instructions so that they can be confident that instructions will first be understood and then followed. As a minimum, the instructions that should be given should consist of four steps:

1. Specifying which monuments will be searched for, and how described.
2. In the event these monuments cannot be found, specifying alternative monuments.

3. Identifying what measurements should be made.
4. Specifying the precision required.

In addition, the surveyor may instruct the subordinate what to do with the evidence and measurements uncovered. Simply because this information is given to the subordinates does not relieve the registered surveyor from ultimate liability if errors are made.

The surveyor should make certain that the field party is furnished with adequate and necessary maps, plats, topography sheets, aerial photographs, deeds, road surveys, and any other information found while researching; a composite map may be furnished, usually prepared in the office, showing the results of the research. Office closure computations relative to the error of closures of descriptions are made, where it is possible to do so from the record data, prior to going into the field, and are furnished to the field personnel. No lot or block in a subdivision, where the subdivision map gives complete dimensional data, should be surveyed without this necessary step. Map blunders or errors can often be resolved before fieldwork; if they cannot, the field crew should be prepared to look for the errors.

Using aerial photographs will often save considerable time, and their use can be a valuable tool for both field and office planning. Stereoscopic examination of the photographs may indicate obstacles to normal measuring procedure, such as extreme slopes, visibility between points, density of vegetation, and possible search areas for corner locations. The necessary control traverse can be laid out on paper with due consideration to the required accuracy, strength of figure, and desired location of control points. Ancient aerial photographs taken at or near the time an adverse use was initiated may be the crucial evidence to either prove or disprove adverse rights. These historical photographs may show that a stream either changed by avulsion or very slowly through accretion or changed in location. Under the *Federal Rules of Evidence* [12], it is not the surveyor's responsibility to worry under which rule the evidence is admissible. Even though aerial photographs are hearsay evidence and technically inadmissible, the attorney may find they fall under one or more exceptions and thus would be admissible. An attorney who understands this evidence may be able to have a number of alternate possible solutions.

The surveyor is able to plan ahead by determining from photographs and plats whose property will need to be entered in order to perform the fieldwork. Evidence of ancient lines (not visible on the ground) may be apparent to a trained individual on photographs, and the field crew can be instructed where to intensify the search for monuments.

Timing is important. Conducting a survey in a business district may require working at nights, holidays, and on Sundays. Surveys next to schoolyards are best made after hours, when the schoolchildren are not present, and measurements anywhere are best performed with the least possible interruption. Other considerations such as the position of the sun, direction of the wind, and stage of the tide must be taken into account for the effective planning of the fieldwork. The surveyor should make realistic estimates of when projects can and will be completed, and all effort should be made to keep the time frame.

Since the responsible surveyor must interpret the meaning of the words in a description and transmit the interpretation to the crew before they go in search of evidence, metes-and-bounds descriptions may require special attention.

Instructions concerning final conclusions cannot be given until sufficient competent evidence is gathered, but instructions in the order of control or the importance of deed elements can be given. The following deed that was furnished to a chief may seem simple and easy, but it should raise many questions as to what to do:

> Hector MacKinnon and Sarah Jane MacKinnon, his wife, both of San Diego County, California for and in consideration of the sum of Twelve Hundred and Fifty Dollars do hereby grant to J. B. Richard of San Diego County, California, all that real property situated in said County of San Diego, State of California, bounded and described as follows:
>
> Beginning at a post 35.95 chains west of the northeast corner of the southeast quarter of Section (26) twenty-six, township 13 south, range 4 west S. B. M.; thence
>
> 1. S 7°55′E, 18.41 chains to post; thence
> 2. S 82°05′W, 13.02 chains to post; thence
> 3. N 7°55′W 20.09 chains to post; thence
> 4. East 13.30 chains to place of beginning and containing (25) twenty-five acres as surveyed April 30th 1893 by Berry McLarsen.

Sample written field instructions to the field party should include the following:

1. The whereabouts of the parties named in the deed and whether there are any living relatives and, if found, an inquiry of where the original posts were located.
2. Whether the starting point is 39.95 chains from the aliquot corner.
3. The posts set by McLarsen.
4. Possession evidence that might prove the original survey position.
5. The east quarter corner of section 26.
6. If the survey of McLarsen is not discoverable from the posts or possession evidence, a solar observation at the east quarter corner run due west 35.95 chains and staking the property.
7. Record measurements in the absence of other acceptable evidence of the original survey.
8. The deed as written will not close mathematically as shown on the traverse sheet furnished.
9. If McLarsen's survey lines are discoverable, tied lines to the east quarter corner.
10. New monuments set for lost or destroyed monuments.

A copy of the written report for the client should be kept in the surveyor's records.

These following were assembled from the various parts of the description. Since the intent of the parties is of paramount importance, the surveyor must try to interpret the description in this light. Of course, if there is any question as to the actual intent between the two parties, the final determination may be left to the courts.

- Is the parcel positively identified by course from an aliquot corner?
- Is the northeast corner of the southeast corner in reality the east quarter corner?
- Does the west bearing mean along the midsection line or a west bearing or a west geodetic bearing?
- Were the individuals in possession of the property or were they just using it at the time of the conveyance?
- Is the east quarter corner an original corner or an obliterated corner or must it be located if it is lost?
- If it is a metes-and-bounds survey and it fails to close mathematically from the recited bearings and distances, can the error(s) be located with certainty?
- Can I find the posts that were set for the deed, or will I have to conduct a retracement to find original corners and then proportion in the lost corners?
- If I find posts, should I replace them with new and better monuments or should I leave them in place?
- What are the minimum technical standards relative to this work, and does the work conform to accepted minimum standards?
- What are the observed or calculated bearings and distances, and how much do they differ from those of the original survey?

These are but a few of the questions that the surveyor must answer before the job is completed. According to law, when a survey in performed and is referred to in a deed, that survey will then control.

11.17 Field Evidence

Collection of evidence in the field consists of three things:

1. Finding monuments or other memorials, both called for and not called for.
2. Tying these together with new measurements.
3. Tying lines of possession to found monuments.

Sometimes this step and the completion step (setting final monuments from conclusions drawn) are performed at the same time, but ordinarily the surveyor determines all of the available evidence before drawing conclusions and setting final monuments.

Of a retracement survey this is perhaps the most interesting phase of the process. The surveyor should have a knowledge of surveying, forestry, soils, be a water specialist and an overall "jack of all trades." The surveyor must have a keen sense of observation and the tenacity of a "bulldog." The surveyor should never give up hopes of finding information.

11.18 Search for Monuments

All retracement surveys must begin at a proven monument or monuments that are tied to an original corner [13]. These must be referenced in the historical records. There are no exceptions. If one were to analyze the causes for surveyors arriving at different conclusions concerning the location of a point or a line, the lack of discovery of correct monuments would often be the underlying cause. The importance of monuments stems from previously stated principles and is summarized in principle 12.

PRINCIPLE 12. *In a description, no one corner, whether monumented or not, is superior to any other corner. Each has equal survey and legal weight in retracing a description.*

In attempting to retrace the call in a description, no single corner or monument is superior or senior to any other corner when it comes to using it to commence a survey or to locate the corners on the ground. It does not matter whether the monument that identifies the corner is first or last in placement in the description; each has equal legal weight. This also applies whether the corner is identified by a natural or an artificial monument; the object of the survey is to locate those that are most certain and positive.

In reading any description, the first task a surveyor should address in the office is to "run" or compute a closure of the recited courses to determine the quality of the description being retraced. In running a closure, blatant errors should be identified when the description fails to close. Through careful analysis of the magnitude of the errors, the surveyor may be able to ascertain the magnitude of the errors and the possible lines where the errors are located. This step cannot be accomplished when the descriptions contain balanced bearings and distances. Many states require the surveyor to indicate the distances and angles obtained from balancing field data. This defeats the job of the retracing surveyor, who is given the legal responsibility to follow, or retrace, the footsteps of the original surveyor. Footsteps cannot be followed when they are no longer footsteps but altered data from the calculation process.

From the standpoint of the retracing surveyor, a description may be found to have errors in the courses and even in the call for monumentation recited. As such, when calculations are made, the numbers will fail to close mathematically. By adding the words "and thence to the point of beginning," the description will legally close, regardless of the mathematical closure.

PRINCIPLE 13. *Except where a senior right is interfered with, record or legal monuments, if called for in a conveyance and if found undisturbed, indicate the true intent of the original parties and, as such, control. If called-for monuments cannot be found or if they are found disturbed, their former position may be identified by competent witness testimony or acceptable physical indicators of boundaries.*

In a retracement, principle 13 emphasizes that, armed with adequate record data, a field search should be made to find the monuments called for in the document or to locate the former position of these monuments. This should be done based on the following:

1. The discovery of the monument itself or accessories to the monument.
2. Other monuments set to perpetuate the position of the original monument and supported by adequate and sufficient writings.
3. Competent witness testimony.
4. Reputation and uncalled-for monuments.

The field search that is conducted is for the evidence of the corner, because it legally exists only as a theoretical position in that it has no physical dimensions. It has only legal position. The field person searching for the corner (evidence) needs to have knowledge of what is there and then what is to be searched for and possess an understanding of the significance of the evidence found. Therefore, a working knowledge of the fundamentals of evidence is essential. Also, one needs to possess "horse sense" or a "sixth sense" to know where to look and to recognize evidence when it is found.

A trace of rust in the soil might mark the position of a former iron pipe; a scar on a tree may indicate the presence of an old hack deep within a tree that was cut there years before by a surveyor marking a survey line; rotted wood may determine the spot of a former wooden stake; fences, hedgerows, ditches, and other boundary barriers indicate likely places to search; signs of cultivation, plow ridges, and tree lines may indicate the directions and possible location of lines. The success or failure of a surveyor is often in direct relation to the ability to discover evidence and how this evidence is used. The surveyor is a professional detective.

According to case law of the various states, uncalled-for monuments do not control a deed, but usually uncalled-for monuments are among the most readily available sources of evidence used by the surveyor. If a corner position that is called for is a "point," its position has been accepted (and located) by several (at least three) surveyors. Not all three agree exactly on the same spot.

The position of these monuments indicates where owners think their lines ought to be, and although they may not be paper title lines, they may identify unwritten ownership lines. The sheer weight of evidence of numerous monuments, all properly interrelated, may prove a location. Sometimes the only evidence available to the surveyor is uncalled-for monuments.

Any conveyance, equally sufficient at the time of its execution, cannot be made insufficient because of subsequent destruction of evidence of the original monuments. It is the duty of surveyors to search, probe, interview witnesses, spade the earth, and examine trees, fences, and other objects, that is, "leave no stone unturned," until they have found the best available evidence. From this evidence the surveyor must then formulate opinions. No rule of time exists; it may take minutes or it may take weeks. Within certain limits, a person can predict how long it will take to measure a given distance, but it is impossible to determine how long it will take to find a corner. This is the uncertainty of boundary surveying and the principal reason many surveys are not susceptible to exact cost estimates. Based on this, principle 14 is self-evident.

PRINCIPLE 14. *The surveyor should hunt and search in the field until the best available evidence is found on which to base the boundary retracement survey. Time should not be a consideration.*

11.19 Uncalled-For Monuments

In locating legal descriptions, a surveyor will often find more uncalled-for monuments on the ground than mentioned in description(s) being retraced. Until recently, few states required the surveyor to record replacement information of original monument locations. In those states where widespread loss of evidence of original monument positions exists, the acceptance of monuments that appear to have been set by a surveyor is quite common, although no chain of history of the monument may exist. In some states the reputation of a monument is the best available evidence, and it becomes controlling for that reason alone. Some states have laws to require the filing of public records of discovered monument positions. In most instances, little administration and enforcement are given to maintaining the intent of the law. The rules of evidence also address this area, in that the hearsay rule provides for exceptions, and as such, this area is addressed as an exception to the hearsay rule by permitting individuals to testify as to ancient boundaries and landmarks about which they have personal knowledge.

Uncalled-for monuments, not referenced in any document, when found, should be identified and noted in the new retracement.

11.20 Importance of Possession

Possession lines should not be ignored. If possession lines do not agree with written or deed lines, the relationship of the written or deed lines to that of possession lines must be shown; possession lines may be the best evidence of ownership lines or they may have no significance whatsoever. Fences, buildings, roads, and other evidence of possession that were constructed on some information or some assumed knowledge of the actual line may serve as adequate and sufficient memorials for the boundary.

Possession is a fact issue for the retracing surveyor but a legal issue for the attorney and the courts. The surveyor should never give legal advice as to possession problems.

PRINCIPLE 15. *Possession may memorialize original survey lines, and, as such, may be the best or the worst evidence of original lines.*

There are two distinct and different types of possession. Possession representing original survey lines marked and surveyed is treated in a vastly different manner from possession not in accordance with original survey lines.

After an original survey is made, if fences and other improvements are constructed in accordance with the original survey, then these fences and improvements may be considered and may be proven to be monuments representing the location of the original surveyor's lines. When all of the original monuments have disappeared, the only remaining evidence of the location of the original survey may be lines of possession that were created when the original survey was still visible.

It must be remembered that a fence is but a fence until it is called for as being the boundary (controlling) line. The surveyor must apply all of the skills of observation and detection in analyzing ancient fences. Was the fence initially placed on posts or was it initially placed on trees? The trees were cut, and then the stumps rotted to leave evidence that must be analyzed as to when the fence was constructed, who constructed it, why it was constructed, and where it was constructed in relation to the line and the corners (see Figure 11.2). Several states have addressed fences.

Ancient fences are used by a surveyor in an attempt to reproduce (follow the footsteps of) an old survey. It was found that such fences were strong evidence of the location of the original lines; if these fences had been standing for years, they should be taken as indicating such lines as against evidence of a survey that ignores such fences and that was based on an assumed starting point [14]!

FIGURE 11-2. Old fence that was attached to living trees. The tree grew around the fence, was then cut, and left to rot. The remains were found buried under leaves.

In Michigan, a court stated that a long-established fence was better evidence of an actual boundary that had been settled by practical location than any new survey that was made after corners of the original survey had disappeared [15].

Possession in accordance with an owner's opinion of where the lines should be, or possession in accordance with an erroneous survey, or possession in accordance with any survey made after the original survey is not possession representing where the original surveyor marked the original lines.

Where a survey is called for, the lines and monuments set by the original surveyor are conclusive and control for resurveys as well as providing one form of evidence for retracement. If possession is proven to be in accordance with the original deed lines, it should be the controlling factor in locating the lines of the original surveyor. But if possession is not related to the original surveys, it may possibly then be treated as an encroachment.

Perhaps the greatest value of possession evidence is as an indicator of where to search; lines of possession usually follow boundary lines, and most monuments are found at the termination of possession. This was discussed by Justice Thomas M. Cooley in the late 1800s in an article entitled "The [Quasi-]Judicial Function of Surveyors" (see Appendix C).

11.21 Testimony

Testimony as to where monuments used to be, especially of long duration that originated at or near the time of the original survey, may be of great value as evidence. Like all other forms of evidence, one must diligently follow the rules of evidence. Human memory often outlives the physical monument itself, but this memory may be subject to the usual flaws attributed to memory. In some states the surveyor has the authority to take statements of living witnesses under oath, which may be of value at a future time. In the absence of this right, a witness's statement can always be subscribed before a notary public. The oath serves to preserve testimony and may deter a person from later changing the former testimony. It may be of value in the event of death or the inability of the witness to appear in court to testify. This depends on the rules of the local court. But before any single line in a description can be located with absolute certainty, the two corners that were placed at each end of the line must be located. However, before all lines in a parcel description are located, all corners described in the original conveyance must be identified and located. The retracing surveyor should not concern himself or herself with the admissibility of such testimony, that is the purview of the attorney to be concerned with. The surveyor should use *any* testimony for "what it is worth."

11.22 Measurements

At least one corner recited in the deed must be proved before measurements can commence. After the discovery of that first proven monument, the processes of searching and measuring are often done simultaneously. Measurements are but one form of evidence and may indicate likely places to search for other monuments. Early in the search for monuments, it may be advantageous to make quick approximate measurements until a suitable starting monument is found. But usually, it may be a waste of time to make approximate measurements, find the corner, and then remeasure precisely. It is often better to measure carefully to find the monument and then tie in the monument and correct the measurement to the true line by computation.

11.23 Uncertainty of Position Caused by Measurements

PRINCIPLE 16. *An original monument found undisturbed usually expresses the intent of the parties of the conveyance, fixes that point as between the parties, and, as such, has no error in position. All restored monuments established by measurements have some error in position.*

A discovered, called-for, and undisturbed original monument accurately determines position. Irrespective of measurements from distant points, the position is unalterable and fixed and has no error of location. If measurements to the point differ from the record, the measurements are in error; the position occupied is correct.

All nonoriginal monuments set by measurements from fixed points have some error in position. Although the error may be insignificant, it is present, and the primary concern of the surveyor is to find starting points that are free from position errors; these are original monuments or the spot formerly occupied by an original monument.

11.24 Permissible Uncertainty of Measurements

PRINCIPLE 17. *The magnitude of permissible uncertainty of measurements is always determined by a court's interpretation.*

Courts are the final authority on interpreting where boundaries called for in old conveyances should be located. It remains a matter of law as to the validity of a description, and if questioned, the judge will determine its validity. This assumed authority includes the right to decide how much measurement uncertainty can be tolerated. The opinion of surveyors as experts may influence the courts' thinking in trial. It is always proper for surveyors to express an opinion on measurement accuracy, but surveyors do not have the final say.

Permissible uncertainty of measurements has varied throughout the years. In the early days, with the use of the compass, directional certainty was poor, and errors in excess of 1 chain in a mile of line were not uncommon. Today, an error of 1 foot in a mile is ordinary accuracy. In the future, an error of 0.10 foot in a mile may be considered excessive for boundary location.

Courts must adjust thinking as new technology results in measurements becoming increasingly more precise. Accuracy standards for resurveys change with the development of new equipment.

A deed stating "N 10° E, 100 feet to an iron pin" has two uncertain elements, and in the event the pin is lost, its location will depend on the bearing and the distance to restore the former pin's position exactly (as closely as the surveyor can measure) with the equipment being used. One surveyor can locate it "N 10°00′00″E, 100.00 feet." The measurement will depend on the equipment that was used to create the position of the original corner and then the equipment that was used to conduct the new retracement.

The rule does not say the distance may be within so many minutes of the direction or within so many feet of 100 feet. We know that the original measurements to the pin had a certain error, in that they were not exact. The knowledge of this does not alter the rule to restore the lost pin at the distance and bearing called for in the description.

In a regular section of land, the section corners and the exterior quarter corners were set by the government surveyor. In the event that the quarter corner of a regular section is lost, the court must declare, as supported by federal laws, being a lost corner, that it be repositioned as prescribed by law, which is "equidistant between two proven section corners." When the corner was originally set, the government would accept as sufficient a position within one chain of that specified. By law, in sectionalized land the government could set rules of accuracy on how close a monument needs to be set to its true position. But this law in no way affects the court's prerogative to specify how the corner is to be relocated in the event that it is lost. The law does not provide for the court to set the lost corner with any flexibility, in that it should be relocated equidistant plus or minus 1 chain; it says "exactly" equidistant. The federal government recognized this principle over the years with passage of the Land Act of 1796 and the Land Act of 1805 and publication of the *Manual of Surveying Instructions*, beginning in 1902 [16].

PRINCIPLE 18. *The error of measurement originally permitted when tying original monuments together is independent of the accuracy required to reestablish an original lost-monument position.*

In restoring a lost corner (not an obliterated corner) to its former reported position, the law, in specifying restoration procedures, assumes that the original reported measurements are free from error. In locating property in accordance with an existing conveyance, the surveyor must be as accurate as necessary to protect a client's interests. If damages occur, the court will decide the responsibility of the surveyor.

One thing is certain: An alleged error of measurement must be proved, and it can be proved only by subsequent measurements between proven corners.

PRINCIPLE 19. *In the absence of the owner's specifying an unusually high precision coupled with an accurate survey, it is presumed that the surveyor will work to that precision consistent with the purpose for which it will be used or the standards of the profession in that locality.*

All surveys describing written conveyances should, as a minimum, be as accurate as those of a reasonably skilled prudent land surveyor under similar circumstances. Obviously, if the land surveyor either knew or suspected that a 14-story building were to be erected adjacent to the property line, the liability would be high. Here, a higher degree of skill would be expected than if the survey were in a rural area.

If a client requests a boundary survey of less accuracy than the surveyor suggests, the surveyor should refuse to conduct the survey. The surveyor has a moral obligation to all adjoiners to the line that is being surveyed. This obligation would not extend to notifying adjoining owners of any problems that may be encountered in the survey.

The boundary surveyor should not reduce accuracy merely because lower fees are accepted. The courts have repeatedly held that cost is not a factor in determining liability; in fact, a surveyor can be held liable for damages resulting from a survey that is done for no cost. Professional registered or licensed surveyors always have the responsibility to instruct their field personnel about the accuracy required and the procedures to be followed. Failure of a subordinate to attain the specified measurement accuracy can never be an excuse to escape liability; surveyors must see that their instructions are understood and followed.

Accuracy of position in surveys, resurveys, and retracements based on the record is difficult to define. It depends not only on attainable measurement accuracies but also on legal interpretation of words used in the description being retraced. A deed reading "the west half of lot two" is difficult to interpret, in that without other information it can present the retracing surveyor with several alternatives (see Figure 11.3). The uncertainty in position arises from legal interpretation, and unless the surveyor can correctly resolve the question of where the easterly line of the west half is, there may be large positional errors of any lines established. A legal question may also present itself: Does the description actually mean that half is by area or by distance?

Surveyors often encounter proportional measurement problems. Suppose that a rectangular block, as originally monumented, contained 12 lots, each 50 × 100 feet. Six lots faced one street. On discovery of the original block corner monuments, the surveyor must divide the frontage into six equal parts. Many types of measuring devices could be used to do this. Monuments set apart one-sixth the length of the block and on a straight line connecting original monuments would be correct in position. A tape too long or too short could be used to divide the block into proportionate parts, but the true unit length of each lot would be unknown. In any proportionate measurement problem, it can happen that the monument as set will be in its correct position, yet the measurements will be inaccurate as compared to a defined unit.

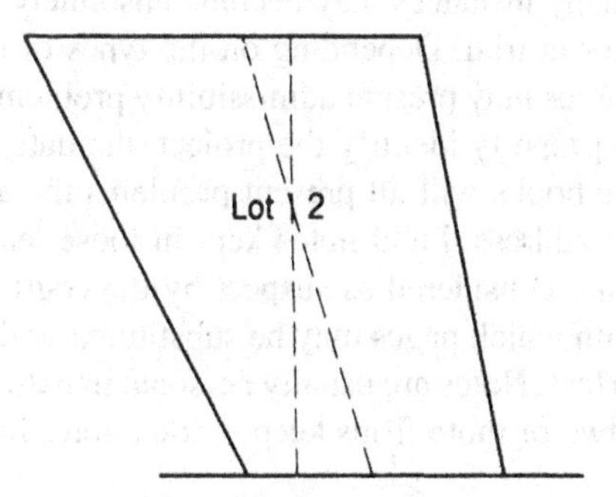

FIGURE 11-3. West half of lot 2.

Assuming that the surveyor correctly interprets a deed and monuments the line described, measurements should be performed with sufficient certainty to prevent damages to the client. The usual reason that an owner wants a survey is as assurance that any improvements will be on their land and not trespassing on an adjoining lot.

Judges are apt to look on errors in terms of positive measurements. If a building were found to be encroaching on the adjacent property by 0.15 feet, the fact that the boundary traverse had an error of closure with a relative error of 1/20,000 or even 1/50,000 would probably fail to impress the judge. The only point that will be considered important to the court is that the building encroaches 0.15 feet.

11.25 Preservation of Recovered Evidence

Found physical evidence, including monuments, must be noted and described in field books or in some other manner to preserve evidence of their location. Photographs are well suited for this purpose, especially when taken systematically and supplemented with a sketch. In the event of litigation, the surveyor, by the evidence that is relied on, must be prepared to prove the location of the property description. Field notes and other records of field evidence are presented to support the surveyor's opinion of boundary location. If photographs are taken, they should be referenced to the survey line whenever possible.

11.26 Field Notes

Field notes may be a record of evidence originally established and to be discovered. They may also be a record of what was done in the field. Each is different, in that the notes that were created as part of the original survey become a part of that survey and, as such, control the position of any monuments set to identify the original corners. Then the notes that were created by the retracing surveyor of the retracement of the original lines merely become evidence of what was done in the field and the information of the original survey that was recovered.

Ordinarily, a sketch represents the results of a boundary survey. Written notes in long-hand, although effective at the moment, may become difficult to interpret at later dates and in many instances may become absolutely impossible to decipher during a retracement or at trial. Depending on the types of notes and their method of compilation, field notes may present admissibility problems for attorneys. Failure by the note keeper to properly identify the project, the date, and crew members as well as erasures in the books will all present problems the attorney who wishes to use them at trial must address. Field notes kept in loose-leaf books or field books that contain erasures are considered as suspect by the courts and present problems because of the ease with which pages may be substituted or data may be altered.

Notes are never perfect. Notes are usually personal in nature. No two surveyors in the same firm or any two or more firms keep written notes in the same form, but all notes, to be effective, should contain certain basic information. Following is a partial checklist of selected information to be included:

1. Date, name, and address of client.
2. Names of party personnel, positions, and duties.
3. Weather condition of wind and observed temperatures.
4. Equipment used, with serial numbers.
5. Legal description or name of parcel being surveyed (e.g., book and page recording, lot and block, subdivision).
6. Work order number and purpose of survey.
7. North arrow with origin of bearings.
8. Description of monuments called for, found, not found, and set.
9. Measurements actually made (e.g., slope distance and vertical angles, temperature, taping, horizontal angles).
10. Corrected distances and angles.
11. Description of monuments set and ties taken to features.
12. Relation of possession to surveyed lines.
13. Outline of parcel surveyed, properly highlighted in pencil.
14. Sketch of parcel staked, showing important features.
15. Oaths of witness evidence (when applicable).
16. Names and addresses of adjoiners, old residents, if available and ascertainable.
17. Reference to any records relied on or called for.

Historical field notes come under the same scrutiny as do modern. They must meet the same challenges of admissibility and relevancy. A surveyor was attempting to introduce some of his own field notes that he had assembled in the 1920s indicating that he had recovered some original GLO cypress bearing trees. The opposing attorney called to the court's attention that the surveyor had erased in the field book. The court refused to admit them into evidence.

As more surveyors use electronic data collectors, the courts will have to examine old principles applying to field notes and records, in that all data are stored electronically and there are no hard copies developed in the field. The question that will be asked is, "Are these business records?"

With the universal acceptance of data collectors for fieldwork, hard-bound field books are being relegated to archives, in that much of the data is now stored electronically. For forensic surveys, it is suggested that field information still be maintained in bound field books, for the legality of the electronic field books has yet to be tested in a court of law. Once the legality of electronic data storage has been tested legally, then a positive statement can be made.

11.27 Assembling Data

After the field search is completed and the discovered monuments have been tied together by measurements, office work such as computations and analysis of results is necessary. Finished plats are often prepared at this point and used for the final staking of the survey.

All discovered evidence, either positive or negative, must be considered during the analysis. In most cases, the field persons can be of great help in deciding where new monuments are to be placed, what is to be shown on the plat, what calls should be used if a new description is to be constructed, and what weight should be given to monuments. Whether used or not, all data should become a permanent part of the file, with sufficient notation and cross-referencing to be of benefit to future surveys.

COMPILATION OF EVIDENCE

11.28 Computations

Computations serve four purposes:

1. To check for the presence of blunders and the quality of measurements
2. To determine where final corners and monuments belong
3. To calculate area
4. To check the precision of the original surveys

If a perimeter is surveyed, the figure must close mathematically. As previously stated, a mathematical closure never proves the absence of errors; it provides some assurance against the presence of blunders and excessive errors. The numbers that are generated in a closure will indicate only the reliability of the measurements; they will not address the correctness of the location of the parcel of land. Systematic errors, such as the incorrect length of a tape, cannot be determined by computations.

Even though area is the least important element in weighing evidence, it is the element that all owners consider in buying land, with road frontage, access, and water or shoreline being also of prime importance to the landowner. In the matter of liability, surveyors have probably paid more damages for errors in area computations than for all other sources. Computation sheets and computer tapes should be preserved and filed with other materials of any survey. They are evidence to express the field measurements and should be free of blunders.

When a surveyor has performed calculations of any manner, it is suggested that a second person perform a separate and individual check of the original calculations, so as to make certain that no blunders were made in the original work.

11.29 Platting

PRINCIPLE 20. *Every property survey should result in the preparation and delivery of a report or plat, whether or not it is to be recorded.*

A plat is the result and summation of the survey. It becomes documentary evidence that the client is entitled to have and to rely on.

Every land survey should require a map properly and correctly representing the survey, showing the information developed by the survey. A given land survey also requires a proper caption, adequate dimensions, and bearings or angles, references to those deeds relied on, and other matters of record pertinent to such survey, including monuments found and monuments set.

Customarily, the original drawing belongs to the surveyor. The client should receive signed, sealed, and numbered copies of the blueprints or blue-line copies of the maps or survey that cannot easily be altered. Modern practice dictates that surveyors have positive control over any maps he or she distributes. This may require that every map distributed be numbered and that surveyors identify the location of each copy. It should be remembered that no client is entitled to the final product until the surveyor has received payment for the survey.

A plat is a special form of a map and must be subject to cartographic discipline. The plat should show land boundaries with all data for description and identification. It should be remembered that the final map is nothing more than a summary of the surveyor's opinions.

11.30 Conclusion

After a surveyor has conducted the necessary research, has completed the field search, and is satisfied that no other evidence can be found, computations can be finalized and the platted positions of known corners and facts be reduced to the plat. Conclusions about the validity of found monuments, where new monuments should be reset, and where boundary lines belong should be addressed and finalized as part of the job. This requires skill and judgment and, at times, might require courage. The boundary surveyor's ability for concluding and expressing an opinion on the location of boundary lines based on the evidence recovered and then interpreted marks the distinction between the professional surveyor and the technician.

To arrive at a proper conclusion, it is essential to understand completely the problem at hand and the object of the survey.

PRESENTATION OF RESULTS

11.31 Final Report and Conference with the Client

Although this step is not always necessary, it is advisable to explain to the client what was done, first in writing by preparing a written report and then possibly in person. The surveyor should talk to the client and explain, on the property if possible, why certain decisions were made and what was done. A written report should be prepared and delivered to the client.

If possession disagrees with title lines, it may be necessary to explain the situation to the client and find out if he or she wants record corners set. If the facts support that the possession has been long and continued, it may be desirable to maintain the status quo and not start a neighborhood squabble by setting monuments in conflict with possession lines.

Title discrepancies discovered by a title search may require early conferences before proceeding further. It might be advisable for the client to seek the counsel of an attorney to correct serious title defects.

During the final conference, the surveyor should take time to explain the value of monuments and point out possible future problems.

If the survey that is performed is a forensic survey, that is, one for litigation, at the request of an attorney, it should be understood at the beginning if a report is required by the attorney. There may be instances when an attorney does not wish a report prepared, as a trial tactic. If this is the situation, then the surveyor should get the request, in writing, that no report is to be prepared. Several of the states require, as part of the minimum technical standards, that a report be prepared on the completion of a survey, and failure to do so may engender a violation complaint.

11.32 Completing the Survey

The final step in a property survey should be the presentation of a plat to the client, a certificate of survey, a surveyor's report, and an invoice for the professional services performed. At times, once the client receives the final plat, the desire to pay the surveyor is of little importance to the landowner, and if the surveyor demands payment, a dishonest client will question the surveyor's work and the surveyor may be in a position of having to defend against a negligence claim. Maps and plats presented should be properly titled, signed, sealed, certified, in accordance with statutes or regulations, and delivered. It is advisable not to deliver any final maps until payment arrangements have been made or until receiving full and final payment. If the surveyor wants to limit the liability, this is the time to do so. Warnings can be given that certain title defects may give future trouble, that certain encroachments may have ripened into a fee title, or that other matters may have adversely affected the land. If a letter is submitted to the client and a copy filed with the job, future disputes and charges of negligence can be averted.

As a protection for the surveyor, it is suggested that only delivered plats be sealed and numbered.

11.33 Forms of Surveying Results

Several states require, by statute, a record of survey that must be filed and recorded. Any survey that leads to or involves a land division usually requires a plat that must be recorded. Even in the absence of statutory requirements, the surveyor should present the client with results in a form that will not only serve the immediate need but also be beneficial to all interested parties in the future. In many of the states that have these recording statutes or requirements, surveyors who have failed to follow the requirements have been fined and have even had their licenses revoked.

A written statement accompanying the plat will include three things:

1. The monuments searched for, discovered, and used for basis of measurements.
2. Records and documents examined for information.
3. Conclusions, results, and written recommendations pertaining to possible unwritten title rights.

Other items, such as photographs of the monuments, measurement data, and monument ties, may be presented to the client.

11.34 The Surveyor's Report

When detailed explanations are to accompany the plat, a report is sometimes in order. The report allows the surveyor to record, at length, differences and discrepancies found, resolutions of problems (with reasons for making specific decisions), recommendations for unresolved problems, and recommendations for further investigations or actions needed by the client, surveyor, or attorney. The report is an ideal device for limiting liability to specific responsibilities, thus reducing the scope of implied warranties. Many times, a surveyor can eliminate unfortunate situations by properly keeping the client informed as to problems encountered and why certain decisions were made.

The requirement that the surveyor prepare a report may be modified, depending on the type of survey that was conducted. Remember, there are two types of surveys, depending on for whom the survey was created. In most instances, a routine survey conducted for a landowner is a utility survey. In this instance, a survey report should be prepared as part of the overall project to preserve evidence and to provide for the surveyor an opportunity to recall information in the event the work is ever questioned in the future.

The client might be an attorney who needs an expert to testify. In this situation, the attorney and his or her client (the surveyor) will enjoy certain privileges as provided by the rules of evidence. When the attorney requests a survey or an analysis of a survey, it should be understood whether a survey report will be furnished by the surveyor. Once a report is prepared, the information may become discoverable. But if no report is requested or prepared, then there is no discoverable information.

PRINCIPLE 21. *The surveyor should conduct each survey as if it will ultimately be contested in court.*

Although there are no assurances that a survey will never be questioned, the surveyor should never take for granted that the utility survey will remain without question indefinitely. Every utility should be conducted to that degree of skill and professionalism in anticipation that at some future time the survey will be questioned by the former client, one who may be in privity with the client, or another surveyor.

REFERENCES

1. *CJS Boundaries,* Vol. 11, Sec. 118. (St. Paul, MN: West Publishing Co., 1938).
2. *Diehl v. Zanger*, 39 Mich. 601 (1878).
3. *Lerned v. Morrill*, 2 N.H. 197 (1820).
4. *Prop. of Kennebuck Purchase v. Tiffany*, 1 Me. 219 (1921).
5. *Bemis v. Bradbeuru*, 2 Me. 462 (1927).
6. *Rivers v. Lozeau*, 539 So. 2d 1147 (Fla. 1989).
7. *Cox v. Hart*, 43 S. Ct. 154, 260 U.S. 427 (1922); *Vaught v. McClymond*, 155 P. 2d 612 (Mont. 1945); *State v. Aucion*, 20 S. 2d 136 (La, 1944).
8. K. E. Hermansen, *Handbook of Annotated Forms for the Surveying Practice* (Hoboken, NJ: Professional Education Systems, 1991). 1993 version by Robillard and Hermansen, Supplement, published by Wiley.
9. *Ski Roundtop Inc. v. Wagerman*, 79 Md. App. 357 (1989).
10. *Cole v. Bradbury*, 29 A. 1097 (1894).
11. *Spaulding v. Abbot*, 55 N.H. 423 (1875).
12. *Federal Rules of Evidence* (St. Paul, MN: West Publishing Co., 1987).
13. *Burke v. Colley*, 495 S.W. 2d 699 (Mo. App. 1973).
14. *James v. Hitchcock*, 309 S.W. 2d 909 (Tex. Civ. App. 1958).
15. *Diehl v. Zanger*, 39 Mich. 601 (1878).
16. Bureau of Land Management, *Manual of Instructions for the Survey of the Public Lands of the United States* (Washington, DC: U.S. Government Printing Office, 1973, 1947, 1930, 1902, 1855).

12

ORIGINAL SURVEYS AND RELATED PLATTING LAWS

12.1 Contents

The creation of new parcels of land by original surveys and their subsequent regulation by governmental agencies are discussed in this chapter.

Land boundaries can be created in two ways:

1. By description in a legal conveyance that usually did not or does not require the benefit of an actual survey.
2. By running lines on the ground by survey, using approved survey methods, by setting monuments at corner points, and then describing the resulting parcel or parcels in documents.

A harmonious parcel is one in which the boundaries described in the written description are superimposed on those lines located in the field by survey. The best results for the benefit of the public are obtained by passage of statutes regulating how land boundaries may be created.

Subdivision law, as the term is commonly used, has a broader meaning than does *platting law*. A *subdivision law* includes regulations for the use of land, how streets must be improved, zoning, and many other restrictions on land. A *platting law* is generally, although not always, thought of as a law regulating the size and shape of parcels, how parcels shall be monumented and measured, who can do the monumenting and measuring, what accuracies must be attained, and how the land shall be platted.

In the early GLO surveys of the 1800s in the United States, the purpose of laws regulating how land could be divided into parcels was to devise a simple means of describing land and ensuring the certainty of location, while creating and providing maps of the entire area surveyed. Currently, subdivision laws usually include regulations on land usage and requirements for improvements.

At one time, the public agencies would accept street dedications without improvements; now it is common practice to require paving, drainage, sewers, water, curbs, sidewalks, and other improvements prior to accepting street dedications. Subdivision laws are becoming stricter and more complex as more land usage restrictions and improvements requirements are included as part of subdivision acts. The purpose of this chapter is to discuss laws pertaining to the creation of land parcels, with emphasis on monumentation and measurements. The preparation of plats is the subject of a later discussion.

Federal laws regulating how the public domain should be subdivided into sections have had a greater impact than any other subdivision laws. At the local level, many states, counties, and municipalities have laws or regulations depicting and regulating how larger parcels can be subdivided into smaller ones.

The principles discussed in this chapter are as follows:

PRINCIPLE 1. *Original surveys to create new land parcels are regulated by legislative laws and rules.*

PRINCIPLE 2. *An original survey does not identify boundaries. It creates them.*

PRINCIPLE 3. *A monument set by the original surveyor and called for by the conveyance has no error of position. It is legally correct, in that only the description may be in error.*

PRINCIPLE 4. *The title to a junior parcel can be no better than the quality of the title of the parent parcel from which the junior parcel was created.*

PRINCIPLE 5. *When retracing the boundaries of parcels of land, the lines and the corners that were created on the ground control over the lines depicted on the plat and the lines described in the description.*

PRINCIPLE 6. *When retracing parcels of land, the priorities of control identified by case law, some statutes, and common law are: actual lines run and marked on the ground; natural monuments (boundaries); artificial monuments; adjoiners; bearings; distances; and area.*

12.2 Regulation of Original Surveys

PRINCIPLE 1. *Original surveys to create new land parcels are regulated by legislative laws and rules.*

Resurveys and retracements are interpretive in nature and are for the purpose of locating parcels described by an existing document. Preparing subdivision plats and performing original surveys are creative in nature and are not interpretative. For original surveys, the governing body usually specifies the size and shape of new land parcels, who may do the surveying and platting, what the land may be used for (including methods and instrumentation), and numerous other regulations. In some areas, the law makes it illegal to convey land other than by lot and block in accordance with a recorded plat.

For a subdivision plat to be valid and have effectiveness, in general, it must be authorized by law and must be in substantial agreement with the law. In this chapter, when we discuss subdivision, the intent is to use the term in the sense that it is a portion of a larger parcel (parent tract) that was divided either by description or survey into two or more smaller parcels. Those creating subdivisions should be certain of their compliance with their enabling act. There are two types of original surveys. First, there are those surveys conducted under federal and some state jurisdictions whereby the original exterior boundaries to which the lands were initially patented or granted by the governmental body were created. This included all public land states (GLO), Georgia, some New England states, and those states that granted state patents or grants to state lands. Second are those surveys that are created, at times under statutes, rules, and regulations, of smaller parcels within larger parcels. The Constitution provides that once land goes into private ownership state laws regulate property.

12.3 Original Boundaries

PRINCIPLE 2. *An original survey does not identify boundaries. It creates them.*

Once statutory authority is granted to the sovereign or the landowner to create boundaries and the survey is conducted, the lines that are run in on the ground in contemplation to create the boundary lines and the corners and any marks made on the lines or monuments created become part of the original survey, including the field notes and plats of the parcels. The rules and regulations that are in effect are for the creation of the parcels. Once a boundary line is created, the presumption is that it was created in accordance with the laws, rules, and regulations in effect at the time. When the survey is accepted by the governmental body, including the courts, no subsequent surveyor has any authority to "recreate" those lines. They remain forever fixed. The responsibility of all subsequent surveyors is to "find" those original lines. This is done by conducting retracements, recovering evidence, interpreting that evidence, and then making decisions.

12.4 Former Presurveying Practices

At one time within the United States, some areas did not have laws regulating the creation of new parcels of land, and in other places the laws were inadequate. In many instances, the better surveyors, without regulatory laws, properly applied sound surveying practices and carefully laid out new additions. However, those with an eye for extracting the last dollar of profit hired the incompetent for a lesser fee and left land parcels in a chaotic condition. To ensure uniformity and certainty of future land location, regulatory measures were required.

Many states and the federal government have created and still have laws regulating land divisions. The failure of these laws in the past has been due to numerous causes, the more important being failure to legislate adequate laws, a failure to enforce laws, and the destruction of monuments after they were set.

The enactment of subdivision laws by the federal government to dispose of the public domain was an example of the best thinking of the time. These laws contemplated and provided for the public domain, with the federal government creating the macro exterior boundaries and local surveyors creating the micro subdivision interior boundaries. One serious flaw was that once land passed into private ownership, federal law had no binding effect on the states as to real property rights. In some states, federal rules for resurveys or retracements have not been adopted. However, most states, either by statute or case law, provide for local surveyors to conduct retracements to retrace originally created boundaries, as well as boundaries created by state laws. Today this conflict still causes serious problems.

12.5 Objective Platting Laws

The object of most platting laws is to specify simple means of identifying land and ensuring certainty of location after land has been identified by references to legally created subdivisions so there can be a consistency of description. Plats are used to identify land, and monuments and measurements are used to locate land. In recent times, land use laws have been tacked onto platting acts. Although some mention is made of land use laws, the purpose of this chapter is to discuss methods of identifying and locating land and creating land parcels.

12.6 Certainty of Land Location

PRINCIPLE 3. *A monument set by the original surveyor and called for by the conveyance has no error of position. It is legally correct, in that only the description may be in error.*

The original surveyor divides a new land parcel, which is usually identified by the lines and the corner monuments that were established.

Federal laws and the courts have given those creating new land divisions a precise method of positioning land boundaries; by merely setting a monument and calling for it, land has an exact location. A parcel of land is certain in a location if an indestructible, unique monument is set and called for at each corner and then found. This principle could also be applied if no monuments were set at the time the parcel was surveyed, and the corners were called points.

Early in the history of our nation, Congress anticipated that subsequent surveyors would attempt to correct and replace the discovered errors of original GLO surveys with better measurements. In an attempt to "legislatively" seal or fix these original boundaries and monuments, the Land Act of February 11, 1805, was enacted, which fixes the original boundaries as they were created. This Act has been so powerful that today most courts simply refer to it as the Act of 1805.

But certainty of location is not necessarily proof of ownership, since a senior right might be interfered with or the subdivider might have had a defective title.

Once a monument and all evidence of its former position disappear, measurements are relied on to replace the monument. The uncertainty of reestablishing a monument in its former position is then dependent on the uncertainty of the originally reported measurements and the uncertainty with which measurements can be made. Original reported measurements cannot be analyzed, checked, or rechecked after monuments are lost; measurements must have been reasonably certain prior to a monument's destruction.

The objectives of platting laws with respect to ensuring future certainty of the location of land parcels require the following:

1. Durable permanent monuments
2. Unique monuments having certainty of identification
3. Frequent setting of monuments
4. Accurate measurements tying monuments together
5. Reference to the monuments in later writings
6. Adequate descriptions identifying a sufficient number of the necessary controlling elements

The responsibility of future certainty of the location of described land parcels lies entirely in the hands of the persons setting the monuments and preparing the descriptions or plats. The only time that location can be assured for future years, without corrective documents, is at the time the conveyance is being made. After a conveyance is made, it becomes a question of interpreting what has been written. It is far easier to correct an ambiguity prior to when a conveyance is being made than it is to try to correct it later.

12.7 Features of Platting Law

Although there are many variations of platting laws, some due to climatic or use conditions and some due to political requirements, the 13 essential features of a platting law are as follows:

1. Definition of when a plat must be made
2. Approval by governing or regulating agencies:
   (a) Planning
   (b) Health (e.g., sewerage, water)
   (c) City engineer (e.g., grades, design)
   (d) Tax assessor (pay back taxes and set new valuation)
   (e) City attorney (approval of form)
   (f) Inspection (usually by city engineer)
   (g) Final approval by the governing body and various environmental agencies

3. Title guarantee
4. Performance bonds
5. Definition of who may prepare maps
6. Dedication and street widths
7. Monumentation
8. Setbacks
9. Easements
10. Measurement data required and accuracies needed
11. Recordation
12. Prohibiting sale of lots prior to recording map
13. Setting monuments at corner positions prior to sale of the lots

12.8 Platting Laws

The number of parcels that may be created or sold prior to the necessity of filing a plat varies with the statutes or ordinances of the states and many local governing bodies. States and/or local agencies usually have specific requirements that must be met before any parcel can be subdivided. These agencies usually do not cooperate well in insisting on a uniformity of regulations. In fact, even adjacent counties may have different requirements.

Before a creating surveyor undertakes the creation of any new subdivision, the surveyor should become familiar with and fully understand the written requirements the governmental agencies demand and require for that particular locality.

12.9 Planning Boards

Within the last 50 years, there has been a marked change in the attitude of people toward land usage. At one time, most people believed they had a constitutional right to do as they wished with their own land; now, in many cities, the opposite is true. Land is designated for a particular use by a governing body, and the land may not be used for purposes inconsistent with the desires of the majority without the consent of the governing body.

The agency charged with planning for land use has been variously designated as the planning commission, planning board, or planning agency. It is this agency's privilege to recommend to the governing board a course action for land usage. Normally, the agency has no legislative authority; by approval of the governing agency is its opinion given force. If a planning board rejects a given plan, the owner has the right of appeal to that governing body.

12.10 Considerations of Title

PRINCIPLE 4. *The title to a junior parcel can be no better than the quality of the title of the parent parcel from which the junior parcel was created.*

Others may have valid claims against the title of a property. This title may appear in the form of senior fee titles, easements or lesser estates, or other encumbrances. The quality of title to the lots of a subdivision will be no better than the title of the parent tract. It is imperative that the relationships be determined and investigated before the development starts. Usually, such matters are referred to an abstractor, title company, or others who guarantee ownership. At times, this responsibility is accepted by the surveyor. If the surveyor undertakes this responsibility, he or she may assume all liabilities resulting from the actions.

Easements are particularly troublesome when land is subdivided. Simply because if an easement falls into nonuse, that does not mean that it is released or abandoned and the title has reverted. The surveyor, attorney, or title examiner must examine the records of all possible utilities or other easement rights to determine any possible impact on the tract. An unrecorded easement may still be valid.

In some jurisdictions, abandonment is the formal proceeding whereby the easement holder gives up his or her rights forever. Other jurisdictions use the word *discontinuance* or *vacation* to mean the same thing, although many times the two are not the same. The main point to remember, or to make clear perhaps, is that mere nonuse of an easement created by grant does not constitute abandonment. Abandonment of an easement necessarily implies nonuser, but nonuser use does not create abandonment, however long continued [1].

The plat that is prepared to show a survey of land boundaries should show all record utilities whether there is evidence of them on the ground or not and all features on the ground that have the slightest possibility of being an easement on the fee title. All easements on the parent title as well as visible uses must be a condition binding on the lots created therefrom.

A New Hampshire court held that a reserved right in a conveyance that is not in its very nature a mere personal and temporary right will always be held to be an easement running with the land, in the absence of some controlling provision to the contrary. Once an easement has become appurtenant to a dominant estate, a conveyance of that estate carries with it the easement belonging to it, whether mentioned in the deed or not [2].

When easements have been discovered on the tract, it must be ascertained if these may exist without greatly hindering the planned subdivision or if there is a possibility of having them relocated to a location that will allow the land development to proceed. In some cases, it is possible to obtain a release from the lessees or easement holders, but often this is impossible because of the restrictive language of the parent document creating the easement. If the easement is to remain in the subdivision tract, the restrictions in the easement must be respected.

In some areas there may be oil and mineral exceptions. Such mineral owners are, in fact, interested parties to the tract, and they must sign releases before these rights are extinguished. They may have a senior right of entry to mine and drill and to otherwise use the land for mining. In many states, these rights may have to be extinguished before land can be subdivided. Some states provide that the owner of the surface can claim any outstanding mineral rights by adverse possession. This is done so that all of the "bundle of rights" can be merged into one common owner.

Covenants and restrictions "run with the land," and the mere act of subdividing does not destroy their effect. A subdivision developed on a tract that was conveyed years earlier, "provided no motor vehicles enter thereupon," caused considerable difficulty before streets could be opened. Restrictions and conditions that have been created in a will can be overcome only by court actions, usually by eminent-domain procedure.

In the city of Laurel, Mississippi, in the early 1940s, the oil rights were sold separately. Then, in 1950, the parcel was sold and the title search failed to identify the previously sold rights. In the late 1950s, the mineral holder started to construct an oil well on one of the lots. Chaos erupted over the drilling noise and the turmoil over the oil. The lot owners had no recourse.

If covenants exist, it must be determined if they are in conflict with the subdivision and if they can be modified.

A fee title to the tract may be clouded because of a prior unsettled lien or tax judgment against the owner or former owner. Fee title does not assure a totally free parcel. *An encumbrance against the parent tract is against the whole, even though only a small portion of the land is subdivided.* Therefore, an owner of a quarter-acre lot that has been carved from 100-acre tract having an encumbrance may find that he or she is in danger of losing the title, or at least some money, when the party who holds a lien or judgment presses or perfects the claim.

12.11 Title Guarantee

To protect the new purchaser of land, it is advisable to require that landowners have some financially responsible body guarantee that if the title is found wanting, they will have financial compensation, either by payment of the decreased value of the land or by defending the title in court.

In those states where title insurance is a common practice, the example format of the following notes usually are endorsed on the face of the map:

> We hereby certify that we are the only owners of, or are interested in, the land embraced within the subdivision to be known as Speer Tract, and we hereby consent to the preparation and recordation of this map consisting of two sheets and described in the caption thereof. I hereby dedicate to public use Midway Drive, the easements for sewer, water, drainage and public utilities, and that portion marked "Reserved for Future Street," together with any and all abutters' rights of access in and to Rosecrans Street and that portion marked "Reserved for Future Street" adjacent and contiguous to Lot L, as shown on this map within this subdivision.
>
> Signed and notarized.
>
> . . . Title Insurance Company, a corporation, hereby certifies that according to Official Records of the County of . . ., state of . . . on the . . . day of 20 . . ., at . . . o'clock, A.M., Mr. So and So were the owners and the only persons interested in and whose consent was necessary to pass a clear title to the land embraced within the subdivision to be known as Speer Tract as shown on this map, consisting of two sheets and particularly described in the caption thereof, except the City of . . ., a municipal corporation, owner of easements which cannot ripen into a fee. In whereof, . . . Title Insurance

Company has caused this instrument to be executed under its corporate name and seal by its proper officers, thereunto duly authorized, the day and year in this certificate first above written.

Signed and notarized.

When the law requires such certification, the initial step in preparing a subdivision is a boundary survey presented to the title-guaranteeing agency for its approval. Such a survey must show monuments found and set, along with existing possession and any unauthorized uses on or beyond the title lines, as well as the new subdivision lines.

12.12 Boundary Survey

In former years, the location of the boundary was the only place the surveyor incurred liability for errors. An interior monument set, even though not in its correct position, may, after it is used, become correct. In recent years, the tendency has been for a developer to produce both house and lot. With delayed staking, the surveyor sets temporary interior stakes for the house location, and if these are wrong, problems ensue.

Boundary monuments of a subdivision along a line adjacent to another owner do not always have the dignity of no error of position. Subdividing land does not automatically erase the rights of the adjoiner; if part of the adjoiner's land is taken, the adjoiner may assert any rights up to and until the time imposed by the statute of limitations.

In the past, errors may have occurred along subdivision boundaries with resultant gaps or overlaps. Within a platted subdivision, wherein one line represents the boundary between adjacent lots, it is not possible to have gaps or overlaps. But along the exterior boundaries, even though the adjoining property is shown as abutting and only one line is drawn, the new subdivision plat does not change the former status of the adjoiner's rights. In resurvey or retracement procedures, lots adjoining boundary lines of the parent tract represent a great hazard to the surveyor and possibly could be a fruitful area for future liability.

12.13 Requirements for Corner Monumentation

All platting laws should require complete and adequate lot monumentation. The requirements should extend beyond that of merely setting lot corner stakes; they should specify the use of permanent identifiable monuments (bounds) at frequent intervals in indestructible locations.

Monuments required may be of two types:

1. Those that are control monuments used to relocate lot corners and exterior boundaries.
2. Those that mark every lot corner within the subdivision.

Control monuments are placed where they are least likely to be destroyed and where they can be conveniently used for future location of the exterior lines of the subdivision. Ordinarily, they are located within a street right-of-way. Boundary monuments have a high propensity for destruction within a subdivision. They are often treated as temporary stakes, and they are often no more than that.

The most important feature of a monumentation law is the requirement for permanent, indestructible, identifiable monuments located at frequent intervals and that are sufficient to enable relocation of any lost lot monument within specified accuracies. These must be set prior to the recording of the plat. Control monuments are usually located at each street intersection, at each point of change of direction of street lines, and at the beginning and end of each curve; they are buried below the frost line or below the street grade line and are situated where the installation of new underground utilities will not destroy them. From the standpoint of ease of computing, monuments set at the intersection of street centerlines are best, but they present traffic hazards.

Better platting laws require that every lot must be monumented prior to the sale of any one lot. In those areas where full improvements are required, the law often requires delayed staking; that is, lot monuments are placed after the curbs, sidewalks, underground utilities, and paving are installed. Some regulations call for lead and disks (with license numbers) in the sidewalks, on an offset, after the installation of sidewalks. Such regulations are of little benefit where sidewalks creep because of sandy soil. In frost-free areas, specifications for lot monuments often include pipes driven substantially in the ground, or even redwood stakes. But in frost areas, iron pins driven below the normal frost line are required.

12.14 Density of Monuments

The distance between control monuments, usually specified by law, should be determined by accuracy considerations and individual needs. Since all original monuments set have no error of position, it is desirable to have frequent monument control. As has been frequently stated, all measurements have some error, and any time it is necessary to measure to relocate a boundary corner, some error will exist. Measurement errors are a function of distance and angle, although not a direct relationship, and it follows that the closer together or more densely located the monuments are, the less uncertainty will exist in relocating land parcels.

In residential areas, where buildings must be set 4 or more feet from the side boundary lines, an error of 2 inches in relocating boundary lines would not be excessive. Permanent monuments located 600 feet apart can be tied together with measurement uncertainties of ±0.06 foot or less; in the event of destruction of a permanent monument, it could be replaced from other permanent monuments with an error of ±0.12 foot. It is logical to require permanent monumentation not further than 600 feet apart in residential areas and preferably at each block intersection. This would ensure certainty of lot location within ±2 inches, even with the destruction of one permanent monument.

In commercial areas, where buildings are permitted to be constructed exactly on the boundary line between two or more parcels, uncertainties of relocation of original positions must be reduced. An error of inch in relocation of positions may be significant. Permanent control monuments not more than 300 feet apart should be required.

Monument density needs depend on the requirement for future certainty of parcel locations. Since future land use is not always foreseeable, it is better to err on the safe side and require more monumentation than is needed at the time of the original survey that created the parcel.

Any monumentation that is created should address seven important elements:

1. Size and composition of monuments
2. Identification of the surveyor setting the monument
3. Maximum distance between the monuments
4. Setting all monuments prior to the sale of the land
5. Responsibilities of the developer, realtor, and surveyor
6. Use of reference monuments
7. The methodology used to create the corner areas

PRINCIPLE 5. *When retracing the boundaries of parcels of land, the lines and the corners that were created on the ground control over the lines depicted on the plat and the lines described in the description.*

This principle, although listed in the middle of this chapter, is perhaps the most important of all principles. From the Land Act of February 11, 1805, through case law throughout the early years of American jurisprudence, courts have held that what the original surveyors created and left on the ground and then described on plats and in field notes, the actual evidence on the ground, became "the footsteps," and the courts would give great evidentiary weight to these footsteps. As only speculation, this principle, perhaps, forecast that the technology and the knowledge the early surveyors possessed and used would not remain the same, and to retrace the original lines with more sophisticated equipment at later times would possibly cause technical and legal problems.

PRINCIPLE 6. *When retracing parcels of land, the priorities of control identified by case law, some statutes, and common law are: actual lines run and marked on the ground; natural monuments (boundaries); artificial monuments; adjoiners; bearings; distances; and area.*

Both the original surveyors who created the boundaries and the retracing surveyors who are given the task to retrace and find the original boundaries have been given both historic and legal guidelines [3] on what elements surveyors should consider. The retracing surveyor has to realize that the only evidence of the boundary line that can be recovered is the only evidence that is created and identified. So, a retracement can only be as good as the evidence that was created and left for the retracing surveyor to recover. The modern surveyor should have knowledge of this priority and should apply it in his/her practice. Not to consider this priority could manifest itself into claims of negligence [3].

12.15 Use of Coordinates in Subdivisions

State plane coordinates, when properly used, are the modern means of ensuring certainty of corner restoration in the event of monument destruction. A legislature can regulate how new land divisions may be made and described, and it is within the scope of law to require compulsory ties to a state-authorized plane coordinate net. The ultimate public advantages far outweigh the complaint of increased initial costs. The system, unless it has been in effect for some time, offers no advantages to the immediate subdivider; it increases the subdivider's cost. But the government should ensure that, in addition to requiring water, sewer improvements, and usage regulations for land, whenever parcels are created, they are of full measure and certain of future identity. The expenses necessary to tie the state plane coordinates are justifiable costs if properly understood and applied.

City or county mapping, major improvement projects, and correlation of a city's activities require a knowledge of distances between subdivided areas. The most accurate and certain method of providing this information is to require that all new parcels of land be located on the same basis of bearing and on the same grid system. Only indifference to a necessary public need and failure of legislative bodies to provide necessary legislation have prevented universal adoption of state plane coordinates in the United States.

State plane coordinates can never replace set local monument control; as long as a called-for monument exists, measurements are secondary evidence. State plane coordinates could be an aid to preserve evidence of monument positions; they are not a means of changing a time-honored system of monument control and land descriptions.

The advantage of a call for a state plane coordinate system is that it increases the number (density) of monuments referred to in a deed. Instead of a newly created parcel having only measurement of relationships to monuments in a local area, the new parcel is related to thousands of monuments in the entire net, if measurements are made to a state plane coordinate system.

To date, little or no land litigation has settled what the courts think of the importance of coordinates; it is conjecture to say what the courts would or would not do. A coordinate system is just as strong as the monuments that mark it on the ground. Without doubt, all coordinate values assigned to a given control monument are in error to some extent. The monument itself is fixed in position, but measurements cannot be exact. If a surveyor measures to a known coordinate point (monument), the tie distance and direction are used to compute the new monument's coordinates. It is the measurement value itself that is the evidence, not the coordinate; it is a measurement tie to a physical monument on the ground or a point that has no monument.

One of the problems of coordinates is the reference system to which the coordinate values are referenced. Reference systems are subject to change, and this defeats the requirement that surveying numbers remain stable for the future surveyors to make reference. With numbers, the need is to have an absolute certainty that the original values will remain stable. The value of tie-out distances to monuments marking a coordinate net is in proportion to the accuracy of the observed measurements.

Poorly made measurements to determine a position in a coordinate net only confuses the record and creates a certainty of change in position when relocating a lost corner. In addition to requiring state plane-coordinate position determination, the law must specify the accuracy of the measurement methods as well as a reference that will never change. To be of real value, errors of position of a state plane-coordinate system created by measurement errors larger than 1 foot in 2 miles could not be tolerated. The ability of the surveyor to replace a coordinate monument or to locate a point that has a coordinate value is determined and related to the precision and accuracy of the original work, the distance from the closest control station on which the values are placed, and the instruments that were used.

State plane coordinates have been recommended by several legal, surveying, and engineering organizations and major governmental bodies, but not courts.

In view of the fact that original surveys are being obliterated each day and property is becoming more valuable, it is imperative that better means be employed to supplement and complement the present system of evidence determination. Following are 17 important advantages of the system of state plane coordinates:

1. Establishes geometric identity.
2. Indicates relative position.
3. Permanently establishes lines.
4. Facilitates the relocation of land.
5. Coordinates may be scaled from government maps (these should not be used as definitive numbers).
6. Self-references for perpetuation.
7. Ties all surveys to a common datum.
8. Eases proportioning mathematically.
9. Often eliminates the need for random lines.
10. Can use other surveys to advantage.
11. Provides a simple means of description.
12. Prevents the accumulation of error due to the shape of the Earth.
13. Permits the ready check on large blunders.
14. Permits easy computations of area in "cutoff" tracts.
15. Is easily adaptable to modern computing methods.
16. Provides an excellent filing system.
17. Permits easy plotting on maps and plats.

In view of the fact that despite being in existence for years, state plane coordinates are not widespread, there are some disadvantages, among which are these:

1. Many surveyors, attorneys, and title officers do not understand the grid projections.
2. Extra cost is involved in additional surveys.
3. Areas do not have sufficient control.

4. Many existing surveys are not of sufficient quality to contribute positions to the system.
5. Positions are dependent on control points that in themselves are subject to destruction.
6. The system to which the coordinates are referenced has no assurance of remaining stable.

12.16 Certification of Survey

In addition to monumentation, a certificate is needed ensuring that the monuments have been accurately set in accordance with the law. Most states require such a certificate to be certified by the surveyor. In some states, the law permits civil engineers to act as surveyors. Such a certificate may take the following form:

> State of . . .
> County of . . .
> I, . . ., hereby certify that I am a Licensed Land Surveyor of the state of . . .; that the survey of the subdivision was made by me or under my direction between . . . and . . . and that the survey is true and complete as shown, that all stakes, monuments, and marks set, together with those found, are of the character and occupy the positions shown thereon, and are sufficient to enable the survey to be retraced.
> Signed.
> Date.

In many instances, monuments set will be destroyed by the construction of improvements. Because of this, many areas have provided delayed staking provisions whereby, after improvements are installed, final stakes must be set. Such a certificate may take this form:

> State of . . .
> County of . . .
> I, . . ., hereby certify that I am a Licensed Land Surveyor of the state of . . .; that the survey of this subdivision was made by me or under my direction between . . . and . . . and that the survey is true and complete as shown. A two- (2) inch pipe twenty-four (24) inches in length will be set at each boundary corner except as shown, and I will set three-quarter-inch (3/4) pipes eighteen (18) inches in length at all lot corners and points of curves along dedicated streets, unless otherwise noted on this map, and each monument will be stamped LS 2554. Said monuments will be set within thirty (30) days after the completion of the required improvements and their acceptance by the city, and such monuments are or will be sufficient to enable the survey to be retraced and will occupy the positions shown thereon.
> Signed.
> Date.

Although some clients wish the surveyor to monument after the signing of the deed and the surveyor may agree to it, this practice is not recommended and should be refrained from. Clients should be encouraged to set all monuments prior to the completion of any deeds.

It is not recommended that monumentation be conducted after the deed is signed, because it defeats the law, in that the deed is sealed at the time of its signing.

12.17 Dedications

Dedications in accordance with statutory law must comply with the letter of the law. There must be an offer to dedicate, an acceptance of the dedication, and a positive description. The form of the dedication may vary from state to state. One form of acceptance is as follows:

> State of . . .
> County of . . .
> I, . . ., city clerk of the city of . . ., State of . . ., hereby certify that the Council of said city has approved this map of Speer Tract, consisting of two sheets, described in the caption thereof, and has accepted on behalf of the public, Shady Pines Drive, as shown on the Speer Tract map, together with any and all abutters' rights of access in and to Rosecrans Street and that portion marked "Reserved for Future Street" adjacent and contiguous to Lot L, and the easements for sewer, water, drainage, and public utilities, all as shown on this map within this subdivision, and hereby rejects that portion marked "Reserved for Future Street" shown on this map within this subdivision. In witness thereof, said council has caused these presents to be executed by the city clerk and attested by its seal this . . . day of . . . 2001.

At times, municipalities desire to have a space reserved for future streets but do not wish to accept the street until such time as the street is improved entirely. Provisions in state laws can be made that "any offer of dedication shall remain open and subject to future acceptance by the city." In the foregoing form above where this law applied, the city rejected the area designated as "Reserved for Future Street" and can at any future date accept the offer of dedication. The only way to remove such an offer to dedicate and the rejection of such is for the city to complete a legal vacation proceeding in the same manner as they are vacated.

12.18 Setbacks

For the purpose of uniformity and to provide for front-yard planting areas, modern zoning laws require minimum setbacks for buildings along front, side, and rear yards. Often such setback requirements are shown on the subdivision map. A blanket requirement that all lots have a given setback is not practical, especially on hillside developments. By mutual agreement between the divider and the planning body, setbacks can be adjusted to fit the topography. On lots rapidly sloping away from a street, excessive setbacks would prevent connections and poor driveways to garages. Setbacks may make some lots unsalable or may prevent the expansion of existing or future improvements. The surveyor must be very careful to make certain that all setbacks are properly indicated and shown on any plats that are prepared. Failure to properly show definitive setbacks may make the surveyor subject to possible negligence claims.

A lot that contains multiple setbacks may end up as being undevelopable or unbuildable. In developing a subdivision, the surveyor should be careful to make certain that a lot should not be encumbered with setbacks that will make it impossible for the lot to be developed.

12.19 Recordation

The certainty of ownership of land sold by lot and block by reference to a map is entirely dependent on the preservation of the map in a place that allows inspection of the map and yet ensures that the map cannot be altered. Recordation of a map in a public office and appointment of a public employee to be responsible for seeing that no map is altered are the best solutions.

Surveyors should take precautions to see that their recorded (or nonrecorded) maps are not altered by third parties. This can be done by assigning a number to each map that is distributed as well as keeping the original copy of the original map in the client's file.

12.20 Examination by Authority

Prior to recordation of a map, the city engineer, the county, the county engineer, the county surveyor, or some identified official must express approval of compliance with improvements and all other elements required by law. Their certification is usually placed on the map, and it will not be recorded without it.

12.21 Presurveys without Recording

In some states the law requires the landowner or the surveyor to record a plat whenever he or she makes an original survey to divide land; in other states the owner may record the plat delivered, at his or her option; and in still other states no plat may be recorded, since no office is delegated to accept plats for recording. Recording a plat is similar, in effect, to recording a deed; it is recorded evidence that may be used in court and to give legal notice of what was the intention of the information indicated on the plat.

Many surveys made for the purpose of dividing land are never recorded. It is unfortunate, because, after the surveyor dies, the evidence of location may well be lost.

12.22 Summary

This chapter is an abbreviated and limited discussion of how subdivisions are made legal. Subdivision laws regulate how and when land may be divided into small parcels. It is within the scope of law to regulate the use of land, how new divisions of land may be created, and who is responsible for monumenting and measuring new parcels. Each state has its own regulations, and the laws are variable from place to place and from year to year. The subject matter presented is general and will not fit any specific area, but it does give general concepts. Whenever a surveyor accepts the responsibility of undertaking work within a subdivision, that surveyor is charged with knowing the full and complete requirements set forth in the statutes and ordinances of the local area of practice. Whenever a surveyor works in several counties or states, a major responsibility is placed on that individual to be knowledgeable of all the laws. The surveyor might find that no two laws or requirements are the same.

The main lesson the student and surveyor should take from this chapter is that the creation of modern subdivisions is highly regulated. To not be totally versed in the specific local requirements required by local governmental bodies could be major source of problems for the surveyor who wishes to work in this area. This means keeping up to date with all of the changes and the most current rules and regulations.

REFERENCES

1. *Adams v. Hodgkins*, 190 Me. 361, 84 A. 530 (1912).
2. *Burcky v. Knowles*, 120 N.H. 244 (1980).
3. *Spainhour v. Huffman*, 377 S.E. 2d 615 (Va. 1989).

13

UNWRITTEN TRANSFERS OF LAND OWNERSHIP

13.1 Introduction

> *Note:* This chapter is not intended to be a source of legal advice for the surveyor. The surveyor should not consider this as legal dicta. It should be used as a source of information only for teaching purposes.

Unrecognized unwritten rights may be a source of trouble and can add unnecessary liability and costs in conducting retracements. The information provided in this chapter is for informational purposes only.

Surveyors always seem to be interested in how they fit into the area of unwritten rights. In reality, unwritten rights are questions of law, usually reserved for attorneys and the courts, not questions of surveying. Surveyors are vital to the unwritten right process by virtue of the fact that they collect and interpret evidence of adverse use and the location of lines of occupation that may be contrary to or conflict with boundary lines described in descriptions.

To be effective in this field, surveyors must understand adverse rights, because their loss or their gain varies, depending on how the surveyor tells the story of the "how" and the "why" of these adverse rights—that is, how he or she interprets collected evidence while doing their retracement. In this chapter, adverse rights will be discussed to provide the retracing surveyor with basic doctrines and information for their effective interpretation and learning.

The Statute of Frauds, originally established in England in 1677, becomes important when dealing with deeds that contain descriptions, because this statute required that any title in land or any recognized interest in land must be transferred and described in writing to be enforceable in a court of law. This statute was originally entitled *A Statute to Prevent Fraud* and originally contained five areas of business where transactions had to have a written document in order to have proof the transaction occurred. As the courts soon found, enforcement of the statute sometimes created a fraud. For this reason, certain exceptions to the law evolved and have been applied to property owners in England and the United States in certain areas of the law, with land being one. As a result of this, the term *written title* or *legal title* probably emerged, in distinction to an occupant of land who was entitled to go to court and through the court obtain an unwritten title through the legal system. Although the term *unwritten title* does not explain or identify the exact nature and may not be the best definition for the exceptions to the Statute of Frauds, it has been used sufficiently to warrant its continuation and acceptance. In all probability, unwritten title may have had its origin through the doctrine of adverse possession that evolved from Roman law. The term *unwritten title* should not and cannot be equated to a written or "legal" marketable title; rather, it identifies rights obtained by law. As far as it is known, those states that recognize English Common Law as the basis of their laws recognize the Statute of Frauds.

Although an unwritten right to the ownership of land is, in theory, good against the written title, it is so only if it can be proved in court and then subsequently adjudicated and granted. Since one cannot be certain of the outcome of any court case, especially when parol evidence is relied on and attorney's costs and court costs often exceed the value of the land in question, it cannot be said that the value of land claimed by occupancy alone is equal to the value of the same land claimed by both written title and occupancy. One duty of the surveyor is to differentiate between boundaries claimed by the different modes of rights to ownership. This would include encroachments on the client's written deed or the client's encroachments on an adjoiner's written deed.

Unwritten ownership rights can be one of two things:

1. A creation of a boundary by agreement or acquiescence.
2. A hostile creation (e.g., adverse possession, estoppel, and acquiescence).

The second creation involves legal means that is looked on by some as a "legal theft" of land and by others as a penalty on the "title" or landowner for not being a good steward of his or her land rights. In recent years, some courts have looked at adverse possession with disfavor.

Agreements to position and to fix a disputed boundary often have the opposite effect. Some individuals believe that a newly created unwritten title will extinguish the written title, yet there is a strong belief existing in the legal field that the original title is never totally extinguished but becomes junior, or inferior, to the newly created adverse title. Land ownership lawfully gained by unwritten means supersedes written title. All of the evidence that one may produce to prove where written titles are located is of no avail against a legal unwritten creation of title. Among the elements proving where boundary lines may lawfully exist, evidence to prove unwritten conveyances must be understood by the surveyor.

This was the decision of the court when a motion was granted striking out the evidence of the surveyor as to the location of the measured line if there was an agreed-upon line. They stated that the location of the measured line was immaterial, and there was no error in striking out the evidence [1].

Determining when possession on the ground, which is in disagreement with the written deed, has ripened into an ownership right is complex and may often depend on parol testimony, which itself may be suspect. A few states delegate to the land surveyor the authority to administer oaths and take oral testimony, but none delegate to the land surveyor the power to determine the validity of property rights. When oral testimony is an indispensable part of determining ownership rights to land, it is often impossible for a surveyor to decide which party has the superior right.

The surveyor must first locate the client's written rights and then relate these to the possible loss or gain of property rights. In the event the client encroaches on an adjoiner or the adjoiner encroaches on the client, the surveyor must fully inform the client of the possible significance of unwritten rights in such a manner as to be protected from possible liability.

In this chapter the following principles concerning adverse rights are discussed:

PRINCIPLE 1. *The surveyor should refrain from giving legal advice to the client relative to land issues.*

PRINCIPLE 2. *A boundary created by unwritten means usually does not carry land title to the described lines.*

PRINCIPLE 3. *Land or land interests lawfully gained by unwritten means extinguish or become superior (senior) to written title rights.*

PRINCIPLE 4. *Written title may or may not precede unwritten title.*

PRINCIPLE 5. *Estoppel by conduct is based on the doctrine that, where a party, through acts, declarations, conduct, admissions, or omissions, misleads another so as to induce that person to injure any rights to the benefit of the person misleading, the party causing the wrong must suffer the loss and is estopped from revealing otherwise.*

PRINCIPLE 6. *If a boundary line existing between adjoining owners is unknown or in dispute and the adjoining owner(s) verbally agree on a line and take possession to that line, that agreed-upon boundary may become the true boundary, regardless of the location of the written property line.*

PRINCIPLE 7. *If the location of the line described in the written document is known to one or both of the parties, another separate line agreed to may not be binding.*

PRINCIPLE 8. *Without a dispute there can be no cause for an agreement.*

PRINCIPLE 9. *Following a dispute over an unknown line, both adjoiners must agree on a common location or method for locating the disputed line.*

PRINCIPLE 10. *Evidence of recognition or acquiescence over a period of time may imply an agreement.*

PRINCIPLE 11. *For a common-law dedication to have force, certain elements must be present.*

PRINCIPLE 12. *Prescription is the doctrine applied to the method of obtaining property rights from long usage.*

PRINCIPLE 13. *Adverse possession is a doctrine used to create title through possession for a statutory period under certain conditions.*

PRINCIPLE 14. *The adverse claimant is charged with the burden of proof if the claimant fails to prove compliance with all adverse possession requirements; the rightful owner is presumed to have never been ousted.*

PRINCIPLE 15. *A land title granted through adverse possession either extinguishes the rightful owner's written title and vests a new title in the former adverse possessor or creates a new and superior title to the original title.*

PRINCIPLE 16. *A survey, although it may reestablish original description lines, does not create or revive rights lost by adverse possession.*

PRINCIPLE 17. *Unless a statute provides otherwise, adverse possession does not run against the United States or a state, county, or city and usually not against quasi-public agencies, individuals with disabilities, and remaindermen.*

PRINCIPLE 18. *If possession is not in agreement with the client's deed, a search of the adjoiner's title must be made to determine the status of possible senior rights and to determine if any changes have occurred in the title's wording at times of subsequent transfers.*

PRINCIPLE 19. *If possession is not in agreement with measurements and if an original survey were made, the surveyor must determine whether the possession represents where the evidence of the footsteps of the original surveyor were. If the possession duplicates the footsteps of the original surveyor, then the possession does not represent an unwritten conveyance right; it is where the written rights belong.*

PRINCIPLE 20. *The surveyor's responsibility is not to determine the validity of adverse claims, but it is to report the facts of possession and use as they relate to the lines described in the description and as they are surveyed and recovered by the retracing surveyor.*

Too many surveyors attempt to solve problems by attempting to give legal advice. The responsibility of the surveyor is to survey the description and then to report the findings as to possession that are in conflict with the lines described to the client or to the client's attorney, and not to give the client legal advice.

PRINCIPLE 1. *The surveyor should refrain from giving legal advice to the client relative to land issues.*

Many surveyors are more knowledgeable and have a better understanding of the law relative to real property than some attorneys and judges, but simply knowing the law does not give them the authority to give legal advice. If the surveyor does give legal advice and the client relies on the advice, then the surveyor can suffer all of the legal consequences that may originate from the bad advice.

PRINCIPLE 2. *A boundary created by unwritten means usually does not carry land title to the described lines.*

Boundaries can be changed or modified by several legal means, but unless the state statutes provided that when these doctrines change or create boundaries they carry with them the change in title to the new boundaries, then the title remains to the original written description. To change title, one must have a deed or a court order.

When one seeks to alter a boundary by unwritten means, the title to the new boundaries must be addressed so that title equals boundary.

13.2 General Concept of Unwritten Conveyances or Rights

A person may gain or lose land rights or modify boundary lines without writings by use of the following:

- Agreement, either expressed or implied, including agreement on practical location, silent recognition and acquiescence, and the doctrine of estoppel, where both agreement and dishonesty enter.
- Adverse relationships (hostile) either expressed or implied (adverse possession).
- Acts of nature (accretions, erosion, and relictions).
- Statutory proceedings.
- Prescription ripening into easements or other rights.

Unwritten Agreements Adjoining owners verbally agreeing on the location of an unknown disputed line are usually bound, by law, to the agreement. In the eyes of the court, they are not transferring land but are merely agreeing on the disputed location of that which is already owned and is uncertain. Technically, a parol agreement is not an unwritten conveyance, since it merely gives definiteness to that which is already conveyed. Usually, the end effect of a parol agreement is the alteration of or making definite the written words of a conveyance. A parol agreement may be considered by some as an unwritten conveyance, although, in the eyes of the courts, it is not to be considered as such.

Not all states honor or recognize oral agreements. This can also be said for many judges who are asked to adjudicate such cases. Some states require all boundary agreements to be in writing to be effective. This is important in that a verbal agreement may die with the parties to the agreement and the future surveyor or landowner will have no knowledge of it.

In many jurisdictions, unwritten agreements must be consummated by occupancy, by practical location, or by recognition and acquiescence to be effective. At law, practical location, recognition, and acquiescence are sometimes erroneously treated as definite types of unwritten conveyances. The more convenient concept considers these as variations of unwritten agreements.

The best way to effectively settle a dispute is to have the disputing parties reach an agreement. Once adjoining owners peaceably settle their boundary problem(s) with a verbal agreement, the court will not usually interfere by declaring some other line to be correct. In many instances, where the court can act appropriately without violating the Statute of Frauds, verbal agreements are recognized as being valid. In some jurisdictions the right of the court to recognize verbal agreements is affirmed by statute law.

For instance, in Illinois, the principle is well established that owners of adjoining tracts may, by parol agreement, settle and permanently establish a boundary line between their lands, which, when followed by possession according to the line so agreed on, is binding and conclusive on them and their grantees. The line is not established by a mutual transfer of title, because that can be done only by properly executed instruments of title, but such settlement determines the location of the existing estate of each and, when followed by possession and occupancy, binds them, not by way of passing title, but as determining the true location of the boundary lines between their lands. Having agreed on the line or on a mode by which it will be determined and having accepted and acquiesced to it by the unequivocal act of taking possession according to the line, they and any subsequent parties in interest are estopped from afterward disputing it [2].

Agreement Deed and Unwritten Agreement An agreement deed is in writing, whereas an unwritten agreement is not. The usual object of an agreement deed is to change the location of that which is known. A legal unwritten agreement fixes the position of that which is unknown or uncertain; it may not be used to alter that which is known.

Adverse Relationships Acquiring title to, or rights in, land by adverse possession includes the idea of hostility (either direct or implied) and can be effective only because of the operation of the statute of limitations. It is an actual taking of one's rights and giving them to another. If a person has been lax and has not asserted his or her rights for a statutory period of time, it is presumed that he or she has given up those rights of ownership. But the law does not look with favor on such methods and has provided many safeguards for the benefit of the rightful owners. In some states, the acquiring of land by adverse occupancy is prohibited by statute law. All states identify the legal requirements necessary to establish these adverse rights.

Acts of Nature A person may lose land by erosion or gain land by reliction or accretion. The initial right to follow receding or encroaching waters is defined or implied by a written conveyance, but the actual land area of a person's holdings is increased or decreased without a change in the writings. The universal belief is that any land that is gained is without any title or description.

Statutory Proceedings A person may have land, or an interest in land, forcibly taken without written consent because of the failure to pay taxes or assessments, eminent domain, or condemnation. In some states, the law provides that certain officials (e.g., surveyors, processioners, and county surveyors) may establish boundary lines between adjoining owners, and if the lines are uncontested for a specified period of time, their location may become conclusive or may become prima facie evidence as to the existence of agreement or true location. The establishment of boundaries under a Land Court System and under the Torrens system is conclusive after the final decree. The taking of title to land for failure to pay taxes or assessments is not discussed in this book.

Prescription Ripening into Easements Since an easement is one of the bundle of rights in land, then long usage of land may not necessarily ripen into an adverse title; at times, the court awards an easement. Use of land for a traveled way or for a pipeline seldom results in a fee right.

Although the requirements for a prescriptive easement and adverse possession are essentially the same, what the court will award is entirely different. With a prescriptive easement, one gets to use the property, but with adverse possession, one gets title to the property. Where adverse possession entitles one to occupy the land, the easement merely permits one to use the land.

Recent Court Developments Recently, court decisions relative to boundaries and title based on prolonged occupancy have seldom resulted in fee title. As original evidence of boundaries becomes more and more difficult to recover, courts are looking at rights of use more than title.

13.3 Unwritten Conveyances and Government

Although there are exceptions, in general, unwritten conveyances (the gaining of rights in land without writings) cannot operate against the federal, state, or local governments, and, in some jurisdictions, against quasi-governmental bodies. In most states, irrespective of how long a person occupies property or uses a road, occupancy gives no rights against the public.

In Texas, most lands were disposed of by the state in sequence by a series of surveys. Vacancies and overlaps between grants occurred. When discovered, these vacancies belonged to the state, and under Texas law, others could claim them.

In the oil fields, vacancy hunters often profited handsomely. One judge humorously referred to a vacancy as "a strip of land completely surrounded by oil and gas wells." New legislation now enables the adjoiner in possession to claim the land, and the discoverer of the vacancy can no longer claim title. This condition also exhibits itself in other nonpublic domain states: Tennessee, North Carolina, and Georgia.

Lands held by the state in a private manner (e.g., road maintenance yards) may be subject to the same rules of unwritten rights, as are applied to private lands. Lands held in trust for the public, except where a statute declares otherwise, are not subject to alienation by unwritten means.

There is a trend that indicates "public" lands may be lost through adverse means if the lands are not being used for the benefit of the general public.

13.4 Unwritten Title Supersedes Written Title

PRINCIPLE 3. *Land or land interests lawfully gained by unwritten means extinguish or become superior (senior) to written title rights.*

Today, two principles exist as to title, and which of the two a court will apply is open for speculation. The first theory is that a new title is created, which will be called a "virgin" title or a title without any documentary foundation. It is an unwritten title that the claimant must ask the court to perfect. This title is superior to the existing written title, but the two titles exist at the same time. The second theory is that the newly created virgin title destroys the existing written title—that is, there is only one title existing at any one time.

The reader may get the impression that we are overusing the term "virgin title." However, as stated in the literature:

> Title acquired by adverse possession is a new and independent title by operation of law and is not in privity in any way with any former title. Generally it is as effective as a formal conveyance by deed or patent from the government or deed from the original owner. In fact, it is a good, actual, absolute, complete, and perfect title in fee simple, carrying all of the remedies attached thereto.
>
> The title acquired can and will pass by deed. After the running of the statute, the adverse possessor has an indefeasible title which can only be divested by conveyance of the land to another, or by a subsequent ouster for the statutory limitation period [3].

But the title can only be affirmed by a court.

The fact that an unwritten right can extinguish a written right is well established, but it is not until after a written right has been identified on the ground that it becomes apparent that an unwritten right might exist. What the surveyor should do about unwritten rights is discussed at the end of this chapter in Section 13.52.

13.5 Written Title

PRINCIPLE 4. *Written title may or may not precede unwritten title.*

Although lands in some instances in the past were acquired by squatter sovereignty, now it is usually true that before acquiring land by prescriptive means, some form of prior written title must exist. The law with respect to unwritten conveyances operates to enlarge or decrease a person's right but not to give him or her a right where none existed. Normally, an unwritten right does not extinguish all of a person's rights; it merely alters the written title rights.

ESTOPPEL

13.6 Introduction

The doctrine of estoppel, a legal doctrine and not a type of unwritten agreement, is a principle at law that sometimes makes unwritten agreements effective. Estoppel is a bar or preclusion to one's alleging or denying a fact because of previous actions. Estoppel by deed precludes a party to a deed from denying, to the detriment of the other party, anything written therein. The written words of a deed are conclusive. Estoppel by record precludes denial of a final judgment of a court. Estoppel by conduct may contribute to unwritten transfers of land rights.

13.7 Estoppel by Conduct

PRINCIPLE 5. *Estoppel by conduct is based on the doctrine that, where a party, through acts, declarations, conduct, admissions, or omissions, misleads another so as to induce that person to injure any rights to the benefit of the person misleading, the party causing the wrong must suffer the loss and is estopped from revealing otherwise.*

The doctrine of estoppel is not limited to land boundary disputes; it has many applications in other areas. Often, the principle of estoppel is recognized and sanctioned by statute. In most jurisdictions it is based on common law. Some courts refer to it as detrimental reliance.

Wherever a party has, by his or her own declaration, act, or omission, intentionally and deliberately led another to believe a particular thing is true and to act on such belief, he or she cannot, in any litigation arising out of such declaration, act, or omission, be permitted to falsify it.

Four considerations are essential for the doctrine of estoppel by conduct to become effective in the classical cases, although these have been liberalized in recent decisions:

1. The party being estopped must know the facts.
2. The party being estopped must intend, or at least act so the other has a right to believe that it intended, that its conduct shall be acted on.
3. The person claiming estoppel must be ignorant of the facts.
4. The person claiming estoppel must have relied on the conduct of the person being estopped and must have been damaged.

All the points must be proven conclusively to have the doctrine of estoppel apply. An example of estoppel pertaining to land occurs if Jones points out to Smith a line he knows to be 20 feet in error. Smith, not knowing where the true line is, accepts the line pointed out and grades and paves a drive for his benefit. After completion, Jones promptly claims the drive as his own, but he is estopped at law from doing so. If the law were otherwise, it would induce people to falsely derive benefits.

Estoppel can arise from long acquiescence in a line, from failure to speak up when a person should speak, and from inferences. As an example of an actual occurrence, a surveyor made an error of 3 feet in locating a lot line for a branch bank. The bank, relying on the survey, built a structure encroaching on the adjoiner. Although these actions were being conducted by both parties, the neighboring landowner sat on a porch and witnessed all of the events. When the bank was completed, the neighboring landowner approached the bank and talked settlement. The results of the court case rested on the fact that the owner knew the correct location of the line and permitted the acts. Thus, the court held that the owner was estopped from claiming otherwise, for there was detrimental reliance on the silence.

A case of acquiescence and agreement by parol occurred in Kentucky, where adjoining owners had established a dividing line by parol agreement and had executed the agreement by taking actual possession of the parts allotted to them respectively up to the agreed-upon line. There is an estoppel to claim otherwise, and the agreement is not within the Statute of Frauds, although in establishing the line each has given up some part of his land to the other [4].

UNWRITTEN AGREEMENT

13.8 Unwritten Agreement

PRINCIPLE 6. *If a boundary line existing between adjoining owners is unknown or in dispute and the adjoining owner(s) verbally agree on a line and take possession to that line, that agreed-upon boundary may become the true boundary, regardless of the location of the written property line.*

Among the many states there is a wide variation of opinion as to what constitutes an unwritten agreement. The rights of innocent third parties (e.g., mortgagee) are protected, and the agreement is binding only between the parties of the agreement and their privies (successors). The requirements of possession vary widely from state to state.

A verbal agreement accompanied by possession and improvement, if founded on good consideration, is not contrary to the Statute of Frauds and will bind the parties and their privies. The agreements between these parties then do not have the effect of conveying from one to another any portion of the land, but merely of determining the boundaries to the land already owned by them. It follows that conveyances by the parties bound by the agreements and their privies giving the same description under which they obtained the title and held the possession would pass a title according to the agreed boundaries [5].

Where the true dividing line between two tracts of land is in doubt and there is a dispute between the adjoining owners as to the exact location of the line, which depends on variable facts or circumstances not susceptible of certain determination, a parol agreement between the adjoining owners establishing a line as the true dividing line is not an exchange of lands and is not within the Statute of Frauds [4].

13.9 Elements of Unwritten Agreement

Most problems come from boundary lines that have been agreed on orally by contiguous owners. Most states have specific requirements that usually include some or all of the following four elements:

1. The true line described by the writings is unknown, uncertain, or unascertainable.
2. The property or boundary line between the contiguous owners is in dispute.
3. The contiguous owners agree to accept the boundary line they establish. In some jurisdictions, possession or acquiescence follows, and the line agreed on is marked.
4. The end points of the line must be monumented.

Although the first three items are uniformly required in all states, the necessity of possession and the amount and kind of possession vary substantially from state to state. The states, in general, may be divided into three classes: (1) those deeming acquiescence or possession unnecessary; (2) those requiring proof of possession or acquiescence for a period at least equal to that required by the statute of limitations; and (3) those requiring that acquiescence or possession for some time (less than that required by the statute of limitations) must be shown. Some states treat these doctrines, agreement, recognition, acquiescence, and practical location as a single subject, all being necessary to establish an unwritten right, whereas other states recognize all four ways as being different means of obtaining unwritten title or rights.

The rule in Illinois is that whenever the boundary line between adjoining lands is unascertained or in dispute, it may be established by parol agreement and possession in pursuance, and the line so established will be binding on the parties and their privies. The effect of such agreement is not to pass title by parol, for such cannot be done, but to fix the location of an unascertained or disputed boundary line. There are but two conditions under which the rule in relation to establishment of boundary lines by agreement applies: where the line is in dispute and where it has not been ascertained [6].

13.10 Described Line Known

PRINCIPLE 7. *If the location of the line described in the written document is known to one or both of the parties, another separate line agreed to may not be binding.*

The Statute of Frauds states that deeds conveying land or interests in land must be in writing. If the true line is known by the parties, a properly written conveyance, based on a consideration and containing words of conveyancing, must be used. The fact that a line is capable of being located by a competent survey does not void an unwritten agreement; the line need only be unknown to the adjoiners and in dispute.

Common law applies the doctrine by holding that to have a boundary by agreement, the boundary must either be in dispute or unascertainable, the dispute being a question as to location between the two parties. As can be seen, courts often misapply doctrines, for in a Rhode Island decision, the court held that an oral agreement was valid—that is, they "would not void" an unwritten agreement that was capable of being ascertained by a survey. The court held that the line had to be unknown *and* in dispute [7].

13.11 Property Line in Dispute

PRINCIPLE 8. *Without a dispute, there can be no cause for an agreement.*

Many authorities cite the necessity of a dispute preceding an agreement. Agreement by both parties is the essential part of conveying title by unwritten agreement. Without a dispute over the location of a division line, adjoining owners have no cause to arrive at an agreement; for this reason, several states consider that a dispute is a necessary part of an unwritten agreement, but not all jurisdictions maintain that an outward dispute is necessary. In some states an uncertainty needs to exist in the minds of the adjoiners; the fact that the true line could be located is not material. If there is an overlap of adjoiner's papers followed by an unwritten agreement, the courts almost always will approve the agreement, but it is not always necessary that there be an overlap.

Numerous authorities have been called to our attention involving disputed boundaries. To settle such disputed boundaries, an agreement resting on parol must be entered into fixing the line, followed by the erection of a fence along the agreed line and acquiescence in that line thereafter.

In one decision there was a total absence of evidence of any agreement to fix a line before the original fence was constructed, and plaintiff's testimony showed that the fence was in existence long before the two pieces were separated and when both were owned by the same person. One of the plaintiff's witnesses testified unequivocally that the fence had been there "50 years anyway" [8].

13.12 Adjoining Owners Must Agree

PRINCIPLE 9. *Following a dispute over an unknown line, both adjoiners must agree on a common location or method for locating the disputed line.*

To agree on a line location, that location must be monumented on the ground: a fence, a tree line, a hedge, or some similar thing. Such agreement may be a fixed distance from a given object, such as an agreed-upon distance from a surveyor's stake.

If the coterminous owners are attempting to establish the true property lines and they err in establishing the true lines, they are not agreeing to a line but merely establishing what they believe to be the true line. In such cases, the owners are not denied the right of claiming their true lines. A mistake in assuming a line to be the true line does not prevent a person from claiming the true line when discovered. Agreeing to accept a particular line, whether it is correct or not, is what is necessary.

In several decisions, the building of a division fence by agreement between adjoining owners, in the mistaken belief that they were placing said fence on the true boundary line between their land, there being no dispute as to the line or contention for a different line, did not preclude one of the parties from insisting on the true line when ascertained [9].

Then an agreement fixing a boundary line under the belief that it is the true line, when in fact it is not, was not binding and may be set aside by either party when the mistake is discovered, unless some principle of estoppel prevents [10].

Finally, a landowner was estopped to assert title to his true boundary where he and those under whom he claimed had, for over 30 years, recognized, by mistake, another line claimed as a true one by the adjoining owner by making partition to it, calling for it in deeds, and pointing it out to others. The adjoining owners, on the faith of such recognition, purchased, cleared, and improved up to that line. It is not material that the landowners were mutually mistaken as to the true line and could have discovered it, at any time, by survey [11].

The element of time, especially if continued beyond the statute of limitation, may change the general rule, as it did in the *Galbraigh* case; recognition and acquiescence then became controlling.

If a line is capable of being established by a survey or the parties agree to have the true line established by a survey and the surveyor errs in the establishment of the true ine, the adjoining owners can claim the true line, since no other than the true line was agreed on. However, if after the error was discovered, there was acquiescence for the statutory time or if there were title by adverse possession, the survey lines may be binding. Where a line is in dispute and unknown and the adjoiners orally agree on a survey and agree that they will be bound by the results of the survey, they are generally bound by the line run.

Where the owners of land order a survey to determine the lines and the surveyor makes and returns an erroneous location, on which they afterward act without knowledge of the error, the owners are not bound by the surveyed line [12].

If one party misrepresents the lines or deceives the other, the agreed-upon lines are not binding except on the person misrepresenting the lines. Fraud cannot be the foundation for any agreement. The element of estoppel may enter here if the person being deceived is induced to harm himself or herself [13].

Owners of adjoining land or those having vested interests therein are the only ones who can establish the disputed line by agreement; innocent third parties are not bound. An agreement between owners but not including the mortgage holder is not binding on the mortgagee.

Merely erecting a fence with no intention of having it represent a boundary line does not prevent a person from claiming a portion of land fenced out, for a person is not compelled to fence in all his or her land.

In such cases, courts will examine all evidence. There is no testimony to show that the parties made any agreement about the boundary line or that such agreement, if made, was executed. In one case, a court stated that one Mrs. Coffman's testimony only went to the extent of showing that she had a survey made and that the adjoining proprietor afterward recognized its correctness and asked permission to move a house situated on the disputed strip. This testimony fell short of showing that the parties made an oral agreement establishing a boundary line that had been in dispute and that the possession of the disputed tracts was taken by Mrs. Coffman by virtue of such agreement. According to her own testimony, she took possession of the disputed tract because the survey that she caused to be made by the county surveyor showed that it belonged to the tract of land owned by her and her husband. The testimony did not establish any agreement between herself and the adjoining property owner Latham as to the establishment of the boundary or that she took possession of the strip in controversy under any such an agreement [14].

How soon an unwritten agreement becomes binding varies somewhat by state. Some say the agreement is binding the moment the agreement is reached; others say it must be followed by acts of dominion, such as erecting a fence. In Maine, two time limits are set. If an unwritten agreement is followed by 20 years of possession, it becomes effective; also, if the parties to the conveyance, soon after the conveyance is formed, mutually establish an unknown line (by survey or other means), the line so established becomes binding, as indicated in the following case.

According to a Maine decision [10], a boundary line may, under certain circumstances, be permanently and irrevocably established by parol agreement of adjoining owners, and a line agreed on by the interested parties and occupied for more than 20 years is conclusive. When the principle of estoppel applies, a shorter period may be sufficient. A line established by agreement of parties, at or near the time of making the conveyance, may be conclusive, even if the occupation is for less than 20 years, as proving the intent of the parties to the conveyance. An agreement fixing a boundary line under the belief that it is the true line, when in fact it is not, is not binding and may be set aside by either party when the mistake is discovered, unless some principle of estoppel prevents this from happening.

In legal writings, the mutual designation of a line is frequently called a practical location. A court in Maine theorized that a practical location at about or near the time of the conveyance indicates the intentions of the parties. This is a strange extension of the meaning of the "intentions of the parties," since in most jurisdictions the intent can only be determined from the writings themselves and the circumstances at the time; subsequent events do not count. Application of this extended meaning is not recommended except where the court has specifically approved it.

Surveyors must be extremely careful not to practice law when they anticipate that unwritten agreements are present. Their responsibility is to collect the facts and evidence and then relate these to the written title lines and then inform the client.

13.13 Practical Location

Practical location by adjoining owners is an actual mutual designation of the location of their dividing line on the ground. Practical location is a legal act of the parties. It is not a means of causing an unwritten conveyance, but it is a practical location, which, along with certain other elements, may be sufficient evidence for an unwritten location. This should be considered as a boundary modification.

The difference between the practical location of a line and an unwritten agreement on a line is that a practical location is a mutual designation by both parties about what they believe to be the true division line, whereas an unwritten agreement on a line is agreeing to accept the line designated, whether or not it is the true line. Agreement is a contract; practical location is an act that may induce a presumption of an agreement. Agreement may be with respect to the mode of establishing a line, whereas practical location includes the act of establishing a line, such as erecting a fence.

Practical location calls for cooperation of both parties—a unilateral action, that is, the action of one party without the approval of the other, has no binding effect. A survey caused by one party, along with notification to the adjoiner, is insufficient to establish practical location, because practical location is dependent on mutual action and mutual agreement to the survey.

In Pennsylvania, the term *consentable line* is used to describe a situation in which parties have agreed to a line by words or actions such as practical location.

One party may be estopped from claiming he or she was a part of a practical location if his or her conduct or behavior has misled the adjoiner to the adjoiner's harm. If this is so, then the doctrine of estoppel may be pleaded.

If a deed given to a buyer is ambiguous and can be interpreted in two or more ways, and if the buyer and seller have mutually erected a common fence, the location of the fence is often considered proof of the parties' intent and is a practical location.

Where two tracts of land owned by adjoining owners interfere, the boundary called for by the title papers of one overlapping that of the other, there being a dispute and controversy as to the superiority of the titles and consequently as to the true dividing line, an agreement between the owners to establish a dividing line between them in pursuance of which such line is established and plainly marked and thereafter recognized by both parties for a considerable time as the true line, is not within the Statute of Frauds and will be upheld [4].

A New York court went on to explain that the doctrine of practical location was originally derived from a long acquiescence by the parties in a line known and understood between them for such a period of time as to be identical with "time immemorial," or "time out of memory." Practical location of a boundary line to be effectual must be an act of the parties either expressed or implied; it must be mutual, so that both parties are equally affected by it. It must be definitely and equally known, understood, and settled. If unknown, uncertain, or disputed, it cannot be a line practically located. Where land is unimproved and uncultivated, mere running of the line through the woods, ex parte, by one of the owners, so long as such line is not settled on and mutually adopted by the adjoining owners as a division line, is an immaterial fact. In such a case, until the adjoining owner shows assent to it, it would amount to a mere expression of the individual opinion of the owner who ran the line. To constitute a practical location of a line or a lot requires the mutual act and acquiescence of all the parties [15].

13.14 Possession Following Agreement

The amount, kind, and extent of possession necessary in conjunction with an unwritten agreement are exceedingly variable, some without reason, and it is usually left up to the courts to determine their validity. In some states, possession must follow for the statutory period. In others, an agreement can become binding just as soon as a legal agreement is consummated, regardless of possession. In some states, a practical location and an agreement to accept the practical location may be binding.

Having agreed on a line, or agreed on a mode by which it shall be determined, and having accepted and acquiesced in it by the unequivocal act of taking possession according to the line, they and their privies are estopped from afterward disputing it. The estoppel arises from the act of the parties in taking possession and occupying their respective tracts to the line thus agreed on and determined. A defendant in a case repudiated the line as it was fixed and retained possession of the strip of land in controversy. It was therefore clear that there had been no practical location of the line by which the parties were estopped. The agreement of the parties was insufficient to authorize recovery [2].

It is generally agreed upon that possession along a line for a period equivalent to the statute of limitations brings about an implied agreement between adjoiners; that is, in the absence of proof of agreement and in the absence of contrary proof (nonagreement), mere possession for the duration of the statute of limitations is sufficient proof (an implied proof) of agreement. Courts have said that lines acquiesced in for a long period of time may be better evidence of the original lines of the original surveyor than are measurements from distant points.

Acquiescence or agreement between public lands and an adjoiner is not possible, since no public official is authorized to acquiesce on behalf of the government. This has caused problems in rural or wooded areas where some government agencies have carried on programs of land line identification and marking where boundary lines were placed off the true line, and the neighboring landowner occupied to the line.

Acquiescence in occupation to a line without agreement (expressed or implied) does not alter written title lines unless such acquiescence is sufficiently long to give adverse rights and then adverse possession becomes the doctrine only if the doctrine is applicable. Various states have different time requirements for establishment of lines by acquiescence, said time being 5, 7, 10, 15, or 20 years.

A code reference was cited in a Georgia decision stating that an unascertained or disputed boundary line between coterminous proprietors may be established (1) by oral agreement if the agreement is accompanied by actual possession to the agreed-upon line or is otherwise duly executed or (2) by acquiescence for seven years by the acts or declarations of the owners of adjoining land as provided in Section 44-5-164 of the code, which states that possession of lands, under written evidence of title, for seven years shall confer good title by prescription, but if such written title is forged or fraudulent and notice thereof is brought home to the claimant before or at the time of the commencement of his or her possession, no prescription shall be based on possession [16].

In Wisconsin, building a division fence on the assumption by both parties that it is on the true line and occupation up to it by both parties for less than 20 years (20 years is the statute of limitations in Wisconsin) do not constitute a binding location of the line, where there was no dispute or uncertainty as to the correct line, no express agreement that the fence should be regarded as such line, and no circumstances such as should estop either party from insisting on the true boundary [17].

By reading selected decisions, one can see that courts have acted rather without direction. The area of unwritten rights is confusing, and a surveyor should be very careful in determining the evidence of adverse possession.

13.15 Recognition and Acquiescence

PRINCIPLE 10. *Evidence of recognition or acquiescence over a period of time may imply an agreement.*

Evidence of recognition and acquiescence often implies an agreement, although none may have existed. Recognition and acquiescence depend on the silence of a party when the party should speak, the declarations of parties, or the inference to be drawn from the conduct of the parties. If the element of estoppel enters, recognition and acquiescence may be effective.

Essentially, establishment of a line by acquiescence is approved by two theories. First, some courts consider acquiescence as evidence of proof of the true line. Second, other courts consider acquiescence as sufficient to give rise to a presumption of a previous agreement between adjacent owners. Often, the theory cited is dependent on the peculiar circumstances of the case being adjudicated.

A long period of acquiescence may raise a rebuttable presumption that a fence is on the true and correct line and often only by contrary evidence can the presumption be overcome. A fence built for mere convenience and admitted by both parties to be not necessarily on the true line can never become a true line by acquiescence. A person acquiescing by mistake in a false boundary that was believed to be on the true line is not necessarily denied the right to claim to the true line, although conduct may estop him from so doing. Acquiescing to a false line is treated as a mistake, and in the absence of an agreement, estoppel, or possession short of the statutory period, a person is not estopped from claiming the true line.

Acquiescence for a long period of time to a boundary line must be with respect to monuments (e.g., fence, hedges) as a boundary, not mere acquiescence in a barrier known to be off from the true boundary. Acquiescence to a monument as a boundary can be deduced from the statements, conduct, or actions of the parties.

Many states now require payment of taxes as a necessary part of the proof for an adverse right claim; in California, the statute requires 10 years of continuous tax payment. Whether this statute was a bar to obtaining an unwritten agreement right was at issue in *Duncan* v. *Peterson* [18]. In 1966, the defendants learned by accident that the line of the fence might not truly reflect the north–south center-section line. They had the land surveyed and confirmed that the true center line was 104 feet east of the fence.

Only two findings were made by the trial court in *Duncan v. Peterson:* (1) that all taxes on the subject property had been paid by defendants and their predecessors and (2) that no taxes had been paid by plaintiffs or their predecessors; also the subject was particularly described by the court when it pointed out the elements *not* necessary were a written agreement or an unascertainable true location. Payment of taxes is not material (for an unwritten agreement).

UNWRITTEN DEDICATION

13.16 Dedication: General

Dedication, the granting of land or rights in land to the public, can be either written or unwritten. For dedication to be effective, there must be a voluntary offer, either expressed or implied, by a donor and an acceptance. Many dedications, made by unwritten means, present difficult problems to the surveyor, since it is customarily the surveyor's habit to mark off those portions free of public road dedications. The decision of whether a person has properly made an unwritten offer to dedicate and whether the public has accepted the offer (by usage) is often outside the responsibility of the surveyor.

13.17 Dedication and Easement

DEFINITION. *Dedication is the giving of land, either as a fee title or as an easement right, by its owner for public, charitable, semipublic, or utility use.*

An easement is an interest in land created by grant or agreement that confers a right on owners (private or public) to some profit, benefit, dominion, or lawful use of the estate of another.

Dedication is the act of giving a right. An easement is the result of one type of giving; it is a right created. A dedication properly consummated may result in either an easement or a fee title, the easement or fee title being held in trust for a particular use.

13.18 Elements of Common-Law Dedications

PRINCIPLE 11. *For a common-law dedication to have force, certain elements must be present.*

Four elements must be present:

1. A donor capable of dedicating land or dedicating rights in land.
2. An identifiable parcel of land to be dedicated.
3. An intention on the part of the donor to dedicate (offer, formal or implied).
4. Acceptance of the dedication (formal or implied).

Consideration is not necessary, since a dedication is a gift, and the advantages gained by public use are sufficient. The existence of an individual grantee capable of taking title is not a necessity, since a dedication is intended for the public or a group as a whole. In most states, a dedication cannot be made to an individual; it must be made to the public, semipublic, churches, or groups. A railroad is usually considered a private corporation and ordinarily cannot receive a dedication.

13.19 Statutory Dedication

Unlike common-law dedication, statutory dedication must comply with the letter of the law. Statutory dedication is regulated by written law and usually requires that the dedicator sign and acknowledge the offer of dedication before a designated officer. Often, designated governing-body officers must sign an acceptance, and the map must be filed in a particular place. The requirements for the dimensioning and designating areas to be dedicated must be complied with.

In some states, statute law specifically provides that an offer to dedicate land is an offer to pass fee title to the land offered for dedication, and in such event fee title passes. In the absence of a statute specifically granting fee title, an easement is granted.

If there is a failure to comply with all the requirements of the law in a statutory dedication, the possibility of common-law dedication is not precluded, since statutory dedication will not operate to void common-law dedication.

Statutory dedications vary considerably from state to state, and it is not the intent to compile all the state statutory laws here.

13.20 Donor of Dedication

Offers of dedication are binding only on those who have the power to alienate property. As in written conveyances, the mortgagee cannot be bound by the acts of the mortgagor. Land owned by several owners can be dedicated only by all parties in interest. A person who has sold part of his or her land or who has contracted to sell part may not dedicate any part sold. The administrator cannot dedicate the estate of others. Corporations may dedicate by proper resolutions of those empowered to act.

13.21 Location and Description of Dedication

The land being dedicated must be capable of positive location; otherwise, the dedication would be void from uncertainty of the description. At common law the location can sometimes be inferred from the acts of the parties, that is, a practical location and need not be an unambiguous, written document. It is only an intent to dedicate that makes a common-law offer of dedication valid; therefore, the land being dedicated need not be described in writing but must be identifiable. A statutory dedication must be described with as much detail as is required for other written conveyances.

13.22 Expressed Intent to Dedicate

At common law it is necessary for the owner only to intend a dedication, and that intent may be by words, acts, or conduct. It is not necessary that the intent be in writing, as is required in statutory dedications. A dedication is a gift, and for this reason the offer to dedicate must be clear, although the offer to dedicate may be inferred from conduct.

Since it is only the intent to dedicate that makes an offer valid, it is not necessary that there be a formal offer. The offer of intent to dedicate may emanate from an oral statement, a writing, or a single or series of acts or a plat. No definite rule exists about what constitutes intent to dedicate.

The act of platting land into blocks and streets presents evidence of intent by inference, but it does not in itself create a right of the public in the streets or alleys. Acceptance must follow. On the selling of a single lot in the plat, a private easement is created for the benefit of that lot and all other lot owners.

The writing of the word *wharf* on a plat is not proof of intent to dedicate, nor is the word *depot*. A wharf is usually subject to private ownership and merely signifies the existence or intent to build a wharf. It has been held an intent to dedicate, and it has also been held as no intent to dedicate where the word *beach* has been written in a blank space adjoining water.

Mere permissive use is not an intent to dedicate. Platting a river shows no intent to dedicate the river for public pier or wharf rights.

13.23 Implied Intent to Dedicate

Long continued acquiescence by the owner may eliminate the necessity of direct expression of intent to dedicate, since intent to dedicate can be inferred from the owner's failure to exercise his or her rights. Mere usage with the consent of the owner does not give the public a right to infer an offer of dedication; the usage must be incompatible with private usage. Setting a fence back to allow a street passageway has been interpreted as an intent to dedicate.

To imply a dedication based on usage, it is usually necessary that the public use the land for a period of time required by the statute of limitations (five years in California). One individual crossing another person's land does not constitute a public right, although it may ripen into a private right. The public is more than one, but does not include five or six individuals.

13.24 Acceptance of Dedication

In any dedication, a competent person must offer or express an intent to offer a dedication, and there must be an acceptance of the offer of dedication by one who has the authority to do so. Acceptance, unlike the offer, does not have to be by a particular grantee. The offer of dedication is to the public, and there are many ways the public can accept a dedication. Acceptance can be by a governing body, the action of maintenance crews, or by usage by the public.

A public body that causes a subdivision (land platted into lots, blocks, and streets) and sells land in accordance with the plat makes an offer and acceptance of dedication on recording the plat. A dedication by deed is effective on recording the deed, since the act of recording is construed to be the act of delivery.

Acceptance of dedication has been construed from such acts as installing public improvements, maintaining a street by city crews, taking possession, appropriating money for streets, issuing of bonds on a street, resolutions of governing bodies, and accepting benefits from the state legislature.

Platting of land into lots, blocks, and streets and recording the same thing do not make it necessary for the city to maintain the streets, since no formal acceptance is made by the city. Liability and maintenance follow acceptance. Formal acceptance must be by those empowered to do so by an order, an ordinance, a vote, legislative action, or simply a plat. Such persons as tax collectors and police cannot accept a dedication.

13.25 Revoking Offer

An offer to dedicate is good until revoked. Prior to acceptance, an offer may be withdrawn or revoked by the person or persons making the offer, but a completed dedication with offer and acceptance is irrevocable. Death of the person offering dedication automatically revokes his or her offer if it has not been accepted. Generally speaking, only the person offering dedication may revoke the offer, but once the dedication is consummated, it usually becomes irrevocable.

13.26 Purpose of Dedication

Dedication is a gift to the public, and the donor may impose any restrictions wished, and the land cannot be applied to any other use. Land donated for a street cannot be used for another purpose such as for buildings. Land dedicated for a park or commons purposes precludes, in some cases, the erection of buildings inconsistent with park purposes. A state dedicating land for park purposes is not precluded from granting the land's use for other purposes. The purpose of a dedication cannot be inconsistent with the city's right of police power or its rights to supervise improvements.

13.27 Effect of Dedication

At common law, the dedication of streets normally gives usage of the land only for the intended purpose or related purposes. Fee title is reserved in the original owner, and the owner may make any usage not inconsistent with the rights of the public in the easement. The fee owner reserves all rights not inconsistent with the purpose of the dedication. But in the matter of dedicating land for churches, schools, and the like, a fee title usually passes. The public can accept a dedication, but it cannot dictate the purpose for which it will be used. On vacation of the dedicated area, the surface rights are united with the subsurface rights.

Statutory dedication operates by way of a grant, and if the statute specifies, a fee title will usually pass. The use usually dictates the quality of the title that reverts. In the absence of a statute stating that a fee title passes, it is presumed that only an easement in streets and alleys is granted, but each individual situation will be decided on its own merits.

Common-law dedications operate by way of estoppel, whereas a statutory dedication operates by way of a grant. In the absence of clear intent to the contrary, a conveyance of land described as bounded by a street or road gives the grantee title to the soil under the road to the middle of the highway if the grantor had such title. Of course, this ownership follows the concept of land ownership from the center of the Earth to the heavens above, subject to an easement in the public. If the street called for does not constitute the boundary for various reasons, then ownership does not extend to the center of it, although there is a presumption and implication that the landowner has an easement to use it.

Where the record owner files a plan or sells lots according to the plan, there is the presumption that a dedication for public use was intended to those roads indicated on the plan for all of the affected lots and the public in general. The surveyor may have problems determining if a dedication exists if the roads were never constructed or opened, for there may have been an abandonment. Dedication can be made only for a public or charitable use; dedications cannot be made for an individual. If usage is sufficiently long that the public rights would be affected by an interruption, estoppel will enter.

After a dedication is accepted, every member of the public has an interest in that dedication. Dedication is not an individual interest but is an interest in common with the general public.

In statutory dedication, the street width is that which is dedicated and includes any part not used.

13.28 Dedication by Plat

A plat made and recorded showing lots, blocks, and streets has an intent to dedicate all streets and alleys shown on it. Each lot sold gives the purchaser a right to use all of the streets and alleys shown. Use by the public for a period of time equivalent to the statute of limitations may give public acceptance. Revoking the offer to dedicate can be done only by all owners of lots.

A village accepted a number of streets and alleys in both the original village and platted additions. But it insisted that an acceptance of some of the streets and alleys of a plat did not constitute an acceptance of the whole. The court held the rule and doctrine to be that an acceptance by the city or village of some of the streets or alleys appearing on a plat is an acceptance of the entire system of streets and alleys [19].

13.29 Prescription

PRINCIPLE 12. *Prescription is the doctrine applied to the method of obtaining property rights from long usage.*

Adverse rights, agreement, and other means of obtaining land result in a fee title, whereas prescriptive usage ordinarily conveys only a right of usage of an easement. A person traveling across a parcel of land for the period of time required by the statute of limitations may acquire an easement right to continue. Dedications are rights acquired by the public; prescription is the means of obtaining dedication or obtaining private easement rights or public easement rights. The public may increase rights they have by prescription.

Encroachment of structures, eaves, power lines, water lines, underground utilities, and the like may ripen into a permanent right of usage. But once the encroachment is removed, the right of usage is sometimes extinguished. Usage rights such as placing a second water line alongside a first line generally cannot be enlarged from time to time.

The right of continuing usage as an easement by the public or by quasi-public agencies is readily recognized by the courts, especially if the usage extends over the period of time required by the statute of limitations. Easements of necessity, that is, rights necessary for the public, or in some instances an individual, are more readily obtained by law that recognizes usage.

13.30 Adverse Possession

PRINCIPLE 13. *Adverse possession is a doctrine used to create title through possession for a statutory period under certain conditions.*

Sometimes the distinction between lands gained by adverse rights and lands acquired by the process of recognition and acquiescence may not be readily discernible, since each might include parts of the other. A feature that usually distinguishes adverse possession from other unwritten methods of obtaining ownership is that possessors must maintain that their possession is their ownership line, irrespective of their written title; adverse rights have the element of hostility; they do not rest on oral claims; they are adverse to the rightful owner.

The doctrine of adverse possession merely maintains a status quo. It is certainly not for the merit of the adverse occupant, since the taking of another's land has no merit. It is probably a penalty against the true owner, since he or she has failed to act on his or her rights within the limitations imposed by law.

13.31 Historical Concepts of Adverse Possession

The doctrine of adverse possession is a concept that evolved both through Roman common law and English common law. Under Roman law, individuals who went into possession of land believed a part of the spirit of that person was transmitted into the parcel of land itself. By virtue of this phenomenon, the person became part of the land and by law became owner of the occupied land. On the other hand, English common law gave title to land superior control over occupancy. A merging of these two historical concepts gives the doctrine that allows an individual to occupy a parcel of land under certain adverse conditions and to obtain a new written title in accordance with the statutes of a given jurisdiction.

The entire concept of adverse possession is one of penalizing the title owner of property for being a poor custodian. It is not one of awarding the adverse possessor, but one of penalizing the owner for permitting adverse use when the owner should have taken legal action to dispossess the adverse possessor.

As mentioned, in the United States, where Torrens titles exist, adverse rights cannot be obtained. In Australia, where the system originated, the law has been changed to permit possession rights under some conditions.

13.32 Statutory Character of Adverse Actions

The means to acquire lands by adverse rights is usually defined by statutes, although on occasion common law may be applied. If a statute exists, it must be strictly complied with.

Often, two means of court actions to acquire title by adverse rights are provided; one is a long possession duration (about 20 years), and the other is a short possession duration (about 7–10 years). For the shorter time requirement, the added elements color of title and payment of taxes are usually imposed.

In some states, especially where title registration is in effect, land cannot be acquired by adverse rights. In those states where some of the titles are registered by a Torrens system and some are not, adverse rights against the Torrens titles, but not other titles, are usually barred.

13.33 Burden of Proof

PRINCIPLE 14. *The adverse claimant is charged with the burden of proof if the claimant fails to prove compliance with all adverse possession requirements; the rightful owner is presumed to have never been ousted.*

In adverse possession actions, to acquire proper title, the burden of proof rests entirely on the person trying to prove perfection of title. Presumptions, except in the matter of good faith, are in favor of the written title owner. Failing to prove conclusive compliance with statutes or other requirements of the law will not remove title from the rightful owner.

In those states where good faith and color of title are required, it is presumed that a person buying paper title, although a defective one, entered in good faith. Hence, where good faith is required, the burden of proof for this item is reversed; the rightful owner must prove bad faith.

13.34 Character of Title Acquired

PRINCIPLE 15. *A land title granted through adverse possession either extinguishes the rightful owner's written title and vests a new title in the former adverse possessor or creates a new and superior title to the original title.*

Once a title is created or perfected by adverse rights, such title is an absolute perfect title in fee, dating back to the time of its inception (original possession). This new title is a "virgin title." It cannot be lost by survey, subsequent litigation, acknowledgment of title in another, or mortgage or liens of the paper title owner. It is a legal title, although not a record title, that is good against all other claims. Title, once acquired by adverse rights, may be removed only by conveyance, another adverse possession ripening into a fee, unwritten agreement, or other lawful means.

Where the evidence showed that the cross-complainants and their grantors had held possession, as the owners of the land since 1840, and that one of such grantors between 1874 and 1880 said to a witness that he did not claim such real estate, the statement could not operate to defeat such person's title, since the prescriptive period had already run prior to the making of such statement and the title had vested.

Theoretically, an adverse title is good against all, provided that it can be proved at law. To be marketable, an adverse title must be adjudicated at law through the courts. Title companies will not ensure unwritten rights (it is what is known as a nonmarketable title), and, in general, lending institutions will not loan money on unwritten titles, mainly because of the possibility of costly litigation to clear a title. Furthermore, a knowledgeable buyer would be unwilling to pay full value for land that is claimed merely on the basis of possession; buyers do not want to purchase a possible litigation. Although it can be stated that an unwritten title that has ripened into a fee right is good against all, it cannot be said that land whose ownership is based on occupancy is of equal value to land whose ownership is based on both writings and possession. The term *virgin title* describes the origin, marketable title described the ability to sell the newly created virgin title.

13.35 Prescriptive Title

Generally, title by prescription is construed to be an easement right as opposed to a fee right by adverse possession. A building erected on another's land by mistake, with no intention of taking title to anything other than the true line, may not confer a fee title, but the right of prescriptive use (easement) may be perfected. Water dripping from eaves and water flooding an upland owner (caused by a dam) have been held to be prescriptive usage rights. Failure to prove adverse rights does not preclude the possibility of an easement right.

Prescriptive rights usually arise from the statute of limitations; if a rightful owner cannot show that he or she has had possession within a given period of time depending on the state statute, then the owner is barred from bringing action for recovery. Although the possessor may not have fulfilled sufficient statutory requirements to bring suit to acquire written title, he or she may rely on the statute of limitations to keep from being ousted.

Under civil (Roman) law, the term *adverse possession* does not exist but is covered by the term *prescription*. An examination of the requirements of the two will indicate that they are both the same. This situation exists today in Louisiana, where prescription is in the code, not adverse possession.

13.36 Effect of Survey on Adverse Rights

PRINCIPLE 16. *A survey, although it may reestablish original description lines, does not create or revive rights lost by adverse possession.*

Although a survey may relocate the original lines created by the original surveyor or it may properly and correctly monument the boundary lines that are described in the written conveyance, it may never revive the rights to land lost by adverse possession. A survey establishing a line between adjacent owners will not revive the right of an original owner against an established boundary (by adverse rights), since all that the survey does is to establish the line and not the title [20].

However, adverse possession cannot change fixed original survey lines [21]. Adverse possession may change the title to real property, but it cannot change the location of the quarter section line [22].

13.37 Against Whom Adverse Rights Do Not Run

PRINCIPLE 17. *Unless a statute provides otherwise, adverse possession does not run against the United States or a state, county, or city, and usually not against quasi-public agencies, individuals with disabilities, and remaindermen.*

The statute of limitations for adverse possession does not operate against the United States; and unless there is a state statute to the contrary, the same is true for the individual states. Even in states where statutory provisions allow adverse rights to run against them, such adverse rights seldom run against lands held in trust for the people. It is usually only lands held in a private manner, such as a road maintenance yard.

By the right of eminent domain, the United States is the absolute and exclusive owner of all public lands that it has not alienated or appropriated. No adverse possession is created by an entry on its public lands [23].

A common-law maxim is, "Once a highway, always a highway." Highways belong to the public and are under the control of the sovereign, either immediately or through the local governmental instrumentalities. The right of the public to the use of the highways is not barred by the statute of limitations. No one can acquire a right to the adverse use of a legally established highway by use, no matter how long such use continues. There can be no such thing as permanent, rightful, private possession of a public street.

The better opinion would seem to be that municipal corporations, in all matters involving private rights, are subject to limitation laws to the same extent as private individuals on those lands that are not used for the public benefit [24].

Testing a case, the county had a perfect right to sell or otherwise dispose of the land at pleasure; the limitation law did not run against it [25].

In Iowa, where the road has been established and continually used, the fact that the fences bordering it were not on the true line and the portion beyond had been occupied by the landowner up to the fence and not been used by the public will not work an estoppel against the public, but the entire width of the highway may be appropriated by the public whenever required for the purposes of travel. The doctrine of acquiescence can have no application to the fixing of a boundary between the abutting owners and the highway, for no one representing the public is authorized to enter into an agreement on or to acquiesce in any particular location. The fee to a street is in the town or city, but always in trust for the public. A municipality cannot sell, convey, or authorize a street for private uses [26].

An example is a Massachusetts statute that provides the following:

> If the boundaries of a public way are known or can be made certain by records or monuments, no length of possession, or occupancy of land within the limits thereof, by the owner or occupant of adjoining land shall give him any title thereto, unless it has been acquired prior to May 26, 1917 [27].

In some states, adverse rights may operate against land held by a railroad; in others, common law or statutes prevent this. Often, an infant, on reaching majority, may oust an adverse claimant, since it was not competent to sue until it did reach majority age. The remainderman usually may not sue as long as the person owning a life estate is alive and is in rightful legal possession; hence, only after the death of the owner of the life estate may the remainderman have the right to oust adverse possessors. There is a difference of opinion about whether adverse rights may operate against school lands; often adverse rights may not.

Torrens registration acts provide by statute that title to registered land may not be attacked by adverse possession. An adverse title perfected prior to registry would be effective, but the Torrens proceedings are essentially quiet title actions, and failure to assert rights during the proceedings could cause loss of adversely held land, especially if title location was part of registration.

In recent years, many governmental bodies are reexamining the concept of adverse possession. It has been an accepted principle that for land that was once in private ownership and subsequently came under government ownership, if adverse possession had matured against the government's predecessor, no title or rights could be obtained by the government, for they had already become lost. Recently, Congress enacted Public Law 92–562, which permits a private party to bring an action at law over property boundaries and surveys but excluding any claim of adverse possession.

If the title to a parcel of land is titled in a governmental authority that is protected from adverse claims, some courts will look at the use for which the authority is using the property. Some courts have found that if the property is being used in a nongovernmental capacity, and not for the "public good," then this agency may be placed in the same capacity as a private individual and they may lose the property through an adverse claim.

On a case-by-case basis, the courts are now raising the standard of quality and quantity of evidence that the adverse possessor must present to prove adverse possession or adverse rights. In earlier decisions, the adverse claimant was required to prove adverse rights only by the preponderance of the evidence; now many courts require evidence that is clear and convincing—a much higher degree.

13.38 Elements of Adverse Rights

Where permitted, an adverse claim will ripen into a new or virgin title when the following acts are complied with continuously and simultaneously for the period of years that is defined by laws of the various states. It should not be assumed that all eleven elements are required for all states; there is much variation in the laws, and surveyors should know the requirements for their states: The usual elements required to meet the legal requirements for adverse possession are the following, or some combination thereof:

1. Actual possession
2. Open possession
3. Notorious possession
4. Claim of title
5. Continuous possession
6. Hostile or adverse possession
7. Exclusive possession
8. Possession as long as required by statute
9. In some states under color of title
10. In some states that all taxes be paid
11. In some states under good faith

Under the U.S. Constitution, the determination of property rights is the exclusive responsibility of each state. Since states are hesitant to address adverse rights in the courts, the courts usually insist on strict adherence to all of the requirements of a given state. Although the statutes of most states spell out the length of time necessary for occupancy to ripen into a fee, none completely stipulates the details of the listed requirements. The requirements are part of the common law or statute law of the respective state.

The real test as to whether possession of real estate beyond the true boundary line will be held adverse is the intentions with which the parties take and hold the possession. It is not merely the existence of a mistake, but the presence or absence of the requisite intentions to claim title. This fixes the character of the entry and determines whether the possession is adverse. Where a fence is believed to be the true boundary and the claim of ownership is up to the located fence, if the intent to claim title exists only on the condition that the fence is on the true line, the intention is not absolute, but conditional, and the possession is not adverse. If, however, in such a case there is a clear intention to claim the land up to the fence, whether it be the correct boundary or not, the possession will be held adverse [28].

13.39 Actual Possession

To acquire land by adverse occupancy, actual possession by some act, such as fencing, cultivating, farming, and possibly including tree farming, and making improvements, is a fundamental necessity. Actual possession is not an argument or contest over possession. The possession cannot be such that the true owner would not be aware of it on occasional visits. The facts necessary to prove actual possession may vary, depending on the circumstances.

An enclosure excluding others from entrance is more conclusive evidence of actual possession; and in some jurisdictions, an enclosure is required by statute. For possession to be effective as evidence, it needs to be so complete, open, and notorious as to charge the owner with constructive notice. The enclosure can be a natural or man-made barrier that is definite and certain of location. Fences in general must be substantial and erected for the purpose of setting off the boundaries of the claimant, and in some jurisdictions must meet legal requirements for sufficiency and legality.

A fence may be sufficient as constituting visible evidence of possession, even though not in good condition. Fences may be the best of evidence or they may be the worst of evidence. In the case of fences, a fence to be legal must first become sufficient as identified by the statutes. This includes first becoming sufficient. Once a fence becomes sufficient and then it becomes legal, if it then ceases to be sufficient, it will always remain legal. The evidentiary problem for the surveyor becomes one of determining when the fence became sufficient [29].

Ordinarily, cutting grass or other naturally growing crops (logging trees) may not be sufficient to show adverse possession; however, if the grass or growth of trees is the best-suited purpose of the land, these facts may have weight as evidence. Recent decisions have modified the forestry aspects by considering tree farming, including the planting of trees in rows, as sufficient indicia for adverse possession.

Cultivation, if intermittent, is not actual possession; but continuous cultivation up to a definite line, especially a barrier, has been accepted as one of the requisites of adverse rights.

Proof of actual possession may be entry denial to the rightful owner or entry denial to the owner's agent or vendee.

Surveying land and the establishment of monuments do not constitute possession. Erecting a fence, planting a hedge, or cultivating to the line of a survey is considered an act of possession. A survey serves to fix bounds and, along with acts of possession, may fix the limits of a claim. The survey of unenclosed land and placing stones at the boundary corners do not constitute taking of possession [30].

An intention to occupy land, however openly proclaimed, is not possession. The intention must be carried into actual execution by such open, unequivocal, and notorious acts of dominion as plainly indicates to the public that the person who performs them has appropriated the land and claims exclusive dominion over it. Anything short of this is not what the law accepts as possession [31].

By actual possession is meant a subjection to the will and dominion of the claimant and is usually evidenced by occupation by a substantial enclosure, by cultivation, or by appropriate use, according to the particular locality and character of the property [32].

13.40 Open and Notorious Possession

Most courts consider "open and notorious" as a single element, but in reality, they should be considered as separate and distinct. Possession must be sufficiently open and notorious such that the true owner by occasional visits can observe the acts of possession. Possession must be adverse; it cannot be secret or with permission. The occupancy must be notorious in such a way that the immediate public is aware of it and that it must be presumed that the true owner is aware of its adverse nature.

The law does not require persons to act or protect their holdings until they are aware or ought to be aware that they need to act; hence, possession must be open and notorious by such acts as fencing, erecting buildings, or farming.

Evidence used to prove open and notorious possession may be acts of possession, construction of buildings, visible use and occupation, a mailbox with name on it, payment of taxes, fencing, erecting "no trespassing" signs, and cultivation.

Surveying and marking of boundaries, payment of taxes, and occasional entries for the purpose of cutting timber are not sufficient to constitute adverse possession [33].

Unloading lumber from cars and piling it on an unenclosed portion of a railroad right-of-way without objection by the railroad company, cutting grass, tying horses out to graze, and similar acts do not constitute adverse possession of the premise, particularly where no notice was ever given to the company that any claim of right was being asserted [34].

The property was in such an obscure place that the city's manifestations of taking possession were not apt to be observed by the owner of the land or by the public [35].

13.41 Claim of Title

Title to land cannot be acquired by squatters' rights. The initial right to possession of land must be in writing, but after the title is acquired, land may be added to or subtracted from by adverse rights. The claim of title may be defective and be only color of title, but, nonetheless, there must be some claim of title.

Like so many of the other elements of adverse possession, what is necessary for a claim of title varies from state to state. In Maine [36], the intention of the possessor to claim adversely is necessary. If a party, by mistake, occupies up to a line that is believed to be the true line, adverse occupancy does not occur. In Connecticut (the majority rule), the mistake is of no importance; the fact that the occupancy occurred in a mistaken belief about the true location of the boundary line is immaterial. In Maine, it is necessary to occupy land with the intention of using it as one's own. In Connecticut, the state of mind of the occupant is not of importance. The unfortunate result of the Maine decision is that it benefits the wrongdoer more than the honest landowner [37].

13.42 Continuous Possession

Possession must be continuous for the period required by statute. Any legal interruption causes the law to operate to restore possession to the true owner. It is presumed by law that possession resides in the true owner; the moment possession is not in the hands of another, it is in the hands of the true owner.

Interruption occurs when there is a break in the continuity of possession. No matter how short the period of interruption, if there is an interruption, the limitation period required by statute must start over. Any admittance on the part of adverse possessors that they do not have title to the land they are occupying ceases hostilities, causes an interruption, and stops the statutory period. Abandonment of the premises for a short period causes interruption. Occupancy temporarily stopped by flood or fire may be renewed after a reasonable period of time. Short periods of vacancy with intent to return do not cause interruption. Renting or leasing adjoining land ceases hostilities between the adjoiners and causes an interruption in possession.

The fact that adverse possession of land is interrupted after it is begun is not material if it is continued for 20 successive years after the interruption [29].

Title by adverse possession can be acquired only by the actual holding and enjoyment of land under a claim of right that is opposed to and inconsistent with any other claim. No possession of land is sufficient to ripen into title unless the holding has been such as to furnish the plaintiff a cause of action for the recovery of the lands every day during the 15-year period. Possession must not only be actual but also must be open, notorious, continuous, adverse, and peaceful for every hour of every day of the whole 15 years. It is incumbent on one asserting title by adverse possession to show affirmatively continuity of possession, and, failing in this, his or her cause must fall. A court looked closely at possession and wrote:

> In this case there are several periods of a year or less in which Sackett and his predecessors did not have actual possession of the land in controversy, or any part of it, during the period of 15 years upon which he relies for the perfection of his title [38].

"Tacking" on of possession by one adverse possessor to another is permitted if there is a privity (e.g., father and son, grandfather and grandson, grantor and grantee) between the parties. Passing title by inheritance does not cause an interruption unless it is followed by abandonment.

In Wisconsin, tacking was limited where the deed read: "beginning 40 rods north of the east 1/4 corner stake; running thence north 40 rods; thence west 80 rods; thence south 40 rods; thence east 80 rods" [39]. *Abland v. Fitzgerald* (1894) concerned a survey that revealed about 1 acre existing between the north line of this description and a 30-year-old fence. The fence marked the line of actual, visible, and exclusive possession. The possessor of the 1 acre claimed adverse rights by tacking on to the possession of his predecessor, but the court ruled that there was an interruption. At the time of conveying title, the predecessor conveyed in accordance with the deed without mentioning the 1 acre he had occupied adversely. A person may sell part of his or her land, and in this case the court ruled that only the part described was sold. Since the 1 acre was not sold, adverse occupancy had to start as of the date of the deed. Because this was less than 20 years, title did not pass. This quotation is from the same case: "It seems that as to land not covered by a deed, a grantee can claim no right founded on the adverse possession of his grantor."

The state of Wyoming also claimed that it acquired title by 10 years of adverse possession in accordance with the rule laid down in that state [40]. This contention could not be sustained. The adverse claim, if it was ever made, did not continue for the requisite period of 10 years. It ceased when the state, through its agent, discontinued to cultivate its land beyond the true boundary, and acknowledged the line established by the U.S. government survey as the true line. Even though the adverse claim commenced to run in 1919, it was interrupted by the state's acts and acknowledgment and could not, during the interruption, ripen into a title. While the state can acquire title to adverse possession, no good reason has been pointed out why it should not be subject to the same rules incident thereto that apply to the case of private individuals [41].

Some states disagree with the rule. In Wisconsin and in other states, the Mother Hubbard clause has been used extensively. A deed with a Mother Hubbard clause usually reads: Lot 1 and all of the adjoining lands possessed by the seller. In Alabama, it is unnecessary to use the Mother Hubbard clause, since the courts have ruled that tacking on can occur even though the area is not specifically described. This state is the exception.

For the purpose of effecting title by adverse possession, where all the traditional elements are present, tacking on the periods of possession by successive possessors is permitted against the coterminous owner seeking to defeat such title unless there is a finding, supported by the evidence, that the claimant's predecessor in title did not intend to convey the disputed strip. The court held that this rule should apply even though the conveying instrument contains no legal description of the property in question and irrespective of the period for which the property was possessed by the present claimant's predecessor in title [42].

13.43 Hostile Possession

Hostile possession must, in most states, be against the interests of the fee owner without the adverse claimant admitting that the land is not his or hers. Hostile possession does not imply ill will but does present a claim on the land to the exclusion of all others. Permissive entry is not hostile entry. As long as the occupant is possessing with the permission of the owner, adverse possession is not operative. The relation of landlord and tenant is not hostile; renting adjoining land stops hostilities and brings about an interruption.

Although hostile possession is almost uniformly required, differences of opinion exist about what is meant by *hostile*. Occasionally, mere recognition and acquiescence for the statutory period are defined by statute as sufficient. An "intent to take title" and an "intent to deliberately take land (theft)" are some of the variations in interpretations.

Most authorities hold that hostility does not need to be present at the inception of possession, but it must be present for a statutory period that may commence either at or after inception.

Where one of two adjacent landowners extend a fence so as to embrace within his or her enclosure lands belonging to the other, but in ignorance of the true boundary line between them, and with no intentions of claiming such extended area, such possession of the land so enclosed is not adverse or hostile to the true owner; but if the fence so extended is believed to be the true boundary line, and one claims ownership to the fence, even though the established division is erroneous, such possession will be adverse and hostile to the owner [43].

The rule is that there must not only be 20 years continuous uninterrupted possession, but such possession must be hostile in its inception and continue as such for the period. It must be visible, exclusive, and notorious and must be acquired and retained under claim of title inconsistent with that of the true owner. All of these elements must concur [44].

The essence of adverse possession is that the holder claims the right to his or her possession, not under but in opposition to the title to which that possession is alleged to be adverse [45].

Hostility is a question of a mental attitude of the claimant, in that adverse possessors must believe that the right of possession is valid and they are entitled to it. This could then be extended to a claim of right. The individuals must believe they have the right to be there because it is theirs.

13.44 Exclusive Possession

Possession cannot be shared with the rightful owner, since the rightful owner can never be deprived of possession. A common entry to two stores does not deprive an owner of possession.

Exclusive possession can be proved by exclusion of the legal owner, by threat of force or legal action, by declarations of the possessor to hold land exclusively, or by refusal to permit the legal owners or his or her agents to enter the land.

The rule is, where the entry is made with the consent of the owner and subservient to his or her claim of title, the law will presume that the continual possession is subordinate to the title of the true owner. The element of hostility to the title of the true owner is an indispensable ingredient of adverse possession [46].

Exclusive not only applies to the rightful owner and adverse possessor, but courts have applied it to an interest claimed by individual possessors. A New Mexico court recently determined that two individuals who were claiming adverse possession as tenants in common did not meet the requirement for exclusive possession or interest. Only one individual could do this [47].

13.45 Statute of Limitations of Possession

Possession must be proven for the length of time required by statute. A person may oust a claimant one day prior to fulfilling the statute of limitations. This ouster can be either a physical ouster or a legal ouster in the form of filing a motion for removal. Originally, the common law of England required that possession extend to time immemorial, or beyond memory. Over a period of time this was gradually shortened to mean 20 years and sometimes less. Practically all states now have a statute of limitations, with the time period varying from 5 or 7 to 20 years. Often within the same state the period of time will vary, depending on the circumstances. In California, the period is either 10 or 20 years, depending on who pays taxes. In Puerto Rico, the Commonwealth has adopted two time elements: 10 years for island residents and 20 years for non-island residents. Those with color of title sometimes need shorter periods of adverse occupancy. An entry based on a judgment or tax lien may require a shorter period of occupancy.

The question of the time element is flexible in the eyes of the court. Although they look at the year requirements, at times they must determine whether the year equals 1 year or 365 days. If the time element is 7 years to prove adverse possession, does the adverse claimant have to prove 7 × 1 year = 7 years or does he or she have to prove 7 × 365 days (1 year) = 2,555 days? If the time is critical, should a leap year be considered as 366 days? The presentation of this critical evidence should be the responsibility of the surveyor. The ability of the surveyor to positively identify actual possession to the exact day is just as critical as being able to prove a "window" of possession based on ancient fences, tree lines, and age of blazes. It is the surveyor's responsibility to be able to place the possession in a required period necessary to meet statutory requirement. The surveyor must be able to work with certainty as to ages of fences, blazes, and all other forms of possession. These are facts about which the surveyor can testify and offer opinions as to, for example, age and about which their extent can be measured.

13.46 Color of Title

Color of title is said to exist when by appearance a written conveyance seems to be good, but in actuality it is not good in that it has legal defects that will make it invalid. Any written instrument, regular on its face and purporting to convey title to described land, is usually sufficient to establish color of title. By statute, some states require that color of title is necessary to obtain an adverse title. The record of a survey of lands does not, in itself, constitute color of title to such lands, but it may be evidence tending to show claim of title [48].

In those states requiring color of title, the conveyance purporting to pass title is regulated by the same laws as for written deeds. That is, the color of title, though not a valid writing, must describe a locatable particular area of land.

A description reads as follows: "Part and parcel of the east half of the north-east quarter of section No. 17, township 40 north, range 14 east of the third principal meridian, being 2½ acres off the north end of the 10 acres conveyed to . . ." [49]. In neither of these deeds (other description was not included) can the particular 2½ acres of land intended to be conveyed be ascertained or located from the face of the deed. The instruments are, because of uncertainty, inefficient as color of title.

The effect of color of title is that if one is in possession, but not of the entire parcel, with color of title and when *all* of the elements have been met, then and only then does adverse possession ripen into title, and the adverse possessor gets whatever is in their description; whereas without color of title, they only get that to which they are in actual possession of. In some jurisdictions, the occupied premises must be enclosed.

13.47 Taxes

In some states, taxes must be paid on the land adversely held. This requirement greatly reduces the possibility of gaining land by adverse means. Sometimes payment of taxes shortens the statutory time required for possession.

Payment of taxes is not proof of possession; it is evidence of a claim of right. Failure to pay taxes is not a necessary element of adverse possession unless required by law. People do not pay taxes willingly; therefore, payment of taxes is evidence of a claim.

13.48 Good Faith

In some jurisdictions adverse possession must be taken and maintained in good faith; but by the majority opinion, good faith need only be with respect to constructive possession. Good faith, if required, means that a person believes that he or she has good title, and takes title in accordance with this belief. A person deliberately taking land, knowing that the title is defective, is not acting in good faith, and where good faith is required, such action is of no avail.

13.49 Actual Possession versus Constructive Possession

The surveyor many times must make a distinction between actual possession and constructive possession, whereas actual possession is a question of the facts, farming, grazing, timbering, then constructive possession become a question of two factors: (1) the ability of the surveyor to place the color of title on the ground; and (2) the surveyor's ability to place the actual possession within the boundaries of the constructive possession. This is mandatory to give legality to the constructive possession.

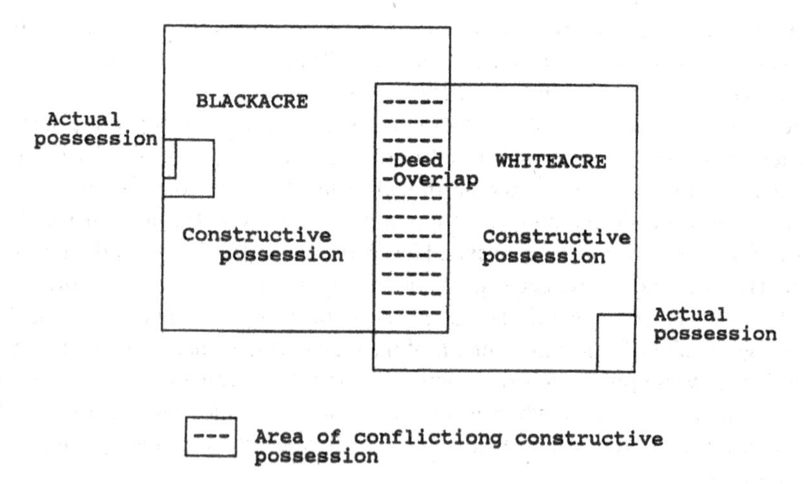

FIGURE 13-1. Each party has a deed to its parcel but is occupying only a small portion. There is no constructive possession in the overlap area.

Once the actual possession is placed within its boundaries, the actual possession constructively extends to the boundaries. Color of title elevates naked possession to constructive possession.

Problems can develop when two individuals are in possession of two separate parcels under the constructive possession theory. This could happen when the color of title of the two parcels overlaps, and the claim is under color of title. Then the courts have determined that adverse possession cannot be granted, for the element of exclusiveness has not been met. Thus, adverse possession fails for the common area (see Figure 13.1).

13.50 Difficulty of Determining Unwritten Rights

As can be deduced from the foregoing, gathering sufficient evidence to prove whether a specific possession is a fee right is extremely difficult and complex. After reading many of the land surveyors' examinations given in various states, it is obvious that surveyors are not questioned in detail on the subject of acquiring ownership by the unwritten processes so far discussed. The logical conclusion is that surveyors are not expected to be experts on the subject; however, it is certainly necessary for them to know that unwritten rights can occur and may be present.

The most difficult of all unwritten rights requirements is to prove that of adverse possession. It is much easier to prove recognition and acquiescence, practical location, or other means of unwritten boundary modification. But most courts make it nearly impossible to prove adverse possession.

A surveyor can assume liability problems for failure to recognize a lawful right obtained by prolonged possession, and he or she can also create problems for failing to monument the deed description when he or she knows that an unwritten right has ripened into a fee right. A few examples clarify these remarks.

A deed, one of several written about 80 years ago in the same locality, reads as follows:

> Commencing at the Southwest corner of theoretical section 17 . . .; thence North along the Westerly line of said section 17 a distance of 10 chains to the true point of beginning; thence continuing along the section line 5 chains; thence at right angles easterly 5 chains; thence southerly parallel with the westerly section line 5 chains; thence 5 chains to the true point of beginning.

A survey (see Figure 13.2) revealed that an angle of 92° existed at the theoretical (Rancho land grant area) section corner and that all fences were constructed parallel with the south line of the section. The deed call of "at right angles" deviated from a 30-year-old fence by 2°. After weighing all evidence, the surveyor decided that the fence would be controlling and then monumented the property with a 92° angle at the corner. The owner, who was not notified of the conditions, erected a house adjoining

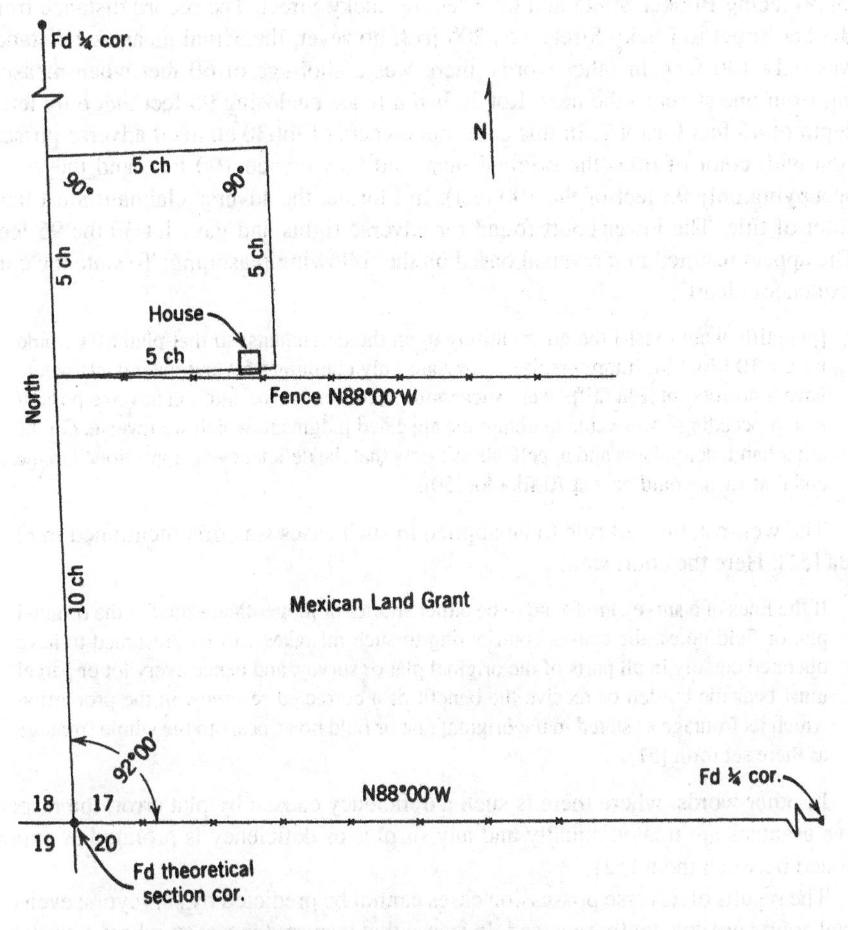

FIGURE 13-2. Using a fence to control a survey.

the south line. On discovery of the facts, the title-insuring agency refused to insure a house loan. The house was several feet over the line when a 90° angle was used, and the surveyor was in trouble. As discussed under evidence, "The written words are conclusive." The deed did not say "parallel with the south line of the section"; it said, "at right angles." If the surveyor had disclosed the true facts to the client, in writing, he would have been without fault; but he, on his own, located the land as though the deed read "parallel with the south section line." To be sure, he could use occupancy rights as a defense, but the fact remains that, win or lose, the client was damaged because of the inability to get a loan. In California, where this particular situation occurred, all house loans are insured by a title company, and the title company will not insure on possession rights; the surveyor should have foreseen this and warned the client of the adverse possibilities.

A Florida case illustrates the difficulty of determining what a court will do in an adverse possession case. Plaintiffs owned lot 30 and defendants owned lot 7; the back line of lot 30 butted against lot 7. The record depth of each lot was 100 feet, with lot 30 facing Booker Street and lot 7 facing Lucky Street. The record distance from Booker Street to Lucky Street was 200 feet; however, the actual measured distance was only 140 feet. In other words, there was a shortage of 60 feet when measuring from one street to the next. Lot 30 had a fence enclosing 95 feet that only left a depth of 45 feet for lot 7. In this case, the owners of lot 30 claimed adverse possession with color of title (the original map said they owned 100 feet, and they were occupying only 95 feet of the 100 feet). In Florida, the adverse claimant must have color of title. The lower court found for adverse rights and gave lot 30 the 95 feet. The appeal resulted in a reversal based on the following reasoning: To state the consequences clearly,

> [plaintiffs would visit] the error entirely upon the defendants, so that plaintiffs would have a 100-foot lot (more precisely, they are only claiming 95′) and defendants would have a 40-foot lot. Plaintiffs were successful in this endeavor and, via adverse possession proceedings, were able to obtain the appealed judgment, which we reverse. On the other hand, defendants and appellants ask only that the deficiency be apportioned to the end that each would have a 70-foot lot [50].

The well-established rule to be applied in such cases was first mentioned in Florida [51]. Here the court said:

> If the lines of a survey are found to be either shorter or longer than stated in the original plat or field notes, the causes contributing to such mistakes will be presumed to have operated equally in all parts of the original plat or survey, and hence every lot or parcel must bear the burden or receive the benefit of a corrected resurvey, in the proportion which its frontage as stated in the original plat or field notes bears to the whole frontage as there set forth [51].

In other words, where there is such a deficiency caused by plat error, the respective grantees are treated equally and any surplus or deficiency is prorated or apportioned between them [52].

The results of adverse possession cases cannot be predicted by surveyors; even the trial courts are constantly reversed. In a case that occurred in a state where a surveyor may be liable to third parties, the surveyor was held responsible for damages [53]. Surveyors in sectionalized land areas know that when the western tier of sections in a township is divided into parts, any deficiency or excess is placed in the last quarter mile. The southwest quarter of one of these sections was short 13 acres. By government survey rules, the east half of the southwest quarter contained 80 acres and the west half contained 67 acres. One party patented the entire southwest quarter and then sold off the west half of the southwest quarter containing 80 acres. Later, the east half of the southwest quarter was sold. About 40 years later, a surveyor was asked to survey the east half of the southwest quarter, and he did so exactly in accordance with the *Manual of Instructions for the Survey of the Public Domain* [54], and this was undoubtedly correct for that description. The surveyor noted the encroachment of a fence on 13 acres, showed it on his plat, and credited the client with 80 acres. The client sold the land on a per-acre basis; a title company insured the land, and in the ensuing litigation the 13 acres were declared to belong to the adjoiner (the west half). The title company paid out over $100,000 for the land and sued the surveyor and won. For correctly locating his client's land in accordance with the client's legal description, the surveyor got into trouble. In other words, the rights of the adjoining landowner should have been looked into; the adverse occupancy should have been the tip-off. It does not pay to certify acreage when someone else is occupying part of the area and may have a right of ownership. The second thing to note about this case is that the surveyor was held liable for damages caused to a third party (those who have a right to rely on the results of a survey), as discussed in Chapter 17.

Here is one other problem involving occupancy rights. A road, called Boundary Street, was offered for dedication and was accepted by the city as part of a subdivision. The road was located along a Rancho Land Grant boundary with control points several miles apart. When constructing improvements in the road, the city discovered that the subdivision had been monumented 5 feet into the adjoiner's land. The city built the road along the true land grant boundary, with the result that a 5-foot strip was taken from the lots on one side of the road and 5 feet was added to the lots on the other side of the road (in a different subdivision). After 40 years of uncontested use of the road, where does the surveyor set the monuments for the lots adjoining the road? Would the surveyor dare to set monuments 5 feet in the road as improved and take a chance on being held liable for a client constructing an improvement in a street that others have obtained by usage? It would be foolish to do so.

13.51 Duties to Unwritten Title Transfers

PRINCIPLE 18. *If possession is not in agreement with the client's deed, a search of the adjoiner's title must be made to determine the status of possible senior rights and to determine if any changes have occurred in the title's wording at times of subsequent transfers.*

The surveyor does not always do the title search; abstract reports are relied on. As previously stated, senior rights control, and it is necessary to know who has senior rights in the event of conflict of deed calls. In the east, deeds commonly can be found that call for one another; the northern deed calls for the other deed, and the southern deed calls for the northern deed. An essential part of the survey is to discover which deed has prior rights. As an example, surveyor 1 has a deed that calls for the adjoiner, and the adjoining landowner's deed also calls for an adjoiner. The surveyor monuments the fence line between the two properties. Surveyor 2 does a search of the records and discovers that some time ago both deeds called for a creek as the boundary line and the fence was built on one side of the creek for mere convenience. The creek still controlled. As part of applying the rules of surveying, a Virginia court held that the failure of the retracing surveyor to understand and then to apply the priority of calls in conducting a retracement was negligence per se in that the retracing surveyor failed to recognize that the identified priority of calls was an absolute legal requirement in retracing boundaries [55].

Although the surveyor does not conduct title searches, in research one usually utilizes many of the same historical documents that the title attorney uses. In most cases the surveyor usually conducts more extensive research into the history of the parcel simply to determine the correct description of the boundaries.

PRINCIPLE 19. *If possession is not in agreement with measurements and if an original survey were made, the surveyor must determine whether the possession represents where the evidence of the footsteps of the original surveyor were. If the possession duplicates the footsteps of the original surveyor, then the possession does not represent an unwritten conveyance right; it is where the written rights belong.*

Determining whether a given possession represents where an original surveyor created the original lines becomes one of evidence, its interpretation, and application for the surveyor. This also requires that the surveyor knows and understands basic principles of unwritten rights, yet no surveyor in America can practice in that area of law that is so vitally important.

One of the basic concepts that the law recognizes is that no individual can lose a greater interest in land than the individual owns. Any interest in land owned by the individual may be lost, but only to the extent of the amount of interest owned. That is, a life tenant can lose only the interest that he or she is entitled to and no more. The adverse claimant cannot claim the rights of any remainderman until that person has a right to the possession of the interest claimed or the property. A person who owns a one-half interest can only lose that one-half interest and no more. When a claimant brings an action for adverse rights, it is important to understand what interest and how much is owned by the individual who is to lose the rights owned.

13.52 Role of the Surveyor in Unwritten Rights

PRINCIPLE 20. *The surveyor's responsibility is not to determine the validity of adverse claims, but it is to report the facts of possession and use as they relate to the lines described in the description and as they are surveyed and recovered by the retracing surveyor.*

After reading this chapter, the reader should realize this area of surveying has more to do with the legal aspects of surveying than the technical aspects. The rule that should be gleaned from this chapter is:

> The surveyor should never give clients legal advice as to unwritten rights and should only give such advice to an attorney if and when asked.

Yet, in order to conduct a retracement, a surveyor should fully understand the requirements of the gaining and/or losing of property and property rights through the various legal doctrines, and then the surveyor should report only the facts to the client or the attorney. The surveyor's responsibility is to collect and report the facts—not to give legal advice.

If the retracing surveyor is engaged to serve as an expert on a boundary issue, it is suggested that definite terms of engagement be made identifying what he or she will be responsible for and who is the client. Because of legal privileges available between the client and the attorney, it is suggested that the terms of the contract be between the attorney and the surveyor expert. In this capacity, the surveyor can advise the attorney on legal issues and not be held legally responsible.

REFERENCES

1. *Doyle v. Bradshaw*, 41 Cal. App. 247, 183 P. 185 (1914).
2. *Berghoefer v. Frazier*, 150 Ill. 577, 37 N.E. 914 (1894).
3. *American Jurisprudence 2d Adverse Possession*, sec. 298 (Rochester, NY: Lawyers Co-Operative Publishing Co., 1962–1977).
4. *Howard v. Howard*, 271 Ky. 773, 113 S.W. 2d 434 (1938).
5. *Smith v. McCorkle*, 105 Mo. 135, 16 S.W. 602 (1891).
6. *Wright v. Hendricks*, 388 III. 431, 58 N.E. 2d 453 (1945).
7. *Aldrich v. Brownell*, 45 R.I. 142, 120 A. 582 (1923).
8. *Wilhelm v. Herron*, 211 Mich. 339, 178 N.W. 769 (1920).
9. *Henderson v. Dennis*, 177 Ill. 547, 53 N.E. 65 (1892).
10. *Bemis v. Bradley*, 126 Me. 462, 139 A. 593 (1927).
11. *Galbraigh v. Lansford*, 87 Tenn. 89, 9 S.W. 365 (1888).
12. *Inhabitants of Wesley v. Sargent*, 38 Me. 315 (1854).
13. *Bailev v. Jones*, 14 Ga. 384 (1853).
14. *Sherrin v. Coffman*, 143 Ark. 8, 219 S.W. 348 (1920).
15. *Adams v. Warner*, 204 N.Y.S. 613 (N.Y. AP. 1924).
16. *Childers v. Dedman*, 157 Ga. 632, 122 S.E. 45 (1924).
17. *Peters v. Reichenbank*, 114 Wisc. 209, 90 N.W. 184 (1902).

18. *Duncan v. Peterson*, 83 Cal. Rptr. 744 (1970).
19. *Village of Lee v. Harris*, 206 Ill. 428, 69 N.E. 230 (1903).
20. *Rennert v. Shirk*, 163 Ind. 542, 72 N.E. 546 (1904).
21. *Grell v. Ganser*, 255 Wisc. 384, 39 N.W. 2d 397 (1949).
22. *Scott v. Williams*, 38 Han. 448.
23. *Cook v. Corbin*, 7 Ill. 652 (1845).
24. *Wolfe v. Sullivan*, 133 Ind. 331, 32 N.E. 1017 (1893).
25. *County of Piatt v. Goodell*, 97 Ill. 84 (1880).
26. *Quinn v. Baage*, 138 Iowa 426, 114 N.W. 205 (1908).
27. *Mass. Statutes*, Chapter 86, sec. 2.
28. *Edwards v. Flemming*, 83 Han. 653. (Handy's Ohio) (1855?).
29. *Shedd v. Alexandar*, 270 Ill. 117, 110 N.E. 327 (1915).
30. *White v. Harris*, 206 Ill. 584, 69 N.E. 519 (1903).
31. *Brumagan v. Bradshaw*, 39 Calif. 24 (1870).
32. *Coryell v. Cain*, 16 Calif. 567 (1860).
33. *Griffith Lumber Co. v. Kirk*, 228 Ky. 310, 4 S.W. 1075 (1929).
34. *Ill. Central RR. v. Haskenwinkle*, 232 Ill. 224, 83 N.E. 815 (1908).
35. *Carrere v. New Orleans*, 162 La. 981, 111 SO. 393 (1926).
36. *Preble v. Maine Central R.R.*, 85 Me. 260, 27 A. 149 (1893).
37. *Lincoln v. Edgecomb*, 31 Me. 345 (1850).
38. *Sackett v. Jeffries*, 182 Ky. 696, 207 S.W. 454 (1919).
39. *Abland v. Fitzgerald*, 87 Wisc. 517, 58 N.W. 745 (1894).
40. *City of Rock Sps. v. Strum*, 39 Wy. 494, 273 P. 908 (1929).
41. *State v. Vanderkoppel*, 45 Wy. 432, 19 P. 2d 955 (1933).
42. *Wilson v. Price*, 356 So. 2d 625 (Ala., 1978).
43. *Hess v. Rudder*, 117 Ala. 525, 23 So. 136 (1897).
44. *Clark v. Clark*, 349 Ill. 642, 183 N.E. 13 (1932).
45. *Mills v. Laings*, 38 Calif. App. 776, 177 P. 493 (1918).
46. *Timmons v. Kidwell*, 138 Ill. 13, 27 N.E. 756 (1891).
47. *Hernandez v. Cabrerra*, 107 N.M. 435, 758 P. 2d 1017 (1988).
48. *Atkinson v. Patterson*, 46 Vt. 750 (1874).
49. *Allmenain v. McHie*, 189 Ill. 308, 59 N.E. 517 (1901).
50. *Madison v. Hayes*, 264 So. 2d 852 (Fl. Ct. App. 4th Cir. 1972).
51. *City of Jacksonville v. Broward*, 120 Fla. 841, 844, 163 So. 229, 230 (1935).
52. *Peerless et al. v. Magoon et al.*, 78 Wisc. 27, 46 N.W. 1047 (1890).
53. *City of Pompono Beach v. Beatty*, 177 So. 2d 261 (Fla. App. 1965).

54. U.S.D.I., Bureau of Land Management, *Manual of Instructions for the Survey of the Public Lands* (Washington, DC: U.S. Government Printing Office, 1973).

55. *Spainhour v. Huffman*, 337 S.E. 2nd 615 (Va. 1989).

ADDITIONAL REFERENCES

Backman, J. A., and Thomas, D. A., *A Practical Guide to Disputes between Adjoining Landowners—Easements*. New York: Mathew Bender, 1991.

Hand, J. P., and Smith, C., *Neighboring Property Owners*. New York: Shepards-McGraw-Hill, 1988.

Wilson, D. A. *Easements and Reversions*. Rancho Cordova, CA: Landmark Enterprises, 1991.

14

GUARANTEES OF TITLE AND LOCATION

14.1 Scope

Ownership of land consists of one of two things:

1. A good written title with the right of possession.
2. Possession with the right to acquire written title as a result of a lawful, unwritten conveyance.

Because most laypeople are incapable of determining when written title defects exist or when others have lawful title rights arising from possession, a need has arisen for experts willing to research, explain, and then guarantee titles to be free of defects.

The professional surveyor certifies to land location and status of what lawyers refer to as encroachments; he or she does not issue a policy of location guarantee but may be held liable for failure to exercise due care. Title companies, attorneys, courts, or the state (Torrens titles), depending on the jurisdiction, guarantee written title adequacy; sometimes, on request, they will guarantee location. How titles are guaranteed and who guarantees written titles are the subjects of this chapter. Such discussion cannot be complete without mention of possession guarantee, since lawful occupancy can defeat paper title.

Land is more than the soil one sees on examination of a parcel. Land consists of a collection of rights, and the rights extend from a point at the center of the earth outward, piercing the land surface and extending into the heavens. Land is not one facet; it includes many different rights. Such rights include but are not necessarily limited to rights of possession, ownership, timber, mineral, water, and aerial. All rights can be subsurface, surface, or aerial.

In its true perspective, ownership of land is never complete; someone else always has a right against the land (e.g., taxing authority, zoning). Ownership is more properly a right of possession and title.

A deed to land is never proof of ownership; it is evidence of ownership. Others may have rights against the land because of taxes, liens, mortgages, easements, judgments, bonds, improvement assessments, incompetency to alienate, restrictions, zoning, and many other items. Because of the numerous hazards, new landowners demand ownership guarantee or at least a statement of ownership condition, along with a guarantee that no rights exist other than those stipulated.

Exposures to written title defects are not visible and cannot be seen; exposure to possession defects can be seen by the purchaser. The demand for title guarantee has been greater than the demand for possession guarantee. In recent years, because of instances of damages resulting from possession matters, an increased demand for guarantee of both possession and title matter exists.

The basic concept of land in the United States is one of the commingling of two ancient doctrines or concepts of law—one with its origin in English common law, in which title to land is paramount, and the other with its origin in Roman common law, in which possession of land is superior.

14.2 Registration of Titles Versus Registration of Ownership

Recording a deed in a public place is a registration of *evidence* of a claim to land ownership. Torrens registration of land includes both registration of title and registration of *ownership*. When an automobile is registered, the name of the owner is registered; when an automobile is sold, the state requires that the new owner must be registered within a definite period of time. The Torrens registration system is similar; new transfers of land must be accompanied by a change in registration of ownership. The difference between the two systems is that one *registers title*, whereas the other registers *ownership*.

14.3 Aids to Title and Location Guarantee

Unwritten title rights are aids to cure location defects. The statute of limitations is of material assistance in maintaining the status quo and preventing spite litigation. These aids to cure defective titles do not include the element of financial guarantee and are not a part of this chapter.

Courts, in general, do not guarantee title; courts declare by decree what are the existing rights between litigants. Parties not involved in the action have their rights unimpaired. But the Massachusetts Land Court actually guarantees ownership of both title and location; an assurance fund is set up to pay damages to those inadvertently harmed by the court's decree.

In Torrens title registration, a similar situation exists, provided the registration includes both written title and location guarantee. However, in many Torrens registrations, location is not a part of the registration, and location is not guaranteed.

14.4 Title and Possession Guarantees

Within the United States, title guarantee, possession guarantee, and statements of title condition guarantees (abstracts) vary from state to state and can be classified as follows:

1. Warranty deed
2. Abstract (guarantee of facts)
3. Abstract with attorney's opinion
4. Written title insurance
5. Written title and possession insurance
6. Torrens title (including Massachusetts Land Court system)
   a. Written ownership of title guarantee
   b. Written ownership guarantee

Warranty Deed The grantor of a warranty deed personally guarantees the title to be free of defects. Such a guarantee is sufficient, provided the grantor has adequate funds to pay in the event of damages if a defect is discovered at a later date. From the grantor's viewpoint, he or she is often unwilling to issue a warranty deed and would rather pay to have others assume the risk of guaranteeing title. Death or financial ruin of the grantor leaves the grantee without guarantee; therefore, the grantee is usually willing to pay for responsible assurance of title.

Patents Sovereigns issuing patents to lands do not guarantee title. A patent is merely a quitclaim to any rights or interests possessed by the sovereign. If the land had been previously patented to others or if the patent is contrary to law, it is without force.

In any patent from a sovereign, legal opinion is needed about whether the patent was properly executed in accordance with the then-existing laws, whether the correct description of the land involved was properly described, and whether any prior patent was issued on the land described.

Chain of Title All land titles originate from a sovereign, whether it be a king or a government, and the sequence of title transfers from the sovereign to the present is called a *chain of title*. Like a chain, the strength of the present title is no stronger than the weakest link. If a defective link exists, the written title may be voidable or faulty from that point on. A person guaranteeing title must, of necessity, examine all transfers back to the creation of the title.

Although all titles originated from the sovereign or governmental utility, it is not always necessary to search the chain back to the original patent. The sovereign, by operation of law, may declare a new start for a title through court decrees or Torrens title actions.

14.5 Abstract of Title

An abstract of title is a statement of publicly recorded facts relating to the chain of title. An abstract is not a complete statement of every detail of past title transfers; it is a summary of the essential facts necessary to pass judgment on the sufficiency of written title.

An abstract is usually completed for a period of time that varies from state to state according to the accepted title standards in effect. In Maine and New Hampshire, abstract search is usually for a period of 40 years or to a warranty deed at 40 years or further back. For title insurance, the search usually extends to at least 60 years. In Massachusetts, the minimum search is for 60 years, and in Georgia it is 50 years. In California, search usually stops at the previously issued title insurance policy.

Abstract Companies In early years it was the custom for attorneys to search public records, extract essential information, and pass opinion on condition of title. With the increase in the volume of title transfers, the problem of searching more numerous records created a demand for those specializing in title abstract information.

Many abstract companies have complete title records that enable them to prepare a title abstract with little outside search. Such systems are indexed by location, grantees, and grantors. The result is a more efficient operation; a few companies serve many people and save the cost of duplicating research efforts.

Sources of Abstract Information Public records (an abstract pertains only to public records) from which the title searcher obtains title history are scattered. There are seven main sources:

1. Public records offices (e.g., recorder of deeds, office of recorder, hall of records, county clerk).
2. Courts (federal, state, and municipal).
3. County offices (e.g., surveyor, engineer, clerk, treasurer, assessor).
4. State offices (e.g., Division of Highways, Division of Water Resources, Division of Beaches and Parks, Department of Mines).
5. State and national archives, where original land patents and records are preserved.
6. Quasi-public bodies (irrigation, sanitation, water, and other taxing utility districts).
7. Educational districts.

Typical Abstract of Title The following is an abstract that was abridged and insertions made when it was believed the changes would not detract from its value as an example of an abstract of title:

1. CAPTION. An abstract of title in the southeast quarter of section 9, Township 22 North, Range 8 East of the Third Principal Meridian, situated in the County of Champaign, State of Illinois.

2. Beginning with Title in the United States Twp. 22 N, R 8 East of the Third P. M. Plat of this title is recorded in the Book of Township Plats, p. 25 in the County Clerk's. . . . (At the beginning of the book is the following certification) . . .
   Office of Land Titles
   St. Louis, Missouri
   June 16, 1869 . . .
   The correctness of plats appearing in this book is certified . . .;
   Albert Siegel, Recorder
   by E. H. Hesse, Deputy
3. A copy of the original township plat showing distances and areas.
4. East half of southeast quarter containing 80 acres entered March 15, 1854b. Robert Houston as appears from Book of Original Entries, County Clerk's Office, Urbana, Illinois:
5. United States PATENT
   to ___ Dated January 10, 1855
   Robert Houston, assignee of ___ Filed July 18, 1870
   Michael Brown Book 23, p. 2
   Military Warrant No. #52232
   Grants east half of Southeast quarter, sec. 9, T. 22N, R 8 E. of the 3rd P. M.—district of lands subject to sale at Danville, Illinois, containing 80 acres according to official plat of said land.
   Signed by President Franklin Pierce by H. E. Baldwin, assistant secretary
   Recorded in Vol. 268, p. 181.
6. United States
   To
   Robert Houston
   The west half of the southeast . . . was entered Mar. 15, 1854, as appears. . . .
7. United States PATENT
   To ____ Dated January 10, 1855
   Robert Houston, assignee of ___ Filed July 18, 1870
   Phebe Dunn. Book 23, p. 4
   Military Warrant #53862
   (Grants the east half of the quarter similarly to the west half as indicated under item 5) Recorded in Vol. 268, p. 180.
8. Robert Houston MORTGAGE
   To ___ Dated March 1, 1855
   School Commissioners for ___ Filed March 3, 185
   the Use of Inhabitants of Twp. ___ Book G, p. 578 of deeds
   22N, R 8E. Secures the sum of $856.00
   payable in one year with interest
   at 10 percent payable semiannually.
   Conveys the south half of sec. 9, Twp. 22N, R 8E of the 3rd PM and other lands.
   Acknowledged March 1, 1855, before Isaac Devore, Justice of the Peace of Champaign County, Illinois.

9. On the margin of the record of the above mortgage appears the following: Be it remembered that on this third day of March, 1856, having received . . .
10. (satisfaction) . . . (I release) . . .
    Robert Houston and Thompson Dickson, Treasurer,
    Eliza M., his Wife, Twp. 22N, R 8E of the 3rd PM.
    to ___ WARRANTY DEED
    Theodore L. Houston, son, Dated April 1, 1872
    of above. Filed May 2, 1872
    Book 33, page 156
    Considerations: $5,000.00
    Conveys the south half of sec. 9, Twp. 22 North, Range 8 East of the Third Principal Meridian.
    Acknowledged April 1, 1872 before James S. Jones, Notary Public of Champaign County, Illinois.
11. Theodore L. Houston, MORTGAGE
    having no wife. Dated August 9, 1876
    to ___ Filed August 10, 1876
    Joshua M. Clevenger. Book 17, p. 608
    Secures loan of $3500.00
    Conveys south half of sec. 9....
    Acknowledged August 9, 1876, before C. H. Yeomans, Notary Public, of Ford County, Illinois.
12. M. Clevenger RELEASE
    to ____ Dated August 12, 1880
    C. Houston. Filed April 28, 1881 Book
    Book 52, p. 630
    Consideration: $1.00
    Releases all interest acquired by a mortgage dated . . . and recorded . . . to premises described therein, to wit: south half of sec. 9 . . .
    Acknowledged August 12, 1880, before George F. Beardsley, N.P.
13. Theodore L. Houston, WARRANTY DEED
    a bachelor. Dated April 22, 1879
    to ___ Filed May 9, 1879
    Laura Houston and ___ Book 56, p. 459
    Ida Houston, both married. Consideration: $8,000.00
    Conveys south half of Sec. 9 . . .
    Acknowledged April 22, 1879 before James S. Jones, N.P.
14. Theodore L. Houston and ___ WARRANTY DEED
    Babie L., wife. Dated August 30, 1880
    to ___ Filed September 10, 1880
    Laura Morden and ___ Book 58, p. 485
    Ida Houston. Consideration: $2000.00
    Conveys south half of sec. 9 . . .

Remark: Second grantor does not acknowledge as wife.
Acknowledged August 30, 1880, before Thomas E. Hayes, Notary Public of Essex County, New Jersey.

15. Affidavit of George F. Beardsley. Subscribed and sworn to
Sept. 18, 1884
Filed September 19, 1884
Misc. Records Book 3, p. 364
(States that the affiant personally knows Ida Houston and Laura Houston as referred to in the deeds preceding and that Ida Houston is the same person as Ida Nelson and Laura Houston is the same person as Laura Morden.)

16. Ida Houston, unmarried. QUITCLAIM DEED
to ___ Dated February 24, 1881
Laura Morden, wife of ___ Filed February 24, 1881
W. J. Morden. Book 41, p. 460
Considerations: $750.00
Conveys interest in the south half of sec. 9 (and other lands) . . .
Acknowledged February 24, 1881, before John L. Pierce, N.P.

17. Affidavit of Geo. F. Beardsley Subscribed and sworn
February 2, 1901
Filed May 7, 1901
Misc. Records Book 8, p. 204
(Affiant states that Laura Morden referred to in item 16 is the same person as Laura Houston.)
Subscribed and sworn to February 2, 1901, before E. S. Clark, N.P.

17–24. These entries, not included, show a mortgage, a release, a trust deed, a release, a mortgage, an assignment, and a release.

25. Laura H. Morden WARRANTY DEED
to ___ Dated July 28, 1906
William J. Morden. Filed July 31, 1906
Book 141, p. 339
Consideration: the natural love and affection she has for her son, William J. Morden, and for his better maintenance and support.
Conveys the south half of sec. 9 . . . with all buildings and improvements thereon.
Acknowledged . . .

26. William J. Morden and ___ WARRANTY DEED
Florence Happer Morden, his wife. Dated June 29, 1922
to ___ Filed July 6, 1922
Stacy C. Mosser, as trustee. Book 186, p. 183
under provisions of trust. Consideration: $1.00 and other
agreement dated June 29, 1922, good and valuable consideration.
Document 158283

27. Stacy C. Mosser, as trustee. TRUSTEE'S DEED
    to ___ Dated May 2, 1939
    William J. Morden. Filed May 11, 1939
    Book 246, p. 227
    Considerations: $1.00 and other.
    Releases all interest for consideration in south half of sec. 9 . . .
28. William J. Morden and ___ DEED IN TRUST
    Myrtle Irene Morden, his wife. Dated December 8, 1952
    to ____ Filed December 27, 1952
    First National Bank of Chicago, Book 471, p. 88
    under trust agreement dated ___ Consideration: $1.00 and
    November 28, 1952 and ___ other good . . .
    known in its records as Trust No. 42675,
    Document 501663
    Conveys south half of sec. 9 (and other lands).
    (Grants full power and authority to improve, manage, subdivide, sell, lease for 198 years or less, grants easements etc. . . .)

29–30. (Terms of the trust agreement)

31. First National Bank of Chicago, TRUSTEE'S DEED
    as trustee for William J. J. Morden. Dated October 8, 1958
    to ___ Filed October 15, 1958
    Irene Morden. Book 607, p. 437
    Cons.: $10.00 and other . . .
    Conveys and quits claim to southeast quarter and southwest quarter of sec. 9 . . . (subject to taxes for 1958).
32. In matter of Hillsbury Drainage Assessment Roll . . . (assessment of $1585.40).
33. NOTE: Property is within limits of Hillsbury Slough Drainage District and subject to taxes and assessments levied by the district.
34. NOTE: Property is located in Champaign County Soil Conservation District and subject to rules, regulations, assessments, etc.
35. Taxes for 1957 on southeast quarter of sec. 9 . . . (have been paid by) . . . the First National Bank of Chicago.

Most offices have developed their own forms for abstracts, yet they essentially include the same information. Although the surveyor does not guarantee the abstract of title, any surveyor who researched for descriptions will, in most instances, include the same documents in researching descriptions as the attorney uses in abstracting titles.

Effect of Abstract of Title An abstract by a reputable abstract company is a guarantee of a statement of facts, and liability is limited to the effect of omission of publicly recorded facts. Issuance of an abstract is not a guarantee that a title is free of defects; someone else passes opinions on the sufficiency of the title, which is a legal question. The tract is merely a history of ownership.

The abstractor's job is to compile a complete statement of all publicly recorded matters affecting the ownership of a particular parcel of land. Who signed a conveyance will be noted, but whether the person had the authority to sign the conveyance is not always noted. If a word is misspelled in a conveyance, the misspelled word is perpetuated and usually is underlined to direct attention to it.

Completion of an abstract reflects the paper history or condition of the real property. The desire of the attorney and the abstractor is to give certainty about the "health" of the written title. Changes in the description of the property through subsequent conveyances disturb these individuals. The most certain title is property that retains the exact description as in the original conveyance. The surveyor, however, is interested in describing the same parcel in relation to boundaries and monuments on the ground today. This discrepancy can be overcome by describing parcels of land relative to the boundaries and monuments on the ground and then making reference in the description to the original and subsequent source of title.

Commencing around 1900, abstracts were the norm for individuals purchasing land. In many rural courthouses, young attorneys fresh from law school would spend hours and weeks "doing abstracts." Today a "complete abstract from the sovereign to the present" is the exception rather than the rule. Many individuals now rely on title insurance as a substitute for a title search. Basically, the issuance of a title policy without an adequate title search is like buying life insurance without a physical examination.

14.6 Abstract and Attorney's Opinions

An abstract by itself is not a guarantee of ownership; someone else assumes the responsibility of proclaiming the sufficiency of the facts stated, usually attorneys. The attorney's opinion does not include possession matters, and unless there is a survey, location is not guaranteed. The opinion of the attorney is limited to *facts revealed by the abstractor* through the abstract. Attorneys' opinions do not include guarantee against fraud, forgeries, or acts of omission or commission not of record. Title policies often do:

> They undertook to furnish an abstract of what appeared from the public records affecting the title to his property, and he relied on their competency and fidelity in this respect: Failure to note the judgment on the sale of the land for taxes, it was held that the party making the abstract was liable for damages to the purchaser [1].

14.7 Title Insurance Policy

Title insurance is a written guarantee that, as of a particular moment, no title defects, other than those stated, exist. The limit of financial liability is stated on the face of the policy and may be increased or decreased in proportion to the fee paid. Unless otherwise stated, a policy is limited to title facts of record, not matters of location or unrecorded documents.

Policies show the amount of the policy, the person or persons whose title is guaranteed, the estate (e.g., fee simple, life estate, lease) being guaranteed, and the description of the real estate covered by the policy.

Title companies are in the risk business; for a fee, they will assume a financial responsibility. If a person examines any title close enough, he or she can almost always find something that will present an uncertainty. In the matter of obtaining Torrens titles through court quiet-title actions, it may take months or even years to clear a small defect that may never materialize. Title companies generally do not delay title transfers because of remote defects; they can and do take calculated risks. For speed and quick service, the title companies are far superior; for this reason, more and more people are relying on them for title matters. Most lending agencies insist on title insurance.

Matters that title policies ordinarily insure against are taxes, bonds, assessments, trust deeds, mortgages, easements (e.g., sewer, water, gas, electricity, oil lines, drainage), liens, insanity, minor children's rights, forgery, false impersonations, wives with community rights, irregularity of conveyance form, omissions, heirs, leases, attachments, judgments, restrictions, reservations, reversal of court decision, senior rights, errors in public records, errors in transcribing and interpreting the public records, attack on the title, and many other items.

Title policies do not insure against everything. No guarantee is made against bankruptcy power, eminent domain, future police power or law, ordinance or regulation limiting the use of property such as zoning ordinances, building ordinances, or other regulations. Defects arising from a person's own violation (fraud or things known by the person to whom the guarantee is issued), matters not of record, such as a mechanic's lien not recorded, possession, and survey are excluded.

Owners' policies are based on the record, and no attempt is made to investigate matters of possession. Mortgage policies, however, are subject to claims of possession and questions of survey. Ordinarily in a mortgage policy, the title company sends out an inspector to determine possession, and if doubt exists, a survey is required.

A title insurance policy is of considerable aid to the surveyor in locating land, since he or she then has a statement that includes senior rights of others and a list of all easements. If there is an omission or error of statements, others assume that responsibility.

Of course, a title abstract of a parcel is only as good as the public records that were reviewed. In those counties that have well-designed research facilities, abstracting is "a piece of cake." But in those courts where the clerks are deficient in maintaining and indexing records, abstracting is "a nightmare." Recently, for example, in an urban county in Georgia, a surveyor was sued for failing to show an easement on his plat when an easement was actually present. The surveyor relied on the information furnished by the title insurance company, which had an abstract performed. The attorney who performed the abstract had conducted his usual abstract in looking at the filed public records. The facts did reveal that an easement that was purchased by the State Highway Department did exist, but it was "missed" by the attorney in his research. The facts were that the county accepted the recorded easement, stamped it in, and placed it on the floor next to a desk in a pile of maps and deeds that were being accumulated for future filing, "when they got time."

There is no question there was a recorded easement. It had been overlooked. Surveyors relying solely on title searches that are furnished may run a great risk in being responsible for something that should have been checked by them. Standards for survey are not the same as standards for title examination. The surveyor conducting research as to title information usually does this to have a good description from which to work.

Wording of Title Policies The wording of title policies varies, depending on the title association. In general, a named person, persons, and/or corporations are insured for a maximum stipulated sum of money against loss or damages sustained because of the following:

- Unmarketability of the described title unless it is caused by conditions listed
- Title to the land being vested otherwise than as stated
- Any defects in title not listed (including easements)
- Defects in previous execution of mortgages, deeds of trust, liens, and the like
- Priority of other claims over mortgage being insured

Among the exceptions listed are

- Taxes, liens, easements, encumbrances, and assessments not shown on the public records
- Rights of persons in possession of land not shown on the public records
- Mining claims, water rights (in the West), claim or title to water, and reservations in patents
- Any rights, interests, claims, or facts that could be ascertained by an inspection of the land or by inquiry of the person in possession or by a correct survey
- Laws, government acts, or regulations (e.g., zoning, restrictions, use of land, setback, and occupancy)

Standard title policies and title guarantees generally contain three standard exceptions: subject to items outside the courthouse; subject to items outside the period of search; and subject to items that would be disclosed by an accurate survey. Some policies may contain a fourth item: subject to implied dedication(s).

14.8 Title and Location Insurance Policy

If a parcel of land is surveyed by a competent surveyor and encroachments are certified, most title companies will extend the coverage to include matters of possession without extra charge. But it must be remembered that if a title company is insuring location as based on a survey, it is its prerogative to accept or reject the results of a surveyor.

14.9 Location Guarantee

Surveyors, because of professional responsibility, must guarantee their work to the extent required by law. Liability is not stipulated as in a title policy but is one of mistakes or omissions caused by failure to do what an ordinary skilled surveyor would do and one of subsequent reliance by the parties involved.

Many surveyors carry liability insurance to protect themselves from loss. This type of insurance is not for the benefit of the client as is a title policy; it is designed to protect the surveyor from the client.

The amount of damages that can be collected from a surveyor is in proportion to the surveyor's assets and liability insurance; it is wise for those using the surveyor's services to inquire into his or her financial condition and insurance coverage.

14.10 Corrective Instruments

In certain matters, title companies will not guarantee title without corrective instruments. In general, the courts of this country take the attitude that "any description by which the property might be identified by a competent surveyor with reasonable certainty, either with or without the aid of extrinsic evidence, is sufficient" [2]. Title examiners and lawyers judge the merits of the conveyance entirely from the instrument itself and matters of public record. Examination with the aid of extrinsic evidence may be sufficient to make an otherwise invalid conveyance valid. Courts may take testimony and arrive at a conclusion binding on the parties, whereas the title examiner does not. It follows, then, that titles that may be legally sufficient to the court may not be to the title examiner.

14.11 Torrens Principle of Title Registration

In essence, the Torrens registration requires a quiet title action for all first registrants, requires registration of title by the state, includes guarantee of sufficiency of title by the state, has optional and sometimes mandatory requirements for location guarantee, and has mandatory requirements that all matters pertaining to the land's title be recorded on the Torrens title. Historically, a system similar to the Torrens titles was practiced as early as the thirteenth century in Bohemia. The present Torrens title registrations, developed by shipping clerk Sir Robert Torrens of Australia, was adapted from the method of titling merchant ships. The system was passed into law in 1858, and Robert Torrens became the first registrar of titles. In recognition, Torrens was knighted. This system was adopted in England, parts of Canada (British Columbia, Alberta, Saskatchewan, and parts of Ontario), Norway, the Philippines, Sweden, Denmark, Puerto Rico, France, and many British colonies. To a limited extent, it has been put into practice in the United States. Illinois, California, Massachusetts, Minnesota, Oregon, New York, North Carolina, Mississippi, Ohio, Nebraska, South

Carolina, Georgia, Tennessee, North Dakota, South Dakota, Washington, Utah, Virginia, Colorado, and Hawaii have passed permissive legislation for such a land registration system. California repealed the act in 1958 because of its disuse. Many other states seldom use Torrens registration.

Characteristics of the Torrens System The validity and sufficiency of a written title are dependent on a continuous chain of valid transfers from the first sovereign's alienation to the present. The chain of title grows with time. One primary purpose of Torrens registration is the elimination of maintaining previous transfer records (chain of title). To fulfill this objective, the written title must have a new starting point, and this is done by a quiet title action and court guarantee of ownership.

Essentially, initial Torrens title registrations are quiet title actions. A responsible party is appointed to examine the sufficiency of title; and if the title is found wanting, a quiet title suit is instituted.

Those registering titles are charged with the responsibility of protecting other valid title rights; titles are not supposed to be registered if defects exist. But in the event a title is registered and others do have a valid claim, the state is financially responsible and pays damages from an assurance fund.

The state guarantees title; this creates a new starting point for the title. If others are found who have rights to the land, they cannot claim possession or usage; they can only claim damages from the state. To compensate those collecting damages, an assurance or indemnity fund is provided, said fund being accumulated by a charge on each title transfer.

As soon as title sufficiency is satisfied, the land is registered in the name of owners, and a certificate of registration is issued to the owners. All new valid transfers or encumbrances on the title must be done by an entry in the register. Provision in the law is made that failure to register transfer or encumbrances operates only between the parties of the transaction but not against future register owners. Thus, the new owner is assured that only items registered can interfere with his or her rights.

Since the Torrens system was first developed in Australia, it was impossible to obtain possession rights against a Torrens title holder. Since that time the policy has eroded, and under some circumstances, land can be obtained by occupancy.

Advantages and Disadvantages of Torrens Because title and ownership are up to date each time a new conveyance appears, searches are unnecessary after the system is in force. The indemnity fund, if properly set up, will cover most claims that arise from clerical errors and omissions. The protection also covers loss from fraud and wrongful possession. After initial registration, there is some saving of time and money to the parties of the conveyance, and the administration realizes some benefit in improving tax rolls.

Since the condition of the title can be determined immediately, the delay and expense of having the title abstracted are eliminated. In some cases, the registered land can be transferred without the need of a new survey unless further subdivision is made. Forgery is quite unlikely, if not impossible, since the certificate of title must be presented before a transfer is effected.

Even though it is theoretically an ideal system, the Torrens system has disadvantages. The cost involved in first entering the properties into registration prevents many landowners from taking advantage of Torrens registration. In many cases, the system has not worked because of the poor ties of the property to the ground. In California, it was a failure and was repealed; loan companies would not loan on Torrens titles because of the insufficiency of the assurance funds.

Torrens land acts prohibit the acquiring of land by adverse possession. However, any state can at any time pass a statute law prohibiting adverse possession acquisition of land for any title; hence this feature of the law is not exclusive to the Torrens system.

The Torrens system is a monopoly by the state; the success or failure of the system is dependent on political whims. The inefficient and incompetent can be placed in charge.

Land Registry in Illinois The first state to enact legislation to permit Torrens type of land registration was Illinois in 1897. After the disastrous Chicago fire of October 8–9, 1871, many land records of Cook County were lost forever, and it was difficult to complete the chain of property title. A statute approved May 1, 1897, made possible the land registration, particularly in the counties having more than 500,000 population. This was a means to quiet title. Cook is the only county with Torrens registration in Illinois.

Land Courts A small number of states have land courts that specialize in land disputes, title matters, and disputes concerning rights and interests in land. These courts seem to make good sense since many states have specialized probate courts, domestic courts, juvenile courts, and the like. But few legislatures have seen the need and wisdom of having land matters determined by individuals trained in land law. Two of the more progressive land courts are in Massachusetts and Hawaii. These courts have their own judges and rules and specialize in land-related matters.

The Massachusetts Land Court In 1898, the Massachusetts Registration Act, a Torrens act, provided for a court of registration with a judge, an associate judge, and a recorder. In 1904, the court was renamed the "land court." All original registration proceedings, as provided by law, were to be handled by the court of registration, sitting in Boston or elsewhere, by adjournment as public convenience demanded. After a decree of the court had been issued and forwarded to the assistant recorder in the district where the land was located, it would be transcribed into a book as an original certificate of title, and a duplicate certificate would be furnished to the owner of the land. A certificate contained five pieces of information:

1. Number of the certificate in that particular registry.
2. Name and address of the owner.
3. Description of the land with reference to a plan filed either with the certificate or some other certificate in the same registry (the locus was fixed in detail on the plan and generalized as a bounding rather than a running description).

4. All appurtenant rights and encumbrances as decreed by the court.
5. On the back, all conveyances, mortgages, easements, liens, and so on, pertaining to the land.

The powers of the court were increased in 1904 and at later dates to include the following 11 items:

1. Writs of entry
2. Petitions to require action to try title to real estate
3. Petitions to determine the validity of encumbrances
4. Petitions to discharge old mortgages
5. Petitions to foreclose tax titles
6. Petitions to establish power or authority to transfer an interest in real estate
7. Petitions to determine the boundaries of flats (along the ocean)
8. Petitions to determine whether equitable restrictions are enforceable
9. Petitions to determine county, city, town, or district boundaries
10. Petitions to determine the validity of zoning laws and the like
11. All equity actions affecting land, except those relating to specific performance of contracts

When a jury trial is demanded, the issues are framed in the land court, and the arguments on these issues are tried before a jury in superior court. With the adjudication of the questions of fact, the work in the land court proceeds, and the ultimate decision is in no way amendable to the machinery of the superior court. A petitioner for registration in the land court has a constitutional right to demand a jury trial, but this right is seldom exercised.

The work of the court is, in general, divided into two classes of cases: contested and uncontested:

> The Land Registration Act necessitates an adjudication by the Court of the boundaries as well as the ownership of the premises, and the boundaries so determined remain definite and fixed.
>
> The procedure for registration of land is as follows: A petition setting forth the claim of the petitioner, accompanied by a plan of the property and a certificate signed by the Assessors of Real Estate of the locus setting forth the abutters at the last date of assessment, is filed with the Land Court. The Court appoints an Examiner, who must be an attorney qualified by the Court, to examine the records and report by filing a narrative, opinion, and abstract of all the records that he feels are necessary to meet the requirements of law. He may be required to furnish supplemental reports if the Court feels that they are necessary. Upon completion of his report citations are prepared and served on the abutters, and all others who may be affected, by registered mail with return receipt. The land is posted by a deputy sheriff with a copy of the citation, and publication is made for three successive weeks in a newspaper that is circulated in the particular vicinity of the locus.

> Thus, all persons are given a chance to be heard and an opportunity to file appearances and answers. In due course the petition reaches the judge's chambers, and after determining the validity of the title by trial or otherwise, he issues an order for a decree setting forth exactly what the petitioner will receive.
>
> A plan is prepared by the Land Court Engineering Department, and a decree drawn in conformance with the order and plan so prepared. This is returned to the judge for his final initialing and, upon final payment of all fees, the order is sent out to the registry district where the assistant recorder issues a Certificate to the petitioner.
>
> The Land Registration Act does not provide for an engineer, but the Honorable Charles Thornton Davis, Judge of the Land Court from 1898 to 1936, recognized the necessity of an engineering department to properly make adjudication of boundary lines upon the earth's surface. Judge Davis established the Engineering Department as an adjunct to the Recorders Department. It is the duty of this department to prepare a "Manual of Instructions for the Survey of Lands and Preparing Plans for the Land Court." The department inspects the plans submitted to see they are made in accordance with the instructions. The engineer for the Court is called upon to interpret engineering features, to conduct investigations, and to fix the boundaries that may be decreed by the judges. The manual is issued for the information and guidance of all engineers and surveyors making surveys and plans for the Land Court. The manual may be amended from time to time, and new editions appear about every 5 years, to keep pace with changes in survey practices.
>
> Each plan made for a petitioner to file for registration must have a statement that the survey has been made in accordance with Land Court instructions over the signature of the engineer or surveyor who has supervised the work. (The term survey applies not only to the actual field work, but also the preparation of the notes, computations and plans [3].)

The Hawaii Land Court Hawaii provides for the registering of title to land through its land court. The rules of the court require that any survey of registered land be under the direct supervision of a registered surveyor who has been approved by the court.

The rules indicate the types of notes that will be kept, the types of plots that will be made, and all requirements that one must consider.

The rules of the court provide for the consolidation of noncontiguous parcels, redesignation of individual lots, and combining lots. Not all lands are filed through the land court, but if it is decided to file a parcel through the land court, the individual must follow the explicit rules to the letter.

The court has 18 pages of rules that surveyors must follow. They are too numerous to list here. It may be that more land courts will be the solution to many of our legal problems.

REFERENCES

1. *Chase v. Heaney*, 70 Ill. 268 (1873).
2. *Smiley v. Fries*, 104 Ill. 416, 416 (Ill. 1882).
3. T. S. Llewellyn, "Massachusetts Land Court," *Surveying and Mapping*, vol. 8, no. 4 (1948), p. 182.

ADDITIONAL REFERENCES

Flick, C. P., *Abstract and Title Practice*, 2nd ed. St. Paul, MN: West Publishing, 1958.

Pitch, L. D., *Abstracts and Titles to Real Property*. Chicago: Callaghan and Co., 1954.

15

USING AND UNDERSTANDING WORDS IN BOUNDARY DESCRIPTIONS

15.1 Introduction

Perhaps in the cycle of creating or retracing, surveyors should start with this chapter, because it is from the original survey that the evidence is created by the surveyor that furnished the information. The retracing survey begins with the evidence the original surveyor created on the ground and then described in subsequent documents.

Descriptions of real property or interests in real property can be mystifying to those individuals who prepare them, to those who are asked to interpret them, and to the courts who are asked to pass on their legality or validity. On numerous occasions attorneys, surveyors, and even the courts misuse the terms *description* and *deed*. Attorneys have asked surveyors, "Did you locate the deed on the ground?" or "Where is the deed located?" meaning, "Did you locate the description or was the description located?"

A distinction must be made between the two. Surveyors work with descriptions; attorneys and the courts work with deeds. The distinction is quite important, in that the deed is the instrument that "carries" within its four corners the description. A deed can be considered valid, voidable, or void, depending on certain legal circumstances—that is, as having been conveyed in fraud or not meeting certain legal requirements set by law.

An Arizona decision defined a deed as follows:

> Formally, the word "deed" was synonymous with the word "act," and, as the law recognized a peculiar sanctity in the use of a seal, and act under seal became something finished, sacred and worthy of great respect. Hence a man's act under seal became his deed, and likewise any instrument under seal became his deed, irrespective of the nature of the transaction involved. But the term "deed" is used now in a more limited sense and as the meaning a written instrument, duly acknowledged by a competent party, conveying title to land [1].

For practical purposes, a "deed" is the legal instrument that carries within it an essential element without which it would be void, and there would be a need for the description.

Since the deed is a legal concept, the description becomes a factual concept as well as a legal concept. Words create the description, and it is the words that must be interpreted, usually by the surveyor, who must place the words that are used in the deed "on the ground."

To the land surveyor, property descriptions are usually prepared by two types of individuals with limited data indicated: (1) by surveyors, who usually show calls for monuments, courses, adjoiners, and other pertinent information; and (2) by others, from attorneys to landowners, who believe that a "do-it-yourself" deed will save them time, trouble, and money.

Surveyors, attorneys, and courts tend to consider the terms deed and description as being synonymous. In reality, they are different. Attorneys and judges have asked surveyor witnesses such questions as, "Have you seen the deed" or, "Have you read the deed?" Essentially they are asking, "Have you read the description?"

The property, or *legal*, description is one element of the deed. For a deed to be valid, it must contain a description that is capable of being located on the ground. If one asks three different surveyors the question, "Can you locate (survey) this description from the words contained in the deed?" you may get three answers: Surveyor 1 says "Yes, I can"; surveyor 2 says, "No, I can't"; and surveyor 3 says, "I just don't know."

Real property law, as well as the Statute of Frauds, requires that conveyances, deeds, of an interest in real property abide by the following:

1. Be in writing.
2. Identify grantors and grantees.
3. Identify the interests being conveyed.
4. Express an intent to convey the interest identified.
5. Identify the location of the land or the interests that are being conveyed with legal certainty.
6. Be signed by the grantor.
7. Be accepted by the grantee.

In this chapter, two specifics of property (deed) descriptions are examined:

1. Writing descriptions.
2. Understanding descriptions that are presented for interpretation and retracing them by placing the description of the land boundaries on the ground.

Through the use of the words in a description, *land descriptions* identify the physical location of the land or the interest conveyed or to be conveyed, whether that interest is a fee title, life estate, easement, royalty, or one of the many other interests that make up the "bundle of property rights," but this area is not the subject matter of this chapter.

The deed or conveyance of a parcel or interest in real property is composed of two basic elements:

1. The words that are used to describe the interest to be conveyed and how much.
2. A description of the land itself.

As a result of legal conflicts, the courts have evolved a common-law ranking of the preference that they will consider to determine if there is a conflict in the wording used in land descriptions. Early in the legal history of descriptions, it was determined that a call for a monument has greater legal force than a recited distance. Topic headings in this chapter are arranged to follow this presumed order of precedence.

The principles discussed in this chapter are as follows:

PRINCIPLE 1. *A deed may be void, ab initio, if the description contained in that deed is incapable of being located by a competent surveyor.*

PRINCIPLE 2. *The description contained on a deed is but one element of the deed, and as such, it should be read in unison with all of the other elements.*

PRINCIPLE 3. *A description is legally sufficient when it relates how to identify a unique and particular area of property relative to monuments on the surface, below, or above the earth or a water body.*

PRINCIPLE 4. *A land description to be legally sufficient in the matter of locatability must identify a particular area or areas to which the interest conveyed attaches.*

PRINCIPLE 5. *If a written conveyance is valid, recorded, and locatable when executed, no later act or event will operate to make the conveyance invalid.*

PRINCIPLE 6. *Junior descriptions should call for the senior description as an adjoiner or should be written with calls identical to that of the senior adjoiner.*

PRINCIPLE 7. *Before attempting to describe land, the scrivener first must carefully determine the intent of the parties and then select words, keeping in mind the presumed order of importance of description elements which will precisely describe that intent.*

PRINCIPLE 8. *A legally sufficient description is one that a competent surveyor can locate on the ground.*

PRINCIPLE 9. *If the intent is to describe land as following a particular survey, that fact must be stated and the survey must be identified and should be specifically called for.*

PRINCIPLE 10. *Other than senior rights and contrary expressed intentions, a call for a monument is presumed to prevail, and if the intent is to limit the boundaries to particular existing monuments, it must be so stated.*

PRINCIPLE 11. *Directions by bearings are relative to whatever the writings define as the basis of bearings; astronomic north is merely one of the many ways to define directions and is sometimes assumed to be the basis of bearings in the absence of other evidence.*

PRINCIPLE 12. *When a monument is called for in a written description, that monument, if it is undisturbed, is controlling over other elements in the description.*

PRINCIPLE 13. *Unless specifically specified, a description of a parcel of land includes all of the land within its boundaries.*

PRINCIPLE 14. *When a new description is created as a result of a retracement, the corners and lines should make reference to the elements recited in the original description.*

PRINCIPLE 15. *No person should use the same description from one deed to another. A new description should be prepared for each subsequent conveyance.*

15.2 Graphic and Written Descriptions

The purpose of a land description is to identify a unique parcel or area of land or interest in land, either on the surface of the Earth or below or above the ground or its waters. There are many accepted and/or legal methods to identify a land interest. Hence, by necessity and possibly tradition, land descriptions take many forms. Many individuals have tried to put personality into some descriptions, to the point that they are humorous. Courts will consider the content, not the form. That is, do the form and the words meet all legal requirements? One such "crazy" deed was tested in an Illinois court and found to be legally sufficient (see Figure 15.1).

There are two methods of identifying a land area or an interest in land:

1. By a graphic representation (plats or maps and other forms of diagrams).
2. By written words.

All graphic descriptions include some written words and also rely on the use of many and various symbols to describe the property. This discussion of description writing is divided into graphic problems (including symbols and words commonly used as part of a graphic presentation) and written descriptions.

Many descriptions are combinations of both graphic representation and written words. Descriptions logically fall into three classes:

1. Those calling for a graphic description (lot and block on a given map).
2. Those describing an area by written words (metes and bounds).
3. Those using combinations of the above.

Once a series of lines close, then an area can be determined and identified. But without a closed figure, balanced mathematically, the area contained in a parcel of land cannot be determined with any certainty.

WARRANTY DEED

I, J. HENRY SHAW, the grantor herein,
Who lives in Beardstown-the County within,
For Seven Hundred Dollars to me paid today
By Charles E. Wyman, do sell and convey-
LOT TWO (2) in BLOCK FORTY (40), said county and town,
Where Illinois River flows placidly down,
And warrant the title forever an aye,
Waiving homestead and mansion to both, a goodby,
And pledging this deed is valid in law,
I add my signature,

(seal) J. HENRY SHAW

dated July 25, 1881

I, SLYVESTER EMMONS, who lives in Beardstown,
A Justice of Peace of fame and renown,
Of County of Cass in Illinois State,
Do certify here that on this same date,
One J. Henry Shaw to me make known,
S he stated he sealed and delivered same
His homestead therein, but left all alone,
Turned his face to the street and his back to his home.

Date August 1, 1881

S. EMMONS, JP

FIGURE 15-1. This deed has been declared by the Illinois courts as valid (filed August 9, 1881, vol. 40, p. 251, Cass County, IL).

15.3 Subdivision Descriptions

The term *subdivision description* should not be confused with the term *aliquot description* used in describing a land parcel of a Public Land Survey. Here, subdivision is used to identify land by reference to a complete parcel of land, which is depicted and represented on a plat. An example of such a description is as follows:

1. Lot 2, Block 27, Hillside Estates, according to Map Number 2761 filed. . . . The term subdivision description used herein refers to entire parcels described by reference to a plat.
2. Section 5, T 3 S, R 2 E, Fifth Principal Meridian.

Descriptions of a part of a parcel, such as "the westerly 100 feet of Lot 7," are herein considered as intermediate types.

Reference to a plat or map in a land description incorporates into the description all the data or facts indicated or depicted on the plat. Knowledge of how to prepare subdivision descriptions is knowledge of how to incorporate the data shown or depicted on specific plats. Subdivision descriptions are "reference descriptions"; they must call for a plat that identifies the land. Both legal as well as evidentiary problems may arise when the surveyor is confronted with an unrecorded plat.

Some subdivision descriptions have both a call for a plat and a metes-and-bounds description. These usually lead to confusion since one may conflict with the other. It is at that time that courts will step in and attempt to rationalize and apply certain rules as to "what controls": The U.S. government, in conveying sectionalized lands, sold the land by reference (e.g., Sec. 10, T 3 S, R 2 E, SBM), and the reference automatically included the field notes. The field notes are a written perimeter description of each section, having all the similarities of metes-and-bounds descriptions. Many states, when alienating their lands, also required a survey, a plat, and field notes. As determined by case law, the call for an aliquot parcel also includes a call for the plat as well as the field notes of the section or township [2].

15.4 Metes-and-Bounds Descriptions

The term *metes-and-bounds description* has a general, not a specific, meaning. Metes means "to measure" or to assign by measure, and bounds means the "boundary" of the land. These words are similar and probably represent the same thought. As used in deeds, "bound" is sometimes used in a restricted sense when referring to a monument as a "stone bound." The classic form of a metes-and-bounds description is perhaps the oldest form of description known.

Bounds means the legal, imaginary lines by which different parcels are divided [3]. Metes and bounds means the boundary lines or limits of a tract [4].

As commonly used by surveyors, the metes-and-bounds description means a complete perimeter description wherein each course is described in sequence and the entire description has a direction of travel around the area described. The distinguishing feature of this type of description, herein called true metes-and-bounds description, is that each course identified must be described one after another in the same direction of travel that would occur as if a person walked around the entire perimeter in either a clockwise or a counterclockwise direction. Once a direction is selected, it must be consistent for the remainder of the description. Clockwise is preferred to counterclockwise. Technically, all written descriptions can be metes-and-bounds descriptions, even those calling for a lot and block by referral. But surveyors and the public do not, by common usage, classify lot-and-block descriptions as metes-and-bounds descriptions.

In writing a metes-and-bounds description, consistency is key. Although each description should be a reflection of the person who wrote it, there are certain basic suggestions that a scrivener should follow:

1. The point of commencement usually should be located off the parcel being described. The point of beginning should be described with a degree of certainty that will permit a person who is trying to locate the parcel to be able to proceed to the exact point.

2. The point of beginning should be located on the parcel proper. And the point of beginning of a newly created parcel should be, if possible, related to the parent tract. If possible, a point of beginning should not be described as a computed point or an unmonumented corner. The point of beginning should be a monumented point.
3. The basis of bearing should be noted, as well as the unit of horizontal measurement used.
4. Each course should stand alone on its own. A metes-and-bounds description should not have the individual courses run together. Each course should be on its own line and should stand alone.
5. If a specific survey is referenced or is controlling, that survey should be called for in the metes-and-bounds description, and it should be identified.
6. The description should be signed by the person who created the description and dated.

Rarely is a metes-and-bounds description written without a reference call. A description reading "Beginning at the southwest corner of section 10; thence," has a reference call, plus a call for the township plat, and its field notes.

A metes-and-bounds description that states, "Along the easterly property line of . . .:" [5] is a bounds call and includes any plat called for by the adjoiner's deed. "According to the map of Hillside Acres" is a reference call for a plat and survey.

Within a metes-and-bounds description, taking the term in its broadest sense, lines defining the limits of the area can be classified as being described:

1. By measurements, usually bearings and distances.
2. By monuments, natural monuments, artificial monuments, adjoiners, and bounds.

A line described by measurement must have a starting point, a distance, a direction, and a terminal point. The term *monument* is used in its broadest meaning.

A call in a description described as "Commencing at a 6" concrete post, with a brass cap stamped 1, Tract 2, and thence one course as follows: North 26 degrees and 10 minutes East for a distance of 226.45 feet to a point has no absolute legal or survey termination, in that both the bearing and the distance are subject to errors. But consider a call in a description described as "Commencing at a 6″ concrete post, with a brass cap stamped 1 Tract 2, and thence one course as follows: North 26 degrees and 10 minutes for a distance of 226.45 feet, to 2, 6-inch square concrete post with a brass cap set and marked 2, Tract 2." In retracing this description, the post may lack survey certainty in bearing and distance, but it possesses legal certainty, in that the retracing surveyor may alter or modify both bearing and distance to have those called for in the description control.

Specific names have evolved to describe certain types of metes-and-bounds descriptions, such as proportional conveyances, bounds conveyances, linear conveyances, strip conveyances, area falling on one side of a described line, and true metes-and-bounds conveyances. Some names, such as "bounds descriptions," have become so commonplace that title men refer to them as a distinct type.

15.5 Bounds Descriptions

A bounds description calls for an adjoiner or adjoining parcel (monument) on each side of the described area; it is a referral description.

The following description states that the area is bounded by two natural monuments or boundaries (the Ohio River and Rose Creek) and two legal monuments (Jones and lot 12):

> Bounded on the north by the Ohio River bounded on the south by Jones; bounded on the west by Rose Creek, and bounded on the east by Lot 12, Acre Park Subdivision.

The lack of necessity of direction of travel around a bounds description distinguishes it from true metes-and-bounds descriptions. This form of description is the easiest to prepare, but it may be the most difficult to retrace. The boundaries are dependent on the location of all of the called-for and identified adjoining parcels.

Many of the descriptions in the northeast U.S. are of the bounds type. A large number may name the abutter or a road, and it becomes necessary to trace the chain of title back to the point where a metes-and-bounds or some other type of controlling description can be found.

Bounds descriptions are best used to describe a remainder (a junior title holder). Bounds descriptions calling for an adjoining junior parcel should never be used. Let the junior parcel call for the senior, but do not let the senior call for the junior. In the past, many scriveners did not understand this principle—unfortunately, many still do not.

15.6 Proportionate Conveyances

Proportionate conveyances are a fractional part of a whole area. Ordinarily, unless a statute exists to the contrary, proportional conveyances are a proportion of the total area. In sectionalized lands, a federal statute defines the method of surveying the southwest quarter; hence, it is not a proportion of area.

Statutory proportional conveyances are those described by fractional parts (southeast quarter of the southeast quarter; section 2) and are not mathematically proportionate parts of the area. Fractional parts conveyed under common law are always proportionate parts of the area; the only exception is the sectionalized lands where statute defines otherwise.

A proportionate conveyance can be a fractional part of either a subdivision description or a metes-and-bounds description. Ordinarily, the term is applied only to whole lots in a subdivision.

The major problem with proportional measurements is in the descriptions of parcels that are not rectangular (e.g., what is the NE half of a square?).

15.7 Area Described as Being on One Side of a Described Line

Areas can be described as being all of a certain described parcel lying on a given side of a line, said line being described by monuments or measurements. Some examples are as follows:

> Lot 12, except the easterly 20 feet.
> Lot 12 lying westerly of the easterly 20 feet.
> All of Smith's land lying southerly of the following described line.
> All of section 10 lying northerly of Alvarado Creek.

15.8 Strip Conveyances

By describing a line and by including a given number of feet on each side or one side of the line, an area is described. The strip conveyance is a special form of the description given in Section 15.7.

Strip descriptions are used for boundaries that are usually long, such as roads, power lines, or utility lines. The center line is usually described and the boundaries usually are described as being perpendicular to the "following described center line." This type of description is not recommended. It is suggested that the perimeter of the parcel be described and the description be a closed parcel.

15.9 Area Conveyances

Area conveyances (e.g., north 5 acres of lot 5) are used to convey a specific quantity of land. To be effective, one dimension of the area must be omitted, so that area becomes the determining quantity to calculate the omitted dimension.

Generally, area conveyances are written without the benefit of a recent survey. The "north 10 acres of lot 12" conveys exactly 10 acres. If there were a recent survey, a better description could be written. This description is uncertain, since one dimension is omitted (direction of the southerly line of the north 10 acres). An infinite number of configurations is possible. Even though the courts have, at times, given definiteness to such descriptions, writing an area description without defining the direction of the dividing line creates an ambiguity. A description such as "the NE half," has usually been held to be invalid in most states, but valid in a very few states.

15.10 Linear Conveyances

Such descriptions as "the easterly 50 feet of Lot 16" are linear conveyances. The wording of the description is short, concise, and exact, provided everyone understands that the "easterly 50 feet" means the easterly 50 feet is measured so as to give the person owning the 50 feet the maximum amount of land. This description is a combination of a "subdivision description" and a linear distance and is just as certain of location as are the lots on the plat called for.

Linear conveyances can be used with metes-and-bounds descriptions, such as "the westerly 50 feet of the following described property: . . ."

15.11 Indispensable Parts of a Description

PRINCIPLE 1. *A deed may be void, ab initio, if the description contained in that deed is incapable of being located by a competent surveyor.*

The distinction between a deed and a description was discussed early in this chapter. No attempt will be made to discuss elements that may cause a deed to be void. But one of the principal reasons a deed may be void is lack of certainty of the description of the parcel of land resulting from errors in the description, the experience of the surveyor, the loss of monuments by the ravages of time, to name but a few. A description may be legally sufficient to pass title to the land, but it may be technically insufficient to be able to locate the parcel on the ground.

A description should reflect the status of the corners and lines as they were at the time of the conveyance. The more remote the description is from the time of the survey or from the time conveyance was executed, the more difficult it will be for the retracing surveyor to locate and place the description on the ground.

If the retracing surveyor is unable to locate the parcel, it may be adjudicated that "nothing passed when the deed was executed," and thus the deed may be void.

PRINCIPLE 2. *The description contained on a deed is but one element of the deed, and as such, it should be read in unison with all of the other elements.*

Since a deed is a legal document, to be effective, numerous legal requirements must be met in order to effectuate the transfer of property interests from one individual to another.

The modern deed is an evolution of generations of common law, and this piece of paper is the modern equivalent of the ancient livery of seisin practiced in early times. Before there were written words to effectuate a transfer of a property right, the two individuals—the landowner who was selling the property, today's grantor, and the person who was buying the property, today's grantee—would go onto the property. The landowner would take a handful of soil or a twig and then physically hand over the soil or the twig to the person to whom the land was being conveyed. This deed or symbolic act would effectuate the transfer of the property. By walking the bounds (boundaries) of the parcel, the grantee could actually see what it was he or she was buying.

Today, there is no longer the requirement that the two individuals go onto the property to have a transfer of land. The entire transaction is accomplished by words on paper. The two individuals no longer walk the boundaries, because the written description contained in the deed now is substituted for the act of walking the boundaries. As such, there are several legal presumptions that have evolved. One is that the grantor is presumed to convey the entire parcel described in the descriptions, and a second is that the grantee is presumed to have read the description of the land and knows what is being purchased.

In early times, communities were closely knit, and it was probable that some town's folks accompanied the parties when they conducted the act of transfer, and since there was no written document, the knowledge of the transfer was made public by personal publication. Today the same is accomplished by recording the deed in the public records, thus giving "constructive" or "legal" notice of the transfer.

PRINCIPLE 3. *A description is legally sufficient when it relates how to identify a unique and particular area of property relative to monuments on the surface, below, or above the earth or a water body [6].*

Descriptions never completely describe a particular area of land; they merely relate how to find it. As yet, a suitable or infallible system whereby land can be identified without reference to monuments has not been developed. Even latitude and longitude are relative to the Earth's poles and the Prime Meridian. Every description must somehow *identify an area relative to fixed monuments*. Although some descriptions seem to be written without a call for a monument, the call, in actuality, is always there. A call of "Lot 1, Block D, according to Map 2201" calls for all the monuments originally set by the surveyor when surveying the area and depicted on map 2201. A call for an adjoiner also is a call for any monuments necessary to locate the adjoiner.

Measurements of bearing and distance are always stated relative to a monument on the ground. A bearing itself may be relative to a monument (north or south pole).

In legal literature and elsewhere, it is often stated that a conveyance must identify a unique area, meaning one, sole, lone, or single area. A conveyance cannot be equally applicable to two areas; it must be unique in this respect. The interest being conveyed can be attached to land being used for some other interest. The right to use an underground location or the air rights to erect a building above a certain elevation may all attach to the same description: It is the location that needs to be unique, not the interest.

Again, a description may overlap another described area; the description as a whole is not always unique with respect to its entirety. A conveyance that overlaps another is not always void or even, at times, voidable; hence, described areas with respect to their entirety are not always wholly unique.

15.12 Objectives When Describing Land

PRINCIPLE 4. *A land description to be legally sufficient in the matter of locatability must identify a particular area or areas to which the interest conveyed attaches.*

It is desirable that a land description should

1. Contain title identity.
2. Not interfere with the senior rights of others.
3. Be so written that either at the present time or at a future date the description can readily be located by a competent surveyor.
4. Not contain words capable of more than one interpretation.

5. Contain measurement data sufficient to describe a geometric area that closes mathematically.
6. Be based on a recent survey.

15.13 Sufficiency of Descriptions

The purpose of a description is to identify a particular area, not to fully describe it; a description is sufficient if it identifies the land. What is legally sufficient and what is desirable are two entirely different things. Although this is the overlying purpose, one should consider the purposes to which land descriptions are applied. Six of the more important are:

1. To identify unequivocally and definitively one and only one unique parcel of land.
2. To identify for legal purposes.
3. To identify for the purposes of describing, recovering, and retracing land boundaries.
4. To identify for cadastral purposes (taxation and valuation).
5. To locate encumbrances positively (e.g., easements).
6. To identify for addresses and indexing.

Before writing any description for a client, the surveyor should:

1. Clearly understand the purpose of the description.
2. Understand that he or she is concerned only with boundaries and their description, not concerned with what constitutes ownership.
3. Prepare a concise, well-written description, with proper spelling and punctuation.
4. Understand what the courts have said relative to the proper wording that is required.
5. Obtain any instructions or directions from the client.

Competent title authors strive to make a description complete within itself, that is, the description recites all senior claimants, describes all monuments, describes the size and shape by measurements, and agrees with existing possession on the ground (a fact that requires a recent survey).

The courts will try to declare a conveyance valid rather than void. Only in cases where the area described by the writings is equally applicable to more than one location or where the deed fails to identify an area does the court reluctantly declare a deed void. How land may be described and what is sufficient are properly regulated by law and, if the law prohibits a certain form of description, it may not be used.

The right to refer to plats or maps (subdivision descriptions) without including a perimeter description is excluded in patenting federal mining claims and in describing certain political boundaries. In some areas building site conveyances by metes-and-bounds descriptions are illegal.

15.14 Title Identity

Title identity is the relationship between a particular description and its adjoiners. Certainty of location can be attained without certainty of title identity, as illustrated in the following description:

Parcel 1

Commencing at a 40-inch blazed pine tree, blazed with 4 hacks on each of four sides, whose East Zone Coordinates are x = _______ and y = _______: thence a grid bearing of South 25 degrees 17 minutes West, a distance of 223.25 feet to the True Point of Beginning, the 38″ blazed pine tree: thence North 301.00 feet to a 60″ oak tree: thence West, 300 feet to a stone mound: thence South 301.00 feet to the Point of Beginning.

This description is certain of location but it is completely unidentifiable relative to adjoiner titles.

Whether the land overlaps another's rights is not stated, nor can it be determined without adequate research and a survey. If this describes land some or all of which is not owned by the grantor, the conveyance, either partially or wholly, is without force. Title-guaranteeing agencies generally will not issue title insurance on the basis of such titles; it is not a marketable title.

When composing a new conveyance, the most desirable description from the title-guaranteeing agencies' viewpoint is one calling for adjoiners without reciting measurement information other than that necessary to complete a perimeter, such as shown here:

Parcel 2

Beginning at the southwest of Thomas Brown's land as recorded in . . ., thence along the southerly line of said Brown's land to the southeasterly thereof: thence in a straight line to the northeasterly of Doris Muller's land as recorded in . . .: thence westerly along the northerly line of said Doris Muller's land to Third Street: thence along Third Street to the point of beginning.

Bounded on the north by the land of Thomas Brown as recorded in . . .; bounded on the west by Third Street; bounded on the south by the land of Doris Muller as recorded in . . .;

Bounded on the east by a straight line connecting the southeasterly of said Thomas Brown's land and the northeasterly of said Doris Muller's land.

In the previous two descriptions, parcel 1 is a true metes-and-bounds description, in that there are calls for courses (bearings and distances) as well as monuments. Once the lines and monuments are located and identified, the person claiming under it takes all of the lands described and contained within it. Once the monuments are found and identified, you have found the parcel. Parcel 2 is a bounding description, in that locating the parcel, one must survey the lines of Brown, Third Street (center line or right-of-way line), and the lands of Muller; then a new line must be run in the field after the two parcels are identified.

In this type of description exposure to damages resulting from discrepancies between monument locations and measurement data is eliminated: size is not stated. But such a description is or should be unsatisfactory to the new owner; it does not disclose on its face the quantity of land purchased. To be satisfactory to a new owner or to a surveyor, a description should be dimensionally complete. Although title identity is not necessarily essential for validity, it is a part of every good description.

To the title insurer, the most important feature of a land description is its stated relationship to adjoiners: Whether it is readily locatable is of little importance. To the surveyor, the most important feature of a land description is its simplicity of location on the ground and its dimensional completeness: Title identity is a bother. It is this difference of viewpoint that causes opinions to clash. Both are essential features of good descriptions.

15.15 Senior Rights

Whether or not a conveyance interferes with the rights of an adjoiner does not necessarily alter the validity of the conveyance, and as such it should not be the concern of the surveyor: but, in instances, it may operate to make a conveyance partially void. If an author inadvertently included portions of adjoining lands, the land inadvertently included is not conveyed, since a person cannot legally convey another's land. Such inclusion does not operate to void the remainder of the conveyance, since the grantee can claim the land remaining after the overlap is excluded. Because a person may be held liable for selling land not his, it is desirable in composing descriptions to exclude the lands of others. From a title insurer's point of view, it is a necessity that others' lands are excluded; the title insurer is liable for such inclusions, and by recent court opinions, in some states, the surveyor may also be liable.

From the surveyor's viewpoint, all the surveyor can and should do is report the facts relied on and recovered, not state legal conclusions.

15.16 Effect of Loss of Evidence on Location

PRINCIPLE 5. *If a written conveyance is valid, recorded, and locatable when executed, no later act or event will operate to make the conveyance invalid.*

The fact that the conveyance can be located at the time it was made is all that is necessary; later destruction of monuments, along with the death of witnesses who knew where the monuments were and resulting confusion over location, cannot operate to make a description nonlocatable or void. The land must be located; it is a matter of evidence and law to determine where the land belongs. Descriptions, when composed, should contain sufficient words to ensure certainty and ease of future location. Unfortunately, since this is not a legal necessity, many (especially title people) compose descriptions without considerations for future location.

Unrecorded deeds are valid as between the parties but generally not as against the rights of innocent parties. Usually, if a party sells land twice, the first recording is recognized. If the second buyer records first and has no knowledge of the first sale, he or she can usually obtain ownership and possession.

Surveyors want descriptions that are dimensionally complete; that is, related accurate bearings and distances or some other accurate equivalent dimension should be recited for all lines. Monuments called for should be permanent or certain for future location.

15.17 Ambiguity

English words often have multiple meanings and, if improperly selected, may cause ambiguity. Words in one part of a description may conflict with words in another part. Ambiguities or conflicts in general do not operate to void a conveyance; they merely require legal interpretation and sometimes will require a legal determination to clarify the meaning of the description. Thus, in the event that North is written instead of South, the deed is not necessarily void, since errors can be corrected or explained if sufficient writings remain in that description to identify the land. Conflicts or ambiguities are eliminated from the writings by competent scriveners.

A common-law principle permits the surveyor, or the court, to assume that the element first recited in the description is presumably correct and the second element is in error. If problems of ambiguities occur, most courts will look to the surveyor to explain these ambiguities. A Florida court placed a unique responsibility on the registered surveyor when it wrote:

> LAND SURVEYORS Although title attorneys and others who regularly work with them develop expertise as to land descriptions, the only professional authorized to locate land lines on the ground is a registered land surveyor. In fact, the definition of a legally sufficient real property description is one that can be located on the ground by a surveyor. However, in the absence of statute, a surveyor is not an official and has no authority to establish boundaries; like an attorney speaking on a legal question, he can only state or express his professional opinion as to surveying questions [7].

15.18 Mathematical and Legal Correctness and Closure

All descriptions have to have a closure, either mathematical or legal. The mathematical relates numbers and the legal relates words. Of the two, the legal is probably the most important, because in a description when the words state, "and thence to the Point of Beginning," regardless what the numbers state, the courts WILL MAKE IT CLOSE. The courts will make North, South, East, West, they will make 10 the number 1,000. But without these "magic" words, they will look at numbers and if the numbers do not come out to N 00.00 and East 00.00, problems will be encountered.

In many existing conveyances, the area described is not mathematically correct and in some instances may not be geometrically correct. This does not in itself invalidate a conveyance, since identity is all that is required to meet the legal requirements.

This fact may be discussed or used by an attorney to argue the validity of the deed. But that is a separate problem.

The question of the validity of a property description is a legal question, but the surveyor contributes professional ability to interpret the description and whether it is capable of being located. If a surveyor can locate the land described and measure and determine its area, the conveyance is sufficient. When there is a call for monuments (including record monuments), it is almost impossible to ensure mathematical and geometric closures without a new accurate survey. Unfortunately, mathematical and dimensional completeness of a description is not a legal necessity; it is merely a desirable feature of a good description.

When a surveyor is asked to retrace an existing description, the first act should be to "run a closure" of the recited courses indicated in the description. The results will indicate to the surveyor if the courses recited in the description are actual field courses, if they are balanced courses, and if there are any blunders or other errors present.

15.19 Description Based on a Survey

Unwritten conveyances or prolonged possession may legally create unwritten rights or may even transfer title to lands without writings. Possession is one of the many necessary elements of ownership. The purpose of a deed is to convey whatever interest or ownership in the land the individual has; without knowledge of possession relative to the described lines, certainty of transfer of ownership is not always possible.

There are no legal requirements that demand a landowner perform a survey to prepare a description to convey a parcel of land. There are many instances where a description can be written without benefit of a survey, but it usually takes expert knowledge to determine when a survey is not necessary. If a property is well monumented by definite monuments, it may be a simple matter to describe it simply calling for the monuments.

Today, especially in urban areas, the demand for title policies that guarantee both title and possession has increased the necessity for both accurate and precise surveys. In many areas, new divisions of land are prohibited without surveys.

15.20 Desirable Qualities of a Scrivener

Competent scriveners of property descriptions, as a minimum, should have or develop the following seven attributes:

1. A knowledge of trigonometry, geometry, and elementary mathematics.
2. Knowledge of the legal meaning of all words and phrases used in land descriptions.
3. An appreciation for the authority of the location of known controlling monuments in the area of the description.

4. Knowledge of the science of measurements and calculations to prove the correctness of measurements.
5. An understanding of senior rights of adjoining owners.
6. The ability to understand and apply the order of importance of conflicting elements.
7. The ability to write English effectively and correctly.

Unfortunately, the activities of scriveners are seldom regulated by law, and many scriveners are incompetent to perform their duties. Those generally considered qualified are surveyors, title company employees, abstractors, and attorneys. Attorneys specializing in title work are eminently knowledgeable of the legal meaning of words and phrases used in land descriptions, and they know how to use proper conveyance forms. But they are often weak in mathematics and the science and techniques of measurements and lack knowledge of possession on the ground.

Unfortunately, such land descriptions, although legal, are often difficult or almost impossible to locate on the ground due to the uncertainty of monument locations. The skilled surveyor should be eminently qualified in the techniques of measurements, mathematics, and monument locations and the legal interpretation of words and phrases, but often he or she is weak in the knowledge of the rights of adjoining landowners. In some states, such as in the sectionalized land areas, the necessity of knowledge of seniority of conveyances is minimal, whereas in other states, such as the metes-and-bounds states, the necessity of seniority knowledge is imperative.

Better land descriptions can be made with the benefit of survey. Others occupying the area described may have occupancy rights or monuments may be in positions other than that indicated by the record.

Subdivision descriptions of land (lot and block by map) are a surveying approach to a description problem. Only the qualified surveyor should prepare such maps, although the preparation of the land description ("Lot 1, Block C of Map 1842") can be done by any person.

Perhaps, the best land descriptions of the metes-and-bounds type are prepared by the teamwork of the surveyor and those engaged in title-guarantee work. The surveyor locates the lines, describes the lines, and certifies as to possession; the title-guaranteeing agency, abstractor, or attorney seeks out possible adjoiner rights that might overlap the property described and corrects legal defects.

15.21 Changing Description Wordings

A land description should reflect the condition of the land at the time the deed was written. Yet in reality, if a description is based on a survey, the description is a reflection of the lines and corners at the time the survey was conducted. If there is a period of time between the completion of the survey and the signing of the deed, then there is always the possibility that minor or even material changes of the ground conditions could happen. Once a description is written and a conveyance is made using that description, no matter how poorly it was written, it is rarely advisable to change it, except by agreement deed between the interested landowners and parties in interest or by a court order. Generally, the object in changing a poor description is to include additional terms or descriptions that may make the future location of the property more certain or to correct errors that were discovered in the original document. But who is to say what may be added without changing the original intent of the description? A deed reading "Commencing at a white oak; thence N 10° E, a distance of 200.00 feet; thence S 80° E a distance of 300.00 feet; thence, . . .:" could be more locatable by changing the form to "commencing at a white oak located N 16° E, 300.00 feet from the southwest corner of Lot 10; thence N 10° E along a rail fence 200.00 feet to an iron pipe and disk stamped LS 2554; thence S 80°20′ W along a rail fence 300.00 feet to a found iron axle; thence . . ." But no title person would insure or accept such changes without a written agreement between the adjoining landowners or between the original parties to the original deed. The new description changes the basis of bearings (along a rail fence), adds a monument (iron axle) not intended in the original deed, and also changes the deflection angle.

A new monument added to a description often changes the intent of the description. Some items may be added to descriptions without altering the original intent and may improve the description. A description reading

> Commencing at a white oak; thence the following courses: N 10° E a distance of 200.00 feet to an old Ford axle; thence S 80° E a distance of 300.00 feet to an old Ford axle; thence . . .

is greatly improved by writing it as

> Commencing at a white oak found to be located N 16° E, 300.00 feet from the southwest corner of Lot 10; thence N 14° 20′E astronomic north (N 10° E magnetic north, 1902), as determined by surveyor Nolan, for a distance of 200.00 feet to an old found Ford axle; thence S 75° 40′ E (S 80° W magnetic north), a distance of 300.51 feet (record 300.00 feet) to an old found Ford axle; thence . . .

Since monuments control over bearing and distance, correcting the errors of measurements does not alter the intent to go to the monuments. But even in harmless instances, title people object to changes and often will not insure changes in wording.

As stated, the object of changing a deed is usually to add evidence that makes location more certain. But such evidence can be perpetuated by other means. The best way to correct errors and discrepancies in a land description is to file a public record of survey disclosing evidence found and new points set. The surveyor discovering an error in the deed wording and filing a public record puts all future surveyors and title people on notice of the facts.

The title-insuring agency, abstractor, or court can judge on the merit of the evidence. Evidence of this generation is often lost to the next generation; hence it is advantageous to have surveyors give public notice of discovered evidence and to give new tie-outs to found old evidence. This method of preserving evidence is superior to arbitrary attempts of one person to change a deed's wording.

15.22 Technique of Writing

The art of writing descriptions is not a simple task. Carelessly used words—or omission of words—can alter the entire meaning of a conveyance. This statement would also apply to the misinterpretation of an abbreviation.

Too many words may cause conflicting statements in a deed, whereas too few words may cause ambiguity and uncertainty. The best deed authors use a minimum of terms that give a clear intent without error, conflict, or ambiguity. The selection of the proper words to use comes from knowledge, experience, and practice. It is not the verbose writer filling many foolscap pages who wins acclaim in writing descriptions; the writer who is applauded is the one who condenses but omits nothing essential, creates no conflicts, and is clear.

15.23 Parts of a Description

Descriptions are divided into four parts: *caption, body, qualifying clauses (including reservations)*, and *augmenting clauses*. Although in writing a particular description the four parts may be intermingled, and it may be difficult to distinguish one from another, it is better practice, when feasible, to keep the four parts distinctly separated. In general, the distinctions between the four parts are as follows:

1. The caption recites a general area or locality and directs attention to a general vicinity.
2. The body pinpoints a particular area in the given locality described in the caption.
3. The qualifying clause takes back part of that given by the body or by the caption.
4. The augmenting clause gives a right of usage of land outside that conveyed (usually easements).

If the description author develops the habit of following this order, fewer mistakes will be made.

An example of parts of a description follows:

(Caption) All that portion of northeast one quarter of Section 30 of Rancho de la Nacion according to Map thereof No. 166, surveyed by Morell and filed in the office of the Recorder of San Diego County, California, being more particularly described as follows.

(Body) One acre of land forming an equal-sided parallelogram at the southeast of said quarter section 30 of said map.

(Qualifying clause) Excepting therefrom the easterly 100.00 feet. *(Augmenting clause)* And granting an easement for road purposes over the westerly 25 feet of the above excepted easterly 100 feet.

This description is not dimensionally complete, since the angle in the southeast of quarter section 30 is not recited; also, no monuments are described. This could be a critical flaw when it comes to being able to retrace this parcel.

Caption A logical arrangement in writing descriptions is to first recite the general area or locality of the land so that attention is directed to a vicinity. The caption or introductory part of a description serves this purpose. A checklist of the items that often, but not always, appear in the captions is as follows:

1. State
2. County (parish)
3. City
4. Subdivision
   (a) Map name and number
   (b) Meridian, township, range, and section
   (c) Land-grant name
   (d) Court map
   (e) Any other identifiable map
5. Recorded conveyance of which instant description is a part
6. Place where record map or recorded conveyance is filed

The wording of the caption may take many forms, either one of the following examples being sufficient:

> All that portion of Section 10, Township 15 South, Range 2 East, San Bernardino Meridian according to the United States Government plat filed January 12, 1885, located in county, state of . . ., more particularly described as follows.

> All that portion of Lot 12, Block 15 according to tract Map No. 16.213 recorded in the office of the County Recorder of Los Angeles County, California, more particularly described as follows.

Body of a Description The body of a description identifies a particular land area within the locality designated by the caption. The body, taken together with the caption, must identify a certain area; otherwise, the conveyance will be void. In addition, the body ought to contain complete dimensional information that renders the intent of the deed more certain and clear.

Monuments called for in a conveyance locate the land. However, monuments are subject to destruction. If a monument is to be replaced after it is destroyed, its measured position must be known before it is destroyed. Often land areas can be described without mention of measurements (calling for monuments alone). Such a description, although legally valid, is unsatisfactory, since no measurements indicate how to replace monuments if lost. The surveyor in describing land is vitally interested in the ease of future locatability; he or she includes writings designed to serve as a means of identification in the event of monument destruction.

Courts, when interpreting the meaning of conflicting elements within a description, presume an order of importance. Explanations of how to write the body of descriptions follow this order of importance.

It is in the body of the description that the surveyor should place the major controlling elements. Having an understanding of what courts will rely on when litigating a boundary issue. The minimum elements that should be included in the body of the description are:

1. If a survey is conducted, call for the lines surveyed.
2. Call for any natural boundaries followed or natural monuments created or found.
3. Identify any artificial monuments found or set.
4. Call for any adjoining parcels necessary and controlling.
5. Call for bearings—with reference meridian
6. Call for distances—with method of creation.
7. Call for area.

Qualifying (Habendum) Clauses This portion of the description may perhaps be considered as being used to forestall future legal problems if the description elements are ever questioned. If elements or prior phrases in the body leave the attorney, the court, or the surveyor with ambiguities, then this clause can be used to eliminate any possible ambiguities. For instance, if a portion of the description calls "and thence along the road," this clause can be used to explain that the intent is to have the boundary as "following the centerline" or it could be used to show the "intent of the line from corner 5 to corner 6 is intended to be the thalwag of the creek as of June 9, 2010." There are no limitations on what is explained in this clause. Such phrases as "the lines between the respective corners" are computed distances and angles from random traverses run between proven corners of the original description.

15.24 Senior Rights and Calls for Adjoining Parcels

PRINCIPLE 6. *Junior descriptions should call for the senior description as an adjoiner or should be written with calls identical to that of the senior adjoiner.*

In the order of importance of conflicting deed elements, except where unwritten titles exist, senior rights rank first. If a junior title is being written, that fact should be noted by inserting a call for the adjoiner.

Inserting a call for a senior claimant serves the purpose of title identity; it states the relationship between the instant property and the neighbor.

Senior deeds should never call for the junior claimant. If Jones's deed reads "bounded on the south by Brown" and Brown's deed reads "bounded on the north by Jones," confusion results. One or the other is senior, yet the fact is not disclosed; it appears that both are senior.

In Figure 15.2, Black is senior and Smith is junior. A description of the new parcel being described and reading "beginning at the southwest corner of Lot 8; thence N

35° E, 200 feet along the line of Smith as recorded in . . ." is wrong in that it implies that Smith has the senior right.

When writing a new description, it is better and probably more certain to start writing it from a survey that is referred to the senior or parent parcel, because this gives an instant reference to the senior parcel.

The necessity of calling for an adjoiner can be avoided if the calls are identical. In Figure 15.2, if Black's description reads "Beginning at the southeast corner of Lot 8, thence West 500.00 feet; thence N 35° E, 665.25 feet; . . ." and Smith's deed, at the time it was written, commenced at the same point and called for both "West 500 feet" and "N 35° E, 665.25 feet," the two parcels would join, and the necessity of calling for Black as an adjoiner could be eliminated. But Smith's deed should not read "beginning at the southwest of Lot 8; thence N 35° E, 665.25 feet; . . ." A surplus or deficiency in lot 8 would cause a gap or overlap. If identical calls are not used, a statement "beginning at the most westerly of (Black's land); thence N 35° E along (Black's land) . . ." is legally satisfactory but not dimensionally complete on the face of the deed itself.

In addition to serving as a means of title identity, a call for an adjoiner serves to prevent a gap or overlap; it makes certain the last described parcel coincides with the first described parcel. Here, again, in describing a new parcel from a part of a senior claimant's land, the description should not call for the junior land, even though the two are intended to be coincident along the common line. Let the junior deed call for the senior.

When the adjoiner is senior, the call for the adjoiner usually takes the form "N 16° E, 1227.32 feet more or less to the southerly line of that land conveyed to Thelma Brown and recorded in . . ." But if the adjoiner is a junior claimant and a new call is being made that will reach the perimeter of the instant property, the call takes the form "N 16° E, 1227.32 feet to the northerly line of this parcel being divided . . ." No call is made for the junior claimant. A call for a lot line within a subdivision is a

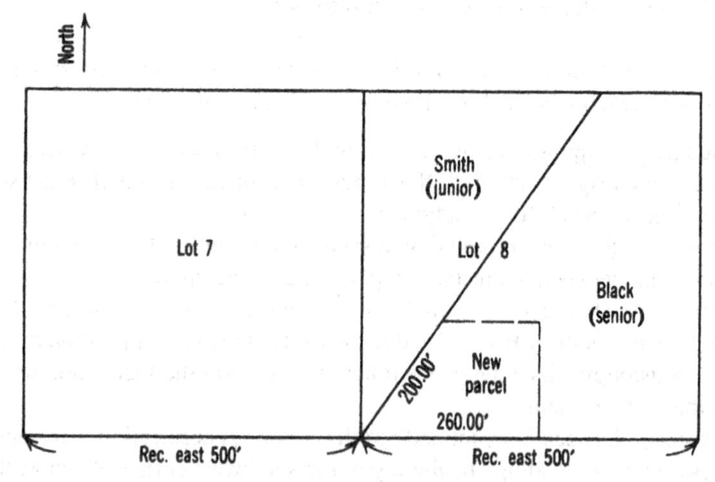

FIGURE 15-2. Calls for junior/senior deeds.

call for a record monument, but it is never a call for a senior claimant. Lot lines are common to both lots, and both lots have equal claims to the line. A call of "N 10° E, 212.12 feet to the north line of Section 12" is considered as calling for a record monument, and the call is presumed to have force over measurements. A call of "to the north line of Section 12 and along the north line of Section 12" is for the purpose of preventing gaps or overlaps, not to disclose seniority, since none exists. It is a useful technique, if properly used.

It has been said by late William C. Wattles, a well-known title authority, that a remainder description should have sufficient title identity to be able to locate the land if all dimensions were omitted. Bounds descriptions are often used for remainders.

15.25 Intent of a Description

The intent of the description or conveyance can either be placed in the body of the description or in the habendum clause at the end.

When preparing a new description, the intent of the scrivener should be to describe the elements (e.g., monuments, line distances, and angles) as they are at the time the survey was conducted and hopefully at the time the description was created and the deed was executed and delivered.

In the order of importance of conflicting deed elements, the intent of the parties to a conveyance, as expressed by the writings, is the paramount consideration of the court, senior rights excepted. In composing a description, the objective of the deed author is to describe exactly and correctly by writings the intent of the grantor. At a later date, all the verbal evidence in the world will not change the intent as expressed by the written words. Also, any ambiguous terms usually will be construed most strongly against the grantor or the individual who wrote the document.

PRINCIPLE 7. *Before attempting to describe land, the scrivener first must carefully determine the intent of the parties and then select words, keeping in mind the presumed order of importance of description elements which will precisely describe that intent.*

PRINCIPLE 8. *A legally sufficient description is one that a competent surveyor can locate on the ground.*

Using the description of the property to be surveyed, the surveyor's responsibility in a retracement is to first put meaning to the words. Then the surveyor must manifest those words into actions and the final placement of certainty on the ground of the parcel described. The final placement depends on the training, experience, and capabilities of the retracing surveyor, as well as the amount of time and costs he or she is willing to put forth.

15.26 Call for a Survey

PRINCIPLE 9. *If the intent is to describe land as following a particular survey, that fact must be stated and the survey must be identified and should be specifically called for.*

A call for a survey can be direct as "according to the survey made by Jones and dated 9 June 1930" or indirect as "according to Map No. 1272 of Lakeview Terrace," wherein the Lakeview Terrace map states "surveyed by Jones." In the Public Land Survey states, a description reading "T 10 S, R 2 E. Sec. 12. SBM" automatically calls for a survey, by case law, and that is the survey that was caused to be made by the government when Section 12 was created.

If a survey is made by a surveyor prior to writing a description and if the owner wants that survey to be a consideration of the conveyance, that survey must be called for in the description; otherwise, it will not be a consideration of the writings. If a public record is filed of the survey called for, all the evidence disclosed on the plat is incorporated into the deed.

Frequently, the platting laws of a state require a survey for all subdivisions. If the law does require a survey for a subdivision, the presumption is that a survey was made, and any call for that subdivision also calls for the presumed survey. It is thus possible to call for a survey by reference to another document without expressly saying that a survey is called for.

In calling for a private, unrecorded survey, the effect of the survey on the interpretation of the conveyance is often dependent on the description wording, such as "beginning at the southwest of Lot 10 as marked by a 2-inch iron pipe with brass disk stamped LS 2554; thence N 10° 05′ E, 200.00 feet to a 2-inch iron pipe with disk stamped LS 2554; thence N 09° 22′ E, 301.54 feet to a spike driven into an oak tree; thence . . . according to the survey made by Brown."

The calls for monuments, bearings, and distances describe the survey and, on recording, make such a matter of public record. Merely reciting bearings and distances without calls for monuments set or discovered deprives a called-for survey of a great amount of legal force. What the surveyor did must then be proved by testimony, and after the surveyor is gone, this is difficult.

Incorporating the data given on a publicly recorded survey into a description is done by merely calling for the survey; public records of surveys are admissible as evidence. Since private, unrecorded surveys are sometimes not admissible without testimony from the surveyor, it is better to incorporate the calls for monuments in the description.

15.27 Call for Monuments

PRINCIPLE 10. *Other than senior rights and contrary expressed intentions, a call for a monument is presumed to prevail, and if the intent is to limit the boundaries to particular existing monuments, it must be so stated.*

Uncalled-for monuments are not a consideration of a description and will have no effect. Recital of bearing and distance that presumably go to a particular monument is not an absolute assurance that the monument will be a part of the boundary; the monument must be called for. Any error of measurement, transposition of figures, or miscalculation can defeat an intent to go to an uncalled-for monument.

A call for a monument, without reciting the particular part of the monument, can cause ambiguity. If there is a possibility of confusion, the better procedure is to spell out the exact intent. Calls for natural monuments (boundaries) have been a continuous source of litigation because of failure to define limits of the monument.

Such wording as "along a river," "along the shore," and "by the stream" are indefinite and should be explained even though various court decisions have specified how to interpret this wording. Instead of saying "along the Susquehanna River," say whatever the limits of private ownership are. "Along a river" can mean along the average high-water mark, along the average low-water mark, along the thread of the stream, along the main channel of the stream, or along the gradient boundary, depending on the circumstances and the state. To clarify this, the exact meaning should be stated.

"Along a highway," "in a highway," "by a highway," and like expressions are usually interpreted to go to the centerline of the highway, that is, if the grantor owns that far. The exact intent should be stated as "along the centerline of the highway," "along the sideline of the highway," or "together with any rights the grantor has in the highway." Although ambiguous statements do not make a conveyance void or even voidable, it is better practice to clarify the terms at the time both parties are consummating a sale.

A call for a tree, a stake, a rock mound, or like object is construed to the central part of the object, and if a different intent is wanted, it must be clearly stated.

Where a deed conveyed land east of a line beginning four rods east of a tree, the distance is to be measured from the center of the tree [8].

15.28 Lines

Invisible lines constitute the perimeter of all described land areas. These invisible lines may be

1. Straight
2. Circular curved
3. Spiral curved
4. Curved to fit the intersection of the Earth's surface with a given elevation (waterlines and contours)
5. Gradient (along rivers)
6. Of geometric relationship (e.g., parallel with)
7. Geodetic (follow latitudes and longitudes)
8. Irregular (along a creek bottom, along a fence, along a road)

Except for those continuous lines described as following the perimeter of a geometric area (e.g., circle, square), a line description forming one side of an area must describe a definite "starting point" and describe a definite "terminus."

Lines are assumed to be straight, except where otherwise qualified. These lines are irregular:

> "Southerly along the ocean and easterly along Boulder Creek."
>
> "In a general northerly direction along the westerly boundary line of that land conveyed to James Case in . . ."
>
> "Northerly and easterly, parallel with and 300 feet distant from the westerly and northerly side line of the railroad right-of-way . . ."

To positively "fix" a line that originally was monumented by corners on each end of the line and that was also described by a bearing and a distance, *both* of the original corners must be recovered. If only one corner was recovered and a bearing and a distance were used to locate the second, there is error in the location of the second due to the measurements. Two original monumented corners must be used to definitely locate a line.

Straight Lines For a local area, a straight line forming one side of an area is the shortest measurable distance between its starting point and its terminus, unless otherwise stated in the writings; it is a horizontal line. Technically, all horizontal lines are curved lines following the curvature of the Earth's surface, but the effect of curvature in local areas is not discernible by plane land survey methods.

In the absence of something in a deed to cause a deviation, a boundary line between two points is presumed to be a straight line [9].

A straight line along the perimeter of an area can have either its beginning, terminus, direction, or all three defined by five elements:

1. Monuments.
2. Direction and distance from a given point.
3. Coordinates as based on a defined grid system.
4. Latitude and longitude as defined by the Earth's poles.
5. Geometric relationship (e.g., parallel with).

A problem may occur on a retracement when the description calls for a straight line between two corners, but the retracing surveyor finds original line marked from the original survey, that departs from a straight line. This decision has bothered surveyors: "What do I do?" Report the facts, and let the court determine the outcome.

Straight Lines Defined by Monuments All land descriptions must somehow be related to monuments. Technically, all directions in land descriptions are defined relative to the direction of monuments, and those monuments can be the polar axis and the Prime Meridian or any two suitably defined points or physical objects. The length of a line can be limited by either monuments or distance.

By defining any two fixed monuments, one as the beginning and the other as the terminus of a line, sufficient information is given to identify one side of an area, and the recital of distance and direction is not legally essential. Because monuments are so frequently disturbed and because distance and direction are aids to replace lost monuments, it is always desirable that measurements be quoted, provided, of course, that the distance and direction quoted are based on correct measurements. Quotations of incorrect measurements create ambiguities and confusion and invite litigation. Quotations of approximate distance and direction do serve the useful purpose of defining a general area in which to search for a monument. But, as so often happens, after a monument is destroyed and witness evidence is not available concerning its original location, the approximate measurements become absolute, a fact not intended.

In past years, the most fruitful source of land litigation has been the discrepancies between lines defined by both monuments and measurements. One of the principal functions of the surveyor is to ensure harmony between monument location and measurements. The reason surveyors so often compose descriptions is that they are capable of determining monument–measurement relationships.

Record monuments (call for adjoiners) can be used to define the direction, distance, termination, or start of a line. "Along Jones' easterly line" defines direction and "to the northeasterly of Jones' land" describes the termination of a line. The recital of record monuments is to prevent gaps and overlaps and is a very desirable feature of a good description. But the practice of reciting the adjoiner without mention of measurements, although sufficient, is undesirable in that owners do not know the quantity of their land.

Straight Lines Defined by Dimension from a Point A straight line forming the side of an area can be defined by a direction and a distance. The starting point can be either a monument, the termination of the preceding line, or a point on a previously defined and described line.

Which is more important, distance or direction, varies from state to state, as well as between federal and state courts. For metes-and-bounds descriptions, the majority of the states hold that direction is more important. Generally speaking, unless a monument is called for, both distance and direction are necessary to determine a bound. To definitively fix a retraced line with absolute certainty, the surveyor must recover both monuments on each end of the described line.

In writing a description, the surveyor should never fall victim to demands from clients or attorneys. Consider the surveyed lines "Thence to a point" or just "251.00 feet, more or less." The first "to a point" is possibly surveyable, "but 251.00 feet, more or less" may render a description void.

15.29 Basis of Bearings for Direction

PRINCIPLE 11. *Directions by bearings are relative to whatever the writings define as the basis of bearings; astronomic north is merely one of the many ways to define directions and is sometimes assumed to be the basis of bearings in the absence of other evidence.*

Direction definition in writings is probably the greatest source of misunderstanding. Too many people assume that directions, as defined by bearings, are always relative to astronomic north, an assumption far from the truth.

Direction is whatever the deed defines it as being; the astronomic definition of direction is merely one of many definitions. Direction can be defined as relative to physical monuments, and since this method entails less expense and time, when describing new parcels, most surveyors resort to this method.

Surveyor Rogers monumented (see Figure 15.3) a new parcel northwesterly of lot 13 as based on the found, original corners of the lot. On the original map, the westerly line was described as north 600 feet and the northerly line was described as west 600 feet. The measured angle at the NW corner as based on the original monuments, was 89° 58′. The new description as written reads:

> Beginning at the northwesterly corner of said Lot 13: thence East along the northerly line of said Lot 13 a distance of 200 feet: thence S 0°02′ E 200 feet, thence West 200 feet to the westerly line of said Lot 13: thence N 0°02′ W along the westerly line of said Lot 13 a distance of 200 feet to the point of beginning.

Translating this literally, the description author is saying,

> This description commences at the northwesterly of lot 13, wherever it may be, and travels along the northerly line of Lot 13 in the same direction traveled by the original surveyor. While no true determination of the direction of this line was made, if, on

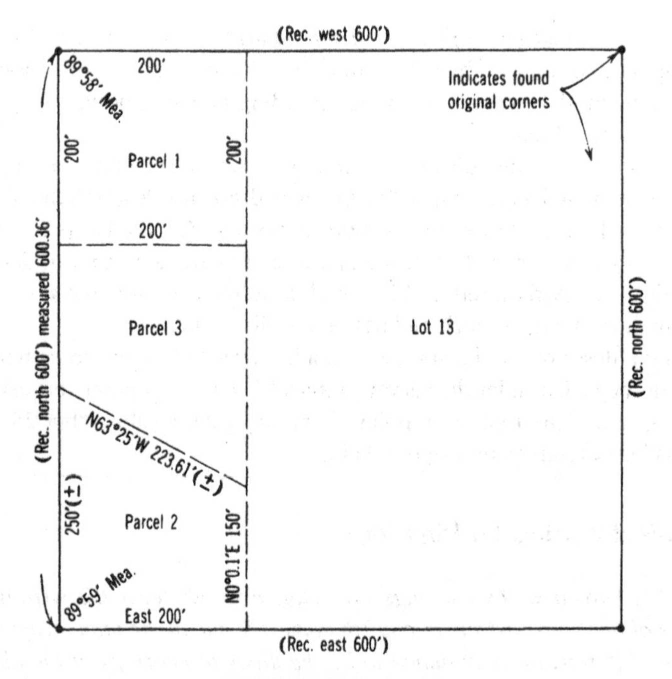

FIGURE 15-3. Lot 13.

actual determination, it is found that this line is not astronomic east, for the purpose of this description, irrespective of the true direction of the northerly line, the direction of the second line is determined by the angle formed by the mathematical difference between east and S 0° 02′ E.

The basis of bearings is the line as run by the original surveyor, not the astronomic east as purported to have been run by the original surveyor. If the original surveyor erred, that error is perpetuated.

If the description is written

> Beginning at the northwesterly of said Lot 13; thence N 89°58′ E, astronomic (record east), along the northerly line of said Lot 13 for a distance of 200 feet; thence.

then the deed author is saying

> The true astronomic direction of the north line of Lot 13 was determined and the original surveyor was found to be in error, but irrespective of this fact, the basis of bearings for this description is N 89° 58′ E along the northerly line of Lot 13 as originally monumented, and other lines will be relative to this basis.

In the event of destruction of the monument at the northeasterly corner of Lot 13, "N 89°58′ E, astronomic" is evidence to aid in reestablishing the lines.

In Figure 15.3 the description of parcel 2 reads:

> Beginning at the southwesterly of Lot 13, thence East 200 feet along the south line of Lot 13; thence N 0°01′ E parallel with the westerly line of Lot 13 a distance of 150 feet: thence N 63°25′ W, a distance of 223.61 feet to the westerly line of Lot 13; thence S 0°01′ W along the westerly line of Lot 13 a distance of 250.00 feet to the point of beginning.

The problem is to write a description of parcel 3. Either north, west, east, N 0°02′ W, or N 0°01′ E can be used as the basis of bearing. Whichever one is used, it should be consistent throughout the description. The body of this description could be written as follows:

> Commencing at the northwesterly corner of Lot 13: thence South along the westerly line of Lot 13 a distance of 200.00 feet to the southeasterly of that certain parcel conveyed to Smith by, said being the True Point of Beginning; thence S 89°58′ E (record east) along the southerly line of said Smith's land, 200 feet to the southeasterly corner thereof: thence South, a distance of 250.18 feet to the northeasterly corner of that certain parcel of land conveyed to Barry by [. . .] thence N 63°26′ W (record N 63°25′ W) along the northeasterly line of said Barry's land 223.61 feet, to the westerly line of Lot 13; thence North along the westerly line of Lot 13 150.30 feet more or less to the true point of beginning.

This description as written is based on an assumed fact: The westerly line of Lot 13 is in fact north as originally called for by the original surveyor of Lot 13. The fact that the line is later proved to be other than astronomic north will not alter the intent to make all lines relative to the assumed north called for.

Basis of bearings may be relative to the following reference meridians (the one the surveyor uses should be indicated):

1. Astronomic north
2. Grid (state plane coordinates) north
3. Local grid north
4. Any two defined monuments
5. Magnetic north
6. Adjoiner's recited bearing on a given line of the adjoining description
7. A previous survey
8. Bearings quoted on a map
9. Arbitrary, assumed

One of the essential functions of the scrivener is to clearly indicate the basis of bearings used. When no basis of bearings is stated or implied, the general presumption of the state will usually be invoked, as in a New Hampshire case: The courses in a deed are to be run according to the magnetic meridian, unless something appears to show that a different reference is intended [10].

Changing the Basis of Bearings In land descriptions, it is assumed that all bearings within the same description are on one basis. Writing new descriptions by assembling the data contained in adjoiner descriptions may result in lines being described by unrelated bearings. Orcutt, Brown, and Jones owned land that was described respectively by astronomic north, grid north, and magnetic north. A description of the remainder reads:

> Beginning at the southeasterly comer of Orcutt's land as described in . . . thence along Orcutt's easterly line, N 12°02′ E 200.00 feet to the southwesterly of Brown's land as described in . . . thence N 89°01′ E along said Brown's land 129.82 feet to the northwesterly corner of Jones' land as described in . . . thence S 0°03′ E along said Jones' land 200.00 feet: thence. . . .

In this particular location the grid and astronomic north varied by 0°31′ (east of north), and the angle as figured by the difference between N 12°02′ E and N 89°01′ E is 0°31′ in error. The difference between N 09°01′ E and N 0°03′ E is incorrect by several degrees because of the magnetic declination.

Lines described both by monuments (physical and record) and bearings always have a potential difference in basis of bearings, and unless the lines are related by accurate angular measurements of a recent survey, ambiguities may result.

It will seem to the retracing surveyor or the expert surveyor who is engaged to testify in court that undue emphasis is given to bearings. Any person who understands bearings will find that there will be instances when frustration is encountered when explanations are fruitless as to the reasons bearings vary so much.

In the retracing process, courts tend to give greater legal weight to bearings in the evidentiary scale, because early in the creation of boundaries, the most-educated person and probably the best-trained individual "ran" the compass. With total stations today, this is not so.

Azimuth Direction by azimuth is seldom used, except in Hawaii, and is not recommended for descriptions. Azimuth is a horizontal, clockwise angle recorded from a fixed direction such as astronomic "north." Azimuths are not divided into quadrants but vary from 0° to 360°. Either north or south may be used as an origin of azimuth. In any description using azimuth, the direction of the origin line must be specified.

Azimuths are not recommended for descriptions, even though they may be legally sufficient.

Deflection Angles Almost all property surveys today are performed by turning angles with a transit or theodolite. Descriptions can be written without converting the angles to bearings. Directional calls for the angles turned must be stated because four possible directions exist for a given angle. Normally, the angle is described from the prolongation of a line as follows:

> Thence 20° 01′ to the right from the prolongation of the last described course, 200.00 feet; thence northeasterly 624.26 feet along a line deflected northerly 16° 02′ from the prolongation of the preceding course; thence southerly 221.02 feet along a line deflected 168° 22′ from the last described course.

In the last instance the angle described is not from the prolongation of a line. The form "thence 20° 01′ to the right 200.02 feet" is used, although not clearly. The "to the right," as commonly used by surveyors, means to the right from the prolongation of the previous course. The longer explanation is better and clearer.

Although the technical methodology does permit this method of describing lines, it is not recommended by the practicing surveyor.

Straight Lines Defined by Coordinates Points on the Earth's surface can be defined relative to a coordinate system, and straight lines can be defined as connecting two coordinate points. Any coordinate system used in a description must have the following two items:

1. A defined point of coordinate origin
2. A direction definition

A deed call for latitude and longitude is a call for a system defined by three monuments: the North Pole, the South Pole, and the Prime Meridian. Utilizing the most modern equipment and exercising the greatest of care will not locate a point by latitude and longitude closer than about 10 feet, and this position is subject to the deflection of the plumb line, thus introducing even greater error. Because of this, latitude and longitude have seldom been used to define private land rights, but the system has often been used to distinguish political boundaries such as between the United States and Canada and between some of the states.

The United States Coast and Geodetic Survey (USCGS), formerly the National Geodetic Survey (NGS) and now the National Oceanic and Atmospheric Administration (NOAA), has established numerous monuments with precise latitude and longitude values assigned, but such values have uncertainty about the true geodetic position, even though their relative agreement is good.

A call of "latitude and longitude based on the control established by the United States Coast and Geodetic Survey" incorporates the USCGS monuments into the deed and makes location more certain.

A suggested description using state plane coordinates follows. For the purposes of land surveying, state plane coordinate values are far superior to geodetic coordinates:

> All that portion of Section 10, T35N, R2E, S.B.M., County of ______, State of ______, according to the U. S. Government Survey of ______, and more particularly described as follows:
>
> Commencing at the southeasterly of said Section 10 as marked by a stone mound having grid coordinates X, Y of Zone 7 of State Coordinate System; thence westerly along the southerly line of said Section 10, S 89°58′ W 200.00 feet to a point having grid coordinates of X, Y of said Zone 7; thence northerly to grid point X, Y of said Zone 7; thence easterly to grid point X, Y.

Today with the advent, introduction, and use of GPSs in surveying, surveyors should realize that such instrumentation can give positions that were never realized in early surveys. Submeter precision can be obtained. Yet these systems may be suitable in creating boundaries, but one must realize that such methodology does not permit the lines to be actually run on the ground.

Directional Calls All perimeter descriptions contain definitions of how to locate a sequence of lines and, except for bounds descriptions, have a direction of travel along the lines described. In land descriptions it is never wise, even in obvious circumstances, to omit directional calls.

Directional calls are not specific but give a general direction such as "northerly" or "southeasterly." The suffix *ly* usually designates a directional call, but should be considered as being approximate:

> Beginning at the *southerly* quarter of Section 10, TI2S, R6E, Meridian; thence *easterly* along the *southerly* line of Section 10 a distance of 200 feet to the Point of Beginning of a 600-foot-radius tangent curve to the *right* from the prolongation of the last described course; thence *easterly*, *southerly*, and *westerly* through a central angle of 180°00′; thence westerly parallel with the *southerly* line of said Section 10 a distance of 200 feet; thence *northerly* to the point of beginning.

In the last call, "northerly to the point of beginning," the northerly is not essential, since "to the point of beginning" defines direction, but it is advisable to insert the directional call even when it is not essential.

A call of "along Red River" has two directions. If the river runs southerly, the call must be either "northerly along Red River" or "southerly along Red River" or "up the Red River" or "down the Red River."

The call "north," "east," "west," or "south" may be either directional or a specific bearing. If the wording is "thence east 200.00 feet to a 2″ iron pipe," the call of east is directional in a generally easterly direction to a 2-inch iron pipe. In the absence of a qualifying term, as "thence east 200.00 feet," the east is specific and is relative to the defined basis of bearing.

Any bearing call or, for that matter, any distance call, that is controlled by a call for a monument is reduced to more or less measurement status. In a call of "N 10°02′ E, 200.00 feet to a 36-inch oak tree," the N 10°02′ E is more properly directional and the 200.00 feet is plus or minus to assist in the discovery of the specific oak tree rather than either being specific calls.

Directional calls in themselves are indefinite as "southerly 300 feet." Directional calls are used with other locative calls that together make a certain boundary as "thence to the beginning of a tangent curve concave northerly." The tangent curve could be either northerly or southerly: The directional call of concave northerly makes the direction of the curve certain.

Lines Defined by Geometric Relationships Use of the expression "parallel with" another line invites caution in certain situations. A line "parallel with a street," as in Figure 15.4, introduces ambiguity at curve returns. The proper form is "parallel with ZX Street and its prolongation." The expression "100 feet northerly from and parallel with a bank of Cedar Creek" is undesirable, as shown in Figure 15.5, and is not recommended. If the creek bank is irregular, the so-called parallel line may not have a similar shape to that of the bank. The upland limits of ownership of Figure 15.5 are determined as of the day of the conveyance, and erosion or accretions cannot alter them. After erosion occurs, uncertainty of the property line may occur. The usage of "parallel with a particular waterline" is not recommended.

Individuals will use the term "parallel with the curve." This term is erroneous in that one cannot have a line parallel with a curve, but the proper term is "concentric with [or to] a curve as described."

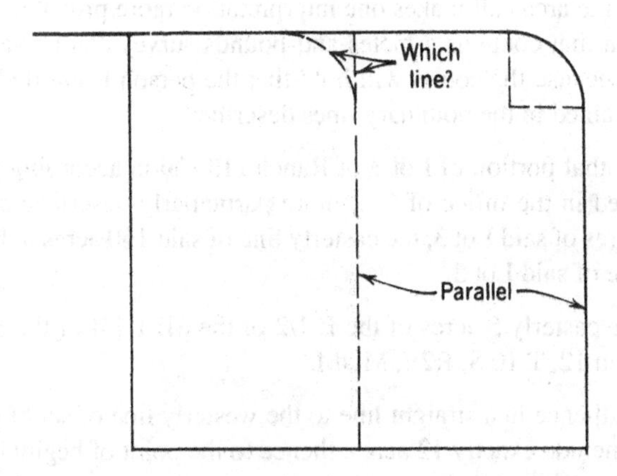

FIGURE 15-4. Parallel lines and corner cutoff.

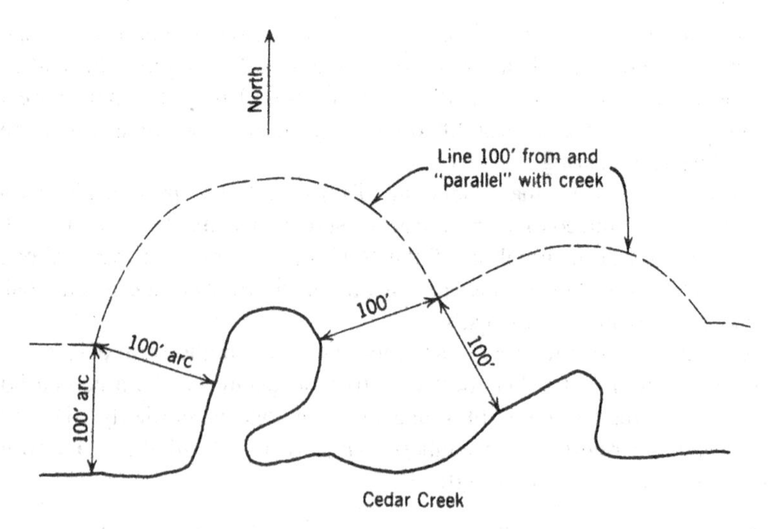

FIGURE 15-5. Lines parallel with a creek.

The form "the westerly 50 feet of Lot 10" is essentially a call for a line parallel with and 50 feet easterly from the westerly line of lot 10 and is a variant of "parallel with." Again, parallel calls with a cutoff radius or waterlines are to be avoided.

15.30 Area

If a description is complete, definite, and certain within itself, a statement that the parcel contains a definite number of acres does not alter the certain description of the parcel that is described within the lines. For all practical purposes, a description map cannot be complete without a recitation of area. Area calls have force only when a perimeter description is dimensionally incomplete or the specific description is ambiguous and the area call makes one interpretation more probable.

A description that contains a metes-and-bounds survey can be valid without the citing of area, because the courts will hold that the person is entitled to ALL of the area that is contained in the boundary lines described.

EXAMPLE 1. All that portion of Lot 3 of Rancho El Cajon according to the partition map thereof filed in the office of . . ., more particularly described as follows: The westerly 160 acres of said Lot 3, the easterly line of said 160 acres to be parallel with the westerly line of said Lot 3.

EXAMPLE 2. The easterly 5 acres of the E 1/2 of the SE 1 1/4 of the SE 1 1/4 of the SE 1/4 of Section 12, T 10 S, R2E, MDM.

EXAMPLE 3. . . . thence in a straight line to the westerly line of said Lot 13 in such a direction as to include exactly 12 acres; thence to the point of beginning.

In these examples, the acreage specified makes the perimeter complete. In the second example, the direction of the westerly line of the easterly 5 acres is not defined and should be. This type of description has been the subject of litigation, and yet a rule for direction determination in all cases has not evolved.

Also in the second example, if one-half were meant instead of 5 acres, it should have been stated. Merely because the original area was 10 acres does not prove that 5 acres is one-half of the land to be found by a modern resurvey. Very rarely does a government section contain its original intended acreage, and it is probable that 5 acres is not one-half of the record 10 acres: "Ten acres of land in Rancho Cuyamaca," is not locatable, and the conveyance is void. On the other hand, the description, "Ten acres at the northwest corner of Lot G of Rancho Cuyamaca," is 10 acres as a parallelogram and is certain and locatable.

Beginning Point and Ending Point In order to have a line, that line must have two end points. In 1687, John Love described a point as follows:

> A point is that which hath neither Length or Breadth, the least thing that can be imagined, and which cannot be divided . . .

Earlier, in 1657, a colleague of Love, Leybourn, described a point as

> A point is that which is void of all magnitude, and is the least thing that by minde and understanding can be imagined, . . .

To be absolute, a point must be adequately identified.

All perimeter descriptions, except bounds descriptions, have a point of beginning. Various words have been used such as "commencing at," "beginning at," or even "starting at." If there is a sequence of calls prior to reaching the area being described, the form used is this:

> Commencing at the southeasterly corner of Lot 10 according to Map . . .; thence S 12° 10′ E, 222.01 feet; thence N 87° 26′ E, 421.07 feet; thence S 10° 22′ E. 100.00 feet *to the true point of beginning*; thence (area described) . . .; *thence the true point of beginning.*

Sometimes the form "to the true point of commencement of the land herein described" is used. Rarely is "to the true starting point of the land herein described" used. Another form used is "Beginning at . . .; thence . . .; thence . . . to the point of beginning."

All closed-perimeter descriptions must go back to the point of beginning. In many instances a described perimeter, if mathematically perfect, will form a closed figure without reciting the last course as "thence to the point of beginning." But in descriptions, regardless of the perfection of the mathematics, a precautionary statement of "to the point of beginning" or an equivalent expression is inserted.

The beginning point and ending point must be described with compatible words: "Beginning at . . . thence to the point of beginning" or "commencing at . . .: thence to the point of commencement" or "To the true point of beginning . . .: thence to the true point of beginning" or "Starting at . . .: thence to the starting point." Descriptions do not *commence at* and then end by going *to the point of beginning*.

For a description of a parcel where the parcel is remote from the commencing point, the proper terminology is "Commencing at a point described as follows: ______, and thence to the Point of Beginning" or "True Point of Beginning."

For a parcel in which the point of commencement and the point of beginning are the same, the proper terminology is "Starting at the Point of Beginning of said parcel."

Distance of Described Line from Starting Point If a principle were stated with respect to the certainty of location of a described area, it would include this: "The uncertainty of location of the land area described is a function of the distance from the nearest known fixed monument." A description reading

> Beginning at the southwest of Rancho El Cajon, thence N 28°02′ E by astronomic bearing 10,282.00 feet to the true point of beginning: thence North 100 feet: thence East 100 feet: thence South 100 feet; thence West 100 feet to the true point of beginning

and surveyed by two surveyors working independently and with better than average care would not be located at the same point within a radius of 1 foot and probably 2 feet. This means that the staking of the description could vary in position, even by competent, careful surveyors, by as much as 4 feet and very likely at least 2 feet. Such calls, unless dictated by necessity created by previous conveyances, should be strictly avoided.

The easiest way to eliminate long calls is the insertion of calls for nearby permanently fixed monuments at the time of the conveyance. If after the words "true point of beginning" in the foregoing description there had been inserted the words "from which a leaded disk stamped LS 2554, set in an 8 × 12 foot boulder, bears N 50° 2′ E, 101.03 feet," the future certainty of identification of the intended location is increased. Of course, such calls can be determined only by those competent to measure—hence all the more reason why surveyors should be the ones to write land descriptions.

Also, to be effective, such calls must be inserted prior to conveyancing.

15.31 Curves

Circular Curved Lines Mathematically defined curves used in legal descriptions are, with minor exceptions, based on circular curves. Curved lines defined by monuments or elevations (e.g., along the mean high-tide line, contour lines) are irregular and can at best be described by approximate measurement information.

To define the location of a circular curve, the following must be specified:

- Location of the starting point of the curve
- Location of the center of the circle of which the curve is a part, or sufficient information so that the center is locatable
- Location of the terminus of the curve

No set rule exists about how a circular curve shall be described, since many forms are available to the scrivener.

The definition of the starting point of a curve can be the termination of the preceding line, a monument, a coordinate, or other suitable method that fixes a definite point on the Earth's surface.

Two points—the location for the center of a circle and the location of one spot on the perimeter of a circle—will define a circle. Curves as used in descriptions are rarely complete circles; hence, the start and end of the curve must also be definite.

Most circular curves used in descriptions are tangent curves; that is, the radial line at the beginning of the curve is at right angles to the preceding straight line; or if the preceding line is a circular curved line and the next curve is circular and tangent, the radial lines of both curves at the point of contact coincide.

The statement "thence east, 100.00 feet to the beginning of a tangent curve" does not locate the direction of the curve since the curve's center can be either north or south of the beginning point of the curve. The direction of the curve can be defined by "thence east 100.00 feet to the beginning of a tangent curve concave northwesterly" or "thence east 100.00 feet to the beginning of a tangent curve whose center bears north."

Compound curves or reverse curves are tangent at the point of change of curvature (see Figure 15.6). The usual methods of describing these curves are "to a point of compound curvature; thence along a curve having a radius of 200 feet . . ."; "to a point of reverse curvature; thence along a curve having a radius of 300.00 feet . . ."

In defining tangent curves, three elements must be given:

1. Two or three dimensions.
2. Direction of curvature.
3. Direction of travel (see Figure 15.7).

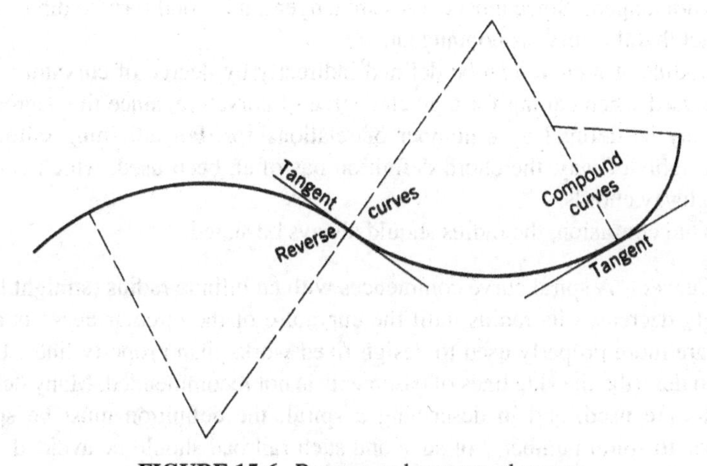

FIGURE 15-6. Reverse and compound curves.

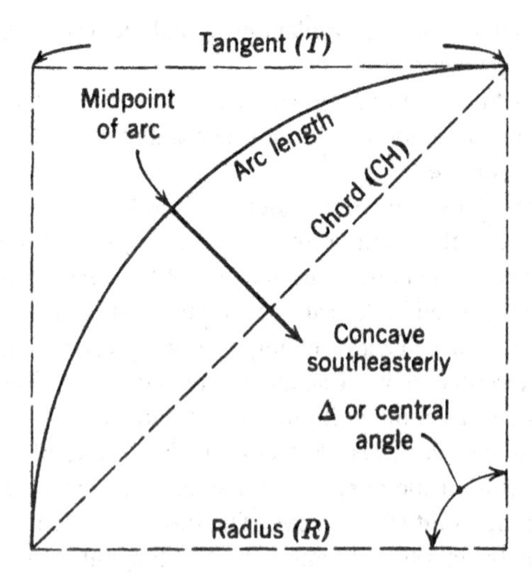

FIGURE 15-7. Direction of curvature.

An example is "to the beginning of a tangent curve of 300.00 foot radius concave northwesterly; thence northeasterly through a central angle of 100°20′ a distance of 54.01 feet . . ." Two mathematical elements of a curve are all that are needed, although three are usually quoted. Radius, central angle (delta), and curve length are more often used than are chord, middle ordinate, tangent, degree of curvature, or external distance.

A nontangent curve is normally defined by a radius bearing as "thence N 89°00′ E to the beginning of a 300.00 foot radius nontangent curve whose center bears N 10° E; thence northeasterly along said curve through a central angle of 3°01′ a distance of . . . feet: thence. . . ." The N 10° E is not at right angles to N 89°00′ E; therefore, the curve is not tangent. Since most curves are tangent, it is good form to direct attention to the fact that the curve is nontangent.

The radius of a curve can be defined indirectly by degree of curvature. Caution must be used when calling for a given degree of curvature, since the degree of curvature may be defined by a number of relationships. For adjoining railroads and railroad rights-of-way, the chord definition has often been used, which is different from highway chords.

To avoid confusion, the radius should always be stated.

Spiral Curves A spiral curve commences with an infinite radius (straight line) and uniformly decreases its radius until the curvature of the circular curve is attained. Spirals are more properly used to design fixed works than property lines. Usage of spirals to describe the side lines of easements is not recommended. Many definitions of spirals are used, and in describing a spiral, the definition must be specified. Reference to spiral number 7 of such and such railroad should be avoided; railroad records are not public records.

Calls of parallel with the centerline of a railroad as it exists are calls for all the spirals and adjustments of curvature. Since tracks have a habit of creeping toward the outside of a curve because of the centrifugal force of the train, certainty of future location is not assured by such calls.

Irregular Curves Many types of undefined irregular curves result from calls for monuments such as lakes, rivers, elevation, arroyos, and the like. Where water floods areas, as behind dams, land boundaries are often defined as all that exists below a certain elevation. In attempting to map land and determine acreage, the surveyor often describes a meander line composed of a sequence of straight lines. Elevation lines (waterlines) are rarely straight, and any attempt by the surveyor to describe a curved line by a series of straight lines is merely an approximation. This, of course, has resulted in the courts declaring meander lines as not having controlling force.

Calling for "along a monument" without dimensional information to locate the monument is unsatisfactory to the tax assessor and the new owner; the amount of land conveyed is uncertain. Specifying measurements without calling for the monument is unsatisfactory, since the intent is to go along the monument. A call of along a monument and meander information (although only approximate) is preferred. When meander information is not given, difficulty may ensue in describing the line leaving the monument. A common form used is "to the Red River, thence westerly along the Red River to a point that is S 10° E from the true point of beginning; thence N 10° W to the true point of beginning." It is never advisable to make the call for a distance along a river the controlling consideration, since erosion or accretion may change the direction of a river. A conveyance reading "thence west 200.00 feet to the Red River; thence northerly along the river 200.00 feet; thence east 200.00 feet; thence to the true point of beginning" is indefinite at future dates. The 200-foot measurement is determined as of the date of deed; but unless it is marked immediately, the location of the northerly line may be made uncertain by the whims of a flood.

15.32 Forms Used to Indicate Superior Call

Occasions sometimes arise whereby the deed author wants to indicate the superiority of one course over another. This is done by emphasizing the more important line, as:

> . . . thence N 89°22′ E, 321.02 feet to a point in the westerly line of the land conveyed to Dr. Maguire, said point being exactly 100.00 feet from the southwesterly of said Dr. Maguire's land; thence S 00°38′ E, 100.00 feet to said Dr. Maguire's southwesterly corner; thence N 15° W, 200.03 feet more or less to a point that bears N 89°00′ E, 200 feet from the true point of beginning; thence S 89°00′ W, 200.00 feet to the true point of beginning.

Each of these forms conveys the intent to make the second course controlling over the first. Repetition of statement produces this purpose.

15.33 Strip Conveyances and Stationing

Along highways, railroads, or other strip conveyances, stations are often used. Each station is 100 feet along the defined centerline of the strip; stations are numbered from some arbitrary starting point.

Stations are usually in reference to a map filed in a public agency's office, and the zero point is determined by the map referred to. The notation 24 + 16.32 means that the point is located 2,400 feet, plus 16.32 feet, from the zero point and is located on the stationed line (usually centerline or construction line). A notation of "right 24 feet from station 24 + 16.32" means that the point is 24 feet to the right from the direction of increasing stations, and the 24 feet are at right angles or radial to the line at station 24 + 16.32:

> Beginning at Station 0 + 16.32 on the centerline of Road Survey number 1632 as filed in the office of the County Surveyor of . . . County, State of . . .; thence N 16°02′ W, 50.00 feet to a point on the sideline of said road survey, said point being the true point of beginning.

Strip conveyances are often described in their entirety and easements obtained for any portion crossing a specific property. The form used is as follows:

> All that portion of land lying 25 feet on each side of the following described centerline: Beginning at the southwesterly corner of Lot 21 of Blane's Subdivision according to Map 2167 filed in the office of the . . . of . . . County, State of . . .; thence N 10°01′ W along the westerly line of said Lot a distance of 21.03 feet to the centerline of the line being described; thence N 87° E, 1222.40 feet to the easterly line of said Lot 21. The side lines of the described strip are to be lengthened or shortened to terminate in the side lines of said Lot 21.

The proper wording is 25 feet on "each side," not 25 feet on "either side." "Either side" is indefinite, since it may be on one side or both sides. Insertion of the words "a distance of" in the above description as "along the westerly line of said lot a distance of 21.03 feet" is necessary, since the wording "along the westerly line of said Lot 21.03 feet" is ambiguous. Lot 21 and Lot 21.03 feet can be confused.

One of the major problems a surveyor may encounter is a description that reads "An easement 40 feet wide of the centerline of the gas line, to be built." In this instance we have nothing but a "blanket easement" until the subject of the easement is located. Then we could also have a description "An easement 40 feet in width across the lands of John Smith." The same principle still applies; until the utility is placed in the ground, the description is still vague and indefinite. The surveyor has an obligation to positively locate the easement, and the actual position is surveyed, as it is placed in the ground, in all three dimensions *X*, *Y*, and *Z* coordinates.

15.34 Abbreviations in Descriptions and Punctuation

The use of abbreviations in descriptions is an accepted practice, but it is not encouraged that surveyors resort to abbreviations when writing descriptions. The reasoning for their use is understandable, in that "way back in the Dark Ages" of surveying and writing descriptions, everything was "writ by hand." Abbreviations saved time and effort. Today this is not so. Most attorneys and surveyors have computers and mechanical means for creating written documents, including descriptions and maps. As such, there is little place for abbreviations in today's professional world. Yes, we have used abbreviations in many of our illustrations in this book, but this has been done to try to duplicate the "real-world" conditions the surveyor will encounter. We do not encourage or recommend abbreviations today.

Certain abbreviations in land descriptions, especially on plats, are in common usage and usually are well understood. A notation of S 23, T12S, R2E, SBM is clear, certain, and identifies section 23 as existing in only one spot. But the description "W on R NE 1/2, Sec. 8, Tp. 5 north, range 4 east" means nothing. The "R" could mean road, ridge, or river [11].

In the description "sections 22 and 28, Tp. 79, R 13, Poweshiek County, Iowa," the court commented:

> It appears to us that there is no uncertainty or indefiniteness in these contractions of words. They are in almost universal use in this State in describing lands, and everybody understands that they mean "township" and "range." It is true the contraction does not state whether the range is east or west, but that was wholly unnecessary, as the land was described as in Poweshiek County, and the courts in this state take judicial notice that all of the land in that County is in range west [12].

Abbreviations, numbers, or characters in common use and understood by most people may be used in legal descriptions. But if doubt exists, it is better to spell the words out. It is recommended that all words be spelled out in their entirety. Plats frequently use abbreviations because of the greater ease of interpretation. The following abbreviations are commonly used in written descriptions:

N:	North	V:	Vara
E:	East	ft or ′	Feet
S:	South or section	in. or ″	Inches
W:	West	delta:	Central angle
Sec:	Section (preferred to S)	Ely:	Easterly
T:	Township	NWly:	Northwesterly
R:	Range or radius	°:	Degree
Ac:	Acres	′:	Minute or feet
ch:	Chain	″:	Second or inches

This list is not complete, and as a general rule, it is better to spell all words out rather than abbreviate any words in written descriptions. On maps it is desirable to abbreviate, since this enables the viewer to more quickly and accurately evaluate the map. The object on a plat is to present a graphic picture by symbols in the clearest and most certain shorthand method. Spelled-out explanations only clutter up a map and make it difficult to understand.

Very often, the certainty of meaning of the abbreviation is dependent on its association with other symbols, numbers, or lines. The letter "R" can be "range" or "radius." When used as R3E or R = 200′, no uncertainty exists about the meaning. By law, abbreviations may be used on tax descriptions, and descriptions of land sold at a tax sale are often abbreviated more than is permissible in a conveyance. For regular conveyancing, tax abbreviations are not advisable, since uncertainty will probably result.

Many states have enacted minimum technical standards that specify the approved abbreviations that surveyors must use in descriptions and on maps. There are instances when the abbreviations specified are different than the ones in general use in other places. In order to not be in violation of the minimum standards, surveyors in those states must use the designated abbreviations. Little is written relative to punctuation in descriptions. The surveyor or person who writes the description, rather than depending on punctuation to identify, should use words, if possible. The elimination of or adding such elements as a comma (,) or a period (.) could be a reason to question the meaning of the description.

An example of a comma could be "a parcel of land described as NW 1/4, NE 1/4 of section 6," describes a 160-acre parcel; without the comma it is "NE 1/4 NW 1/4," which is a 40-acre parcel.

On a similar note, a line was described as "N 14 degrees 14 feet and 27 inches W." In attempting to make the deed void, extrinsic evidence was introduced that indicated the original description was originally written as "N 14 14′ 27″ W."

The answer to this problem is obvious.

15.35 Easements

An appurtenant easement—that is, an easement necessary for the enjoyment of the land—passes automatically whether recited or not. Irrespective of this, it is always good practice to include a description of the easement as "Lot 12, Block 40, Horton's Addition . . . together with an easement for road and utility purposes over the easterly 20 feet of Lot 4." The best way to describe appurtenant easements is to separate the fee conveyance and the easement into two distinct parts.

The intent of a fee or easement conveyance may be reversed by a description "Lot 12, reserving therefrom the westerly 20 feet for road purposes." Ordinarily, the grantor reserves the fee subject to a 20-foot easement. If it is intended to convey the fee and reserve an easement, the form "Lot 12, reserving therefrom an easement for road purposes over the westerly 20 feet" is proper.

In writing descriptions for easements, the surveyor and the attorney should give considerable thought to their creation and descriptions. Courts will give strict application to the words, and as such, errors in writing descriptions can be devastating. All easements should be described with clarity and specificity. Courts will not enlarge an easement for "ingress and egress" to include utilities. Nor will they permit additional use of a "pedestrian" easement to be used by animals.

The retracement of easements is particularly troublesome in that a written easement has three descriptions that should be identified: the dominant estate, the servient estate, and the easement lines themselves.

15.36 Exceptions in Descriptions

Exceptions in descriptions are particularly prone to ambiguities. A description reading "Lots 10 and 11, except the south 20 feet thereof" may mean an exception for both lots or an exception for only Lot 11.

Double exceptions are doubly apt to be used in error. "Lot 13 except the east 12 feet, except the south 10 feet" may mean "Lot 13 except the east 12 feet and also except the south 10 feet of all of Lot 13," or it may mean "Lot 13 except the east 12 feet," except that the south 10 feet of the east 12 feet is not conveyed. This description is greatly improved by writing "Lot 13, except the east 12 feet of Lot 13 and also except the south 10 feet of all of Lot 13." Exceptions with respect to roads and rights-of-way may inadvertently be erroneously worded as, for example, "Lot 13 except M Street as opened" can be construed to mean retention of fee title to M Street. Is a fee or easement exception intended? A description reading "Lot 13, excepting an easement over the westerly 20 feet of Lot 13 as granted to . . ." is clear.

In considering exceptions, usually, but not always, the surveyor may have to look in some other historical documents to find the description of the original line of the easement.

15.37 Whole Descriptions

A whole description is made up of words, and the possible number of words that can be used in a description is infinitely variable. It would be impossible to compile a complete list of all description variations.

A whole description identifies only a particular area. It can be simple, short, and to the point, or it can be long, complex, and difficult to understand. Which one to use is the choice of the scrivener. Compare these two descriptions:

1. Lot 1, Block 4 Highland Addition to Colton, Map 1304, Riverside County, California.
2. All that portion of land located in Highland Addition to Colton according to Map 1304 as recorded in the office of the Recorder, Riverside County, California, and more particularly, described as follows: Beginning at the southeasterly of said Lot 1 as shown on said Map 1304; thence along the southerly line of Lot 1, also being the northerly line of Hope Avenue to the southwesterly of said Lot 1; thence northerly along the westerly line of said Lot 1 to the northwesterly of said Lot 1; thence easterly along the northerly line of said Lot 1 to the northeasterly of said Lot 1; thence southerly along the easterly line of said Lot 1 to the point of beginning, also including reversionary rights in Hope Avenue.

Description 2 says nothing more than is contained in description 1. The desirable feature of a whole description is brevity without loss of clarity or identity (either now or in the future). Land can never be completely described. Many volumes could be written describing the exact size, shape, and contents of every grain making up a small area of land. The deed author need not feel it his or her duty to do such writing, but at the same time, a description cannot be so short as to cause ambiguity.

How to write whole descriptions brief, concise, exact, and without error is the sum total of a large body of knowledge; it includes mathematics, law, and English customs. It cannot be completely discussed in limited manner; nor is it desirable to do so.

15.38 Whole Descriptions by Referral

The most complete descriptions requiring a minimum of words yet conveying a maximum of information are reference calls for a given map. A map is a shorthand notation devised by people to convey, by symbols, a large amount of information without the necessity of reading many words. If a map is complete, it will completely describe a parcel of land. The only knowledge needed by the deed author is how to call for the map, as "Lot 12. Block C, West Highland Addition, according to Map 1604 as recorded in the office of the Recorder, County of . . ., State of . . ." Not all states have or have had platting laws. In the past there were many parcels described as "Lot 12, Block C, West Highland Addition to the City of . . . according to the attached plat." The strength of this type of description resides entirely in the plat; and if the plat is poor, not showing monuments found or set, the description is poor.

Reference to sectionalized lands by township plat is similar in form. Because a large number of these descriptions are used, abbreviations acceptable to everyone have evolved. Undoubtedly, sectionalized land descriptions give more information with fewer words than any other possible description, for example, Sec. 2, T 2 S, R 4 E, Mt. Diablo Meridian (MDM). Not only does this describe a unique area by law, it also calls for all the original surveyor's field notes and monuments.

15.39 True Metes-and-Bounds Descriptions

The variations in writing a metes-and-bounds description are illustrated in Figure 15.8. In describing parcel A by a perimeter description, at least 12 ways of beginning can be used; a good scrivener would soon eliminate all but two. When the description is being prepared, the location of the intersection of Frog Creek with either the road or the railroad would be relatively certain; but because of erosion or accretion, it would be highly undesirable to cause a land description to be dependent on a changing point.

An exact quantitative perimeter description of this property cannot be written without a survey locating the railroad, the creek, and all of Lot 10. The owner of Lot 10 sold the following parcels:

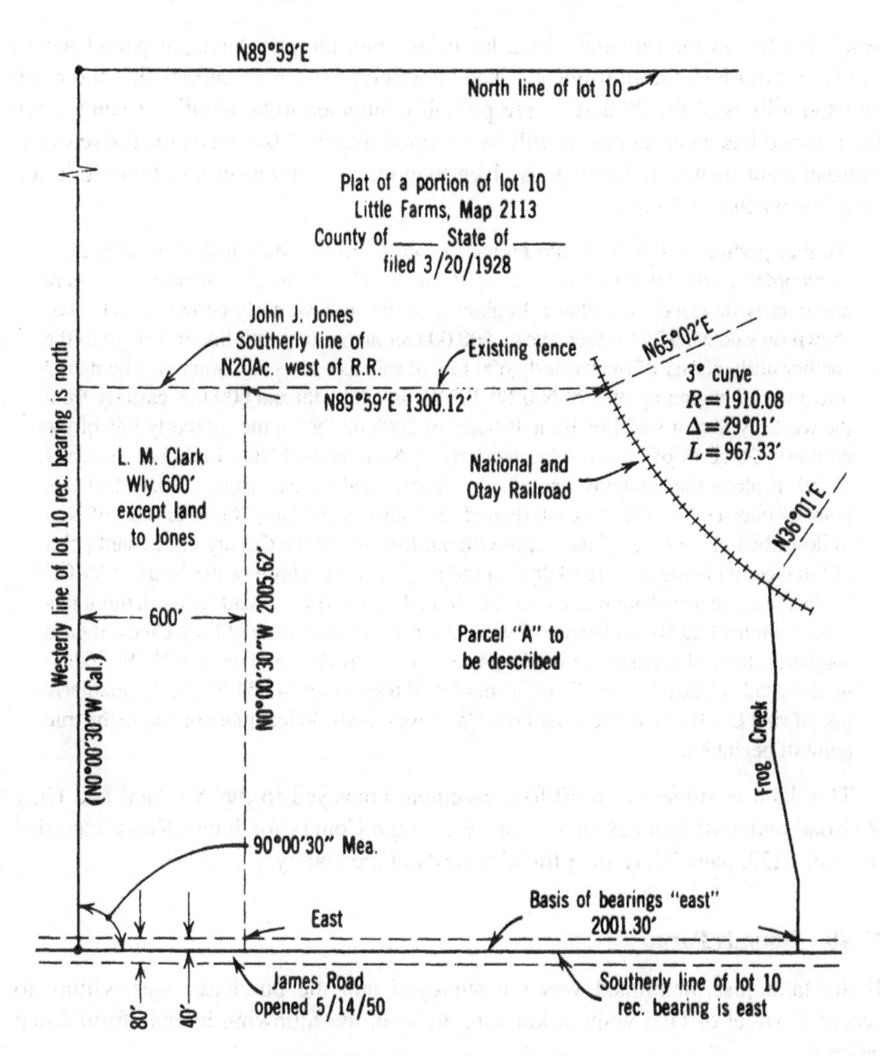

FIGURE 15-8. Lot 10, Little Farms.

> To John J. Jones the northerly 20 acres of Lot 10 lying westerly of the Nation and Otay Railroad. To L. M. Clark the westerly 600 feet of Lot 10, excepting therefrom that portion sold to John J. Jones. To the National and Otay Railroad an easement described in book . . . page . . . of Official Records.

The description to Jones is defective; the direction of the southerly line is not given. More serious is whether "the northerly 20 acres" means the northerly 20 acres of all of Lot 10 or 20 acres entirely westerly of the railroad. The fence indicates that 20 acres is occupied, and it will be presumed that Jones did acquire 20 acres. Also, since this leaves the remainder as a lesser amount, the new buyer of parcel A will not be buying into future litigation. The "northerly 20 acres" implies that the north and south lines of the 20 acres were probably intended to be parallel. Even though the railroad has an easement, it will be assumed that the 20 acres is exclusive of the railroad right-of-way (advantage is given to grantee). After a survey, the description could be written as follows:

> All that portion of Lot 10, Little Farms, according to the Map thereof number 2113 as recorded in the Office of the County Recorder, County of . . ., State of . . ., more particularly described as follows: Beginning at the southwesterly corner of Lot 10 as shown on said Map 2113; thence east 600.00 feet along the south line of Lot 10 to the southeasterly corner of the westerly 600 feet of said Lot 10: said corner also being the true point of beginning; thence N 0°00′30″ W, parallel with and 600 feet easterly from the westerly line of said Lot 10, a distance of 2005.62 feet to the southerly line of the northerly 20 acres of Lot 10 lying westerly of National and Otay Railroad; thence N 89°59′ E along the southerly line of said 20 acres and its extension, 1300.12 feet to a point of intersection with the centerline of the National and Otay Railroad right-of-way as described in book . . ., Page . . . of Official Records of the County of . . . said point of intersection being on a 1910.08 foot radius (3°) curve whose center bears N 65°02′ E; thence southerly along said curve and along the centerline of said railroad, through a central angle of 29°01′ a distance of 967.3 3 feet to the centerline of Frog Creek; thence southerly along the centerline of Frog Creek the following course; S 4°01′ E, 300.00 feet; S 2°59′ W, 261.50 feet: S 10°31′ E, 469.20 feet; south 245.54 feet to the southerly line of said Lot 10; thence leaving Frog Creek west 2001.30 feet more or less to the true point of beginning.

This land is subject to a 40-foot easement conveyed to the National and Otay Railroad and a 40-foot easement conveyed to the County for James Road, recorded in book 1232, page 23, of the official records of the county.

15.40 Bounds Form

If the land just mentioned was not surveyed and the purchaser was willing to accept a parcel of land without knowing its size, the following bounds form could be used:

> All that portion of Lot 10, Little Farms, according to the Map thereof number 2113 as filed in the Office of the County Recorder, County of . . ., State of . . ., bounded as follows:
>
> On the East by Frog Creek.
>
> On the Northeast by the centerline of the National and Otay Railroad.
>
> On the North by the southerly line of the northerly 20 acres of Lot 10 lying westerly of the National and Otay Railroad.
>
> On the West by the easterly line of the westerly 600 feet of Lot 10.
>
> On the South by the southerly line of Lot 10.

The ambiguity of the 20 acres is left just as it was found. Only the courts or the owners can change it. If corrective action is wanted, the two owners can come to an agreement, or litigation can be instituted.

15.41 Exception Form

Writing a description by exception is similar to the bounds form; instead of saying "bounded by the following" the form "excepting the following" is used:

> Lot 10, Little Farms, according to the Map thereof number 2113 as filed in the Office of the County Recorder, County of . . ., State of . . ., excepting the following described parcels:
>
> (1) All land lying east of Frog Creek.
> (2) All land lying northeasterly of the centerline of the National and Otay Railroad.
> (3) The northerly 20 acres of Lot 10 lying westerly of the National and Otay Railroad.
> (4) The westerly 600 feet of Lot 10.

The bounds form and the exception form are complete and identify the land, but neither gives quantitative data.

The advantage of the metes-and-bounds form given in Section 15.40 is that it gives definiteness to the meaning of the 20-acre parcel, and it gives quantities by which any other surveyor can retrace what the boundaries are. If the surveyor erred in the measurements, the monument calls are superior.

15.42 Monument Calls

PRINCIPLE 12. *When a monument is called for in a written description, that monument, if it is undisturbed, is controlling over other elements in the description.*

The metes-and-bounds description in Section 15.38 has one obvious defect: It does not recite the monuments set by the surveyor. At each point of change of direction, the description should state what was found or set; this would enable retracement with greater certainty.

Many times surveyors, in retracing a parcel description, will find physical monuments in the close proximity of deeds and on closer examination will discover that these monuments were not called for in the description being surveyed. The decision must be made as to what control these monuments have. To be totally controlling, these found monuments must be called for in the description; otherwise, they are nothing more than evidence and may identify property rights but not the parcel described.

PRINCIPLE 13. *Unless specifically specified, a description of a parcel of land includes all of the land within its boundaries.*

Any description of a parcel of land includes only the amount of land that is identified or described to be within the boundaries of the description. If a clear and positive description of the land is present, the only land conveyed is the area that falls within the described lines. The recitation of the area in the description, unless it is considered as being the controlling element, has no effect and cannot alter or change the exterior lines described in the conveyance.

15.43 Use of Coordinates in Descriptions

The metes-and-bounds description in Section 15.39, along with its monument calls, does not completely ensure the ability of a future surveyor to correctly locate the parcel. Monuments can be destroyed and can be replaced, provided their position is known prior to destruction. Ordinarily, when a conflict in calls exists, distance and direction will be superior to coordinates. This principle holds because coordinates, a product of distance and direction, are more subject to error. Nevertheless, the recitation of state plane coordinates in the description and on the plat provides additional evidence helpful to those interested in the location of the property.

In metes-and-bounds descriptions, grid north may serve as a reference for bearings, provided it is properly identified in the body of the description and is recognized by law in the jurisdiction. The corners can be identified by their x and y values (see sample description that follows).

It has been said that a call for coordinates is a call for all points of the system. An example of a body of a description by metes and bounds using state plane coordinates is as follows:

> Beginning at a drill hole in a stone bound that is set in a stone wall on the north line of Farm Road at the southwest corner of land of Peter Prince and at the southeast of land hereby conveyed, the coordinates of which monument referred to the Illinois State Coordinate System, West Zone, are $x = 617{,}603.29$, $y = 1{,}316{,}042{,}17$; thence on an azimuth of 81°39′ 30″ a distance of 123.39 feet along the northerly line of Farm Road to an iron pin at the southwest of the tract hereby conveyed . . . Zero azimuth is grid south in the Illinois State Coordinate System, West Zone . . .

When state plane coordinates are used in describing parcels within the U.S. Rectangular System, it will be apparent how the section or part of the section was divided. An example of such a description follows:

> A tract of land lying in Jackson County, State of Alabama, on the left side of the Tennessee River, in the south half (S 1/2) of the Northwest Quarter (NW 1/4) of Section Three (Sec. 3), Township Six South (T6S), Range Five East (R5E), and more particularly described as follows: Beginning at a fence corner at the southwest of the Northwest Quarter (NW 1/4) of Section Three, a point accepted by landowners as being the true center of the section whose (Coordinates N 1,470,588; E 416,239), said being north six degrees twenty-four minutes west (N 6°24′ W) twenty-six hundred (2600) feet from the southwest corner of Section Three (3) (N 1,468,004; E 1,416,529), and a corner to the land of T. E. Morgan; thence to Morgan's line, the west line of Section Three (3), and a fence line, north five degrees thirty-three minutes west (N 5°33′ W) thirteen hundred and four (1304) feet to a fence (N 1,471,886); E 416,113. . . .

The coordinates referred to in this description are for the Alabama Mercator (East) Coordinate System as established by the U.S. Coast and Geodetic Survey in 1934. The Central Meridian for this coordinate system is longitude eighty-five degrees fifty minutes no seconds (85°50′00″).

When state plane coordinates are required on plats of subdivisions, these values serve to perpetuate the entire subdivision. Such ties are now required by ordinances in some areas.

Few realize that the accuracy of the coordinates is dependent on the precision of the instrumentation and the individuals who created them and accuracy of the control station(s) from which the traverses were run. Of the course, this also includes the distances the computed points are removed from the control stations. The further the surveyor has to run from the station, the greater will be the error of the computed points. Some states limit by statute the distance that a surveyor is permitted to run a traverse.

In spite of predictions, the law is not ready for a complete conversion and acceptance of state plane coordinates. There are a few surveyors (and some college professors) who advocate a complete abandonment of monuments, replacing them with coordinate values. These people are not in touch with reality. We must consider that the landowner has an absolute right to have his or her lands monumented and to be able to see and touch the monuments that define the property. This is far more important than making some clerk of the court happy with a flawless record. In the future, measurements may take on a relationship that bearings and distances may have a more controlling aspect to serve a meaningful purpose.

As an example, assume that two identical subdivisions exist: one a plat with bearings and distances identified on it and the other a plat with corners identified only by coordinate positions. Then apply the probability that there is one error only on each plat. In the real world, these errors may remain undetected for many years. On the first plat the draftsman transposed the distance on a line from the actual distance of 129.72 feet to 127.92 feet. On the second plat the draftsman also transposed the *X* coordinate of 512,129.72 and recorded it as 512,127.92 feet. This is the same amount of error. Then in each case on a retracement, it was determined that the corner was a lost corner.

The solutions in each case are entirely different. In the first case, the surveyor notices the discrepancy and runs closures of all the lots that are affected by that corner and the entire perimeter: The error of closure may indicate the transposition. The cause may be ascertained and reported. Checks can be made on the erroneous distance, and its location can be placed, with a great degree of certainty, at the fence.

In the second instance, the story is much different. The data indicating the lots do not show an error. All of the references are to the same erroneous point. Based only on the numbers, the surveyor locates the point 1.80 feet from an existing fence, which according to the adjoining owner was placed there when the monument was still in existence. Now, one landowner wants the entire 1.80 feet in his description and the other owner refuses to accept that "coordinated" point because she knows it is wrong. The landowner cares little that the coordinates were established by someone with a PhD in geodesy or that the closure met the requirements of the USCGS (now NOAA) for first-order control.

Periodically, the point of reference for geodetic references changes. Whenever coordinates are used in descriptions, the datum of reference should always be noted.

The law has not addressed the validity of using coordinates in descriptions and until the courts have spoken, these writing are only speculation as to what the "crystal ball" will tell.

15.44 Checklist for Descriptions

Because of the variable nature of land descriptions, no checklist can be thorough and complete. To be of value, a checklist should be brief, and brevity causes omissions. The following list will serve as a reminder for the more important considerations included in descriptions. Such a checklist can be an aid to any person who assumes the responsibility of preparing a description that will be inserted into a deed:

1. General descriptions (caption with title identity)
   a. State
   b. City, county, or parish
   c. Subdivision: name, number, date, and place of recording
   d. Court plat: case number, date, title, or other information necessary for identity
   e. Recorded document: book and page, where filed, date, title, and other identity information
   f. Township plat: date of recording (where there is more than one plat), range, township, section, portion of section
   g. Land grant: name, date, court case number, and other identity information
2. Seniority of deeds and record monuments: a call for all adjoining senior or superior deeds or use
   a. Exact wording of the adjoiner deed when describing the line along a senior deed
   b. Call for property lines of equal rights (e.g., lot lines, section lines)
   c. Proper identity of senior deed or property lines of equal rights
3. Call for a survey
   a. Indirect: a call for a plat that calls for a survey (identify the plat)
   b. Direct call for a survey: identify where plat and field notes are recorded, date, and surveyor
4. Point of commencement
   a. Certainty for present and future identity
   b. Compatibility with previous deeds (Is it the same point as used in previous deeds, and is the call the same?)

5. Point of beginning
   a. Correct measurements from point of commencement
   b. Certainty of identity: set monument, found monument, tie-out measurements from other monuments, all identified and described
6. Call for physical monuments found or set
   a. Described for present and future identity: size, shape, material, marks on monuments (blaze, cross, license number, etc., particular means of identifying)
   b. Reference ties from other monuments by bearing and distance or other measurements
7. Call for natural monuments
   a. General identity: Atlantic Ocean, Columbia River, Grand Canyon
   b. Locative position on monument: thread of stream, mean high-tide line, along the top of the bank, centerline of Keeney Street, etc.
8. Directions
   a. Definition of basis of bearings or basis of direction (e.g., astronomic north, magnetic north, assumed, grid)
   b. All bearings on same basis
   c. Directional calls along a monument (northerly along Pine Creek, southerly along Lake Superior, easterly along the centerline of Cabrillo Road)
   d. Directional calls along geometric or irregular lines (southerly along a circular curve of . . ., westerly parallel with Milar Road; southerly, westerly, and northerly along the 330-foot elevation line; etc.)
   e. Bearings: correct quadrant
   f. Azimuth: definition of reference meridian (south or north)
   g. Grid: definition of reference meridian
9. Distance: consistent units and definition of units where necessary
10. Curves: A minimum of two elements needed; a third for a check
    a. Circular: radius, central angle, chord, arc length, radial direction, direction of concave side, compatibility of parts where more than two curve elements are given
    b. Spiral: complete definition of basis of spiral
    c. Elevation: definition of elevation datum
    d. Irregular define line (along centerline of Rose Creek, along the average low-water mark)
11. Coordinates
    a. Origin or coordinates
    b. Basis of direction: a definite and identifiable reference
12. Area—more or less except where it is controlling with one dimension omitted: Gross or net area? To what degree of accuracy?

13. Calculations
    a. All bearings and distances on same basis; figure that closes mathematically
    b. Consistency of parts (e.g., length of arc, central angle, radius)
    c. All lines and parts defined by consistent dimensions
14. "Of" conveyances
    a. Westerly 50′ of lot 2 is 50′ at 90°
    b. Westerly 1/2—define direction of dividing line
    c. Northerly 10 acres—define dividing line
    d. Sectionalized lands W 1/2 of NW 1/4 and W 80 Ac of NW 1/4 are not the same
    e. In all except sectionalized lands, a fractional part is a fraction of area
15. Strip conveyances
    a. Extend and shorten terminal lines
    b. Each side of centerline not "either"
    c. Define station and origin
    d. Make certain no gaps overlap
16. Intent: Does the description properly express the intent of the seller? Did he or she intend to sell by a survey, monuments, dimensions, or area? Does the entire description express this intent?
17. Final steps: date and signature by person who wrote the description; sealed and delivered to client

PRINCIPLE 14. *When a new description is created as a result of a retracement, the corners and lines should make reference to the elements recited in the original description.*

When a historic description is retraced and a new description is created from new field survey, in order to make the connection, rather than by referencing the document, a better way is to reference the individual elements of the original description. This indicates that the retracing surveyor found the points recited.

An example would be this:

> To Corner 5, originally Corner 2, found a 6 in. by 4 inch heart pine stake, partially burned, 9 inches above the ground with ax mark on one side. Originally described as a pine stob, 6 in. by 6. In, marked with 3 hacks. I buried the pine stake and placed a 5 inch by 5 inch concrete monument with brass cap over the stake.

PRINCIPLE 15. *No person should use the same description from one deed to another. A new description should be prepared for each subsequent conveyance.*

In the hopes of saving money and for expediency, many surveyors are asked to copy historic descriptions to be placed in a currently written description. This practice should be discouraged, in that the purpose of creating a new description is to describe the parcel including the lines and corners at the time the new deed is written and not at the time the original deed was created.

The new description should reflect the status of the corners and lines on the ground at the time of the creation of the current deed. An attorney came into a surveyor's office with a deed that was created about 1898 and wanted the surveyor to write a new description for the deed. The original deed referenced corners, lines, and an element as "thence 25.9 to "station tree."

The surveyor asked to do a new survey, but the attorney did not want to spend the money, and then asked the surveyor, "What is the station tree?" The surveyor explained this was a tree that was on the original line and was marked, and if you found the tree, you found the line. Wisely, the surveyor did not write the new description.

15.45 Summary

If a competent expert or layperson can locate land from the identification given in a land description, the description is sufficient. But legally sufficient descriptions are not always satisfactory descriptions. Good descriptions, in addition to identifying land, describe the geometric shape and size of the land identified, recite title identity, give sufficient locative calls to ensure future location, are based on a recent survey, and are clear, unambiguous, precise, brief, and certain. Proficiency in description writing is the result of a combined knowledge of mathematics and particular knowledge of the legal meaning of description words and phrases.

A well-written land description is a work of art that will withstand the test of time. Accurate and precise original surveys, properly punctuated sentences in proper business and technical English, coupled with adequate monumentation, will be a tribute to both the surveyor and the attorney for generations to come.

REFERENCES

1. *Test Oil Co. v. La Tourette*, 19 Ok. 214, 91 P. 1025 (1907).
2. *Heath v. Wallace*, 11 S. Ct. 380, 138 U.S. 573 (Cal. 1891).
3. *Walton v. Tifft*, 14 Barb. 216.
4. *Moore v. Walsh*, 37 R.I. 436, 93 A. 355 (1915).
5. *Duff v. Fordson Coal Co.*, 298 Ky. 411, 182 S.W. 2d 955 (1944).
6. *Town of Brookhaven v. Dinos*, 431 N.Y.S. 2d 567, 76 A.D. 2d 555, Affd N.Y.S. 2d 151, 54 N.Y. 2d 911, 429 N.E. 2d 830 (1980).
7. *Rivers v. Lozeau*, Fla. App. 5 Dist., 539 So. 2d 1147 (1989).
8. *Coombs v. West*, 115 Me. 489, 99 A. 445 (1916).
9. *Leigh v. LaPierre*, 113 N.H. 633, 312 A. 2d 699 (1973).
10. *Wells v. Jackson Iron Mfg.*, 47 N.H. 235 (1886).
11. *Kims v. Rolfe*, 177 Ark. 52.
12. *Ottumwa, CF & St. Paul RR v. McWilliams*, 71 Iowa 164, 32 N.W. 315 (1887).

ADDITIONAL REFERENCES

Brown, C. M., Robillard, W. G., and Wilson, D. A., *Boundary Control and Legal Principles*, 3rd ed. New York: John Wiley & Sons, 1986.

Patton, R. G., and Patton, C. G., *Patton on Titles*. St. Paul, MN: West Publishing, 1938.

Robillard, W. G., and Bouman, L., *Clark on Surveying and Boundaries*, 6th ed. Charlottesville, VA: Michie Co., 1993.

Thompson, G. W., *Thompson on Real Property*. Indianapolis, IN: Bobbs- Merrill, 2004.

Wattles, G. H., *Writing Legal Descriptions*. Orange, CA: Gurdon H. Wattles Publications, 1976.

Wattles, W. C., *Land Survey Descriptions*. Los Angeles, CA: Title Insurance and Trust Company, 1956.

Wilson, D. A., *Deed Descriptions I Have Known But Could Have Done Without*. Rancho Cordova, CA: Landmark Enterprises, 1982.

16

PROFESSIONAL LIABILITY

16.1 Introduction

Professional land surveyors are unique because they have the obligations of diligence and competence to each of their clients, but the health, safety, and welfare of the public are paramount concerns that rise above all other duties. Regardless of the type of work that each individual surveyor practices, whether locating the boundaries of small lots or designing multimillion-dollar skyscrapers, all surveyors should understand the effect of today's action on tomorrow's liability because the courts have imposed on land surveyors the certainty that they will be held liable for costs resulting from professional negligence. And unlike engineers or doctors, some states still have no statute of limitations for causes of actions against professional land surveyors. This greatly increases the risks of professional negligence for surveyors; the costs of completing surveys in these states should follow the risks involved in stamping such survey plats.

Civil courts, the licensing boards of professional land surveyors of each state, and the state statutes governing occupations all exist for one purpose: *the protection of the public's welfare*. All professionals should keep in mind that licensing statutes are for the public's good, and professionals exist to serve the public, not themselves. If it were not for the threat of future liability, some professionals who are willing to do substandard work for less money would soon lower the standing of all professionals within that community. Those surveyors who carelessly fail to protect or inform the client in all survey matters may find themselves burdened with defending alleged negligence and the costs of defending against claims of negligence.

This chapter will discuss the following principles relative to professional land surveying negligence:

PRINCIPLE 1. *The discovery rule as applied to land surveyors means that the statute of limitations for liability commences to run from the time the alleged error is discovered or after a client should have known or became aware of an error.*

PRINCIPLE 2. *Surveyors are expected to perform with the same degree of care and skill as a fictitious "reasonable person" within the same profession.*

PRINCIPLE 3. *The surveyor is liable for damages arising from expressed guarantees and from damages arising from implied guarantees expected of the profession.*

PRINCIPLE 4. *Unless there is a law permitting the surveyor to trespass when necessary to do a survey, the surveyor can be held liable for damages of trespass, either compensatory or punitive.*

PRINCIPLE 5. *To avoid liability, the professional surveyor should do more than the ordinarily prudent surveyor would do under similar circumstances.*

PRINCIPLE 6. *To prove professional negligence, one must show four elements: (1) a standard, or duty, exists within the profession; (2) the surveyor breached that standard, or duty; (3) that breach was the proximate cause of the damages; and (4) the actual damages resulted.*

PRINCIPLE 7. *If a surveyor can be proven negligent in the performance of his or her duties, punitive damages may be awarded in addition to actual and contract (compensatory) damages.*

PRINCIPLE 8. *A surveyor should not contract to do work for which he or she is not qualified, educated, and experienced to accomplish.*

PRINCIPLE 9. *A surveyor can be sued either in tort or in contract. With a written contract, the terms of the contract become the controlling elements.*

PRINCIPLE 10. *The use of another surveyor's work that is found to be substandard does not relieve the second surveyor from liability for the substandard work of the first surveyor. The liability will attach to the surveyor who uses any work product that is erroneous.*

A client employing a professional land surveyor expects a certain, minimum level of expertise. For that reason, laymen often "call around" for surveying quotes in an attempt to find the "same product for the lowest price." While this type of shopping is frustrating to the professional, it makes sense to the client because of their inability to appreciate that some land surveyors may provide a higher standard of expertise for the same job.

As a professional land surveyor, there is only one rule: Bring to the client the level of expertise required by state statute, board rules, and state case law. While one survey may be more accurate, or contain more research before a boundary conclusion, the primary concern is whether the survey plat adequately meets the minimum standards as required under the law. Surveyors who are able to foresee possible problems through the surveying process are valuable assets to their clients.

Surveyor's liability may be imposed by a variety of laws and standards. Primary sources include state statute or common law. Typically, state statutes in the form of the "acts" and state board of land surveying "rules" govern the practice of professional surveying in each of the 50 states. In the event that a statute or rule is ambiguous, the courts will intervene in order to interpret the meaning of the clauses that are in dispute.

Unfortunately, the statutes enacted by legislatures and the resulting court interpretations found within the various states may be quite diverse; the requirements to become a surveyor, survey plat elements, and continuing education standards often vary greatly from state to state. Therefore, few broad statements can be made regarding the duty of a professional land surveyor and his or her professional liability. In many states the courts have recently reversed previous decisions, possibly because of a change in the court's makeup. On the other hand, negligence theories such as *res ipsa loquitor,* balancing tests, and four elements of liability that a plaintiff must prove in order to hold a professional liability for negligence rarely vary between states.

Liability results when a surveyor fails to do what he or she purports to do. The surveyor must also know and understand the statute of limitations in his or her state, whether based on statute or current court decisions. Liability rules and their application are continually being redefined and reevaluated by the courts. Many times, these liability claims result because of a communication failure, in that the surveyor fails to keep the client informed of the progress or the situation unfolding. The surveyor should keep the client informed at all times throughout the project, not only to reduce his or her professional liability, but also because it is good business practice. Happy clients pay their invoices without argument and recommend the surveyor to family and friends

When any statute is enacted, it cannot be applied *ex post facto,* or prior to the date of the statute's enactment or adoption. When a court's opinion is rendered or handed down, it says in effect: "This is the law as it is commencing now." For example, in Kentucky, the board of registration recently attempted to reprimand a surveyor for surveys that did not meet the minimum standards the state of Kentucky had passed. The surveys the board cited as failing to meet the minimum standards were conducted prior to the enactment of the statute that gave the board the authority to "police" the surveying profession. The statute began to run on completion of the work. Surveyors were thus assured that after six years, any errors or omissions were forgiven. In another case, in 1976, a court decided that the time the statute of limitations begins to toll is six years after the discovery of an error [1]. As a result of this decision, the jeopardy of Colorado surveyors immediately changed from six years to the discovery rule. Exposure to all of the old skeletons in the survey closet was resurrected by that court decision. However, those bad surveys that had been tried by the courts could not be reintroduced. Currently, not all states have adopted the discovery rule, although it is believed that in the interest in making clients "whole" for bad surveys, most will eventually do so.

16.2 The Discovery Rule

PRINCIPLE 1. *The discovery rule as applied to land surveyors means that the statute of limitations for liability commences to run from the time the alleged error is discovered or after a client should have known or became aware of an error.*

The courts are reevaluating those individuals who hold themselves out to the consuming public as professionals. The development of this discovery rule has generally been by court interpretation, not legislative enactment. One of the first decisions applied to the medical profession, ruling that doctors could be held negligent for such actions as leaving a sponge in a patient, which promulgated the practice to include a metallic thread in all sponges so that the thread itself can be detected in X-rays. An early decision that applied to land surveyors was in Illinois in 1969 [2]. Two of the questions raised were certification and discovery. Soon the same rule spread to several other states.

Recently some state courts have ruled against the discovery rule, and in New York it was the opinion of the court that, if the discovery rule should apply, it should be by statute law, not court interpretation (common law). Courts now applying the discovery rule in recent or somewhat recent times are Delaware (1973), New York (1977), and Tennessee (1962). If in a given state the supreme court or a court of appeals has not recently handed down a decision on the discovery rule, the surveyor is uncertain of his or her position; however, some guidance might be obtained from the courts' attitude toward other professions.

A recent decision in the state of Washington [3] illustrates a modern court's thinking in applying the discovery rule. In 1962, an appellant surveyor orally agreed to survey and stake boundary lines for several lots that respondents owned on Mercer Island in King County. The parties were aware of surveying problems in the area, and when the work was completed, it physically appeared that the lines laid down did not correspond to the general pattern of other properties in the vicinity. Respondents Fischer built a house on one of the lots, sold it, and then bought it back to settle the suit for encroachment.

Two questions are presented: first, whether there was sufficient evidence to support the findings of the trial court as to liability; second, whether the court erred in deciding that a three-year statute of limitation did not bar respondents' claim because their action against appellants did not accrue until they discovered or had reasonable grounds to discover the error.

Expert witnesses were called by each side to give testimony concerning the accuracy of the survey and the methods employed by the appellant surveyor in conducting it. There was substantial evidence in the record that the appellant surveyor was negligent in his conduct of the survey and that this negligence was a proximate cause of the loss.

The controlling question is whether the action, which was based on a negligent breach of duty, was begun within the time limited by law. Until recently, the judicial resolution of this question has been that the action "accrues" when the breach of duty occurs and the start of the running of the statute of limitations is not postponed by the fact that actual or substantial damages do not occur until a later date.

Until 1969, the same rule and reasoning have been uniformly applied in legal and medical malpractice cases [4]. Then in 1969, in *Ruth v. Dight*, a specific exception was made in a situation in which a surgeon left a sponge in a patient and the discovery of it was made more than 20 years later. Previous medical malpractice decisions were overruled, and the "discovery rule" was adopted, whereby the statute of limitations commenced to run when the patient discovered, or in the exercise of reasonable care should have discovered, the injury. The discovery rule has been discussed and applied in subsequent medical malpractice cases [5].

The question of when a cause of action accrues in the case of the negligence of a land surveyor had not been decided in this state. The legislature in 1967 limited the accrual of a cause of action against a surveyor, among others, under certain circumstances, to six years from the date of substantial completion or termination of improvements to real estate but left with the courts the determination of when during the six-year period the action accrues.

Although *Ruth v. Dight* and subsequent decisions applying the discovery rule involved medical malpractice, the reasons given for the rule are as persuasive in this case where a landowner, intending to build and knowing that there were boundary location problems, obtained, for pay, the services of a professional land surveyor to locate his property.

The court saw no distinction between the medical and other professions, insofar as the application of the discovery rule was concerned. Although the damages resulting from medical malpractice are more personal in character, the pecuniary loss caused by malpractice in the other professions can be as great or greater. In this case, it was illogical to charge respondents with the obligation of retaining the services of another surveyor to prove out the stakes that appellant surveyor placed. That would be the only way that respondents could discover the error; unless, of course, the second surveyor was in error, calling for a third survey, and so on ad infinitum.

The injustice caused by the strict definition of "accrual" can be avoided by the application of the discovery rule. Of course, the legislature has the power to provide special statutory periods for malpractice cases [6].

Because of our state system of government, opinions rendered in one state jurisdiction are not binding on another state. If the case itself is one of first impression within a state, courts of other states will look to the decision for advice. The only legal decisions binding on all state court jurisdictions are those rendered by the U.S. Supreme Court on a subject granted to the federal government by the Constitution. By contrast, opinions rendered by a state court of appeals or state supreme court are primary law for courts of all levels within that particular state. Thus, any decision rendered, either of joining a previous decision or completely reversing "old law," will affect future decisions rendered relative to liabilities of professions.

In Colorado, the discovery rule was expanded even more; the statute of limitations commences running from the time damages are awarded, not from the time the person should have been aware that there were damages [1].

The facts are that James and Florene Doyle sought recovery of damages resulting from an allegedly negligent survey made by the defendant surveyor. Following a trial to the court, it found that the defendant was negligent and that plaintiffs had been substantially damaged as a result of such negligence. However, the court dismissed the action on the ground that the action was barred by the operation of two statutes of limitations. The plaintiffs appealed that dismissal.

The sole issue before the court was whether the action was barred by a statute of limitations. The appeals court held that it was not and reversed the judgment of the trial court.

In 1960, the Doyles had engaged the surveyor to prepare a boundary survey of a parcel of mountain property that they were considering buying. The surveyor knew that the Doyles had selected a site for a home they proposed to build and would buy the property if the site was actually within the plot's boundaries. In September 1960, the lot was surveyed and the surveyor staked the boundaries and delivered a certificate and plat to the Doyles. According to this survey, the desired home site was within the boundaries of the plot. In reliance on the survey, the Doyles bought the land and began building their house in November 1960. The house was not completed until 1970.

The land adjacent to the Doyle plot on the north is owned by the United States, being part of the Pike National Forest. In 1964, government agents began negotiating with the Doyles for permission to widen an access road to the National Forest, which road crossed the Doyle land. During the discussions, the agents advised the Doyles that there was a possibility that their house was on government land. In November 1964, the Bureau of Land Management (BLM) began a "dependent survey" of the boundary line. The survey was completed in June 1965 and was officially accepted and approved by the BLM on February 1, 1968. According to this survey, the Doyle house was on government land. In October 1968, the United States brought suit against the Doyles in the U.S. district court for trespass and obtained a judgment that gave the Doyles 90 days to remove the house. On appeal, the judgment was affirmed [7].

Following this, the Doyles moved the house onto their own land. The cost of this removal constituted the bulk of the damages sought in the 1973 action. The trial court found that "plaintiffs suffered $14,721.46 loss and damages, and this loss and damages were proximately caused by defendant's negligence." On disputed evidence, the court further found that on completion of the survey in 1965, the government advised the Doyles that their house was, in fact, on government land. The court concluded that the six-year statute of limitations began to run in 1965, the date of the notice to the Doyles of the government's claim.

Here plaintiffs suffered no damages until the adverse claim of the government was determined to be valid and plaintiffs were found to be trespassers, with the consequent damages resulting from their being required to remove their house from government land. The statute of limitations did not begin to run until the judgment became final in 1972. The judgment of the trial court was reversed and the cause was remanded with direction to enter judgment in favor of plaintiffs for the amount of loss and damages [1].

16.3 Privity of Contract

DEFINITION. *In an action for damages, privity of contract means only those who were parties to the contract may bring suit.*

Along with the development of the concept of the discovery rule has been the erosion of the concept of privity of contract. The old rule, approved by most states, was that only those with privity of contract could sue in the event of damages. On the sale of a property, unless the new buyer was part of the contract of survey, he or she was barred from claiming damages from the surveyor. The newer court opinions in several states is this: Those in privity of contract and those to whom the surveyor can reasonably foresee as having a right to rely on the surveyor's work can sue for damages. This opens a Pandora's box. Now title companies or lawyers who relied on the survey in issuing insurance or opinions, the contractor who built in accordance with the monuments, the next buyer who relied on the monuments, and any others who had a right to rely on the monuments, if damaged, may be able to sue.

The impact of the discovery rule, coupled with the erosion of privity, has had an astronomical effect on the cost of errors and omission insurance. The fee of one California firm jumped from $100 per year in 1947 to $7,500 in 1965, and the deductible changed from $0 to $3,000 per occurrence.

To date, California, Pennsylvania, and Illinois have adopted expansion of the concept of enlarged privity, and it is probable that many more states will do the same. In Tennessee, the rule of necessity of privity of contract was apparently approved and the discovery rule not adopted [8].

Defendants made the survey of a parcel of land "for a former owner" in the year 1934. The record of the survey described the parcel by metes and bounds and showed it contained 2.3 acres, more or less. Twenty-four years later, in 1958, the plaintiffs purchased the parcel, relying on the accuracy of the survey and description.

In May 1960, the plaintiffs learned of the errors in the survey, which were that their west line is 19.6 feet shorter, their north line 0.8 foot shorter, and their east line 13.1 feet shorter than shown in the 1934 survey. The area of plaintiffs' lot was less than that shown by the survey, and they are entitled to damages for the deficiency in the area.

The question was whether the defendants, in surveying the lot for the then owner in 1934, owed a duty of care to any remote purchaser, not in privity, who might purchase it 24 years later.

Earlier the court stated, "It is true the old rule was that there was no duty of care upon a defendant to a plaintiff not in privity. But it can hardly be said that such a general rule any longer exists" [9].

The court also stated:

> On principle and authority, we think the rule of liability cannot be extended to a case like that before us. If these surveyors could be held liable to such an unforeseeable and remote purchaser 24 years after the survey, they might, with equal reason, be held liable to any and all purchasers to the end of time. We think no duty so broad and no liability so limitless should be imposed [9].

16.4 Standard of Care

PRINCIPLE 2. *Surveyors are expected to perform with the same degree of care and skill as a fictitious "reasonable person" within the same profession.*

By virtue of their license, surveyors are not restricted in the areas in which they may practice. A surveyor licensed in his or her respective state has the ability to practice in all areas of land surveying. The professional may perform property surveys, GPS observations, first-order leveling, LiDAR scanning, in some states, drainage and percolation studies, all under the umbrella of a professional land surveying license. While it is legally permissible to undertake all of these different land surveys, once a surveyor undertakes any project outside of his or her area of surveying expertise, the courts will assume that the surveyor will practice with competence and diligence when doing so. That means that the surveyor will exhibit a degree of skill, care, and ability equal to or above the minimum requirements of that type of surveying. No quarter will be given to surveyors who conduct certain classes of surveys only incidental to their general practice.

Usually, when a professional's work is questioned, a court will require testimony of an expert who practices within the same field as the professional. This expert testimony is used by the court to establish a reasonably degree of care owed by the professional to his or her client. Essentially, the expert informs the court as to the standard of practice within the particular profession through the presentation of an expert report and testimony at depositions and/or the trial. This expert testimony, whether written or oral, is based on standard primary sources, such as court cases and state statutes, as well as secondary sources, such as textbooks, journal articles, or a continuing education presentation created by fellow nationally accepted professionals at various live and online conferences. The expert witness is expected to set a certain level of competence using such objective sources as are described above and then explain to the courts whether the professional surveyor in question has met those standards.

In a court claiming damages against a surveyor, it is proper to introduce evidence and testimony from other surveyors about what an ordinary skilled surveyor prudently is expected to do during the performance of a particular type of survey. But in rare situations, expert testimony is not required to establish professional negligence.

In a recent Louisiana decision, the court decided that the plaintiff did not have to introduce such evidence since it was obvious to all that the conduct of the surveyor was unprofessional and below reasonable standards. The Lawyers Title Insurance Company sued a surveyor to recover damages that it had to pay out because of reliance on a surveyor's work [10]. The basic facts are as follows: Lawyers Title Insurance Company (Appellant) subrogee of its insured appealed the dismissal of its subrogated demand against defendant surveyor and the surveyor's employee for reimbursement of a claim paid pursuant to a policy of title insurance issued by the appellant covering a parcel of land purchased on the strength of a survey by the surveyor. The survey erroneously located drainage facilities on the land, which in reality were on the neighboring land. Relying on the survey, the owner planned its drainage to utilize the facilities indicated as being on its property. The error was discovered after he had installed the drainage. The work installed had to be removed and new facilities constructed. The trial court found that the surveyor made a mistake but that a mistake does not necessarily constitute accountable negligence and rejected the appellant's demand on the grounds that the appellant failed to produce expert testimony to establish the surveyor's negligence by showing that the surveyor failed to exercise the degree of skill and care exercised by other surveyors in the vicinity. The appeals court reversed and rendered judgment for the appellant.

The court stated that surveyors are expected to perform with the same degree of care and skill exercised by others in the profession in the same general area. Ordinarily, proof or lack of such skill and care or proof of failure to exercise such skill and care in a given instance rested on the plaintiff. They deemed it reasonable, however, to exempt from the general rule those instances when the conduct of a surveyor may be so unprofessional, so clearly improper, and so manifestly below reasonable standards dictated by ordinary intelligence as to constitute a prima facie case of either a lack of the degree of skill and care exercised by others in the same general vicinity or failure to reasonably exercise such skill and care. They found that the omission of a visible drainage structure from a surveyor's plat falls within the exception. No profession may, by adopting its own standards of performance, method of operation, or paragons of care, insulate itself from liability for conduct that ordinary reason and logic characterize as faulty or negligent.

In *Lawyers Title Insurance Co. v. Hoges*, the court ordered, adjudged, and decreed that the judgment of the trial court be reversed and set aside and judgment rendered herein in favor of plaintiff Lawyers Title Insurance Company and against the defendant surveyors for all damages [10].

The older rule, used by some courts today in cases involving damages against a land surveyor, is best expressed as the prudent surveyor rule [11].

The question was that of the negligence of a civil engineer. The jury was instructed that an engineer is bound to exercise that degree of care that a skilled civil engineer of ordinary prudence would have exercised under similar circumstances. This being the rule, it was important that the jury should know what such ordinarily prudent engineer would do under the circumstances of this case. Might he or she simply examine the land records and muniments of title and observe the fixed monuments and evidences of present occupancy and ownership, or was he or she bound to scour the neighborhood to learn whether there had been adverse occupancy and claims of ownership of any part of the premises covered by his or her client's deed by others than the latter's predecessors in the title appearing of record? Would good engineering practice require that he or she examine a city engineer's map, called the "1870 map," the failure to examine which was claimed to constitute negligence on the part of the defendant? The jury would not know, unless informed by evidence, and such evidence was admissible. The tendency in courts today is to lean more toward the opinion in *Lawyers Title Insurance Co. v. Hoges* rather than the rule just cited. In other words, surveyors cannot always escape liability merely because they can introduce evidence showing standards of practice below what would be expected of the professions. In some courts, some things are patently negligent without proof in the form of testimony.

Certainly, the surveyor can expect to be held to the standards adopted by the state association in existence at the time, and this applies regardless of whether the surveyor is a member of the association. If a state has not adopted a standard of care, the court may inquire about what the standard of care is in other state associations and adopt them for criteria.

The fact that a surveyor can be held liable for negligence and lack of skill is well established; the only question is what constitutes negligence and lack of skill. The elastic definition—the surveyor must exercise that degree of care that a surveyor of ordinary skill and prudence would exercise under similar circumstances—will undoubtedly continue as the court's rule.

However, it must be remembered that with improved equipment, better dissemination of knowledge of boundary laws, and improved scientific methods, the exact parameters that govern the standard of care owed will change along with these technological innovations. What was an acceptable minimum measurement accuracy, or closure, 50 years ago would certainly not be tolerated today because of the advances in measurement science. Furthermore, the standard of care required 50 years from now will be quite different from what is expected today.

There are few national standards for land boundary surveys. National standards that do exist include the BLM *Manual*, ALTA/NSPS Specifications, and various publications from federal agencies concerning how data is collected, handled, and disseminated. The American Land Title Association standards, published in conjunction with the National Society of Professional Surveyors, probably comes the closest to national standards for property surveys. Additionally, the National Geodetic Survey publishes standards for horizontal and vertical surveys, and these are applied by many courts. For land surveys in the public land states, the *Manual of Surveying Instructions* may be considered a "national" standard for retracements. Although this was developed for the surveyor of currently owned public lands, several states have adopted this manual as their state standard in retracing privately owned lands as well, either by statute or case law reference [12]. Today, most courts require an affidavit from a second surveyor, indicating the nature of the negligence.

Taking the exercise of proper care and skill a step further, it may be said that the standard expected of any professional is that of "due diligence." Claims for the erroneous designation of boundaries or in locating improvements are obvious, but some expectations might not be so obvious. Under some circumstances, for instance, surveyors can be held liable for failing to note the existence of easements on land being surveyed. Surveyors are not required to guarantee the discovery of *all* easements but may be required to report items that a reasonable visual inspection of the property would disclose.

In the case of *Jarred v. Seifert* [13], a surveyor was hired by a developer to measure boundaries for future construction and to set stakes accordingly. While setting the stakes, the surveyor discovered a manhole on the site. The surveying firm failed to search the records further, neglected to report the manhole, and failed to relocate the stakes. When construction was nearly completed, it was discovered that a building corner encroached on the sewer easement. The owner was forced to pay for relocation of the sewer line and sued the surveyor to recover his expenses.

In upholding a jury verdict against the surveyor for negligence in failing to note the existence of the manhole, the court held the surveyor responsible for failing to use reasonable diligence in inspecting the property for the existence of easements. If easements are discovered, the surveyor has a duty to bring them to the owner's attention.

With the advances in technology and the fact that this technology is readily available to any person, many professionals believe that possessing and using this new advanced technology will save them from claims of negligence. This is far from so. The purpose of retracement is to determine from the evidence that was created, the evidence found, and the evidence argued, "I am on the correct parcel of land." This was emphasized as early as 1912 by A. C. Mulford when he stated:

> When it comes to a question of stability of property and peace of the community, it is more important to have a somewhat faulty measurement of a spot where the line truly exists than it is to have an extremely accurate measurement of a place where the line does not exist at all [14].

16.5 Certificates of Merit

In about half the states, before a property owner can file a civil suit against a surveyor, that plaintiff must obtain a certificate of merit. The purpose of the certificate of merit is to limit or eliminate frivolous lawsuits against professionals, such as doctors, nurses, engineers, architects, and land surveyors. Basically, another professional (i.e., a licensed surveyor) must write a report that states the defendant-surveyor acted negligently. If you are an expert in your area of expertise—by education, experience, or both—why not help the public by authoring these reports?

A certificate of merit is essentially an expert report by a land surveyor that (1) describes the standard of care owed by a hypothetical professional surveyor to his or her client(s) based on state statutory law, state case law, and land surveying texts; (2) explains what work the current surveyor-defendant did for the client and the methods used to complete such survey; and (3) provides an expert opinion as to whether the surveyor-defendant was professionally negligent.

In California, the state's Code of Civil Procedure sets forth the requirements of a certificate of merit. According to 411.35:

> (a) In every action, including a cross-complaint for damages or indemnity, arising out of the professional negligence of a person holding a valid . . . land surveyor's license issued pursuant to Chapter 15 (commencing with Section 8700) of Division 3 of the Business and Professions Code on or before the date of service of the complaint or cross-complaint on any defendant or cross-defendant, the attorney for the plaintiff or cross-complainant shall file and serve the certificate specified by subdivision (b).
>
> Furthermore, (b)(1) states:
>
> That the attorney has reviewed the facts of the case, that the attorney has consulted with and received an opinion from at least one architect, professional engineer, or land surveyor who is licensed to practice and practices in this state or any other state, or who teaches at an accredited college or university and is licensed to practice in this state or any other state, in the same discipline as the defendant or cross-defendant and who the

attorney reasonably believes is knowledgeable in the relevant issues involved in the particular action, and that the attorney has concluded on the basis of this review and consultation that there is reasonable and meritorious cause for the filing of this action. The person consulted may not be a party to the litigation. The person consulted shall render his or her opinion that the named defendant or cross-defendant was negligent or was not negligent in the performance of the applicable professional services. [15]

16.6 Negligence Versus Breach of Contract

PRINCIPLE 3. *The surveyor is liable for damages arising from expressed guarantees and from damages arising from implied guarantees expected of the profession.*

Historically, all negligence actions result from breach of contract. In all states a surveyor is required to do what he or she contracts for and may be held liable for failure to fulfill a contract correctly. In some states, double or triple damages can be claimed in addition to contractual damages if the work is done in a negligent manner. The extra damage is called *punitive damages*, which are usually awarded as a punishment by the courts. In states where punitive damages are allowed, the attorney usually adds the charge of negligence in addition to the real or compensatory damage.

According to the fourth edition of *Black's Law Dictionary*, negligence is "omission to do something which a reasonable man, guided by those ordinary considerations which ordinarily regulate human affairs, would do, or the doing of something which a reasonable and prudent man would not do" [16]. To collect punitive damages in some states, as in New Jersey, malice must be shown. It seems that the tendency of courts is to become more liberal in what they consider to be negligence.

In an action for trespass to plaintiffs' real property in Rahway, New Jersey, resulting from the wrongful erection of a fence thereon, the jury awarded plaintiffs $900 as compensatory damages against all defendants [17]. In addition, the jury allowed punitive damages totaling $4300. The sum of $300 was assessed against the neighbors who ordered the fence erected, $1,000 against the defendant contractor who erected the fence, and $8,000 against the defendant surveyor, whose erroneous staking of the boundary line between the Smith and La Bruno properties caused the Smiths' fence to be erected on the La Bruno property.

This case would thus seem to indicate that one of two factors must be found before punitive damages can be awarded in a suit for trespass to real property:

1. Actual malice, which is nothing more or less than intentional wrongdoing, an evil-minded act.
2. An act accompanied by a wanton and willful disregard of the rights of another.

Clearly, each case must be governed by its own particular facts. Accordingly, all facts must be examined as to each of the defendants to decide the validity of the award of punitive damages in each case.

The court first considered the surveyor whose initial mistake of wrongfully placing the stakes on the La Bruno property started the chain of causation that resulted in the damage and litigation. The surveyor committed a trespass, and this was the mistake. Honest belief or professional neglect is no defense to a trespass. A trespass by mistake or through carelessness will ordinarily not justify an award of punitive damages, because the element of mistake or neglect negates any intentional wrongdoing. A mistake of 1 foot in the boundary line does not, per se, bespeak willful and wanton disregard of the rights of another.

The surveyor, however, aggravated his original trespass or mistake by a stubborn refusal to examine his prior survey and correct his alleged error when the matter was called to his attention by La Bruno originally and later by the Smiths when they received a letter from La Bruno's attorney. Even when La Bruno put it to him that he had to be wrong one time or the other, he said, "I'll cross that bridge when I come to it." The surveyor was fully aware from his personal inspection, when he placed the stakes, that the erection of the fence by the Smiths along that boundary line would damage the La Bruno walk, patio, and flower bed. As a surveyor, he knew or should have known that the Smiths and their contractor would probably rely on his professional skill and judgment in erecting the fence. He demonstrated a willful and wanton disregard of the property rights of the plaintiffs that was reasonably calculated to aggravate his original mistaken trespass by the consequent trespasses of the Smiths and Cohill. The jury could justifiably find the surveyor's overall conduct and attitude, as they did, a proper basis for the award of punitive damages against him. He probably gave the jury the impression that he was a man who would rather see property destroyed by his error than admit that he had made the error.

Turning next to the contractor, the court found no justification for the award of punitive damages. There was no evidence of any actual or express malice on his part.

Accordingly, the appeals court dismissed all punitive damages against the others and held only the surveyor liable [17].

16.7 Expressed and Implied Guarantees

PRINCIPLE 3. *The surveyor is liable for damages arising from expressed guarantees and from damages arising from implied guarantees expected of the profession.*

Advertising oneself as a land surveyor carries with it implied guarantees that will be enforced by the court, and these guarantees will be imposed by the court as based on fairly high standards set by the profession. Such questions as these can be inquired into. What are the standards set by the state surveyor's society, the ACSM, and the ASCE? What accuracy does the governing agency require for original (subdivision) surveys? Although at one time the rule was that surveyors could testify about the degree of skill required in their particular area and thus establish minimum standards, it appears that this relaxed rule is on its way out.

At times, surveyors foolishly make positive guarantees such as, "This plat of survey carries our absolute guarantee of accuracy" [17]. Such stipulation invites liability beyond what is required by law and should be avoided. This situation was addressed in the Illinois decision previously mentioned [2]. The surveyor's statement of "absolute guarantee" held him responsible when it was discovered that his survey was inaccurate.

Surveyors are, of course, liable to adopt the work of others without checking it. This includes but is not limited to the acceptance of plats, corners, and surveys of others. In Louisiana, a surveyor copied the acreage figure from a previous survey and was held liable for damages from its inaccuracy [18]. A surveyor was employed by the plaintiff's attorney to "confirm" an earlier survey of a tract of land that he had agreed to purchase but had not yet acquired by act of sale. The plaintiff did not specifically request an acreage computation in connection with it, but the earlier survey, which was subsequently confirmed by the surveyor, included a written notation that the tract contained 13.16 acres. As a matter of fact, the surveyor did not actually confirm the acreage figure but relied on it. The surveyor (or his employee) simply assumed that the acreage figure on the earlier survey was correct since all the measurements of the boundary lines on the earlier survey were confirmed for accuracy when they were located and measured. However, the original mathematical computation of the acreage figure had been incorrectly made.

The parties agreed that, after the sale, a correct computation showed that the tract contained only 11.26 acres, not 13.16 acres, as shown on the surveys, although the lines were all correctly measured in both surveys. The surveyor contended that he had relied, to his detriment, on the acreage information confirmed by the survey. Since the sale by which he acquired the tract was based on a dollar-per-acre price, he claimed injury in the amount of $6,840, the alleged "overpayment" computed at $3,600 per acre.

The appeals court did not find error in the trial court's determination that the surveyor was liable. The inclusion of the incorrect acreage figure may not, at its inception, have been anything more than a mistake of judgment on the part of the surveyor. However, it was ultimately transmitted, duly certified as correct, by the surveyor, and the owner reasonably relied on it. The loss was fairly and directly a result from a faulty and defective exercise of care and skill on his part.

A similar result was found in Maryland in 1972 [19], when a surveyor reported 22.075 acres when in fact there was only 19.58 acres. Judgment was for $3,339.93, plus costs, as entered against the surveyor on the basis of $1,250 per acre for land not acquired.

Any time a surveyor adopts a monument proclaimed by someone else as being correct or adopts measurement data or computations as being that surveyor's product, the surveyor assumes full liability for the correctness of such. This can be likened to endorsing a check or cosigning a note; everyone whose signature is on the check is liable for any damages. Recommendations are not to copy data or assume something to be true unless you can prove such to be correct, for if you do, you assume any and all liability for any errors.

16.8 Trespass Damages

PRINCIPLE 4. *Unless there is a law permitting the surveyor to trespass when necessary to do a survey, the surveyor can be held liable for damages of trespass, either compensatory or punitive.*

Most surveyors assume the right to go onto the land of others. Unless there is a law giving the surveyors this right of entry, such right does not exist. Even if a law does give a right of entry, the surveyor at no time can damage the vegetation existing on the land. Brush cutting or line-of-sight clearing may result in a damage claim. At common law any unauthorized trespass into the invisible bounds of a parcel of land is a common-law tort, or civil wrong. When it is determined that the trespass was unauthorized, a court will consider the person's ability to pay when it awards damages. In Indiana, a total of $420 compensatory damages and $100,000 punitive damages were awarded to the landowner as a result of a power company's survey crew illegally removing corn to clear a line of sight [20]. This reward was made even though the power company was in the process of establishing what it wanted to take under eminent domain.

16.9 Avoiding Liability

PRINCIPLE 5. *To avoid liability, the professional surveyor should do more than the ordinarily prudent surveyor would do under similar circumstances.*

With present-day thinking on liability to third parties and the application of the discovery rule, the surveyor must take precautions to avoid paying avoidable damages.

Liability can result from improperly locating the written document, failure to inform the client of the possibility of unwritten title rights, and errors in computations. The remainder of the book is devoted to the subject of how to locate written deeds on the ground. The major cause of damages in computations is in computation of acreage. Every client can claim the purchase price is based on a price per acre; the courts are full of damage suits based on an error in acreage computations.

The most disturbing feature of current liability laws pertains to third persons. Title companies have a right under the third-party rule, when applicable, to rely on the results of the survey. When they ensure title as based on a survey and damages result, they have a right to seek restitution from the surveyor. In effect, the surveyor assumes some of the risk in title insurance business. This makes it necessary for the surveyor to present the results of his or her survey in such a fashion so that all third parties are apprised of possible title problems such as unwritten rights as a result of prolonged occupancy.

If a surveyor sets monuments to include land occupied by an adjoiner, there is always the possibility of the adjoiner having an unwritten title right. Unless this is made clear to the client, the client may be prone to do something that will lead to damages, and the possibility exists that the surveyor will be held responsible. In the event of an encroachment on the client's written deed, the surveyor should be certain that the client understands the significance of the encroachment and that the information is presented in a manner that third parties are appraised of it. This means that the surveyor should present a plat clearly showing all encroachments and noting the possibility of occupancy rights in relation to deed or title lines. If an unwritten title right or the probability of such exists, it is foolish to include the occupied area in the acreage of the client as shown by his writings.

In the case of *Western Title Guaranty Co. v. Murray & McCormick, Inc.* [21], the survey company was sued for failing to disclose the significance of a potential fence encroachment, or more accurately, the difference between the record line and a fence, or line of occupation—a difference of 230 feet, amounting overall to 14 acres.

The surveyor correctly located the deed line according to all the rules, those based on court decisions and those in the BLM Manual [12] (this was a Public Lands State). However, 230 feet *inside* the record line there was a fence, which the surveyor also located and showed on the plat. The title company relied on the survey and insured the parcel to the record line, and subsequent litigation resulted in the fence line being the boundary between the two ownerships, not the record line. The title company then was sued, lost, and ultimately sued the surveyor. Whether the title company or the surveyor or both were truly at fault is immaterial here; the point is that litigation ensued and an action was brought against the survey firm.

What the surveyor did not do but should have done was to make everyone aware of the encroachment or report the difference between the record title and the occupation (possession). While a surveyor does not pass on the ultimate ownership of the land, he or she is obligated to report a potential conflict in ownership whenever apparent. A question raised at the beginning may prevent erroneous assumptions and conclusions by others at a later time. Some surveyors report all the facts, show record and physical evidence, then let someone else worry about what it all means. Many times, that is not sufficient. At the very least, if the surveyor is retained to "show the boundary" and all he or she does is show the evidence, good, bad, indifferent, conflicting and confusing, without commentary, leaving conclusions to be formed by someone else, he or she has not fulfilled the contractual obligation. That in itself may be grounds for a lawsuit.

The most important third parties that the surveyor has to be aware of are future buyers, title companies, and contractors. The surveyor who omits damaging information from a plat merely because a client does not want it disclosed may be inviting possible damage suits from third parties. The day of privity of contract between the surveyor and client is over; the surveyor is obligated to disclose, for the possible benefit of third parties, all information that may lead to damages. In short, surveyors now have an obligation to anyone they can reasonably foresee as relying on their work.

16.10 Elements of Liability

PRINCIPLE 6. *To prove professional negligence, one must show four elements: (1) a standard, or duty, exists in the profession; (2) the surveyor breached that standard, or duty; (3) that breach was the proximate cause of the damages; and (4) actual damages resulted.*

In 1912, A. C. Mulford stated, "The surveyor is isolated in his calling and therein lies his responsibility" [22]. Liability or negligence is complicated and defies exact definition and application. Simply because a surveyor conducts his or her work to a high degree of skill does not preclude that his or her actions may be legally questioned. Negligence differs from an intentional wrong (tort) in that the negligent surveyor in all probability did not intend to do wrong.

16.11 Damages

PRINCIPLE 7. *If a surveyor can be proven negligent in the performance of his or her duties, punitive damages may be awarded in addition to actual and contract (compensatory) damages.*

To show negligence, all four requirements listed in principle 6 must be met, not just one or two of them. If negligence is in fact demonstrated, the court (often the jury) may award punitive damages in addition to compensatory damages. Punitive damages are intended to punish the negligent individual and to convey a message to others that what was done was not a simple matter but rather cause for great concern. Compensatory damages are intended to compensate victims for their loss(es). In a given situation, punitive damages may be several times greater than the compensatory damages, depending on the severity of the offense.

If a court determines a surveyor is negligent, as a matter of law, there may be instances that a court may grant summary judgment in a negligence claim.

PRINCIPLE 8. *A surveyor should not contract to do work for which he or she is not qualified, educated, and experienced to accomplish.*

Like all other professionals a surveyor should only practice in those areas in which he or she is qualified. Even though a professional license authorizes one to work in all aspects of the profession, the surveyor should only practice in those in which he or she is qualified. If one does conduct work in an area other than his or her specialty area, that individual will be held to the standard that most qualified persons in that area possesses.

PRINCIPLE 9. *A surveyor can be sued either in tort or in contract. With a written contract, the terms of the contract become the controlling elements.*

Studies indicate that two-thirds of the claims of negligence brought against surveyors are initiated by their own clients for whom they have conducted work. If the surveyor is working under a written contract, then the terms of the contract will be examined and will be controlling in claims of negligence.

PRINCIPLE 10. *The use of another surveyor's work that is found to be substandard does not relieve the second surveyor from liability for the substandard work of the first surveyor. The liability will attach to the surveyor who uses any work product that is erroneous.*

Surveyors will contact colleagues and will often use recorded data or maps in doing retracements, not knowing the validity of the previous work.

16.12 Liability of an Abstractor

Some surveyors also do abstracting of title documents in addition to their other duties, while others rely on abstractors or abstracts of title for their research information. While this type of service exists, surveyors should be aware that the minimum standards of practice for abstractors and those for surveyors may not be the same.

For instance, as previously discussed in Chapter 14, title examination and abstracting are different from records research for survey and boundary purposes. While many of the same records are likely to be reviewed, there are departures, and in general, survey research will demand a wider scope of investigation in most cases.

While they may be all-inclusive under the right circumstances, generally a standard title examination or certification is confined to the record, and many title reports and title insurance policies exclude certain items. The standard search constitutes the following:

1. It is confined to one parcel, the subject locus.
2. It reviews only those items on the public record.
3. It contains only those items within the period of search where a limited search period is designated in the title standards (which may be as short as 35 or 40 years).
4. It covers only nonsurvey items, since title issues are the focus.

Therefore, the examination of title is subject to those items that may be shown by an accurate survey, which could include any or all of these exceptions.

Ordinarily, an abstractor's duty is simply to examine the records and to compile therefrom and incorporate in the abstract a statement of the facts appearing of record that affect title to the property; he or she is not required to show matters that are not of record in the absence of any agreement to investigate and determine whether there are matters affecting the title that are not of record or to show all the facts and circumstances that might affect the title, such as possession and who were the legal heirs of the deceased owner, where the administration is in a different jurisdiction [23].

The liability of an abstractor is a contractual one, and the general rule is that he or she cannot be held liable in damages to a third person for his or her negligence in the preparation of the abstract even though such third person relied on an abstract prepared for the abstractor's employer [24].

In the case of *Crawford v. Gray & Associates* [25], both lawyer and surveyor were successfully sued by the client for failure to disclose a right of way adjoining an existing highway because the client had to remove the part of his store that encroached on the right of way.

The lawyer searched the title and gave a written opinion of good and valid title based on his examination of local public records for "at least the past sixty years." The surveyor's plat indicated no easements for rights of way. Before the building was completed, the state highway department informed the client that the state had a 50-foot right of way on each side of the highway which was being encroached upon by the building. A 9 × 12-foot section had to be removed and the client sued both the lawyer and the surveyor. The court, in finding for the client, noted the following:

1. The lawyer's title opinion letter, which stated that it was based on an examination of the local public records, was false.

2. He would have discovered the right of way if he had actually examined those records.

The court also went on to find the surveyor negligent since he had not exercised "reasonable skill" in the performance of his survey.

REFERENCES

1. *Doyle v. Lynn*, 37 Colo. 214; 574 P.2d 257 (Colo., 1973).
2. *Rozney v. Marnul*, 43 Ill. 2d 54 250 N.E. 2d 656 (1969).
3. *Jundahl v. Barnett*, 468 P. 2d 1164 (Wash. 1971).
4. *Ruth v. Dight*, 75 Wash. 2d 660, 456 P. 2d 351 (1969).
5. *Frazer v. Weeks*, 76 Wn. 2d 819, 456 P. 2d 451 (1969).
6. *Kundahl v. Barnett*, 80 Wash. 2d 1001, 468 P. 2d 1164 (1071).
7. *U.S. v. Doyle*, 468 F. 2d 633 (Colo. 1973).
8. *Howell v. Betts*, 211 Tenn. 134, 362 S.W. 2d 924 (1962).
9. *Burkett v. Studebaker Bros. Mfg. Co.*, 126 Tenn. 467, 150 S.W. 421 (1900).
10. *Lawyers Title Ins. v. Hoges*, 358 So. 2d 964. (La. App. 1978).
11. *Ferris v. Sperry*, 85 Conn. 377, 82 A. 972 (1912).
12. Bureau of Land Management, "Technical Bulletin 6," in *Manual of Surveying Instructions* (Washington, DC: U.S.D.I. Bureau of Land Management, 1973).
13. *Jarred v. Seifert*, 22 Wash App. 476, 591 P. 2d 809 (1979).
14. A. C. Mulford, *Boundaries and Landmarks* (New York: D. Van Nostrand Company, 1912. Reprinted 2008).
15. Code of Civil Procedure, available at: https://leginfo.legislature.ca.gov/faces/codesTOC-Selected.xhtml?tocCode=CCP&tocTitle=+Code+of+Civil+Procedure+-+CCP
16. *Black's Law Dictionary*, 4th ed. (St. Paul, MN: West Publishing, 1951).
17. *La Bruno v. Lawrence*, 166 A. 2d 822 (N.J. Sup. Ct. 1960).
18. *Jenkens v. Krebs*, 322 So. 2d 426 (La. 4 Cir. 1975).
19. *Downs v. Reighard*, 265 Md. 344, 289 A. 2d 299 (1972).
20. *Indian & Mich. Ele. Co. v. Stevenson*, 363 N.E, 2d 1254 (Ind. 1977).
21. *Western Title Guaranty Co. v. Murray & McCormick, Inc.*, Placer County, California Superior Court, case 36933.
22. Mulford, *Boundaries and Landmarks*, p. 88.
23. "Abstracts of Title," in *American Jurisprudence* (2d), vol. 1, sec. (Rochester, NY: Lawyers Co-operative Publishing Co.), p. 6.
24. Ibid., p. 15.
25. *Crawford v. Gray & Associates*, 934 So. 2d 734 (La. App. 1986).

ADDITIONAL REFERENCES

Brown, C. M. "Land Surveyor's Liability to Unwritten Rights." *ACSM Surveying and Mapping*, vol. 39 (1979): 119–123.

Felder, F. E. "Surveyor's Failure to Exercise Due Care in Making Survey." *Proof of Facts (2d)*, vol. 11 (1976): 397–409.

"Liability of Surveyor or Civil Engineer." In *American Jurisprudence (2d)*, vol. 58, Occupations, Trades and Professions, sec. 78. Rochester, NY: Lawyers Co-operative Publishing Co.

Madson, T. S., II, and Munro, R. J. "Understanding Your Professional Liability as a Land Surveyor." Paper presented at Land Surveyor's seminar, Gainesville, FL, 1979.

Ytreberg, D. E. "Surveyor's Liability for Mistake in, or Misrepresentation as to Accuracy of, Survey of Real Property." *American Law Review (2d)*, vol. 35 (1971), pp. 504–511.

17

PROFESSIONAL STATURE

17.1 Contents

When asked, surveyors identify with, believe they are, and espouse to be professionals, but seldom do we see in print what constitutes professional standards. The discussion in this chapter in no way expresses what professional behavior should be. Facts and stated opinions were gained from discussions with varied individuals, reading and researching available literature, references relating to other professions and professionals, as well as references to case law from the various states.

In this chapter, we will discuss some personal philosophy, some personal viewpoints, and some discussions of the past, as well as some concerns for the present and some hopes for the future.

Today, surveyors are caught in the dilemma of attempting to portray themselves as a hardy, rugged, outdoors-oriented people who are seeking acceptance as professionals in areas that require a coat-and-tie atmosphere and relationships. They are trying to fit into a modern society with a definition of a professional while attempting to maintain a sense of individualism that history has placed on them.

17.2 Definition of Profession or Professional

An exact definition of the word *profession* is elusive; each authority gives it new and varied meanings. In a narrow historical sense created by English common law, there were three basic professions known as "the professions": theology, law, and medicine. Today, those professions are limited to include doctors, clergymen, and attorneys; in a broad sense, professions include professional boxers, baseball players, football players, and many others also witnessed by personal claims as well as telephone directories. To expand on this, it is suggested that the students turn to the business pages of their local telephone book. Count the number of listings that start with the word *professional*. The following three court cases may serve to define the term.

Almost a century ago, a court held that a "beauty specialist" was exempt from a special tax on "Professionals." The judge rationalized that the beauty specialist was not a profession because "the skill required of the taxpayer in the use of her hands and the handling of tools and machinery predominated over the intellectual and mental requirement" [1].

In the decision *Aulen v. Triumph Explosives, Inc.*, the U.S. District Court identified six elements that were needed to identify whether a person's work was professional:

1. The work is predominantly intellectual and varied in character as opposed to routine mental, manual, mechanical, or physical.
2. The work requires constant exercise of discretion and judgment in its performance.
3. The work is of such a character that the output produced or the result accomplished cannot be standardized in relation to a given period of time.
4. The "subprofessional" work does not exceed 50 percent of the time involved (paraphrased).
5. Knowledge of an advanced type in a field of science or learning is customarily acquired by prolonged course or specialized intellectual instruction and study, as distinguished from general academic education. It is also distinguished from an *apprenticeship* and from training, the result of which depends primarily on the invention, imagination, or talent . . . (emphasis added).
6. The work is predominately original and creative in character in a recognized field of artistic endeavor [2].

Very generally, the term *profession* is employed as referring to a calling in which one professes to have acquired some special knowledge, used by way of instructing, guiding, or advising others or of serving them in some art.

> Formerly theology, law, and medicine were specifically known as "the professions": but, as the applications of science and learning are extended to other departments of affair, other vocations also received the name. The word "profession" is a practice dealing with affairs as distinguished from mere study or investigation; and an application of such knowledge for others as a vocation, as distinguished from the pursuits for its own purposes [1].

> The word "profession" in its larger meaning, means occupation, that is, if not industrial, mechanical, agricultural, or the like, to whatever one devotes oneself; the business which one professes to understand and follow. In a restricted sense it only applies to the learned professions [3].

> A "profession" is not a money-getting business. It has no element of commercialism in it. True, the professional man seeks to live by what he earns, but his main purpose and

> desire is to be of service to those who seek his aid and to the community of which he is a necessary part. In some instances, where the recipient is able to respond, seemingly large fees may be paid, but to others unable to pay adequately, or at all, the professional service is usually cheerfully rendered [4].

One may look at the base verb, *to profess*. Originally, the professional had a moral dimension. The original description was given to those individuals who "professed" or gave their word to devote their lives to a higher order or to serve mankind. Those to whom this could be said were teachers ("professors"), doctors, clergy, and lawyers. Theoretically, these individuals put service to humanity ahead of making money. Gradually, as more and more education was required for some of the tradespeople, the word was expanded to include dentists, engineers, and accountants.

Most dictionaries state that a profession is a "calling." Today we are caught in the dichotomy of considering a professional as an individual doing full-time what most people do part-time.

Thus, we have professional ball players, professional pest exterminators, and professional boxers. If money is the criterion, then a professional is one who performs a job or produces a product that is worth the price being charged for it. Pride and workmanship go hand in hand. Perhaps the main attributes of a true professional are enthusiasm, a sense of vocation or calling, and inspiration to do a good job. A professional job is readily distinguishable from a mediocre one. Being a professional requires being calm in times of adversity and showing confidence in the specialty field practiced. A professional knows that any subject cannot be completely mastered. Regardless of the years in practice, a professional is always learning. A professional must constantly renew knowledge and expertise through constant reading and conferences. A professional is always available to perform, whether ready or not.

Because of dereliction of many professionals, many states have attempted to regulate professionals through legislation.

Being professional is charging an adequate price for a good product and not in undercutting prices that will inevitably produce an inferior product. Surveyors who fail to charge an adequate price or who undercut prices know full well that a suitable product cannot be produced for the price indicated. The true professional is not that individual who has been certified by a licensing body, but that person on whom the name has been conferred by clients and peers.

The possession of a license does not make one a professional.

In this chapter, the following principles will be discussed:

PRINCIPLE 1. *Professional stature cannot be attained by self-proclamation; it must be earned, and others must bestow the title on the profession.*

PRINCIPLE 2. *Superior and distinct education in a field of knowledge is one essential feature of a profession.*

PRINCIPLE 3. *The surveyor has the following obligations to the public: (1) to see that the client's boundaries are properly monumented without subtracting from the rights of the adjoiner; (2) not to initiate boundary disputes; (3) not to aid in unauthorized surveying practice; (4) to see that those licensed to survey are properly qualified by character, ability, and training; (5) to see that those who prove unworthy of their privileges have those privileges deprived; (6) to agree not to attempt to practice in any professional field in which one is not proficient; and (7) to produce a quality product regardless of the price.*

PRINCIPLE 4. *No surveyor should permit his or her name to be used in aid of or to make possible the unauthorized practice of surveying by any agency, person, or corporation.*

PRINCIPLE 5. *The surveyor does not attempt to injure falsely or maliciously, directly or indirectly, the professional reputation, prospects, or business of another surveyor.*

PRINCIPLE 6. *In the event a property surveyor discovers an error or disagrees with the work of another property surveyor, it is the duty of that surveyor to inform the other surveyor of such fact.*

PRINCIPLE 7. *It is unprofessional to review the work of another property surveyor or work for the same client, except with the knowledge or consent of such property surveyor or unless the connection of such property surveyor with the work has been terminated.*

PRINCIPLE 8. *It is unprofessional to act in any manner or engage in any practice that will tend to bring discredit on the honor or dignity of the surveying profession.*

PRINCIPLE 9. *No division of fees for surveying services is proper, except with another surveyor who by his or her license is permitted to do property surveying work.*

PRINCIPLE 10. *It is unprofessional to take employment on a contingency basis.*

PRINCIPLE 11. *Violation of professional or ethical standards may be considered as "some evidence" of malpractice.*

PRINCIPLE 12. *Professionals may have differences of opinion. Simply a difference of opinion will not constitute negligence.*

PRINCIPLE 13. *Having a valid license does not make a professional. The license requirements are governmental requirements in order to practice surveying.*

PRINCIPLE 14. *The foundation of a profession is having a solid basis of education commensurate to other professions.*

PRINCIPLE 15. *A true professional merges practical experience with formal education to attain the degree of sophistication needed to serve the client base.*

17.3 Attaining Professional Stature

PRINCIPLE 1. *Professional stature cannot be attained by self-proclamation; it must be earned, and others must bestow the title on the profession.*

The lazy say, "Give me the prize without the training, the wages without work, a profession's prestige without a profession's skill." Fortunately, professional stature is something that must be earned. It cannot merely be claimed.

Many attempt to filch good names. Industry, public agencies, and others have found that by bestowing a fancy title on the lowly laborer, they can keep him or her better satisfied at a lower rate of pay. The boy who cuts brush and carries the surveyor's stake bag is certain of his importance if he is called an engineering aide rather than an axman. Mere claiming of a good name is not proof of right to the name. A person's actions, behavior, and conduct are far more significant proof.

Some occupations inherently carry a historical connotation of profession. The professions that everyone recognizes and acclaim as professions, that is, the doctor, clergy, and attorney, need not use the title professional doctor, professional attorney, or professional clergy; all consider them professional. The title was bestowed because of the ethics, behavior, and standing of the members of these groups in the community.

Legislating a title "professional surveyor" or "professional engineer" does not in itself prove that the activity the bearer of the title engages in is a profession. If surveyors wish to acquire and maintain a professional reputation, it can be earned only through the average standing of the entire profession and not by any single individual.

Within land boundary surveying practices, there are many workers, each of whom is different in capabilities and personalities. A significant failure is to properly distinguish between technical-level work and professional-level work and responsibilities. On highway work, the engineer in charge has surveyors who make measurements to determine the shape of the ground. The surveyors are merely measuring the ground as it exists and recording the facts as they are. To be sure, the surveyor must have superior ability in knowing how to use instruments and how to make measurements, but these acts are technical in nature and not professional. They do not include road and drainage design, nor do they utilize the measurements they produce.

Again, the engineer may tell surveyors to grade stake a road in accordance with a given plan. Since no design or judgment is involved, it is a purely technical matter. To some engineers, surveyors are technicians who carry out orders. And often, surveyors are just that. But the engineer frequently overlooks the fact that a professional property surveyor has both technical attributes and professional attributes.

In surveying, if a person is successful at a task, he or she will be forced to leave it. The chain person who is good soon advances to an instrument person. With study, the instrument person becomes a chief of party. With further study, licensing is obtained and the individual tries to emerge as a professional, although still lacking the basic requirement of education that is typical and required of many of the other learned professions. As a person advances with experience and promotion, he or she becomes a manipulator of people. If a person is to succeed as a professional, it will soon be evident that a person qualified at one level must craft routines to others and desert, at least to some degree, those craft companions who practice below the expected level. As professionalism is attained, ideas and people are the determining factors, not surveys.

Usually, when a young surveyor attains registration, the individual seeks to commence practicing surveying under the newly obtained license, little realizing that in self-employment, added liabilities attach to the work performed.

17.4 Attributes of a Profession

If the surveyor wants to be considered in a learned profession, the individual must seek the attributes of the learned professions. Ten of these attributes are as follows:

1. Having a unique and superior education in a specific field of knowledge.
2. Providing a service to the public in having the ability to persuade.
3. Placing oneself in a position of trust.
4. Conducting practice within a code of ethics.
5. Desiring to gain high eminence with financial return of secondary importance.
6. Using independent judgment and accepting liability.
7. Providing services to those unable to pay.
8. Charging fees to those able to pay, such fees being dependent on the services rendered rather than labor or product.
9. Becoming a persuasive and effective communicator.
10. Seeking continued education to maintain professional competency.

17.5 Superior Education in a Field of Knowledge

PRINCIPLE 2. *Superior and distinct education in a field of knowledge is one essential feature of a profession.*

Formal education is not necessarily the only means of acquiring knowledge, since knowledge can be attained by experience and self-effort. At one time many attorneys were admitted to the bar without the advantage of a college degree, but today that practice has been eliminated. Before an attorney is admitted to the bar, he or she must demonstrate the possession of superior education and knowledge.

For years within the surveying profession there have been major discussions as to the "formal" requirements for a formal education: college degree or no degree? This text simply states that a college degree is only the beginning of opening of the professional world to the young student. Most states now require a basic college degree in order to qualify for licensing; this does not indicate that a college degree makes one a surveyor. It is the foundation, and, coupled with experience, field training, business practice, and many other areas, all are part of the profession.

Learned professions depend heavily on formal education and curricula leading to a professional degree. A variety of surveying degrees are now offered in a few university-level colleges and in a number of two-year associate programs. As of this writing, land surveyors as a whole do not have the formal educational standing possessed by the higher professions. If education is the development of the thinking faculties of the individual and not training, land surveyors may deserve higher standing than formal education indicates, but so far, this has not been recognized by the public, yet each year more legislatures impose the requirement of a four-year college before one can be certified "to sit" for the professional licensing examination.

Education and knowledge are personal, and each individual acquires it on an individual basis. The major deterrents to the development of surveying as a learned profession are the current low requirements for the right to practice. Until such time as the knowledge requirements for registration are raised, surveyors, in the minds of others, will be relegated to an inferior professional position.

Although formal college education is a requirement to enter the profession, there must be continuing education requirements after registration is attained to maintain a degree of proficiency. One can never learn all there is to learn about a profession, no matter how much one studies.

In a recent decision in Florida, the court of appeals held that "a 'profession' is any vocation requiring a minimum of a four-year college degree before licensing is possible." It went on to state, "There is no requirement that the four year degree itself be in a field of study specifically related to the vocation in question" [5].

As of 1999, Florida now requires a minimum education of a four-year college degree.

17.6 Position of Trust

Knowledge in itself does not make a person a professional; one must use the knowledge to aid, assist, teach, and benefit others. The professional has a call of duty beyond the fee and beyond other selfish interests. Application of knowledge to assist and aid in the affairs of others is an essential part of the definition of a profession. A professional has clients or students who directly benefit from the application of this knowledge.

Professionals are often delegated an exclusive franchise for the purpose of protecting the public from the unqualified. In exchange for this exclusive franchise, the professional has moral obligations to the public; this individual is in a position of trust. In monumenting the exterior lines for a client, the surveyor must also determine the adjoiners' boundaries. In most surveys the adjoiner accepts the results without question. If an error is made, someone usually suffers a loss. As an obligation to the public, the surveyor should never do anything to compromise a client or an adjacent landowner.

The surveyor must protect *all* parties, but the contractual obligations are to the client.

17.7 Gaining Eminence

Professional eminence is earned because of superior ability to apply knowledge for the benefit of others. It is not derived from the ability to earn money or from the notoriety of undesirable publicity.

To be a successful professional surveyor, one must have more than a narrow technical education. Technical education has to do with things. Employees at a lower professional scale deal with things; professionals deal with people, situations, and ideas. The fundamental concept in human relationships is that it is not sufficient to be right; a person must be able to persuade. Things cannot be persuaded; humans can be persuaded.

All the technical knowledge in the world is of little value unless a person can also convey this knowledge to others in a convincing manner.

If surveyors are going to participate in activities outside their technical fields and furnish leadership in the broader affairs of the community, they must possess the ability to persuade and the knowledge to determine the right courses or direction in which to lead. The manipulation of things, the manipulation of the transit, and the manipulation of computers do not constitute professional standing; these are but the tools of the profession.

17.8 Independent Judgment and Liability

The lay person seeks help or aid in solving problems beyond his or her capabilities. The knowledge and experience of a professional, combined with the ability to reason and apply sound judgment, are why services are sought and purchased. The professional, after an initial consultation, determines the facts and uses independent judgment in arriving at a solution to a problem. Because professionals usually charge a fee for their advice, and as they profess to have superior knowledge, they assume a liability for failure to exercise proper care. The only products the professional surveyor sells are time, knowledge, and advice.

One of the greatest deterrents to substandard professional practice is the fear of liability. Although liability to an individual may be considered a disadvantage, it is an advantage to a profession as a whole. Without the threat of liability, those willing to accept professional prestige and yet to do substandard work for less money would soon ruin any professional standing gained by the majority. Pecuniary punishment of the careless and indifferent soon improves lagging professional attitudes.

Liability can result in several instances, first under a contract basis or under tort.

17.9 Services to Those Unable to Pay

Any learned professional is considered to have a moral duty to all regardless of the ability to pay. Doctors are obligated to serve the sick, the clergy to serve those in trouble, and attorneys to defend the criminal. However, fees are charged, and those able to pay must pay. This is as it should be. Seldom do property surveyors display this type of understanding or obligation. Property surveying is not an urgent necessity; if it is not done today, it can be done tomorrow. If a person cannot pay today, the survey can wait until tomorrow.

17.10 Fees

Money in itself should not enter into the definition of a profession. It is accepted that professionals must somehow gain a livelihood, and services are not denied to those in dire need.

The most suitable means of obtaining a livelihood seem to be based on a suitable fee charged commensurate with the service rendered. True professional fees are not dependent on the physical labor or force applied but should be based on the complexity of the situation and the resultant decision rendered. Personal knowledge gained through experience and education creates the demand for the service rather than the size of the muscle in the arm. Businesses compete on the basis of lowest prices; professions should not [6].

Although money does not enter into the definition of a profession, it does have a profound influence on what others think of a profession. A group that shows by its outward appearances that it is not successful in handling its own financial affairs and lacks sound business judgment can hardly instill confidence in the public.

17.11 Ethics

DEFINITION. *Ethics includes that branch of moral science concerned with the duties members of a profession owe to their public, their professional associates, and their clients.*

Because of its ambiguity, *ethics* cannot be easily defined. The foregoing definition, adapted from *Bouvier's Law Dictionary* [7], expresses the intent and purpose of ethics.

From the early Greeks and Romans we obtained a basis of what ethics is, but since a person's ethics are personal, it is difficult if not impossible to accurately define ethics. We can all tell unethical behavior if we see it, but we are hard pressed to define it beforehand.

Without a firm foundation of ethics, one cannot be a professional.

It is a fact with which everyone is familiar that an individual may strictly observe the laws of the land and yet be an undesirable citizen and a poor neighbor. The idea that each individual can and should establish for himself or herself rules of conduct for such relations as are not covered by law and without reference to experience or opinions of others seems equally as absurd as would a similar attempt to establish principles of law. Laws must be established by the majority action of a legislative body, and rules of professional conduct must be based on the concurrent opinions of the members of a profession [7].

Rules or codes of ethics, as adopted by any profession, should be considered as general guides of conduct and behavior.

Advocating observances to ethics is not sufficient; the surveyor's personal example is far more important. It is not sufficient that the surveyor alone feels that he or she has honesty and integrity; the public, clients, and fellow practitioners must also believe it. The proof of observance of ethics lies in integrity as seen by the public, clients, and fellow practitioners.

Integrity is required if a practitioner is to be recognized as professional and ethical. People should be answerable for their mistakes, and should be ready to correct them. Integrity is defined by *Webster's Dictionary* as "the quality or state of being of sound moral principle; uprightness, honesty, and sincerity." Groups are judged by their individual members; professions are judged by the actions of the individuals they comprise. The actions of a few can cast doubt on an entire profession, resulting in a lack of confidence and respect by members of the public.

If a profession is to maintain a respected position in the community, it must look beyond the "club" of the law to ethical standards that prohibit the doing of what the law does not forbid.

Perhaps one of the most difficult ethical positions to be placed in as professional is reporting on the substandard work of a fellow surveyor. Many times in doing retracement work, a surveyor will follow another surveyor or will end up having problems with the other person's work. In cases of claims of malpractice, the law requires a statement from another surveyor stating what the standard is and what were the deviations from the standard. This is where the statement "to protect the public" enters. If one does make a statement as to the substandard work of a second surveyor, the surveyor should be totally committed as to his or her facts and opinion.

17.12 Surveyor's Obligations to the Public

PRINCIPLE 3. *The surveyor has the following obligations to the public: (1) to see that the client's boundaries are properly monumented without subtracting from the rights of the adjoiner; (2) not to initiate boundary disputes; (3) not to aid in unauthorized surveying practice; (4) to see that those licensed to survey are properly qualified by character, ability, and training; (5) to see that those who prove unworthy of their privileges have those privileges deprived; (6) to agree not to attempt to practice in any professional field in which one is not proficient; and (7) to produce a quality product regardless of the price.*

17.13 Obligations in Monumenting Boundaries

When conducting a boundary survey for a client, the boundary surveyor also identifies the boundary of an adjoiner, since each boundary has at least two owners for each boundary line run. A reason for giving surveyors the exclusive privilege of marking boundaries is to prevent the unskilled from monumenting lines that encroach on the bona fide rights of an adjoiner. As an obligation to the public, the surveyor should not in any way assist a client in acquiring rights to land that are not his or her to enjoy legally. Although some surveyors consider themselves as advocates, they sometimes disregard this advice.

17.14 Provoking Litigation

Provoking litigation, according to common law, is a crime known as *maintenance*. If the offender in a land-boundary case is a surveyor, he or she is doubly at fault. A surveyor may act as an arbitrator and try to smooth over a difficult boundary situation, but the surveyor should not stir up or fester litigation as a solution to the problem, especially when he or she would collect an expert fee as part of the litigation.

Because the surveyor is morally and ethically bound to protect the bona fide rights of the adjoiner, while holding the bona fide rights of the client paramount, he or she should not hesitate to point out the rights of the adjoiner. If there is a long-continued possession and title has probably passed by acts of possession, and if knowledge of such facts would tend to prevent the client from entering in litigation, the surveyor should not hesitate to disclose such facts. However, the surveyor should also suggest that attorneys are the proper parties to render a legal opinion and prepare legal documents on such matters. At no time should a surveyor discuss the legal ramifications of a case with a client.

17.15 Aiding Unauthorized Surveying Practice

PRINCIPLE 4. *No surveyor should permit his or her name to be used in aid of or to make possible the unauthorized practice of surveying by any agency, person, or corporation.*

What is unauthorized practice of surveying must ultimately be resolved by the courts, and ethics committees should be bound by their findings.

The common practice of some surveyors of selling signatures for a fee—that is, a surveyor signing a surveyor's certificate and certifying to the correctness of a survey where the work was not performed by the registrant or employees or direct subordinates of the registrant—is the most flagrant violation of the intent of a licensing law.

A lay person may be properly hired by a surveyor, provided that his or her services do not constitute the practice of surveying and that compensation is not a proportion of the fee. Having a lay person be a partner in charge of a corporation practicing surveying is a violation of ethics.

17.16 Qualifications of Surveyors

Applicants for registration are required to furnish a list of professionals as references. It is the duty of those replying to a board of registration, in response to questions of a person's qualifications, to disclose all unfavorable as well as favorable qualifications of the applicant. Friendship, family relationships, sympathy, or any other reason should not influence a statement.

Likewise, surveyors and surveying organizations must take the initiative to comment on the quality of the questions being asked on surveyors' examinations and to call attention to infractions of ethical considerations or differences in their technical capabilities. A most difficult position in which to place a surveyor is using the individual as a recommendation without first asking approval.

17.17 Surveyors of Questionable Capabilities

Occasionally, there are those who seek licensing who are not qualified. They will ask for recommendations from friends or acquaintances. These individuals feel an obligation to give a favorable recommendation, knowing full well that the applicant is not capable of doing professional work. If more people would give fair recommendations before a license is obtained, there would be fewer problems from those registered. If it is determined that those who are licensed prove themselves by their conduct unworthy of licensing, they should have this privilege removed. Surveyors are better able than lay people to appraise the qualifications of other surveyors. If any surveyor is frequently negligent, that fact will be noticed by other surveyors, who, as a group, can prefer charges.

17.18 Surveyor's Obligations to the Client

The surveyor, when performing a given service for a client, assumes certain ethical obligations in addition to liabilities. But these obligations to the client may not supersede or interfere with the surveyor's obligations to associates or to the public. He or she should serve clients faithfully but should refuse to do that which is illegal and unethical and that which violates a duty of responsibility to others. The surveyor advises clients about what is right and proper. However, if a client insists otherwise, the surveyor must withdraw, with notification being given to the client in writing.

Regardless of the fee charged, the surveyor is obligated to perform a correct survey within specified accuracies.

At times a property owner may indicate that an inaccurate or approximate property-line survey will suffice, but a surveyor should not accept such a commission. Future owners, not knowing the circumstances under which the monuments were established, will be misled. Approximately located monuments could be the basis for fraud or deceit on the part of the property owner or the client. Most people assume that surveyors' monuments are located correctly; hence the mere finding of a property corner may be the cause of a misconception and lead to costly litigation.

Like all other professions, communications between the surveyor and client are confidential and should remain so. But the surveyor may not be a party to an illegal act or fraud, and communications concerning illegal acts or frauds are not confidential. At law the surveyor is not given the right to withhold privileged communications. But unless required by law to disclose the business of a client, communications are confidential.

17.19 Surveyor's Obligations to Other Surveyors and the Profession

A profession is distinguished by the fairness and courteousness of one practitioner to another and the unwillingness to encroach on the clients of another. Businesses aggressively compete for customers; professions do not. Members of a profession value the esteem of associates and the prestige of their calling, especially so for those of mature age. But those who steal another's clients do not induce cordial reception or pleasant relationship, as it ought to exist, among surveyors.

Surveyors not only have obligations to one another but also to the profession as a whole.

In general, surveyors should not criticize another's work in public; take advantage of a salaried position to compete unfairly with other surveyors; or become associated with other surveyors who do not conform to ethical practices. They should cooperate with other surveyors concerning information of mutual interest of benefit and support their professional organizations.

17.20 Professional Reputation

PRINCIPLE 5. *The surveyor does not attempt to injure falsely or maliciously, directly or indirectly, the professional reputation, prospects, or business of another surveyor.*

Confidence and respect for a profession are gained by praise of one member of the profession for another. Constant sniping between professional people can only degrade the profession.

This ethical rule prohibits the surveyor from falsely or maliciously harming the reputation of another. This does not prohibit the right of any surveyor to give proper advice to those seeking relief from negligent surveyors. Surveyors should expose, with substantial evidence, at the proper time and place, dishonest conduct in their profession and should not hesitate to accept employment that will assist a client who has been wronged. But it is distinctly bad taste and poor manners to accept the word of the client without first checking with one's professional associates. Many times, those doing the accusing are biased in the presentation of their side of the story. If it is determined that writings or written statements were made maliciously, and they are proven to be false, this could be considered as libel per se. Under this doctrine, all the claimant has to prove is that the statement was made and that it affected his or her ability to conduct business.

17.21 Employment

At one time it was considered unprofessional to supplant another land surveyor after definite steps were taken toward employment, and it was unprofessional to compete with another surveyor for employment by reducing usual charges. Since the time the federal government brought antitrust actions against various professional groups and won, now it is illegal to include such statements in rules of ethics. Under present laws the surveyor may not take into consideration minimum prices adopted by group actions.

17.22 Discovered Errors

PRINCIPLE 6. *In the event a property surveyor discovers an error or disagrees with the work of another property surveyor, it is the duty of that surveyor to inform the other surveyor of such fact.*

Surveys are not for the purpose of creating disputes between neighbors. If the adjoining property has been surveyed by another surveyor and the two surveys are not in agreement, the matter should be discussed between the surveyors prior to announcing that an error exists. Sometimes evidence found on the first survey may indicate that a different principle should be used in a later survey. Of course, if a surveyor has made a genuine blunder, other surveyors should not honor the mistake, but the first surveyor should be given an opportunity to review and prove the correctness of the survey, if possible.

This rule applies in the conduct of *utility* surveys, but when conducting a *forensic* survey, no contact should be made with the opposing surveyor without obtaining permission of the attorney.

17.23 Review of Another's Work

PRINCIPLE 7. *It is unprofessional to review the work of another property surveyor or for the same client, except with the knowledge or consent of such property boundary surveyor or unless the connection of such property surveyor with the work has been terminated.*

Consulting with another's client is considered to be, or has the appearance of being, an attempt to supplant the other engineer or surveyor. If a request is made to review the work of another, the person making the request should be informed of the ethics involved, and the original surveyor should be promptly notified of the facts. Even in the event that the work of the original surveyor appears to be fraudulent or neglectful and the original surveyor probably will be charged with misconduct, it is the duty of the second surveyor to communicate with the original surveyor and give him or her the opportunity to reply. This probably will not apply in the event either surveyor is named as a party in litigation. In such an event, the second surveyor should only communicate with the attorney who is employed to defend the original surveyor's work.

This situation may arise when one surveyor is asked to review another surveyor's work or work product as to technical compliance and personal qualifications. Unfortunately, this will place the surveyor being reviewed in a serious and unfortunate position.

17.24 Discredit to the Profession

PRINCIPLE 8. *It is unprofessional to act in any manner or engage in any practice that will tend to bring discredit on the honor or dignity of the surveying profession.*

A surveyor should not aid a client in perpetrating a fraud or assist the client in any illegal act. The professional surveyor should not, in personal appearance or manner of conduct before the public, other persons, or other professionals, bring personal discredit on the individual surveyor or any other surveyor.

17.25 Fees

PRINCIPLE 9. *No division of fees for surveying services is proper, except with another surveyor who by his or her license is permitted to do property surveying work.*

PRINCIPLE 10. *It is unprofessional to take employment on a contingency basis.*

The concept that a surveyor should not accept payment for services is primarily directed to the surveyor who is involved in litigation. Being paid by a client, with payment being rendered on the final results of the impending litigation, is an absolute *no.* Since the result of having won or lost the dispute dictates the amount of remuneration, this arrangement will jeopardize the integrity of the surveyor and the necessity of remaining an unbiased witness.

In determining the amount of the fee, it is proper to consider the following:

1. The time and labor required, the novelty and difficulty of the questions involved, and the skill required to properly conduct the survey.
2. The customary charges for a similar service.
3. The amount of liability involved and the benefits resulting to the client from the services.
4. The contingency or the certainty of the compensation.
5. The character of the employment, whether casual or for an established and constant client.

No single consideration in itself is controlling.

Recommended minimum-fee schedules can never constitute a binding agreement between surveyors. These have been considered as violating antitrust laws, which prohibit such price fixing, and a conviction may lead to civil and possible criminal penalties. This will be expanded to include prices recommended by state organizations or their functioning chapters. If such an organization or unit adopts such a minimum-fee schedule, all members may be included as possible litigants by the government.

17.26 Advertising

Professions usually refrain from advertising in self-laudatory language or in any other manner derogatory to the dignity of the profession. Professionals earn their stature; they do not try to win acclaim by advertising like a used-car dealer. Recent court decisions now bar professional societies from adopting a *no* advertising policy in their codes of ethics.

17.27 Ethical and Professional Standards

PRINCIPLE 11. *Violation of professional or ethical standards may be considered as "some evidence" of malpractice.*

All professions work under some ethical and/or professional standards, either written or unwritten. These standards serve several purposes: They give the professional a platform from which they can service their clients and the general public and they provide the consuming public, including the respective boards of licensure, standards against which a professionals conduct can be evaluated.

Whenever a client or a licensing board attacks a surveyor for failing to meet published or unpublished standards, reference must be made to standards that are adopted or are accepted as being the ones the surveyor is alleged to have breached. It has been pointed out that the first two elements of negligence are having a standard in place and breaching that standard. The courts have held that once a surveyor is proved to have breached a standard that should have been followed, simply being able to prove the standard was breached can be used to show negligence.

PRINCIPLE 12. *Professionals may have differences of opinions. Simply a difference of opinion will not constitute negligence.*

As professionals mature in judgment and practice, each realizes that two professionals may have differences of opinion as to what evidence indicates. As has been pointed out, if all professionals recovered the same evidence and then evaluated that evidence in a similar manner, differences in opinion are minimized.

For one surveyor to state that a second surveyor was negligent simply because there was a difference of opinion will not suffice. Professionals are entitled to differences of opinion. This is why patients seek a second opinion from a doctor or clients seek separate opinions from two attorneys. As professionals, surveyors are entitled to differ, yet when determining a boundary, all evidence should lead to the same boundary line. Failure to properly apply the rules of evidence may cause the surveyor to be negligent "as a matter of law" [8].

There will be instances when two surveyors will place different values on the same evidence, and as such there will be different opinions.

PRINCIPLE 13. *Having a valid license does not make a professional. The license requirements are governmental requirements in order to practice surveying.*

As pointed out earlier in this chapter, a formal degree in surveying or a related field should be an absolute requirement. A degree does not assure that the individual meets the requirements to be registered surveyor, but it does indicate the person possesses the tenacity and the willingness to use this education as a foundation for his or her profession.

PRINCIPLE 14. *The foundation of a profession is having a solid basis of education commensurate to other professions.*

PRINCIPLE 15. *A true professional merges practical experience with formal education to attain the degree of sophistication needed to serve the client base.*

Living is a learning experience. The young surveyor is fortunate in that each decade has the potential of new and greater possibilities for the chosen profession. As the closing words in this text, we could not say it better than did A. C. Mulford did in his publication, *Boundaries and Landmarks:*

> Yet it seems to me that to a man of an active mind and high ideals the profession is singularly suited; for to the reasonable certainty of a modest income must be added the intellectual satisfaction of problems solved, as sense of knowledge and power increasing with the years, the respect of the community, the consciousness of responsibility met and work well done. It is a profession for men [and women] who believe that a man is measured by his work, not by the purse, and to such I commend it [9].

17.28 Professional Standing of Surveyors

What the surveyor thinks of himself or herself is not proof of professional stature. It is what others proclaim the surveyor as being that determines his or her standing and is the criterion.

In response to liability litigation, courts have taken a firm stand that all surveyors are liable in the same manner as are other professional people. This is proof of professional standing, but it is not proof of degree of eminence of standing, nor is it prima facie evidence that everyone thinks surveyors are professionals.

One thing is certain: Professional standing can never be attained by self-proclamation; if a person wants professional standing, it must be earned! Others must bestow that title.

REFERENCES

1. *State v. Cohn*, 184 La. 53, 165 So. 2d 449 (1936).
2. *Aulen v. Triumph Explosives, Inc.*, 588 F. Supp. 4 (Md. 1944).
3. *Geise v. Pa. Fire Ins. Co.*, 107 S.W. 555 (Tex. Civ. App. 1908).
4. *Stiner v. Yelle*, 174 Wash. 402, 25 P. 2d 91 (1933).

5. *Garden v. Frier*, 602 So. 2d 1273 (Fla. 1992).
6. *Royal Bank of Canada Newsletter*, vol. 71, no. 6 (1990).
7. *Bouvier's Law Dictionary* (Kansas City, Mo.; Vernon Law Book; St. Paul, MN: West Publishing, 1914).
8. *Spainhour v. Huffman & Associates*, 237 Va. 340, 377 S.E. 2d 615 (1989).
9. A. C. Mulford, *Boundaries and Landmarks*, 1912, from a reprint by Walt Robillard, Atlanta, GA.

ADDITIONAL REFERENCES

Brown, C. M. "The Professional Status of Land Surveyors." *Surveying and Mapping, vol.* 21, no. 1 (1961).

Drinker, H. S. *Legal Ethics*. New York: Columbia University Press, 1953.

18

THE SURVEYOR IN COURT

18.1 Introduction

As a professional soldier trains and studies to avoid war, so should the professional surveyor train and study to avoid litigation. Hopefully, a surveyor can practice his or her entire career without entering the doors of a courtroom, other than as an observer. Yet in today's litigious environment, this will probably be very unlikely. Surveyors have increasingly become targets of litigation. In all probability, land surveyors will become involved in litigation in at least one or more areas. Whether the land surveyor is employed by a company or by a government agency, the individual in professional practice can expect to appear in court either as a litigant or as an expert witness. Litigation may encompass several areas of the law. Surveyors asked to testify about land ownership will usually be questioned about measurements or evidence of work that was accomplished in the ordinary course of their duties. Surveyors are experts in gathering, interpreting, and evaluating survey evidence, but the uninformed general public may also expect them to be experts in all aspects of surveying. This cannot be.

In today's litigious environment, surveyors may find that they are parties to the litigation in negligence or malpractice actions for products they prepared, either property surveys or measurements conducted for clients.

In today's legal system, some surveyors may find themselves called to testify as experts either for the parties to the litigation or on behalf of the court. This specialized area of surveying, namely, expert testimony, is a relatively new field that has developed to meet current needs. Expert testimony is a complex and highly specialized field of surveying that may be open to few practicing surveyors who excel in certain areas of the profession. In some states, the court, on its own motion, is privileged to call on the surveyor as an *amicus curiae* to act as an expert in the determination of property lines and other survey matters. Attorneys may ask surveyors to act as expert investigators or witnesses in boundary disputes or to evaluate the prior work of other surveyors.

Court procedure and witness behavior, particularly that relevant to the surveyor, attorney, and the court, are discussed in this chapter.

Courts are not creative institutions and do not legislate or pass laws; they merely attempt to interpret what legislatures have lawfully decreed as proper or legal, but not necessarily just. However, courts may create what can be called *judicial legislation,* for want of a better word. That is, they may "judicially legislate" certain decisions. In early case decisions, the courts in a particular jurisdiction may have interpreted the word *gray* to mean white. Then the old justices are replaced by a newer group, and when this question is raised again, they will interpret the word *gray* to mean black. A recent Georgia decision such as this was rendered on the question of a line tree—that is, a tree through which a property (deed) line between two lots passes. Heretofore, the Georgia courts had accepted common law dating from Roman law prior to AD 300 and English law before 1600 that a line tree is owned by both parties as "tenants in common" [1]. In a far-reaching modern decision, Georgia Superior Court held that a line tree was tantamount to a "party wall" in that each party owned their respective parts and the exact portion on their side of the line and owed the other party a right of easement of support to the remaining portion. The decision of the trial court was affirmed by the Supreme Court of Georgia, stating that it was time to change the law [2].

In this chapter the following principles will be discussed:

PRINCIPLE 1. *A surveyor should always be open to any evidence, pro or con, on all sides of the issue in question, even after he or she has made an opinion. Failure to do so will result in the surveyor becoming an advocate for the client, rather than being an unbiased witness.*

PRINCIPLE 2. *The surveyor is not an advocate but merely the reporter of facts.*

PRINCIPLE 3. *When a portion of a witness's testimony is found to be less than truthful, then the entire testimony may be discredited.*

Although the Rules of Civil Procedure vary in each state, certain principles apply to all civil litigation.

18.2 Pretrial Discovery

Today, a surveyor's first experience in litigation or court may be in the form of a deposition. Depositions are used by attorneys as one form of pretrial discovery.

At one time, attorneys rarely used depositions but relied on the trial testimony. Today, attorneys are taught in law school that the use of depositions is acceptable and efficient. They are not taught that depositions are expensive to the parties and can be troublesome for the surveyor. A deposition is a method of pretrial discovery that is transcribed by a court reporter, usually at the office of one of the attorneys. The witness (deponent) is placed under oath and is asked questions by the attorney requesting the deposition, usually the opposing attorney, who will use the information obtained as bait for a "legal fishing expedition." Unlike at trial, the witness is usually required to answer all the questions with few exceptions. The attorney for the surveyor being deposed is then permitted to cross-examine the witness. The witness (deponent) does not have to be a party to the action, but any information or statements gathered can be used in court as "evidence" for or against the party. The surveyor or witness (deponent) may also be deposed by his or her own attorney. This may be for several purposes. The main purpose is to "preserve" testimony in the event the surveyor is unavailable for trial, and then if the case is particularly strong, it may be used to force the opposing party into settlement.

18.3 Initial Court Appearance

In the first appearance before the court, the surveyor should be guided by the attorney. The judge usually will not permit random testimony. Unless the surveyor has thoroughly discussed the case with the attorney, problems can be expected. To keep from being disappointed by the questions asked by the attorney, the surveyor (witness) should understand the mode or procedures that will be used to extract his or her testimony. Many attorneys are concerned about the cost of litigation and will not employ the expert until the time of trial. This is a serious mistake on both the parts of the attorney and the expert who accepts the job.

Many witnesses make a serious mistake in trying to "win" or convince the opposing attorney of the sanctity of their position. One should always realize that the opposing attorney is there for one purpose—to win by any manner or means possible, not to change. Court cases should be won or lost on the merit of the evidence presented and on the foundation of law involved. The surveyor's only function in court should be to present, discuss, and explain evidence; the surveyor usually cannot present law without the permission of the court. In presenting evidence, the surveyor will be permitted to answer only questions asked by the attorneys or the judge. Thus, it is important that the attorney and the surveyor have a preconsultation to make certain the right questions will be asked. They must become a team, with each member of equal importance.

No person can predict the outcome of litigation. It may be concluded by the surveyor witness that a certain monument is an original, but the judge will decide that the evidence of a witness, whom the surveyor had good cause not to trust or gave little weight to, fixes the corner location. The surveyor may conclude that a certain principle of law applies, yet the judge, along with the aid of a persuasive lawyer, may conclude differently.

Surveyors cannot and do not decide how litigation should be conducted; they are not experts on the law of boundaries and evidence, at least in the eyes of the court. The role of the surveyor should be one of a consultant and as a witness.

18.4 Duties of a Surveyor in Court

PRINCIPLE 1. *A surveyor should always be open to any evidence, pro or con, on all sides of the issue in question, even after he or she has made an opinion. Failure to do so will result in the surveyor becoming an advocate for the client, rather than being an unbiased witness.*

Perhaps the most difficult job of the surveyor is to remain an unbiased witness. As an expert witness, the surveyor is considered free to testify as to facts and his or her opinions, yet in a boundary dispute in which the surveyor's survey work is being used as the "thrust" for the litigation, in reality the surveyor is biased to see that his or her viewpoints expressed on the evidence prevail. All surveys are but *opinions*. They are not proof. In a recent trial decision in Georgia, three different surveyors placed the same lot line in three different positions. The court, after becoming frustrated as to the conflicting results, requested the local surveying society to advise it on which one of the three surveys was correct. The court then rendered its decision. Neither the court nor society totally understood that the ultimate question was not one of a lot line location but which party had superior title.

In litigation involving land boundaries, the surveyor may be either a lay or an expert witness. The surveyor's major function is as a witness describing known and perceived facts or as an expert witness describing techniques and then expressing opinions within a special field of surveying. The actual relationship is not always definitive, for whether a surveyor will act as a lay witness or be qualified as an expert witness is at the discretion of the judge. Depending on each surveyor's abilities and capabilities, a unique relationship with the attorney is developed, and as such the surveyor witness may also be asked to accompany the attorney during the litigation and serve as an adviser during the course of the trial.

In the practice of surveying, the client usually asks the surveyor for solutions to a particular boundary problem. The surveyor seeks to determine a solution based on the evidence, as it is perceived. Acting as the sole judge of the quality and weight of evidence and of the conclusions presented, recommendations are made to the client. But in court, the surveyor presents the facts, and as such, becomes an advocate for that particular position espoused. The court ultimately decides on the relevancy, admissibility, and quality of evidence and the conclusions produced. Thus, in court the surveyor must change his or her mode of thinking and must always remember that the problem may have other solutions that are not familiar to the surveyor. The surveyor is merely a witness presenting the facts to the judge or the jury for determination. This concept was once discussed in a Michigan case. The decision was as follows:

> The maps which plaintiff introduced did not give the data necessary to measure the rights of parties on any theory. They gave no means of getting at just where the thread of the stream was supposed to be, or the shore conformation, except in a limited range, and were in other respects deficient.

> We have had occasion, in several instances, to point out that a surveyor cannot be allowed, under any circumstances, to fix private rights or lines by any theory of his own. Before a surveyor's evidence can be received at all, it must be connected with a starting point or other places called for by the grants under which the parties claim. His duty is neither more or less than to measure geometrically in accordance with those data, and his science goes no further. It is not his business to decide the questions of law, or to pass upon facts that belong to the tribunal dealing with the decision of facts. His testimony, as a man of science, is never receivable except in connection with the data from which he surveys, and if he runs lines they are of no value unless the data are established from which they are run, and those must be distinctly proven, or there is nothing to enable anyone to judge what is the proper result [3].

18.5 Conduct on the Stand

PRINCIPLE 2. *The surveyor is not an advocate, but is merely the reporter of facts.*

The surveyor's conduct on the witness stand should always reflect an attitude of a reporter of facts. At no time should a witness give the impression of being an advocate; that is the role of the attorney. Once a surveyor is placed in the position of an advocate, the effectiveness of treating facts and evidence impartially will be lost or seriously questioned. His or her effectiveness is impaired. Effective witnesses are those who answer questions fully, factually, correctly, and honestly, both on direct and cross-examination. No witness is under any obligation to volunteer any information to the opposition party. Harmful facts should be presented just as straightforward and with as much dignity as the facts relied on to make the case. The surveyor's attorney should be apprised of both supporting and detrimental facts. There should be no surprises as to facts or evidence. To convey the idea of attempted concealment of facts is a fatal mistake and could lead to the discrediting of all previous and subsequent testimony. Since the surveyor's duty is to testify as to the facts and opinions concerning these facts, attorneys sometimes "opinion shop" for a witness to fit their facts. The surveyor may find that an attorney will ask for opinions before asking for facts. This can be very dangerous to the young surveyor. Once the facts are changed, the opinions may also change. At all times the surveyor should realize he or she is playing in the attorney's area of expertise.

18.6 Object of Litigation

Generally speaking, each side believes its position, evidence, and law are correct. Based on these presumptions, each hires attorneys and experts to assist in proving the case in chief. The objective for the attorney is to win. The objective for the surveyor, as an expert, is to assist the client in winning—provided, and only provided, that the testimony is absolutely honest. The surveyor who is given an exclusive franchise to perform property–boundary line surveys is also given the responsibility of absolute honesty.

18.7 Surveyor-Attorney Relationship

The attorney is the advocate and the final interpreter of the facts, plus the final judge as to how the facts will be presented in court. The attorney is in full and complete command of the case. The surveyor should be in on all planning, and, often in land boundary litigation, may often become the major defense or attack for the client. Unlike the attorney, the surveyor is not there to win for the glory of winning. The surveyor is there to present survey evidence recovered and then to defend an opinion that resulted from the interpretation of that evidence. If the surveyor cannot or does not wish to do this, then the surveyor should inform the attorney at the earliest moment and withdraw, if it becomes necessary.

Any engagement between the attorney and the expert witness should be reduced to writing by executing a written contract, including such elements as when the contract becomes effective, the amount of compensation to be paid and how will it be collected, the exact role the expert will play, and will reports be required. It should be understood that the contract is discoverable by the opposing party at any time, and as such, what is revealed should be limited.

No contract can anticipate every situation that may be encountered, and as such provisions should be made for collecting the fee, should it become necessary. The contract should add an element of certainty and predictability to the expert–attorney relationship.

Since the contract is discoverable, no contract should even attempt to outline how the expert should do the job or what results the attorney wants the expert to conclude or draw.

In a Georgia appeals decision, the attorney was dissatisfied with the testimony of the expert during the course of the trial. Upon presentation of the final bill by the expert, the attorney refused to pay the expert's bill. The expert filed suit against the attorney based on a breach of contract as well as tort. The jury awarded the expert his full contract fee, as well as punitive damages for bad faith on the part of the attorney, in the six-figure category.

18.8 Court Trials

All the procedures and methods used in court trials, although interesting, are not of vital importance to the surveyor. Since the surveyor is a witness, it is important that each knows what constitutes proper conduct and demeanor. It is also very important that each witness knows what fees can and should be charged. All trial procedure, rules of evidence, and methodology are usually prescribed in the states by statutes and in the federal courts under rules of procedure.

18.9 Pretrials

Pretrial motions, as practiced in some states, are for the purpose of eliminating unnecessary court time and usually are an attempt to reach an agreement on the issues in a case. The attorneys and the judge, in conference, discuss the issues and agree on which points of evidence will be taken and the witnesses that will be used are identified. At this time, the judge or either party's attorney may ask for an expert witness to be appointed or named.

18.10 Types of Court Trials

State courts are not uniform. Most are divided into inferior, superior, and appeal. Inferior courts, so called, are police courts, small claims courts, night courts, traffic courts, municipal courts, and other courts wherein transcripts of proceedings are not usually kept. Decisions are usually instant. Land litigation is generally assigned to superior courts, and a transcript of the entire trial is kept by a court reporter. Usually, the cost of the transcript is paid for by each of the parties sharing the costs.

Trial court is a term used to designate the court wherein facts of a case are presented. Usually there is no appeal from the trial court's decision relative to what the facts are, but there can be appeals from questions of law. As an example, Jones and Brown are in a dispute over their boundary as located along a creek. The trial court will decide from the facts presented by the witnesses where the creek is located, and usually there can be no appeal from these facts. But whether the limit of ownership is on the bank, along the gradient line, or at the thread of the stream is a question of law that may be, if wrongly applied, appealed. Any irregularities of trial procedure, inadmissibility of evidence, and similar questions of law are appealable. Appeals are based on the facts presented in the trial court, and, unless a new trial is ordered during the appeal, no new facts may be presented or permitted at the appeal. Published court cases are confined to appellate court decisions and, sometimes, courts immediately inferior to the Supreme Court. In some lower courts, the judge may request the decision to be published if, in his or her opinion, the nature of the case is so significant as to "make new law."

When researching case law, usually only final courts of appeal records will be in state or federal reports.

In most instances, the surveyor has little function beyond the trial court; the surveyor's testimony and usefulness are usually confined to presenting facts. Since only questions of law may be appealed, a surveyor does not appear before the appellate or higher court. The special group of surveyors who serve as consultants to attorneys may be used to help research and to draft appeals, but this field of expertise is open to only a very few highly trained and specialized surveyors.

Surveyors may be used in preparing affidavits for trial courts and for use in those few cases that are appealed. Although new evidence cannot be presented above the trial court level, a wise attorney will seek the help of the expert witness in preparing and stating facts should a case be appealed to a higher court. This ensures that no salient facts are omitted.

18.11 Oath, Questions, and Answers

Prior to testifying, whether at a deposition or at trial, a witness is placed under oath to "tell the truth, the whole truth, and nothing but the truth." Some of the more modern courts will ask the witness to "affirm." This oath relates only to questions asked and not to information known to the witness. Being under this oath does not mean that the witness has to volunteer any known information, nor does it give liberty for the witness to ramble along on any subject that is thought to be relevant or that may have a bearing on the case. The truth applies only to answering questions as presented. If the answer given to a question results in a half-truth or could leave the court with a false impression and the witness feels that further information should be given, yet the information is not called for by the question, the witness may ask permission of the court to clarify any answer. The reason testimony in general is limited to the answering of questions is that the court wants to know in advance whether the subject matter the witness is to talk about is admissible evidence. Also, the opposing side then has an opportunity to object if the question asked is improper or flawed.

18.12 Direct Examination and Cross-examination

Although many individuals consider these two forms of examination as different, the experienced trial attorney will use both direct examination and cross-examination to prove the case. The examination of a witness at the start of a trial is first by the plaintiff or moving party. The individual testimony is called direct examination; the examination of the same witness, on the same matters covered in direct examination, by the adverse party is called cross-examination. The direct examination must be completed before the cross-examination begins, unless the court directs otherwise. The exact manner of presentation is usually identified in the rules of the court.

If a party does not have an opportunity for the cross-examination of a witness, then the direct testimony will probably be completely excluded.

In direct examination the surveyor's usual function is to answer questions directed at introducing certain evidence to tell "the story." The kinds, types, sources, and nature of evidence and the relationships of other evidence (usually measurements) are the foundation of the testimony.

If a question is not understood, the witness may ask for clarification, and without the clarification, the witness is not bound to answer. The question must be understood by the witness, not the attorney. After a question is asked and is understood, prompt answers should be made; long delays and hedging create the impression of indecision. As a trial tactic and as an aid to the attorney, on cross-examination the witness should permit a sufficient delay in giving an answer so as to permit the client's attorney an opportunity to object, if such an objection is required. It should be realized that all of the questions and the resulting answers are being recorded by the court recorder. Answers of "from this point to that point" mean nothing on the record; answers of "from point C on plaintiff's exhibit A to the centerline intersection of Main and Johnson Streets" are positive and clear. Body expressions such as nods of yes or no cannot be recorded. At times an attorney or the judge asks the witness to place a symbol on an exhibit. This can be done, and the witness should clarify the action by saying, "I am now placing an A at the point previously referred to." The record must state what was done. This gives any court of review the opportunity to relate the transcribed record to the trial evidence. The importance of this action was expressed in a Colorado case when the appellate court stated:

> We call attention to the fact that in the record there are more than 50 instances where the witness indicated points on exhibits without any mark whatever thereon to enable us to ascertain therefrom the point to which the witness was directing attention, and in several instances the exhibit about which the witness was testifying was not even identified. This laxity on the part of the trial court and counsel made it impossible for us to give consideration to such evidence [4].

Your attorney's rule to you should be "do not volunteer information; answer only questions asked."

In cross-examination, the opposing attorney has several objectives in mind, including these three main objectives:

1. To discredit testimony in controversy with his or her claims.
2. To use the testimony to his or her advantage.
3. If possible, to cause the witness to lose credibility in the eyes of the jury or court.

If the witness is testifying to facts, the truth will prevail. Facts are facts. These are difficult to discredit. All factual questions must be answered, and if some fact does help the other side, it should be of no concern to the surveyor. An attorney who is astute in trial procedure will not leave adverse facts to chance but will usually explain these adverse facts in the course of the direct testimony, and then the jury is not surprised if they are revealed. If an answer cannot be avoided, a prompt, normal answer leaves less impression in the mind of the jury, but it should not be so prompt that the attorney does not have time to object. A trained trial attorney can use cross-examination to help prove the case.

Further, on cross-examination, the witness must be prepared for rude treatment, twisted half-truths, and attempts to discredit abilities and qualifications by the opposing attorney. Through all of the theatrics of the cross-examining attorney, the witness should not fall prey by "losing one's temper."

18.13 Direct and Leading Questions

Court rules are universal as to the form of questions that may be asked of the witness. On direct testimony, only direct questions may be asked of the witness. An attorney who has not developed a "road map" of what is trying to be proved will usually have a difficult time. Some attorneys will write out questions beforehand, while others will "wing" it. Here, the witness can answer only the question asked. No explanations are permitted.

On cross-examination, leading questions are permitted. Any question suggesting to the witness the answer that the examining party desires is called a *leading* or *suggestive question*. As stated, on direct examination leading questions are not allowed, except in the sound discretion of the court and under very special circumstances, making it appear that the interests of justice require it.

On cross-examination, the opposing party may cross-examine the witness about any facts stated in the direct examination or connected therewith, and in so doing, leading questions may be asked. If the opposing party examines the witness about other matters, such examination is to be subject to the same rules as a direct examination and may be objected to and excluded. With an expert witness, the more modern rule does permit cross-examination into any related matter that may be considered of importance. As a rule of thumb, the surveyor or witness should pause before answering any question.

18.14 Hearsay

Boundary litigation, survey matters, and related areas are full of hearsay evidence. Hearsay can include the field books in which the surveys are recorded to statements made to the survey crew while they were conducting the survey. Evidence not obtained from the personal knowledge of the witness but from the mere repetition of what has been heard or what others have said is called *hearsay evidence*.

Evidence based on rumor, common talk, or the statements of other people is considered hearsay and, with few exceptions, is inadmissible. In surveying practice, the surveyor often bases opinions on hearsay facts, such as monuments found and commonly reported by many as being in a correct position or statements made in old field books. Obviously, a surveyor could not be present in person when all the original surveys were made; hence, opinions of where property lines belong may be based entirely on hearsay evidence of where the original surveyor set the original monuments that created the parcels. Expert opinions are admissible even though they may be based on inadmissible hearsay evidence. This was reinforced in two 1980 decisions by the Supreme Court of Georgia. By Code, an expert witness is permitted to testify as to his or her opinion on the facts presented by other witnesses.

In one case an expert witness, such as the surveyor, was permitted to testify as to his opinion on the ultimate issue in the case without invading the province of the jury so long as the subject was an appropriate one for opinion evidence. Also, an expert was permitted to base his opinion on hearsay and was allowed to testify as to the basis for his findings. When an expert's testimony is based on hearsay, the lack of personal knowledge on the part of the expert does not mandate the exclusion of the opinion but, rather, presents a question for the jury as to the weight that should be assigned the opinion. The evidence should go to the jury for whatever it might be worth [5].

Similarly, documentary evidence illustrative of oral testimony and authenticated by oral testimony is admissible. Further, where the admissibility of such evidence is doubtful, it should be admitted and its weight left to the jury [6]. This was also raised when the appellants' third enumeration of error, that the testimony and plat of appellees' surveyor were based on hearsay and should be excluded, was also without merit. An expert, such as a surveyor, was able to base his opinion on hearsay. The presence of hearsay did not mandate the exclusion of the testimony; rather, the weight given the testimony was a question for the jury [7]. Accordingly, it was held that the third enumeration of error was without merit.

18.15 The Jury

The jury system is very ancient in our legal history. Either one of the parties may request a jury trial, or a jury trial is assured, and the parties may waive the jury. All questions of fact are for the jury, if one is set, and when the trial is by jury, all evidence is for their determination. As stated, a jury trial may be requested by either side. If there is a jury trial, the jury decides questions of fact. For instance, if a government section corner is lost and there is a dispute about which of several monuments represents the true and correct corner, the jury would listen to the evidence presented by the surveyors and then would decide which monument represents the true corner. If a fence has been in existence for a period of time and the question is raised about how long the fence has been in existence, the jury would decide based on the testimony and the evidence. But if there is a question of the legal interpretation of the meaning of a deed term, or the legal question, then the judge would decide.

Since most boundary disputes involve conflicting and technical discussions, few attorneys desire jury trials in boundary litigation; hence, the judge must then decide questions of fact (evidence) as well as the application of the law.

18.16 Witnesses

Witnesses may be placed into two categories: lay and expert. In our legal system, each serves a specific purpose. In most boundary litigation, the attorney usually insists that the witness engaged be an expert in the field of surveying. In most instances, the attorney should evaluate what is it that is to be accomplished with a witness, and then once it is determined, what is needed to use the witness for that purpose. The attorney should then weigh whether the witness will be a lay witness or an expert witness. Each plays a specific and unique part in the scenario of the trial.

18.17 Lay Witness

Lay witnesses are usually used only in the capacity in which testimony may be only to the facts that are within their knowledge or perceived by the five senses. Ordinarily, they are not permitted to express opinions.

As with most rules, there are many exceptions. In fact, practically all of the discussion of the rules centers on instances that are exceptions.

Often, there is no way of extracting the facts from a lay witness, other than by allowing them to express an opinion on things derived from their own perceptions. A statement that there was an "old" fence between the two properties is an opinion. "Hot" or "cold" may only be an opinion, especially when a person says how hot or cold a thing was. A lay witness may testify that a car was going "fast" or "slow" or that an individual appeared "drunk." The lay witness testifies about subject matter readily understood by the court. Preventing the lay witness from stating opinions, especially on subjects regarding which he or she is not qualified, tends to prevent fraud and perjury and is one of the strongest safeguards of personal rights.

Specifically, the California Code (sec. 1845 CCP) is this:

> Witnesses can testify to only those facts that they know of their own knowledge—that is, facts that are derived from their own perceptions—except in those few express cases in which their opinions or inferences or declarations of others are admissible.

And, of course, a witness is presumed to speak the truth. This presumption, however, may be repelled by the manner in which the witness testified, by the character of the witness's testimony, by evidence affecting his or her character for truth, honesty, or integrity and veracity, by his or her motives, or by contradictory evidence. And the jury members are the exclusive judges of the credibility of the witness.

In analyzing lay testimony, conclusions and opinions usually are inadmissible, except when they are specifically derived from personal observation of the facts in issue and when no better evidence is available to the jury.

Like all other evidence, for lay testimony to be admissible, the judge must find that the testimony is based on personal observation and that the resulting opinion will be helpful to a clear understanding of the witness's testimony or the determination of the fact in issue. This last area includes areas in which normal people usually form opinions: "He looked sick" or "He was as drunk as a hoot owl" or "The fence was as old as Moses." Statements of this sort convey to the jury an impression by a manner of speech.

18.18 Expert Witness

In most instances in which a surveyor is called as a witness, he or she will usually appear as an expert witness. Before a witness is permitted to appear as an expert, the judge determines the witness's qualifications in relation to the ultimate issues. In arriving at his or her decision to either permit or exclude the expert witness, the judge must determine that the subject matter to be presented goes beyond the everyday knowledge of surveyors of ordinary experience and education. *Not all surveyors will qualify for the expert witness category.* A surveyor whose practice is predominantly in urban subdivision layout would not qualify as an expert in rural section retracements. If a situation should occur in which a surveyor was determined to be unqualified, he or she could possibly be held negligent for practicing in a field in which he or she is unqualified.

Any opinion expressed by the expert surveyor will go beyond what is normally expected. The court must consider if the witness is specially qualified by examining special knowledge and skills and the variety and depth of experience. It is the responsibility of the proponent who wants to introduce the expert to persuade the judge that the expert possesses the necessary qualifications. If the opposing party objects, the burden then rests on the proponent to convince the judge about the witness's capabilities. Factors used to convince the judge may include, but are not limited to, the witness's training and education, experience, familiarity with standard references and authorities in the field, membership in professional associations and societies, and general association within his or her profession.

A land surveyor testified that he had run out the lines of lots surveyed by a former surveyor and was familiar with his mode of marking corners, and the witness then testified to certain marks on certain alleged corners as having been made by the former surveyor. It was held that his belief that the marks were those made by the former surveyor was not evidence to be received by the jury as the opinion of an expert, but was merely the testimony of a witness to the effect within his knowledge and was to be credited by the jury only so far as they believed him able, from his personal knowledge, to identify the marks in question [8].

Ordinarily, the surveyor, as an expert, confines testimony to measurements, monuments found, and the interrelationship of monuments found. The case of *Curtis* v. *Donnelly* illustrates the evidence given:

> P. W. Warner, County Surveyor of Marion County testified he made and recorded a survey of Block 3, and record of his plat and survey was introduced in evidence. He testified that he dug into the earth and round the original corner stone in Block 2, which is 66 feet south of the southwest corner of Block 3, that he located the section line, that he measured the distance from the corner of Block 3, established as 66 feet from the corner of Block 2 to the section line; that he found a gain of 17 inches and of 12 inches; that he gave all the gain to the end lot; that he found certain corner stones and not others; that he did find the section corner stones; that fence located was 9 feet off at one end and 19½ feet off at the other. In rebuttal, E. C. T., another surveyor, testified that he measured the south side of Block 3 on the south line thereof beginning 14 inches east of the east side of the concrete walk on the east side of Green Street, which he treated as the southwest corner of said Block 3, and that from that point he measured east, and after allowing to each lot owner of said seven lots the number of feet as shown by the original survey of said block that belonged to each lot owner, and he found that the fence corresponded to the measurements. This is the full extent of his survey. He gave as his reason for beginning 14 inches east of the concrete walk in question, that he understood that the walk was built 14 inches west of the line of Green Street. The man who constructed the walk testified that it was built 10 or 11 feet west of the property line [9].

In survey matters, those not licensed may sometimes testify about boundary surveys. This could be somewhat confusing, in that many attorneys place great credence on the fact that an individual is or is not registered in that specific state, or even registered at all. Most surveyors will practice their entire lifetime in a single community or state, but there are other surveyors who are quite mobile, in that they are available to work and to testify in many states. This is especially true with experts. Attorneys are always asking, "Are you registered in this state?" The usual reply is, "Yes." But with experts who are authors or educators, many times they are not registered in the state in which they are to testify. The opposing attorney often will attempt to place great weight on the fact that the individual is not registered in the state in which he or she is testifying.

In Colorado, the court held that inasmuch as Furlong was not a licensed surveyor, he could not testify. The statute is 4696, C.L. 1921, which makes it unlawful for any person "to practice or to offer to practice engineering or land surveying in this state, unless such person has been duly licensed under the provisions of this act." Such a statute may or may not have any application in a case. In this instance, the testimony of Furlong was that he was engaged in land surveying in Minnesota, that he, at the time of his survey, was a clerk in the post office, and that he made the survey for his father who was one of the owners of the property. Such a survey was held as not practicing surveying. The appellate court held that to practice a profession is to hold oneself out as following that "profession as a calling, as one's usual business" [10].

This may be applied to "full-time experts," or those whose practice it is to testify on a regular basis. Expert testimony may not be considered as surveying. The authors have no knowledge of any state law that identifies land surveying as including "expert testimony."

18.19 Appointment of Expert Witnesses

Either side may call its own expert witness. In some jurisdictions the court may appoint an expert. Rule 706 of the *Federal Rules of Evidence* states:

> (a) Appointment. The court may on its own motion or on the motion of any party enter an order to show cause that expert witness should not be appointed, and may request the parties to submit nominations. The court may appoint any expert witness agreed upon by the parties, and may appoint expert witness of its own selection. The expert witness shall not be appointed by the court unless he consents to act. A witness so appointed shall be informed of his duties by the court in writing, a copy of which shall be filed with the clerk, or at a conference at which the parties shall have opportunity to participate. A witness so appointed shall advise the parties of his findings; if any, his deposition may be taken by any party. He shall be subject to cross-examination by each party, including the party calling him as a witness [11].

Many states have similar rules or statutes concerning expert witnesses. A California statute states as follows:

> Whenever it shall be made to appear to any court or judge thereof, either before or during the trial of any action or proceeding, civil or criminal, pending before such court, that expert evidence is, or will be required by the court or any party to such action or proceeding, such court or judge may, on motion of any party, or on motion of such court or judge, appoint one or more experts to investigate and testify at the trial of such action or proceeding relative to the matter or matters as to which such expert evidence is, or will be required, and such court or judge may fix the compensation of such expert or experts for such services, if any, as such experts may have rendered, in addition to his or their services as a witness or witnesses, at such amount or amounts as to the court or judge may seem reasonable [12].

The practice of the parties hiring their own expert witnesses has been severely criticized. There is an implication of bias toward the party calling the expert, especially when a sizable fee is involved. A precommitment or pre-consultation with a party in itself presents the flavor of bias. A party may interview many experts before finding one that is favorable to his or her cause and thus not present a true picture to the court. It is indeed unfortunate that many are often tarred by the same brush. Such ideas are an injustice to the conscientious expert whose only concern is the truth of the principle involved.

Today there is a trend toward liberalizing the rules governing the qualification of expert witnesses. The ultimate determination is, "Would the testimony be likely to assist the jury?"

The ultimate discretion as to the qualification of the expert is in the sole authority of the trial judge.

18.20 Duties of Expert Witnesses

In summary, an expert witness has seven duties:

1. Answer all questions clearly and intelligently.
2. Be absolutely unbiased and honest.
3. Have expert knowledge of the particular subject.
4. Be prepared to discuss the opinions of other authorities and state why you agree or disagree with them.
5. Limit testimony to things and opinions that you can defend before experts in the particular field.
6. Remain unbiased and open to information on all sides of the problem.
7. Remain current in the field through education and supplemental study and reading.

The opinions of an expert are generally only arguments on behalf of a litigant. Hence an expert's testimony is often valuable only because of the reasons and facts given to support his or her conclusions. Rejection of conclusions unsupported by facts or reasons can be expected.

Testimony that cannot be understood by the jury or judge is practically without value. A statement that "I ran a closure around the property using the deed calls and found an error of closure of two feet" is simple to the surveyor but obscure jargon to the average person. It is doubtful that the judge or jury would know that the traverse consists of an office calculation and not a field measurement. "Error of closure" certainly needs clarification for most people. A judge became a judge because he or she was an attorney, not a surveyor. One of the assets of a good expert witness is the use of simple, clear English. As a witness, the surveyor should not try to awe the jury with ability to "elucidate" with "ostentatious" terminology. The expert should be able to explain in simple terms.

At no time should surveyors place themselves in the position where they are obligated to take a particular side in a testimony. When the expert is on the stand, he or she should make every effort to be rid of any bias or prejudice resulting from who is paying the fee or who has previously consulted with him or her.

But this does not mean that the surveyor is to avoid expressing positive, sincere convictions. The surveyor is on the stand because he or she is supposed to know what to do under the circumstances. The surveyor must present the facts and opinions as he or she sees them and present arguments or reasons so clearly that anyone can understand and believe them. A person who is biased or prejudiced tends to deliberately slant reasons without sound foundation. A person who has a firm conviction based on sound reasons should prove it. If a person's testimony is doubtful, he or she should have avoided appearing as an expert in the first place.

No one can expect to remember in detail all the facts to be presented without refreshing one's memory. Surveyors take field notes and develop evidence in writing. Maps are examined and deeds are studied. Prior to taking the stand, the surveyor should review all the data; otherwise, embarrassment may result from a searching cross-examination. He or she should be prepared in particular to explain the error of the contrary opinion.

18.21 Opinion Evidence

As previously stated, the lay witness is limited to testimony concerning facts and may sometimes give opinions on things that are easy for everyone to understand. The expert may give opinions on subjects that are beyond the knowledge of average people. But such right to give opinions is not unlimited.

The expert witness's opinions may be based on such elements as the following three:

1. Facts or data made known to him or her outside the court (field books of prior surveyors, plats, and descriptions).
2. Facts or data from personal observations (distances between monuments and descriptions of found monuments).
3. Facts or data made known to the witness during the trial (testimony of other witnesses, assumed facts to hypothetical questions, and offered evidence).

Usually, an expert surveyor may be required to disclose facts or data prior to offering an opinion concerning the facts.

Federal Rules of Evidence Rule 703 states:

> The facts or data in the particular case upon which the expert bases an opinion or inference may be those perceived by or made known to him at or before the hearing. If of a type reasonably relied upon in the particular field in forming opinions or inferences upon the subject, the facts or data need not be admissible in evidence [13].

At times, courts may hold evidence on which expert opinions are based as inadmissible. As long as the facts are of the nature on which experts usually rely in that field in forming opinions, the facts themselves need not be admissible in evidence. At other times, it is impossible for the expert to frame a positive answer and must express the testimony in terms of probability. There is no legal requirement that an expert witness be positive or even reasonably certain in formulating opinion. This testimony goes only to the "weight" or evidence and not to the admissibility [14].

No witness can give opinions on the ultimate fact that is being tried. Permitting an expert to tell the members of the jury what they must decide is usurping their exclusive rights. If the north quarter-corner location is in dispute and any of three stone mounds might be the right one, the surveyor cannot tell the judge or jury which of the mounds to select. He can walk all around the subject and even answer hypothetical questions that almost give direct answers on the solution. He could describe in detail the shape and size of the mounds and describe any special markings found. If leaves were found under one of the mounds, indicating recent construction, such facts could be emphasized. What was found in other similar locations and whether the original surveyor consistently set a certain type of monument could be discussed. Needless to say, the examining attorney would have to be well versed on the facts observed by the surveyor: otherwise, he or she would not know which questions to ask. The surveyor is more or less limited in the responses to the questions asked.

A Mississippi court ruled on the question of whether the surveyor could be asked, "Where is the true line?" The court replied:

> This was not a matter about which they could give their opinion. It was a matter capable of being fully stated to the jury. The witnesses had surveyed the land and had fully described all the facts as found by them on the ground. It was the province of the jury to conclude from all the facts proved whether the line was where contended for by appellants or where insisted upon by appellees. The matter sought to be shown by the opinion of the witnesses was the very question in issue between the parties and was one for the jury alone to determine [15].

No expert opinions may be given to the members of jury if they are capable of forming their own opinion. The purpose of giving expert opinions is to advise the members of the jury on matters beyond their knowledge, not on matters within their knowledge.

Fortunately, the expert witness does not have to decide when he or she can or cannot give opinions. The judge will decide. A question is presented to the witness, and if the other side objects, the judge will rule on whether the question can be answered.

18.22 Hypothetical Questions

All attorneys thrive on hypothetical questions to opposing experts. Of course, the "hypo" usually contains only those statements that will require a response favorable to that particular attorney. A hypothetical question is a question put to an expert witness containing a recital of facts assumed to have been proved or proof of which is offered in the case and requiring the opinion of an expert witness thereon. Hypothetical questions have caused considerable criticism in many courts, but most jurisdictions still permit them.

Because witnesses cannot express an opinion on the ultimate fact in issue, hypothetical questions are asked. "Hypothetical" implies assumed without proof. Each side thinks or hopes that it can prove certain facts to be true. But neither side can be certain of what the jury will declare as being true. So, a hypothetical question such as "Assuming that this, this, and this are true, could you express an opinion as to whether a person so injured could continue doing surveying work?" is asked. The facts assumed to be true are the facts presented, and the facts must be in the record of the court when the questions are asked, or the examiner must have made an offer of proof. If the jury finds that the facts are not as assumed, the opinion of the expert is without effect. Several hypothetical questions may be asked by adding extra, assumed facts or subtracting assumptions. Hypothetical questions were frequently used for doctors in personal injury cases. Today in complex land disputes or negligence litigation, the surveyor will find such questions presented to him or her. The most favorable reply to a hypothetical question in which extraneous facts are included is, "Those are not the facts in this case, and I am unable to answer that."

18.23 Cause and Effect

An expert may testify about what might have caused an event, but it is better not to say what did cause it. Thus, a surveyor might testify that the improper location of a house might be caused by measuring from an incorrect monument located 10 feet from the correct monument. He or she should not say that it was the cause: that is for the jury to decide. To say that a 2 × 2-inch pine stake located at a proper corner would completely disappear in 30 years because of decomposition, termites, and so on, would be improper. To say that at all locations where pine stakes were originally set, none was found that was over 30 years old, and that those older than 10 years were in a bad state of decomposition, would be proper.

18.24 Textbooks and Treatises

Accepted textbooks and treatises are often used by surveyors and attorneys to support their cases. There may be instances when both attorneys will rely on the same books for both the plaintiff's case in chief and the defendant's case in chief. In reality, textbooks and treatises are considered as hearsay evidence. Because the author of the books cannot be cross-examined and the author did not write the books under oath, the books are, by common law, excluded as evidence; however, there are exceptions that vary from state to state.

An exception is identified in Rule 803 (18) of the *Federal Rules of Evidence* as follows:

> To the extent called to the attention of the expert witness on cross-examination or relied upon by him in direct examination, statements contained in published treatises, periodicals, or pamphlets on a subject of history, medicine, or other science or art, established as a reliable authority by the testimony or admission of the witness or by other expert testimony or by judicial notice. If admitted, the statements may be read into evidence but may not be received as exhibits [13].

In most states, the right to use books of science or art and published maps or charts is permitted under limited conditions by statutes and under common law:

> Evidence of this sort is confined in a great measure to ancient facts that do not presuppose better evidence in existence . . . The work of a living author, who is within the reach of the process of the court, can hardly be deemed of this nature. Such evidence is only admissible to prove facts of a general and public nature, and not those which concern individual or mere local communities. Such facts include the meaning of words

> which may be proved by ordinary dictionaries and authenticated books of general literary history, and facts in the exact sciences founded upon conclusions reached from certain and constant data by processes too intricate to be elucidated by witnesses when on examination. Thus, mortality tables for estimating the probable duration of the life of a party at a given age, chronological tables, tables of weights, measures and currency, annuity tables, interest tables and the like, are admissible to prove facts of general notoriety and interest in connection with such subject as may be involved in the trial of a cause [16].

In some states, the expert is allowed to read what other experts have written on the question being considered, especially if it is a matter of science. Since the judge is an expert on law, it is seldom wise to try to tell the judge what the law is. Stating, "It is my understanding that the surveyor must accept monuments in preference to measurements," is much better than stating, "The law is the surveyor must accept monuments in preference to measurements." Stating that, "I was following code Section 23:726 in making my determination," is usually acceptable.

Under *Federal Rules of Evidence* Rule 803 (18), if an expert has relied on statements from scientific texts standard in the profession or other learned treatises, such statements may be read into evidence under a special exception to the hearsay rule, either by the party calling the expert or by the adverse party in cross-examination. It is accepted that an expert may be subjected to rebuttal testimony by other experts. Thus, it is always possible to show on cross-examination that the texts the expert relied on in forming his or her opinion in reality do not support the expert; other portions of the book may have contrary thoughts expressed.

The *Federal Rules of Evidence* permit cross-examination of any expert witness about opposing views expressed in any text, whether the expert relied on it or not [Rule 803 (18)]. These same rules provide that the opposing statements from the text may be read into evidence under a special exception to the hearsay rule and thus may be used as proof of contrary views. A surveyor who wants to rely on texts to reinforce his or her opinions should be knowledgeable about all texts within the specific field.

Since the admissibility of reading texts in court is variable, the surveyor should obtain advice from an attorney in the state about what is permitted.

If a surveyor does use a text or treatise in his or her preparation and if permission is granted to quote from it during testimony, this simple act will open up the entire book for questioning on cross-examination.

18.25 Use of Photographs

Like electronic distance measuring equipment and theodolites, it is well established that photographs are tools that an expert can use in court. Photographs, both terrestrial and aerial, of the site to show the conditions existing, if proved to be correct, are admissible into evidence. The use of aerial photographs has been rejected in some cases and admitted in others. Overall, the courts have been quite receptive to aerial photographs as evidence, perhaps because of their uniqueness.

A picture taken directly below the airplane has a certain amount of distortion to the untrained person because of the "explosion effect," and straight lines appear bent where changes in elevation occur. In a photograph of a forest directly under the camera, images of the center trees are vertical, but due to radial displacement of the lens, all vertical trees away from the center appear to slope outward as would occur in a bomb explosion. For these and other reasons, the courts have held aerial photographs as not being accurate, especially when other evidence gives a better picture.

These errors have not deterred attorneys and surveyors from attempting to make "precise measurements" during the course of testimony. Many individuals who use aerial photographs as evidence or in their testimony do not have a sound fundamental training and understanding in their use. Aerial photographs should be used with caution.

Aerial photographs are valuable tools as evidence in boundary disputes to show items such as the age of fences, adverse possession, or lack thereof, changes in roads, and water courses. It is highly recommended that if the attorney and surveyor wish to use such photographs the witness first be qualified as an expert in aerial photograph use and interpretation.

Today, videotaped evidence is being used and computer-generated animations are being produced to aid testimony in major cases. One element often overlooked is that animation is nothing more than the creator's "opinion" of what happened and as such is hearsay.

A surveyor who wishes to use aerial photographs as an aid to testimony or as evidence should be very careful that no traps are set by opposing attorneys during qualification. Remember, a *surveyor who uses aerial photographs is not a photogrammatrist.*

18.26 Power to Compel an Expert to Testify

The right of an expert to refuse to testify about matters of opinion, without compensation, is traced back to English common law. A decision in 1843 stated that the right of the expert to demand compensation prior to testifying was upheld [17].

For the usual witness fee, a lay or expert witness is required to testify to facts seen or observed by the witness. If a surveyor observes certain facts or sets certain monuments, he or she may be compelled to testify about these facts. But when a party selects an expert to render an opinion on a peculiar subject, the expert can refuse until satisfactory arrangements are made for compensation.

Many jurisdictions hold that an expert must testify to facts within his or her knowledge, even though special study, learning, or skill was required to determine them. In a New Jersey decision, a civil engineer was denied the right to recover compensation for services as an expert witness. He was called as an expert by the party who employed him to make a survey. Since he attended court under a subpoena, he was bound to testify about what he knew, however he acquired the knowledge [18]. Then, in a Texas decision, a medical expert was compelled to report on the findings of a postmortem examination without extra compensation [19].

But there is nothing in the law that compels an expert witness to make a free preliminary investigation to prepare for expressing an opinion. A surveyor who is asked to perform certain surveys as a preparation for litigation cannot be compelled to do so without compensation. If a deed is presented to a surveyor in court and the surveyor is asked where it is located on the ground, he or she may refuse to read the document and give an opinion. But if the witness has already read the document and has already formed an opinion, he or she probably would be bound to express the opinion.

A District of Columbia court observed that it was an obligation of an expert to serve on payment of a reasonable fee [20].

In summary, surveyors in many jurisdictions are bound to testify about the results of surveys performed in the past. They are not bound, without compensation, to express opinions on things that require professional preparation prior to expressing the opinion.

It is interesting to note that an expert witness may not refuse to answer a question on the grounds that it will result in civil liability. Usually, the only ground for refusal to answer a question is that the answer may cause the witness to self-incriminate in a crime. Refusal to answer a question because it may cause embarrassment, disgrace, or monetary loss from civil liability is not an excuse. The privilege of refusal to answer a question applies only to crime [21].

18.27 Cross-Examination of Expert Witness

PRINCIPLE 2. *The surveyor is not an advocate but merely the reporter of facts.*

Once direct testimony is concluded, few surveyors are prepared for the cross-examination by the opposing party. It is the duty and responsibility of the advocate attorney to discredit the opposing expert in the eyes of the court and to strengthen a client's case. An expert can be cross-examined like any other witness, and the areas of examination may include the expert's qualifications, experience, honesty, prejudice, affiliation with the client, and common knowledge of the field, including publications and people.

The courts hold that cross-examination is the most effective manner of testing credibility and authenticity of testimony, and to ensure due process to parties in litigation, cross-examination is a right. Further, in cross-examination, leading questions and innuendos may be used. But cross-examination does have some restrictions: the cross-examiner may not utilize misleading and compound questions, cannot be argumentative, cannot assume facts not in evidence, and, above all, cannot require answers to questions in areas in which the witness is not qualified.

On cross-examination and direct examination, it is of utmost importance that the expert surveyor be completely honest in the testimony, for if it is determined that a portion of his or her testimony lacks credibility, this could jeopardize the entire testimony of the witness. As Mark Twain once said, "If you tell the truth, you will never have to lie."

When the cross-examination has attempted to impeach or discredit an expert witness, the party for whom the witness is testifying is permitted to redirect in an attempt to "rehabilitate" the witness. Basically, this involves restoring the credibility of the witness before the court.

18.28 Expert Witness Fees

PRINCIPLE 3. *When a portion of a witness's testimony is found to be less than truthful, then the entire testimony may be discredited.*

With any witness, the key words in regard to testimony, whether in court or in court documents, are honesty and credibility. As with any witness, honesty and integrity are expected, but there are those few witnesses who slant testimony or who are less than honest in statements made under oath. If it is ever ascertained that the witness's testimony is less than truthful, than the present testimony, as well as past testimony, as well as future testimony, will always be subject to questioning.

This goes not only to factual testimony but to professional opinions as well. If the surveyor opines an opinion as to a certain position, but at a later date the same surveyor, for the same or for a different client, offers an opinion contrary to the prior opinion, the integrity is questioned and the surveyor's effectiveness is seriously impinged. So, the key ideas in such a situation are, be consistent and be truthful at all times.

The surveyor's fee for expert testimony is a contractual arrangement between the surveyor and the party engaging the surveyor.

The courts cannot compel a person to perform work. If a measurement or observation is required, the right of a litigant to hire an expert is recognized. Before commencing the work, the surveyor should have a clear understanding of what the fee will be. If the work is performed, and there is no understanding about the amount of the fee, the surveyor may be compelled to testify about what he or she knows without the benefit of a fee. It is advisable to include within the fee all expenses that will be incurred in court appearances, since a person is compelled to appear in court, whether there is a fee or not. The only collectible part of the contract is the work for preparation to appear in court.

The compensation of expert witnesses cannot be dependent on the outcome of the litigation, since it furnishes a powerful motive for exaggeration, misrepresentation, or suppression.

The surveyor should be aware that witnesses who are to be called to give expert testimony involving special knowledge and skill, and often requiring examination and study on a particular branch of science, are, from the necessities of the case, justified in demanding and receiving compensation for their time and labor devoted to the investigation of the particular science about which they are to testify. This practice is usually allowed from the necessities of the case, and the inability of courts and juries to determine questions without the benefit of such expert knowledge. Such agreements, however, can never be valid where the amount to be paid is to depend on the testimony that is to be given and where the right of compensation depends on the result of the litigation [22].

Since 1968, a California law has regulated the procedure used to pay for expert witnesses. A person who is not a party to an action and who is required to testify before any court or tribunal or in the taking of a deposition in any civil action or proceeding solely about an expert opinion that he or she holds on the basis of special knowledge, skill, experience, training, or education and who is qualified as an expert witness will receive reasonable compensation for the entire time required to travel to and from the place where the court or other tribunal or, in the taking of a deposition, the place of taking such deposition are located and while he or she is required to remain at such place pursuant to subpoena. The court may fix the compensation or it can be by agreement with the attorney. The surveyor should ask for a subpoena under this law; it makes payment of a fee more certain.

18.29 Amount of Expert Witness Fee Is Subject to Cross-Examination

Cross-examination may extract from an expert the fact that he or she is to receive a fee and how much the fee is. Usually, the surveyor feels uneasy stating that compensation is being received. There is nothing wrong in being compensated for professional time. It is the time that is being sold and not the testimony, for if not, the surveyor's testimony may not be usable.

In an actual decision, the appellate court in Illinois stated that an engineer who testified for the petitioner may be asked, on cross-examination, if he had not been promised "considerable money" if the proceeding went through, as it is always competent to ask a witness, on cross-examination, by whom he is employed and whether he is paid, in order to show his interest [23].

But then in another state, in a park condemnation proceeding, it was error not to allow a question asked of a real estate agent, who had estimated values for the city, as to whether he was interested in the same land that would be benefited by the park [24].

18.30 Discovery by Interrogatories, Depositions, Admissions, or Production of Documents

In recent years the process of discovery has been added to the preliminary features of court trials; each attorney is given the privilege of asking the opposing attorney to produce certain documents and to ask for admissions, depositions of witnesses, and answers to certain questions of parties while under oath. In cases involving land boundaries the process of discovery is implemented by four devices:

1. Interrogatory is a list of questions directed to a party for answers under oath.
2. Requests for admissions are in writing; they ask a person to acknowledge the truth of certain facts or to authenticate certain materials.
3. Requests for the production of certain documents for the purpose of inspection and study.
4. Requests for depositions of witnesses who may appear at the trial. Deposition hearings are conducted with a court reporter and attorneys present.

Frequently, the surveyor is asked to give a deposition, and since it is recorded under oath, it is imperative that consistent answers be given at the trial. Before any trial testimony is given, a deposed surveyor should, as a matter of routine preparation, review his or her prior deposition(s).

The attorney is allowed privileged information; he or she does not have to reveal information that is considered privileged. When a surveyor works directly for an attorney and develops information directly for the attorney, the information may fall into the privileged class; during the deposition the surveyor should not answer the questions without a slight pause, thus giving the attorney time to object. If the information is privileged, the attorney will object.

Also, if the question is one of opinion, an objection will be made. Such questions as, "In your opinion is . . .?" and "Do you believe that . . .?" need not be answered, since they are a matter of opinion. Questions of facts must be answered. An answer to the question, "What did you find at the northwest corner of Wilson's property?" may be, "I found a 1-inch iron pipe at the corner of Wilson's fence." The answer does not limit that you think the fence is the corner of Wilson's property; it may be that a measurement controls. Be certain that the question and any possible implications of the question are understood before you answer.

When being served a subpoena for either a deposition or testimony at trial, it may be a subpoena *duces tecum*, which means, "Bring with you the requested items." The request may include field books, maps, reports, or other information furnished the client or anything bearing on the court case. A simple deposition without the request for documents or exhibits only requires the surveyor to appear.

The surveyor may be asked to assist the attorney in the preparation of questions for discovery and may be called on to answer questions, as may any other witness or party to the litigation.

18.31 A Survey Must Be Done by a Surveyor

This area is still open for considerable discussion and confusion between the courts and state registration boards. Those states that require surveys to be conducted "under the direct supervision" of a registered surveyor usually interpret this to mean that the surveyor does not have to be in the field when the survey work is actually performed. The surveyor can give direction and guidance and be available for decision making.

By contrast, the courts have taken an entirely different approach to the "direct supervision" requirement. A California court held that the surveyor who testifies in a case must be the one who made the survey. The court noted: "it was error to permit a witness to testify that a certain arch protruded over the lot line where said witness testified that the survey was not made by him personally, but by persons in his employ" [25].

An Alabama court also held that "direct supervision" meant exactly that, namely, the registered surveyor would be in the field at all times in supervision of the work of employees [26].

18.32 Preparation for Testimony

Perhaps the Boy Scout motto "Be prepared" is the best advice. It is assumed that all sources of information have been investigated, monuments have been searched for, calculations have been made, pictures have been taken, and the trial date is approaching.

Between the time of gathering evidence and the time of trial, several months to a year or more have passed. A thorough and complete review of all data, evidence, and acts is the first order of business for "refreshing one's memory." No one can answer questions promptly and certainly without the benefit of review. Some trivial fact, a fact sufficient for the opposing lawyer to use as a discredit, will escape momentary memory. Re-walking the survey lines, re-examination of monuments and notations of peculiar markings, observation of topography and plants, and conditions, offenses, and possession are all cross-examination questions. Particular attention should be paid to the apparent age of possession. These questions may be for the purpose of confirming personal presence on the site.

The most important exhibits of the surveyor are plats, pictures, and monument material. Plats are for different conditions to place emphasis on different features. Working sketches of the entire area, an enlargement of some detail, or overlays are all used to advantage.

Seldom is a boundary dispute or title problem litigated without a profusion of maps. New maps, old maps, maps from questionable sources, and unidentified maps may all find their way into the surveyor's testimony. There may be maps that were never seen before the trial, and there may be second-generation copies of an original map. Maps are nothing more than opinions and pictorial representations of the results of work, either by a surveyor or even the client. Maps must be large enough for the judge and jury to see and must be clear enough to be understood. Important points are given letters or numbers to facilitate ready reference in testimony. Numbering all corners, such as Corner No.1, and so on, is an aid to testimony.

Nothing should be put on a map that cannot be authenticated: It is a map of fieldwork actually performed and observed. The opposing attorney delights in and is very adept at discrediting and embarrassing a witness by revealing defects in a map.

Maps may be created simply to show a composite of other maps, yet they are all referred to as "maps."

The map should be complete with respect to what was done and sufficient to prove the points. It should show all measurements, points found, the names of the surveyor and survey crew, date of survey, legend, scale, north arrow, and all pertinent facts.

Maps, plats, and diagrams explanatory of locations may be introduced in evidence in connection with the testimony of a witness and as an aid to explain testimony. Such drawings are not produced as evidence within themselves but are to enable the jury to understand and apply the testimony in the case. The surveyor had testified fully as to the manner in which he or she made the survey. It was therefore competent for the appellant to offer to produce before the jury, in connection with the testimony of the surveyor and for the purpose of enabling the surveyor to point out such places to the jury—a map or plan of a subdivision and of the adjoining subdivisions, so marked to indicate the places where the stakes in question were found; in addition, evidence tending to show that such map or plat was a correct representation of the subdivision and correctly indicated the points where such stakes were found was proper for the consideration of the jury together with such map or plat [27].

Maps may be used in conjunction with other evidence. If pictures are taken, they should be numbered on the map and the position and direction of the camera indicated. In testifying, the statement, "I stood at Point Q on plaintiff's Exhibit Land and faced the corner in the direction indicated by the arrow and took picture #5 to reveal the condition of the corner as I found it," is clear and concise. Equally as important, the same facts should be labeled on the picture. Fumbling around and trying to identify where you were do not improve the opinion that the jury or the judge has of the witness.

Detailed drawings are cross-referenced to the plot plan by a general note, "See Detail P," and on the detail, reference is made to the plot plan as, "Detail or corner #11 on the plot plan."

The object of testimony is to prove a point, not to just assume it. In a 1932 bench trial, the trial judge ordered a new trial; evidence was insufficient for him to reach a conclusion. The dispute was over the location of the west one-quarter corner of a section in Missouri. Two surveyors testified. One testified as follows that he determined the southwest corner of section 3, from which he commenced his survey:

Q: I understand that you never established any section corner down there.

A: It was not lost and was not necessary to establish it.

Q: You never found it?

A: I found the exact point where the corner should be and was from reference points.

Q: Where did you start your survey?

A: At the intersection of the two fences that are known to be on the section line, and also the intersection of the two roads that my notes in the office show were located on the section line by Pleas Stubblefield in 1894.

Q: You never found that corner stone?

A: I located that corner at the intersection of the two fence lines that are known to be on the section line, and the two roads known to be on the section line.

Q: You just assumed that they were?

A: It was not necessary to survey, and there is no better way to get about it to establish a corner than the way I did. There is no other way to get the corner than the theoretical way I did.

Q: But you never made any survey to determine that. You determined that by looking at the ground?

A: That was all that was necessary [28].

He established the quarter section corner from which he ran the line between the northwest quarter and the southwest quarter 39 chains and 10 links north of the southwest corner of section 3. He said:

A: I gave each section of land the measurement shown on the plats of the United States Government—which was by proportionate measurement—I gave the half mile on the south side the forty acres government measure which was a little different from my chaining, and the north half mile thirty-eight chains and thirty-eight chains and ninety links, which is figured by the government, which showed that their chain would equal ninety-seven point seven five six links as against my chain which would equal one hundred links, and at the point figured by proportionate measurement I drove an iron pin and measured into a number of trees for reference and recorded the survey.

Q: In other words, each half mile bears its percentage of the loss, rather than one half-mile taking all the loss?

A: Yes, sir, it does [28].

His chainman testified concerning the determination of the southwest corner from which the last survey started as follows:

Q: Did you find any evidence of a corner being there?

A: We didn't find no corner but we found the fences. We had two fences and two roads to get the point of beginning.

Q: You just assumed there was a corner there because of those two fences there and two corners.

A: There was two roads that seemed to have been there quite a while.

Q: There was no evidence of that corner outside of these fences and roads.

A: We didn't find no rock.

Q: Did you look for one?

A: Yes, sir. The road had been cleared around in a curve.

Q: You never measured that to any other corner to establish this corner did you?

A: No [28].

The plaintiff's other surveyor, a former county surveyor, had made the survey of 1902. He said he made a record of his survey but he had lost the book and testified from recollection. He said that the west side of section 3 measured about 175 links short of what the government field notes gave it. He said, "We didn't find the section corner; however, on the west side of the section." Concerning the corner from which he started he testified as follows:

Q: You don't know whether you started from the corner or not?

A: I started from the recognized corner. It was fifteen feet west of what is now Kimball's fence, and it was in the center of the road west from there and south from there.

Q: You didn't try to establish the fact that was a corner by any further surveying?

A: I run that line. I started from the corner. We dug and found the corner.

Q: Do you know whether you found the established corner or not?

A: No, sir.

Q: You didn't go out to establish a corner?

A: We don't do that. When we find a corner that is not disputed we don't want any trouble.

Q: You assume everybody was abiding by that.

A: Yes, sir [28].

Asked as to how he then determined the boundary line between the southwest quarter and the northwest quarter, he said: "Just divided what belonged to each quarter as the shortage. I divided the shortage equitably. There was a shortage in the government survey. I determined that because there was a shortage between the distance across the section that the government field notes said should be there and the actual measurement."

The court went on to say:

> However, we think defendant is correct in his contention that plaintiff's evidence was insufficient to re-establish the lost quarter-section corner and to show in which quarter section the land in controversy was located. It may be that the point at which plaintiff's surveyors commenced to survey is the southwest corner of section 3, but plaintiff did not prove that it was. The testimony of these surveyors, which is herein above set out shows that the original corner monument could not be found and that they assumed that the place from which they started was the section corner from the roads and fences. If they had other information upon which to base this assumption, it was not introduced in evidence. Their testimony is replete with conclusions of fact or of law rather than evidence in regard to facts. If the roads were on the section lines, the section corner might have been either on the north side of the road, the south side of the road, or in the middle of the road. Evidence of a survey, which is not definitely shown to have commenced from a corner established by the government, or, if lost, re-established in accordance with our statutes, is of no probative force. *Clark v. McAtee*, 227 Mo. 152,127 S. W. 37; Id., 253 Mo. 196, 161 S. W. 698: *Atkins v. Adams,* 256 Mo. 2, 164 S. W. 603: *Nelson v. Cowles* (Mo. Sup.), 193 S. W. 579: *Wright Lumber Co. v. Ripley County*, 270 Mo. 121, 192 S. W. 996. There is not even one word of evidence in this record that the northwest corner monument of section 3 was found or that this corner was otherwise properly established. A surveyor's testimony "is never receivable except in connection with the data from which he surveys, and if he runs lines they are of no value unless the data are established from which they are run, and those must be distinctly proven, or there is nothing to enable anyone to judge what is the proper result." *Jones v. Lee*, 77 Mich. 35.43 N. W. 855, 857. See also, *Robells v. Lynch*, 15 Mo. App. 456; *Dolphin v. Klann*, 246 Mo. 477, 151 S.W. 956. Plaintiff, on another trial, may be able to show a survey between definitely established corners which would prove that the land sued for or some part of it is in the northwest quarter. Furthermore, on another trial, the court should have the benefit of the United States government survey [28].

The appearance of a surveyor in court is actually a team effort with the attorney. Prior to litigation, it is important that surveyors identify with their attorney their strengths and weaknesses, manner of testimony, and frank opinions. Neither the attorney nor the surveyor should be surprised by the other during the course of the trial. It is imperative that each thoroughly understands the manner of presentation, the evidence, and any problems that could possibly present themselves.

18.33 View of Site by the Court

At the discretion of the court, either party may request a view of the premises or parts thereof to better relate the testimony to the court. The view may be taken at the beginning of, during, or at the end of the trial, or more than just once.

The attorneys may point out certain elements or points to the judge or jury, but at no time may any of the witnesses talk, unless asked a question by the judge. The court reporter is usually present to record what is said. Ordinarily, the surveyor accompanies the court by being available to answer questions regarding particular items of evidence.

REFERENCES

1. Anonymous, King's Bench [1 Roll. R 255], 1622.
2. *Willis v. Maloof*, 184 Ga. App. 349, 362 S.E. 2d 512 (1987).
3. *Jones v. Lee*, 77 Mich. 43, 43 N.W. 855 (1889).
4. *Everet v. Lantz*, 126 Colo. 504, 252 P. 2d 103 (1952).
5. *Napier v. Little*, 137 Ga. 242; St. Hwy Dept. v. Peters, 121 Ga. App. 167 (1911).
6. *Savannah Ice Del. Co. v. Ayres*, 127 Ga. App. 560 (1972); *King v. Browning*, 246 Ga. 47 (1980).
7. *Cheek v. Wainwright*, 246 Ga. 171 269 S.E. 2d 443 (1980).
8. *Barron v. Cobleigh*, 11 N.H. 557 (1841).
9. *Curtis v. Donnelly*, 273 Ill. 79, 112 N.E. 293 (1916).
10. *Beaverbrook Resort Co. v. Stevens*, 76 Colo. 131, 230 P. 121 (1924).
11. *Federal Rules of Evidence* (St. Paul, MN: West Publishing, 1974), p. 105.
12. California Civil Procedure, Calif., sec. 1871.
13. *Federal Rules of Evidence*.
14. *Finda v. Boltón*, 6 NJLJ 240 (1883).
15. *Gichner v. Antonio, Title Co.*, 410 F. 2d 238 (Miss. 1969).
16. *Kirby Lumber Co. v. Adams*, 127 Tex. 376, 93 S.W. 2d 382 (1936).
17. *Gallager v. Market St. Ry. Co.*, 67 Calif. 13 (1885).
18. *Finda v. Bolton*, 6 NJLJ 240 (1883).
19. *Webb v. Page*, 1 Car. and K. (England 1843).
20. *Summers v. State*, 5 Tex. Ct. App. 365 (1897).
21. *United States v. Cooper*, 21 D.C. 491 (1893).
22. *Corpus Juris Secundum*, vol. 98, Brooklyn, NY: American Law Book Co., 1957, p. 98.
23. Schapiro, 144 App. Div. 1, 128 Y. Supp. 852 (1911).

24. *West Skokie Drainage Dist. v. Dawson*, 243 Ill. 175. 90 N.E. 377 (1909).

25. *Hermance v. Blackburn*, 206 Calif. 653, 257 P. 783 (1929).

26. *Lawler v. Hare*, 587 So. 2d 387 (Ala. 1991).

27. *Justen v. Schaaf*, 175 Ill. 49, 51 N.E. 695 (1898).

28. *Cordell v. Sanders*, 52 S.W. 2d 834 (Mo. 1932).

ADDITIONAL REFERENCES

Briscoe, J. *Surveying the Courtroom.* Rancho Cordova, CA: Landmark Enterprises, 1984.

Kaplan, J., and Waltz, J. R. *Evidence.* Gardena, CA: Gilbert Law Summaries, 1979.

Mundo, A. L. *The Expert Witness.* Los Angeles: Parker and Baird, 1938.

Stem, S. T. *Introduction to Civil Litigation.* St. Paul, MN: West Publishing, 1977.

The Consulting Engineer as an Expert Witness. Burlingame, CA: Consulting Engineers Association of California, 1972.

19

THE SURVEYOR, THE LAW, AND EVIDENCE

A ProfessionaL Relationship

The student surveyor has now been exposed to various elements and concepts of evidence relating to surveying. Yet, a client who has a boundary problem wants to know the kind of result that can be expected if he or she must litigate a dispute based on the evidence recovered and the evidence that is presented by the attorney, through the surveyor, at a trial.

Justice Oliver Wendell Holmes stated: "Law is a prediction of what the courts will do." In this final chapter, we will attempt to predict what one might expect if the evidence of a boundary dispute is litigated.

Using prior court decisions from several states, both metes-and-bounds and GLO, we will present how the courts discussed evidence and the subsequent rules these courts gave to both the surveyor and the attorney as to what each could possibly expect.

The surveyor must realize that these decisions in no way guarantee future decisions, but knowledge of how the courts have addressed prior decisions may aid in applying evidence and the rules thereof to present problems.

These decisions are presented chronologically so that the student may appreciate the evolution of how the surveyor fits into this complex area of evidence, boundaries, and the law. Preceding each decision are salient points we think the student should glean from each case. Also, in each decision, the important points, as we see them, have been highlighted. But the student should decide for himself or herself if these points are important or whether there are other points that should be highlighted.

19.1 *Bryan v Beckley*, 16 Ky. 91 (1809)

This is one of the earliest decisions relative to boundaries in the United States. The principles set forth seem like instructions an experienced party chief gives a young surveyor on his or her first day at work.

The surveyor will notice the emphasis placed on the compass and its relation to retracements, as well as the importance of chaining to the surveyor and his or her everyday work-related measurements.

November 18, 1809, Decided
PRIOR HISTORY: From the Fayette Circuit Court.
DISPOSITION: Decree reversed and set aside, and cause remanded.
HEADNOTES: To restore lost lines and corners.
The principles of the decision in this cause are declared correct.
The principles contained in a former decree directing the court below, cannot be changed.
When the cause returns by appeal from the execution of that decree in the court below.
Where those principles fall short, others apposite may be applied.

1. Course and distance not to be departed from, but in cases of necessity. Distances must first yield.
2. Allowances to be made for variation of the needle. Unevenness of ground to be allowed for.
3. Mistake in distance originally committed in one line, could have affected only the opposite. A mistake in one course, not to be presumed to have affected any other course.
4. Court bound to take notice that there is magnetic variation from the true meridian.
5. Surveyors generally took their courses from the magnetic meridian. Partial mistakes and inaccuracies never to vitiate a claim to land—Vide *Helm v. Small*, Hard. 369, *Morrison v. Coghill's legatees*, Pr. Dec. 382.Horizontal measure to be attained, being the basis of the art of surveying.
6. Existing lines and corners to govern, however variant from the courses called for.
7. When visible and actual landmarks fail, resort is to be had to courses and distances.
8. Departure from distance not indulged, further than necessary.
9. The order of the courses and distances in certificate of survey, not to govern; it is no evidence that the same order was observed in executing the survey.
10. A complainant ought to recover only as far as his claim is reasonably certain. Where it is uncertain whether land was or was not included originally, it ought not to be decreed.

JUDGES: CH. J. BIBB. OPINION BY: BIBB
OPINION: OPINION OF THE COURT, BY CH. J. BIBB.
THE present cause comes before this court upon a writ of error to the decree of the Fayette circuit court, subsidiary to the former opinion of this court and the consequent decree, whereby the circuit court was directed to ascertain and restore the lost lines and corners of Beckley's military survey.

That opinion and decree need not be herein recited; it will be found under the title of *Beckley v. Bryan and Ransdale,* in the *Printed Decisions*, p. 107.

The case was this: Beckley was complainant in chancery, claiming under a survey made in 1774 by virtue of a warrant for military services; his survey was confirmed by the statute of Virginia of 1779; but Bryan and Ransdale has obtained letters patent, containing a grant for a part of that survey, founded on posterior rights, but elder in date than the letters patent granting the land to Beckley. The military survey purported to be of 2,000 acres, but contained much surplus; the grant to Beckley described the land by reference to four courses and distances, and as connecting four corners, designated by marked trees. Two lines could not be traced by any visible marks, and two of the corners could not be found. The two corners found were connected by the line between them, and another line from one of those corners was visible; but the distance on that line was not terminated by the existence of the corner trees called for; it was in fact, and so named in the grant to Beckley, the longer line of an adjoining survey. So that the proposition formerly before this court, was, one line of the survey being ascertained and visible on the land by its terminating corner trees, and another line from one of those corners being ascertained as to the visible direction thereof, how shall the remaining two corners and lines be ascertained, restored, and connected with those actually existing, so as to close the survey by the courses and distances specified in the grant? The court, in the former opinion and decree, after reasoning upon the case and giving the directions to be observed by the inferior court, drew this corollary: "From his northwestwardly corner, an elm, buckeye and ash, extend a line south, twenty degrees west, four hundred and sixty poles; from his eastwardly corner, a white walnut and hoopwood, extend a line southwest, with a line styled in his grant, Wm. Preston's five hundred poles; the extremities of those two extended lines will be the lost corners; and then connect those corners with a line running *parallel* to the line which connects the two first mentioned corners and the survey will be closed."

In attempting to carry that decree into execution, new difficulties have occurred in the court below, by a disclosure and development of *facts* (by the report of survey) not before exhibited to this court. The most material facts, newly disclosed, are these: That an actual mistake has happened in surveying the ground originally, by running the given line before mentioned, variant in fact from the course intended and certified in the original plat and certificate of survey; lastly, a greater excess of the given line beyond the distance specified therefor in the grant, than was represented formerly to this court. Before the former opinion of this court was given, the attention of the parties was attracted to objects of greater importance and of greater collision; upon going on the ground to prepare for executing that opinion, they were particularly directed to the points above stated. In adjudicating upon these subjects, it is distinctly to be understood that we feel ourselves bound to adhere to the principles of the former decision, as well because we are fully persuaded that they are correct in themselves, as because, if they were not orthodox, we have no lawful power to change or oppugn their application to this controversy. So far as they apply, we disclaim all and every authority to counteract them by any other. If the case in its new features, can be adjusted according to the *principles* of the former decision, taken in their true spirit and effect, then assuredly it must be so decided; where those principles fall short, *others apposite* and *coadjuvant* may be applied.

For restoring and renewing lost lines and corners, the following principles are contained in the former opinion, viz:1st. "That nothing but necessity will justify a departure, either from course or distance." 2d. "When a departure from either course or distance becomes necessary, that the *distances* ought to *yield*." 3d. That in all cases where lost lines and corners are to be renewed, due allowances must be made for the variation of the magnetic needle from the true meridian. 4th. That proper allowances are to be made on each line for the unevenness of the ground over which it passes. 5th. That a mistake in distance committed in the original survey, on one line, could have affected the opposite line only, "and the presumption, in violation of the length of the other lost lines, cannot be carried further." From which may be deduced as a rational inference this 6th principle follows. 6th principle: That a mistake in one course, evidenced by applying the patent to the ground, cannot be applied, or be made by presumption, to affect any other course named; because the courses were taken from the quartered compass, by the magnetic needle, and therefore a mistake in one course, does not necessarily or probably argue a mistake in running any other course.

Having thus premised, we come to the decree of the circuit court now complained of, as being contrary to the former decree of this court, and to the facts established by the surveyor's report, upon which the said decree complained of was predicated.

The surveyor's report, confirmed and decreed by the circuit court, has exhibited Beckley's grant as bounded and closed by the lines, courses, distances and corners following, and represented in the annexed diagram [Figure 19.1]: E. elm, buckeye and ash, extant, as likewise expressed, line A to B, south 43 degrees west, 500 poles.

Patent course and distance, named for that line, south 45 degrees west, 500 poles; corner called for at B, not extant.

B to C, north 69 1/2 degrees west, 793 poles, being parallel to EA; course and distance expressed in the grant north 70 degrees west, 600 poles; corner named at C, not extant.

E to C, south 20 1/2 degrees west, at right angles to EA, which line the surveyor reports extended 460 poles only, falls seventeen poles short of intersecting the line B C. The patent course and distance of E C, south 20 degrees west, 460 poles.

E to A, south 69 1/2 degrees east 984 poles; patent course and distance, south 70 degrees east 800 poles.

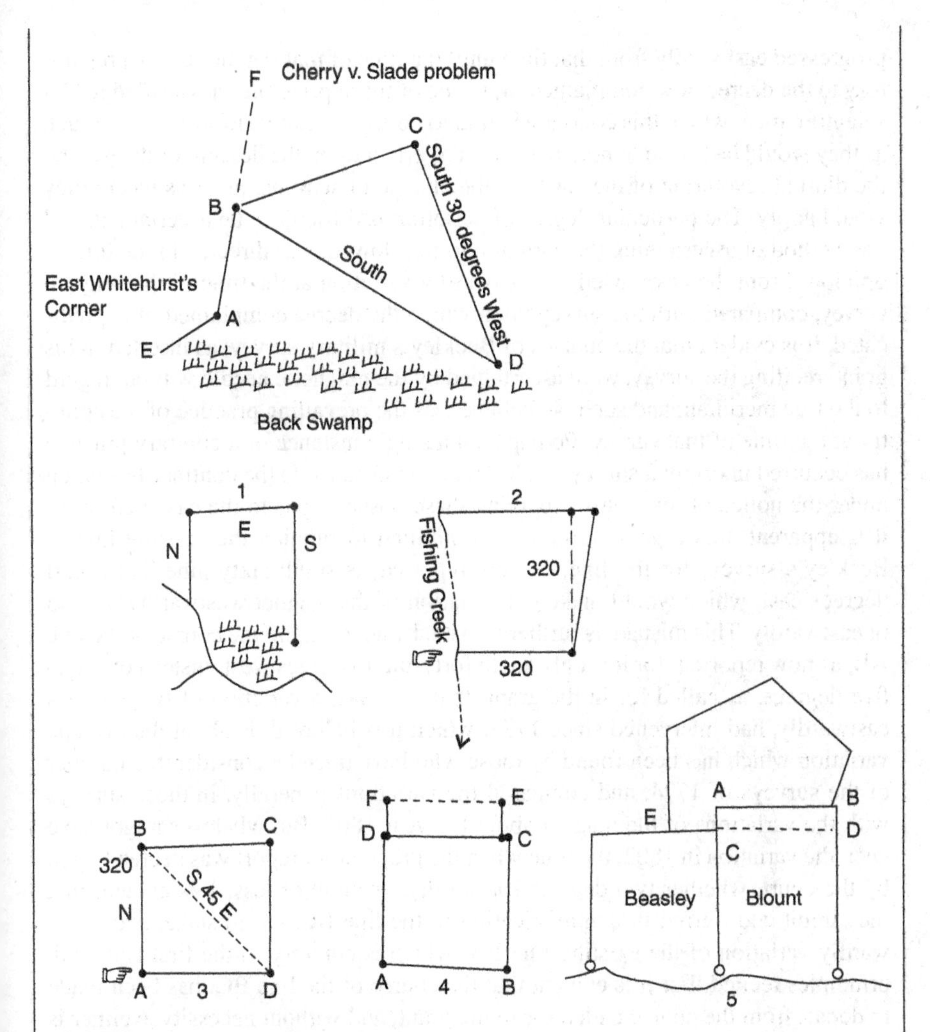

FIGURE 19-1. Case studies' diagrams.

The errors assigned in this decree, are:

1st. "The inferior court has erred in not pursuing the decree of the court of appeal, which they have not done either in its spirit or in its letter."

2d. "The inferior court erred in extending the line EC, beyond the distance prescribed for that line by the court of appeals."

3d. "In not having allowed on the line (EC) two degrees east, variation of the magnetic needle from the true meridian, that being the variation, or in not allowing some other variation to the east."

The variation of the magnetic meridian from the true meridian is recognized by the statutes, and by the former opinion of this court. That such variation was eastwardly of the true meridian at the time of the original survey, (in 1774,) that it had progressed eastwardly from that time until the time of making the survey preparatory to the decree now complained of, is one of those principles acknowledged by scientific men, which this court are bound to notice as relative to surveys, as much as they would be bound to notice the laws of gravitation, the descent of the waters, the diurnal revolution of the earth, or the change of seasons, in cases where they would apply. The particular degree of variation is difficult to be ascertained, and the method of ascertaining the variation to be allowed, was directed in the former opinion. From the acknowledged eastwardly variation at the time of the original survey, compared with the survey upon which the decree complained of is predicated, it is evident that the courses of Beckley's military survey, as inserted in his grant, reciting the survey, were ascertained by the magnetic needle without regard to the true meridian; and such we believe was the prevailing practice of the country at the time of that survey. Perhaps not a single instance of a contrary practice has occurred in original surveys; at least, not an instance to the contrary has fallen under the notice of this court. Applying these observations to the case before us, it is apparent, that a gross error was committed in running the existing line of Beckley's survey; for the line, as now reported, is south sixty-nine and a half degrees east, which would make the variation of the magnet westwardly instead of eastwardly. This mistake is further proved, by adverting to the course of the line AB, as now reported; for it is only south forty-three degrees west, instead of forty-five degrees, as called for in the grant, that is to say, a variation of two degrees eastwardly, had intervened since 1774, which it is believed, is about the average variation which has been found by those who have traced a considerable number of the surveys of 1774, and compared the variations generally, in those surveys with the variations of the magnet about the year 1802. But whatever might have been the variation in 1802, the time when the preparatory report was decreed upon by the court, whether two degrees eastwardly, or more or less, it is evident that the circuit court erred in communicating to the line BC, the mistake, and westwardly variation of the existing line EA, which is contrary to the first and sixth principles recited. For it is evident that the course of the line BC, has been made to depart from the course called for in the grant, and without necessity. Neither is this departure owing to a *due* allowance for the magnetic variation; but an *undue* and improper westwardly variation has been given to it, instead of an eastwardly. The circuit court have no doubt fallen into this error by attending to the direction, that this line should be parallel to the given or existing line, without examining that direction in conjunction with the principles of the decree; but nakedly and abstractedly, as well from those principles as from the additional light afforded by the report to which that direction has been applied. According to the grant, those lines appeared parallel; the court of appeals, not informed of the mistake committed in the *existing* line, directed the corresponding line to be renewed by running a parallel. But it is evident from the first, second, third and fifth principles extracted from that decree, that if the court had been informed of the mistake existing in that line, they could not have directed a parallel to it, without doing violence upon the very principles which they meant to preserve unimpaired. The mind cannot for a moment assent to the proposition that by any rational construction of the former opinion, the court have intended, by introducing the word "parallel," to produce a departure from the course of that line as expressed in the grant, after due allowance was made for the variation of the magnet. If any intention of departing from course, except from *inevitable necessity* had been intended, the direction to connect the extremities of the two lines, the one of five hundred poles, and the other of four hundred and sixty poles, would have been nugatory; for a line run from those extremities to connect them, would have closed the survey. But so to connect those lines did not comport with the views and intention of the court. Such a connection might have, and in fact would have, produced a departure from the patent course expressed for that line. For it is a truth, that the courses and distances named in the grant, without any variation, without mistake in any one distance, without any mistake in course, and tried upon a perfect plane, will not close the survey. To guard against such an event, and against mistakes, the court, therefore, added that these extremities were to be connected by a line "parallel to the line which connects the two first mentioned corners;" which provision was only tantamount to a direction to observe the *patent* course, for in the patent those lines were apparently parallel, each to the other. The opinion does not notice any mistake or departure in the given line. The very decretal order declares that Beckley was to have a decree for so much of Bryan's settlement survey, as should be found to be within Beckley's military survey when its lost lines and corners are run and fixed conformably to the *exception* contained in the aforegoing opinion; and that exception expressly declares, that the two lost lines are to be run from the corners extant, and extended on the courses and to the distances called for in the plat and certificate of survey, making a proper allowance on each line for the unevenness of the ground over which it passes, "and also to preserve the *course of the other lost line, as called for* in the same." Hence it may be affirmed with confidence, that a departure from *course* is repelled by every part of the former decision. The transfusing a mistake in one course into another course expressed in the patent, and thereby producing a departure from that other course, as is done by the decree complained of, was contrary to the manifest spirit and intention of the former decree of this court in the premises.

For the same reasons, the decree of the circuit court is erroneous, in establishing the line of four hundred and sixty poles at right angles to the existing line; since thereby a similar departure will be produced from the corresponding course of that line of four hundred and sixty poles; which supports the first, as well as the third numbers of the assignment of error.

Upon the second member of the assignment, it must be remarked, that although the surveyor has reported that the line of four hundred and sixty poles fell short by seventeen poles of intersecting the parallel line which he had run; yet by so reporting, he has thereby convicted his work of an error somewhere. That error may have been in some one or more of the courses, or in one or the other of the distances, or in the allowances made for unevenness of ground. Because if no such error had been committed, taking the courses and distances as reported by himself throughout, the survey will close by accurate delineation on a plane surface, or according to calculation by latitude and departure.

This is but another memento, that *partial* mistake or inaccuracies should never be suffered to vitiate a claim to land; and induces a remark, that reasonable accuracy can never be attained in completing this business, without good instruments, careful chain-carriers, and such allowance, or other safeguard, for unevenness of surface as will be equivalent to horizontal admeasurement, upon which the art and rules of surveying are founded.

It further appears that the decree of the court of appeals heretofore made in the premises, has not been carried into execution by the decree complained of, in this; that *due* allowance has not been made for the variation of the magnetic meridian. The circuit court have in fact communicated to every line to be renewed, the mistake which had been committed in the given line, and thus have closed the survey upon a mistake, and have thereby contravened the decree of this court, by departing from the courses of the grant in the lines to be renewed.

We would have it understood, that in alluding to courses of the grant as to the *lost* lines, we mean they shall be run with proper allowances for the variation of the magnet since the date of the original survey, to be ascertained in the mode pointed out in the former decree, so as to close the survey, and fix the lost lines and corners where they were originally, according to the survey recited in the grant. The ancient lines and corners now *lost*, are the objects to be attained by the best evidence the nature of the case now affords, to wit, the description as certified by the surveyor and recited in the grant.

It is further to be understood, as to those lines which are extant, or whose bearings are ascertained by existing corners, they are to govern, however variant from the courses called for. It is to the lost lines and corners that these rules are to be applied; where the ancient boundaries are visible and identified, resort to courses and to distances is unnecessary. Visible and actual boundaries, as rules to the govern property of men, are far preferable to ideal lines and corners; but when these actual landmarks fail, we must resort to the next best evidences, courses, and distances, as producing a reasonable degree of certainty, and a necessary security against the acts of the fraudulent and depraved, and against time and the elements.

It now appears that the courses and distances expressed in the grant, when extended from the remaining corners of this survey, will not close, as well on account of some inaccuracy in the original certificate of survey, when tested by calculation, or on a plane, even according to the courses and distances in the grant expressed, as on account of the particular mistake before mentioned as to the given line. The case is provided for in the last sentence of the sixth section of the former opinion as printed, (printed decisions, p. 110). The length of one or the other of the lines of five hundred poles, or four hundred and sixty poles, that is to say of AB, or of EC, must be departed from, and then the length of the one which is made to yield must be determined by calculation. We say the one or the other, because there is no necessity for both to yield, and to make them both do so, would be a departure from the spirit of the first and the fifth principles, and to the second case put, in the section just alluded to. A departure from distances even, ought to be indulged further than necessary. A departure in the distances of both lines, is not necessary; either lengthening the line of four hundred and sixty poles, or shortening that of five hundred poles, will do; and the section directs a lengthening *or* shortening; but not to do both, unless upon necessity. The former opinion excludes the idea of deciding the matter by going around from the beginning called for in the grant, in the progressive order of the courses and distances therein recited; because, by the fourth section of the opinion, it is clearly to be understood, that the order of the courses in the certificate of survey, is no evidence that the same order was observed in executing the survey. But there is a principle of equity which seems to be decisive on this subject; namely, that a complainant in equity ought not to have relief further than his claim is reasonably certain. Now it is evident, that if from the extremity of the line of four hundred and sixty poles (from C) a line is extended the patent *course* to intersect the line of five hundred poles (AB, shortened) and from the extremity (B) of the line of five hundred poles, a line be extended the patent course to intersect the line of four hundred poles (EC) extended, the claim to all the land between those two lines is *in dubio*, the land may have been excluded or included by the grant. And this argument would apply also to an extension *and* diminution of the distances on the lines EC and AB in equal proportions. But as to all the lands which would be included by retaining the line of four hundred and sixty poles in length, and reducing the line of five hundred poles by calculation to close the survey, the grant would be certain in every part, and for that only the claim should be established. It is, therefore, decreed and ordered, that the said decree of the circuit court be reversed and set aside, and the cause remanded, to have the former *decree* of this court carried into execution according to the principles thereof, in the said *opinion* contained, and as now explained in the aforegoing opinion; which is ordered to be certified.

And it is further decreed and ordered, that the defendant in error pay to the plaintiffs their costs in this behalf expended.

19.2 *Cherry v. Slade*, 7 N.C. 3 Mur. 82 (1819)

This early nineteenth-century decision is important for the opportunity it presents to the surveyor to use imagination in attempting to solve it. It identifies the methodology a surveyor should use in attempting to solve a complex boundary problem. Unfortunately, students ask, "What was the solution?"

Interestingly, at the time this decision was handed down, there was no answer. The answer is, "The decision gives no solution." It only offers guidelines a surveyor should use in retracing land boundaries.

The primary one being as follows: One should find the steps (evidence) of the original surveyor who created the boundary and then use this evidence as control for the parcel. In reading this decision, the student should attempt to ascertain the number of solutions that are possible using various elements in the description, eliminating those elements that are uncertain, and then attempting to put that description "on the ground."

The rules that have been established by the decisions of the courts, for settling questions relative to the boundary of lands, have grown out of the peculiar situation and circumstances of the country; and have been molded to meet the exigencies of men, and the demands of justice.

There are at least four possible solutions to this problem. How many can you find?

May, 1819, Decided

* * *

PRIOR HISTORY: In this case the Jury found a special verdict, so much of which as relates to the point which was sent up for the opinion of this Court, was as follows:

"Hislop's patent calls for Whitehurst's corner described in the plat of survey as at A, and thence along Ward's line 80 poles. "Ward's line is from C to D." Thence south on his line 320 poles to the back swamp. The line from C to D is south 30° west; and D is on the back swamp, and is the corner of Ward's patent, which covers the land to the east of the line from C to D [see Figure 19.1(b)]. If the 80 poles from Whitehurst's corner at A be run east, the course and distance will lead to B: and the south course called for in the patent from the termination of this line, will lead to D. If the east course called for in the patent, from Whitehurst's corner at A be continued, it will lead to F. and not touch Ward's line. If the Court be of opinion that the line from Whitehurst's corner at A, should be continued east 80 poles, and thence to Ward's line in the course nearest the description in the patent; or that it should run directly to Ward's line along the course nearest to that called for in the patent, the Jury find for the Plaintiff, and assess his damages to $ 345. But if the Court be of opinion that the line from Whitehurst's corner at A, should run east 80 poles, and thence south, though not on or with Ward's line, the jury find for the Defendant. It does not appear that Ward had any lands, except those covered by the patent referred to in this case."

DISPOSITION: New trial granted.

NOTES: The Court will award a *venire facias de novo*, where the jury, in a special verdict, find the evidence and not facts. The rules which have been established by the decisions of the Courts, for settling questions relative to the boundary of lands, have grown out of the peculiar situation and circumstances of the country; and have been molded to meet the exigencies of men, and the demands of justice.— These rules are,

That whenever a natural boundary is called for in a patent or deed, the line is to terminate at it, however wide of the course called for, it may be: or however short or beyond the distance.

Whenever it can be proved that there was a line actually run by the surveyor, was marked and a corner made, the party claiming under the patent or deed, shall hold accordingly, notwithstanding a mistaken description of the land in the patent or deed.

When the lines or courses of an adjoining tract are called for in a deed or patent, the lines shall be extended to them, without regard to distance: Provided those lines and courses be sufficiently established, and no other

departure be permitted from the words of the patent or deed, than such as necessity enforces, or a true construction renders necessary.

Where there are no natural boundaries called for, no marked trees or corners to be found, nor the places where they once stood ascertained and identified by evidence; or where no lines or courses of an adjacent tract are called for; in all such cases, we are of necessity confined to the courses and distances described in the patent or deed: for however fallacious such guides may be, there are none other left for the location.

JUDGES: TAYLOR, Chief Justice. HENDERSON, Judge. HALL, Judge, concurring.
OPINION BY: TAYLOR; HENDERSON

OPINION: TAYLOR, Chief Justice.—The land claimed by the Plaintiff under Hislop's patent, is described by the letters, A, B, C, D, and E. "Beginning at Whitehurst's corner at the letter A; thence east along Ward's line 80 poles," thence "south on his line 320 poles to the back swamp." If the lines first called for, be extended according to the course and distance in the patent, they would run from A to B, and thence to D, and leave out the land claimed by the Plaintiff; but in so running, they would not be on or with Ward's line; they would depart from the boundaries called for in the patent, for the sake of preserving the course and distance. If the course of the first line, viz. east, be pursued, it will lead to F, and will not strike Ward's line, which is intersected by running a course south 50 [degrees] east, at the letter C: and the question is, whether after running the first line 80 poles, east, it shall diverge in the course nearest to that called for, for the purpose of meeting Ward's line, and "of running on it" 320 poles to the back swamp, although the course of Ward's line is south 90 [degrees] west, instead of south, as in the patent.

The decisions which have taken place in this State on questions of boundary, have grown out of the peculiar situation and circumstances of the country, and have, beyond the memory of any persons now alive, been molded to meet the exigencies of men and the demands of justice, where the mode of appropriating an almost uninhabitable forest, had involved land titles in extreme confusion and uncertainty. In many cases, surveys were no[t] otherwise made than upon paper; and in many others, when an actual survey was made, the purchasers from the Lord's Proprietors were in danger of losing their lands by an inaccurate description of them, the omission of whole lines, and the mistake of courses and appropriated by a general description of courses and distances, without natural boundaries or marked lines, cannot be identified after the lapse of a considerable interval of time. If a beginning tree only were marked, the property continually revolves around it, and never can be ascertained; for no person can pronounce what course must now be run in order to ascertain a line, said to be run in a certain direction an hundred years ago, from the uncertainty in the variation of the compass, and from carelessness or the want of skill in measurement. It is easy to conceive, therefore, how utterly impossible it would have been, to render any thing like justice to claimants under old patents, if the land described in them were to be allotted only according to the courses and distances, to the neglect of natural boundaries, marked lines, and the well established lines and corners of adjoining tracts. Hence, certain rules have been laid down and repeatedly sanctioned by adjudications, which, in their application, have been found effectual for the just determination of almost every case that has arisen, and which have been considered for so great a length or time as part or the law of the country, that they ought not be abrogated by any power short of that of the legislature. These rules are,

That whenever a natural boundary is called for in a patent or deed, the line is to terminate at it, however wide of the course called for it may be, or however short or beyond the distance specified.

The course and distance may be incorrect, from any one of the numerous causes likely to generate error on such a subject; but a natural boundary is fixed and permanent, and its being called for in the deed or patent, marks, beyond controversy, the intention of the party to select that land from the unappropriated mass. In confirmation of this rule, many cases have been decided, only a few of which have been reported; but as some of them are fully up to the rule, and have been uniformly acquiesced in, it may be useful to bring forward the principal features of them.

In *Sandifer v. Foster*,[1] Gee's patent began on the mouth of dividing run, thence north, thence east, thence south to a white oak, thence along the river to the beginning. This white oak stood half a mile from the river, and if the line were run thence to the beginning, a large part of the land described in the Plaintiff's grant would be left out of Gee's patent.[2] It was decided that the river must be considered the boundary of Gee's patent. In *Pollock v. Harris*,[3] a swamp, a pocosin and a marsh, are severally called for in the patent, as the termination of lines, which if run according to the courses and distances did not extend to them. The natural boundaries were held by the Court to be the proper terminations of the lines. To these cases may be added *Witherspoon and Wife v. Black*,[4] *Hammond v. McGlaugon*, and *Swaine v. Bell Bethune*.[5]

Whenever it can be proved that there was a line actually run by the surveyor, was marked and a corner made, the party claiming under the patent or deed, shall hold accordingly, notwithstanding a mistaken description of the land in the patent or deed.

I understand the first decision of *Bradford v. Hill*, to be an authority for this rule; for although the Court directed the Jury to find according to the courses and distances called for in the deed, it was in the absence of proofs tending to establish the old marked line leading from Pollock's to Bryant's corner. The boundaries in the patent were, "beginning . . . on fishing Creek, thence east 320 poles to Pollock's Corner thence north the same number of poles to Bryant's, thence along Bryant's lines west 320 poles to the creek"—Bryant's corner being four degrees to the east of north from Pollock's corner, the line from Pollock's corner intersected Bryant's line considerably to the west of Bryant's corner. It was proved that there was an old marked line leading from Pollock's to Bryant's corner, but that in running by the compass north 54 [degrees] east, which was the general course of that line, it would be sometimes on the one side and sometimes on the other of that run by the compass, *whence it was taken by the Jury to have been run by some person after the survey.* The triangle formed by the said north line, part of Bryant's line, and a line from Pollock's corner to Bryant's corner, included the land in dispute. It was decided by the Court that the courses and distances in the deed must be adhered to, because the line from Pollock's to Bryant's corner, was not proved to have been run by the surveyor; but that in cases of evident mistake by the surveyor, parol evidence was admissible, though it ought to be admitted with caution.

The same case under the name of *Burton v. Christie*,[6] came on for trial before Judge Moore, when the only additional fact proved was, that some ancient deeds were bounded by the old marked line from Pollock's to Bryant's corner. The Judge directed the Jury to establish that line, if they believed that to be the one intended. In *Eaton v. Person*, the deed called for a course and distance, which carried the second line through the body of the land, leaving out a triangular piece, included in the second and third lines really run; but the second and third lines really run, as well as the corners, were marked and proved, by persons present at the survey for the patentee; upon which evidence the claimant under the patentee recovered.[7]

In *Person v. Roundtree*,[8] the latter entered a tract of land, lying in Granville County, upon Shocco Creek, which was run out, "beginning at a tree on the bank of "Shocco Creek, running south a certain number of poles" to a corner, thence north a certain number of poles to a "corner on the creek, thence up the creek to the beginning." By a mistake, either in the surveyor or secretary who filled up the grant the courses were reversed, placing the land on the opposite side of the creek to that on which it was really surveyed, so *that the grant did not cover any of the land surveyed.*—Roundtree settled on the land surveyed, which was afterwards entered by Person, who obtained a deed from Lord Granville, and brought an ejectment against Roundtree, who proved the lines of the survey and possession under his grant. The Court decided that Roundtree was entitled to the land intended to be granted, and which was surveyed; and that he should not be prejudiced by the mistake of the surveyor or secretary.

It was decided by Judge Haywood, in the case of *Beatty*, that if a course and distance be called for, and there be a marked line and corner, variant from the course, which is proved to be the line made by the surveyor as a boundary, that marked line shall be preserved.[9] In *Stanton v. Bains*,[10] the Plaintiff claimed under Askill, who patented a tract of land in 1740, extending, as he alleged, to a line distinguished in the plat by the name of the dotted line. The courses and distances expressed in the patent, extended not so far, but only to a line distinguished in the plat as the black line. The Defendant entered this intermediate tract in the year 1784, and took possession; whereupon the Plaintiff brought suit. The Court permitted evidence to be given, that the dotted line, which was a marked one, had, for a long time since the year 1740, been reputed the line of Askill's tract. The patent called for a gum standing in Robert's line; this gum was found at the termination of the dotted line. It next called for two lines of Robert's tract; the dotted line was upon these two lines. It called for Hoskins' corner; the dotted line went to that corner; and there was nothing in favor of the black line, but course and distance.

But there was no witness who could prove positively that the dotted line was the line of Askill's tract.

In *Blount v. Benbury*, Judge Hall says, there have been many decisions in this country which warrant a departure from the line described in the deed or patent, to follow a marked line, which the Jury have good reason to believe was the true one.[11]

The circumstances of that case, of which I have a fuller note than any published, afford a striking confirmation of the rule. The question arose on the title of a piece of land, which lay between two parallel lines, AB and CD. The latter line was contended for by the Defendant as the line of J. Blount's patent; in which case the Defendant was not in the possession of the Plaintiff's land: but if the line AB was the line of J. Blount's patent, the Defendant was in possession. The patent under which the Defendant claimed, Beasley's line and Blount's line at E, south 85 [degrees] east, as one of the boundaries.

The person under whom the Defendant claimed, in his deed dated in 1785, called for Blount's line, and at the same time, marked as such, the line from C to D. The principal question was, whether the line thus marked should be the boundary of the deed, or whether Blount's, wherever that should be ascertained to be, should be so considered.

It is to be remembered, that the line from A to B was an old marked line. It was decided by the Court, that Blount's line wherever it was, should be the boundary: that although the patent calls for Beasley's line and Blount's line south 85 [degrees] east for one boundary; still the Jury might consider Beasley's line the boundary as far as it went, and the line AB, which was 51 poles to the north of it: and that line was consequently established.[12]

In *Johnson v. House*,[13] Person surveyed the land for the patentee, under whom House claimed, and extended the line in question 160 poles, and marked and cornered it, as also the next line: but upon calculation, he found he had included 712 acres, instead of 640, and he cut off the land in question, by drawing from 80 poles instead of 160. But he returned a plat to the Secretary's office, mentioning the corner red oak, marked at the end of 160 poles, and the corner white oak, marked at the end of the next line drawn from thence. The plat having been returned with those corners, although mentioned to stand at the distance of 80 poles instead of 160, they were taken, notwithstanding the distance mentioned in the plat, to be the true corners. The corner marked at the end of 80 poles, was a white oak instead of a red oak, called for in the patent.

These cases, and many others which have occurred since, sufficiently prove the existence of the rule, which, if it had not been adhered to, would, in every case cited, have deprived the true owner of part, or the whole of his land.

Where the lines or corners of an adjoining tract are called for in a deed or patent, the lines shall be extended to them, without regard to distance, provided those lines and corners be sufficiently established, and that no other departure be permitted from the words of the patent or deed, than such as necessity enforces, or a true construction renders necessary.

This rule is founded upon the same reasons with the preceding ones, the design of all being to ascertain the location originally made; and calling for a well-known line of another tract, denotes the intention of the party, with equal strength, to calling for a natural boundary, so long as that line can be proved.

In *Miller v. White*,[14] the Plaintiff claimed under a patent to Nathan Bryan, beginning at a corner tree, thence south 80 [degrees], east 40 poles, to Walter Lane's line. There was no actual survey; the 40 poles were completed before arriving at Lane's line. The second line was with Lane's line to his corner, a certain course and distance; but that distance would not reach the corner. Supposing the line to be drawn from the point of intersection of the first line with Lane's line, if the first line extended to Lane's line, and if the second line went with Lane's line to his corner, then the land claimed by the Plaintiff was within Bryan's grant. If Judge Johnson was of opinion that the line should extend to Lane's line.

In *Smith v.* Murphey,[15] the Defendant produced in evidence two deeds: the third course of the latter deed called for 42 poles to a corner standing on the other tract. Forty-two poles were completed before arriving at the first tract. If the last line of the second tract should be drawn from the point where the 42 poles were completed, the land which the Plaintiff had obtained a grant for, was not within any of the Defendant's deeds; but if the line be extended beyond the 42 poles to an intersection with the lines of the other tract, then the land claimed by the Plaintiff was covered by the Defendant's second deed. It was held by the Court that the line should be extended to that of the other tract.

Where there are no natural boundaries called for, no marked trees or corners to be found, nor the places where they once stood can be ascertained and identified by evidence, or where no lines or corners of an adjacent tract are called for, in all such cases, we are, of necessity, confined to the courses and distances described in the patent or deed; for, however fallacious such guides may be, there are none others left for the location.

A case recently decided may seem, upon a hasty glance, to limit the application of the third rule, and in some degree to shake its authority; but an accurate attention to the circumstances of that case, will show that the decision and the rule accord with each other. The words of the Plaintiff's patent were, "Beginning at a pine in or near" his own line, and runs south 240 poles to a stake in William Hooks' line, thence with or near his line, north 73 [degrees] east, 400 poles to a stake, thence north 305 poles to a "pine, thence to the beginning." Hooks had three lines in the direction of the first line of the Plaintiff's patent, the two first of which formed an acute angle, and the corner of a patent which was ten years older than the Plaintiff's: the third line of Hooks formed one side of a patent which was seven years younger than the Plaintiff's. The question in the case was, whether Hook's line which first presented itself as one side of the angle, that which formed the other side of the angle, or the still more distant line of the new patent, should form the *terminus ad quem* of the Plaintiff's patent. The claim of the last line was rejected without hesitation, because, being made several years after the Plaintiff's location, it could not have been called for in the survey, although there was evidence of Plaintiff's declarations that the patent was not surveyed till after the grant issued. The Court decided in favor of the first line, as the establishment of that formed the least departure from the words of the patent, and as the course from that line was north 73 [degrees] east, as called for in the Plaintiff's patent; and would also run with or near Hook's line, which would not be the case, if the first line of the Plaintiff's patent were extended to Hooks' second or third line. Upon this case it is to be observed, that it is an express authority to show that where the line of another tract is called for, the line calling for it shall be extended thither, and in the terms of the rule, with no greater departure from the words of the patent than is necessary to satisfy them. The words of the Plaintiff's patent were satisfied by stopping at the first line, and would have been departed from, if the line had been extended to the second. It was also a part of the case, that no actual survey of the Plaintiff's land had been made till after the patent issued.[16]

It may also be thought that the second rule and the cases which support it, are broken in upon by the late decision in *Herring v Wigg's*;[17] but such an inference is not authorized by the very peculiar circumstances of that case. Michael Herring was the owner of a patent covering the whole land in dispute; and in 1778, conveyed to Keetley, under whom the Defendant claimed, "beginning at a" pine tree of Jacob Herring's and George Graham's land, and running with George Graham's line, and the same" course continued to a corner including 100 acres of land, "running north course to the patent line," Twenty-seven years afterwards, Michael Herring conveyed to his son the Plaintiff, by deed, which after describing several lines, called for "Richard Keetley's corner, a pine, thence" with Rutley's line south 98 poles to a pine, standing by "the side of Graddy Herring's fence." The Plaintiff proposed to prove by parol evidence that a marked line from L. to K. was the true line of the Keetley deed; and he did prove that at the time of the conveyance to him (the Plaintiff) Michael Herring and himself actually ran to K, and thence to L, where there had stood the pine by Graddy Herring's fence; and that Keetley had sent his son to shew this line. The Court were of opinion, that no parol evidence was admissible to show a line in contradiction to the deed, which would give less than 100 acres; that as Keetley's corner was admitted to be at U, a line running thence could not alter the location, which was fixed by a prior deed from Michael Herring, to which permanency was affixed by registration; that as to the line in question the deed did not purport to be bounded by a tree or a stake, but was to be run in such a way as to include quantity. It was, therefore, a boundary, which could not be mistaken, altered, or changed by memory or reputation, but would always speak in the same tone of decisive notoriety. With respect to the circumstances of the land's being run off before Michael Herring executed the deed to the Plaintiff, of the conduct of Keetley in recognizing the line K-L as the true one, they were held not to affect Keetley's right, or to change that which was originally made the dividing-line by the deed from Herring to Keetley.

This case does not affect the general rule; because the question in it was not where Keetley's line was originally fixed, but whether, when the survey was made and the lines established so as to include 100 acres, the posterior circumstances should have the effect of changing the line, so as to include a less quantity.

The right of the case before us depends upon the application of the third rule. The patent calls for Whitehurst's corner, which is ascertained to be at the letter A, and this is sufficient authority for running the course and distance of the next line from it, notwithstanding the unaccountable insertion of the words "along Ward's line."—After running out the 80 poles, the words are "thence south on Ward's line 320 poles to the Back Swamp."—Here we are presented with a choice of difficulties. If we run according to the course, we reach the Back Swamp, but we do not run on Ward's line [see Figure 19.1]. On the other hand, if we continue the first line in an eastern course, beyond the 80 poles, Ward's line will never be touched. It is therefore, less a question of construction respecting the patent than of fact to be ascertained upon evidence to the Jury, "whether the line described as *Ward's* was the line" originally called for, and according to which the land was located; and if Ward's line be established by proof, whether the second line in Hislop's patent was run from B to C, or from B. to D. I am of opinion these facts ought to be enquired into by the Jury; for which purpose there must be a new trial.

HENDERSON, Judge.—I think a *venire facias de novo*, should be awarded, because the Jury, instead of finding the facts have only found the evidence. That the line C. D. is Ward's line or a line of a tract of land belonging to Ward, is matter of *evidence*. That it is the line of Ward called for in Hislop's patent, is a question of *fact*, for the Jury to find from the evidence: and this fact may depend upon a variety of circumstances, all proper for the consideration of a Jury. This error has become too common, from confounding the evidence with the facts. A line, when once established to be the one called for, no matter by what evidence (if it be legal evidence,) whether it be artificial or natural, will certainly control course and distance, as the more certain description. A natural boundary, such as a water course, is designated from other water courses by its name, or by its situation, or by some other mark. One of those means of identifying the water course cannot control all the rest, if those other means are more strong and certain. A name, for instance, is the most common mean[s] of designating it; and this in general is sufficiently certain: but it cannot control every other description; and where there are two descriptions incompatible with each other, that which is the most certain must prevail. Cases might be put, where it must be evident that the parties were mistaken in the name, and therefore the name must yield to some other description more consistent with the apparent intent of the parties. It is true, that in cases of water courses or other natural boundaries, and in some cases of artificial boundaries, which are of much notoriety and have therefore obtained well known names, the other descriptions must be very strong: but if they be sufficiently so, the name must give away, and be accounted for from the misapprehension or mistake of the parties. This doctrine was fully illustrated in the famous suit relative to the *Cattail branch*, between Bullock's heirs and Littlejohn, which was more than once in this Court, and finally decided on the Circuit, to the entire satisfaction of the bench and the bar. Hislop's patent begins "at *A*, Whitehurst's corner, thence along Ward's line east 80 poles, thence south *on his line* 320 *poles*, to the back swamp." Ward's line mentioned in the case, is almost 240 poles from Whitehurst's corner, running in a southwestern direction, and would not be intersected by an east line from the beginning; and is the boundary of a tract of land of Ward's lying entirely east of that line. Of course, it cannot be the line of Ward called for from A to B or the first course in the patent. It is therefore almost certain, that it is not "the Ward's line" called for in the second course of the patent, and therefore would not control the course and distance, which is from A to B, and from B to D; leaving out the triangle B, C, D. But this is a question of fact, and the evidence should have been submitted to the Jury.

I presume the Chief Justice has correctly examined all the cases stated in his opinion; but I have not had an opportunity of looking into them; nor do I deem it necessary to do so, in order to illustrate my views of the points arising in this case—I am of opinion that there should be a new trial.

CONCUR BY: HALL, Judge—In this case I concur in the opinion delivered by Judge Henderson.

I think it is unnecessary to take any different view of the case than that taken by Judge Henderson—I concur in the opinion that there should be a new trial. New trial granted.

Riley v. Griffin, 16 Ga. 141 (1854) This decision was one of the first of many written by Chief Justice Lumpkin. Justice Lumpkin identified numerous rules of surveys and of law that apply today. It should be noted that one of the reference decisions referred to is the previous decision listed, *Cherry v. Slade*.

Rather than list these rules, we refer you to the decision that follows.

August 1854, Decided.

PRIOR HISTORY: Ejectment, in Bibb Superior Court. Tried before Judge POWERS, May Term, 1854.

This action was brought by the defendants in error, against the plaintiff in error, to recover a lot of land in the County of Bibb.

The plaintiffs in the action, introduced a grant from the State to Lewis L. Griffin, one of the lessors of the plaintiff, dated in 1836, to fractional Lot No. 3, in said county, and introduced testimony, showing that the grant covered the premises in dispute, and that defendant was in possession, at the time of bringing the writ.

Defendant introduced a deed from Willis Wilder to O. H. Prince, dated in 1835, and a deed from Prince to Wm. Riley, here intestate, dated in 1845, for part of fractional Lot No. 2, Macon Reserve. Defendant introduced testimony, going to show that this deed covered the premises in dispute, (the lots adjoining) and also to show possession and claim of ownership, since 1836. Defendant also introduced Warren B. Riley, the son of intestate, whose testimony was as follows:

"I have known the premises in dispute, since 1835 or 1836. My father was in possession of the premises, since 1835 or 1836, first as the agent of Maj. O. H. Prince, and after the death of Prince, he bought the land at Prince's sale, and remained in possession until his death; and his widow and administratrix has been possession ever since, and is now.

"During the lifetime of Prince, father always held possession as his property, using it as such; and after he bought it, used it as his own ever since, cultivating it and exercising acts of ownership over it, as his own. I know the lines well; I know the line is the old original line; have frequently seen the old blazes on the trees, which had all the appearance of the original Surveyor's mark. The owners of the adjoining lands set it up as the original line, and were governed by it, and no dispute about it until Adams bought the adjoining land."

If this land is not covered by the deeds from Wilder to Prince, and from Prince's administrator, (W. Poe) to Wm. Riley, then there is no land, whatever, that can be claimed under the deeds.

All the parties acquiesced in agreeing to the line, as fixed and known, until Adams bought, and then Brantly (Adam's father-in-law) made a fuss about the lines, in a Justice's Court. Riley was in possession of the premises in dispute occupying it and using it as his own, when Adams bought fractional Lot No. three, which they claim to cover the place sued for."

After argument, the Court charged the Jury as follows:

"This is a question of boundary. If you believe, from the evidence, that the premises in dispute are covered by the grant from the State to Lewis L. Griffin, then you will find for the plaintiff.

And if you believe, from the evidence, that defendant went into possession of the premises in dispute, under a mistaken apprehension, as to the real boundary, when, in fact, the premises were not covered by his deed, possession and user of it, as his own, under such misapprehension, for seven years, does not confer title, as against the true owner; and so, if you believe that plaintiff's grant covers the premises in dispute, you will find for plaintiff; otherwise, you will find for defendant."

Defendant's Counsel requested the Court to charge the Jury, that if Riley, and those under whom he claims, had been in the actual possession of the premises in dispute, for more than 7 years before suit brought, claiming it as his own, by himself and those under whom he claimed, and cultivating it and exercising acts of ownership over it—That such possession under the case made by the evidence, (if the Jury believed the testimony) conferred a good title in defendant, and the Jury would find for defendant, if they believed the evidence, as to the possession. The Court refused so to charge, but did charge, that a purchase of Lot No. 2, and going into possession of Lot No. 3, conferred no title to No. 3; and 7 years' open, adverse possession of No. 3, did not convey title to No. 3, against the true owner, because not bona fide, under claim of right, and adverse to the true owner. The party making such a mistake, does so at his peril—if such mistake, in your opinion, from the testimony, was made.

The Jury found, for the plaintiff, the premises in dispute.

Which said charge, and refusals to charge, defendant, by her Counsel, then and there excepted, and now assign the same for error.

DISPOSITION: Judgment reversed.

SYLLABUS: The rules set by the appellate court are as follows:

1. A possession which is the result of ignorance, inadvertence, misapprehension or mistake, will not work a disseizin.
2. Marked trees, as actually run, must control the line, which courses and distances would indicate.

3. If nothing exists to control the call for courses and distances, the land must be bounded by the courses and distances of the grant, according to the Magnetic Meridian: but courses and distances must yield to natural objects.
4. All lands are supposed to be actually surveyed: and the intention of the grant is, to convey the land according to that actual survey.
5. If marked trees and marked corners are found, distances must be lengthened or shortened, and courses varied so as to conform to those objects.
6. Where the calls of a deed or other instrument, are for natural, as well as known artificial objects, both courses and distances, when inconsistent, must be disregarded. And this rule is supposed to prevail, in most of the States of this Union.
7. Whenever a natural boundary is called for in a grant or deed, the line is to determine at it: however wide of the course called for, it may be, or however short, or beyond the distance specified.
8. Whenever it can be proved that there was a line actually run by the surveyor, or was marked, and a corner made, the party claiming under the grant or deed, shall hold accordingly, notwithstanding a mistaken description of the land in the grant or deed.
9. When the lines or courses of an adjoining tract are called for in a deed or grant, the lines shall be extended to them, without regard to distances, provided these lines and courses be sufficiently established.
10. When there are no natural boundaries called for, no marked trees or courses to be found, nor the places where they once stood, ascertained and identified by evidence; or where no lines or courses of an adjacent tract are called for, in all such cases, Courts are of necessity confined to the courses and distances described in the grant or deed.
11. Courses and distances occupy the lowest, instead of the highest grade, in the scale of evidence, as to the identification of land.
12. Any natural object, and the more prominent and permanent the object, the more controlling as a *locator*, when distinctly called for and satisfactorily proved, becomes a landmark not to be rejected, because the certainty which it affords, excludes the probability of mistake.
13. Courses and distances, depending for their correctness on a great variety of circumstances, are constantly liable to be incorrect; difference in the instrument used, and in the care of surveyors and their assistants, lead to different results.
14. In ascertaining boundaries, the locations of the original surveyor, so far as they can be found, are to be resorted to; and where they vary from the proprietor's plan, the locations actually made, will control the plan.
15. Whenever, in a conveyance, the deed refers to monuments, actually erected as the boundaries of the land, it is well settled that these monuments must prevail, whatever mistakes the deed may contain, as to courses and distances.
16. Courses and distances are pointers and guides, rather to ascertain the natural objects of boundaries.

17. Where a given line is exceeded in a grant, according to the courses and distances, evidence may be given of long occupation under it, to prove the boundaries.
18. Boundaries and courses may be proved by hearsay, from the actual necessity of the case.
19. Where a line has been run and agreed on by the co-terminous proprietors, and acquiesced in and possession held to it, for eighteen or twenty years, the parties and those claiming under them, are bound by it, no matter when, nor by whom, the line was run.

COUNSEL: STUBBS & HILL; WHITTLE, for plaintiffs in error. RUTHERFORD; POE; NISBET & POE, for defendants.
JUDGES: Lumpkin, J., delivering the opinion.
OPINION BY: LUMPKIN
OPINION: By the Court.—LUMPKIN, J., delivering the opinion.

We recognize the doctrine, that a possession which is the result of ignorance, inadvertence, misapprehension, or, in other words, mistake, will not work a disseizin—as, for instance, A has a grant to Lot No. 2; he is a stranger in the country, and calls upon some one residing in the vicinity of his land, who points out No. 3 instead of No. 2. A, acting upon this mistake, and not intending to occupy any other land than that which his grants covers, enters upon No. 3 and lives on it as his own for more than seven years. An occupancy, under such circumstances, would not, we apprehend constitute adverse possession. *Brown vs. Gray,* (3 *Greenleaf's R.* 126.)

It is the intention to claim title which makes the possession of the holder adverse; and this is the doctrine upon which the decision in every case proceeds. If it be clear, therefore, that there is no such intention, there can be no pretence of an adverse possession. (*Angell on Limit* 402, 412.) If one be the owner of a tract of land, and at the same time the agent of the owner of an adjoining tract, he cannot avail himself of the Statute, to support his title to a part of the land of his principal, of which he had taken possession upon a misapprehension of the boundary. *Cornegis vs. Carley*, (3 Watts 280). Whether there be any evidence to justify the charge, that the defendant's occupancy, in this case, may have been the result of mistake, is somewhat questionable.

But there is another portion of the charge, which requires more consideration. The Court instructed the Jury, that if Griffin's grant covered the premises in dispute, then the verdict must be for the plaintiff.

Is this proposition necessarily correct? We think not.

Let us refer, for a moment, to the testimony of young Riley and Jonathan Wilder. Warren B. Riley swears, that he has known the premises in dispute since 1835 or 1836. His father has been in possession since that time, first, as the agent of Major O. H. Prince, and afterwards, in his own right—he having become the purchaser of the property, when sold as the estate of Major Prince, by Col. Poe, the administrator. That either as agent of Prince, or in his own right he had always held the land, exercising acts of ownership over it, by cultivating it, &c. This witness testifies that he knows the lines well, and that the line to which his father claimed was the old original line. He has frequently seen the old blazes and trees which had all the appearance of the original Surveyor's marks. That the owners of the adjoining lands set it up, as the original line, and were governed by it, and there was no dispute about it until Adams bought the adjoining land. He further stated that all the parties, that is, those residing on the contiguous tracts, acquiesced in the line as fixed, until Adams bought; and then Brantley, Adams' father-in-law, made a fuss about the line in a Justice's Court.

Jonathan Wilder swore that he acted as the agent of his uncle, Willis Wilder, who owned the land before Major Prince bought it. That at the time it was sold, the blazes made by the surveyor who run the land, were plain on the trees; and that he followed the original marks. That Riley's fence is nearly on the line as run round by witness. That it is a little over at the corner, as well as he can recollect, judging from his eye and from memory; subsequent examination has confirmed him in this opinion. He knows he is not mistaken as to the lines, because he followed the original Surveyor's, marks, then fresh and plain on the trees, which were then standing, very few, if any, having been cut down. Prince and Riley, together have been in possession of the land, for the last eighteen or twenty years, claiming it as their own, under Willis Wilder. There never was any dispute about the boundary while witness controlled it, as the agent of his uncle. Witness has known the place, ever since the original survey was made, and before that time; has often seen the original Surveyor's marks, and could trace the original lines by them, and did so.

Now, it would seem, according to the proof, that when Lot No. 2 was originally surveyed, the lines may not have been run straight, according to courses and distances. But still, if the Surveyor marked these as the true lines it is quite clear that the owner of No. 2 will hold to these boundaries. Marked trees, or the line, as actually run, must control the line which courses and distances would indicate. If nothing exists to control the call for course and distance, the land must be bounded by the courses and distances of the grant, according to the Magnetic Meridian; for it is the practice, undoubtedly, of Surveyors, to express, in their plots and certificates of survey, the courses which are designated by the needle. But it is a general principle, that the course and distance must yield to natural objects called for in the grant.

All lands are supposed to be actually surveyed; and the intention of the grant is, to convey the land, according to that actual survey.

Consequently, if marked trees and marked corners be found, distances must be lengthened or shortened, and courses varied, so as to conform to these objects. *McIver's Lessee vs. Walker* (9 Cranch 173). Where the calls of a deed or other instrument, are for natural, as well as known artificial objections, both courses and distances, when inconsistent, must be disregarded. And this rule, says Mr. Justice *Washington*, is supposed to prevail in most of the States. *McPherson vs. Foster* (4 Wash. C. C. 45). Whenever a natural boundary is called for in a grant or deed, the line is to determine at it, however wide of the course called for it may be, or however short, or beyond the distance specified.

And whenever it can be proved that there was a line actually run by the surveyor, was marked and a corner made, the party claiming under the grant or deed, shall hold accordingly, notwithstanding a mistaken description of the land in the grant or deed.

When the lines or courses of an adjoining tract, are called for in a deed or grant, the lines shall be extended to them, without regard to distance, provided these lines and courses be sufficiently established.

Where there are no natural boundaries called for—no marked trees or corners to be found, nor the places where they once stood, ascertained and identified by evidence; or where no lines or courses of an adjacent tract are called for—in all such cases, Courts are, of necessity, confined to the courses and distances prescribed in the grant or deed; for however fallacious such guides may be, there are none other left for the location. *Cherry vs. Slade's Adm'r* (3 Mur. 82). The foregoing rules, Chief Justice Taylor remarked, had grown out of the peculiar exigencies of the country, and were molded by experience, to meet the demands of Justice.

And thus, it will be seen that courses and distances occupy the lowest grade, instead of the highest, in the scale of evidence, as to the identity of land. And it is reasonable that this should be so; for any natural object, when called for distinctly, and satisfactorily proved and the more prominent and permanent the object, the more controlling as a *locator becomes* a landmark not to be rejected, because the certainty which it affords excludes the probability of mistake:

While course and distance, depending, for their correctness, on a great variety of circumstances, are constantly liable to be incorrect. Difference in the instrument used, and in the care of Surveyors and heir assistants, lead to different results (*Lessee of McCay vs. Gallway,* 3 Ham. 282. *Thornberry vs. Churchill and Wife*, 4 Monroe's Ken. *R.* 32. *McNeil 1 vs. Massey*, 3 Hawk 91. *Beard's Lessee vs. Tullot*, 1 Cook 142. *Preston's Heirs vs. Bowmar*, 6 Wheat. 580).

This doctrine is found scattered, broadcast, throughout the authorities; and I had supposed to be too well understood and established to require to be discussed at this day.

In *Brewer vs. Gay*, (3 Greenleaf's R. 126,) it was held, that in ascertaining the boundaries of lots of land, where a township has been laid out, the locations of the original Surveyor, so far as they can be found, are to be resorted to; and where they vary from the proprietor's plan, the locations actually made will control the plan.

So, in *Dodge vs. Smith.* (2 N.H. 303, 1) the Supreme Court say, "whenever, in a conveyance, the deed refers to monuments actually erected, as the boundaries of the land, it is well settled that those monuments must prevail, whatever mistakes the deed may contain, as to the courses and distances." The same principle was decided in *Brand vs. Dawny* (20 Marten's Lon. Rep. 159).

In *Doe vs. Paine & Sawyer*, (4 Hawk's N. Rep. 64), the Court refers to courses and distances, as pointers or guides merely, to ascertain the natural objects of boundary. So, also, it has been held, that where a given line is exceeded in a grant, according to the courses and distances, evidence may be received, of long occupation under it, to prove the boundaries. (*Makepeace vs. Bancroft*, 12 Mass. 469. *Sargent vs. Towne*, 10 Mass. 303. *Baker vs. Sanderson*, 3 Pick. 354. *Livingston vs. Ten Broeck*. 16 Johns. 23). *Vide Davies' Abrid. of Am. L. vol.* 3, p. 307, where some of the early cases decided in Massachusetts, upon this subject, are collected. And, as landmarks are frequently formed of perishable materials, which pass away with the generation in which they are made; and are often destroyed, as in the case before the Court, by the improvement of the country and other causes, the boundary and corners may be proved by hearsay, from the necessity of the case (*Nicholls vs. Parker*, 14 East. 331. 12 East. 62. 1 M. & S. 679. 1 T. R. 466. 5 T. R. 26. 2 Ves. 512. 6 Peters 341. 4 Day's Conn. R. 265. 1 Harris & McHenry. 84. 368, 531. 2 Hayw. 349. 1 Yates 28. 6 Binney 59). So then, notwithstanding Mr. Wood run out Lot No. 3, according to the courses and distances designated in the plot accompanying the grant, and conceding that the lines, thus run, would cover the premises in dispute; still, if the testimony of Wilder and Riley be true, as to the original lines of Lot No. 2, actually run and marked by a surveyor, as a question of law, the plaintiff was not entitled to recover, nor were the Jury bound so to find. Again: Suppose the line sworn to, is not that which was marked by the original Surveyor; still, if it were agreed on by the coterminous proprietors, and acquiesced in, and possession to it held for eighteen or twenty years, the parties, and those claiming under them, would be bound by it, no matter when nor by whom the line was run and chopped. (*Brown vs. McKinney*, 9 Wharton's R. 567. *Burrell vs. Burrell*, 11 Mass. 294.) There is a small piece of ground which stands in this predicament; it is included in Riley's fence, but outside of the old boundary line. That being so, the title to this parcel of ground will not depend on the actual line run, as proven by the witnesses, because it is not included in it; still, it is enclosed by Riley's fence. Here, then, is *possessio pedis*. If this fence has been built seven years, then this strip of land will be covered by actual occupancy; otherwise, the plaintiff's right to that will not have been lost or taken away.

Judgment reversed.

19.3 *Stewart v. Carleton*, 31 Mich. 270 (1875)

This is one of two decisions discussed in which Justice Thomas Cooley participated (see Appendix C for a discussion of the Cooley doctrine).

Whereas the earlier decision primarily addressed the surveying aspects of evidence, this and the following cases address the legal aspects of the profession.

In this decision, Justice Cooley identifies how far a surveyor can go in making legal decisions as to the determination of private rights.

January 26, 1875, Heard.
January 29, 1875, Decided.
PRIOR HISTORY: Appeal in Chancery from St. Clair Circuit.
DISPOSITION: Decree affirmed, with costs.

When there has been an honest difficulty in determining the lines between two neighboring proprietors, and they have actually agreed by parol upon a certain boundary as a true one, and have occupied accordingly with visible monuments or divisions, the agreement long acquiesced in, shall not be disturbed, although the time has not been sufficient to establish an adverse possession: Campbell, J., in *Smith v. Hamilton*, 20 Mich., 433; see, also, *Joyce v. Williams*, 26 Mich. (second edition), 332 and note 1; *Fahey: v. Marsh*, 40 Mich., 236; *Bunce v. Bidwell*, 43 Mich., 542; *Bums v. Martin*, 45 Mich., 22 and cases cited on p. 24; compare *Hayes v. Livingston*, 34 Mich., 384.

A long practical acquiescence by the parties concerned in supposed boundary lines should be regarded as such an agreement upon the lines as to be conclusive, even if originally located erroneously: *Diehl v. Zanger*, 39 Mich., 601; *Dupont v. Starring*, 42 Mich., 492.[18]

But where adjacent land-owners occupy and define their possession under mutual mistake, neither doubting the rights of the other, and neither attempting either to fix or disturb the line, improvements not interfered with do not operate as an estopped. Estoppels when allowed must be based on something else than the silent permission of improvements by a party who has not acted in bad faith, or done any act which the other party had a right to regard as meant to govern his conduct: *Cronin v. Gore*, 38 Mich., 381, citing and distinguishing the principal case and *Smith v. Hamilton* and *Joyce v. Williams*, cited *supra*.

The mere building of a division fence by adjacent proprietors and occupancy up to the same by them, in the absence of any agreement binding upon the parties will not prevent either from thereafter contesting the correctness thereof; *Chapman v. Crooks*, 41 Mich., 595. A street that has been established and recognized and used for more than ten years, has become practically located, and cannot be changed unless according to charter and statutory conditions: *Pratt v. Lewis*. 39 Mich., 7.

Where streets have been opened in supposed conformity to a plat, and have been long acquiesced in as opened, they should be accepted as fixed monuments in locating lots or blocks continuous thereto or fronting thereon: *Van Den Brooks v. Correon*. 48 Mich., 283; *Twogood v. Hoyt*, 42 Mich., 609.

***Evidence of Boundaries: Questions of fact.* The question of the location of a section line or a starting point is one of fact for the jury, and not one of theory, to be determined finally upon the opinion of surveyors as experts.**[19]

In this case, the only evidence of the location of a section line, which is made a starting point in the description of premises in a deed, is the opinion of a surveyor, based on his examination of records, deeds and abstracts and other documents from which he protracted his maps and plans, and not upon any *data* made known by testimony in the cause; it would be clearly inadmissible to disturb a tangible and established boundary on any such evidence.

COUNSEL: Frank Whipple and A. E. Chadwick, for complainant. Atkinson Bros., for defendant.
JUDGES: Campbell, J. Graves, Ch. J., and Cooley, J., concurred. Christiancy, J., did not sit in this case.
OPINION BY: Campbell
OPINION: The object of the bill in this case is to rectify the boundaries, and remove a cloud on the title of a portion of a lot of land owned by complainant in Port Huron.

The entire lot of complainant is a parcel of land formerly owned by Beard and Haynes, and the easterly end slopes downward to a piece of somewhat lower ground, concerning which the difficulty seems to have arisen. In January, 1856, Beard and Haynes agreed to sell to the Port Huron & Milwaukee Railroad Company the eastern part of a larger tract of land which they then owned (and of which the premises in dispute are a part), and in the contract the western boundary of the railroad tract was to be bounded "*by the slope of the hill*." A fence was built then or previously, which ran along the base of the slope, and which complainant claims was the line of the tract in question. In July, 1856, Beard and Haynes, at the request of the company, and in pursuance of the contract, executed a deed, which they were assured by the company's agent and engineer was in conformity with the contract, whereby the western boundary was fixed at four hundred and twenty feet from the easterly line of section fifteen, of which the tract was a part. The reason given for mentioning distances was, that the company preferred an exact measurement; and the parties, when they conveyed, were assured that the actual survey put the line where they had agreed it should be, at the fence, or a trifle east of it.

This fence continued undisturbed, and coincided in direction with the line of railroad lands bought of adjoining owners.

In August, 1856, the land west of the fence was conveyed to complainant's grantors, and in the deed was described, not by metes and bounds, but by adjoining property, and the easterly line was defined as "*the line of lands sold by said Beard and Haynes to the Port Huron & Milwaukee Railroad Company*."

In October, 1871, Carleton, the defendant, obtained a deed from the Port Huron & Lake Michigan Railroad Company (which appears to have purchased the property of the Port Huron & Milwaukee Railroad), and in that deed the land he bought was described as bounded "On the west by the west line of said railroad company's lands; the same being a line parallel with, and four hundred and twenty feet distant west from the east line of said section fifteen." Since that time Carleton has asserted ownership over the tract in dispute, by various acts not acquiesced in by complainant, but has not fenced it separately, nor defined his own exact boundary by any very clear lines, and it does not appear to us that there has been any complete ouster.

Upon the testimony in the case, the facts do not seem to us to be left in any doubt. There can be no question but that all parties assumed and acted on the assumption, that the fence was the boundary, and that no one ever questioned it until defendant made his purchase. The railroad company from whom he bought, did not suppose there was any controversy about lines, and had no actual knowledge on the subject at all. The line had been acquiesced in as properly located, for fifteen years before defendant purchased. And although defendant undertakes to make out that, before he purchased, complainant was fully informed of the true boundaries, and admitted the error, his cross-examination, in connection with other proofs, furnishes a complete refutation of any such notion, and shows defendant to have made the purchase with full knowledge of the real state of things, and not in good faith.

In this view of the case, we think defendant is in no better position than the original railroad company would have been, and subject to any remedy which could have been enforced against them. There is nothing to show any fault in complainant for not discovering an error which he had no reason to suspect, and no means of correcting without a survey, which he had reason to suppose had been correctly made already. Under all these circumstances, we think the case comes within familiar equitable doctrines, and does not depend on the statutory remedy to quiet title. The landmarks which have been recognized and acted on so long, ought not to be disturbed, even if there had been some variance from the liens intended. But here the line so fixed was the line actually agreed upon; and if the deed varied from the contract, it was by mutual mistake or fraud; and we do not think any fraud was designed. If it had been, the equity would be still stronger. Unless against an honest purchaser without notice, there is no reason for refusing relief, and defendant is not on that relation.

We have thus far proceeded, as perhaps we are bound to proceed, on the assumption made by all the parties, that a mistake was actually made. And we affirm the decree on that basis.

But the facts require some reference to the testimony of boundary, which seems to us to have been introduced on a somewhat dangerous theory. It appears to have been supposed that surveyors are competent not only to testify to measurements and distances, but also to pass judgment themselves and on information of their own choosing upon the position of lines and starting points. **This is not the only case in which we have encountered such evidence on important private rights; and surveyors seem to have the idea that they may act entirely on their own judgment in determining important private and public rights.**

This is a very dangerous error. The law recognizes them as useful assistants in doing the mechanical work of measurement and calculation, and it also allows such credit to their judgment as belongs to any experience which may give it value in cases where better means of information do not exist. But the determination of facts belongs exclusively to courts and juries. Where a section line or other starting point actually exists, is always a question of

fact, and not of theory, and cannot be left to the opinion of an expert for final decision. And where, as is generally the case in an old community, boundaries and possessions have been fixed by long use and acquiescence, it would be contrary to all reason and justice to have them interfered with on any abstract notion of science. The freaks of opinionated surveyors have led to much needless and vexatious litigation and disturbance, and it is much to be desired that they should be confined to their legitimate place as witnesses on fact, and not on opinions, which lie beyond the domain of science.

If we examine the testimony in the present case carefully, we find that there is no legal evidence, properly receivable, which shows that complainant's boundary is not strictly correct. The presumption is quite as strong in favor of Gray's survey, as of Carleton's work, and there is nothing except assumption to show where the east line of section fifteen is to be found. Defendant professes to base his opinion on an examination of records, deeds and abstracts and other documents, from which he protracted his maps and plans. But neither he, nor any other witness, testifies from any *data* made known by testimony in this cause. Whatever may be the probabilities of correctness, this is not the proper way of proving boundaries. And we should not be justified in holding complainant's line disproved, even if not fixed by the conduct and acquiescence of the parties.

The decree must be affirmed, with costs. Graves, Ch. J., and Cooley, J., concurred. Christiancy J., did not sit in this case.

19.4 *Diehl v. Zanger*, 39 Mich. 601 (1878)

In the previous decision, Justice Cooley addressed possession, resurveys, fences, acquiescence, and erroneous surveys. In this decision, you will see some survey philosophy that will be found in the next decision discussed in Section 19.5, *Cragin v. Powell*, a U.S. Supreme Court decision, 10 years later.

Once again, Justice Cooley instructs the surveyor as to the duties that should be expected in a retracement.

October 25, 1878, Submitted October 31, 1878, Decided.
PRIOR HISTORY: Error to Superior Court of Detroit. Defendants bring error.
DISPOSITION: Judgment reversed with costs, and a new trial ordered. In defense to an action of ejectment based upon an alleged mistake in the original survey, evidence is admissible that the existing boundaries had been defined for more than twenty years by buildings, fences, and harmonious occupancy.

A re-survey, made after the monuments of the original survey have disappeared, is for the purpose of determining where they were, and not where they ought to have been.

A long-established fence is better evidence of actual boundaries settled by practical location than any survey made after the monuments of the original survey have disappeared.

Long practical acquiescence in a boundary, between the parties concerned, may constitute such an agreement on it as to be conclusive, even if it had been erroneously located.

COUNSEL: Henry M. Campbell and Alfred Russell for plaintiff in error. Long acquiescence and practical location are controlling proof of boundaries. *Baldwin v. Brown*, 16 N.Y., 359; *Reed v. Farr*, 35 N.Y., 113; *Bower v. Earl*, 18 Mich., 367.
JUDGES: Graves, J. Campbell, C. J. and Cooley, J., concurred. Marston, J., did not sit in this case.
OPINION BY: Graves
OPINION: Graves, J. This ejectment was brought by defendants in error for a strip of land four feet and seven inches wide, and claimed in the case as being the westerly four feet and seven inches of the easterly half of lot thirty-nine of the subdivision of out-lot one hundred and eighty-two of the Rivard farm in the city of Detroit.

February 7, 1873, plaintiffs in error owned this lot thirty-nine (39), and at that time, by deed, granted to defendants in error the easterly half according to the recorded plat, "together with the dwelling house being thereon," and about a month later they went into possession under the grant and have since occupied accordingly. At the time of the grant a fence which still remains was standing at the east between defendants in error and the adjoining proprietor of lot thirty-eight. The occupancy on each side was bounded by the fence, and defendants in error have continued to hold under their deed from that fence twenty-four feet westerly and to the fence dividing the lot into halves, including a house and barn situated on the parcel.

Defendants in error alleged and gave evidence tending to show that the fence between their occupancy and that east on lot thirty-eight was four or five feet too far east; that they did not know there was any mistake about boundaries; and that the question had not been mentioned between them and plaintiffs in error until a short time before this suit was commenced.

They further gave evidence tending to show that according to a recent survey made by a surveyor of long experience in Detroit, all the lots in the subdivision of out-lot one hundred and eighty-two are out of the way; that the fences and buildings on all the lots are incorrectly located; that the owner of the lot on Prospect Street,—which street bounds the subdivision on the east,—has encroached on said street between four and five feet, and that the owners of lots west of that lot, as they actually occupy, respectively encroach on the owner next east to the same extent, and that for three or four blocks the lines of the lots along Prospect Street encroach on the street between four and five feet. They also gave evidence tending to show that according to this survey the true line would run through the dwelling house on the westerly half of lot thirty-nine where plaintiffs in error reside.

The plaintiffs in error submitted evidence conducing to show, among other things, that lots thirty-nine and forty, as well as other contiguous lots in the subdivision, had for twenty years and upwards been identified and defined in their position and extent upon the ground by buildings, fences and harmonious occupancy, and that at the very time of the grant to defendants in error the physical evidences of recognized and long admitted bounds which plaintiffs in error contend for, were visible and apparent to everybody. There was no conflicting evidence in regard to these facts. They were not disputed, and there is no evidence that the practical locations and proprietary and possessory recognitions ever deviated until after the remarkable results of the late survey. The evidence was abundantly sufficient to require the case to be submitted to the jury on the defense. If believed, it was competent for the jury to find that the east line of lot thirty-nine had become fixed and established at the place marked by the fence.

The court, however, refused to regard this evidence as having any force, and in effect withheld it.

He instructed the jury to find for defendants in error. The matter of defense the court thus refused to recognize is within the doctrine which has been expounded and fully approved by this court in several cases. *Smith v. Hamilton*. 20 Mich. 433. *Joyce v. Williams*. 26 Mich. *Stewart v. Carleton*. 31 Mich. 270. *Pratt v. Lewis*, ante p. 7. And in some of them the impropriety of disregarding the various landmarks which time and actual occupancy and improvements and the behests of usage and general acquiescence have produced, in deference to the disordering achievements of some modern survey, has been distinctly adverted to and explained. This record contains enough to justify such observations. The requests of plaintiffs in error should have been given. The judgment is reversed with costs, and a new trial ordered.

CONCUR BY: Cooley

CONCUR: Campbell, C. J. and Cooley, J., concurred. Cooley, J. **The judge below took this case from the jury, instructing them that on the facts the plaintiffs were entitled to recover. We think, on the other hand, that if he had submitted the case to the jury on the facts, it would have been their plain duty to return a verdict for defendants.**

The controversy concerns part of a lot on a plat of a subdivision of a part of out-lot 182, Rivard farm, surveyed by Thomas Campau in 1850, and recorded in 1851. There are forty-eight of these lots in the subdivision. Whether they have all been sold off and improved by the purchasers we are not informed, but it appears from the record that many of them have been. It also appears that there has been a practical location of a street on one side [of] the plat and of other streets across it, and also of the lot lines. The lot the boundary of which is in dispute in this case has been fenced in for twenty years by fences on the supposed lines, and it does not appear that the lines have been disputed until recently. The adjoining lots have also been claimed, occupied, and improved according to the practical location of the lines.

This litigation grows out of a new survey recently made by the city surveyor. This officer after searching for the original stakes and finding none, has proceeded to take measurements according to the original plat, and to drive stakes of his own. According to this survey the practical location of the whole plat is wrong, and all the lines should be moved between four and five feet to the east. The surveyor testifies with positiveness and apparently without the least hesitation that "the fences and buildings on all the lots are not correctly located"

and there is of course an opportunity for forty-eight suits at law and probably many more than that. When an officer proposes thus dogmatically to unsettle the landmarks of a whole community, it becomes of the highest importance to know what has been the basis of his opinion. The record in this case fails to give any explanation, but the reasonable inference is that the surveyor has reached his conclusion by first satisfying himself what was the initial point of Mr. Campau's survey, and then proceeding to survey out the plat anew with that as his starting point. Of course by this method if no mistake is made, there is no difficulty in ascertaining with positive certainty where, according to Mr. Campau's plat, the original street and lot lines ought to have been located; and apparently the surveyor has assumed that that was all he had to do.

Nothing is better understood than that few of our early plats will stand the test of a careful and accurate survey without disclosing errors. This is as true of the government surveys as of any others, and if all the lines were now subject to correction on new surveys, the confusion of lines and titles that would follow would cause consternation in many communities. Indeed, the mischiefs that must follow would be simply incalculable, and the visitation of the surveyor might well be set down as a great public calamity.

But no law can sanction this course. The surveyor has mistaken entirely the point to which his attention should have been directed. The question is not how an entirely accurate survey would locate these lots, but how the original stakes located them. No rule in real estate law is more inflexible than that monuments control course and distance,—a rule that we have frequent occasion to apply in the case of public surveys, where its propriety, justice and necessity are never questioned. But its application in other cases is quite as proper, and quite as necessary to the protection of substantial rights. The city surveyor should, therefore, have directed his attention to the ascertainment of the actual location of the original landmarks set by Mr. Campau, and if those were discovered, they must govern. If they are no longer discoverable, the question is where they were located; and upon that question the best possible evidence is usually to be found in the practical location of the lines, made at a time when the original monuments were presumably in existence and probably well known. *Stewart v. Carleton*, 31 Mich. 270. As between old boundary fences, and any survey made after the monuments have disappeared, the fences are by far the better evidence of what the lines of a lot actually are, and it would have been surprising if the jury in this case, if left to their own judgment, had not so regarded them.

But another view should have been equally conclusive in this case. The long practical acquiescence of the parties concerned, in supposed boundary lines, should be regarded as such an agreement upon them as to be conclusive even if originally located erroneously. We had occasion to apply this doctrine in *Joyce v. Williams*, 26 Mich. 332, and need not enlarge upon it here. See also *Smith v. Hamilton*, 20 Mich. 433; *Stewart v. Carleton*, supra.

I agree with my brethren that the case should be sent back for a new trial. Campbell, C. J., concurred.

Marston, J., did not sit in this case.

19.5 *Cragin v. Powell*, 128 U.S. 691 (1888)

This decision found its way to the U.S. Supreme Court after a long legal road through the state courts, then to the federal courts and back to the state courts, and finally to the U.S. Supreme Court.

Powell was a surveyor who attempted to "correct" an original GLO survey that he found to be in error by some 20 chains.

The state courts and the federal district courts said that was acceptable, but the U.S. Supreme Court disagreed. In fact, no court, including the U.S. Supreme Court, has the authority to correct an original survey once it had been approved.

The Supreme Court went on to discuss the sanctity of the original surveys that were conducted under federal laws. In addition, it chastised the surveyor who had attempted to benefit from the alleged error.

This decision is perhaps one of the most quoted decisions relative to the public lands. It set the tone for all future retracements.

Argued and submitted October 26, 1888. December 17, 1888, Decided

DISTRICT OF LOUISIANA.

THIS was a proceeding under a local statute of Louisiana for the purpose of ascertaining the boundary line between coterminous proprietors. The case is stated in the opinion of the court.

1. The power to make and correct surveys of the public lands belongs to the political department of the government, and while the lands are subject to the supervision of the General Land Office, its decisions in such cases are unassailable by the courts, except by a direct proceeding.
2. The courts have power to protect the private rights of a party who has purchased lands in good faith from the government against the interferences or appropriations of corrective resurveys, made by the Land Department, subsequent to such purchase.
3. Where the appellee, a surveyor, while employed by appellant to make a survey of his plantation, thought he discovered an error in the public lands claimed by appellant, by which it would appear that such lands were not in fact situated by Bayou Four Points as officially surveyed, and induced a third party to obtain the patent for the land which he, the appellee, then purchased from him, knowing that it had been possessed and cultivated by the appellant for a long period of years, a court of equity will not readily enforce an advantage thus obtained.

SYLLABUS: When lands are granted according to an official plat of their survey, the plat, with its notes, lines, descriptions, and landmarks, becomes as much a part of the grant or deed by which they are conveyed, and, so far as limits are concerned, controls as much as if such descriptive features were written out on the face of the deed or grant.

It is not within the province of a Circuit Court of the United States or of this court to consider and determine whether an official survey duly made, with a plat thereof filed in the District Land Office, is erroneous; but, with an exception referred to in the opinion, the correction of errors in such surveys has devolved from the earliest days upon the Commissioner of the General Land Office, under the supervision of his official superior, and his decisions are unassailable by the courts, except in a direct proceeding instituted for that purpose.

When the General Land Office had once made and approved a governmental survey of public lands, the plats, maps, field notes and certificates having been filed in the proper office, and has sold or disposed of such lands, the courts have power to protect the private rights of a party who has purchased in good faith from the government, against the interferences or appropriations of subsequent corrective resurveys made by the Land Office.

One who acquires land knowing that it covers a portion of a tract claimed by another will be held either not to mean to acquire the tract of the other, or will be considered to be watching for the accidental mistake of others, and preparing to take advantage of them, and as such not entitled to receive aid from a court of equity.

COUNSEL: Mr. J. D. Rouse, with whom was Mr. William Grant on the brief, for appellant. Mr. J. S. Whitaker, for appellees, submitted on his brief.
OPINION BY: LAMAR
OPINION: MR. JUSTICE LAMAR delivered the opinion of the court.
The appellees, Christian L. Powell, Joseph O. Ayo, and Ludger Gaidry, on the 1st of November, 1880, brought an action of boundary in the state court against the appellant, George D. Cragin, praying for a judgment of the court to fix the boundaries between certain lands, the property of appellees, and the contiguous lands belonging to appellant, and that he be ordered to deliver to appellees possession of the lands claimed and set forth in their petition.

On the 12th of July, 1880, the cause was removed into the Circuit Court of the United States, on the ground of diverse citizenship.

The answer of appellant sets up that he and his grantors, who had acquired the lands from the original patentees, had been, in public, peaceable and continuous possession of the lands included in his deed by well-defined boundaries for more than thirty years, and without notice of the claims of any person whatsoever, and that it is unnecessary to fix or establish any boundaries as prayed in the petition.

On the 2d of May, 1881, on motion of counsel for appellees, the court appointed a surveyor, for the purpose of ascertaining and fixing the boundary lines between the properties of the respective parties litigant, and ordered him to report his proceedings within reasonable time. By mutual consent of parties, Benjamin McLeran was selected by the court as such surveyor.

On June 6, 1881, McLeran filed his report of the survey made by him, and its results. From this report it appears that the township and sections in which the lands of the parties are located were officially surveyed in 1837 by one G. W. Connelly, as part of the public domain, and that the plat of such survey was filed in the United States Land Office of the district; that he considered this survey of Connelly so incorrect, and the traces of its lines and corners so difficult to identify, that he was unable to locate any proper line between the lands in question, except upon the basis of a resurvey of the entire township, in accordance with certain corrective resurveys of adjoining townships, which had been made in 1850, and succeeding years, by one Joseph Gorlinski, a deputy United States surveyor. In this view, and guided by the theory of these corrective surveys, McLeran proceeded to run a line which he considered the proper boundary between the lands in question, and recommended its adoption to the court "as substantially such a line as would have been run had the whole township been resurveyed at the time when Deputy Surveyor Gorlinski was resurveying the adjoining townships." With this filed two maps, No. 1, a map of his own survey, and No. 2, a map designed to exhibit the discrepancies between the Connelly survey, and the survey of Joseph Gorlinski and that of McLeran himself. These discrepancies are: (1) By the Gorlinski and the McLeran surveys the township lacked half a mile of being six miles square, the eastern tier of sections thereof losing fully one-half acre of the area given by them in the official plat, which official survey establishes a full township as prescribed by law; (2) By Connelly's plat "a bayou, known as Bayou Four Points," is located on appellant's lands, whilst by McLeran's map that bayou is located on the lands of appellees. In his supplemental report McLeran says, "it appears that Bayou Four Points was erroneously reported by the original survey." The report also says: "The ridges on either side of the bayous are composed of a rich, black, loamy soil, . . . and when put under cultivation become the best sugar-producing lands in the South. The far greater portion of the township consists of a marsh, . . . worthless for cultivation."

The line recommended by McLeran places the lands of the appellees where those of the appellant are located by the official survey, and thus gives to the former the rich ridges along the bayous now in the possession of the latter.

The appellant was required to show cause by the 19th of November, 1881, why the report of McLeran should not be approved and homologated as being a true and correct survey in the premises. Thereupon the court, upon motion of the appellant, and against the opposition of the appellees, ordered that the cause be placed on the equity docket and proceed as in equity. Opposition to the report was afterwards duly filed, alleging that, if approved, the appellant would be deprived of lands to which he held title through mesne conveyances from United States patents, and of which he and his grantors had held possession for thirty years and upwards.

An amended answer by appellant and replication by appellees having been filed, the cause was put at issue. The court, upon the pleadings and evidence, confirmed the report of the surveyor, and rendered a decree fixing the boundary line between the two estates according to the prayer of the original petition.

The primary object of the action of boundary, under the Civil Code of Louisiana, is to determine and fix the boundary between contiguous estates of the respective proprietors. The provision of the code in article 845, and other provisions under title 5 of the code, that the limits must be fixed according to the titles of the parties, are held by the Supreme Court of Louisiana to apply to cases in which neither party disputes the title of his antagonist. *Sprigg v. Hooper*, 9 Rob. La. 248, 253; *Zeringue v. Harang*, 17 Louisiana, 349; *Blanc v. Cousin*, 8 La. Ann. 71. The title to the property is not allowed to be litigated in this action, whose purpose is to fix a line or boundary between adjoining claims. When, therefore, in the course of the proceedings in this case, the surveyor appointed to survey and fix a boundary between the respective properties of the parties made a report, alleging mistakes in the official survey, and recommending a fine, the effect of which, if adopted, would eject the appellant from the lands held by him under a claim of valid title, the court below ordered the case to be placed upon the equity side of the docket, thus bringing, it was supposed, within its equitable cognizance the essential rights of the parties, unaffected by the special limitations governing the action of boundary.

To determine the grounds upon which this court is asked to reverse the decree of the court below, it is necessary to advert in some detail to the fact as shown by the record.

In 1844, the United States issued to one Bach patents to certain portions of sections 10, 15 and 22 of township 20 south, range 17 east, in the southeast district west of the river, according to the official plat of survey and said lands returned to the General Land Office by the surveyor general.

The appellant is the owner of the lands thus patented to Bach; and for many years he, and those under whom he claims, have been in possession of the lands, which, according to the official survey, were embraced in said patents.

In April, 1878, one Samuel Wolf purchased from the State of Louisiana portions of the same sections, 10, 15 and 22, and also portions of sections 14 and 23 of the same township, all adjoining the lands of the appellant. These lands last described were given to the State as swamp lands, under the act of the 20th of March, 1849, and were noted as such this property to Powell, one of the appellees, who in May, 1880, sold an undivided half to the other two appellees; and in the same year they brought this action of boundary.

In support of the decree of the court below, it is urged by counsel for appellees that "there is nothing in the patents or title on record to show, by word or otherwise, any distinct calls, designating their location; nothing given descriptive of the property, except the township, the section and the range; nothing to describe the lands patented or conveyed, either as high lands, swamp or overflowed lands, or as having upon them any water course or bayou." He admits, however, that the plat in evidence contains upon its face the names of certain bayous, as "Bayou Cailliou," "Grassy," "Sale," and others; but says "that the original patents and conveyances, apart from the plat, are silent upon the subject, except that the defendant's title calls for land on Bayou Grand Cailliou."

In this view, which seems to be the one on which the court below must have acted, the learned counsel is mistaken. It is a well-settled principle that when lands are granted according to an official plat of the survey of such lands, the plat itself, with all its notes, lines, descriptions, and landmarks, becomes as much a part of the grant or deed by which they are conveyed, and controls so far as limits are concerned, as if such descriptive features were written out upon the face of the deed or the grant itself.

The patent of the State of Louisiana to Wolf was of the east half of southeast quarter of section 10, east half of east half of section 15, etc., "containing 635 58/100 acres tidal overflow according to the official plat of the survey of said lands in the state land office." By that plat the portions of the sections patented to Wolf were noted as tidal overflow; and as such they had been certified to the State by the General Land Office and the Interior Department. By the same plat Bayou Four Points was noted as running through those portions of sections 10, 15 and 22, which had been patented to Bach, who doubtless entered them, and obtained patents for them, because of high lands so noted on this bayou to the character of the land, whether as swamp or high land. The statutes of the United States make it the duty of the surveyor general to note "all water courses over which the line he runs may pass; and also the quality of the lands." Rev. Stat. §2395, subdiv. 7.

And they provide that a copy of the plat of survey shall be kept for public information in the office of the surveyor general, in the offices where the lands are to be sold, and also in the office of the Commissioner of Public Lands. They further provide that "the boundary lines actually run and marked in the surveys returned by the surveyor general shall be established as the proper boundary lines of the sections or subdivisions for which they were intended, and the length of such lines, as returned, shall be held and considered as the true length thereof." Rev. Stat. §2396, subdiv. 2.

The surveyor, McLeran, insists not only in his original report of his survey, but also in his second explanatory report, and in his oral evidence, that this governmental survey is incorrect; some of it more incorrect than the rest, but especially erroneous in the length of its lines and in the location of Bayou Four Points on the portions of the sections patented to the appellees. The plat, he reports, is totally inconsistent with that of the governmental survey, and should have been rejected by the court below. Whether the official survey made by Connelly is erroneous, or should give way to the extent of its discrepancies to the survey reported by McLeran, is a question which was not within the province of the court below, nor is it the province of this court to consider and determine. The mistakes and abuses which have crept into the official surveys of the public domain form a fruitful theme of complaint in the political branches of the government. The correction of these mistakes and abuses has not been delegated to the judiciary, except as provided by the act of June 14, 1860, 12 Stat. 33, c. 128, in relation to Mexican Land claims, which was repealed in 1864, 13 Stat. 332, c. 194, §8. From the earliest days matters appertaining to the survey of public or private lands have devolved upon the Commissioner of the General Land Office, under the supervision of the Secretary of the Interior. Rev. Stat. §453. The Commissioner, in the exercise of his superintendence over surveyors general, and of all subordinate officers of his bureau, is clothed with large powers of control to prevent the consequences of inadvertence, mistakes, irregularity and fraud in their operations. Rev. Stat. §2478; *Bell v. Hearne*, 19 How. 252 and 262. Under the authority of specific appropriations by Congress, for the purpose, the resurveys of public lands have become an extensive branch of the business of the General Land Office. In 1848, the surveyor general of Louisiana urgently recommended a resurvey of certain townships in the district of Louisiana fronting on Bayou Cailliou, in Terre Bone, which had been surveyed by G. G. Connelly and other named surveyors. It was in accordance with this recommendation that Gorlinski made the resurveys above referred to. But the Commissioner of the General land Office very soon put an end to this system of resurveys, and in a letter to the surveyor general, which throws no little light upon the subject, he says:

"The making of resurveys or corrective surveys of townships once proclaimed for sale is always at the hazard of interfering with private rights, and thereby introducing new complications. A resurvey, properly considered, is but a retracing, with a view to determine and establish lines and boundaries of an original survey, . . . but the principle of retracing has been frequently departed from, where a resurvey (so called) has been made and new lines and boundaries have often been introduced, mischievously conflicting with the old, and thereby affecting the areas of tracts which the United States had previously sold and otherwise disposed of."

It will be perceived that McLeran's survey not only disregards the old original survey making new lines and boundaries, but does so in contravention of the order from the Land Office that those resurveys should not be extended into this township.

That the power to make and correct surveys of the public lands belongs to the political department of the government and that, whilst the lands are subject to the supervision of the General Land Office, the decisions of that bureau in all such cases, like that of other special tribunals upon matters within their exclusive jurisdiction, are unassailable by the courts, except by a direct proceeding; and that the latter have no concurrent or original power to make similar corrections, if not an elementary principle of our land law, is settled by such a mass of decisions of this court that its mere statement is sufficient. *Steel v. Smelting Co.*, 106 U.S. 447, 454-5, and cases cited in that opinion; *United States v. San Jacinto Tin Co.*, 10 Sawyer, 639, affirmed in 125 U.S. 273; *United States v. Flint*, 4 Sawyer, 42, affirmed in *United States v. Throckmorton*, 98 U.S. 61; *Henshaw v. Bissell*, 18 Wall. 255; *Stanford v. Taylor*, 18 How. 409; *Haydel v. Dufresne*, 17 How. 23; *West v. Cochran*, 18 How. 403; *Jackson v. Clark*, 1 Pet. 628; *Niswanger v. Saunders*, 1 Wall. 424; *Snyder v. Sickles*, 98 U.S. 203; *Frasher v. O'Connor*, 115 U.S. 102; *Gazzam v. Phillips*, 20 How. 372; *Pollard v. Dwight*, 4 Cranch. 421; *Taylor v. Brown*, 5 Cranch. 234; *McIver v. Walker*, 9 Cranch. 173, 177; *Craig v. Radford*, 3 Wheat. 594; and *Ellicott v. Pearl*, 10 Pet. 412. 1888 U.S. 2264, 15; 32 1. Ed. 566, in the field better done and divisions more equitably made than the department of public lands could do. 17 How. 30. It is conceded that this power of supervision and correction by the Commissioner of the General Land Office is subject to necessary and decided limitations. Nor is it denied that, when the Land Department has once made and approved a governmental survey of public lands (the plats, maps, field notes and certificates all having been filed in the proper office,) and has sold or disposed of such lands, the courts have power to protect the private rights of a party who has purchased, in good faith, from the government against the interferences or appropriations of corrective resurveys made by that department subsequently to such disposition or sale. But there is nothing in the circumstances of this case which brings it within any such limitations. The appellee, Powell, is a surveyor, who, in the year 1877, while employed by appellant to make a survey of his plantation, thought he discovered an error in the public lands, whereby it would appear that his lands were not in fact situated on Bayou Four Points. From his own evidence it is shown that he induced Wolf to obtain the patent from the State of Louisiana for the land which he, the said appellee, purchased from him. When he purchased this land from Wolf he knew that the tracts to which he was laying claim had been possessed and cultivated by the appellant for a long period of years. An advantage thus obtained, a court of equity will not readily enforce. As was said in *Taylor v. Brown*, 5 Cranch, 234, 256:

"The terms of the subsequent location prove that the locator considered himself as comprehending Taylor's previous entry within his location. . . . He either did not mean to acquire the land within Taylor's entry, or he is to be considered as a man watching for the accidental mistakes of others, and preparing to take advantage of them. What is gained at law by a person of this description, equity will not take from him; but it does not follow that equity will aid his views."

For the reasons above stated, the decree of the Circuit Court is reversed, with directions to dismiss the petition.

19.6 *United States v. Doyle*, 468 F.2d 633 (1972)

What originally started as a trespass action ultimately ended up with a very basic question: Is the corner a lost corner or an obliterated corner?

The facts are undisputed. For a number of years, the United States had used a corner as being the corner between them and Doyle. The United States had posted signs that stated, "National Forest Land Behind This Sign." When a correct survey was subsequently conducted by a contract private surveyor, it was determined that the corner that had been used could not be proved by reference to the original GLO notes, and thus the surveyor considered it as a "lost" corner and as such located it by law, proportioning between two original section corners. The Doyles protested, stating that the United States had used the corner, and as such it should hold as being the true and correct corner between the two landowners.

The trial court held that the United States was not bound to the posted line, because the quarter corner was a lost corner and as such would be positioned "equidistant" between the two sections corners, as identified by law [Land Act of Feb. 11, 1805] and not as established by agreement or any other doctrine.

The trial court referenced the definition of a lost corner and an obliterated corner, as identified by the Bureau of Land Management in its official publications. It should be noted that the quantum proof of evidence identified is that of "beyond a reasonable doubt." Recent case law has held that no governmental agency can establish the measure of proof needed in a civil evidentiary situation, and as such, case law has now set the criteria of proof for proving corners as not being "beyond a reasonable doubt," as discussed in the *BLM Manual,* but as being that required by civil requirements of the "preponderance of the evidence."

U. S. COURT OF APPEALS FOR THE TENTH CIRCUIT November 6, 1972, Affirmed.

JUDGES: Murrah, Breitenstein and Holloway, Circuit Judges.
OPINION: HOLLOWAY, Circuit Judge.

This is an action brought by the Government alleging occupancy trespass by defendants of a portion of the Pike National Forest and seeking injunctive relief against trespass and for removal of improvements from the property in dispute. The case was tried to the court. The trial court determined the boundary dispute in the Government's favor on the basis of a dependent resurvey. Injunctive relief was granted and this appeal followed.

The Government owns land in the Pike National Forest including the SW $\frac{1}{4}$ SE $\frac{1}{4}$, Section 11 Township 8 South, Range 71 West of the 6th Principal Meridian in Jefferson County, Colorado. The defendants are the owners of the north 250 feet of the east 700 feet of the NW $\frac{1}{4}$ NE $\frac{1}{4}$, Section 12, adjoining to the south.

The dispute here concerns the north boundary line of the defendants' property which is formed by the section line between the described portions of Sections 1 and 12 as it runs along the north of their property. According to the Government, the correct section line lies south of the property line claimed by the defendants. The defendants say that that the true line is about 124 feet north on one end and 147 feet north on the other end of the section line that the Government claims to be correct.

The Government's position is that the correct section line and therefore the north property line of the defendants' property should follow a resurvey by the single proportionate measurement method made by a Government surveyor, Mr. Brinker, in 1965.

It says that a stone marker which was described in the original 1872 survey performed by a Mr. Oakes is lost and that the loss of this marker makes the quarter corner at the northwest corner of the NE $\frac{1}{4}$ of Section 12 a lost corner. Therefore, the Government maintains that the resurvey made between the northeast and northwest corners of Section 12, and establishing a straight line between them, was the proper basis for locating the true section line and the quarter corner which was located at the midpoint of that line. The defendants, on the other hand, essentially argue that collateral evidence consisting of Forest Service signs, tree blazes and testimony sufficiently established as correct the boundary relied on by them. They say that a determination that a corner is lost is disfavored and that the trial court applied the incorrect criteria and burden of proof in making its determination that the corner was lost and erred in accepting the boundary established by the 1965 resurvey.

The trial court found that a stone relied on by the defendants was not the actual quarter section marker, and that this corner was lost; that the tree blazes were too recent to be relied on; that the dependent resurvey was a proper basis for determining the boundary; and that, therefore, defendants were in trespass on the disputed strip of land.[20] We are satisfied that the record supports findings by the trial court that the original quarter corner monument was lost and that a stone claimed by defendants to be the marker was not the original monument.[21] We find, from all of the evidence, that the stone claimed by defendants to be the quarter corner marker is not the original monument for the quarter corner. There was no testimony from any witness claiming to have knowledge of the original location. The tree blazes relied on by defendants were of too recent origin to be of significant evidentiary value; the blazed line nearest the claimed monument is some 80 to 90 feet south of the stone. The markings on the stone are not discernible. The claimed location of the monument cannot be confirmed by the field notes of the original survey.

"We therefore conclude that the original quarter corner is lost and its original location cannot be identified. In such circumstances, the accepted method of establishing the lost quarter corner is to locate a straight line between the section corners and find a point equidistant therefrom; such point then is the relocated quarter corner. *Vaught v. McClymond*, 116 Mont. 542, 155 P.2d 612 (1945); *Littlejohn v. Fink*, 109 Neb. 282, 190 N.W. 1020 (1922); Sec. 376, *Manual of Surveying Instructions*, supra. This method was followed by plaintiff in the 1965 dependent resurvey. The quarter corner so relocated and fixed by the brass-capped rod set in the ground is found to be the correct quarter corner.

"We consider this question later in this opinion. On appeal the defendants argue two propositions. First, they say that the trial court erred in failing to sustain their motion to dismiss under Rule 41 Fed.R.Civ.P. at the close of the Government's case. Secondly, they argue that the trial court used the wrong criteria in determining the facts and that it erred and imposed a burden on the defendants to establish conclusively the original boundary line. We will discuss the facts in more detail in treating these issues.

The first proposition of the defendants is without merit. As it was entitled to do under Rule 41 Fed.R.Civ.P., the trial court declined to rule on the motion to dismiss when the Government rested or to render any judgment until the close of all the evidence. The defendants chose not to stand on their motion but offered their proof. In these circumstances the defendants may not claim error by the refusal of the trial court to grant the motion made when the Government rested. *A. & N. Club v. Great American Insurance Co.*, 404 F.2d 100 (6th Cir.). The defendants argue that the Government failed to prove its case or establish the property line, which the plaintiff must do in a trespass case. The merits of this point are considered below in judging the sufficiency of the record as a whole to support the judgment. See *Bogk v. Gassert*, 149 U.S. 17, 23, 13 S. Ct. 738,37 L. Ed. 631; *Dindo v. Grand Union Co.*, 331 F.2d 138 (2d Cir.). However, no error is shown by the trial court's action in refusing to sustain the motion to dismiss. The second issue concerns whether the trial court erred in finding that the quarter corner was lost and in accepting the location of it and the boundary line established by the 1965 resurvey. And this issue involves also consideration of defendants' contentions that the incorrect criteria and burden of proof were used by the trial court in determining the facts. If the court properly found that the corner was lost and that the resurvey boundary should be accepted, there is no question as to the accuracy of the new line or the location of the quarter corner at its midpoint.

The guiding legal principles are not in dispute. Where there is no controlling federal legislation or rule of law, questions involving ownership of land are determined under state law, even where the Government is a party. *Mason v. United States*, 260 U.S. 545,558,43 S. Ct. 200, 67 L. Ed. 396; *United States v. Williams*, 441 F.2d 637,643 (5th Cir.); *Standard Oil Co. of California v. United States*, 107 F.2d 402,415 (9th Cir.). The rule is recognized implicitly by the federal statute permitting resurveys. See 43 U.S.C.A. §772.[22]

The original survey as it was actually run on the ground controls. *United States v. State Investment Co.*, 264 U.S. 206,212,44 S. Ct. 289,68 L. Ed. 639; *Ashley v. Hill*, 150 Colo. 563, 375 P.2d 337, 339. It does not matter that the boundary was incorrect as originally established. A precisely accurate resurvey cannot defeat ownership rights flowing from the original grant and the boundaries originally marked off. *United States v. Lane*, 260 U.S. 662,665, 666,43 S. Ct. 236,67 L. Ed. 448; *Everett v. Lantz*, 126 Colo. 504,252 P.2d 103, 108. The conclusiveness of an inaccurate original survey is not affected by the fact that it will set awry the shapes of sections and subdivisions. See *Vaught v. McClymond*, 116 Mont. 542, 155 P.2d 612, 620; *Mason v. Braught*, 33 S.D.559, 146 N.W. 687. The actual location of a disputed boundary line is usually a question of fact. *Gaines v. City of Sterling*, 140 Colo. 63, 342 P.2d 651. ". . . The generally accepted rule is that a subsequent resurvey is evidence, although not conclusive evidence, of the location of the original line." *United States v. Hudspeth*, 384 F.2d 638, 688 n. 7 (9th Cir.); accord, see *Ben Realty Co. v. Gothberg*, 56 Wyo. 294, 109 P .2d 455, 458, 459. And in its trespass action the burden of proving good title to the land rests on the Government. *Yakes v. Williams*, 129 Colo. 427,270 P.2d 765; see also *Cone v. West Virginia Pulp & Paper* Co., 330 U.S. 212,67 S. Ct. 752,91 L. Ed. 849.The procedures for restoration of lost or obliterated corners are well established. They are stated by the cases cited below and by the supplemental manual on *Restoration of Lost or Obliterated Corners and Subdivisions of Sections* of the Bureau of Land Management (1963 ed.).[23] The supplemental manual sets forth practices and contains explanatory and advisory comments. Practice 1 of the supplemental manual recognizes that an existent corner is one whose position can be identified by verifying evidence of the monument, the accessories, by reference to the field notes, or "where the point can be located by an acceptable supplemental survey record, some physical evidence, or testimony." Practice 2 recognizes that an obliterated corner is one at whose point there are no remaining traces of the monument, or its accessories, but whose location has been perpetuated, or the point for which may be recovered beyond a reasonable doubt, by the acts and testimony of the interested landowners, competent surveyors, or other qualified local authorities, or witnesses, or by some acceptable record evidence. Practice 3 states that a lost corner is one whose position cannot be determined, beyond reasonable doubt, either from traces of the original marks or from acceptable evidence or testimony bearing on the original position, and whose location can be restored only by reference to one or more interdependent corners.

The authorities recognize that for corners to be lost "they must be so completely lost that they cannot be replaced by reference to any existing data or other sources of information." *Mason v. Braught*, supra, 146 N.W. at 689, 690. Before courses and distances can determine the boundary, all means for ascertaining the location of the lost monuments must first be exhausted. *Buckley v. Laird.* 158 Mont. 483, 493 P. 2d 1070′ 1075 (Mont.); Clark, *Surveying and Boundaries* § 335, at 365 (Grimes ed., 1959); see advisory comments of the supplemental manual, supra at 10. The means to be used include collateral evidence such as boundary fences that have been maintained, and they should not be disregarded by the surveyor. *Wilson v. Stork*, 171 Wis. 561, 177 N.W. 878, 880. Artificial monuments such as roads, poles, fences, and improvements may not be ignored. *Buckley v. Laird*, supra, 493 P.2d at 1073; *Dittrich v. Ubi*, 216 Minn. 396, 13 N.W.2d 384, 390. And the surveyor should consider information from owners and former residents of property in the area. See *Buckley v. Laird*, supra, 493 P.2d at 1073–1076. "It is so much more satisfactory to so locate the corner than regard it as 'lost' and locate by 'proportionate' measurement." Clark, supra § 335 or 365.

Measured against these standards we believe the trial court's acceptance of the resurvey is supported by sufficient evidence of substantial compliance with proper procedures. The sufficiency of the resurvey investigation is not free from doubt since the record shows little contact with owners, who are one proper source of relevant collateral evidence.[24] However, the record shows that the surveyors searched the records of the Bureau of Land Management before the Brinker survey. The Government surveyor approving the resurvey, Mr. Teller,[25] said that prior surveys were also sought, but none were located, by requests to the county clerk, recorder and surveyor for plats filed as surveys of record. The special instructions for this resurvey recite that inquiry was made of incumbent and past surveyors and Rangers and that no factual data was discovered that would assist in recovery of the missing quarter section corner. And the original quarter corner marker placed during the *1872 survey was not located. The trial court found that this monument was not* located, after a thorough search of the area by the surveyor and his crew, and the record supports this finding.

We are satisfied that the record supports the trial court's finding that the corner was lost. On the entire evidence we are not left with a definite and firm conviction that a mistake was committed. *Zenith Radio Corp. v. Hazeltine Research, Inc.*, 395 U.S. 100, 123, 89 S. Ct. 1562, 23 L. Ed. 2d 129. We likewise are persuaded that the record sustains acceptance of the validity of the resurvey, and the accuracy of the location of the lost quarter by the resurvey. We

cannot agree that the trial court erred in the criteria used in determining that the corner was lost. Nor do we feel that the court erred and placed the burden of proof on the defendants to establish the boundary conclusively themselves. The record does not show that this was done and such error is not presumed. *Smith v. Crouse*, 413 F.2d 979, 980 (10th Cir.).

The defendants argue that the court erred in concluding that the corner was lost, disregarding acceptable collateral evidence. They say the overall line of the original survey was about 192 feet longer than the straight line established by the proportionate measurement method; that the blazed tree line and Forest Service signs showed a boundary relied on since at least 1942; and that a former owner of property along the line, including the Doyle property, testified about markers and monuments which had previously existed along the blazed tree line. This proof was all weighed by the trial court. It was found that the irregular line of blazed trees was of too recent origin to be of significant evidentiary value and that it was some 80 to 90 feet south of the stone relied on by the defendants as the lost quarter corner monument. And on conflicting testimony by the surveyors the court found that this stone was not the lost marker. Applying the criteria from the authorities we have discussed, we are satisfied the findings and conclusions of the trial court are sustained.

Affirmed.

19.7 *Rivers v. Lozeau*, 539 So. 2d 1147, Fla. App. 14 Fla. L. Weekly 523 (1989)

The final decision discussed is an important decision for several reasons: First, it identifies the role of the modern surveyor and the responsibilities in surveying land boundaries. Second, it identifies, with exactness, a retracement and a resurvey. Finally, it chastises the lawyers who take shortcuts in writing land descriptions.

Your attention is called to note 33. This statement by the court identified the absolute responsibility that is placed on the individual who writes land descriptions to call for monuments that are placed as a result of a survey.

February 23, 1989, Filed
Review Denied June 9, 1989.
Appeal from the Circuit Court for Marion County, Raymond T. McNeal, Judge.

FINAL DISPOSITION: Reversed and remanded.
PROCEDURAL POSTURE: Both appellant and appellee landowners sought review from the Circuit Court for Marion County (Florida), which ordered that land in a boundary line dispute be evenly split between them.

FACTS: A dispute arose between landowners involving a boundary line bordering land owned by the U.S. Government. An earlier landowner had retained a surveyor to establish land lines dividing it into parts eventually owned by appellants and appellees. Later dependent surveys by the U.S. Government showed that the surveyor had failed to locate the original government survey lines. In an action by appellees in ejectment and for declaratory judgment, the trial court ordered the property to be split between both parties and both appealed. On appeal, the court reversed and remanded, finding that appellees had acquired title to the disputed section established by the original government surveyors. Neither the title nor boundaries to a deeded parcel moved about in time based on where a particular surveyor might erroneously believe the correct location of the true boundary line to be.

FINAL DECISION: An order requiring land to be split was reversed and remanded because appellees had acquired title based on the true boundary line established by the original government surveyors which did not move in time based on a particular surveyor's erroneous belief of the correct location.

The basic rules gleaned from the decision are as follows:

1. Real property descriptions are controlled by the descriptions of their boundary lines which are themselves controlled by the terminal points or corners as established on the ground by the original surveyor creating those lines.
2. When there is an inconsistency between the description of a corner, i.e., a line terminal point, in field notes and plats subsequently made and recorded and the original monument evidencing that corner on the ground, the original monument on the ground controls.
3. Although title attorneys and others who regularly work with them develop expertise as to land descriptions, the only professional authorized to locate land lines on the ground is a registered land surveyor . . . In fact, the definition of a legally sufficient real property description is one that can be located on the ground by a surveyor. However, in the absence of statute, a surveyor is not an official and has no authority to establish boundaries.
4. In a "retracement" survey, not a resurvey, the second and each succeeding surveyor is a following or tracing surveyor and his sole duty, function and power is to locate on the ground the boundaries corners and boundary line or lines established by the original survey; he cannot establish a new corner or new line terminal point, nor may he correct errors of the original surveyor. He must only track the footsteps of the original surveyor. The following surveyor, rather than being the creator of the boundary line, is only its discoverer and is only that when he correctly locates it.
5. The owner of a parcel of land, being the grantee under a patent or deed, or devisee under a will or the heir of a prior owner, has not authority or power to establish the boundaries of the land he owns; he has only the power to establish the division or boundary line between parcels when he owns the land on both sides of the boundary line he is establishing. In short, an original surveyor can establish an original boundary line only for an owner who owns the land on both sides of the line that is being established and that line becomes an authentic original line only when the owner makes a conveyance based on a description of the surveyed line and has good legal title to the land described in his conveyance.

6. The approved and accepted boundary lines established by the federal government surveyors are unchangeable and control all references in deeds and other documents describing parcels of land by reference to the federal government of sections, townships, and ranges.
7. Neither the title to land nor the boundaries to a deeded parcel move about from time to time based on where someone, including a particular surveyor, might erroneously believe the correct location of the true boundary line to be.

COUNSEL: Bryce W. Ackerman of Savage, Krim, Simons, Fuller & Ackerman, P. A., Ocala, for Appellants/Cross-Appellees.
H. Randolph Klein of Klein & Klein, Ocala, for Appellees/Cross-Appellants.
JUDGES: Cowart, J. Cobb, J., and Glickstein, H. S., Associate Judge, concur.
OPINION BY: COWART
FINAL OPINION: This is a land boundary line dispute case.
THE FACTS: The controversy in this case involves the correct location of the line between two parcels of land lying within the 40-acre quarter-quarter section described as the Southeast 1/4 of the Southwest 1/4 of Section 15, Township14 South, Range 24 East, in Marion County, Florida.

In 1964, Joseph Rizzo and his wife owned that portion of this quarter-quarter section that is in question. The U.S. Forestry Service owns the land to the north. At that time, the Rizzos retained a surveyor, Moorhead Engineering, to survey their land and to establish certain internal land lines dividing it into parts. Moorhead undertook to locate and monument Rizzos' external boundary lines and corners and to establish and monument the terminal points of certain internal division lines.

In 1969, the Rizzos conveyed to Marcus E. Brown and wife by deed containing the following land description:

The north 400.00 feet of SE 1/4 of SW 1/4 of Section 15, Township 14 South, Range 24 East, Marion County, Florida.

The west, north, and east lines of the Brown parcel followed the outer or external boundary lines of the property owned by the Rizzos. The south line of the Brown parcel did not follow any internal line established by the Moorhead survey. Mr. Rizzo showed Marcus Brown the monuments Moorhead had set as being the north corners of this quarter-quarter section and certain other Moorhead monuments which the Rizzos told Marcus Brown were 33 feet south of the south line of the parcel the Rizzos conveyed to Brown. Later in 1977 or 1978, Marcus Brown measured 33 feet north of the Moorhead monuments shown him by Mr. Rizzo and placed a metal rod at the point Mr. Rizzo had told him was his south boundary line. Marcus Brown conveyed this property by the same description to George Brown who conveyed by the same description to appellees Raymond S. Lozeau and his wife.

In 1975, the Rizzos conveyed a parcel of their remaining land to Paul W. Adams and wife, which parcel was described by reference to the boundary lines of this quarter-quarter section with the north line of the property conveyed being described as thence N 89 53′ 01″ E. along a line 400.00 feet south of and parallel to the North line of said SE 1/4 of SW 1/4 a distance of 1327.04 feet to a point on the East line of said 1/4 of SW 1/4.

Using substantially the same land description, the Adamses conveyed to Daniel E. Reader and wife, who conveyed to appellants Harold J. Rivers and wife.

In 1982, the U.S. Bureau of Land Management did a "dependent resurvey" of the lands of the U.S. Forestry Service which retraced the lines of the original government survey and identified, restored, and remonumented the original position of the corners of the original U. S. government survey.[26] This remonumenting of the original government survey, along with a 1986 survey by Whit Holley Britt, made obvious to all the true location of the north line of this quarter-quarter section on the ground and that the Moorhead monuments intended to denote that lien were actually located 28.71 feet north of the true location of that line as it was originally established by the official U.S. government survey and reestablished by the 1982 government "dependent survey." Appellees Lozeaus brought this action in ejectment and for declaratory judgment against the appellants Riverses who had possession of the south 28.71 feet of the north 400 feet measured from the north line of the quarter-quarter section according to the U.S. government (and Britt) surveys. The Lozeaus argued that they acquired legal title to the disputed land by virtue of the 1969 deed from Rizzo to Marcus Brown and the successive conveyances to them. The Riverses argued that Moorhead was the original surveyor and that his monuments on the ground controlled the location of the land subsequently conveyed by Rizzo, notwithstanding that "later" surveys, i.e., the government survey of 1982 and the 1986 Britt survey, may show the Moorhead monuments to have been in error.[27] After a non-jury trial, the trial court found that the property descriptions of the parties overlapped and ordered that the exact dimensions of the overlap be established and the overlapping property be split evenly between the plaintiffs and defendants. The Riverses appealed and the Lozeaus cross-appealed.

LAND DESCRIPTIONS Since time immemorial, parcels of land have been identified and described by reference to a series of lines or "calls" or "courses" that connect to completely encircle the perimeter or boundaries of a particular parcel. A particular property description may consist entirely of descriptions of original lines that compose it or it may, in whole or in part, refer to other sources which themselves show or describe previously surveyed and existing lines or calls. An individual line or call in a property description usually, but not always,[28] refers to an imaginary straight line customarily described in several ways: (1) by reference to its length, (2) by reference to its terminal points (commonly called "corners" or "angles"), (3) by reference to its angle with regard to true north, magnetic north, or to one or more other lines. A property description composed of descriptions of its constituent boundary lines or calls is known as a "metes-and-bounds" description. Of the ways that boundary lines are described, the reference to terminal points is the strongest and controls when inconsistent with other references.[29] In effect, real property descriptions are controlled by the descriptions of their boundary lines which are themselves controlled by the terminal points or corners as established on the ground by the original surveyor creating those lines.

A property description that refers to, and adopts by reference, the description of a boundary line is *DEPENDENT* upon the proper location of the adopted line, which is dependent upon the location of the terminal points of the adopted line, which are dependent on their location on the ground as established by the original surveyor creating that adopted line

LAND SURVEYORS—Although title attorneys and others who regularly work with them develop expertise as to land descriptions, the only professional authorized to locate land lines on the ground is a registered land surveyor.[30] In fact, the definition of a legally sufficient real property description is one that can be located on the ground by a surveyor. However, in the absence of statute, a surveyor is not an official and has no authority to establish boundaries; like an attorney speaking on a legal question, he can only state or express his professional opinion as to surveying questions. In working for a client, a surveyor basically performs two distinctly different roles or functions: First, the surveyor can, in the first instance, lay out or establish boundary lines within an original division of a tract of land which has theretofore existed as one unit or parcel. In performing this function, he is known as the "original surveyor" and when his survey results in a property description used by the owner to transfer title to property[31] that survey has a certain special authority in that the monuments set by the original surveyor on the ground control over discrepancies within the total parcel description and, more importantly, control over all subsequent surveys attempting to locate the same line. Second, a surveyor can be retained to locate on the ground a boundary line which has theretofore been established. When he does this, he "traces the footsteps" of the "original surveyor" in locating existing boundaries. Correctly stated, this is "retracement" survey, not a resurvey, and in performing this function, the second and each succeeding surveyor is a "following" or "tracing" surveyor and his sole duty, function, and power are to locate on the ground the boundaries corners and boundary line or lines established by the original survey; he cannot establish a new corner or new line terminal point, nor may he correct errors of the original surveyor. He must only track the footsteps of the original surveyor. The following surveyor, rather than being the creator of the boundary line, is only its discoverer and is only that when he correctly locates it.[32]

ORIGINAL LAND LINES When there is a boundary dispute caused by an ambiguity in the property description in a deed, it is often stated that the courts seek to effectuate the intent of the parties. This is not an accurate notion. The intent of the parties to a contract for the sale and purchase of land, both the buyer and the seller, may be relevant to a dispute concerning that contract, but in a real sense, the grantee in a deed is not a party to the deed, he does not sign it and his intent as to the quality of the legal title he receives and as to the location and extent of the land legally conveyed by the deed is quite immaterial as to those matters. The owner of a parcel of land, being the grantee under a patent or deed, or devisee under a will or the heir of a prior owner, has no authority or power to establish the boundaries of the land he owns; he has only the power to establish the division or boundary line between parcels when he owns the land on both sides of the boundary line he is establishing. In short, an original surveyor can establish an original boundary line only for an owner who owns the land on both sides of the line that is being established and that line becomes an authentic original line *only when the owner makes a conveyance based on a description of the surveyed line*[33] and has good legal title to the land described in his conveyance.

UNITED STATES AS ORIGINAL OWNER AND ITS CADASTRAL ENGINEER. Subject only to certain rights or individuals under Spanish grants, the United States became the owner of all land now in the State of Florida by virtue of a treaty with Spain dated Feb. 22, 1819 and ratified Feb. 22, 1821 and, as original governmental owner, caused Florida to be surveyed in accordance with a rectangular system of surveys of public lands adopted by Acts of Congress. The permanent seat of government having been established at Tallahassee, an initial point of reference was located nearby through which a north-south guideline was run according to the true meridian and a base (township) line was run east-west on a true parallel of latitude.[34] North-south range lines, six miles apart and parallel to the Tallahassee Principal Meridian, were run throughout the state except where impracticable because of navigable waters, etc.

Likewise, East-West township lines, six miles apart and parallel to the base line, were also run throughout the state to form normal townships six miles square, each of which were divided into thirty-six square sections, one mile long on each side containing as nearly as may be, 640 acres each. These sections were numbered respectively, beginning with the number one, in the northeast corner and proceeding west (left) and east (right) alternately through the townships with progressive numbers. Sections were divided into squares of quarter sections containing 160 acres. The quarter-quarter section corners are placed on the line connecting the section and quarter-section corners, and midway between them.

Although theoretically conceived and invisible, these lines are not merely theoretical concepts but are real lines, actually run and marked on the ground with terminal points monumented by surveyors acting under the authority of the cadastral engineer of the Bureau of Land Management. The approved and accepted boundary lines established by the federal government surveyors are unchangeable and control all references in deeds and other documents describing parcels of land by reference to the federal government of sections, townships and ranges.

THE LAW APPLIED TO THE FACTS OF THIS CASE In establishing the internal lines within Rizzo's subdivision, Moorhead acted as an "original surveyor" but in attempting to locate and monument Rizzo's external boundary lines which are described by reference to the federal rectangle system of surveying, Moorhead was a "following surveyor" and not only failed to properly find the northern boundary of this quarter-quarter section where it was located by the original government surveyor (and also re-established by an authorized federal government resurvey) but to evidence his erroneous opinion as to the true line, the Moorhead surveyor placed monuments 28.71 feet north of the true north line of this quarter-quarter section. From the time the federal government granted this quarter-quarter section to the original grantee down to the Rizzos, the title conveyed was to a tract of land located according to the original government survey and by the deed from the Rizzos to Brown, and subsequent deeds, the Lozeaus acquired title to the north 400 feet of this quarter-quarter section according to the true boundary line established by the original government surveyors. This is true regardless of the fact that Mr. Rizzo showed Marcus Brown the erroneous monuments set by the Moorhead surveyors and regardless of where anyone erroneously thought or believed the correct location of this land boundary line to be. Neither the title to land nor the boundaries to a deeded parcel move about from time to time based on where someone, including a particular surveyor, might erroneously believe the correct location of the true boundary line to be. In 1975, the Rizzos conveyed to appellant Rivers' predecessor in title property the northern boundary of which is defined as being 400 feet south of, and parallel to, the north line of this quarter-quarter section. Regardless of any assertion that this conveyance was made relying on the Moorhead survey, the description itself does not describe the line in question by reference to the surveyor monuments set by the Moorhead surveyor. On the contrary, that description adopts by reference the true north line of this quarter-quarter section which is necessarily controlled by the location of that line as established by the original government survey. Even if the description in the subsequent deed is considered to overlap the south 28 feet of the property previously conveyed by the Rizzos to Lozeaus' predecessor in title (which it does not), it is quite immaterial because, at the time of the conveyance to Paul W. Adams, Mr. and Mrs. Rizzo did not own that south 28 feet, they having previously conveyed legal title to it to Marcus Brown, Lozeaus' predecessor in title. All else argued in this case is immaterial. The Lozeaus are entitled to prevail in this controversy. All legal theories that could change the result in this case, such as those relating to adverse possession, title by acquiescence, estoppel, lack of legal title, etc., were neither asserted, nor argued, nor material in this case. This case is reversed and remanded with instructions that the trial court enter a judgment in favor of the appellees Lozeau and wife, in accordance with the land description as controlled by the official U.S. government survey.[35]

REVERSED and REMANDED

19.8 Epilogue

After reading these decisions, it is hoped that the student of surveying, as well as the practicing surveyor, will have gained insight and an appreciation of evidence and its importance in professional practice. The surveyor is an important link in the legal chain of evidence from its initial creation to its subsequent recovery in order to ascertain lines created decades ago.

Also, today's surveyor must not only be attuned to the modern technological aspects of measurements for ascertaining boundary lines but also be able to, as Justice Cooley so aptly stated over 100 years ago, perform quasi-judicial functions.

Evidence, boundaries, land titles, and law combined together to identify and locate land boundaries and property rights, ownership lines, rights, and interests, as well as possession lines all make up the legal aspects of the surveying profession. The modern surveyor, as a locator of boundaries created by individuals some time in the past, must be familiar enough with each aspect and how they relate to one another to perform his or her function, that of being the impartial identifier and the locator of land boundaries and land parcels. Without one, the other cannot exist, and thus the surveyor is a vital element in the total picture of accomplishing these tasks.

Notes

Note: All cases presented in this chapter are from lexis.com with verbal approval. Footnotes have been adapted and bold emphasis added by authors.

1. Hayw. 237.
2. Diagram 1.
3. 1 Hayw. 254.
4. 2 Id. 496.
5. 2 Id. 67, 139.
6. Diagram 2, Taylor 118.
7. Diagram 3, 1 Hayw. 22.
8. 1 Hayw. 378.
9. 1 Hayw. 376.
10. Diagram 4, 1 Hayw. 283.
11. 2 Hayw. 354.
12. Diagram 5.
13. 2 Hayw. 301.
14. 2 Hayw. 160.
15. 2 Hayw. 183, Tayl. 301.
16. 1 N.C. Term. Rep. 1.
17. 9 Id. 34.
18. Editor's note: This is unusual in that Diehl was decided in 1878, three years subsequent.
19. All bounds and starting points are questions of fact to be determined by testimony, and surveyors have no more authority than any other men to determine boundaries on their own notion (*Cronin v. Gore*, 38 Mich. 381; *Case v. Trapp*, 48 Mich. 59).
20. The pertinent portions of the trial court's findings are as follows: The original monument placed at the 1/4 corner in the 1872 survey was not found, after a thorough search of the area by this surveyor and his crew. No accessories to the monument were found, and no record of other surveys pertinent to the involved sections was found in the county records.
21. We feel that these findings do not necessarily lead to the finding that the corner was "lost" rather than obliterated and capable of being reestablished by other sources of information.
22. 43 U.S.C.A. §772 provides in pertinent part: "The Secretary of the Interior may, as of March 3, 1909, in his discretion cause to be made, as he may deem wise under the rectangular system on that date 468 F.2d 633, *636; 1972 U.S. App. provided by law, such resurveys or retracements of the surveys of public lands as, after full investigation, he may deem essential to properly mark the boundaries of the public lands remaining undisposed of: Provided, That no such resurvey or retracement shall be so executed as to impair the bona fide rights or claims of any claimant, entryman, or owner of lands affected by such resurvey or retracement . . ."

23. This manual is a supplement to the *Manual of Survey Instructions* (1947) of the Bureau of Land Management. The courts have recognized the manual as a proper statement of surveying principles. See *Reel v. Walter*, 131 Mont. 382, 309 P. 2d 1027 (1957).
24. A government surveyor testified that they were aware that the defendants, "among other people," had a deed originating from the sixteenth corner of the line between sections 1 and 12. While the surveyors should investigate and consider collateral evidence during the resurvey, we note that here information from the Doyles' surveyor was received later, shortly before trial, in a conference in the field and considered by the government surveyor, including the stone that he submitted for the quarter corner monument.
25. When the resurvey was commenced, Mr. Teller was Chief of Cadastral Surveys at the Colorado office of the Bureau of Land Management. On July 1, 1965, he was promoted to the position of Chief of the Bureau of Land Management, Denver Service Center, Cadastral Surveying Branch.
26. This is only a reestablishment of the true position of the original survey by retracement. See Jon S. Grimes, *Clark on Surveying and Boundaries*, 4th ed. sec. 650, "Dependent Surveys" (Indianapolis: Bobbs-Merrill Co., 1976), p. 956.
27. See *Akin v. Godwin*, 49 So. 2d 604 (Fla. 1950); *Willis v. Campbell*, 500 So. 2d 300 (Fla. 1st DCA 1986); *Zwakhals v. Senft*, 206 So. 2d 62 (Fla. 4th DCA 1968); *City of Pompano Beach v. Beatty*, 177 So. 2d 261 Fla. 2d DCA 1965); and *Froscher v. Fuchs*, 130 So. 2d 300 (Fla. 3d DCA 1961).
28. Property descriptions sometimes refer to irregular natural lines capable of identification, such as the banks, shores, and high and low marks of bodies of water such as oceans, lakes, rivers, and streams and to the midtread of streams, the face of cliffs, the ridge of mountains, etc.
29. In a similar manner, when there is an inconsistency between the description of a corner (a line terminal point) in field notes and plats subsequently made and recorded and the original monument evidencing that corner on the ground, the original monument on the ground controls. See *Tyson v. Edwards*, 433 So. 2d 549 (Fla. 5th DCA 1983), rev. denied, 441 S. 2d 633 (Fla. 1983).
30. See sec. 472.005(3), Florida Statutes.
31. This is a most important qualification.
32. See *Clark on Surveying and Boundaries*, 4th ed., Chap. 14, "Tracking a Survey" (Grimes, 1976), p. 339.
33. Neither the 1969 deed from the Rizzos to Marcus Brown nor the 1975 deed from the Rizzos to Paul W. Adams contains property descriptions of lines bounded by monuments set by surveyor Moorhead in 1964. This would be an entirely different case if the land descriptions in question described lines "commencing at (or running to) a concrete monument set in 1964 by surveyor Moorhead, etc."
34. See sec. 258.08, Florida Statutes and *Fla. Stat. Annot.*, vol. 1, p. 119 (St. Paul, MN: West Publishing, 1961). Unfortunately, this helpful material has been omitted from the 1988 edition of this volume of F.S.A.
35. Notwithstanding that Rizzo and Brown both may have subjectively believed or intended Rizzo's deed to Brown to convey the land between the erroneous Moorhead monuments, because the deed described land by reference to the U.S. government survey it conveyed the legal title to the north 400 feet of this quarter-quarter section as measured from the true location of the original government survey. To the extent that Rizzo's deed conveyed legal title to land Rizzo did not intend to convey, Rizzo's remedy would have been to have brought a reformation suit in equity to have his deed reformed to describe the correct parcel by a correct description. Of course, the resulting litigation can easily be visualized:

Rizzo would claim that he and his grantee Marcus Brown intended Rizzo's deed to convey land only south to a point 33 feet north of one of Moorhead's monuments and his deed should be reformed accordingly. Brown would admit that was true but would then claim that the parties also obviously intended that Brown was to obtain property 400 feet wide from north to south and that Brown should either keep the 400 feet described in the deed or be entitled to obtain money damages from Rizzo or to rescind the transaction because of Rizzo's misrepresentation that he owned the erroneous Moorhead monument located 28.71 feet north of Rizzo's true line and Rizzo did not own that northern 28.71 feet. These contentions, which never matured, existed only between the original parties and do not inure to any subsequent good faith purchasers who took legal title to their parcels according to the land descriptions contained therein, and the equitable and legal rights between Rizzo and Brown, being personal to them, are immaterial in litigation between subsequent owners.

APPENDIX A

THE SURVEYOR'S REPORT

The document in this appendix is a suggested outline for a report of survey that the professional should prepare to preserve information and to provide the client with an analysis of the work performed.

It is suggested that such a report be prepared for each utility survey but that in a forensic survey it be prepared at the request of the attorney.

The surveyor should provide the client with a copy of the report, but the surveyor should also place a copy in the project files.

Surveyor's Report

____

[client's name]

I. *General Information*

Survey Firm: The survey was performed by ________________________ ____[surveying firm], located at _________________________________[address], ____________________________ [telephone number], and was performed by ________________________________ [surveyor], _____________________[license number], on ______________________[date].

Client: The survey was performed on behalf of ___________________________ [client's name] residing at ____ __________________________________[address], ______________________________ [telephone number].

Project: The property is located along ____________________________ [road, stream, physical feature, etc.] in ________________________________ [municipality], county of ____, state of ______.

II. *General Survey Information*

Purpose: Between ______________ [date] and ______________ [date], a ______________ [type of survey: retracement, subdivision, etc.] was performed to locate the ______________ [record boundaries, possession boundaries, ownership boundaries, easements, topography, etc.].

Specifications: The survey complied with

(1) accuracy specifications of ______________ [order, class, etc.] promulgated by ______________ [A.C.S.M./state society/etc.];

(2) requirements fixed by ______________ [governing agency/etc.].

Record: The survey is recorded in ______________ [computer file name/book volume/etc.].

Azimuth Basis: The basis of azimuth was ______________ [true/magnetic/grid/etc.], determined by observing ______________ [sun/Polaris/etc.], and calculated using the ______________ [hour angle/altitude/etc.] method while observing at ______________ [latitude, longitude, station] on ______________ [date] using ______________ [number direct/reverse] positions.

Declinations: Based on this survey and prior records for the property [add specific information as follows]

(1) a ______________ [date] magnetic declination of ______________ [amount] was calculated;

(2) a grid declination of ______________ [amount] was calculated based on ______________ [projection system].

Tolerances and Calculations: The relative error of closure was calculated and found to be 1: ______________ [amount] and adjusted using ______________ [compass rule/least squares/etc.].

Conduct of the Survey and Actions Taken: On ______________ [date] ______________ [a record search took place, a field reconnaissance took place, letters were sent to adjoining landowners, a field survey was conducted, computations were completed, a plat of the survey was prepared, a review occurred, etc.].

III. *Comments, Problems, Analysis, and Results*

Comments: ______________ [gaps, encroachments, fences, measurements, monuments, etc.].

Problems Encountered: During the course of the survey, the following problems were noted: ______________ [research: missing records; reconnaissance: construction; field survey: deep snow; computations: deed closures, etc.].

Analysis: ______________ [corner number, boundary, etc.] was established ______________ [discussion: procedure, method, evidence, see corner form, etc.].

Results: Based on the analysis and best available evidence, ________________________________ [description: monuments set, markings made, actions taken, etc.].

IV. *Revised Parcel Description*

Based on the survey, the following revised description was prepared: ________________ ________________ [property description], ________________ [appurtenant easements/rights: "together with"], ________________ ________________ [existing easements/rights: subject to], ________________ [fee excluded: excepting], ________________ [rights excluded/retained: reserving], ________________ [history: being], ________________ ________________ [reference: surveyor, plan, date].

Seal

____
[surveyor's signature]

____
[surveyor's name]

____
[license number]

Appendices: (include as required)

A. Contract for services
B. ________ [source] survey specifications
C. Record research table
D. Measurement comparison table
E. Monument comparison table
F. Traverse information
   1. Solar azimuth computations
   2. Traverse stations established
   3. Field measurements obtained
   4. Reduced measurements
   5. List of final coordinates
   6. Road easement apportionment
   7. Map (hard copy)
G. Error estimation (standard deviation)
H. Curve data
I. Reference monuments established
J. Areas (easements, fee, encroachments, etc.)
K. Miscellaneous maps and documents (unrecorded)
L. Time sheet

AGREEMENT FOR SERVICES

Heading: This agreement made ______________________ [date] between ______________________ [client's name] of ______________________ [street, route, etc.], city of ______________________, county of ______________________, state of ______________________, hereafter called client, and ______________________ [surveyor's name], of ______________________ [street, route, etc.], city of ______________________, county of ______________________, state of ______________________, hereafter called surveyor.

I. *Recital*

Client requests surveying services in the form of ______________________ [list type or types of surveys: retracement survey, subdivision, mapping, control, mining, etc.], hereafter called the project, for property owned by ______________________ [owner's name], located ______________________ [specific location: intersection Mud Run and Route 94] in ________ [municipality], county of ________, state of ______________________ and shown as ______________________ [lot, block, subdivision, tax parcel number, etc.] in ______________________ [deed/plat/etc.], book ______________________ [volume number], page ______________________ [number].

Client and surveyor in consideration of the mutual promises set forth and agree as follows:

II. *Responsibilities of the Surveyor*

General: Surveyor agrees to perform professional services in connection with the project hereafter stated. Surveyor will provide those services on behalf of client for those phases of the project to which this agreement applies and will give consultation and advice to client during the performance of the service to include the following: ______________________ ______________________. [Specify the type of work, documents to be prepared and delivered, and other services. The following may be used for guides.]

(1) Surveyor will serve as client's professional representative in designing and supervising the subdivision of client's property according to the limitations set forth in this agreement.

(2) Surveyor agrees to survey and locate the boundaries according the best available evidence and agrees to provide an unbiased opinion on the location of the record, possession, and ownership boundaries. For this purpose, surveyor will research public records to the extent possible or necessary. Surveyor will notify those adjoining landowners (as indicated by the tax records) of the survey. Surveyor will perform a survey according to ______________ [class of standards], promulgated by ______________ [state society/A.C.S.M./state law/etc.].

Monumentation: Record corners that are set, established, or replaced will be monumented by ______________________ [pin/#5 rebar/1″ pipe/4″ diameter concrete post/etc.] or witnessed by ______________________ ______________________ [pin/pipe/etc.]. All monuments will be appropriately marked ______________________ [with surveyor's name and license number/"reference"/etc.] and set ______________________ [flush/protruding/below ground/etc.].

In the event the survey indicates possible title problems or that monumenting the record corner will cause a dispute, surveyor will inform client in writing and will defer setting any monument that may cause conflict until written instructions are received from client.

Easements: Surveyor ______________________ [agrees/is not obligated] to ______________________ [show, monument, etc.] the location of recorded easements ______________________ [and/or] visible easements that may appear within the boundary of the project site.

Survey Plans: Surveyor will provide a survey plat. The plat will meet the requirements of the ______________ [agency/municipality: Plat Act of the State of Georgia, etc.] and will contain, as a minimum, the following: ______________________ ______________________ [record, possession, ownership boundaries, adjoiners, bearings, distances, monumentation, etc.].

Subdivision Plans: Surveyor will provide those plans required by ______________ ______________[governing body] according to specifications and standards listed in ______________________ [governing ordinance].

Completion: The survey will be considered complete when the monuments marking or referencing the corners have been set.

Description: Surveyor will provide client with a description of the project and ______________ ______________________ [parcel, lots, easements, etc.].

Release and Note: Upon final payment of surveyor's fee, as provided under the terms of this agreement, surveyor agrees to provide client with a statement that all fees have been paid and to release any mechanics' or other liens placed on client's property by surveyor.

III. *Additional Services of Surveyor*

General: If authorized in writing by client, surveyor will furnish the following additional services: ______________________.
[List any services, other than those previously identified.]

Report: Surveyor will prepare a survey report depicting and describing the general history of the survey and the project. Surveyor will provide client with a copy of the survey report. The report will include as a minimum the following: ______________________ ______________________ [problems, analysis, decisions, etc.].

Dissemination: All information obtained as a result of the survey will be considered confidential and surveyor will not provide any information without the express written permission of client or his or her authorized representative.

Recording: Surveyor will record appropriate plats, plans, or documents with public officials or agencies.

Prints: Surveyor will provide client ______________________ [specify number] copy[ies] of the survey plat. If requested, surveyor will provide client with a reproducible print.

Other Documents: In addition to the plat and description, surveyor will provide client with copies of the following: ______________________ ______________________ [copy of other documents, field notes, etc.].

Witness: Surveyor agrees to serve as an expert witness in any litigation or other adjudicatory proceeding involving the project that is subject to this agreement.

IV. *Responsibilities of the Client*

General: Client will be responsible for the following:
________________. ________________ [Insert any of the following as applicable.]

Permission: Client by this agreement agrees to obtain reasonable access across public and private lands to reach the project site and will obtain permission for surveyor to enter upon the project to do those tasks necessary to provide professional services under this agreement. Client is responsible for obtaining approval, consent, or permission from those parties that have any interest and could otherwise prevent completion of the project in a reasonable and timely manner.

Indemnification: Client agrees to indemnify surveyor against reasonable damage of client's property that reasonably occurred during the course of surveyor's services on behalf of client.

Assistance: Client will assist surveyor by placing at surveyor's disposal all available information pertinent to the services that are to be provided by surveyor including ______________ ____________[potential conflicts, previous survey documents, deeds, correspondence, plats, easements, other encumbrances, etc.].

Representation: In the event surveyor is acting on behalf of several owners or parties, client will designate in writing a person to act as representative for the various parties or co-owners. Such person shall have complete authority to give instructions, receive information, and make decisions. The first name listed as "client" will be the designated representative in the event there is a failure to designate a particular representative.

Notice: Client will give ______________________________ [prompt/X number of days] notice to surveyor whenever client observes or otherwise becomes aware of any problem concerning the project or surveyor's work.

V. *Duration of Service*

Period: Surveyor shall provide client with services that will [give period of service]

(1) Start ____________________ [before/around/etc.] ____________________ [date] and be completed ________________ [before/around/etc.] ________________ [date]. Documents will be submitted to client for review within ____________________ [amount] days following ______________________________ [the completion of the survey/receipt of the final payment/etc.].
(2) Terminate upon ______________________________ [delivery of a plat/approval of the final plan/completion of the survey/etc.].
(3) Continue at the pleasure of surveyor or client.
(4) Terminate on ____________ [date].

Retainer: A retainer of $ ____________________________ [amount] shall be required and [list reasons and time of transfer].

(1) The retainer shall be due upon signing this agreement.
(2) The retainer will be forfeited by client should client breach this agreement.

(3) The retainer, plus ____________________ [amount] percent interest compounded ____________________ [daily/monthly/etc.] shall be returned to client should surveyor breach this agreement.
(4) The retainer shall be returned to client without interest should the project be terminated for any reason beyond the control of either client or surveyor.
(5) The retainer without interest shall be applied to the final payment at the completion of the project by surveyor or the termination of surveyor's services, whichever is sooner.

Basic Services: For basic services (Section I), client agrees to pay surveyor [indicate method or amount of payment]
(1) a lump-sum fee of ______________ [amount] dollars;
(2) cost of the survey plus ______________ [amount] percent;
(3) an hourly rate as set out in the fee table ____________________ [estimated/guaranteed] not to exceed ____________________ [amount] dollars.

Supplemental Charges: In addition, client agrees to pay surveyor the actual ____________________ [cost of monuments, recording fees, paint, transportation, phone costs, etc.].

Additional Services: For additional services (Section II), client agrees to pay [indicate method or amount of payment]
(1) according to the amounts listed in the fee schedule;
(2) cost plus ______________ [amount] percent;
(3) a rate of ____________ [amount] dollars per hour.

Interest: All amounts billed are net in ____________________ [amount] days. Client agrees to pay surveyor ____________________ [amount] percent interest on all outstanding amounts due and payable in excess of ____________________ [number] days.

Unanticipated Termination: In the event services are terminated for any reason prior to the completion of the project, client shall pay surveyor for reasonable services performed by surveyor on behalf of client and completed prior to the termination of surveyor's services.

Payment Schedule: Periodic payments for services performed shall be made according to the following:
(1) Boundary retracement ____________________ [schedule]:
  (a) upon 90 percent completion of the research ____________________ [percent lump sum/number of hours times hourly rate];
  (b) upon completion of the reconnaissance ____________________ [percent lump sum/number of hours hourly rate];
  (c) upon completion of the field survey ____________________ [percent lump sum/number of hours times hourly rate];
  (d) when the corners and boundaries are marked ____________________ [percent lump sum/number of hours times hourly rate];
  (e) upon the delivery of documents [percent lump sum/number of hours times hourly rate] minus the retainer fee of ____________________ [amount] dollars.
(2) Subdivision ____________________ [schedule]:
  (a) upon estimate of the cost and preparation of estimate ____________________ [percent lump sum/number of hours times hourly rate];
  (b) upon completion of the sketch plan ____________________ [percent lump sum/number of hours times hourly rate];

(c) upon completion of the preliminary plan ______________________
[percent lump sum/number of hours times hourly rate];
(d) upon completion of all improvements ______________________
[percent lump sum/number of hours times hourly rate];
(e) upon completion of the final plan ______________________
[percent lump sum/number of hours times hourly rate] minus the retainer fee of ______________________ [amount] dollars.

3. ______________ [weekly/monthly/etc.] at a rate of ______________ [percent lump sum/number of hours times hourly rate] to begin on ______________ [date/start of the project/etc.] until ______________ [date/dollars paid/project completed/etc.], at which time the balance is due and payable.

VI. *General*

Modification: Client or surveyor at any time during the course of this agreement may request a modification to this agreement. This request will in no way affect or make void this agreement, but the fair and reasonable value of the services agreed upon shall be added to or deducted from the amount specified in this agreement as the case may be. This agreement shall be held to be completed when the work is finished according to the original agreement, as amended by such changes.

Termination: In the event of substantial failure to perform in accordance with the terms of this agreement by any party, this agreement may be terminated by the other party upon ______________ [amount] days' written notice. If this agreement is so terminated, surveyor shall be paid as provided in Section V.

Ownership of Documents: All documents, including the original drawings, field notes, and other data gathered by surveyor remain the property of surveyor. Surveyor ______________ [shall/shall not] have the right to provide the information to other surveyors upon their request. Client, at his or her expense, may obtain a set of reproducible record prints of drawings and copies of other documents in consideration of which client will use them solely in connection with the project and not for the purpose of making subsequent extensions or enlargements of the project. Client will not sell, publish, or display them publicly or use them for new projects without the written permission of surveyor.

Arbitration: All claims, disputes, and other matters in question, arising out of or relating to this agreement or the breach of this agreement, shall be decided by ______________ [person/firm/procedure/etc.] according to ______________ [industry/profession/etc.] rules of arbitration that are currently in effect. This agreement to arbitrate shall be specifically enforceable under the prevailing arbitration law.

Notice of Arbitration: Demand for arbitration shall be in writing and sent or delivered to the other party to this agreement and to the ______________ [person/firm/etc.] responsible for arbitrating. The demand shall be made within ______________ [reasonable time/X number of days/etc.] after the claim, dispute, or other matter in question has arisen. In no event will arbitration be allowed after the appropriate statute of limitations has passed.

Award or Judgment: The award or judgment by the arbitrators shall be final, and the judgment may be entered in any court having proper jurisdiction.

Insurance: Surveyor shall secure and maintain such insurance as required ____________ [by law, to protect client] from ________________________________ [claims under the workmen's compensation acts, claims for bodily injury, death, property damage, etc.] that might arise from surveyor's performance of services under this agreement.

Successors and Assigns: Each party to this agreement binds himself or herself, his or her partners, successors, executors, administrators, and assigns to the other party of this agreement and to the partners, successors, executors, administrators, and assigns of such other party in respect to all covenants in this agreement. Except as noted, neither party shall assign, sublet, or transfer his or her interest in this agreement without the written consent of the other party. Nothing in this document shall be construed as creating any personal liability on any officer or agent of any public body which may be party to the document, nor shall it be construed as giving any rights or benefits to anyone other than the parties to this agreement.

Choice of Law: The laws of ____________________________ [state/commonwealth] will apply concerning the interpretation and performance of this agreement.

VII. *Fees and Fee Schedule*

Estimates: Surveyor has no control over the difficulty in locating records, condition of the terrain, density of the vegetation, severity of the weather, and availability of original monuments. Therefore, any estimates of the project cost that were previously provided to client were made on the basis of experience and similar services performed in the past. Surveyor cannot and does not guarantee that the actual costs will not vary from estimates previously named provided, however, that surveyor will notify client as soon as surveyor is aware the fee charged will exceed the estimated cost.

Fees: ____________ [See sec 4.3.]

VIII. *Special Provisions*

The provisions and the exhibits annexed to the agreement represent the entire agreement between the parties. This agreement, the exhibits, and any special provisions to which this agreement is subject may only be altered, amended, or rescinded by a properly executed written agreement entered subsequent to this agreement. This agreement shall be subject to the following special provisions: ____

____

____________ [special provisions].

In witness, the parties have executed this agreement at ______________________ ____________________________ [place of execution], this, the ______________________ day of ____ [month], in the year ____ [year].

____

[surveyor's signature]

____

[client's signature]

APPENDIX B

WOODEN EVIDENCE

This appendix covers the use of certain aspects of evidence.

Sometimes a surveyor must make the distinction between the various species of oaks. When the surveyor may have only fragments of stumps or portions of tree trunks, the surveyor will not be able to say which of the various red or white oaks the wood is, but the surveyor may make the distinction that it is one of the white or one of the red oaks.

The second excerpt in the appendix provides the surveyor with the ability to make the distinction between that broad classification of hardwoods and softwoods. One will not be able to make the distinction between a red oak and a hickory, but the distinction can be made that it is either a hardwood species or a softwood species.

CHARCOAL IDENTIFICATION [1]

Charcoal is largely unaffected by fungi or other wood-destroying organisms. Consequently, it will persist in soil for great periods of time and often turns up in archaeological diggings and paleobotanical studies. The identification of such charcoal is often desirable because it may provide a record of paleoclimatic changes or have significance as evidence of ancient cultural practices. Identification is possible because the anatomical features of the wood remain intact during the carbonization process.

The extreme brittleness of charcoal makes it a difficult material to work with, and the usual wood-sectioning procedures are totally unsatisfactory. Charcoal can be broken across the grain or split along it, but thin sections cannot be cut with a knife because the knife edge, no matter how sharp, simply shatters the cell walls into tiny fragments. Consequently, to identify charcoal, one must work with pieces too thick for the transmitted light system used in ordinary microscopy.

The ideal identification method requires a microscope with incident light attachments. This system provides light from above, which reflects off the specimen surface into the microscope. The anatomical details of a solid material can, therefore, be satisfactorily viewed. Very often, however, such equipment is not available because of its rather high cost. If this is the case, many charcoal specimens can still be identified with a good hand lens and a reference collection of charcoal specimens of known identity.

A charcoal reference collection is relatively simple to prepare. The first step is to accumulate a collection of woods of known identity and then convert them to charcoal. Wood specimens should be large enough so that after charcoaling they can be split and broken to show sharp structural details of the cross, radial, and tangential surfaces.

Charcoaling can be accomplished by any method of heating the wood to a temperature of about 800°F to 900°F in the absence of air. A simple system is to immerse the wood specimens in sand in a covered ceramic crucible and heat with a Bunsen burner. After 15 to 20 minutes, when smoke emission has diminished, the heat can be removed. The crucible should be allowed to cool before the charcoal is removed, or the specimen may begin to burn. The known charcoal specimens can then be stored in labeled containers, such as small jars, for future reference.

A basic knowledge of wood structure and anatomy is necessary to use a charcoal reference collection. A good text for the novice is Volume I of the *Textbook of Wood Technology* [2].

The first step in studying any charcoal specimen is to break it across the grain and examine the broken surface with a 10X or 14X hand lens in good light. Such details as pore size and distribution, parenchyma patterns, and ray widths will be easily discernible from a cleanly broken cross-section. If the specimen being studied is unknown, then the reference collection of charcoal specimens is utilized to compare the known with the unknown. In this regard, it is most helpful to have specimens representing different growth rates in the reference collection. Many specimens can be identified from the cross-section alone, and in some cases it may not be practical to go further. The charcoal of such woods as oak, hickory, and ash can readily be distinguished from cross-sections.

IDENTIFICATION OF OAK WOODS [3]

Over 50 species of native oaks assume proportions of trees, and about 25 are used for lumber. After cutting into lumber, there is no means known to the Forest Products Laboratory by which the wood can be identified as to exact species. By examining the wood alone, however, it is easy to separate the oaks into two groups—the white oak group and the red oak group. For most purposes, fortunately, further classification is not necessary. The oaks all average about the same in strength, but the heartwood of the species of the white oak group is much more durable under conditions favorable to decay than that of the species in the red oak group.

The white oak group includes true white oak, post oak, bur oak, swamp white oak, swamp chestnut oak, chestnut oak, overcup oak, and Oregon white oak. The red oak group includes true red oak, pin oak, black oak, turkey oak, southern red oak, swamp red oak, blackjack oak, water oak, willow oak, and laurel oak.

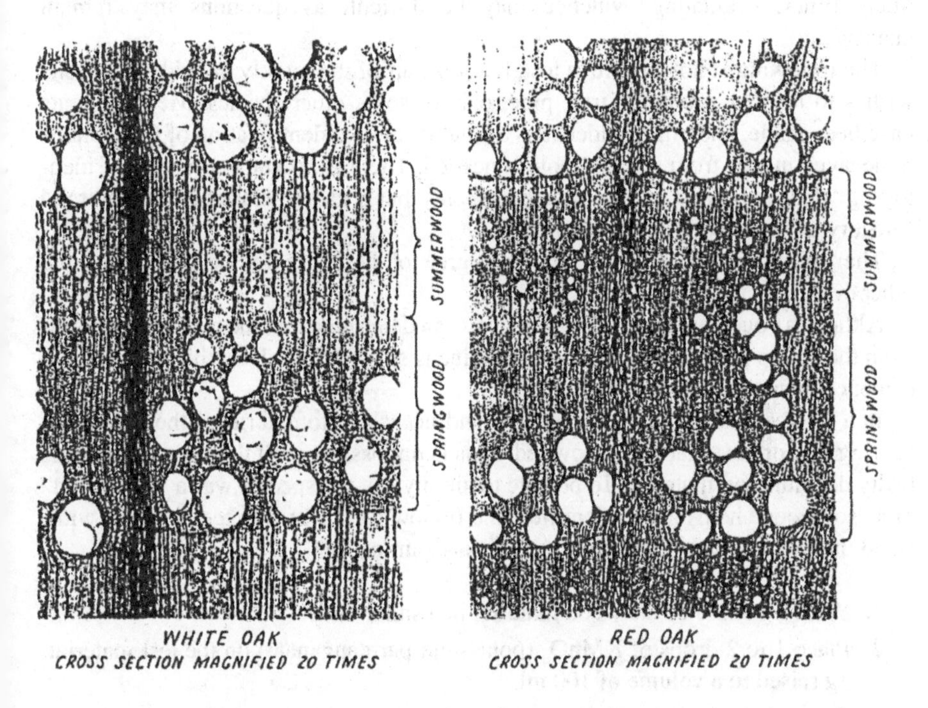

WHITE OAK
CROSS SECTION MAGNIFIED 20 TIMES

RED OAK
CROSS SECTION MAGNIFIED 20 TIMES

The color of the wood is a ready but not absolutely reliable means of distinguishing the wood of the white oak group from that of the red oak group. The wood of the latter group usually has a distinctly reddish tinge, especially near the knots. The wood of the white oak group is generally a grayish brown, but occasionally a reddish tinge is found in white oak lumber.

For more accurate identification, it is necessary to examine the pores of the wood, which appear as tiny holes on a smoothly cut end surface. They vary in size throughout each growth ring, being larger in the springwood, where they are visible to the eye, decreasing in size abruptly toward the summerwood. The large pores in the springwood of the heartwood and inner sapwood of the woods of the white oak group are usually plugged up with a froth-like growth called tyloses, and those of the red oak group are open. This feature, however, is not so reliable for classification as the character of the much smaller pores in the summerwood.

To attain certainty as to whether a piece of oak belongs to the white or red oak group, cut the end grain smoothly with a sharp knife across several growth rings of average width. With a hand lens, examine the small pores in the dense summerwood. If the pores are plainly visible as minute rounded openings, and can readily be counted, the wood belongs to the red oak group. If the pores in the summerwood are very small, somewhat angular, and so numerous that it would be exceedingly difficult to count them, the wood belongs to the white oak group.

THE INFALLIBLE TEST FOR HARDWOOD OR SOFTWOOD

Many times, evaluating evidence may be difficult as questions may remain unanswered.

The following is not a positive test, but it is part of the totality of evidence, if one wishes to identify a bearing tree, pine, oak, or some other species. Measurements have been made, and other evidence is available, but an identification of the remains of decayed matter from a stump hole is needed. There is no known method of identifying "decayed pulp" into its specific species, but it can be classified into a broad category of "hardwood" or "softwood."

The ability to do this may add a degree of certainty to the totality of the other evidence.

Although you cannot say it is a pine, you can determine that the material removed from the stump hole is softwood and that pine is softwood. At least this is a bit more evidence than what was originally there.

In a retracement, it is often desirable and necessary to determine the species of bearing tree or post that was recovered. This is impossible if all the surveyor finds is badly decomposed material. To be able to simply say the species was a "hardwood" or a "softwood" helps in the identification. Following is a proven test that will withstand the examination of the best legal cross-examination:

1. Place a sample of the wood material in a small dish.
2. Place 1 to 2 drops of $KMnO_4$ (potassium permanganate) on the test material: 1g raised to a volume of 100 ml.
3. Leave the solution on the material for 2 to 3 minutes.
4. Remove the excess.
5. Add two drops of 12 percent HCl (hydrochloric acid) solution 12 ml commercial raised to 100 ml.
6. Let stand until decolorized.
7. Add 1 to 2 ml of concentrated NH_4OH (ammonium hydroxide).

The sample will turn the following colors:

Hardwood—bright red to violet.
Softwood—coffee brown to dull brown.

A second "quick" test is to place chips of wood into the following solutions for a period of 10 seconds each:

1. Undiluted household bleach
2. Diluted hydrochloric acid, equal volumes concentrated HCl and H_2O
3. Concentrated ammonium hydroxide

The samples will turn the following colors:

Hardwood—a red-purple color
Softwood—a yellow brown color

Caution! Use rubber gloves and safety glasses, and do it in an open area.

RING COUNT OF DECAYED WOOD MADE EASIER THROUGH FREEZING [4]

Freezing small, rotten wood samples to provide thin sections for investigation is not a new technique. Yet quite by accident, this method was recently "rediscovered" in the Oregon-Washington Cascade Range. A large central portion of a 400-year-old Douglas fir that had been felled for analysis had a brown cubical butt rot. To count rings of old-growth trees is difficult under the best of conditions, but to count such rings in rotten or partially decayed traverse sections is nearly impossible. The decayed wood tissue crumbled easily; and even though the thinnest, sharpest razor blade available was used, sequential radial measurements and ring counts could not be made. This work was done during a freezing rainstorm. The next morning, the rotten stump and breast-high-cut butt sections were frozen. Razor knife cuts clearly revealed the rings that had given so much trouble previously. Thereafter, when decay was encountered, a 4- to 6-inch thick rectangular piece was taken instead of the normal inch-thick piece. Sandwiched between quarter-inch plywood, the piece was taken to the office; there it was submerged in water for several hours and frozen. The process changed wet/green dimensions but little.

REFERENCES

1. This section is from R. C. Koeppen, U.S. Forest Service Research Note FPL-0217, by Forest Products Laboratory, Forest Service (Madison, WI: U.S. Department of Agriculture, 1972).
2. J. Panshin and C. de Zeeuw, *Textbook of Wood Technology,* vol. I: *Structure, Identification, Uses and Properties of the Commercial Woods of the United States and Canada*, 3rd ed. (New York: McGraw-Hill Book Co., 1970), 705 pp.
3. This section is a reprint of Technical Note No. 125 of the same title, which was reissued in April 1961. The original note was written in February 1921.
4. Francis R. Smith, Clark E. Smith, and John E. Firth, research foresters, U.S. Forest Service, Corvallis, Oregon, and at Oregon State University, Corvallis.

APPENDIX C

THE (QUASI-)JUDICIAL FUNCTION OF SURVEYORS[1]

THOMAS M. COOLEY

When a man has had a training in one of the exact sciences, where every problem within its purview is supposed to be susceptible of accurate solution, he is likely to be not a little impatient when he is told that, under some circumstances, he must recognize inaccuracies, and govern his action by facts which lead him away from the results which theoretically he ought to reach. Observation warrants us in saying that this remark may frequently be made of surveyors.[1]

In the State of Michigan, all our lands are supposed to have been surveyed once or more, and permanent monuments fixed to determine the boundaries of those who should become proprietors. The United States, as original owner, caused them all to be surveyed once by sworn officers, and as the plan of subdivision was simple, and was uniform over a large extent of territory, there should have been, with due care, few or no mistakes; and long rows of monuments should have been perfect guides to the place of any one that chanced to be missing. The truth, unfortunately, is that the lines were very carelessly run, the monuments inaccurately placed; and, as the record witnesses to these were many times wanting in permanency, it is often the case that when the monument was not correctly placed, it is impossible to determine by the record, by the aid of anything on the ground, where it was located. The incorrect record of course becomes worse than useless when the witnesses it refers to have disappeared.

It is, perhaps, generally supposed that our town plats were more accurately surveyed, as indeed they should have been, for in general there can have been no difficulty in making them sufficiently perfect for all practical purposes. Many of them, however, were laid out in the woods; some of them by proprietors themselves, without either chain or compass, and some by imperfectly trained surveyors, who, when land was cheap, did not appreciate the importance of having correct lines to determine boundaries when land should become dear. The fact probably is that town surveys are quite as inaccurate as those made under authority of the general government.

RECOVERING LOST CORNERS

It is now upwards of 50 years since a major part of the public surveys in what is now the State of Michigan were made under authority of the United States. Of the lands south of Lansing, it is now 40 years since the major part were sold and the work of improvements begun. A generation has passed away since they were converted into cultivated farms, and few if any of the original corner and quarter stakes now remain.

The corner and quarter stakes were often nothing but green sticks driven into the ground. Stones might be put around or over these if they were handy, but often they were not, and the witness trees must be relied upon after the stake was gone. Too often the first settlers were careless in fixing their lines with accuracy while monuments remained, and an irregular brush fence, or something equally untrustworthy, may have been relied upon to keep in mind where the blazed line once was. A fire running through this might sweep it away, and if nothing was substituted in its place, the adjoining proprietors might in a few years be found disputing over their lines, and perhaps rushing into litigation, as soon as they had occasion to cultivate the land along the boundary.

If now the disputing parties call in a surveyor, it is not likely that any one summoned would doubt or question that his duty was to find, if possible, the place of the original stakes which determined the boundary line between the proprietors. However erroneous may have been the original survey, the monuments that were set must nevertheless govern, even though the effect be to make one half-quarter section 90 acres and the one adjoining, 70; for parties buy, or are supposed to buy, in reference to these monuments, and are entitled to what is within their lines, and no more, be it more or less. While the witness trees remain, there can generally be no difficulty in determining the locality of the stakes.

When the witness trees are gone, so that there is no longer record evidence of the monuments, it is remarkable how many there are who mistake altogether the duty that now devolves upon the surveyor. It is by no means uncommon that we find men, whose theoretical education is thought to make them experts, who think that when the monuments are gone, the only thing to be done is to place new monuments where the old ones should have been, and would have been if placed correctly. This is a serious mistake. The problem is now the same that it was before: to ascertain by the best lights of which the case admits, where the original lines were. The mistake above alluded to is supposed to have found expression in our legislation; though it is possible that the real intent of the act to which we shall refer is not what is commonly supposed.

An act passed in 1869 (Compiled Laws, 593) amending the laws respecting the duties and powers of county surveyors, after providing for the case of corners which can be identified by the original field notes or other unquestionable testimony, directs as follows:

Second. Extinct interior section corners must be reestablished at the intersection of two right lines joining the nearest known points on the original section lines east and west and north and south of it. Third. Any extinct quarter-section corner, except on fractional lines, must be reestablished equidistant and in a right line between the section corners; in all other cases at its proportionate distance between the nearest original corners on the same line.

The corners thus determined, the surveyors are required to perpetuate by noting bearing trees when timber is near. To estimate properly this legislation, we must start with the admitted and unquestionable fact that each purchaser from government bought such land as was within the original boundaries, and unquestionably owned it up to the time when the monuments became extinct. If the monument was set for an interior section corner, but did not happen to be "at the intersection of two right lines joining the nearest known points on the original section lines east and west and north and south of it," it nevertheless determined the extent of his possessions, and he gained or lost according as the mistake did or did not favor him.

EXTINCT CORNERS

It will probably be admitted that no man loses title to his land or any part thereof merely because the evidences become lost or uncertain. It may become more difficult for him to establish it as against an adverse claimant, but theoretically the right remains; and it remains as a potential fact so long as he can present better evidence than any other person. And it may often happen that notwithstanding the loss of all trace of a section corner or quarter stake, there will still be evidence from which any surveyor will be able to determine with almost absolute certainty where the original boundary was between the government subdivisions.

There are two senses in which the word extinct may be used in this connection: One, the sense of physical disappearance; the other, the sense of loss of all reliable evidence. If the statute speaks of extinct corners in the former sense, it is plain that a serious mistake was made in supposing that surveyors could be clothed with authority to establish new corners by an arbitrary rule in such cases. As well might the statute declare that, if a man loses his deed, he shall lose his land altogether.

But if by extinct corner is meant one in respect to the actual location of which all reliable evidence is lost, then the following remarks are pertinent:

1. There would undoubtedly be a presumption in such a case that the corner was correctly fixed by the government surveyor where the field notes indicated it to be.
2. But this is only a presumption, and may be overcome by any satisfactory evidence showing that in fact it was placed elsewhere.
3. No statute can confer upon a county surveyor the power to "establish" corners, and thereby bind the parties concerned. Nor is this a question merely of conflict between State and Federal law; it is a question of property right. The original surveys must govern, and the laws under which they were made

govern, because the land was brought in reference to them; and any legislation, whether State or Federal that should have the effect to change these, would be inoperative, because of the disturbance to vested rights.

4. In any case of disputed lines, unless the parties concerned settle the controversy by agreement, the determination of it is necessarily a judicial act, and it must proceed upon evidence and give full opportunity for a hearing. No arbitrary rules of survey or of evidence can be laid down whereby it can be adjudged.

THE FACTS OF POSSESSION

The general duty of a surveyor in such a case is plain enough. He is not to assume that a monument is lost until after he has thoroughly sifted the evidence and found himself unable to trace it. Even then he should hesitate long before doing anything to the disturbance of settled possessions. Occupation, especially if long continued, often affords very satisfactory evidence of the original boundary when no other is attainable; and the surveyor should inquire when it originated, how, and why the lines were then located as they were, and whether a claim of title has always accompanied the possession, and give all the facts due force as evidence. Unfortunately, it is known that surveyors sometimes, in supposed obedience to the State statute, disregard all evidences of occupation and claim of title and plunge whole neighborhoods into quarrels and litigation by assuming to "establish" corners at points with which the previous occupation cannot harmonize. It is often the case that, where one or more corners are found to be extinct, all parties concerned have acquiesced in lines which were traced by the guidance of some other corner or landmark, which may or may not have been trustworthy; but to bring these lines into discredit, when the people concerned do not question them, not only breeds trouble in the neighborhood, but it must often subject the surveyor himself to annoyance and perhaps discredit, since in a legal controversy the law as well as common sense must declare that a supposed boundary line long acquiesced in is better evidence of where the real line should be than any survey made after the original monuments have disappeared. (*Stewart v. Carleton*, 31 Mich. Reports, 270; *Diehl v. Zanger*, 39 Mich. Reports, 601.) And county surveyors, no more than any others, can conclude parties by their surveys.

The mischiefs of overlooking the facts of possession most often appear in cities and villages. In towns the block and lot stakes soon disappear; there are no witness trees, and no monuments to govern except such as have been put in their places, or where their places were supposed to be. The streets are likely to be soon marked off by fences, and the lots in a block will be measured off from these, without looking farther. Now it may perhaps be known in a particular case that a certain monument still remaining was the starting point in the original survey of the town plat; or a surveyor settling in the town may take some central point as the point of departure in his surveys and, assuming the original plat to be accurate, he will then undertake to find all streets and all lots by course and distance according to the plat, measuring and estimating from his point of departure. This procedure might unsettle every line and every monument existing by acquiescence in the town; it would be very likely to change the lines of streets, and raise controversies everywhere. Yet this is what is sometimes done; the surveyor himself being the first person to raise the disturbing questions.

Suppose, for example, a particular village street has been located by acquiescence and used for many years, and the proprietors in a certain block have laid off their lots in reference to this practical location. Two lot owners quarrel, and one of them calls in a surveyor, that he may make sure his neighbor shall not get an inch of land from him. This surveyor undertakes to make his survey accurate, whether the original was so or not, and the first result is, he notifies the lot owners that there is error in the street line, and that all fences should be moved, say, 1 foot to the east. Perhaps he goes on to drive stakes through the block according to this conclusion. Of course, if he is right in doing this, all lines in the village will be unsettled; but we will limit our attention to the single block. It is not likely that the lot owners generally will allow the new survey to unsettle their possessions, but there is always a probability of finding someone disposed to do so. We shall then have a lawsuit; and with what result?

FIXING LINES BY ACQUIESCENCE

It is a common error that lines do not become fixed by acquiescence in a less time than 20 years. In fact, by statute, road lines may become conclusively fixed in 10 years; and there is no particular time that shall be required to conclude private owners, where it appears that they have accepted a particular line as their boundary, and all concerned have cultivated and claimed up to it. Public policy requires that such lines be not lightly disturbed, or disturbed at all after the lapse of any considerable time. The litigant, therefore, who in such a case pins his faith on the surveyor is likely to suffer for his reliance, and the surveyor himself to be mortified by a result that seems to impeach his judgment.

Of course, nothing in what has been said can require a surveyor to conceal his own judgment, or to report the facts one way when he believes them to be another. He has no right to mislead, and he may rightfully express his opinion that an original monument was at one place, when at the same time he is satisfied that acquiescence has fixed the rights of parties as if it were at another. But he would do mischief if he were to attempt to "establish" monuments which he knew would tend to disturb settled rights; the farthest he has a right to go, as an officer of the law, is to express his opinion where the monument should be, at the same time that he imparts the information to those who employ him and who might otherwise be misled, that the same authority that makes him an officer and entrusts him to make surveys, also allows parties to settle their own boundary lines, and considers acquiescence in a particular line or monument, for any considerable period, as strong, if not conclusive, evidence of such settlement. The peace of the community absolutely requires this rule. It is not long since, that in one of the leading cities of the State, an attempt was made to move houses two or three rods into the street, on the ground that a survey under which the street had been located for many years had been found on a more recent survey to be erroneous.

THE DUTY OF THE SURVEYOR

From the foregoing, it will appear that the duty of the surveyor where boundaries are in dispute must be varied by the circumstances.

1. He is to search for original monuments, or for the places where they were originally located, and allow these to control if he finds them, unless he has reason to believe that agreements of the parties, express or implied, have rendered them unimportant. By monuments, in the case of government surveys, we mean, of course, the corner and quarter stakes. Blazed lines or marked trees on the lines are not monuments; they are merely guides or finger posts, if we may use the expression, to inform us with more or less accuracy where the monuments may be found.
2. If the original monuments are no longer discoverable, the question of location becomes one of evidence merely. It is merely idle for any State statute to direct a surveyor to locate or "establish" a corner, as the place of the original monument, according to some inflexible rule. The surveyor, on the other hand, must inquire into all the facts, giving due prominence to the acts of parties concerned, and always keeping in mind, first, that neither his opinion nor his survey can be conclusive upon parties concerned, and, second, that courts and juries may be required to follow after the surveyor over the same ground, and, that it is exceedingly desirable that he govern his action by the same lights and the same rules that will theirs.

It is always possible, when corners are extinct, that the surveyor may usefully act as a mediator between parties and assist in preventing legal controversies by settling doubtful lines. Unless he is made for this purpose an arbitrator by legal submission, the parties, of course, even if they consent to follow his judgment, cannot, on the basis of mere consent, be compelled to do so; but if he brings about an agreement, and they carry it into effect by actually conforming their occupation to his lines, the action will conclude them. Of course, it is desirable that all such agreements be reduced to writing, but this is not absolutely indispensable if they are carried into effect without.

MEANDER LINES

The subject of meander lines is taken up with some reluctance because it is believed the general rules are familiar. Nevertheless, it is often found that surveyors misapprehend them, or err in their application; and as other interesting topics are somewhat connected with this, a little time devoted to it will probably not be altogether lost. These are lines traced along the shores of lakes, ponds, and considerable rivers, as the measures of quantity when sections are made fractional by such waters. These have determined the price to be paid when government lands were bought, and perhaps the impression still lingers in some minds that the meander lines are boundary lines, and that all in front of them remains unsold. Of course, this is erroneous. There was never any doubt that, except on the large navigable rivers, the boundary of the owners of the banks is the middle line of the river; and while some courts have held that this was the rule on all fresh-water streams, large and small, others have held to the doctrine that the title to the bed of the stream below low-water mark is in the state, while conceding to the owners of the banks all riparian rights. The practical difference is not very important. In this State, the rule that the centerline is the boundary line is applied to all our great rivers, including the Detroit, varied somewhat by the circumstance of there being a distinct channel for navigation, in some cases, with the stream in the main shallow, and also sometimes by the existence of islands.

The troublesome questions for surveyors present themselves when the boundary line between two contiguous estates is to be continued from the meander line to the centerline of the river. Of course, the original survey supposes that each purchaser of land on the stream has a waterfront of the length shown by the field notes and it is presumable that he bought this particular land because of that fact. In many cases it now happens that the meander line is left some distance from the shore by the gradual change of course of the stream, or diminution of the flow of water. Now the dividing line between two government subdivisions might strike the meander line at right angles, or obliquely; and, in some cases, if it were continued in the same direction to the centerline of the river, might cut off from the water one of the subdivisions entirely, or at least cut if off from any privilege of navigation or other valuable use of the water, while the other might have a waterfront much greater than the length of a line crossing it at right angles to its side lines. The effect might be that, of two government subdivisions of equal size and cost, one would be of great value as waterfront property, and the other comparatively valueless. A rule which would produce this result would not be just, and it has not been recognized in the law.

Nevertheless, it is not easy to determine what ought to be the correct rule for every case. If the river has a straight course, or one nearly so, every man's equities will be preserved by this rule: Extend the line of division between the two parcels from the meander line to the centerline of the river, as nearly as possible at right angles to the general course of the river at that point. This will preserve to each man the waterfront which the field notes indicated, except as changes in the water may have affected it, and the only inconvenience will be that the division line between different subdivisions is likely to be more or less deflected where it strikes the meander line.

This is the legal rule, and is not limited to government surveys, but applies as well to water lots which appear as such on town plats. (*Bay City Gas Light Co. v. The Industrial Works*, 28 Mich. Reports, 182.) It often happens, therefore, that the lines of city lots bounded on navigable streams are deflected as they strike the bank, or the line where the bank was when the town was first laid out.

IRREGULAR WATERCOURSES

When the stream is very crooked, and especially if there arc short bends, so that the foregoing rule is incapable of strict application, it is sometimes very difficult to determine what shall be done; and in many cases the surveyor may be under the necessity of working out a rule for himself. Of course, his action cannot be conclusive; but if he adopts one that follows, as nearly as the circumstances will admit, the general rule above indicated, so as to divide as near as may be the bed of the stream among the adjoining owners in proportion to their lines upon the shore, his division, being that of an expert, made upon the ground, and with all available lights, is likely to be adopted as law for the case. Judicial decisions, into which the surveyor would find it prudent to look under such circumstances, will throw light upon his duties and may constitute a sufficient guide when peculiar cases arise. Each riparian lot owner ought to have a line on the legal boundary, namely, the center line of the stream, proportioned to the length of his line on the shore, and the problem in each case is how this is to be given him. Alluvion—when a river imperceptibly changes its course—will be apportioned by the same rules.

The existence of islands in a stream when the middle line constitutes a boundary, will not affect the apportionment unless the islands were surveyed out as government subdivisions in the original admeasurement. Wherever that was the case, the purchaser of the island divides the bed of the stream on each side with the owner of the bank, and his rights also extend above and below the solid ground, and are limited by the peculiarities of the bed and the channel. If an island was not surveyed as a government subdivision previous to the sale of the bank, it is, of course, impossible to do this for the purposes of government sale afterward, for the reason that the rights of the bank owners are fixed by their purchase; when making that, they have a right to understand that all land between the meander lines, not separately surveyed and sold, will pass with the shore in the government sale and, having this right, anything which their purchase would include under it cannot afterward be taken from them. It is believed, however, that the Federal courts would not recognize the applicability of this rule to large navigable rivers, such as those uniting the Great Lakes.

On all the little lakes of the State which are mere expansions near their mouths of the rivers passing through them—such as the Muskegon, Pere, Marquette, and Manistee—the same rule of bed ownership has been judicially applied that is applied to the rivers themselves; and the division lines are extended under the water in the same way. (*Rice v. Ruddiman*, 10 Mich., 125.) If such a lake were circular, the lines would converge to the center; if oblong or irregular, there might be a line in the middle on which they would terminate whose course would bear some relation to that of the shore. But it can seldom be important to follow the division line very far under the water, since all private rights are subject to the public rights of navigation and other use, and any private use of the lands inconsistent with these would be a nuisance, and punishable as such. It is sometimes important, however, to run the lines out for considerable distance in order to determine where one may lawfully moor vessels or rafts for the winter or cut ice. The ice crop that forms over a man's land of course belongs to him. (*Lorman v. Benson*, 8 Mich., 18; *People's Ice Co. v. Steamer Excelsior*, recently decided.)

MEANDER LINES AND RIPARIAN RIGHTS

What is said above will show how unfounded is the notion, which is sometimes advanced, that a riparian proprietor on a meandered river may lawfully raise the water in the stream without liability to the proprietors above, provided he does not raise it so that it overflows the meander line. The real fact is that the meander line has nothing to do with such a case, and an action will lie whenever he sets back the water upon the proprietor above, whether the overflow be below the meander lines or above them.

As regards the lakes and ponds of the State, one may easily raise questions that it would be impossible for him to settle. Let us suggest a few questions, some of which are easily answered, and some not:

1. To whom belongs the land under these bodies of water, where they are not mere expansions of a stream flowing through them?
2. What public rights exist in them?
3. If there are islands in them which were not surveyed out and sold by the United States, can this be done now?

Others will be suggested by the answers given to these.

It seems obvious that the rules of private ownership which are applied to rivers cannot be applied to the Great Lakes. Perhaps it should be held that the boundary is at low water mark, but improvements beyond this would only become unlawful when they became nuisances. Islands in the Great Lakes would belong to the United States until sold, and might be surveyed and measured for sale at any time. The right to take fish in the lakes, or to cut ice, is public like the right of navigation, but is to be exercised in such manner as not to interfere with the rights of shore owners. But so far as these public rights can be the subject of ownership, they belong to the State, not to the United States, and so, it is believed, does the bed of a lake also. (*Pollord v. Hagan*, 31 Howard's U.S. Reports.) But such rights are not generally considered proper subjects of sale, but like the right to make use of the public highways, they are held by the State in trust for all the people.

What is said of the large lakes may perhaps be said also of the interior lakes of the State, such, for example, as Houghton, Higgins, Cheboygan, Burt's Mullet, Whitemore, and many others. But there are many little lakes or ponds which are gradually disappearing, and the shore proprietorship advances *pari passu* as the waters recede. If these are of any considerable size—say, even a mile across—there may be questions of conflicting rights which no adjudication hitherto made could settle. Let, any surveyor, for example, take the case of a pond of irregular form, occupying a square mile or more of territory, and undertake to determine the rights of the shore proprietors to its bed when it shall totally disappear, and he will find he is in the midst of problems such as probably he has never grappled with or reflected upon before. But the general rules for the extension of shore lines, which have already been laid down, should govern such cases, or at least should serve as guides in their settlement.

Where a pond is so small as to be included within the lines of a private purchase from the government, it is not believed the public have any rights in it whatever. Where it is not so included, it is believed they have rights of fishery, rights to take ice and water, and rights of navigation for business and pleasure. This is the common belief, and probably the just one. Shore rights must not be so exercised as to disturb these, and the States may pass all proper laws for their protection. It would be easy with suitable legislation to preserve these little bodies of water as permanent places of resort for the pleasure and recreation of the people, and there ought to be such legislation.

If the State should be recognized as owner of the beds of these small lakes and ponds, it would not be owner for the purpose of selling. It would be owner only as trustee for the public use; and a sale would be inconsistent with the right of the bank owners to make use of the water in its natural condition in connection with their estates. Some of them might be made salable lands by draining; but the State would not drain, even for this purpose, against the will of the shore owners, unless their rights were appropriated and paid for.

Upon many questions that might arise between the State as owner of the bed of a little lake and the shore owners, it would be presumptuous to express an opinion now, and fortunately the occasion does not require it.

QUASI-JUDICIAL CAPACITY OF SURVEYORS

I have thus indicated a few of the questions with which surveyors may now and then have occasion to deal, and to which they should bring good sense and sound judgment. Surveyors are not and cannot be judicial officers, but in a great many cases they act in a quasi-judicial capacity with the acquiescence of parties concerned; and it is important for them to know by what rules they are to be guided in the discharge of their judicial functions. What I have said cannot contribute much to their enlightenment, but I trust will not be wholly without value.

Note

[1] This lecture was prepared by Thomas Cooley (Chief Justice, Supreme Court of Michigan, 1864–1885) as an aid to surveyors and engineers. The student will find sound philosophy and advice that can be applied to survey situations of today. The student should remember this article was written approximately 15 to 20 years after the original GLO surveys were completed in Michigan.

APPENDIX D

GEODAESIA

JOHN LOVE

GEODÆSIA:

OR, THE

ART

OF

SURVEYING

AND

Measuring of Land,

Made EASIE.

SHEWING,

By Plain and Practical Rules, How to Survey, Protract, Cast up, Reduce or Divide any Piece of Land whatsoever; with New Tables for the ease of the Surveyor in Reducing the Measures of Land.

MOREOVER,

A more Facile and Sure Way of Surveying by the Chain, than has hitherto been Taught.

AS ALSO,

How to Lay-out New Lands in America, or elsewhere: And how to make a Perfect Map of a River's Mouth or Harbour; with several other Things never yet Publish'd in our Language.

By JOHN LOVE, Philomath.

Oculus mentis excæcatus & defossus, per sola Mathematica studia instauratur & excitatur, ut res ipsas cernere queat, & à rerum nudis simulachris ad veritatem, à tenebris ad lucem, à materiæ spelunca & vinculis, ad incorporeas, & invisibiles essentias sese erigere. Plato de Repub.

LONDON:
Printed for JOHN TAYLOR, at the *Ship* in S. *Paul's Church-Yard*, MDCLXXXVIII.

THE PREFACE TO THE READER.

WHat would be more ridiculous, than for me to go about to Praiſe an Art that all Mankind know they cannot live Peaceably without? It is near hand as ancient (no doubt on't) as the World: For how could Men ſet down to Plant, without knowing ſome Diſtinction and Bounds of their Land? But (Neceſſity being the Mother of Invention) we find the *Egyptians*, by reaſon of the *Nyles* over-flowing, which either waſht away all their Bound-Marks, or cover'd them over with Mud, brought this Meaſuring of Land firſt into an Art, and Honoured much the Profeſſors of it. The great Uſefulneſs, as well as the pleaſant and delightful Studie, and wholſom Exerciſe of which, tempted ſo many to apply themſelves thereto, that at length in *Egypt* (as in *Bermudas* now) every Ruſtick could Meaſure his own Land.

From *Egypt*, this Art was brought into *Greece*, by *Thales*, and was for a long time called *Geometry*; but that being too comprehenſive a Name for the Menſuration of a Superficies only, it was afterwards called *Geodæſia*; and what Honour it ſtill continued to have among the Antients, needs no better Proof than *Plato*'s ἀ γεωμέτρητος οὐδεὶς ἐσοίτω. And not only *Plato*, but moſt, if not all the Learned Men of thoſe times, refuſed to admit any into their Schools, that had not been firſt entred in the *Mathematicks*, eſpecially *Geometry* and *Arithmetick*. And we may ſee, the great Monuments of Learning built on theſe Foundations, continuing unſhaken to this day, ſufficiently demonſtrate the Wiſdom of the Deſigners, in chuſing *Geometry* for their Ground-Plot.

Since which the *Romans* have had ſuch an Opinion of this ſort of Learning, that they concluded that Man to be incapable of Commanding a Legion, that had not at leaſt ſo much *Geometry* in him, as to know how to Meaſure a Field. Nor did they indeed either reſpect Prieſt or Phyſitian, that had not ſome Inſight in the *Mathematicks*.

Nor can we complain of any failure of Reſpect given to this Excellent Science, by our Modern Worthies, many Noblemen, Clergymen, and Gentlemen affecting the Study thereof: So that we may ſafely ſay, none but Unadviſed Men ever did, or do now ſpeak evil of it.

Beſides the many Profits this Art brings to Man, it is a Study ſo pleaſant, and affords ſuch Wholſom and Innocent Exerciſe, that we ſeldom find a Man that has once entred himſelf into the Study of *Geometry* or *Geodæſia*, can ever after wholly lay it aſide; ſo natural it is to the Minds of Men, ſo pleaſingly inſinuating, that the *Pythagoreans* thought the Mathematicks to be only a Reminiſcience, or calling again to mind things formerly learned.

But no longer to light Candles to ſee the Sun by, let me come to my buſineſs, which is to ſpeak ſomething concerning the following Book; and if you ask, why I write a Book of this nature, ſince we have ſo many very good ones already in our own Language? I anſwer, becauſe I cannot find in thoſe Books, many things, of great conſequence, to be underſtood by the Surveyor. I have ſeen Young men, in *America*, often nonplus'd ſo, that their Books would not help them forward, particularly in *Carolina*, about Laying out Lands, when a certain quantity of Acres has been given to be laid out five or ſix times as broad as long. This I know is to be laught at by a Mathematician; yet to ſuch as have no more of this Learning, than to know how to Meaſure a Field, it ſeems a Difficult Queſtion: And to what Book already Printed of Surveying ſhall they repair to, to be reſolved?

Alſo concerning the *Extraction of the Square Root*; I wonder that it has been ſo much neglected by the Teachers of this Art, it being a Rule of ſuch abſolute neceſſity for the Surveyor to be acquainted with. I have taught it here as plainly as I could deviſe, and that according to the Old way, verily believing it to be the Beſt, uſing fewer Figures, and once well learned, charging leſs the Memory than the other way.

Moreover, the Sounding the Entrance of a River, or Harbour, is a Matter of great Import, not only to Seamen, but to all ſuch as Seamen live by; I have therefore done my endeavour to teach the Young Artiſt how to do it, and draw a fair Draught thereof.

Many more things have I added, ſuch as I thought to be New, and Wanting; for which I refer you to the Book it ſelf.

As for the Method, I have choſe that which I thought to be the eaſieſt for a Learner; adviſing him firſt to learn ſome Arithmetick, and after teaching him how to Extract the Square-Root. But I would not have any Neophyte discouraged, if he find the *First* Chapter too hard for him; for let him rather skip it, and go to the *Second* and *Third* Chapters, which he will find so easie and delightful, that I am persuaded he will be encouraged to conquer the Difficulty of learning that one Rule in the *First* Chapter.

From *Arithmetick*, I have proceeded on to teach so much *Geometry* as the Art of *Surveying* requires. In the next place I have shewed by what Measures Land is Surveyed, and made several Tables for the Reducing one sort of Measure into another.

From which I come to the Description of Instruments, and how to Use them; wherein I have chiefly insisted on the Semi-circle, it being the best that I know of.

The *Sixth* Chapter teacheth how to apply all the foregoing Matters together, in the Practical Surveying of any Field, Wood, *&c.* divers Ways, by divers Instruments; and how to lay down the ſame upon Paper. Alſo at the end of this Chapter I have largely inſiſted on, and by new and eaſie ways, taught Surveying by the Chain only.

The *Seventh*, *Eighth*, *Ninth*, *Tenth* and *Eleventh* Chapters, teach how to caſt up the Contents of any Plot of Land; How to lay out New Lands; How to Survey a Mannor, County or Country: Alſo, how to Reduce, Divide Lands, *Cum multis aliis.*

The *Twelfth* Chapter conſiſts wholly of *Trigonometry.*

The *Thirteenth* Chapter is of Heights and Diſtances, including amongſt other things, how to make a Map of a River or Harbour. Alſo how to convey Water from a Spring-head, to any appointed Place, or the like.

Laſtly, At the end of the Book, I have a Table of Northing or Southing, Eaſting or Weſting; or(if you pleaſe to call it ſo) A Table of Difference of Latitude and departure from the Meridian, with Directions for the Uſe thereof. Alſo a Table of Sines and Tangents,and a Table of Logarithms.

I have taken Example from Mr. *Holwell* to make the Table of *Sines* and *Tangents*, but to every Fifth Minute, that being nigh enough in all ſenſe and reaſon for the Surveyor's Uſe; for there is no Man, with the beſt Inſtrument that was ever yet made, can take an Angle in the Field nigher, if ſo nigh, as to Five Minutes.

All which I commend to the Ingenious Reader, wiſhing he may find Benefit thereby, and deſiring his favourable Reception thereof accordingly. I conclude,

READER,

Your Humble Servant,

J. L.

CHAP. V.

Of Inſtruments and their Uſe.

And firſt of the Chain.

THere are ſeveral ſorts of Chains, as Mr. *Rathborne*'s of two Perch long: Others, of one Perch long, ſome have had them 100 Feet in length: But that which is moſt in uſe among Surveyors (as being indeed the beſt) is Mr. *Gunter*'s, which is 4 Pole long, containing 100 Links, each Link being $7\frac{92}{100}$ Inches: The Deſcription of which Chain, and how to reduce it into any other Meaſure, you have at large in the foregoing Chapter of Meaſures. In this place I ſhall only give you ſome few Directions for the uſe of it in Meaſuring Lines.

Take care that they which carry the Chain, deviate not from a ſtreight Line; which you may do by ſtanding at your Inſtrument, and looking through the Sights: If you ſee them between you and the Mark obſerved, they are in a ſtreight Line, otherwiſe not. But without all this trouble, they may carry the Chain true enough, if he that follows the Chain always cauſeth him that goeth before to be in a direct line between himſelf, and the place they are going to, ſo as that the Foreman may always cover the Mark from him that goes behind. If they ſwerve from the Line, they will make it longer than really it is; a ſtreight Line being the neareſt diſtance that can be between any two places.

Beſure that they which carry the Chain, miſtake not a Chain either over or under in their Account, for if they ſhould, the Error would be very conſiderable; as ſuppoſe you was to meaſure a Field that you knew to be exactly Square, and therefore need meaſure but one Side of it; if the Chain-Carriers ſhould miſtake but one Chain, and tell you the Side was but 9 Chains, when it was really 10, you would make of the Field but 8 Acres and 16 Perches, when it ſhould be 10 Acres juſt. And if in ſo ſmall a Line ſuch a great Error may ariſe, what may be in a greater, you may eaſily imagine. But the uſual way to prevent ſuch Miſtakes is, to be provided with 10 ſmall Sticks ſharp at one End, to ſtick into the Ground; and let him that goes before take all into his Hand at ſetting out, and at the End of every Chain ſtick down one, which let him that follows take up; when the 10 Sticks are done, be ſure they have gone 10 Chains; then if the Line be longer, let them change the Sticks, and proceed as before, keeping in Memory how often they change: They may either Change at the end of 10 Chains, then the hindmoſt Man muſt give the foremoſt all his Sticks; or which is better, at the end of 11 Chains, and then the laſt Man muſt give the firſt but 9 Sticks, keeping one to himſelf. At every Change count the Sticks, for fear leſt you have dropt one, which ſometimes happens.

If you find the Chain too long for your uſe, as for ſome Lands it is, eſpecially in *America*, you may then take the half of the Chain, and meaſure as before, remembring ſtill when you put down the Lines in your Field Book, that you ſet down but the half of the Chains, and the odd Links, as if a Line meaſured by the little Chain be 11 Chains 25 Links, you muſt ſet down 5 Chains 75 Links, and then in plotting and caſting up it will be the ſame as if you had meaſured by the whole Chain.

At the end of every 10 Links, you may, if you find it convenient, have a Ring, a piece of Braſs, or a Ragg, for your more ready reckoning the odd Links.

When you put down in your Field-Book the length of any Line, you may ſet it thus, if you pleaſe, with a Stop between the Chains and Links, as 15 Chains 15 Links 15.15. or without, as thus 1515, it will be all one in the caſting up.

Of Inſtruments for the taking of an Angle in the Field.

There are but two material things (towards the meaſuring of a piece of Land) to be done in the Field; the one is to meaſure the Lines (which I have ſhewed you how to do by the Chain) and the other to take the quantity of an Angle included by theſe Lines; for which there are almoſt as many Inſtruments as there are Surveyors. Such among the reſt as have got the greateſt eſteem in the World, are, the Plain Table for ſmall Incloſures, the Semicircle for Champaign Grounds, The Circumferentor, the Theodolite, *&c.* To deſcribe theſe to you, their Parts, how to put them together, take them aſunder, *&c.* is like teaching the Art of Fencing by Book, one Hours uſe of them, or but looking on them in the Inſtrument-maker's Shop, will better deſcribe them to you than the reading one hundred Sheets of

Paper concerning them. Let it ſuffice that the only uſe of them all is no more (or chiefly at moſt) but this ; *viz.*

To take the Quantity of an Angle.

As ſuppoſe A B and A C are two Hedges or other Fences of a Field, the Chain ſerves to meaſure the

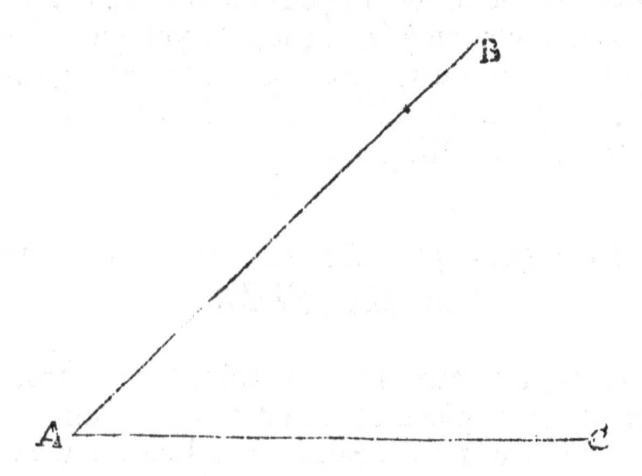

length of the Sides A B or A C, and theſe Inſtruments we are ſpeaking of are to take the Angle A. And firſt by the

Plain Table.

Place the Table (already fitted for the Work, with a Sheet of Paper upon it) as nigh to the Angle A as you can, the North End of the Needle hanging directly over the *Flower de Luce* ; then make a Mark upon the Sheet of Paper at any convenient place for the Angle A, and lay the Edge of the Index to the Mark, turning it about, till through the Sights you

APPENDIX E

LAND ACTS THAT CREATED THE PUBLIC LANDS

The student should become familiar with three major land acts that created the Public Land Survey (PLS) (also called the General Land Office (GLO) System) System in the United States:

- the Act of 1785
- the Act of 1786
- the Act of 1805

This appendix contains the three major land laws under which the PLS or GLO system was created and the basic principles were identified. Each act is separate and distinct and, taken as a whole, they set the established rules that identify the system and as such should be understood by the student and the practicing surveyor.

The acts are summarized next.

1785

The Act of 1785 is famous, not for what it said but for what it did not state. This act created the system of six-mile square townships that supposedly crossed at right angles, an impossible request on a curved surface. Later, this would be recognized and correction lines would be directed.

The law provided that all lines would be measured with a chain. It failed to state the length of the chain.

It also provided that numbering of the "lots" (sections in 1796) would start in the southeast corner of the township.

1796

The Land Act of 1796 plugged some of the loopholes that were discovered in the Act of 1785:

It specified the chain to be "2 perches in length." This equated to a chain 33 feet long.

It stated that lines would be to the true meridian.

It set the method of section numbering starting in the northeast corner.

1805

The Act of February 11, 1805, legally, was perhaps the most far-reaching of all of the land acts.

The major provisions that are still in effect today are summarized as follows:

The original surveys are without error, in that the distances and the courses recited in the field notes and noted on the plat are the "true distances," regardless of what the modern surveyor remeasures with modern equipment.

It provided for "legal, but unmonumented" corners, primarily the center quarter corner and the one-sixteenth corners.

It identified that the contents of the sections, as returned on the plat, will be the true contents of the sections.

It held that the contents of the half sections and the quarter sections, the contents which have not been returned, shall be considered as containing one-half or one-fourth, respectively. This provision coined the term "aliquot."

The text (in whole or in part) of these land acts follows.

An Ordinance for Ascertaining the Mode of Disposing the Lands in the Western Territory (Passed May 20, 1785) [1]

Be it ordained by the United States in Congress assembled, that the territory ceded by individual states to the United States, which had been purchased of the Indian inhabitants, shall be disposed of in the following manner:

A surveyor from each state shall be appointed by Congress or a committee of the states, who shall take an oath for the faithful discharge of his duty, before the geographer of the United States, who is hereby empowered and directed to administer the same; and the like oath shall be administered to each chain carrier, by the surveyor under whom he acts.

The geographer, under whose direction the surveyors shall act, shall occasionally form such regulations for their conduct, as he shall deem necessary; shall have authority to suspend them for misconduct in office, and shall make report of the same to Congress, to the committee of the states; and he shall make report in case of sickness, death, or resignation of any surveyor.

The surveyors, as they are respectively qualified shall proceed to divide the said territory into township of 6 miles square, by lines running due north and south and others crossing these at right angles, as near as may be, unless where the boundaries of the late Indian purchases may render the same impracticable, and then they shall depart from this rule no further than such particular circumstances may require. And each surveyor shall be allowed and **paid at the rate of two dollars for every mile,** in length, he shall run, including the wages of chain carriers, markers, and every other expense attending the same.

The first line, running due north and south as aforesaid, shall begin on the river Ohio, at a point that shall be found to be due north from the western termination of a line, which has been run as the southern boundary of the state of Pennsylvania; and the first line, running east and west, shall begin at the same point, and shall extend throughout the whole territory; provided, that nothing herein shall be construed, as fixing the western boundary of the state of Pennsylvania. The geographer shall designate the townships, or fractional parts of townships, by numbers progressively from south to north; always beginning each range with No. 1; and the ranges shall be distinguished by their progressive numbers to the westward. The first range, extending from the Ohio to the lake Erie, being marked. No. 1. The geographer shall personally attend to the running of the first east and west line; and shall take the latitude of the extremes of the first north and south line, and of the mouths of the principal rivers.

The lines shall be measured with a chain; shall be plainly marked by chaps on the trees, and exactly described on a plat; whereon shall be noted by the surveyor, as their proper distances, all mines, salt-springs, salt-licks, and mill-seats, that shall come to his knowledge; and all water-courses, mountains, and other remarkable and permanent things, over and near which such lines shall pass, and also the quality of the lands.

The plats of the townships respectively, shall be marked by subdivisions into lots of one mile square, or 640 acres, in the same direction as the external lines, and numbered from 1 to 36; always beginning the succeeding range of the lots with the number next to that with which the preceding one concluded. And where, from the causes before-mentioned, only a fractional part of a township shall be surveyed, the lots, protracted thereon, shall bear the same numbers as if the township had been entire. And the surveyors, in running the external lines of the townships, shall, at the interval of every mile, mark corners for the lots which are adjacent, always designating the same in a different manner from those of the townships.

The geographer and surveyors shall pay the utmost attention to the variation of the magnetic needle; and shall run and note all lines by the true meridian, certifying, with every plat, what was the variation at the times of running the lines thereon noted.

As soon as seven ranges of townships, and fractional parts of townships, in the direction from south to north, shall have been surveyed, the geographer shall transmit plats thereof to the board of treasury, who shall record the same, with the report, in well-bound books to be kept for that purpose. And the geographer shall make similar returns, from time to time, of every seven ranges as they may be surveyed. The secretary at war shall have recourse thereto, and shall take by lot therefrom, a number of townships, and fractional parts of townships, as well from those to be sold entire, as from those to be sold in lots, as will be equal to one-seventh part of the whole of such seven ranges, as nearly as may be, for the use of the late continental army; and he shall make a similar draught, from to time to time, until a sufficient quantity is drawn to satisfy the same, to be applied in manner hereinafter directed. The board of treasury shall from time to time, cause the remaining numbers, as well those to be sold entire, as those to be sold in lots, to be drawn for, in the name of the thirteen states respectively, according to the quotas in the last preceding requisition on all the states; provided, that in case more land than its proportion is allotted for sale in any state, at any distribution, a deduction be made therefore at the next.

The board of treasury shall transmit a copy of the original plats, previously noting thereon, the townships, and fractional parts of townships, which shall have fallen to the several states, by the distribution aforesaid, to the commissioners of the loan-office of the several states, who, after giving notice of not less than two nor more than six months, by causing advertisements to be posted up at the courthouses, or other noted places in every county, and to be inserted in one newspaper, published in the states of their residence respectively, shall proceed to sell the townships, or fractional parts of townships, at public vendue; in the following manner, viz: The township, or fractional part of a township, No. 1, in the second range, shall be sold by lots; and No. 2, in the same range, entire; and so in alternate order through the whole of the second range; and the third range shall be sold in the same manner as the first, and the fourth in the same manner as the second; and thus alternately throughout all the ranges; provided, that none of the lands, within the said territory, be sold under the price of one dollar the acre, to be paid in specie, or loan-office certificates, reduced to specie value, by the scale of depreciation, or certificates of liquidated debts of the United States, including interest, beside the expense of the survey and other charges thereon, which are hereby rated at 36 dollars the township, in specie, or certificates as aforesaid, and so in the same proportion for a fractional part of a township, or of a lot, to be paid at the time of sales; on failure of which payment, the said lands shall again be offered for sale.

There shall be reserved for the United States out of every township the four lots, being numbered 8, 11, 26, 29, and out of every fractional part of a township, so many lots of the same numbers as shall be found thereon, for future sale. **There shall be reserved the lot No. 16, of every township, for the maintenance of public schools, within the said township;** also one-third part of all gold, silver, lead, and copper mines, to be sold, or otherwise disposed of as Congress shall hereafter direct. [Here follow the terms of the deed to be given when a township or a lot is sold.]

Which deeds shall be recorded in proper books, by the commissioner of the loan office, and shall be certified to have been recorded, previously to their being delivered to the purchaser, and shall be good and valid to convey the lands in the same described.

The commissioners of the loan offices, respectively, shall transmit to the board of treasury every three months, an account of the townships, fractional parts of townships, and lots committed to their charge; specifying therein the names of the persons to whom sold, and the sums of money or certificates received for the same; and shall cause all certificates by them received, to be struck through with a circular punch; and shall be duly charged in the books of the treasury, with the amount of the money or certificates, distinguishing the same, by them received as aforesaid.

If any township, or fractional part of a township or lot, remains unsold for 18 months after the plat shall have been received, by the commissioners of the loan office, the same shall be returned to the board of treasury, and shall be sold in such manner as Congress may hereafter direct.

And whereas Congress, by their resolutions of September 16th and 18th; in the year 1776, and the 12th of August, 1780, stipulated grants of land to certain officers and soldiers of the late continental army, and by the resolution of the 22nd September, 1780, stipulated grants of land to certain officers in the hospital department of the late continental army; for complying therefore with such engagements, Be it ordained, That the secretary at war, from the returns in his office, or such other sufficient evidence as the nature of the case may admit, determine who are objects of the above resolutions and engagements, and the quantity of land to which such persons or their representatives are respectively entitled, and cause the townships, or fractional parts of townships, hereinbefore reserved for the use of the late continental army, to be drawn for in such manner as he shall deem expedient, to answer the purpose of an impartial distribution. He shall, from time to time, transmit certificates to the commissioners of the loan offices of the different states, to the lines of which the military claimants have respectively belonged, specifying the name and rank of the party, the terms of his engagement and time of his service, and the division, brigade, regiment or company to which he belonged, the quantity of land he is entitled to, and the township, or fractional part of a township, and range out of which is portion is to be taken.

The commissioners of the loan offices shall execute deeds for such undivided proportions in manner and form herein before-mentioned, varying only in such a degree as to make the same conformable to the certificate from the secretary at war.

Where any military claimants of bounty in lands shall not have belonged to the line of any particular state, similar certificates shall be sent to the board of treasury, who shall execute deeds to the parties for the same.

The secretary at war, from the proper returns, shall transmit to the board of treasury, a certificate, specifying the name and rank of the several claimants of the hospital department of the late continental army, together with the quantity of land each claimant is entitled to, and the township, or fractional part of a township, and range out of which his portion is to be taken; and thereupon the board of treasury shall proceed to execute deeds to such claimants.

The board of treasury, and the commissioners of the loan offices in the states, shall, within 18 months, return receipts to the secretary at war, for all deeds which have been delivered, as also all the original deeds which remain in their hands for want of applicants, having been first recorded; which deeds so returned, shall be preserved in the office, until the parties or their representatives require the same.

And be it further ordained, That three townships adjacent to Lake Erie be reserved, to be hereafter disposed of in Congress, for the use of the officers, men, and others, refugees from Canada, and the refugees from Nova Scotia, who are or may be entitled to grants of land under resolutions or Congress now existing or which may hereafter be made respecting them, and for such other purposes as Congress may hereafter direct.

And be it further ordained, That the towns of Gnadenhutten, Schoenbrun, and Salem, on the Muskingum, and so much of the lands adjoining to the said towns, with the buildings and improvements thereon, shall be reserved for the sole use of the Christian Indians, who were formerly settled there, or the remains of that society, as may, in the judgment of the geographer, be sufficient for them to cultivate.

Saving and reserving always, to all officers and soldiers entitled to lands on the northwest side of the Ohio, by donation or bounty from the commonwealth of Virginia, and to all persons claiming under them, all rights to which they are so entitled, under the deed of cession executed by the delegates for the state of Virginia on the first day of March, 1784, and the Act of Congress accepting the same: and to the end, that the said rights may be fully and effectually secured, according to the true intent and meaning of the said deed of cession and act aforesaid, Be it ordained, that no part of the land included between the rivers called Little Miami and Scioto, on the northwest side of the river Ohio, be sold, or in any manner alienated, until there shall first have been laid off and appropriated for the said officers and soldiers, and persons claiming under them, the lands they are entitled to, agreeably to the said deed of cession and Act of Congress accepting the same.

Done by the United States in Congress assembled, the 20th day of May, in the year of our Lord, 1785, and of our sovereignty and independence the ninth.

RICHARD H. LEE, President.

CHARLES THOMPSON, Secretary.

Resolution of May 12, 1786 [2]

Whereas the ordinance for ascertaining the mode of disposing of lands in the Western territory directs … That the geographer and surveyors shall pay the utmost attention to the variation of the magnetic needle, and shall run and note all lines by the true meridian, certifying, with every plat, what was the variation at the times of running the lines thereon noted, which direction will greatly delay the survey of said territory:

Resolved, … That the above-recited clause in the said ordinance be, and the same hereby is, repealed.

Land Act of May 18, 1796

An act providing for the sale of the lands, of the United States, in the territory northwest of the River Ohio, and above the mouth of Kentucky River.

Sec. 1. Be it enacted, ... That a surveyor general shall be appointed, whose duty it shall be to engage a sufficient number of skilled surveyors, as his deputies; whom he shall cause, without delay, to survey and mark the unascertained outlines of the lands lying northwest of the river Ohio, and above the mouth of the river Kentucky, in which the titles of the Indian tribes have been extinguished, and to divide the same in the manner hereinafter directed; he shall have authority to frame regulations and instructions for the government of his deputies, to administer the necessary oaths, upon their appointment, and to remove them for negligence or misconduct in office.

Sec. 2. Be it further enacted, ... That the part of said lands which has not been already conveyed by letters patent, or divided, in pursuance of an ordinance in Congress, passed on the twentieth of May, one thousand seven hundred and eighty-five, or which has not been heretofore, and during the present session of Congress, may not be appropriated for satisfying military land bounties, and for other purposes, shall be divided by north and south lines, run According to the true meridian, and by others crossing them at right angles, so as to form townships of six miles square, unless where the line of the late Indian purchase, or of tracts of land heretofore surveyed or patented, or the course of navigable rivers, may render it impracticable; and then this rule shall be departed from no further than such particular circumstances may require. The corners of the townships shall be marked with progressive numbers, from the beginning; each distance of a mile between the said corners shall be also distinctly marked with marks different from those of the corners. One half of said townships, taking them alternately, shall be subdivided into sections, containing, as nearly as may be, six hundred and forty acres each, by running through the same, each way, parallel lines, at the end of every two miles, and by marking a corner, on each of the said lines, at the end of every mile; the sections shall be numbered, respectively, beginning with the number one, in the north-east section, and proceeding west and east alternately, through the township, with progressive numbers, till the thirty-sixth be completed. And it shall be the duty of the deputy surveyors to cause to be marked on a tree near each corner made as aforesaid, and within the section, the number of such section, and over it the number of the township within which such section may be; and the said deputies shall carefully note, in their respective field books, the names of the corner trees marked, and the numbers so made. The fractional parts of townships shall be divided into sections, in manner aforesaid, and the fractions of sections shall be annexed to, and sold with, the adjacent entire sections. All lines shall be plainly marked upon trees, and **measured with chains containing two perches of sixteen feet and one-half each, subdivided into twenty-five equal links, and the chain shall be adjusted to a standard** to be kept for that purpose. Every surveyor shall note in his field book the true situations of all mines, salt licks, salt springs, and mill seats, which shall come to his knowledge; all water-courses over which the line he runs shall pass; and also the quality of the lands. These field books shall be returned to the surveyor general who shall therefrom cause a description of the whole land surveyed, to be made out and transmitted to the officers who may superintend the sales. He shall also cause a fair plat to be made of the townships and fractional parts of townships contained in the said lands, describing the subdivisions thereof, and the marks of the corners. This plat shall be recorded in books to be kept for that purpose; a copy thereof shall be kept open at the surveyor general's office for public information; and other copies sent to the places of sale, and to the Secretary of the Treasury.

Sec. 3. Be it further enacted, ... That a salt spring lying upon a creek which empties into the Scioto River, on the east side, together with as many continuous sections as shall be equal to one township, and every other salt spring which may be discovered, together with the section of one mile square which includes it, and also four sections at the centre of every township, containing each one mile square, shall be reserved for the future disposal of the United States; but there shall be no reservations, except for salt springs, in fractional townships, where the fraction is less than three-fourths of a township.

Sec. 4. Be it further enacted, ... That whenever seven ranges of townships shall have been surveyed below the Great Miami, or between the Scioto River and the Ohio Company's purchase, or between the southern boundary of the Connecticut claims and the ranges already laid off, beginning upon the Ohio River and extending westward, and the plats thereof made and transmitted, in conformity to the provisions of this act, the said sections of six hundred and forty acres (excluding those hereby reserved) shall be offered for sale, at public vendue, under the direction of the Governor or secretary of the Western Territory, and the surveyor general; such of them as lie below the Great Miami shall be sold at Cincinnati; those of them which lie between the Scioto and the Ohio Company's purchase, at Pittsburg; and those between the Connecticut claim and the seven ranges, at Pittsburg. And the townships remaining undivided shall be offered for sale, in the same manner, at the seat of Government of the United States under the direction of the Secretary of the Treasury in tracts of one quarter of a township, lying at the corners thereof, excluding the four central sections and the other reservations before mentioned; Provided, always, ... That no part of the lands directed by this act to be offered for sale shall be sold for less than two dollars per acre.

Sec. 5. Be it further enacted, ... That the Secretary of the Treasury, after receiving the aforesaid plats, shall forthwith give notice, in one newspaper in each of the United States, and of the territories northwest and south of the river Ohio, of the time of sale; which shall, in no case, be less than two months from the date of the notice; and the sales at the different places shall not commence, within less than one month of each other. And when the Governor of the Western Territory, or Secretary of the Treasury, shall find it necessary to adjourn or suspend the sales under their direction, respectively, for more than three days at any one time, notice shall be given in the public newspaper, of such suspension, and at what time the sales will recommence.

Sec. 6. Be it further enacted, ... That immediately after the passing of this act the Secretary of the Treasury shall, in the manner hereinbefore directed, advertise for sale the lands remaining unsold in the seven ranges of townships, which were surveyed in pursuance of an ordinance of Congress, passed the twentieth of May, one thousand seven hundred and eighty-five, including the lands drawn for the army by the late Secretary of War, and also those heretofore sold, but not paid for: the townships which by the said ordinance, are directed to be sold entire, shall be offered for sale at public venue in Philadelphia, under the direction of the Secretary of the Treasury, in quarter townships, reserving the four centre sections, according to the directions of this act. The townships, which, by the said ordinance, are directed to be sold in sections, shall be offered for sale at public vendue, in Pittsburg, under the direction of the Governor or secretary of the Western Territory, and such person as the President may specially appoint for that purpose, by sections of one mile square each, reserving the four centre sections, as aforesaid: and all fractional townships shall also be sold in sections, at Pittsburg, in the manner and under the regulations provided by this act, for the sale of fractional townships: provided, always, … That nothing in this act shall authorize the sale of those lots which have been heretofore reserved in the townships already sold.

Sec. 7. Be it further enacted, … That the highest bidder for any tract of land, sold by virtue of this act, shall deposit at the time of sale, one twentieth part of the amount of the purchase money; to be forfeited if a moiety of the sum bid, including the said twentieth part, is not paid, within thirty days, to the Treasury of the United States or to such person as shall be appointed by the President of the United States, to attend the places of sale for that purpose; and upon payment of a moiety of the purchase money within thirty days the purchaser shall have one year's credit for the residue; and shall receive from the Secretary of the Treasury, or the Governor of the Western Territory (as the case may be), a certificate describing the land sold, the sum paid on account, the balance remaining due, the time when such balance becomes payable; and that the whole land sold will be forfeited if the said balance is not then paid; but that if it shall be duly discharged, the purchaser, or his assignee or other legal representative, shall be entitled to a patent for the said lands: and, on payment of the said balance to the Treasury, within the specified time, and producing to the Secretary of State a receipt for the same, upon the aforesaid certificate, the President of the United States is hereby authorized to grant a patent for the lands to the said purchaser, his heirs or assigns. And all patents shall be countersigned by the Secretary of State, and recorded in his office. But if there should be a failure in any payment, the sale shall be void, all the money theretofore paid on account of the purchase shall be forfeited to the United States, and the lands thus sold shall be again disposed of, in the same manner as if a sale had never been made: Provided, nevertheless, that should any purchaser make payment of the whole purchase money at the time when the payment of the first moiety is directed to be made, he shall be entitled to a reduction of ten per centum of the part for which a credit is hereby directed to be given, and his patent shall be immediately issued.

Sec. 8. Be it further enacted, … That the Secretary of the Treasury and the Governor of the Territory northwest of the river Ohio, shall, respectively, cause books to be kept, in which shall be regularly entered an account of the dates of all the sales made, the situation and numbers of the lots sold, the price at which each was struck off, the money deposited at the time of sale, and the dates of the certificates granted to the different purchasers. The Governor or Secretary of the said Territory shall, at every suspension or adjournment, for more than three days, of the sales under their direction, transmit to the Secretary of the Treasury a copy of the said books, certified to have been duly examined and compared with the original. And all tracts sold under this act shall be noted upon the general plat, after the certificate has been granted to the purchaser.

Sec. 9. And be it further enacted, … That all navigable rivers within the territory to be disposed of by virtue of this act, shall be deemed to be and remain public highways. And that in all cases where the opposite banks of any stream, not navigable, shall belong to different persons, the stream and the bed thereof shall become common to both.

Sec. 10. And be it further enacted, … That the surveyor general shall receive, for his compensation, two thousand dollars per annum; and that the President of the United States may fix the compensation of the assistant surveyors, chain-carriers and axe-men: Provided, … That the whole expense of surveying and marking the lines shall not exceed three dollars per mile, for every mile that shall be actually run or surveyed.

Sec. 11. And be it further enacted, … That the following fees shall be paid for the services to be done under this act, to the Treasurer of the United States, or to the receiver in the Western Territory, as the case may be; for each certificate for a tract containing a quarter of a township, twenty dollars; for a certificate for a tract containing six hundred and forty acres, six dollars; and for each patent for a quarter of a township, twenty dollars; for a section of six hundred and forty acres, six dollars: and the said fees shall be accounted for by the receivers, respectively.

Sec. 12. And be it further enacted, … That the surveyor general, assistant surveyors, and chain-carriers, shall, before they enter on the several duties to be performed under this act, severally take an oath or affirmation, faithfully to perform the same. And the person to be appointed to receive the money on sales in the Western Territory, before he shall receive any money under this act, shall give bond, with sufficient security, for the faithful discharge of his trust; that, for receiving, safekeeping, and conveying to the Treasury, the money he may receive, he shall be entitled to a compensation, to be hereafter fixed.

Act of February 11, 1805, 2 Stat. 313 [3]

Chap. XIV—An Act concerning the mode of surveying the Public Lands of the United States.

(a) Be it enacted by the Senate and House of Representatives of the United States of America in Congress assembled, That the surveyor-General shall cause all those lands north of the river Ohio, which by virtue of the act, entitled "An act providing for the sale of lands of the United States, in the territory northwest of the river Ohio, and above the mouth of the Kentucky river," were subdivided, by running through the townships, parallel lines each way, at the end of every two miles, and by marking a corner on each of the said lines, at the end of every mile; to be subdivided into sections, by running straight lines from the mile corners thus marked, to the opposite corresponding corners, and by marking on each of the said lines, intermediate corners as nearly as possible equidistant from the corners of the sections on the same. **And the said surveyor-general shall also cause the boundaries of all the half sections, which had been purchased before previous to the first day of July last, and**

on which the surveying fees had been paid, according to law, by the purchaser, to be surveyed and marked, marked by running straight lines from the half-mile corners, heretofore marked, to the opposite corresponding corners; and intermediate corners shall, at the same time, be marked on each of the said dividing lines, as nearly as possible equidistant from the corners of the half section on the same line: Provided, that the whole expense of surveying and marking the lines, shall not exceed three dollars for every mile which has not been surveyed, and which shall be actually run, surveyed, and marked by virtue of this section. And the expense of making the subdivisions, directed by this section, shall be defrayed out of to be paid, the monies appropriated, or which may be hereafter appropriated, for completing the surveys of the public lands of the United States.

Sec. 2. And be it further enacted, **That the boundaries and contents of the several sections, half sections, and quarter sections of the public lands of the United States, shall be ascertained to conformity with the following principles, any act be ascertained, or acts to the contrary notwithstanding:**

1st. **All corners marked in the surveys, returned by the surveyor-general, or by the surveyor of the land south of the state of Tennessee, respectively, shall be established as the proper corners of sections, or subdivision of sections, which they were intended to designate; and the corners of half and quarter sections, not marked on the said surveys,** shall be placed as nearly as possible equidistant from those two corners which stand on the same line.

2d. **The boundary lines, actually run and marked in the surveys returned by surveyor-general, or by the surveyor of the land south of the state of Tennessee, respectively, shall be established as the proper boundary lines of the sections, or subdivisions, for which they were intended, and the length of such lines, as returned by either of the surveyors aforesaid, shall be considered as the true length thereof.** And the boundary lines, which shall not have been actually run, and marked aforesaid, shall be ascertained, by running straight lines from the established corners to the opposite corresponding corners; but in those portions of the fractional townships, where no such opposite corresponding corners have been or can be fixed, the said boundary lines shall be ascertained, by running from the established corners, due north and south, or east and west lines, as the case may be, to the water-course, Indian boundary line, or other external boundary of such fractional township.

3d. Each section, or subdivision of section, the contents whereof shall have been, or by virtue of returned the first section of this act, shall be returned by the surveyor-general, or by the surveyor of the public lands south of the state of Tennessee, respectively, shall be held and considered as containing the exact quantity, expressed in such return or returns, and the half sections and quarter sections, the contents whereof shall not have been thus returned, shall be held and considered as containing the one half, or one fourth part respectively, of the returned contents of the section of which they make part.

Sec. 3. And be it further enacted, That so much of this act entitled "An act making provision for disposal of lands in the Indiana territory, and for other purposes,... **as provides the mode of ascertaining the true contents of sections or subdivisions of sections, and prevents the issue of final certificates, unless the said contents**

shall have been ascertained, and a plot certified by the district surveyor, lodged with the register, be, and the same is hereby repealed."

APPROVED, February 11, 1805.

Notes

[1] This is the text of the Land Ordinance of 1785 as finally approved by Congress.

[2] This resolution, amending the Ordinance of May 20, 1785, is given in part.

[3] The provisions of this Act of Congress, now codified in Title 43 of the United States Code, are still the statute law of the land and brought the public land surveys to the basic system still in use, even though some of the provisions are obsolete. Reprinted from U.S. Statutes at Large 2 (bold emphasis added).

INDEX

30-foot-rule (Texas), 276

A

Abstract, 443
  companies, 444
  information,
    sources, 444
  opinion, 449
Abstractor, liability, 529–531
Abstractor's plat, 207f
Accessory, meaning, 326
Accidental errors, 164
Accomplice, testimony, 43
Accretion bank, 228
Accuracy, 162–163, 181–182
Acquiescence, 400, 413–414, 653
Acts of nature, 403
Actual possession, 425–426
  constructive possession,
    contrast, 431–432
Adjoiner deeds, examination, 22
Adjoining owners, agreements, 408–410
Adjoining parcels, calls, 479–481
Admissibility, 37–38. *See also* Evidence
Adverse actions, statutory character, 420
Adverse claimant, burden of proof, 400, 420
Adverse possession, 3, 400, 419
  historical concepts, 420
  maintenance, good
    faith, 431
Adverse relationships, 402–403
Adverse rights, 422–424
  elements, 424–425
Advertising, avoidance, 548
Aerial photographs, 331f, 332f, 334f
  maps, contrast, 159
  usage, 331–336, 361, 570
Agreement deed, 402
Agreement, validity (dispute), 57
Agrimensores, 153
Agrimensores, competency examination, 7
Alaska, land acquisition, 310–311
Aliquot corner, 326
Aliquot description, 463–464
Ambiguity, 57, 473
Ambiguous deeds/description, 53, 67
American Congress on Surveying and Mapping (ACSM), 5, 61
American Land Title Association (ALTA)
  contracts, 61
  standards, 522

American Title Association policy, basis, 200
Amicus curiae (usage), 551
Amplitude ratio method, 223–224, 223f
Ancient deeds, admissibility, 68
Ancient documents
  rule, 13, 68
  words, meaning, 53, 74
Ancient fences, usage, 134–135
Ancient survey plats/documents, 53, 68
Angles
  distances, relationship, 156f
  measurement errors, 147
Angular measurements, significant numbers (usage), 158
Annual rings, 114f, 120f
Appurtenant easement, 500
Arbitrary plat (arb plat), 207–208
Area, 492–494
  concept, 256–257
  conveyances, 467
  description, 467
  uncertainties, 179–180, 180f
Area of confusion, 347
Arpent, 145, 283t
Assessors, tax plats (relationship), 209–210
Astronomic north (true north), usage, 173–175
Attorney General v. Chambers, 218
Attorneys
  opinions, 449
  privileged information allowance, 574
  surveyors, relationship, 556
Augmenting clauses (description component), 477
Aulen v. Triumph Explosives, Inc., 534
Austin, Moses, 273–274
Azimuth, 489

B

Barkscribed beech tree, 110f
Barney v. Keokuk, 214–215
Bar scale, expression, 194
Baseline, 266, 303
Basic point, determination, 228
Battle of Hastings, 250
Bay City Gas Light Co. v. The Industrial Works, 655
Beach ridge, 228
Bearing references, determination, 150–151
Bearings, 195–196
  astronomic/true north, relationship, 173–175
  basis, 485–494
    change, 488–489
  directions, 462
  legal presumption, 141
Bearing trees, 136f, 289f, 290, 327
  evidence, 93, 123
  scribing, exposure, 109f
Beech tree, barkscribing, 110f
Beginning point, usage, 493–494
Bench mark, 326
Bering, Vitus, 310
Berkeley, John, 265–267
Best available evidence, 18, 24, 28, 47
Best evidence, term (usage), 24
Beyond a reasonable doubt, phrase (usage), 19, 24, 34, 37, 84
Blanchard, Joseph, 264
Blaze, display, 120f
Body (description component), 477, 478–479
Boone, Daniel, 245
Borax Consolidated Ltd. v. City of Los Angeles, 217–219
Borden, John P., 279
Boundaries, 45–46, 218–219
  creation
    original surveys, impact, 340, 343–346, 380
    unwritten means, 399, 401
  defining, written documents (usage), 342–343
  description
    short-cut methods, 256–257
    words, usage/understanding, 459
  gradient boundary, 280–281
  lines, 26, 134–135, 399
  location procedures, 339
  monumenting, obligations, 543
  outer boundaries, display, 202

problem, survey (performing), 20
proof, burden, 30
proving, fences evidence (usage), 134–135
question/litigation, 3
questions, categories, 14
retracement, 380, 79, 88
surveying, 5
tidal boundaries, evidence/procedures, 217–224
Boundaries and Landmarks (Mulford), 8
Boundary corners, 18, 161
Boundary monuments, Old Testament accounts, 249
Boundary surveyor
activities/authority/responsibility, 6–7, 340–341, 349–350
research responsibilities, 354–358
Boundary surveys, 5, 387
documents, 358–359
preparation, 518
regulation, legal authority, 349
Bounds conveyances, 465, 466
Bounds form, 504–505
Bouquet, Henry, 254
Breach of contract, negligence (contrast), 524–525
Brown, Caleb, 263
Bryan v. Beckley, 582–589
case study diagrams, 585f
Burden of proof, 25–26, 30–33, 400, 420–421
Bureau of Land Management (BLM), 73, 303
dependent survey, initiation, 518
Burned pine stump, charcoal discoloration, 117f
Burt, William A., 153

C

Cadastration, 5
Cadastre, 5
Calculations, evidence role, 139
California, survey history, 281–282
Candia, plan (preparation), 263f
Caption (description component), 477, 478
Carolina, surveys, 269
Case law, impact, 28
Case study river, 232f, 233f
Certificates of merit, plaintiff collection, 523–524
Chain, 148f
feet, conversion, 171f
usage, 148–149
Charcoal identification, 643–644
Cherry v. Slade, 315, 589–604
Cinque Bambini Partnership v. State of Mississippi, 216
Circular curved lines, usage, 494–496
Circumferenter, 254, 255f
Circumstantial evidence (indirect evidence), 25, 27
Claiborne, William, 253
Clark, George Rogers, 254
Class, property size, 185
Clear and convincing evidence, phrase (usage), 19, 24, 36–37
Clients
agreements/disagreements, 353–354
conferences, 352, 375–376
contact, 351–354
contracts, 353
final report, 375–376
surveyor obligations, 544–545
surveyor legal advice, avoidance, 399, 401
Clinton, James, 267
Closed-perimeter descriptions, usage, 493
Closing corner, construction, 297
Closure, 473–474
error, calculation, 155
lines, 305
relative error, stating, 165
Collateral facts, admission, 39
Colonial method, 235
Colvin, Verplanck, 266
Common-law dedications, elements, 414–415, 418
Common-law presumptions, 40
Common report, 130–133
Competent parol evidence, summary, 125–126
Compilation plats, 206–208

Compound confusion, 347
Compound curves, 495, 495f
Compound of hiatus, 347
Computations, purposes, 374
Computer-aided design (CAD) systems/scanning, usage, 160
Conclusive evidence, 27, 53
Conclusive presumptions, 40
Concrete monument, witness stake, 107f
Connecticut, surveys, 264–265
Constant accidental errors, 164
Constructive possession, actual possession (contrast), 431–432
Continuous possession, 427–428
Contract privity, concept, 3
Contrary proof, usage, 94
Control corner, 326
Conveyance, 42, 198, 320, 365
  composition, 471
  intent (determination), evidence (usage), 69–70
  party intent, expression, 341, 369
  party intentions, 53
  planned/indiscriminate land conveyancing, 244–245
  survey, 19, 92
  words, interpretation, 69
Cook, James, 310
Coordinates
  creation, 78, 85–86
  usage, 390–392, 489–490
Coordinate value, computation, 79, 88
Corners, 78–79, 82–87, 151
  acceptance, 526
  accessories, 306
  control, 79, 86
  corner monuments, placement, 305
  cutoff, 491f
  history, chain, 122
  marking lines, 305
  monumentation, requirements, 387–388
  position, errors, 177
  posts, usage, 299
  proving, 79, 124–128
  proximity (proof), measurement evidence (usage), 170–171
  replacement, 79, 88
  superiority, absence, 341, 364
  validity, proof, 170–171
Corpus, The, 7
Correction lines (standard parallels), 303
Corrective instruments, 452
Corroborative evidence, 27
Council of New England, 260
County surveyors, records (prima facie evidence), 123–124
Courtroom
  cross-examination, FRE permission, 569
  direct/leading questions, 559–560
  discovery, 573–574
  hypothetical questions, usage, 567–568
  initial appearance, 553
  jury, impact, 561
  oaths/questions/answers, 557–558
  opinion evidence, usage, 566–567
  photographs, usage, 569–570
  pretrials, 556–557
  procedure, 552
  site viewing, 579
  surveyor
    appearance, 551
    conduct, 555
    direct examination/cross-examination, 558–559
    duties, 554–555
  trials, 556
    types, 557
Covenants, impact, 386
Cragin v. Powell, 142, 315
Crawford v. Gray & Associates, 530–531
Creation control retracements, standards, 82–83
Cross-examination, FRE permission, 569
Cubit, usage, 247
Cultural improvements, 199
Curtis v. Donnelly, 563
Curvature, direction, 496
Curves
  starting point, definition, 495
  usage, 494–497
Cypress, overgrowth, 115f

D
Damages, 529
  action, 519
  recovery, 518
  surveyor liability,
    514, 525–526
Data
  assembly, 374
  collectors, 320
De Agrorum Qualitate
    (Frontinus), 246
Decayed wood (ring
    count), freezing
    (usage), 647
Dedication by
    plat, 418–419
Dedications, 393
  acceptance, 417
  donor, 415
  easements,
    relationship, 414
  effect, 418
  expressed intent, 416
  implied intent, 416–417
  location/
    description, 416
  purpose, 417
Deeds, 52–61, 70–72
  conflict, 141, 175–177
  descriptions, 461, 468
  legal definition, 460
  voiding, 461, 468
Deflection angles, 489
De Jure Maris, 218
Delaware, surveys, 269
Dependent survey, BLM
    initiation, 518
Deposition,
    usage, 552–553
Described line
    known, 407–408
Described line, starting
    point (distance), 494
Description lines, rees-
    tablishment, 400, 422
Descriptions
  abbreviations, 498–500
  checklist, 508–511
  classes, 462
  components, 468–469,
    477–479
  coordinates,
    usage, 506–508
  creation, retracement
    (impact), 510
  exceptions, 501
  intent, 481
  legal sufficiency,
    461, 469, 481
  objectives, 469–470
  sufficiency, 470–471
  survey basis, 474
  surveyor preparation,
    470
  wordings,
    change, 475–476
  writing by
    exception, 505
De Witt, Simeon, 289
Diagrams, 190
  evidence, 575
Diehl v. Zanger, 652
Digges, Thomas, 218
"Dignity of Calls," 134
Dimensional
    data, 196–198
Dimensioning, 197f
Direct evidence, 27
Direction, 151–154
  bearings,
    basis, 485–494
  definition, 489
  distance, contrast, 176
  indication, 195
Directional
    calls, 490–491
Disagreements, 29
Discontinuance
    (vacation), 385
Discovered errors, 546
Discovery rule
  application,
    514, 516–518
  impact, 519
Disputable presumptions
    (rebuttable presump-
    tions), 24, 40
Distances, 143–144,
    197–198
  angles, relationship,
    156f
  direction, contrast, 176
  historical determina-
    tion, 147–150
  horizontal measure-
    ment, 141, 172–173
  measurement,
    errors, 147
  relative error, 157
  scaling, impossibility,
    191, 200–201
  uncertainty, calcula-
    tion (factors/
    elements), 166–167
Disturbed evidence,
    108–109
Division fence,
    building, 413
Dixon, Jeremiah, 268
Documents
  depositions/admissions/
    production,
    573–574
  intent, 69–70
  research, 354–360
  usage, decision, 340
Domesday Book, 250
Donation Tract, 272
Double corners, permis-
    sion, 297, 298
Double meridian distance
    (DMD), 180, 258
Drawings, 190, 192
Duces tecum, 574
Due diligence, 522

Due north, phrase (meaning), 66
Duncan, James, 313
Duncan v. Peterson, 414
Dutch West India Company, charter, 265

E

Early states, surveys, 259–271
Easements, 451, 500–501
  conveyance, intent, 500
  dedications, relationship, 414
  location, 341, 359–360
  prescription ripening, 403
East, survey systems, 257–274
Education, superiority/ distinction, 535, 538–539
Egypt, measurements, 247f
Electronic distance measurement (EDM), 86, 87, 145, 149, 178, 185
Elevation, datum, 196
Ellicott, Andrew, 244, 267
Ellicott v. Pearl, 82
Eminence, gaining, 540
Empresario system, 277–278
Encroachment, 347
Encroachments, indication, 203
Ending point, usage, 493–494
England
  common law, 54, 258
  historical measurements, 248f
  surveying system, 4
Equal rights, 70–72
Equity, 236
Errors, 177–178
  altitude, impact, 169f, 178f
  classification/ types, 163–164
Ervey, John, 281
Escarpment (scarp), 228, 232–233
Estate, term (usage), 249
Estoppel, 405–406
Estoppel by conduct, 405–406
Ethical standards, violation, 548–549
Ethics, 541–542
Evidence, 15–23
  admissibility, 39–41
  area, knowledge, 46
  change, impact, 23
  classifications, 26–27
  compilation, 374–375
  conclusions/proofs, 25–26
  conclusiveness, 38–39
  credibility, 88–90
  disappearance, 321–322
  disturbed evidence, 108–109
  estimation, 43
  finding, surveyor duties, 47–48
  historical knowledge, usage, 243
  laws, 15, 28–30, 46
  loss, impact, 472–473
  monuments, value, 99
  perpetuation, 323–328
  possession evidence, 93, 129–130
  preponderance, 34–36
  preservation
    aerial/terrestrial photographs, usage, 331–336
    state plane coordinates, usage, 337–338
  prima facie evidence, 123–124
  proof, 323. *See also* Proof of evidence.
  question, 3
  recording/ preserving, 319
  recovery, percentage, 322f
  retracements, relationship, 95–98
  retracing surveyor recovery, 323
  rules, codification (development), 14
  scope, 18–23
  significance, 59
  substitute, judicial notice (usage), 43–44
  surveyor
    openness, 552
    relationship, 581
  surveyor role, 16, 18
  surveyor usage, 29
  technology, relationship/application, 77, 80
  totality, 22, 77, 321
  types, 25, 27–28, 91, 227–228
  value, 30, 42–43
  variation, 24
Exception form, usage, 505
Exclusive possession, 429–430
Expert testimony, compelling (power), 570–571
Expert witnesses
  appointment, 564–565

cause and effect, 568
cross-examination, 571
duties, 565–566
fees, 572–573
fees, cross-examination, 573
opinions, basis, 566
surveyor qualification, 562–564
Expressed guarantees, 525–526
Extinct corners, 651–652
Extrinsic ambiguity, explanation, 57
Extrinsic evidence, 24, 27, 41–42
introduction, impossibility, 58
usage, 53

F
Fairfax, Thomas, 254
Federal Rules, application (absence), 34
Federal Rules of Evidence (FRE), 13, 19–20, 130, 209, 336, 361, 564, 566, 568–569
Federal Rules of Procedure (FRP), mastery, 13
Federal Test, artificial improvement considerations, 215
Federal Title Test, 215
Fences, 425–428
construction, 410
evidence, usage, 134–135
fence brace, usage, 289f
tree attachment, 367f
usage, 433f
Feudal land system, 250–251
Fiefs/feuds, 250
Field notes, 372–373
plats, conflicts, 210–211
survey plot, 333f
written evidence, 64–65
Fields
evidence, 363
hunt/search, 366
instructions, sample, 362
term, usage, 92
Fieldwork, 360–374
computations, 186–187
intent, 363
planning, 360–363
preparation, 339–340
First-order leveling, 520
First Principal Meridian, 293, 294
First surveys, 346
Following the footsteps (phrase), 2
Forensic survey, 352
Forensic survey, utility survey (contrast), 19, 45–46
Foundation survey, redates (consideration), 203
Found evidence, basis, 84
Found monument
chronological history, incompleteness, 320
history, 198–199
Free surveys, 345
French grants, 272–273
French surveying system, 4
Fry, Joshua, 253

G
Gallatin, Albert, 296
Gates, John F., 32
General Instructions, 300–301
General Land Office (GLO), 36, 324, 581
bearing trees, 136f, 289f, 327
notes, copy, 105f
plat, surveyor usage, 40
reorganization, 300–301, 303
states, 141
surveys, 32, 278, 329, 344
system, 284–286
Geodaesia (Love), 80–81, 143, 158, 250, 256, 659–672
Geographer of the United States, role, 289
Geographic information system (GIS), usage, 6–7, 11, 84
Geometers, usage, 246
Georgia
land divisions, 271f
land patents, granting, 313–314
surveys, 269–270
GLO. *See* General Land Office
Global positioning system (GPS), 149, 337
observations, 520
reading, 86
three-unit GPS system, usage, 87
usage, 6, 11–12, 159, 178
Going forward, burden, 31
Gore, 347, 348
Government lots, 305
Gradient boundary, 280–281

Grantor/grantee,
 intent, 60
Graphic descriptions,
 462–463
Great Mahele (1846), 310
Greece, measurements,
 249
Grid systems, usage, 152
Groma, usage, 286, 288
Guide meridians, 304
Gunter,
 Edmund, 254, 256
Gunter's chain, 254, 255f

H

Habendum clauses
 (description component), 479
Hale, Matthew, 218
Half-tide level, 219–220
Handbook of Annotated Forms for the Surveying Practice (Hermansen), 353
Hardwood, test, 646–647
Harvey survey, 32
Hawaiian Housing Authority v. Midkiff, 315
Hawaii Land Court,
 impact, 456
Hawaii, land ownership, 310
Haydel v. Dufresne, 82
Head right concept, 259
Headright grants, 309
Hearsay, 130–
 133, 560–561
 evidence, 560
Hearsay evidence, 104
Hearsay rule, 30
Hiatus, 347
High water (HW), 225
Historical knowledge,
 usage, 243
Historical sequence,
 proof, 104
Historic private
 surveys, 98
Holland Land
 Purchase, 266
Holmes, Oliver
 Wendell, 581
Homesteading,
 impact, 309
Hostile possession,
 428–429
Hunt v. Feese et al., 54–55
Hutchins, Thomas,
 253–254, 289–291,
 293

I

Illegality/fraud, establishment, 57
Illinois, land registry, 454
Image, perception, 321f
Impartiality, importance/
 necessity, 19–23
Implied guarantees, 525–526
Increment borer,
 usage, 119f
Independent survey,
 conducting, 45
Indirect evidence, 27
Indispensable
 evidence, 26
Inferences, 24
 evidence, substitute, 41
Infrared (IR)/color infrared (CIR) photography, usage, 335
In Re Ashford
 decision, 218
Institutes of Justinian, 214, 217
Integrity, definition, 542
Intent, evidence, 92
Intention, application, 60
Invisible lines, 483
Irregular curves,
 usage, 497
Irregular watercourses,
 655–656

J

Jarred v. Seifert, 522–523
Jefferson, Thomas/
 Peter, 253, 286
Judging effect, 42–43
Judicial knowledge, 43
Judicial notice,
 25, 43–44, 58
Junior deeds, calls, 480f
Junior descriptions,
 461, 479–481
Junior parcel title, quality, 380, 384–386
Junior rights, 70–72
Jury, impact, 562

K

Kelley's Creek and Northwestern Railroad v. United States, 227

L

Lakes, boundaries, 234–235
Land
 acquisition, homesteading (impact), 309
 area, identification
 method, 462
 boundaries
 creation
 methods, 379
 involvement, 573
 surveys, national
 standards, 522
 conveyance, 42, 345
 courts
 powers, increase, 455
 specialization, 454–456

data systems, 11
dedication
(Dutch rule), 4
description, 461, 481
objectives, 469–470
short-cut methods,
256–257
division, 340, 349
grants, resurvey,
282–283
interest
gaining, unwritten
means
(usage), 399, 404
identification
method, 462
location, 359, 382–383
occupation,
intention, 426
ownership
(Hawaii), 310
parcel
description, 505–506
retracing, 380, 389
survey, 22–23
registry (Illinois), 454
sources, 287t
Spanish origin, 274
surveying, biblical
references, 247–249
surveyors
attorney cross-
examination, 20
concerns, 235
surveys
acceptability, 201
procedural
network, 17f
taxes, payment, 431
tenure cases, state
law basis, 34
title (granting),
adverse possession
(usage), 400, 421
Land Act (1785),
284, 291–292,
349, 673–674
Land Act (1796), 144,
284, 291–292, 294,
370, 678–682674
Land Act
(1800), 292, 296
Land Act (1805), 142,
284, 292, 311, 370,
389, 674, 682–684
Land information system
(LIS), usage, 6–7, 11
Land ownership, 3,
47, 359, 441
certainty, 394
precolonial surveys,
245–246
unwritten transfers, 397
Landowners, testimony,
53, 72–73
Land Registration
Act, 455–456
Land surveying, 5, 6, 425
Las Siete Partidas, 217
Latent, term (usage), 42
Latitude and longi-
tude call, 490
Law
presumptions, 40
surveyor, relationship,
581
Laws of evidence,
15, 28–30, 48
*Lawyers Title Insurance
Co. v. Hoges,* 521
League, measurement
unit, 145
Legal closure,
presence, 473–474
Legal history,
importance, 315
Length, units
(Southwest), 283
Length, units (variations),
145, 146t, 147
Liability, 513
avoidance,
514, 527–528
elements, 528
results, 515
LiDAR scanning, 520
Limiting closure, 182
Linear conveyances,
465, 467–468
Lines, 483–485
corners creation, 79, 86
creation, 78, 82–83
defining
geometric
relationships,
impact, 491–492
points/corners,
usage, 79, 87
fixing, acquiescence
(usage), 653
irregularity, 484
location, impact,
399, 407–408
marks, identifi-
cation, 329
position, errors, 177
re-creation, 381
Line trees, evi-
dence, 93, 123
Litigation
object, 555
provocation, 543
Livery of seisin
(delivery of posses-
sion), 250–251
Location
evidence, loss
(impact), 472–473
guarantees, 441, 442, 452
insurance policy,
451–452
surveys, 343–351

Long lake, method, 235, 236, 236f, 237f
Lorman v. Benson, 656
Lost corners
  creation, 78
  recovery, 650–651
  settings, technology (usage), 85
Lost tracts, identification, 335
Louisiana Territory, land grants, 273–274
Love, John, 80–81, 143, 148, 158, 250, 253, 256, 493, 659
Lowest qualified bank, 228
Low water (LW), 225
Luttes decision, 217

M

Magnetic field, discovery/movement, 151–152
Magnetic meridian, reference, 173–175
Mailbox rule, 41
Maine, land boundaries (court decisions), 312–313, 410
Mansfield, Jared, 294, 296
Manual of Instructions for the Survey of the Public Lands of the United States, 140, 160, 228–229, 300–303, 435
  Manual of Instructions (1973), 307
  Surveying Instructions (1947), 303–306
Maps, 189–191, 200, 209
  evidence, 575
Marshall, John, 174
Martin v. Waddell, 214
Maryland, surveys, 268–269
Mason, Charles, 268
Mason-Dixon line, 291
Massachusetts Bay Company, charter, 260
Massachusetts Land Court, providing, 454–456
Masters v. Pollie, 250
Material evidence, disclosure, 325
Material facts, 37f, 38, 583
Materiality, 37
Mathematical closure, presence, 473–474
Meander corners, 305
Meander lines, 654–655
Meander rights, 656–658
Mean higher high water (MHHW), 220
Mean high water (MHW), 219–220, 222, 225
Mean lower low water (MLLW), 220
Mean low water (MLW), 219–220, 222, 225
Mean range (MR), 222, 225
Mean tide level (MTL), 219–220, 222, 225
Measurements, 368–369
  alternatives, 371f
  comparisons, 142–143
  creation, 88
  data, 470
  definitions, examples, 283t
  errors, 140, 162, 370
  evidence, 139, 171–177
  historical application, 154
  history, 247
  precision, 370–372, 570
  principles, 140–141
  refinement, 200
  types, 142–143
  uncertainties, 140, 369–372
    magnitude, determination, 341
  units (Southwest), 283
Measuring devices, 140, 144–145
Mendenhall Order, 145
Meridian observations
  ambient conditions, impact, 169
  attending observations, 169–170
  instrumentation, usage, 169
  reliability, 168–169
Meridians, 152, 303
  assumption/establishment, 195
  determination, celestial observations (usage), 158–159
Metes-and-bounds, 581
  conveyances, 466
  descriptions, 71, 152, 464–466, 472, 506
    writing, variation, 502–504, 503f
  states, 141, 148, 175
  surveys, 93, 346–347
  system, 244, 284
  true descriptions, 502–504
Mexican land grants, 277
Mexican/Spanish land grant systems, 4
Midcall v. Hawaiian Housing Authority, 310

Millicam, Robert, 251–252
Minimum technical standards, 194
Modic Indian Wars, 244
Monumentation, 486f
creation, 389
Monumented positions, preservation (measurement types), 320
Monuments
calls, 482–483, 505–506
control/evidence, 92
density, 388–389
description, 462
disclosure, 198–199
disturbing/moving, probability/possibility, 104
evidence, 99–102
history, chain, 122
indispensable evidence, 103
land, inclusion, 527
location, 123–124, 335–336
marks, identification, 329
measurements, refinement, 186
original position, evidence, 122
party intent, 341, 369
physical characteristics, 104–108
position error, absence, 382–383
proving, witness evidence (usage), 124–125
proximity (proof), measurement evidence (usage), 170–171
record, 325
search, 364–366
testimony, 368
validity, proof, 170–171
value (evidence basis), 99
Moore, John M., 301
Morse v. Emery, 62
Mortgage policies, 450
Mulford, A.C., 8, 162, 523, 528

N

Napoleonic Code, 4
National Council of Examiners for Engineering and Surveying (NCEES), 5
National forest lands, classification, 308–309
National Geodetic Survey (NGS), 336
National Land System, 258
National Society of Professional Surveyors (NSPS), 5, 522
National Tidal Datum Epoch, 221
Natural bridge, topographic call, 106f
Natural levees, 228
Negligence
breach of contract, contrast, 524–525
proving, 528
theories, 515
New England
ordinance, example, 262f
rock walls, property line demarcation, 264f
surveys, 260–264
New evidence, weight, 44–45
New Hampshire, surveys, 265
New Jersey, surveys, 267
New World, property surveys, 253–256
New York, surveys, 265–267
Nixon, J.A., 278
Nonpublic trust waters, boundaries, 233–235
Nontangent curve, defining, 496
Nontidal waters
boundaries, 226–227
navigability, 215
North, reference point, 150
North/south line, establishment, 36
Northwest Ordinance, 285
Note possession, 26

O

Oak woods, identification, 644–645
Oaths, 328–329
Objects, passing calls, 93
Obliterated corner, qualification, 325
Observed directions, errors, 167–168
Official plat, 208
reference, 191
Ohio Company purchase, 272
Ohio lands
donation tract, 272
French grants, 272–273
northwestern Ohio, 273
subdivision, 272–273
Symmes purchase, 273
United States military district, 273
Western Reserve, 273

Oklahoma v. Texas, 280–281
Old evidence, weight, 44–45
Old photographs, usage, 332–334
Open and notorious possession, 426
Opinion evidence, usage, 566–567
Oral evidence, 25
Ordenanza de Intendentes, 277
Ordinance of 1785, 286, 288, 674–678
Ordinance of 1796, 291–292
Ordinary high water line, 226
Ordinary high water mark, 226, 231–233
Oregon Manual (1851), 301
Original boundaries
  impact, 381
  retracement, 340, 346–348
Original corner monuments (proof), evidence (usage), 104
Original corners, 84
  control, 80–81
  positional tolerance location, 83
  weight, 93
Original documents, copies (recording), 329–330
Original evidence, superiority, 85
Original lines
  evidence, representation, 129
  examination, 79
  running (Tiffin method), 298f
Original measurements, 140–143
Original monuments
  control, 102
  identification problems, 108
  position, 122, 130–133
  value, 92, 108
Original plats, 205–211
  platting, laws, 206
  purpose, 205–206
Original surveys, 81–84, 340, 344–348, 379
  bearing/line trees, evidence (consideration), 123
  evidence, 320
  limitation, 88–90
  lines, memorialization, 341, 366–368
  regulation, 380–381
Original surveys, resurveys, 340
Outer boundaries, display, 202
Overlap, 347, 348
Ownership
  lines, 354
  registration, 442
  sovereignty, relationship, 249

P

Pace, measurement unit, 253
Parallel lines, 491f, 492f
Parcel
  absolute location, 92
  deed/conveyance, elements, 461
  deed, party ownership, 432f
Parol evidence, 27–28, 38
  admissibility, 66
  summary, 125–126
  usage, 65, 130–133
Partial evidence, 27
Passing calls, 126–128
Patents, 443
  ambiguities, 42, 67
Pedestrian easement, usage, 500
Pennsylvania, surveys, 268
Pen point, thickness (impact), 200
People's Ice Co. v. Steamer Excelsior, 656
Perennial stream, term (usage), 275
Peters, John, 313
Photogrammetry, 160–161
Photographs
  evidence/availability, 159–162
  explosion effect, 569–570
  special photography, 335
  usage, 331–336, 569–570
Pine trees, examples, 110f, 117f, 118f, 120f
Pipe and brass cap, 306f
Planned/indiscriminate land conveyancing, 244–245
Planning boards, 384
Plans, 189–190
Plats, 63f, 375
  acceptance, 526
  bearings, basis, 195–196
  contents, 204–205
  dimensional data, 196–198
  dimensioning, 197f
  direction, indication, 195

distances, 191
elevation datum, 196
evidence, 189, 210, 330, 575
features, 191
field notes, conflicts, 210–211
information, omission, 200–201
monuments, disclosure, 198–199
preparation, 300
public/private rights-of-way, display, 202
reference, 52
calls, 61–62
retracement survey, results, 193f
scale, 194
surveyor completion, 58
survey results, 200
title, 192–194
types, 190–191
writings as evidence, contrast, 62–63
Platting, 374–375
laws, 206, 379–380, 382–384, 388
Plots, 189
Plymouth Company, organization, 260
Point of confusion, 347–348
Points, usage, 79, 87
Pollard's Lessee v Hagan, 214
Pollard v. Hagan, 657
Position
adjustment, 186
closing error, 182
errors, 177
uncertainty, measurements (impact), 369
Possession
addition, 427
client deed, disagreement, 400, 435–436
evidence, 93, 129–130
facts, 652–653
guarantees, 443
importance, 366–368
insurance, 443
measurements, disagreement, 400, 436
reputation, 129
results, 375–377
ripening, determination, 399
statute of limitations, 430
survey, 347
Practical location, 53, 67, 411–412
Precision, 162–163
Preponderance of evidence. *See* Evidence
Prescription (doctrine), 419
Prescriptive title, 421–422
Presumption, 24, 40
evidence, comparison, 41
Presurveying practices, 381–382
Presurveys, recording (absence), 394
Pretrial, 556–557
discovery, 552–553
Prima facie, 27
case, making, 31
Primary evidence, 27
correlation, 77
Priority of Calls, 256
Prior surveys, usage, 288–289
Private plats, 208–209
Private surveys, records, 98, 330–331
Privity of contract, 519
Probative evidence, 38
Profession
advertising, avoidance, 548
attributes, 538
definition, 533–537
discredit, 547
foundation, 536, 549
surveyor obligations, 545
Professional
definition, 533–537
ethics, 541–542
experience/education, merger, 549
fees, 541
independent judgment/liability, 540
license validity, relationship, 549
moral duty, 541
opinions, differences, 548
trust, position, 539–540
Professional liability, 513
Professional reputation, 545
Professional standards, violation, 536, 548–549
Professional stature, 533
attainment, 537–538
nonattainment, 535, 537–538
Proof of evidence, degree/quantum, 33–34
Property
adjoinder description, 55–56
descriptions, identification, 55–57
monumenting, 35
New World surveys, 253–256

Property *(continued)*
rights, 244, 468
size, 185
surveying, boundary surveying/boundary line (comparison), 5
surveys, 341, 374–375, 520
value, 184
Property boundaries
location specifications, 184
term, 2
Property lines, 2
extension method, 235
method, prolongation, 237–238, 238f
rock wall demarcation, 264f
Proportional conveyances, 465
Proportionate acreage method, 235
Proportionate conveyances, 466
Proportionate medial line method, 235, 237
Proportionate thread of the stream method, 236
Protraction diagrams, preparation, 307–308
Proximity (proof), measurement evidence (usage), 170–171
Public domain (PD), 284–286
absence, 259
land grants, resurvey, 282–283
land sources, 287t
tracts, obtaining, 310
Public land
creation, land acts (impact), 673–684
survey monument, 326
surveys, 308
Public Land Survey System (PLSS), 142, 284, 324
surveys, recovery, 358
Public, surveyor obligations, 536, 542
Public title interest, determination, 214–216
Public trust doctrine, 214
Public trust waters, boundaries, 216–219
Public welfare, protection, 513
Punctuation, abbreviations, 498–500
Punitive damages, 524–525
factors, 524
Putnam, Rufus, 294, 296

Q

Qualifying clauses (description component), 477, 479

R

Rabbittown Creek, 55–56
Rancho de la Nación, redwood post set, 106f
Random errors, 164
Range (R), 225
lines, creation, 32
Ratio, expression, 194
Raw recovered evidence, 26
Reading uncertainty, 168
Real evidence, 25
Real property
conveyance/description, 140, 144–145
interest, deed/conveyance, 461
law, requirements, 460
Reasonable doubt, 37. *See also* Beyond a reasonable doubt
Rebuttable presumption. *See* Disputable presumptions
Reckoning, 152
Recognition, evidence, 400, 413–414
Recordation, 394
Records, research, 354–360
Recovered evidence, preservation, 372
Rectangular System. *See* U.S. Rectangular System
Redwood posts, 106f, 107f
Reference calls, 61–62
Reference descriptions, 464
Reference meridians, bearings basis (relationship), 488
Reference systems, change, 390–391
Registered surveyor, survey examination, 33
Relevancy, 37–38
Relevant evidence, Rule 401 definition, 37
Relocated lost corners/boundaries, determination, 77
Remainder, size, 70–71
Representative fraction, expression, 194
Reputation, 130–133
Research
reasons, 355–356
responsibilities, 354–358
Residual accidental errors, 166–167

Residual errors, 164
Res ipsa loquitor, 515
Restoration of Lost or Obliterated Corners, 307
Resurveys, 282–283, 340, 344–346, 348
Retracements, 346–348
  evidence, relationship, 95–98
  survey, 193f, 337
  surveyor, duty, 243
Retracing surveyor
  equipment/standards examination, 83
  responsibility, 81
  witness function, 16
Reverse curves, 495, 495f
Revoking offer, 417
Rhode Island, surveys, 265
Rice v. Ruddiman, 656
Richfield Oil Co. (case), 66
Right of trespass laws, 328
Riley v. Griffin, 315
Ring counts, 116, 118–122
Riparian allocation methods, 235–238
Riparian evidence, 334–335
Riparian rights, 656–658
  allocation, common law methods, 235
Rittenhouse, David, 290
River. *See* Case study river
Rivers v. Lozeau, 315, 328, 346, 623–629
Robert survey, 32
Rod, measurement unit, 252, 283
Rope stretchers, 246, 246f
Round lake method, 235, 236, 236f
Rule of thumb calculation, 179–180
Rules of engagement, understanding, 19–20
Rules of evidence, codification (development), 14
Rules of Evidence for U.S. District Courts and Magistrates, 14
Ruth v. Dight, 517

S

Salinas Avenue (Del Mar Heights), scarfed redwood post, 107f
Scale (scaling), 194, 197–198
Scarfed redwood posts, 107f
Scribing, exposure, 109f–110f
Scrivener
  qualities, 474–475
  role, 464–465
Search areas/locations (determination), measurements (usage), 23
Secondary evidence, 27, 77
Second-generation corner monument, chronological written history (availability), 320, 323
Second Principal Meridian, 293, 294
Sectionalized lands (survey), true/astronomical north (basis), 174
Section lines
  measurement, chain (usage), 144
  running, 305
Section post, notches (usage), 300f
Sections, subdivisions, 305
Seisin, livery (delivery of possession), 251–252
Self-calibration, accomplishment, 144–145
Senior deeds, calls, 480f
Senior rights, 26, 41, 70–72, 461, 469, 482–483
  adjoining parcels, relationship, 479–481
  alteration, impossibility, 53, 70
  interference, 341, 472
Senior rights, interference, 364–366
Setbacks, requirement, 393–394
Seven Ranges
  deputy instructions, 296
  meridians/baselines, 293–294, 295f
  survey, 288–291
Signal error, 168
Significant figures, consistency/relationships, 155–157, 156t
Significant numbers, usage, 158
Simultaneous descriptions, examples, 71
Slope distance, EDM measurement, 153
Softwood, test, 646–647
Soil composition, changes, 229
Solar compass, development, 153
Solar transit, usage, 306
South, survey systems, 257–274

Southwest
California, survey history, 281–282
early settlements, 277
empresario system, 277–278
gradient boundary, 280–281
lands, Spanish origin, 274
measurements/length units, 283
Mexican land grants, 277
minerals, title passage (absence), 275
Native Americans, effect, 281
Ordenanza de Intendentes, 277
Ordinance of 1785, 284, 286
property ownership, history, 274
public domain, 282–286, 287t
road beds, 276
sovereign, suits, 278
Spanish/Mexican grants, 275, 279–280
Spanish water laws, 275–276
surveyor instructions, 278–279
titles, seniority, 274–275
Sovereignty, ownership (relationship), 249
Spanish/Mexican grants, 275, 279–280
Spanish water laws, 275–276
Spiral curves, usage, 496–497
Spoken evidence (words), value, 72–74
Spoken words, value, 51–53
Squatter's rights, 268
Stadia, usage, 149
Standard of care, 520–523
Standard parallels (correction lines), 303
Standards, effect, 78, 84
State plane coordinates, 390–392, 506–507
States
early states, surveys, 259–271
public domain, absence, 259f
Stationing, strip conveyances (relationship), 498
Stature, 533
Statute
enactment, 515
law, impact, 28
Statute of Frauds, 3, 54, 252–253, 397, 408, 411
Statute to Prevent Fraud, A, 398
Statutory dedication, 415, 418
Statutory proceedings, 403
Stewart v. Carleton, 604–618, 652
Storyteller, witness role, 15
Straight lines, 484–485
defining, coordinates (usage), 489–490
Streams, boundaries/threads, 233–234
Strip conveyances, 465, 467
stationing, relationship, 498
Subdivision
coordinates, usage, 390–392
descriptions, 463–464
law, 379–380, 382
lot, retracement, 63
survey, 345
unit, 305
Subject matter, 23–24
Subject matter, arrangement, 343–344
Superior call (indication), forms (usage), 497
Surface measurement, usage, 172–173
Surrounding, term (usage), 57
Surveying, 1–5, 10f
compass, 299f
fees, division (impropriety), 536, 547–548
history/epochs, 244
results, dorms, 376–377
unauthorized practice, aiding, 536, 543
Surveyor General, appointment, 294, 296
Surveyors, 4–10, 45–48
accuracy, defining, 140
attorneys, relationship, 556
capabilities, questions, 544
certificates, 204
chain-to-feet conversion, 171f

conduct, 555
county surveyors, records (prima facie evidence), 123–124
courtroom
appearance, 551
duties, 554–555
oaths/questions/answers, 557–558
cross-examination, 558–559
direct examination, 558–559
direct/leading questions, 559–560
disagreement, causes, 44–45
discovered errors, 546
document reliance, 22
duty, 654
education, importance, 9–10
employment, 546
evidence, relationship, 581
expert witness qualification, 562–564
fact finding/conclusions, 350–351
failures, 515
footsteps, retracing, 149–150, 154
functions, 343–348
law, relationship, 581
lawsuits, 529
legal advice, avoidance, 399, 401
liability, 514, 525–528
negligence proof, 529
obligations, 536, 542, 544–545
performance, 514, 520–523
professional standing, 549
qualifications, 544
quasi-judicial function, 649, 658
records (monument location), 123–124
report, 377, 633–641
responsibility, 400, 436–437
retracing surveyor, witness role, 16
substandard work, 529
textbooks/treatises, 568–569
work
contract, avoidance, 529
review, 546–547
usage, 514
Surveys, 4–6
calls, 481
certification, 392–393
completion, 376
computations, 154–155
conducting, subject matter/systematic procedures (arrangement), 342–343
control, fence (usage), 433f
documents, 53, 68
evidence, 91–94
intent, 99
location, 358–359
original surveys, 81–83
plats, 53, 68, 189–180, 190–191
precision/accuracy, 162–163
problem, example, 182f
procedures, court reports, 311–314
public record, filing (insufficiency), 325–326
questions, categories, 14
results, plats (impact), 200
specifications, purpose, 180–181
surveyor completion, 574
systems, 257–274
Symbols, 194
Symmes Purchase, 273
Systematic errors, 163–164

T

Tachometric methods, 149
Taped distances, errors, 166–167
Tax maps, usage, 191, 209–210
Tax plats, assessors (relationship), 209–210
Technical words, interpretation, 66
Technology
application, 80
development, impact, 7–8
principles, 78–80
usage, 85, 88–90
Tennessee, surveys, 270–271
Tent survey (campfire survey), 40
Terms and Definitions of Surveying and Associated Terms (NSPS/ASCE), 5–6
Terrestrial photographs, usage, 331–336

Testimony, 368
discrediting, 559
limitations, 126
preparation, 574–579
Texas, 30-foot-rule, 276
Textbook of Wood Technology, 644
Theoretical uncertainty (TU), 166, 183, 184f
Theoretical uncertainty analysis (TUA), 164–166
Third Principal Meridian, 294
Thread of the stream, 234, 236
Tidal boundaries
amplitude ratio method, 223–224
evidence/procedures, 219–224
standard method, 222–223
Tidal datum, 219–225, 220f
Tidal epoch, 219–222
Tidal waters, 216–219, 225–226
Tide level (TL), 225
Tiffin, Edward, 296–300, 298f, 305
Timber Culture Act, 308
Titles
abstract, 444–449
aids, 442
associations, title survey plats, 201–203
chain, 443
character, acquisition, 421
claim, 426–427
color, 430–431
companies, risk business role, 450
concept, 257
considerations, 384–386
defects, 347f
evidence, research, 22
guarantees, 386–387, 441, 443
identity, 199, 469, 471–472
insurance policy, 449–452
origin, 257–258
policies, wording, 451
proof, 134–135
registration, 442
Torrens principle, 452–456
seniority, 274–275
Toise, 145
Topo calls, 126–128
Topographic call, display, 105f, 106f
Torrens principle, 452–456
advantages/disadvantages, 453–454
Torrens, Robert, 452
Torrens system, characteristics, 453
Torrens titles, 420, 441, 443
Total stations, 153, 185
Townships, exteriors/subdivisions, 304–305
Tracts, 304f, 310, 335
Trans-Appalachian states, surveys, 270
Transit lines, 266
Transit/theodolite, terms (usage), 153
Traverse
closures, 181t
errors/uncertainties, 179f
Traversing,
errors, 177–178
Treaty of Guadalupe-Hidalgo, 274, 277, 282
Trees
annual rings, 114f
burned pine stump, charcoal discoloration, 117f
characteristics, 116–122
core sample, 120f
identification, 111–116
increment borer, usage, 119f
ring counts, 116, 118–122
scribing, 109f–110f
species/characters, 116
stump particulars, 116
Trespass, 348
damages, 526–527
Trial court, 557
True metes-and-bounds descriptions, 502–504
True north (astronomic north), usage, 173–175
Trunnion error, 178f

U

Umiat Principal Meridian, establishment, 307
Uncalled-for monuments, 366, 482–483
Uncertainty, 179–180, 347
computation, 167t
expression, 183
Under construction survey, redates (consideration), 203

United New Netherlands Company, charter, 265
United States military district, 273
United States v. Doyle, 618–623
United States v. Flint, 82
United States v. San Jacinto Tin Co., 82
Unknown line, dispute, 408–410
Unwritten agreement, 401–402, 406–414
  elements, 407
  possession, relationship, 412–413
Unwritten conveyances
  concept, 401–403
  government relationship, 403–404
Unwritten dedication, 414–437
Unwritten ownership rights, 398
Unwritten rights
  concept, 401–403
  determination, difficulty, 432–435
  surveyor, role, 436–437
Unwritten title transfers,
  duties, 435–436
  written title, relationship, 399, 404
U.S. Geological Survey (USGS) topographic maps, usage, 56
U.S. Rectangular System, 272, 282, 286, 291–292
  boundaries, 296
  organizational structure, 289
  Protraction Program, 307
U.S. territories, acquisition, 288f
U.S. v. Doyle, 315
U.S. v. Joder Cameron, 227
U.S. v. Parker, 227
Utility survey, 19, 45–46, 352

V

Vara, usage, 41, 145, 147, 283
Verbal evidence, discussion, 52
Video presentations, 337
Virginia
  military district, 272
  surveys, 259–260
Virgin title, 404

W

Warranty deed, 443
Washington, George, 253–254
Washington, John, 254
Water boundaries
  botanical evidence, 230–231
  changes, 238
  evidence, 213, 227–228
  geomorphological features, 228–229
  hydrological evidence, 231
  nonpublic trust waters, boundaries, 233–235
  nontidal waters, boundaries, 226–227
  ordinary high water mark, determination, 231–233
  public title interest, determination, 214–216
  public title waters, boundaries, 216–219
  soil composition, changes, 229
  tidal boundaries, 219–226
  tidal datum determination, 224–225
Waters, riparian doctrines (impact), 108
Weeks Law, 308–309
Wentworth, John, 263–264
Western Reserve, 273
Western Title Guaranty Co. v. Murray & McCormick, Inc., 528
Whiskered decision, 44
Whole descriptions, usage, 501–502
Witness. *See* Expert witnesses
  behavior, 552
  credibility, loss, 559
  testimony, 552
  usage, 561–562
Witness evidence, 104, 124–125, 328–329
Witness monuments, 104
Witness objects, 93, 109–111
Wooden evidence, 643
Words
  definitions, court reliance, 69
  evidence, 51

Words *(continued)*
translation, 51, 53, 73–74
value, 72–74
Work, review, 546–547
Writings
ambiguities, extrinsic evidence (usage), 65–66
evidence, 54–55
exceptions, 57
interpretation, rules, 59
reference calls, 61–62
testimony, relationship, 126
Writings as evidence, plats (contrast), 62–63
Writings, referral, 52
Writing, technique, 477
Written conveyance, 52–53, 71–72
validity, 461, 472–473
Written descriptions, 462–463
abbreviations, 499
Written document, 52, 55–56
Written evidence, 25, 64–65
Written ownership guarantee, 443
Written titles, 404–405
classification, 47
defects, exposures, 442
insurance, 443
unwritten title, relationship, 399, 404
Written words, 51–53, 57–59